Springer Series in Statistics

Advisors:
P. Bickel, P. Diggle, S.E. Feinberg, U. Gather,
S. Zeger

More information about this series at http://www.springer.com/series/692

Mark J. van der Laan · Sherri Rose

Targeted Learning in Data Science

Causal Inference for Complex Longitudinal Studies

 Springer

Mark J. van der Laan
Division of Biostatistics and
Department of Statistics
University of California, Berkeley
Berkeley, CA, USA

Sherri Rose
Department of Health Care Policy
Harvard Medical School
Boston, MA, USA

ISSN 0172-7397   ISSN 2197-568X   (electronic)
Springer Series in Statistics
ISBN 978-3-030-09736-3   ISBN 978-3-319-65304-4   (eBook)
https://doi.org/10.1007/978-3-319-65304-4

Printed on acid-free paper

This Springer imprint is published by Springer Nature
The registered company is Springer International Publishing AG
The registered company address is: Gewerbestrasse 11, 6330 Cham, Switzerland

To Martine, Laura, Lars, and Robin
To Burke, Ani, and Soozie

Foreword

The authors are to be congratulated for their new book *Targeted Learning in Data Science*. The book is a welcome addition to the literature, merging two exciting fields. It is a sequel text that builds on their earlier highly successful general treatment of machine learning and causal inference in *Targeted Learning* (van der Laan and Rose 2011). Given that both targeted learning and data science are rapidly evolving fields with fuzzy boundaries, to narrow the scope, we write from the perspective of data scientists in the biomedical sciences and the manner in which targeted learning can help unify the foundations in this area, particularly, how targeted learning addresses a key divide in causal foundations by connecting mechanistic modeling and randomization-based inference.

To elaborate, in mechanistic modeling, practitioners seek to build an accurate model of the underlying data-generating process. In other words, the entire distribution of the observed data unit or perhaps only relevant portions of this distribution are modeled. For example, the conditional distribution of an outcome given treatment type and baseline covariates may be explicitly modeled. This is at the heart of traditional likelihood-based modeling, including common Bayesian approaches. In mechanistic modeling, causality is often thought of informally, if at all, with a primary focus instead on modeling or prediction. Historically, parsimony guides the analysis, often via linearity assumptions on conditional mean models. More recently, machine learning approaches have allowed the creation of more accurate but computationally elaborate models. This has motivated some to characterize approaches to mechanistic modeling as being either model-based or algorithmic—see, for example, Breiman et al. (2001). Model-based mechanistic modeling consists of describing the data-generating process using relatively few parameters. Even despite its inherent risk of model misspecification, this approach may be appealing since in several cases, parsimonious models, even if wrong, provide useful descriptions of key features of the data-generating process. Model-based approaches can suffer when subject-level inference is of interest, since the simplifications made for the sake of parsimony are generally incompatible with rich data-generating processes and may not accurately capture population heterogeneity. In contrast, complex prediction algorithms often do facilitate accurate subject-level inference for use in individual decision-making. However, they generally cannot be described parsimoniously and do not readily allow for inference on useful population summaries. Nevertheless, irrespective of whether algorithmic prediction or parsimonious statistical models are used, a model is built for the data-generating mechanism.

In contrast, many statistical approaches to causal inference focus on ancillary randomization, or ideas based on ancillary randomization, to estimate marginal effects. The average treatment effect (ATE) is often the target of interest. The ATE is a desirable estimand, as it summarizes the causal impact of a policy, such as what the average benefit of treating patients with a particular drug would be. Research in causal inference has focused on reducing the assumptions that more mechanistic, model-based approaches would require to estimate the ATE or related causal estimands. In such approaches, formally incorporating the randomization scheme, or models of treatment assignment, is the price required to avoid the onerous modeling assumptions to obtain causal estimates out of traditional methods.

It is fair to say that both algorithmic and machine learning approaches to mechanistic models and robust causal marginal estimates have been revolutions in statistics, dominating much of the zeitgeist of late. These two approaches both have desirable aspects, yet little in common. *A core appeal of targeted learning is the formal unification of these approaches.* Targeted learning of an ATE requires estimation of both the conditional mean outcome (i.e., outcome regression) and the treatment assignment distribution (i.e., propensity score), but also yields robust estimation of marginal causal effects through the use of targeted minimum loss-based estimation (TMLE). By performing this unification, targeted learning builds up a theory of causal inference based on underlying mechanistic models (as discussed, for example, in the foreword to van der Laan and Rose 2011, by Judea Pearl). This underlying framework will be satisfying to data scientists, who tend to think more mechanistically than researchers in causal inference, who base their foundations on notions of experimentation. A key aspect of targeted learning is that while it is based upon the estimation of underlying mechanistic models, it does not require knowing or postulating simplistic models or model classes to be effective.

While the framework of targeted learning unifies these two disparate areas, the specific implementation of the framework recommended in the text offers somewhat of a free lunch similar to modeling treatment assignment probabilities (propensity scores). Specifically, the mere act of modeling relationships between a treatment and covariates is informative. This, of course, applies to targeted learning if the target requires modeling treatment assignment. Similarly, the exercise of building up several conditional outcome models in the pursuit of targeted learning will often be extremely illuminating about the data and setting. In other words, targeted learning does not eliminate the nontargeted, informal learning that is so valuable in model-based statistics. The authors explicitly encourage a very broad approach to the conditional modeling and suggest model stacking/super learning as a method for blending estimates. This approach, within the context of targeted learning, shatters the divide between model-based and algorithmic approaches to mechanistic modeling, rendering the distinction between the two practically moot. It also eliminates concern over inference after model selection, a key issue in more mechanistic approaches. All relevant models and algorithms can contribute to the stacked prediction model in an objective, data-driven but a priori specified manner. This fact, in conjunction with use of the targeting step of the TMLE procedure, essentially eliminates concerns over biased error rates due to model selection.

Another benefit is obtained by forcing researchers to actually specify the marginal estimand of interest (the target). This has the mathematical benefit that TMLE can improve estimation by modifying the output of the mechanistic modeling step to focus on the target of interest. This is in contrast to regular maximum likelihood, for example, wherein a global, nontargeted assessment of fit is optimized. Targeted learning has the practical benefits of focusing the discussion and prompting an a priori specification of hypotheses. As a metaphor, in targeted learning, one cannot shoot an arrow and paint a bullseye around where it lands, since the target must be pre-specified. It is worth emphasizing that the effect of interest need not be formally causal, merely any global summary of the data-generating mechanism. This is important in the context of neuroimaging and neuroscience (some of our areas of interest), as scientists are often interested in "treatments," such as disease status or brain morphometry, that are not conceptually assignable.

Speaking of our areas of interest, the incorporation of the mechanistic and machine learning aspects of targeted learning is highly appealing in neuroscience and neuroimaging data science, as mechanistic and model-based causal approaches dominate. Many of the most popular techniques are model based: dynamic causal modeling, Granger causality, and structural equation models (Friston et al. 2013; Zhang et al. 2015; Chén et al. 2017; Friston et al. 2003; Penny et al. 2004). "Generative modeling" is a phrase that is used frequently in the neuroimaging literature to (positively) motivate an approach. One could conjecture that the goals of a mechanistic understanding of the brain and its disorders predispose the field toward more mechanistic approaches to observational modeling.[1] However, much less attention has been paid to causal inference and excessive focus is placed on the single-final-model based statistics that the authors rightly criticize. It is interesting to note that our mechanistic understanding of the brain has led to artificial neural networks and modern artificial intelligence through deep learning. These flexible approaches contain large swaths of traditional statistical modeling as special cases and have come to dominate data science, especially in tech-related industries. One could envision applications of targeted learning to existing artificial intelligence systems to perform on-the-fly causal analyses in lieu of formal time-consuming A/B tests.

---

[1] Interestingly, the reverse, utilizing ideas from statistics creates hypothetical models of neural organization, also appears to be true. Recent attempts at characterizing the brain as intrinsically Bayesian represent exactly such a case (Knill and Pouget 2004).

To summarize, this book will serve as a bridge for existing data scientists wanting to engage in causal analyses and targeted estimates of marginal effects. It will also help the causal statistical community understand key issues and applications in data science.

Baltimore, MD, USA   Brian Caffo
Seattle, WA, USA   Marco Carone
December 2017

Preface

This book builds on and is a sequel to our book *Targeted Learning: Causal Inference for Observational and Experimental Studies* (2011). Since the publication of this first book on machine learning for causal inference, various important advances in targeted learning have been made. We decided that it was important to publish a second book that incorporated these recent developments. Additionally, we properly position the role of targeted learning methodology within the burgeoning field of data science.

This textbook for scholars in statistics, data science, and public health deals with the practical challenges that come with big, complex, and dynamic data. It presents a scientific roadmap to translate real-world data science applications into formal statistical estimation problems by using the general template of targeted maximum likelihood estimators. These targeted machine learning algorithms estimate quantities of interest while still providing valid inference. Targeted learning methods within data science are a critical component for solving scientific problems in the modern age. The techniques can answer complex questions, including optimal rules for assigning treatment based on longitudinal data with time-dependent confounding, as well as other estimands in dependent data structures, such as networks. *Targeted Learning in Data Science* contains demonstrations with software packages and real data sets that present a case that targeted learning is crucial for the next generation of statisticians and data scientists.

Key features of *Targeted Learning in Data Science*:

1. Represents essential data analysis tools for answering complex big data questions based on real world data
2. Machine learning estimators that provide inference with data science
3. Introductory chapters present an accessible explanation of causal inference and targeted learning for complex longitudinal data
4. Filled with real world applications and demonstrations of (a) the translation of the real world application into a statistical estimation problem, and (b) the targeted statistical learning methodology to answer scientific questions of interest based on real data
5. Demonstrates targeted learning from experiments in which the data on the different experimental units are dependent, such as those described by a network
6. Deals with the practical challenges that come with big, complex, and dynamic data while maintaining strong theoretical foundation

Outline

Similar to our last book, *Targeted Learning in Data Science* is special as it contains contributions from invited authors, yet is not a traditional edited text. As the authors, we have again spent substantial time reworking each chapter to have consistent notation, style, and a familiar road map. This led to a second cohesive book on targeted learning that reads as one text.

Part I—Targeted Learning in Data Science: Introduction

In Chap. 1, we provide the motivation for targeted learning and a general overview of its roadmap involving (1) data as a random variable; (2) a statistical model representing the true knowledge about the data experiment; (3) translation of the scientific question into a statistical target parameter and (4) targeted minimum loss-based estimation (TMLE) accompanied with inference. In Chap. 1 we also define our running longitudinal example inspired by a "when to treat" application in HIV research.

In Chap. 2, we review causal models for longitudinal data and their utility in defining a formal causal quantity/query representing the answer to the scientific question of interest. We focus on causal quantities defined in terms of counterfactual means of an outcome under a certain intervention rule. We present the g-computation estimand that identifies the causal quantity as a function of the data distribution, under the sequential randomization and positivity assumption. At this point, we have defined the statistical model and estimand and thereby the statistical estimation problem: under the causal model and identifiability assumptions, the estimand equals the desired causal quantity, but either way it has a valid statistical interpretation of interest.

The last two chapters in Part I focus explicitly on estimation. Chapter 3 presents the sequential super learning approach to estimate the prediction of a counterfactual outcome as a function of baseline covariates, an object of independent interest. Since the expectation of this prediction function equals the (marginal) mean of the counterfactual outcome, this sequential learning also provides the initial estimator in the definition of the TMLE presented in Chap. 4. Chapter 4 presents the TMLE of the counterfactual mean outcome in our running example. It demonstrates the general roadmap for computing a TMLE in terms of the efficient influence curve, a local least favorable submodel that uses the initial estimator as an off-set and loss function whose score spans the efficient influence curve, and an iterative algorithm iteratively updating the current estimator with the maximum likelihood estimator of least the favorable submodel that uses as off-set the current estimator. Chapter 4 also demonstrates the general analysis of TMLE and formal inference in terms of its influence curve.

Part II—Additional Core Topics

Part II concerns theoretical and methodological developments for the general TMLE. There are many estimation problem for which the TMLE involves iteratively updating the initial estimator by iteratively maximizing the log-likelihood (or, more generally, minimizing an empirical mean of a loss function) along the local least favorable submodel through the current update. The iterative nature of such a TMLE can result in unstable finite sample TMLE, especially in the case that the data provides sparse information for the target parameter. In Chap. 5, we develop a general one-step TMLE based on a so called universal (canonical) least favorable submodel that is uniquely characterized by the canonical gradient/efficient influence curve. We also develop this one-step TMLE for a multivariate target parameter, and even for an infinite dimensional target parameter. The philosophy of this TMLE strongly suggest that this one-step TMLE is more robust and stable than an iterative TMLE. An example in survival analysis is worked out and simulations are used to demonstrate the theory.

Chapter 6 presents a new general estimator of a parameter defined as the minimizer of an expectation of a loss-function such as a conditional mean or conditional density. We refer to this estimator as the highly adaptive lasso (HAL) estimator since it can be implemented by minimizing the empirical risk over very high dimensional linear combination of indicator basis functions under the constraint that the sum of the absolute value of the coefficients is bounded by some constant, which itself is selected with cross-validation. We show that this estimator is guaranteed to converge to its true counterpart at a faster rate than $n^{-1/4}$ in sample size n, even for complete nonparametric models and high-dimensional data structures.

In Chap. 7, we show that a TMLE that uses the HAL estimator as initial estimator, or uses a super learner whose library contains this HAL estimator, is asymptotically efficient under very weak regularity conditions, as long as the strong positivity assumption holds. The formulation of a TMLE relies on the computation of the efficient influence curve. There are many estimation problems in which this object only exists in an implicit form and is extremely hard to compute according to this implicit form. In Chap. 8 we present a machine-learning based estimator of an efficient influence curve which avoid the need for the analytic computation of the efficient influence curve, and by using the HAL estimator we are guaranteed to estimate it accurately. In particular, we show that this allows us to construct TMLE that could not be formulated previously due to the immense complexity of its efficient influence curve. This is demonstrated with various censored data examples (e.g., interval censored data and bivariate right-censored data).

In Chap. 9 we present a general class of data-adaptive target parameters, which allows a statistician to mine the data to determine the target parameter of interest while obtaining valid confidence intervals. Specifically, we present a cross-validated TMLE (CV-TMLE) of this data-adaptive target parameter, develop the formal asymptotics theorem, and demonstrate this CV-TMLE in a variable importance analysis for continuous variables. Following this, in Chap. 10, we propose a general class of collaborative TMLE (C-TMLE) for targeted selection of the nuisance parameter estimator among a continuum of candidate estimators. We show that it is theoretically superior to a TMLE that estimates the nuisance parameter with an estimator (e.g., super learner) that is optimized for estimation of the nuisance parameter itself. C-TMLEs are of enormous practical importance due to their ability to protect the TMLE against using fits of the nuisance parameter that are harmful for the performance of the TMLE (e.g., a fit of a propensity score that includes instrumental variables unknown to the user).

Part III—Randomized Trials

Part III is concerned with TMLE for randomized controlled trials (RCTs), including cluster randomized controlled trials (CRTs). Chapter 11 develops a TMLE of the locus-specific causal effect of vaccination on time to HIV infection due to a virus that matches or mismatches the vaccination at this locus. Results of such an analysis allows one to evaluate the effectiveness of the vaccination at various loci and thereby directs future improvements of the vaccination. The TMLE utilizes baseline covariates and time-dependent covariates to gain efficiency and to allow for informative censoring. The method is demonstrated on an HIV vaccination RCT.

Chapter 12 considers the TMLE of the sample average treatment effect, which is defined as the sample average of the individual causal effects over the individuals in the actual sample. Robust statistical inference is studied in detail and the methods are evaluated with simulations. It is demonstrated theoretically and practically that the sample average treatment effect can be estimated at greater precision than the population average treatment effect. It is also argued that the sample average treatment effect has the advantage that it does not require viewing the sample as a random sample from some target population. The importance of the latter is demonstrated with CRTs involving comparing treated and nontreated communities that represented a certain geographic region in East Africa that cannot be viewed as a random sample.

Chapter 13 presents a novel data-adaptive TMLE for a CRTs in the common situation that the number of communities is small (e.g., 30). Remarkably, this TMLE still uses super learning to estimate the outcome regression with a library of simple candidate targeted regressions adjusting for one or two potentially important covariates. The cross-validation criterion that is used to select the best estimator among the candidate regression estimators is aimed at minimizing the variance of the TMLE. The superior practical performance of this data-adaptive TMLE relative to a simple marginal estimator is demonstrated with a simulation study. This chapter is important in that it demonstrates the key concepts of TMLE also apply to very small sample sizes.

Part IV—Observational Data

Part IV concerns the analysis of observational studies with TMLE. Chapter 14 develops TMLE of the causal effect of stochastic interventions for a single time-point data structure (W, A, Y). It focuses on stochastic interventions that depend on the unknown treatment mechanism. Chapters 15 and 16 represent powerful applications of the longitudinal TMLE to complex observational longitudinal studies. Chapter 15 evaluates the causal effect of different breast feeding regimens on child development in the PROBIT study, also dealing with cluster sampling. Chapter 16 evaluates different intensification rules for controlling glucose level on long-term time-to-event outcomes in diabetes patients, based on a large Kaiser Permanente database. Chapter 17 concerns causal mediation analysis in longitudinal studies. Specifically, it develops a novel TMLE for estimation of the causal natural direct or indirect effect of a point treatment on a survival outcome controlling for a time-dependent mediator, while allowing for right-censoring affected by time-dependent covariates.

Part V—Online Learning

Part V concerns the development of scalable online super learning and scalable online TMLE for online time-series dependent data. As a special case, it includes i.i.d. data that is ordered artificially, in which case one can run the online estimators over various orderings of the data set. Chapter 18 presents an online super learner for time-series dependent data and develops its optimality by establishing an oracle inequality for the online cross-validation selector. The statistical model assumes Markov dependence and stationarity, but leaves the stationary datagenerating mechanism unspecified. Chapter 19 develops online TMLE for timeseries dependent data of the causal impact of stochastic or deterministic interventions on certain treatment nodes in the time-series on a future outcome. These online estimators can be applied to a single time series, in which case the asymptotics are based on the number of time points. The models and methods developed in this part will generate much future research due to the importance of online data, scalability, and time series dependence. In particular, in the era of precision medicine, the option to learn from experiments and data collected on a single subject is of great importance.

Part VI—Networks

Part VI concerns TMLE of the causal effect of stochastic or deterministic interventions on an average outcome for a finite (large) population of interconnected units in which the network structure is observed over time: i.e., for each unit, we know at each time point the set of friends it potentially depends upon. Chapter 20 introduces a causal and statistical model for longitudinal network data and develops a TMLE of the desired causal effect when one observes on each unit baseline covariates W, a treatment A, and an outcome Y_t at various time points t. Chapter 21 focuses on the special case that one only observes the outcome at one point in time t, presumable shortly after A. In this special case it is shown that the TMLE exists in closed form and it is supported by an R package. Many interesting issues are discussed in detail, and simulations are presented evaluating the practical performance of the TMLE. Chapter 21 extends much of the causal inference literature for the point treatment data structure (W, A, Y) for i.i.d. data by allowing that (1) the outcome Y of a unit is affected by the treatment A and baseline covariates W of its friends, and (2) that the treatment of the unit is affected by the baseline covariates of its friends. Formal asymptotic theory establishing the asymptotic normality of the TMLE is reviewed as well. We note that both Parts V and VI develop TMLE and its theory for sample size one problems where one only observes a single realization of a complex experiment involving possibly a single unit over many time points or many interdependent units at a finite set of time points.

Part VII—Optimal Dynamic Rules

The three chapters in Part VII concern estimation of the dynamic treatment allocation rule that optimizes the mean outcome. The chapter focus on the case that one observes on each subject baseline covariates W, binary treatment A, and outcome Y, but generalizations have been worked out in accompanying articles. Chapter 22 develops a super learner of the optimal rule, a TMLE of the counterfactual mean under the optimal rule, and a TMLE of the data-adaptive target parameter defined as the counterfactual mean under the estimate of the optimal rule. Chapter 23 extends this work to the optimal rule under resource constraints.

Chapter 24 proposes a group sequential adaptive randomized design in which the randomization probabilities for the next group of subjects are based on an estimate of the optimal treatment rule based on the data collected on the previously recruited subjects. In this manner, one simultaneously learns the optimal rule and allocates treatment according to the best estimate of the optimal rule. In this type of group sequential targeted adaptive design, a novel TMLE is developed for the counterfactual mean under the optimal rule. It is shown that this TMLE is asymptotically consistent and normally distributed, under the single assumption that one succeeds in learning the optimal rule. If the latter does not hold, one still obtains valid inference for the data-adaptive target parameter defined as the counterfactual mean under the estimate of the optimal rule: that is, just as one always obtains valid inference for the average treatment effect in an RCT, we preserve this guarantee for the much more complex causal question: *"What is the counterfactual mean outcome under this estimate of an optimal rule we learned based on the data?"* The problem tackled in Chap. 24 is a low dimensional version of the long standing multiple bandit problem.

Part VIII—Special Topics

Part VIII dives into some important special topics in the field of targeted learning. Targeted learning has largely focused on pathway differentiable target parameters. Chapter 25 studies the estimation of a nonpathwise differentiable target parameter such as a density at a point. It approximates the target parameter with a family of pathwise differentiable parameters indexed by a bandwidth. Subsequently, it develops a CV-TMLE and a selector for the bandwidth. In addition, it demonstrates that the resulting CV-TMLE is asymptotically normally distributed at an unknown adaptive rate that depends on the underlying smoothness of the data density. It also develops asymptotically valid confidence intervals. This chapter opens up a general approach for targeted learning of nonpathwise differentiable target parameters while still providing formal statistical inference.

Chapter 26 reviews the theory of higher-order influence functions for higher-order pathwise differentiable target parameters and demonstrates that the TMLE framework easily allows the construction of higher-order TMLE. The benefit of higher-order pathwise differentiability is that it allows one to develop estimators based on higher-order Taylor expansions so that one only needs to assume that a higher-order remainder is negligible, instead of having to assume that a second-order remainder is negligible. However, unfortunately, most target parameters for realistic models are only first order pathwise differentiable. If the target parameter is only first-order pathwise differentiable, then we show that a second-order TMLE based on an *approximate and tuned* second-order influence function can yield significant finite sample improvements relative to the regular TMLE that only targets the first order efficient influence function.

Sensitivity analysis has as its goal to set an upper bound for the difference between the estimate of the estimand and the causal quantity of interest. It naturally concerns statistical bias due to using a biased estimator of the estimand and identifiability bias due to violation of causal assumptions that were needed to identify the causal quantity from the observe data distribution. Many sensitivity analysis are made confusing by using biased estimators (e.g., regression in parametric models). Therefore, one wants to use estimators such as TMLE based on highly adaptive super learners to provide maximal guarantee for honest statistical inference concerning the estimand. Given that this is achieved, there is still need for sensitivity analysis with respect to the nontestable assumptions. Once again, many methods proposed in the literature utilize parametric sensitivity models so that the interpretation of the sensitivity parameters (whose bounds are presumably provided by external knowledge) completely depend on the correctness of these models. Therefore, in order to make sensitivity analysis transparent and helpful it is important to use a well-defined sensitivity parameter. Chapter 27 presents such a nonparametric sensitivity analysis approach so that the sensitivity parameter is nonparametrically interpretable. A real case study is used to demonstrate the power and transparency of this approach.

The nonparametric bootstrap generally fails to consistently estimate the sample distribution of an estimator when the estimator uses machine learning, such as the typical TMLE for realistic statistical models. Since the bootstrap picks up second-order variability of the estimator that is not captured by first order asymptotics, it is important that the bootstrap is also an option for the TMLE. Chapter 28 proposes a targeted bootstrap method for estimation of the limit distribution of an asymptotically linear estimator. The targeted bootstrap estimates the sampling distribution of the estimator by resampling from a TMLE estimator P_n^* of the data distribution P_0 that targets the variance of the influence curve of the estimator. This general approach is demonstrated for the TMLE of a counterfactual mean for the point treatment data structure (W, A, Y). The failure of the nonparametric bootstrap and the superior performance of this targeted bootstrap is evaluated in a simulation study.

Chapter 29 considers the fast computation of (inefficient) TMLE by replacing the TMLE based on all the data by a TMLE based on a controlled random sample of much smaller size from the database. The sampling probabilities are allowed to be a function of a measurement that is available for all and easy to compute. It works out the optimal sampling probabilities that maximize efficiency of the TMLE. One can now consider group sequential designs where one adjusts the sampling probabilities based on past data so that the design minimizes the variance of the TMLE. This general approach can be used to scale TMLE to large data sets.

Finally, Chap. 30 presents a historical philosophical view on the books topics. In his essay "The Predicament of Truth: On Statistics, Causality, Physics and the Philosophy of Science" the author discusses some main implications of recent developments in data science, for statistics and epistemology. He shows that these developments give rise to a specific, uncomfortable vision on science, which seems hardly adequate for both statistics and the philosophy of science. He then shows that, given this antistatistical and antiepistemic stance, improving the relation between statistics and philosophy of science could be considered a matter of well-understood self-interest, but it appears that such a liaison is highly problematic. It would seem that nowadays preoccupations with truth advance along at least two distinct lines with separate roles for epistemology and research methodology, thus inducing a rigid and regrettable demarcation, which also applies to other epistemological key issues, including causality. The role and significance of targeted learning in this debate is analyzed in detail and a few initial steps toward a philosophy of data science are made.

Appendix

Lengthy proofs of fundamental results are deferred to our Appendix. Specifically, in Sect. A.1 we present the general analysis of the CV-TMLE for data-adaptive target parameters (Chap. 9). Section A.2 establishes three fundamental results for mediation analysis (Chap. 17). In Sect. A.3, we provide the proof of the oracle inequality of the online super learner for time-series dependent data (Chap. 18). Lastly, in Sect. A.4, we provide first order Taylor expansions of causal target parameters based on their canonical gradient for time series data, which provide the basis for the analysis of the TMLE (Chap. 19).

Acknowledgments

This book reflects the greater effort of a large team of people, without whom it would not have been completed. We thank the following students and colleagues, many of whom wrote a chapter in this book, for their contributions to and discussions on these methods, as well as their vital collaborative efforts: Samrachana Adhikari (Harvard), Laura B. Balzer (UMass), Oliver Bembom (Pandora), David Benkeser (Emory), Savannah Bergquist (Harvard), Aurélien Bibaut (Berkeley), Brian Caffo (Johns Hopkins), Wilson Cai (Berkeley), Marco Carone (Washington), Wilson Cai (Berkeley), Antoine Chambaz (Université Paris Descartes), Jeremy Coyle (Berkeley), Victor De Gruttola (Harvard), Irina Degtiar (Harvard), Iván Díaz (Weill Cornell), Francesa Dominici (Harvard), Randall Ellis (BU), Ani Eloyan (Brown), Bruce Fireman (Kaiser Permanente), Peter Gilbert (Washington), Jeff Goldsmith (Columbia), Susan Gruber (Harvard), Laura Hatfield (Harvard), Joe Hogan (Brown), Alan E. Hubbard (Berkeley), Nicholas P. Jewell (Berkeley), Emilien Joly (Université Paris Nanterre), Cheng Ju (Berkeley), Chris J. Kennedy (Berkeley), Tim Layton (Harvard), Erin LeDell (h2o), Sam Lendle (Pandora), Alexander R. Luedtke (Fred Hutch), Xavier Mary (Université Paris Nanterre), Alex McDowell (Harvard), Thomas McGuire (Harvard), Caleb Miles (Berkeley), Andrew Mirelman (York), Ellen Montz (Harvard), Erica E.M. Moodie (McGill), Kelly L. Moore (Netflix), Ian Nason (Harvard), Romain S. Neugebauer (Kaiser Permanente), Sharon-Lise Normand (Harvard), Patrick J. O'Connor (HealthPartners), Elizabeth L. Ogburn (Johns Hopkins), Judea Pearl (UCLA), Maya L. Petersen (Berkeley), Robert W. Platt (McGill), Eric C. Polley (Mayo Clinic), Kristin E. Porter (MDRC), A.J. Rosellini (BU), Michael Rosenblum (Johns Hopkins), Daniel B. Rubin (FDA), Stephanie Sapp (Google), Julie A. Schmittdiel (Kaiser Permanente), Mireille E. Schnitzer (Université de Montréal), Jasjeet S. Sekhon (Berkeley), Taki Shinohara

(Penn), Akritee Shrestha (Wayfair), Michael J. Silverberg (Kaiser Permanente), Oleg Sofrygin (Berkeley), Jake Spertus (Harvard), Richard J.C.M. Starmans (Universiteit Utrecht), Ori M. Stitelman (Dstillery), Ira B. Tager (Berkeley), Boriska Toth (Berkeley), Catherine Tuglus (Amgen), Hui Wang (VA), Yue Wang (Novartis), C. William Wester (Vanderbilt), Wenjing Zheng (Netflix), and Corwin Zigler (Harvard).

Berkeley, CA, USA   Mark J. van der Laan
Boston, MA, USA   Sherri Rose
March 2017

Contents

Part V Online Learning

Contributors

Mark J. van der Laan
Division of Biostatistics and Department of Statistics, University of California, Berkeley, Berkeley, CA, USA

Sherri Rose
Department of Health Care Policy, Harvard Medical School, Boston, MA, USA

Laura B. Balzer
Department of Biostatistics and Epidemiology, School of Public Health and Health Sciences, University of Massachusetts - Amherst, Amherst, MA, USA

David Benkeser
Department of Biostatistics and Bioinformatics, Rollins School of Public Health, Emory University, Atlanta, GA, USA

Aurélien Bibaut
Division of Biostatistics, University of California, Berkele, Berkeley, CA, USA

Wilson Cai
Division of Biostatistics, University of California, Berkeley, Berkeley, CA, USA

Marco Carone
Department of Biostatistics, University of Washington, Seattle, WA, USA

Antoine Chambaz
MAP5 (UMR CNRS 8145), Université Paris Descartes, Paris, France

Jeremy Coyle
Division of Biostatistics, University of California, Berkeley, Berkeley, CA, USA

Iván Díaz
Division of Biostatistics and Epidemiology, Department of Healthcare Policy and Research, Weill Cornell Medical College, Cornell University, New York, NY, USA

Peter Gilbert
Vaccine and Infectious Disease Division, Fred Hutchinson Cancer Research Center, Seattle, WA, USA

Susan Gruber
Department of Population Medicine, Harvard Medical School, Harvard Pilgrim Health Care Institute, Boston, MA, USA

Alan E. Hubbard
Division of Biostatistics, University of California, Berkeley, Berkeley, CA, USA

Emilien Joly
Modal'X, Université Paris Nanterre, Nanterre, France

Cheng Ju
Division of Biostatistics, University of California, Berkeley, Berkeley, CA, USA

Chris J. Kennedy
Division of Biostatistics, University of California, Berkeley, Berkeley, CA, USA

Sam Lendle
Pandora Media Inc, Oakland, CA, USA

Alexander R. Luedtke
Vaccine and Infectious Disease Division, Fred Hutchinson Cancer Research Center, Seattle, WA, USA

Xavier Mary
Modal'X, Université Paris Nanterre, Nanterre, France

Erica E. M. Moodie
Department of Epidemiology, Biostatistics, and Occupational Health, McGill University, Montreal, QC, Canada

Romain S. Neugebauer
Kaiser Permanente Division of Research, Oakland, CA, USA

Patrick J. O'Connor
HealthPartners Institute, Bloomington, MN, USA

Elizabeth L. Ogburn
Department of Biostatistics, Johns Hopkins Bloomberg School of Public Health, Baltimore, MD, USA

Maya L. Petersen
Division of Epidemiology and Division of Biostatistics, University of California, Berkeley, Berkeley, CA, USA

Robert W. Platt
Department of Epidemiology, Biostatistics, and Occupational Health, McGill University, Montreal, QC, Canada

Julie A. Schmittdiel
Kaiser Permanente Division of Research, Oakland, CA, USA

Mireille E. Schnitzer
Faculté de pharmacie, Université de Montréal, Montreal, QC, Canada

Oleg Sofrygin
Division of Biostatistics, University of California, Berkeley, Berkeley, CA, USA

Richard J. C. M. Starmans
Department of Information and Computing Sciences, Buys Ballot Laboratory, Universiteit Utrecht, Utrecht, The Netherlands

Wenjing Zheng
Netflix, Los Gatos, CA, USA

Abbreviations and Notation

Frequently used abbreviations and notation are listed here.

A	Treatment or exposure
A-IPCW	Augmented inverse probability of censoring-weighted/weighting
A-IPW	Augmented inverse probability weighted/weighting
C	Censoring
CRT	Community randomized trial
i.i.d.	Independent and identically distributed
IPCW	Inverse probability of censoring-weighted/weighting
IPW	Inverse probability of weighted/weighting
LTMLE	Longitudinal targeted maximum likelihood estimation/estimator
MLE	Maximum likelihood substitution estimator of the g-formula *Not to be confused with nonsubstitution estimators using maximum likelihood estimation. MLE is also known as g-computation*
MSE	Mean squared error
O	Observed ordered data structure
P	Possible data-generating distribution
p	Possible density of data-generating distribution P_0
P_0	True data-generating distribution; $O \sim P_0$
p_0	True density of data-generating distribution P_0
P_n	Empirical probability distribution; places probability $1/n$ on each observed $O_i, i \ldots, n$
RCT	Randomized controlled trial
SCM	Structural causal model
SE	Standard error
SL	Super learner
TMLE	Targeted maximum likelihood estimation/estimator
W	Vector of covariates
Y	Outcome
Y_1, Y_0	Counterfactual outcomes with binary A

Uppercase letters represent random variables and lowercase letters are a specific value for that variable. If O is discrete, $p_0(o) = P_0(O = o)$ is the probability that O equals the value o, and if O is continuous, $p_0(o)$ denotes the Lebesgue density of P_0 at o. For simplicity and the sake of presentation, we will often treat O as discrete so that we can refer to $P_0(O = o)$ as a probability. For a simple example, suppose our data structure is $O = (W, A, Y) \sim P_0$ and O is discrete. For each possible value (w, a, y), $p_0(w, a, y)$ denotes the probability that (W, A, Y) equals (w, a, y).

$\mathcal{M}$	Statistical model; the set of possible probability distributions for P_0
$P_0 \in \mathcal{M}$	P_0 is known to be an element of the statistical model $\mathcal{M}$

In this text we often use the term *semiparametric* to include both nonparametric and semiparametric. When semiparametric excludes nonparametric, and we make additional assumptions, this will be explicit. A *statistical model* can be augmented with additional nonstatistical (e.g., causal) assumptions providing enriched interpretation, often represented as $\{P_\theta : \theta \in \Theta\}$ for some parameterization $\theta \to P_\theta$. We refer to this as a *model* (e.g., the probability distribution of the observed data $O = (W, A, Y = Y_A)$ could be represented as a missing data structure on counterfactual outcomes Y_0, Y_1 with missingness variable A, so that the probability distribution of O is indexed by the probability distribution of (W, Y_0, Y_1) and the conditional distribution of treatment A, given (W, Y_0, Y_1)).

$X = (X_j : j)$	Set of endogenous variables, $j = 1, \ldots, J$
$U = (U_{X_j} : j)$	Set of exogenous variables
$P_{U,X}$	Probability distribution for (U, X)
$p_{U,X}$	Density for (U, X)
$Pa(X_j)$	Parents of X_j among X
f_{X_j}	A function of $Pa(X_j)$ and an endogenous U_{X_j} for X_j
$f = (f_{X_j} : j)$	Collection of f_{X_j} functions that define the SCM
$\mathcal{M}^F$	Collection of possible $P_{U,X}$ as described by the SCM; includes nontestable assumptions based on real knowledge; $\mathcal{M}$ augmented with additional nonstatistical assumptions known to hold
$\mathcal{M}^{F*}$	Model under possible additional causal assumptions required for identifiability of target parameter
$P \to \Psi(P)$	Target parameter as mapping from a P to its value
$\Psi(P_0)$	True target parameter
$\hat{\Psi}(P_n)$	Estimator as a mapping from empirical distribution P_n to its value
$\psi_0 = \Psi(P_0)$	True target parameter value
ψ_n	Estimate of ψ_0

Consider $O = (L_0, A_0, \ldots, L_K, A_K, L_{K+1}) \sim P_0$.

L_k	Possibly time-varying covariate at $t = k$; alternate notation $L(k)$
A_k	Time-varying intervention node at $t = k$ that can include both treatment and censoring
$Pa(L_k)$	$=(\bar{A}_{k-1}, \bar{L}_{k-1})$
$Pa(A_k)$	$=(\bar{A}_{k-1}, \bar{L}_k)$

P_{0,L_k} True conditional probability distribution of L_k, given $Pa(L_k)$, under P_0

P_{L_k} Conditional probability distribution of L_k, given $Pa(L_k)$, under P

P_{n,L_k} Estimate of conditional probability distribution P_{0,L_k} of L_k

P_{0,A_k} True conditional probability distribution of A_k, given $Pa(A_k)$, under P_0

P_{A_k} Conditional probability distribution of A_k, given $Pa(A_k)$, under P

P_{n,A_k} Conditional probability distribution of A_k, given $Pa(A_k)$, under estimator P_n of P_0

ϵ Fluctuation parameter

ϵ_n Estimate of ϵ

$\{P_\epsilon : \epsilon\} \subset \mathcal{M}$ Submodel through P

H^* Clever covariate

H_n^* Estimate of H^*

$D(\psi)(O)$ Estimating function of the data structure O and parameters; shorthand $D(\psi)$

$D_0^*(O)$ Efficient influence curve; canonical gradient; alternate notation $D^*(P_0)(O)$, $D^*(P_0)$ or $D^*(O)$

$IC_0(O)$ Influence curve of an estimator at P_0, representing a function of O

$IC_n(O)$ Estimate of influence curve

We focus on the general data structure $O = (L_0, A_0, \ldots, L_K, A_K, L_{K+1}) \sim P_0$ in many chapters, introduced on the previous page. In this setting, the following specific notation definitions apply:

L_0 Baseline covariates

$\bar{L}$ $= (L_0, \ldots, L_{K+1})$

$\bar{A}$ $= (A_0, \ldots, A_K)$

L_d Counterfactual outcome for regime d

d_0 Optimal regime depending on P_0

Q_{0,L_k} True conditional probability distribution of L_k

Q_{L_k} Possible conditional probability distribution of L_k

Q_{n,L_k} Estimate of Q_{0,L_k}

Q $= (Q_{L_0}, \ldots, Q_{L_{K+1}})$

$L(O, Q)$ Example of a loss function where it is a function of O and Q; alternate notation $L(Q)(O)$ or $L(Q)$

$\{Q_\epsilon : \epsilon\}$ Submodel through Q

$\bar{Q}_{L_k}$ Conditional mean of the probability distribution of L_k

G_{0,A_k} True conditional probability distribution of A_k

G_{A_k} Possible conditional probability distribution of A_k

G_{n,A_k} Estimate of G_{0,A_k}

$\bar{Q}_n^0$ Initial estimate of $\bar{Q}_0$

$\bar{Q}_n^1$ First updated estimate of $\bar{Q}_0$

$\bar{Q}_n^k$ kth updated estimate of $\bar{Q}_0$

$\bar{Q}_n^*$ Targeted estimate of $\bar{Q}_0$ in TMLE procedure

$\Psi(Q_0)$ Alternate notation for true target parameter when it only depends on P_0 through Q_0

$\Psi(Q_n^*)$ Targeted estimator of parameter

Pf Expectation of $f(O)$ under P, e.g., $P_0 L(Q) = \int L(Q)(o) dP_0(o)$

Part I
Targeted Learning in Data Science:
Introduction

Chapter 1
Research Questions in Data Science

Sherri Rose and Mark J. van der Laan

The types of research questions we face in medicine, technology, and business continue to increase in their complexity with our growing ability to obtain novel forms of data. Much of the data in both observational and experimental studies is gathered over lengthy periods of time with multiple measures collected at intermediate time points. Some of these data are streaming (such as posts on Twitter), images, DNA sequences, or electronic health records. Statistical learning methods must be developed and adapted for these new challenges.

In 2010, Google Flu Trends was touted as an shining example of collective intelligence. Researchers claimed that their ability to predict flu in over two dozen countries by using millions of user search terms had an accuracy of 97% and identified a flu spike 2 weeks earlier than the Centers for Disease Control and Prevention. However, it was soon discovered that their techniques were frequently overpredicting flu, aggregating across multiple illnesses, and had substantial problems related to overfitting to the data. The initiative is not currently publicly active.

The $1 million Netflix Prize made a similarly large splash in the media and data science communities by offering a large cash award to the team that improved their movie recommendation algorithm. The winning team developed an algorithm that made the Netflix recommendation system 10% better. However, Netflix never implemented the winning team's algorithm due to the engineering complexity involved in deploying it. This Prize continues to be lauded as a prime example of the promise of big data and collaborative teams in data science despite this failure.

S. Rose (✉)
Department of Health Care Policy, Harvard Medical School, 180 Longwood Ave, Boston, MA 02115, USA
e-mail: rose@hcp.med.harvard.edu

M. J. van der Laan
Division of Biostatistics and Department of Statistics, University of California, Berkeley, 101 Haviland Hall, #7358, Berkeley, CA 94720, USA
e-mail: laan@berkeley.edu

© Springer International Publishing AG 2018
M.J. van der Laan, S. Rose, *Targeted Learning in Data Science*,
Springer Series in Statistics, https://doi.org/10.1007/978-3-319-65304-4_1

Data science is moving toward analytic systems that can take large data sets and estimate quantities of interest both quickly and robustly, incorporating advances from the fields of statistics, machine learning, and computer science. These two recent examples demonstrate that underpinning a big data system of this nature must be a methodological grounding in statistical theory *combined* with computational implementation that is fast, flexible, and feasible. This text on targeted learning is aligned with these goals and describes empirical techniques suitable for big data to estimate a number of complex parameters while remaining computationally feasible.

1.1 Learning from (Big) Data

Targeted learning focuses on efficient machine-learning-based substitution estimators of parameters that are defined as features of the probability distribution of the data, while additionally providing inference via bootstrapping or influence curves. Targeted learning is a broad framework that includes targeted maximum likelihood estimators (TMLEs) for effect estimation questions and super learning, an ensembled machine learning technique, for prediction. TMLEs build on the literature in loss-based estimation for infinite-dimensional parameters in order to target lower-dimensional parameters, such as effect parameters. These estimators are constructed such that the remaining bias for the effect target feature is removed. Super learning is completely integrated into the estimation of the relevant components of the TMLE algorithm. TMLEs have many desirable statistical properties, including being double robust, well-defined substitution estimators. Targeted learning uniquely solves the enormous challenge of combining data-adaptive estimation with formal statistical inference.

There has been a concerted effort in the scientific community to address issues that can impact the soundness of research, including the design of experimental and nonexperimental studies and the statistical analyses used to evaluate these studies. Targeted learning contributes critically to this area by focusing on prespecified analytic plans and algorithms that make realistic assumptions in more flexible nonparametric or semiparametric statistical models. The goal is to take our knowledge about the data and underlying data-generating mechanism to precisely describe our observational unit and model, while accurately translating the research question into a statistical estimation problem. Targeted statistical learning machines then take our data and knowledge as inputs into the system, while using rigorous a priori specified evaluation benchmarks and estimators grounded in theory to produce interpretable policy-relevant results.

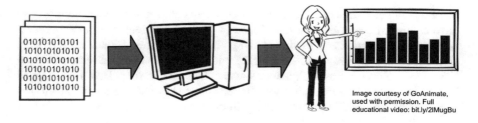

Image courtesy of GoAnimate,
used with permission. Full
educational video: bit.ly/2IMugBu

*This is all computationally efficient and practice focused. The idea being that
we take the theoretically optimal and make sure it translates into a fast and
user-friendly tool.*

Over the last decade, targeted learning has been established as a reliable framework for constructing effect estimators and prediction functions. The continued development of targeted learning has led to new solutions for existing problems in
many data structures in addition to discoveries in varied applied areas. This has
included work in randomized controlled trials, parameters defined by a marginal
structural model, case-control studies, collaborative TMLE, missing and censored
data, longitudinal data, effect modification, comparative effectiveness research, aging, cancer, occupational exposures, plan payment risk adjustment, and HIV, as well
as others. In many cases, these studies compared targeted learning techniques to
standard approaches, demonstrating improved performance in simulations and real-
world applications.

1.2 Traditional Approaches to Estimation Fail

While there are many methods available for classic cross-sectional studies, such as
traditional parametric regression and several off-the-shelf statistical machine learning techniques, there is a dearth of methodology for the complex longitudinal studies
found in data science disciplines. These methods fail and their assumptions break
down in cross-sectional studies, and this is exacerbated when applied to complex
data types, such as those involving time-dependent treatments, networks, or streaming data. This book aims to fill this gap, by presenting targeted learning methods tailored to handle such difficult questions. The first *Targeted Learning* book addressed
the previously mentioned methods for cross-sectional studies and the first estimators
for TMLE in longitudinal studies, demonstrating the advantages of targeted learning approaches for many data structures. We present additional novel advances here,
hence the subtitle *Causal Inference for Complex Longitudinal Studies.*

The general wisdom has also been that statistical inference was not possible in the
context of data-adaptive (i.e., machine-learning-based) estimation in nonparametric

or semiparametric models. Let's make this statement more concrete. Suppose we have computed a machine-learning-based fit of the conditional mean of a clinical outcome as a function of a treatment and patient characteristics in an observational study. We can use an ensemble learner for this; one that combines a library of algorithms and relies on cross-validation, such as the super learner. This fit is mapped into an estimate of the treatment-specific mean by (1) evaluating the predicted outcome under the specified treatment condition and (2) averaging these predictions across all n subjects in the sample.

Historically, the default approach has not been to use machine learning; instead estimating the regression with a maximum likelihood estimator based on a parametric regression model. Under this setting, the resulting treatment-specific mean is a simple function of the maximum likelihood estimator of the unknown regression coefficients. As a consequence, if the regression model is correctly specified, this maximum likelihood estimator of the treatment-specific mean is asymptotically linear. (This means that the maximum likelihood estimator minus the true treatment-specific mean equals an empirical mean of its influence curve up to a negligible remainder.) As a result, it is approximately normally distributed with mean the true treatment-specific mean and variance equal to the variance of the influence curve divided by the sample size. Confidence intervals are constructed analogue to confidence intervals based on sample means. However, in practice, we know that this parametric model is misspecified, and therefore the maximum likelihood estimator is normally distributed, but biased, and the 95% CIs will have asymptotic coverage equal to zero.

If we use a machine learning algorithm, as initially proposed above, then the estimator of the treatment-specific mean will generally not be normally distributed and will have a bias that is larger than $1/\sqrt{n}$. Because of this, the difference between the estimator and its true value, standardized by $\sqrt{n}$, converges to infinity! Since the sampling distribution of the estimator is generally not well approximated by a specified distribution (such as a normal distribution), statistical inference based on such a limit distribution is not an option.

Remarkably, a minor targeted modification of the machine-learning-based fit may make the resulting estimator of the treatment-specific mean asymptotically linear with influence curve equal to the efficient influence curve. Thus, this minor modification maps an initial estimator (of the data distribution, or its relevant part, such as the regression function in our example) for which its substitution estimator of the target parameter is generally overly biased and not normally distributed into an updated estimator for which the substitution estimator is approximately unbiased and has a normal limit distribution with minimal variance.

1.3 Targeted Learning in Practice

There are high-profile examples of the benefits of targeted learning in varied real-world analyses. For example, data scientists at Pandora implemented targeted learning to discover that streaming music spins increase music sales by 2.3% for new

Table 1.1 Top five targeted learning effect estimates and estimates from parametric regression

Medical condition	Targeted learning	Parametric regression
Multiple sclerosis	67,011	30,715
Congestive heart failure	19,904	4131
Lung, brain, and severe cancers	19,233	24,528
Major depressive and bipolar	15,398	3498
Chronic hepatitis	10,530	5539

music and 2.7% for catalog music. This study was discussed in *Billboard Magazine*. One particular area where targeted learning has been used with frequency is in health care. How can targeted learning improve health care? In work published in *Lancet Respiratory Medicine*, investigators developed a super learner for intensive care units to predict mortality with improved performance over severity scores. The algorithm is available in a user-interface online for implementation by clinicians. In another study, published in *World Psychiatry*, a novel function for predicting PTSD after traumatic events was generated. This algorithm had extraordinary performance, placing 96% of PTSD outcomes in the top 10% of predicted values. In a recent computational health economics analysis using a large health record claims database, the impact of individual medical conditions on total health care spending was examined. Targeted learning estimators for effect estimation ranked the medical condition categories based on their contributions to total health care spending, controlling for demographic information and other medical conditions. The impact of medical conditions on health care spending has largely been examined in parametric regression formulas for plan payment risk adjustment and aggregated means without confounder adjustment. The results of this study demonstrated that multiple sclerosis, congestive heart failure, severe cancers, major depressive and bipolar disorders, and chronic hepatitis are the most costly medical conditions (see Table 1.1). In contrast, parametric regression formulas for plan payment risk adjustment differed nontrivially both in the size of effect estimates and relative ranks. If current risk-adjustment formulas are not accurately capturing the incremental effects of medical conditions, selection incentives to health insurers may remain. We refer to Sect. 1.6 for additional references to earlier work.

1.4 The Statistical Estimation Problem

We present a simplified in vitro fertilization (IVF) example here to introduce longitudinal statistical estimation problems where we estimate the probability of success (i.e., live birth resulting from embryo transfer) of a program of at most two IVF cycles, controlling for time-dependent confounders. Infertility is a global public health issue and various treatments are available. IVF is an increasingly common treatment method, but accurately assessing the success of IVF programs has proven challenging since they consist of multiple cycles.

1.4.1 Data

Consider vectors of covariates L_t, for each time t ($t = 0, \ldots, T + 1$). Baseline covariates are denoted by L_0. We focus on a specific data structure for our IVF study, for illustrative purposes, where we have interventions only at two sequential time points. This data structure is a simple extension from cross-sectional data (the case with an intervention at a single time point).

For our data structure, covariates at each time point t are $\bar{L} = (L_0, L_1, L_2)$, and $T = 1$. The set of covariates $\bar{L}$ is also referred to as the set of states in the sequential decision process literature, although subscript notation, such as $L_{1:2}$ is also used to indicate a specific subset of the covariate set. The covariates at L_0 include maternal age, IVF unit, number of oocytes harvested, number of embryos transferred or frozen, and indicators of pregnancy. L_1 and L_2 encode whether the IVF cycle was successful or not. The set of interventions (also called "actions") is denoted by $\bar{A} = (A_0, A_1)$, where the random variable A_t is the intervention at time t. In the IVF study, each A_t in $\bar{A}$ will be binary and indicates whether the IVF cycle was attempted. By convention, if $A_t = 0$ then $L_{t+1} = 0$ and if $L_1 = 1$ then $L_2 = 1$.

One can then represent the data on one randomly sampled subject as a time-ordered data structure:

$$O = (L_0, A_0, L_1, A_1, L_2),$$

where it is assumed that L_t occurs before A_t. We denote the final measurement L_2 by Y, which represents the outcome of interest. We consider the case of Y being binary valued, for simplicity. The sample is comprised of n i.i.d. draws $O_1, \ldots, O_n$ of the random variable O. Realizations of these random variables are denoted $o_1, \ldots, o_n$. The probability distribution of O can be factorized according to the time-ordering of the data:

$$p(O) = p(L_0) \times p(L_1 \mid A_0, L_0) \times p(Y \mid A_1, L_1, A_0, L_0)$$
$$\times p(A_0 \mid L_0) \times p(A_1 \mid L_1, A_0, L_0).$$

1.4.2 Model and Parameter

We assume a nonparametric statistical model $\mathcal{M}$, which contains the possible set of probability distributions, for the observed data-generating distribution. The parameter of interest, the probability of success (i.e., live birth resulting from embryo transfer) of a program of at most two IVF cycles, can be written as

$$\Psi(P_0) = E_{P_0}\Big(\sum P_0(Y = 1 \mid A_{0:1} = 1, L_{0:1} = l_{0:1})$$

$$\times P_0(L_1 = 1 \mid A_0 = 1, L_0 = l_0)\Big).$$

Under causal identifiability assumptions we discuss in Chap. 2, the causal parameter can be written

$$P(Y_{(1,1)} = 1),$$

and is equal to $\Psi(P_0)$, where $Y_{(1,1)}$ is the counterfactual outcome under the intervention $A_{0:1} = 1$.

1.4.3 Targeted Minimum Loss-Based Estimators

A targeted *minimum loss-based* estimator (TMLE) can be established for this research question to estimate $P(Y_{(1,1)} = 1)$. The TMLE framework is an incredibly general system defined by a loss function, initial estimator, and least favorable submodel through the initial (or current) estimator. Precisely, it requires:

1. A target parameter defined as a mapping from a (typically) infinite dimensional parameter of the probability distribution of the unit data structure into the parameter space,
2. Deriving the efficient influence curve of the pathwise derivative of the target parameter,
3. Stipulating a loss function,
4. Specifying a least favorable submodel through an initial (or current) estimator of the parameter such that the linear span of the loss-based score when the fluctuation is zero includes the efficient influence curve, and
5. An algorithm for the iterative minimization of the loss-specific empirical risk over the fluctuation parameters of the least favorable parametric submodel and updating of the initial (or current) estimator.

The iterative minimization will be carried out until the maximum likelihood estimators of the fluctuation parameters are close to zero. By the generalized loss-based score condition on the submodel and loss function, the resulting TMLE of the infinite dimensional parameter solves the efficient score equation. This gives us the basis for the double robustness and asymptotic efficiency of the corresponding substitution estimator of the target parameter. Targeted *maximum likelihood* estimators are one type of TMLE. In Chap. 5, it is shown that one can always select a least favorable submodel so that a single minimization of the loss-specific empirical risk suffices to solve the efficient score equation. These special types of local least favorable submodels are called universal least favorable submodels.

1.4.4 Other Common Estimation Problems

Rule-Specific Mean. When studying clinical questions in longitudinal observational data, it is often of interest to evaluate treatment rules that extend over time, referred to as dynamic rules (sometimes regimes or regimens). The larger mathematical sciences context of dynamic rules falls within sequential decision theory, where a key component is that the best intervention decision for a specific time point may differ when one considers an immediate outcome versus a long-term outcome. This type of decision making is common in statistics, especially for medical and public health questions, and therefore methods to estimate the optimal dynamic rule are of considerable importance. The rules considered may also have complex classifiers. For example, consider the question of when to start antiretroviral treatment among therapy-naive HIV-infected individuals in the United States. Here, we wish to consider a set of prespecified thresholds for CD4 count (e.g., 200–500 cells/mm^3 in intervals of 50), where falling below the threshold indicates one should start antiretroviral treatment. Each dynamic rule will be indexed by both a CD4 threshold.

Let $\mathcal{D} = (d_1, \ldots, d_K)$ be the set of dynamic rules we consider. Each dynamic rule d_k encodes a time sequence $d_{k,t}$ ($t = 1, \ldots, T$) of rules, where $d_{k,t}$ represents the function mapping a patient's previously measured covariates $\bar{L}_t$ to the treatment $a(t)$ that should be followed at time t. We suppress the subscripts on an individual rule d for notational simplicity when removing it does not cause ambiguity. An individual is said to be following rule d through time t if the interventions received are the interventions indicated by rule d. Let Y^d be the (counterfactual) value that would have been observed had the subject been set to follow rule d at all time points. Our goal is to determine the best dynamic rule d^*, i.e., the rule in $\mathcal{D}$ that maximizes the expected value of the potential outcome Y^d. We first consider the problem of estimating, for each rule $d \in \mathcal{D}$, the population mean of Y had everyone followed rule d, i.e., $E(Y^d)$. Since Y is binary in our example data structure, this is equivalent to the probability that $Y^d = 1$. Under a set of strong identifiability assumptions, including the assumption of no unmeasured confounders, $E(Y^d)$ can be represented as a function of the observed data-generating distribution P, using the g-computation formula:

$$\Psi^d(P) = \sum_{l_0} \sum_{l_1} P(Y = 1 \mid A_1 = d(\bar{l}_1), \bar{L}_1 = \bar{l}_1, A_0 = d(l_0))$$
$$\times P(L_1 = l_1 \mid A_0 = d(l_0), L_0 = l_0) P(L_0 = l_0).$$

We provided sufficient detail here on the data structure and statistical estimation problem for dynamic rules as we describe this study as a worked example in several places in Chaps. 2–4

Community Randomized Trials. In community randomized trials we are interested in the setting that interventions are assigned to a small number of communities, with covariate and outcome data collected on a random sample of individuals from each of the communities. Drawing on our previous work in biased sampling, we can identify interesting parameters, such as causal contrasts, in these trials. This data structure is described in more detail in Chap. 13.

Networks. Network data is increasingly common. Starting with disease transmission and then email and now exploding with social networks, we regularly observe populations of causally connected units according to a network. The data structure is typically longitudinal, with time-dependent exposures and covariates. These data structures and parameters are described in more detail in Chaps. 20 and 21.

1.5 Roadmap for Targeted Learning

The first four chapters of this book provide critical foundational material on targeted learning for longitudinal data, including the targeted learning road map and prediction and causal inference estimation problems (Fig. 1.1). These first chapters are guided introductions to main concepts through the focus on the data structure $O = (L_0, A_0, L_1, A_1, L_2) \sim P_0$, a nonparametric statistical model $\mathcal{M}$, and the causal parameter $\Psi^d(P)$. This initial chapter motivated the need for new methods to handle complex longitudinal data science problems and introduce the data, model, and target parameter. The road map for targeted learning will be further explained in Chaps. 2–4.

Defining the Model and Target Parameter. A structural causal model (SCM) is a model for underlying counterfactual outcome data, representing the data one would generate in an ideal experiment. This translates knowledge about the data-generating process into causal assumptions. The SCM also generates the observed data O and allows us to determine what additional assumptions are needed in order to obtain identifiability of the causal effect from the observed data.

Super Learning for Prediction. We need flexible estimators able to learn from complex data, and we introduce ensemble super learning for longitudinal structures. Super learning can be integrated within effect estimation or used as a standalone tool for prediction problems. Some previous knowledge of cross-validation and machine learning will be beneficial, such as Chap. 3 from the first *Targeted Learning* book.

TMLE. With TMLE we are able to target (causal) effect parameters by making an optimal bias–variance tradeoff for the parameter of interest, instead of the overall probability distribution P_0. These estimators have many desirable statistical properties, and in some cases, are the only available estimators for certain complex parameters.

This brief teaser is provided as a guidepost to the upcoming chapters so readers can anticipate where we are headed, how the roadmap fits together, and why the material is presented in this chronology.

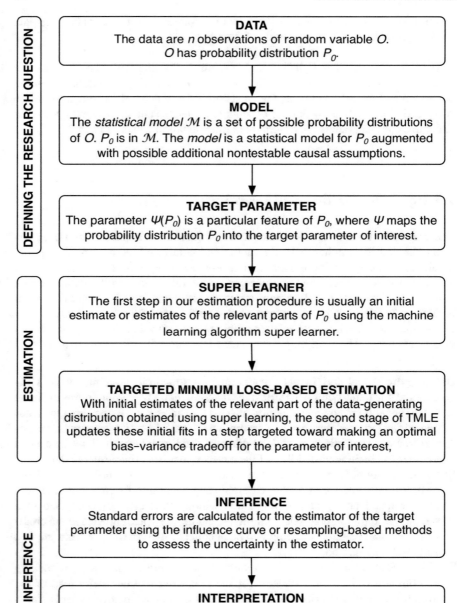

Fig. 1.1 Road map for targeted learning

1.6 Notes and Further Reading

We motivated this first chapter by presenting the challenges of real-world data science in Google Flu Trends and the Netflix Prize. More information about the history of Google Flu Trends can be found in *Wired* magazine (Lazer and Kennedy 2015) and the Netflix Prize background is described in *Forbes* (Holiday 2012). Crucial foundational material on targeted learning can be found in the first *Targeted Learning* book published seven years ago (van der Laan and Rose 2011).

In Sect. 1.1 we discuss the many areas where targeted learning methods have led to new solutions for existing problems. Explicit citations for these areas follow: randomized controlled trials (Rubin and van der Laan 2008; Moore and van der Laan 2009a,b,c; Rosenblum and van der Laan 2009), parameters defined by a marginal structural model (Rosenblum and van der Laan 2010a), case-control studies (van der Laan 2008a; Rose and van der Laan 2008, 2009, 2014a,b), collaborative TMLE (van der Laan and Gruber 2010; Gruber and van der Laan 2010a; Stitelman and van der Laan 2010), missing and censored data (Stitelman and van der Laan 2010; Rose and van der Laan 2011), effect modification (Polley and van der Laan 2009; Stitelman and van der Laan 2011), longitudinal data (van der Laan 2010a; van der Laan and Gruber 2012), networks (van der Laan 2014a), community-based interventions (van der Laan 2010c), comparative effectiveness research (Neugebauer et al. 2014a; Kunz et al. 2017), variable importance for biomarkers and genomics (Bembom et al. 2009; Wang et al. 2011a,b, 2014; Tuglus and van der Laan 2011; Wang and van der Laan 2011), aging (Bembom and van der Laan 2007; Rose 2013), cancer (Polley and van der Laan 2009), occupational exposures (Chambaz et al. 2014), health economics (Rose 2016; Rose et al. 2017; Shrestha et al. 2018), and HIV (Rosenblum et al. 2009). The paper by Rose et al. (2017) is also an example of the targeted learning framework in algorithmic fairness, accountability, and transparency. They demonstrated how insurers could use drug claims with ensemble machine learning to identify 'unprofitable' enrollees, despite protections for preexisting conditions, and then target them for disenrollment. The increasingly pervasive use of algorithms in society has broad risks, for example, because there are typically biases imbedded within the data. For a short introduction to algorithmic fairness with respect to criminal justice reform we refer readers to Lum (2017).

The computational health economics project summarized in Sect. 1.3 is discussed in detail elsewhere (Rose 2018). The papers described in Sect. 1.3 are Peoples (2014); Pirracchio et al. (2015); Kessler et al. (2014). A tutorial on TMLE in a point treatment setting with continuous outcome, geared toward an applied public health audience, has also been published (Schuler and Rose 2017). For a binary outcome, see Chap. 4 of *Targeted Learning* (2011).

We covered the problem of inference in the context of machine learning, and how targeted learning can address these shortcomings, in Sect. 1.2. We provide additional background here, specifically regarding how the bootstrap can fail for the purposes of inference with machine learning. The nonparametric bootstrap estimates the sampling distribution of an estimator with the sampling distribution of the estimator when sampling from the empirical distribution. That is, one estimates the data distribution with the empirical distribution and one hopes that the convergence of the empirical distribution to the true data distribution translate into convergence of the sampling distribution of the estimator. It makes sense that this method would work well if the sampling distribution of the estimator only depends on smooth functions of the data distribution, but that it can be expected to fail when it depends in an essential way on the density of the data distribution (which is poorly estimated by a discrete empirical distribution). Indeed, the nonparametric bootstrap is a valid consistent method for estimating the sampling distribution if the estimator is a nicely (so called Hadamard or compact) differentiable functional of the empirical probability distribution (Gill 1989; van der Vaart and Wellner 1996). On the other hand, estimators that rely on smoothing, model selection, cross-validation or other forms of machine learning are not Hadamard differentiable functionals of the data at all, so that the nonparametric bootstrap can be expected to be inconsistent.

Chapter 2
Defining the Model and Parameter

Sherri Rose and Mark J. van der Laan

We are often interested in the estimation of a causal effect in data science, as well as an assessment of the uncertainty for our estimator. In Chap. 1, we described the road map we follow to estimate causal effects in complex data types for realistic research questions. This chapter details the formal definition of the model and target parameter, which will vary depending on your research question. However, the concepts presented here will be carried throughout the book for multiple parameters, and the template is general.

We encourage readers to familiarize themselves with basic concepts in causal inference prior to reading this chapter, such as Judea Pearl's text *Causality*, published in a second edition (Pearl 2009a). Chapter 2 of the first *Targeted Learning* book summarizes key material from Pearl's book for point treatment data structures for an average treatment effect parameter and is a shorter piece of background material compared to Pearl's book. We do not repeat all of that basic content here, but do provide the material needed for tackling the complex causal questions we wish to target. A crucial component for readers to take away from this chapter is that causal inference requires both a causal model to define the causal effect as a target parameter of the distribution of the data *and* robust semiparametric efficient estimation. This book focuses almost exclusively on the latter, estimation, while the work of Pearl the former.

S. Rose (✉)
Department of Health Care Policy, Harvard Medical School, 180 Longwood Ave, Boston, MA 02115, USA
e-mail: rose@hcp.med.harvard.edu

M. J. van der Laan
Division of Biostatistics and Department of Statistics, University of California, Berkeley, 101 Haviland Hall, #7358, Berkeley, CA 94720, USA
e-mail: laan@berkeley.edu

© Springer International Publishing AG 2018
M.J. van der Laan, S. Rose, *Targeted Learning in Data Science*,
Springer Series in Statistics, https://doi.org/10.1007/978-3-319-65304-4_2

Key Definitions and Notation:

- Statistical model $\mathcal{M}$ is a collection of possible probability distributions P.
- P_0 is the true distribution of O.
- Definition of a target parameter requires specification of a mapping Ψ applied to P_0. Ψ maps any $P \in \mathcal{M}$ into a vector of numbers $\Psi(P)$. We write the mapping as $\Psi : \mathcal{M} \to \mathbb{R}^d$ for a d-dimensional parameter.
- ψ_0 is the evaluation of $\Psi(P_0)$, i.e., the true value of our parameter.

The statistical estimation problem is to map the observed data $O_1, \ldots, O_n$ into an estimator of $\Psi(P_0)$ that incorporates the knowledge that $P_0 \in \mathcal{M}$, accompanied by an assessment of the uncertainty in the estimator.

(*See Chap. 2 of* Targeted Learning (2011) *for additional background.*)

The data $O_1, \ldots, O_n$ consist of n i.i.d. copies of a random variable O with true probability distribution P_0. For our data structures from Chap. 1, such as

$$O = (L_0, A_0, L_1, A_1, L_2 = Y),$$

with vector of covariates $L_{0:1}$, vector of interventions $A_{0:1}$, and outcome $L_2 = Y$, uppercase letters represent random variables and lowercase letters are a specific value for that variable. With all discrete variables, $P_0(L_0 = l_0, A_0 = a_0, L_1 = l_1, A_1 = a_1, Y = y)$ assigns a probability to any possible outcome (l_0, a_0, l_1, a_1, y) for $O = (L_0, A_0, L_1, A_1, Y)$.

We will now move forward to define a *model* that is augmented by nontestable causal assumptions, building on the underlying minimal assumptions of our *statistical model*. This allows us to define a parameter of interest that can be interpreted causally, as well as determine the necessary assumptions for establishing identifiability of the causal parameter from the distribution of the observed data. Lastly, we commit to a statistical model and target parameter.

2.1 Defining the Structural Causal Model

We describe a set of endogenous variables $X = (X_j : j)$, where these endogenous variables are those where our structural causal model (SCM) will define them as a deterministic function of other endogenous variables and exogenous error. The exogenous variables are given by $U = (U_{X_j} : j)$ and are never observed. For each X_j we specify the parents $Pa(X_j)$ of X_j among the other X variables. The endogenous variables X often include the observables O, but may also include nonobservables. We make the assumption that each X_j is a function of $Pa(X_j)$ and an exogenous U_{X_j}:

$$X_j = f_{X_j}(Pa(X_j), U_{X_j}), \ j = 1 \ldots, J.$$

The functions f_{X_j}, together with the joint distribution of U, specify the data-generating distribution of (U, X) as they describe a deterministic system of structural equations that deterministically maps a realization of U into a realization of X. These functions f_{X_j} and the joint distribution of U may be unspecified, or we may have subject-matter knowledge that informs our willingness to specify them in a more restrictive way. It is unlikely our knowledge will support a fully parametric SCM.

> **A SCM Is a Statistical Model for the Random Variable** (U, X). The set of possible data-generating distributions of (U, X) is defined by varying:
>
> 1. the collection of functions $f = (f_{X_j} : j)$ over all permitted forms, and
> 2. the distribution of the errors U over all possible error distributions.

In our IVF study and when to start HIV treatment study discussed in Chap. 1, we have $j = 1, \ldots, J$, where $J = 5$ and all variables in X observed. Thus, $X = (X_1, X_2, X_3, X_4, X_5)$. We then rewrite X as $X = (L_0, A_0, L_1, A_1, Y)$ with $X_1 = L_0$, $X_2 = A_0$, $X_3 = L_1$, $X_4 = A_1$, $X_5 = Y$. The vectors L_0 and L_1 may contain both binary and continuous variables, with A_0, A_1, and Y binary for both examples. It is important to explicitly remark on the time ordering in the generation of these variables:

$$L_0 \to A_0 \to L_1 \to A_1 \to Y.$$

Focusing on our IVF study, baseline variables L_0 are measured, including maternal age, IVF unit, number of oocytes harvested, number of embryos transferred or frozen, and indicators of pregnancy. A_0 occurs next, sequentially, and establishes whether an IVF cycle was attempted. L_1 then follows, indicating whether the IVF cycle was successful. A_1 occurs after L_1, and indicates whether a second IVF cycle was attempted, followed by Y, whether any IVF cycle attempted at L_1 was successful. Recall that, by convention, if $A_t = 0$ then $L_{t+1} = 0$ and if $L_1 = 1$ then $Y = 1$.

Thus, we have the functions $f = (f_{L_0}, f_{A_0}, f_{L_1}, f_{A_1}, f_Y)$ and the exogenous variables $U = (U_{L_0}, U_{A_0}, U_{L_1}, U_{A_1}, U_Y)$. Our structural equation models are given as

$$
\begin{aligned}
L_0 &= f_{L_0}(U_{L_0}), \\
A_0 &= f_{A_0}(L_0, U_{A_0}), \\
L_1 &= f_{L_1}(L_0, A_0, U_{L_1}), \\
A_1 &= f_{A_1}(L_0, A_0, L_1, U_{A_0}), \\
Y &= f_Y(L_0, A_0, L_1, A_1, U_Y).
\end{aligned}
\tag{2.1}
$$

We choose not to make restrictive assumptions about the functional form of f_{L_0}, f_{A_0}, f_{L_1}, f_{A_1}, and f_Y; the functions f are nonparametric. We will additionally assume in this study that there are *no unmeasured confounders*. Thus, to be explicit, in (2.1), we assume that the data were generated by:

1. Drawing U from probability distribution P_U ensuring that $U_{A_{0:1}}$ is independent of U_Y, given $L_{0:1}$,
2. Generating L_0 as a deterministic function of U_{L_0},
3. Generating A_0 as a deterministic function of L_0 and U_{A_0},
4. Generating L_1 as a deterministic function of L_0, A_0, and U_{L_1},
5. Generating A_1 as a deterministic function of L_0, A_0, L_1, and U_{A_1},
6. Generating Y as a deterministic function of L_0, A_0, L_1, A_1, and U_Y.

We make the assumption that our *observed* data structure $O = (L_0, A_0, L_1, A_1, Y)$ is a realization of the endogenous variables (L_0, A_0, L_1, A_1, Y) generated by the structural equations in this system and defines the SCM for O.

Our SCM represents a set of *nontestable causal assumptions* made regarding our belief about how the data were generated. As discussed in our earlier treatment of SCMs for single-time-point interventions in the introductory material for *Targeted Learning* (2011), the SCM for O also involves defining the relationship between the random variable (U, X) and O, such that the SCM for the full data implies a parameterization of the probability distribution of O in terms of f and P_U of U. Each possible probability distribution $P_{U,X}$ of (U, X) in the SCM for the full data is indexed by a choice of error distribution P_U and a set of deterministic functions $(f_{X_j} : j)$ and implies a probability distribution $P(P_{U,X})$ of O. Thus, the SCM for the full data implies a parameterization of the true probability distribution of O in terms of a true probability distribution of (U, X), so that the statistical model $\mathcal{M}$ for P_0 of O can be represented as $\mathcal{M} = \{P(P_{U,X}) : P_{U,X}\}$, with $P_{U,X}$ varying over all probability distributions of (U, X) allowed in the SCM. If this $\mathcal{M}$ is nonparametric, none of the causal assumptions encoded by our SCM are testable in the observed data.

If subjects had instead been randomized to IVF treatment, our structural equation models might be given as

$$
\begin{aligned}
L_0 &= f_{L_0}(U_{L_0}), \\
A_0 &= f_{A_0}(U_{A_0}), \\
L_1 &= f_{L_1}(L_0, A_0, U_{L_1}), \\
A_1 &= f_{A_1}(A_0, U_{A_0}), \\
Y &= f_Y(L_0, A_0, L_1, A_1, U_Y).
\end{aligned} \tag{2.2}
$$

In (2.2), A_0 is evaluated only as a deterministic function of U_{A_0}. A_1 is a deterministic function of only A_0 and U_{A_0}. We hypothesized a randomization design that did depend on previous treatment, but not other exogenous variables. While this text largely focuses on observational studies without randomization, the targeted learning framework is quite general, and we have three chapters on randomized trials in Part III.

2.2 Causal Graphs

In the previous section, we described SCMs as a systematic way to assign values
to a set of variables from random input, define required causal assumptions, and
assess the identifiability of the causal target parameter. Causal graphs are another
popular way to represent some of the assumptions encoded in our SCM. All the
causal graphs in this book are directed acyclic graphs, with only one arrow on the
edges that connect the nodes and no closed loops. However, with longitudinal data,
networks, and other complex data structures, this representation can become visually complicated quickly. The nonparametric structural equations in the previous
section do not have this drawback, and may be preferable in some settings. Additionally, causal graphs are not specific for the target parameter of interest, and
therefore identifiability assumptions from the causal graph may be stronger than
necessary.

We start by presenting a possible causal graph for (2.1) in Fig. 2.1, where we
make causal assumptions by defining $Pa(X_j)$ for each X_j and the joint distribution
P_U. The relationships given in f guide the connection of all $Pa(X_j)$ to X_j with a

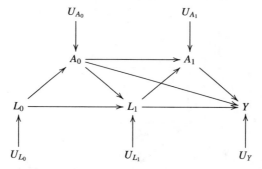

Fig. 2.1 A possible causal graph for (2.1)

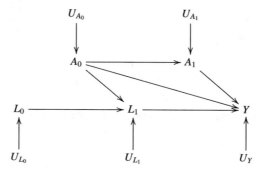

Fig. 2.2 A possible causal graph for (2.2)

directed arrow into X_j and a directed arrow into each X_j from each U_{X_j}. These assumptions are ideally made with our subject matter knowledge related to the scientific question of interest. A direct effect, such as that between L_0 and L_1, is illustrated by a directed arrow between two nodes. When we are uncertain whether there is a direct effect between two variables, our default is to include it, as the explicit absence of the arrow signals that a direct effect is known not to exist. For example, a possible causal graph for our SCM (2.2) is displayed in Fig. 2.2. Here there is an explicit absence of a directed arrow between L_0 and A_0 as well as between L_1 and A_1 due to our a priori knowledge regarding the randomization represented in the SCM.

Both Figs. 2.1 and 2.2 do not include any arrows between the endogenous errors $U = (U_{L_0}, U_{A_0}, U_{L_1}, U_{A_1}, U_Y)$. This indicates that a strong assumption about the joint independence of the endogenous error has been encoded as an assumption in (2.1) and (2.2). However, it is unlikely that this assumption is one we can make in practice. When we wish to reflect relationships between the endogenous variables U, they are represented using dashed double-headed arrows. If we make no assumptions about the distribution of P_U for (2.1), a causal graph would be given as drawn in Fig. 2.3. This figure has many so-called backdoor paths between our treatment nodes and the outcome Y. In order to isolate our causal effect of interest, we must block all unblocked backdoor paths from $A_{0:1}$ to Y.

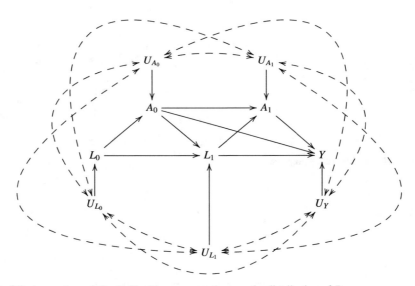

Fig. 2.3 A causal graph for (2.1) with no assumptions on the distribution of P_U

2.3 Defining the Causal Target Parameter

With an SCM for our data-generating mechanism, we now move toward defining the causal targeted parameter. To be very precise, we define this causal target parameter as a parameter of the distribution of the full-data (U, X) in the SCM. Formally, we denote the SCM for the full-data (U, X) by $\mathcal{M}^F$, a collection of possible $P_{U,X}$ as described by the SCM. $\mathcal{M}^F$, a model for the full data, is a collection of possible distributions for the underlying data (U, X). Ψ^F is a mapping applied to a $P_{U,X}$ giving $\Psi^F(P_{U,X})$ as the target parameter of $P_{U,X}$. This mapping needs to be defined for each $P_{U,X}$ that is a possible distribution of (U, X), given our assumptions encoded by the posited SCM. In this way, we state $\Psi^F : \mathcal{M}^F \to \mathbb{R}^d$, where $\mathbb{R}^d$ indicates that our parameter is a vector of d real numbers. The SCM $\mathcal{M}^F$ consists of the distributions indexed by the deterministic function $f = (f_{X_j} : j)$ and distribution P_U of U, where f and this joint distribution P_U are identifiable from the distribution of the full-data (U, X). Thus, the target parameter can also be represented as a function of f and the joint distribution of U.

For our IVF study with observed data $O = (L_0, A_0, L_1, A_1, Y)$ and SCM given in (2.1) with no assumptions about the distribution P_U. We can define $Y_{a_{0:1}} = f_Y(L_0, a_0, L_1, a_1, U_Y)$ as a random variable corresponding with intervention $A_{0:1} = a_{0:1}$ in the SCM. The marginal probability distribution of $Y_{a_{0:1}}$ is given by

$$P_{U,X}(Y_{a_{0:1}} = y) = P_{U,X}(f_Y(L_0, a_0, L_1, a_1, U_Y) = y).$$

Recall our statistical parameter of interest given in Chap. 1

$$\Psi(P_0) = E_{P_0}\Big(\sum P_0(Y = 1 \mid A_{0:1} = 1, L_{0:1} = l_{0:1})$$
$$\times P_0(L_1 = 1 \mid A_0 = 1, L_0 = l_0)\Big).$$

As we will discuss later, under a randomization and positivity assumption his statistical parameter equals the causal target parameter of the distribution of (U, X) given by

$$\Psi^F(P_{U,X}) = P_{U,X}(Y_{(1,1)} = 1).$$

2.3.1 Interventions

One can intervene upon our system defined by the SCM by setting the intervention nodes $A_{0:1}$ equal to some values $a_{0:1} \in \mathcal{A}$, where $\mathcal{A}$ is the set of possible values for our exposure IVF treatment. Intervening allows us to describe the data that would be generated by the system at the levels of our intervention variables. This is a critical concept because we define our causal target parameter as a parameter of the distribution of the data (U, X) under an intervention on the structural equations in f. The intervention defines a random variable that is a function of (U, X), so that the target parameter is $\Psi^F(P_{U,X})$.

In our IVF study, we can intervene on $A_{0:1}$ in order to observe what would have happened under specific exposures to IVF treatment. Notably, intervening on the SCM, changing the functions f_{X_j} for the intervention variables, does not change the other functions in f. For our SCM in (2.1) we can intervene on f_{A_0} and f_{A_1} and set both $a_0 = 1$ and $a_1 = 1$:

$$
\begin{aligned}
L_0 &= f_{L_0}(U_{L_0}), \\
A_0 &= 1, \\
L_1 &= f_{L_1}(L_0, 1, U_{L_1}), \\
A_1 &= 1, \\
Y_{(1,1)} &= f_Y(L_0, 1, L_1, 1, U_Y).
\end{aligned}
$$

The intervention defines a random variable that is a function of (U, X), namely, $Y_{a_{0:1}} = Y_{a_{0:1}}(U)$ for $a_0 = 1$ and $a_1 = 1$. Our target parameter is a parameter of the postintervention distribution, which is the probability distribution of the (X, U) under an intervention. Thus, the SCM for the full data allows us to define the random variable $Y_{a_{0:1}} = f_Y(L_0, a_0, L_1, a_1, U_Y)$ for each $a_{0:1}$, where $Y_{a_{0:1}}$ represents the outcome that would have been observed under this system for a particular subject under exposure $a_{0:1}$.

2.3.2 Counterfactuals

The "ideal experiment" where we observe each subject under all possible trajectories of exposure is not possible. Each subject only contributes one Y, the one observed under the exposure they experienced. Above, we intervened on $A_{0:1}$ to set $a_0 = 1$ and $a_1 = 1$ in order to generate the outcome for each subject under the condition that they received two rounds of IVF treatment. Recall that $Y_{a_{0:1}}$ represents the outcome that would have been observed under this system for a particular subject under exposure $a_{0:1}$ and we have $(Y_{a_{0:1}} : a_{0:1})$, with $a_{0:1} \in \mathcal{A}$. For each realization u in our study, which corresponds with an individual randomly drawn from the target population, we generate counterfactual outcomes $Y_{(1,1)}(u)$ by intervening on (2.1). The counterfactual outcomes are implied by our SCM; they are consequences of it. That is, $Y_{(1,1)}(u) = f_Y(L_0, 1, L_1, 1, u_Y)$. The random counterfactual $Y_{(1,1)}(U)$ is random through the probability distribution of U.

2.3.3 Establishing Identifiability

Are the assumptions we have already made enough to express the causal parameter of interest as a parameter of the probability distribution P_0 of the observed data? We want to be able to write $\Psi^F(P_{U,X,0})$ as $\Psi(P_0)$ for some parameter mapping Ψ, where we remind the reader that the SCM also specifies how the distribution P_0 of the observed data structure O is implied by the true distribution $P_{U,X,0}$ of (U, X). Since the true probability distribution of (U, X) can be any element in the SCM $\mathcal{M}^F$, and each such choice $P_{U,X}$ implies a probability distribution $P(P_{U,X})$ of O, this requires that we show that $\Psi^F(P_{U,X}) = \Psi(P(P_{U,X}))$ for all $P_{U,X} \in \mathcal{M}^F$.

This step involves establishing possible additional assumptions on the distribution of U, or sometimes also on the deterministic functions f, so that we can identify the target parameter from the observed data distribution. Thus, for each probability distribution of the underlying data (U, X) satisfying the SCM with these possible additional assumptions on P_U, we have $\Psi^F(P_{U,X}) = \Psi(P(P_{U,X}))$ for some Ψ. O is implied by the distribution of (U, X), such as $O = X$ or $O \subset X$, and $P = P(P_{X,U})$, where $P(P_{U,X})$ is a distribution of O implied by $P_{U,X}$.

Let us denote the resulting full-data SCM by $\mathcal{M}^{F*} \subset \mathcal{M}^F$ to make clear that possible additional assumptions were made that were driven purely by the identifiability problem, not necessarily reflecting reality. To be explicit, $\mathcal{M}^F$ is the full-data SCM under the assumptions based on real knowledge, and $\mathcal{M}^{F*}$ is the full-data SCM under possible additional causal assumptions required for the identifiability of our target parameter. We now have that for each $P_{U,X} \in \mathcal{M}^{F*}$, $\Psi^F(P_{U,X}) = \Psi(P)$, with $P = P(P_{U,X})$ the distribution of O implied by $P_{U,X}$ (whereas P_0 is the true distribution of O implied by the true distribution $P_{U,X,0}$).

Theorems exist that are helpful to establish such a desired identifiability result. For a particular intervention d on the A nodes, and for a given realization u, the SCM generates deterministically a corresponding value for $L_{1:2}$. We denote the resulting realization by $L_d(u)$ and note that $L_d(u)$ is implied by f and u. If $O = X$, and the distribution of U is such that, at each time point t, A_t is independent of L_d, given $Pa(A_t)$, then the g-formula expresses the distribution of L_d in terms of the distribution of O:

$$P(L_d = l) = \prod_{t=1}^{T} P(L_t = l_t \mid Pa_d(L_t)) = Pa_d(l_t)),$$

where $Pa_d(L_t)$ are the parents of L_t with the intervention nodes among these parent nodes deterministically set by intervention d. This so-called sequential randomization assumption can be established for a particular independence structure of U by verifying the backdoor path criterion on the corresponding causal graph implied by the SCM and this independence structure on U. The backdoor path criterion states that for each A_t, each backdoor path from A_t to an L_t node that is realized after A_t is blocked by one of the other L_t nodes.

In this manner, one might be able to generate a number of independence structures on the distribution of U that provide the desired identifiability result. That is, the resulting model for U that provides the desired identifiability might be represented as a union of models for U that assume a specific independence structure. The sequential randomization assumption is also referred to as the *no unmeasured confounders assumption*. We define confounders as those variables in X one must observe in O in order to establish the identifiability of the target parameter. We note that different such subsets of X may provide a desired identifiability result. If we return to our IVF study example and the structural equation models found in (2.1), the union of several independence structures allows for the identifiability of our causal target parameter by meeting the backdoor path criterion. The independence structure in Fig. 2.3 does not meet the backdoor path criterion.

2.3.4 Commit to a Statistical Model and Target Parameter

The identifiability result provides us with a purely statistical target parameter $\Psi(P_0)$ on the distribution P_0 of O. The full-data model $\mathcal{M}^{F*}$ implies a statistical observed data model $\mathcal{M} = \{P(P_{X,U}) : P_{X,U} \in \mathcal{M}^{F*}\}$ for the distribution $P_0 = P(P_{U,X,0})$ of O. This now defines a target parameter $\Psi : \mathcal{M} \to \mathbb{R}^d$. The statistical observed data model for the distribution of O might be the same for $\mathcal{M}^F$ and $\mathcal{M}^{F*}$. If not, then one might consider extending the Ψ to the larger statistical observed data model implied by $\mathcal{M}^F$, such as possibly a fully nonparametric model allowing for all probability distributions. In this way, if the more restricted SCM holds, our target parameter would still estimate the target parameter, but one now also allows the data to contradict the more restricted SCM based on additional doubtful assumptions.

For our IVF study, our corresponding statistical parameter $\Psi(P_0)$ is given

$$\Psi^F(P_{U,X,0}) = P(Y_{(1,1)} = 1)$$

$$= E_{P_0}\Big(\sum P_0(Y = 1 \mid A_{0:1} = 1, L_{0:1} = l_{0:1})$$

$$\times P_0(L_1 = 1 \mid A_0 = 1, L_0 = l_0)\Big) \equiv \Psi(P_0).$$

This identifiability result for the causal effect as a parameter of the distribution P_0 of O required making the sequential randomization assumption. This assumption might have been included in the original SCM $\mathcal{M}^F$, but, if one knows there are unmeasured confounders, then the model $\mathcal{M}^{F*}$ would be more restrictive by enforcing a randomization assumption that we believe to be incorrect.

Another required assumption is that of sequential positivity. In our IVF study, this means that the probability of IVF treatment at each of our two time points is nonzero, given covariate history. Without this assumption, the probabilities of $L_{1:2}$ in $\Psi(P_0)$ are not well defined. However, the positivity assumption is a more general name for the condition that is necessary for the target parameter $\Psi(P_0)$ to be well defined, and it often requires the censoring or treatment mechanism to have certain support.

2.3.5 Interpretation of Target Parameter

We may not have knowledge that supports the causal assumptions in the SCM, and be unwilling to rely on these additional assumptions. By assuming that the time ordering of observed variables $L_{0:1}$, $A_{0:1}$, and Y is correct:

$$L_0 \rightarrow A_0 \rightarrow L_1 \rightarrow A_1 \rightarrow Y,$$

our target parameters still represent an interesting and well-defined effect and can be interpreted as a variable importance measure

> The observed data parameter $\Psi(P_0)$ can be interpreted in two possibly distinct ways:
>
> 1. $\Psi(P_0)$ with $P_0 \in \mathcal{M}$ augmented with the truly reliable additional non-statistical assumptions that are known to hold (e.g., $\mathcal{M}^F$). This may involve bounding the deviation of $\Psi(P_0)$ from the desired target causal effect $\Psi^F(P_{U,X,0})$ under a realistic causal model $\mathcal{M}^F$ that is not sufficient for the identifiability of this causal effect.
> 2. The truly causal parameter $\Psi^F(P_{U,X}) = \Psi(P_0)$ under the more restricted SCM $\mathcal{M}^{F*}$, thereby now including all causal assumptions that are needed to make the desired causal effect identifiable from the probability distribution P_0 of O.

2.4 Notes and Further Reading

We refer readers to the in-depth presentation of SCMs found in Pearl (2009a). This chapter builds and relies on Chaps. 2 and 24 of *Targeted Learning* (2011). Some content reappears from Chap. 2, with permission. The g-formula for identifying the distribution of counterfactuals from the observed data distribution, under the sequential randomization assumption, was originally published in Robins (1986).

Chapter 3
Sequential Super Learning

Sherri Rose and Mark J. van der Laan

Suppose a doctor is interested in predicting the individual outcomes for a group of patients under two particular drug regimens at two time points in the future. She is therefore asking, what would happen to each of these patient's future outcomes at these time points if I were to enforce drug regimen 1 or drug regimen 2? Which treatment will be better for the patients' efficacy outcomes? Which treatment will be better for the patients' safety outcomes? Prediction problems can be longitudinal in nature, generally, and we frequently wish to understand what the mean outcome of patients with certain characteristics would be months or years in the future. Often, this is under the setting where we would hypothetically assign a particular treatment "rule" to the patients.

This chapter discusses sequential super learning for longitudinal data problems, and provides us with a framework for prediction at any time point. In Chap. 3 of *Targeted Learning* (2011), we introduced super learning for prediction in a simple data structure $O = (W, A, Y) \sim P_0$ and estimated the conditional mean $E[Y \mid A, W]$. Now, we are examining longitudinal data structures and more complex prediction questions. These types of questions in longitudinal data have a natural analog to our $E[Y \mid A, W]$ prediction question: $E(Y^d \mid W) = E(Y|A = d(W), W)$. With $O = (W, A, Y)$, there was only one observed outcome after baseline and one intervention node, making the prediction problem very simple: a single regression (estimated with super learning). We introduce the concept of *sequential prediction* and generalize the concepts and methodology presented in Chap. 3 of *Targeted Learning*

S. Rose (✉)
Department of Health Care Policy, Harvard Medical School, 180 Longwood Ave, Boston, MA 02115, USA
e-mail: rose@hcp.med.harvard.edu

M. J. van der Laan
Division of Biostatistics and Department of Statistics, University of California, Berkeley, 101 Haviland Hall, #7358, Berkeley, CA 94720, USA
e-mail: laan@berkeley.edu

© Springer International Publishing AG 2018

M.J. van der Laan, S. Rose, *Targeted Learning in Data Science*,
Springer Series in Statistics, https://doi.org/10.1007/978-3-319-65304-4_3

(2011) for counterfactual estimation of conditional means under Y^d, given baseline covariates, with multiple time point dynamic interventions d.

For these multiple time point interventions, $E(Y^d \mid L_0)$ can be estimated with sequential regressions, and these sequential regressions will also be critical components for the estimation of $E(Y^d)$ (and other parameters) with LTMLE. We could simply average over our estimator of $E(Y^d \mid L_0)$ to estimate $E(Y^d)$. However, in the next chapter we will discuss augmenting the sequential regression estimator of $E(Y^d)$ using targeted at each step in the LTMLE.

Therefore, we keep in mind that prediction questions are frequently the scientific goal, and sequential super learning is the appropriate stand-alone tool for these problems. However, we highlight that we are also interested in estimating a target parameter of the probability distribution of the data, and this will often be a target parameter that can be interpreted as a causal effect. In these settings we will implement an LTMLE. An integral component of this estimation procedure in research questions involving longitudinal data are sequential regression estimates of the relevant parts Q of P_0 that are needed to evaluate the target parameter. This step is presented in Chap. 3, with sequential super learning, and LTMLE will be presented in Chap. 4. Thus, this chapter focuses on the estimation of conditional means within the road map for targeted learning that are useful for both prediction and causal effect questions.

Effect Estimation vs. Prediction in Longitudinal Data

Both causal effect and prediction research questions in longitudinal data are inherently *estimation* questions. In the first, we are interested in estimating the causal effect of an intervention or dynamic process or other longitudinal effect question. For prediction, we are interested in generating a function to input the variables $(A_{0:t}, L_{0:t})$ and predict a value for the outcome, possibly under a dynamic process. These are clearly distinct research questions despite being frequently conflated. LTMLE involves prediction steps within the procedure, thus understanding the sequential super learner for prediction in longitudinal data is a core concept for both research questions.

3.1 Background: Ensemble Learning

As introduced in Chap. 1, ensemble learning has been developed for various data types and research questions, as well as applied in many substantive areas. The core prediction framework is an "ensembling" super (machine) learning approach that leverages cross-validation to produce an optimal weighted average of multiple algorithms. This solution solves the critical challenge for prediction: many algorithms are available, from decision trees to penalized regressions to neural networks, and any individual algorithm may have disparate performance in a given data set.

The historical issue has been: How do we know beforehand which algorithm will perform the best in our data? Even in similar data types, we may find that a logistic regression in a misspecified parametric model outperforms a decision tree in one study, but the decision tree outperforms the logistic regression in another study. There are many such examples in the data science literature.

Given our nonparametric statistical model, we may initially be drawn to nonparametric methods that smooth over the data without overfitting. However, a simple nonparametric estimator, such as local averaging, still requires partitioning the covariate space to define the smoothness of this regression estimator. Even with optimally selected partitions, a logistic regression in a misspecified parametric statistical model may outperform this nonparametric estimator if the true underlying data-generating distribution is very smooth.

These considerations led to substantial statistical work in ensembling, and ultimately the super learner. Super learning protects against a selecting a poorly performing single algorithm a priori. Instead, we consider many algorithms, and need not worry that our local averaging will be outperformed by a logistic regression as we include both, and many others. This is due to the fact that super learning constructs a prediction function that is the optimal weighted average of all considered algorithms, based on an a priori specified loss function.

Notable Applications

- *Publicly Available Data in R*: In 13 publicly available data sets from R, all with small sample sizes (ranging from 200 to 654) and a small number of covariates (ranging from 3 to 18), super learner outperformed each single algorithm studied. This study was notable for demonstrating in real data that the benefits of super learning do not require large samples and many covariates. Parametric linear regression was only the 8th best algorithm overall, of the single algorithms considered, out of 20 (Polley and van der Laan 2010; Polley et al. 2011).
- *Mortality Risk Scores*: This study generated an improved function for predicting mortality in an elderly population with super learning. The work was also notable for demonstrating that a small carefully collected cohort study one-tenth the size of a large health database (both in terms of subjects and number of covariates) generated a more accurate prediction function. Thus, the manuscript was an early contribution to the literature revealing the limitations of large electronic health databases to answer targeted scientific research questions (Rose 2013).
- *Mortality Risk Scores in ICUs*: Developing risk scores for mortality in intensive care units is a difficult problem, and previous scoring systems did not perform well in validation studies. A super learner developed for this problem had extraordinary performance, with area under the receiver operating characteristic curve of 94% (Pirracchio et al. 2015).
- *HIV RNA Monitoring*: This study demonstrated that implementing super learning with electronic health record data on medication adherence may be useful for identifying patients at a high risk of virologic failure (Petersen et al. 2015).
- *Plan Payment Risk Adjustment*: Current methods for establishing payment to health plans are fully parametric. The results of recent super learning work for

this problem indicate that a simplified risk adjustment formula selected via this nonparametric framework maintains much of the efficiency of a traditional larger formula. This could impact health insurers' ability to manipulate the system through aggressive diagnostic upcoding or fraud (Rose 2016).

The super learner has established, desirable statistical properties, discussed in detail in Chap. 3 of *Targeted Learning* (2011). Briefly, in realistic scenarios where none of the candidate algorithms in the super learner achieves the rate of convergence of an a priori correctly specified parametric statistical model, the super learner performs asymptotically as well (not only in rate, but also up to the constant) as the best choice among the possible weighted combinations. We restate this formally with the finite sample oracle inequality from van der Laan and Dudoit (2003):

Finite Sample Oracle Inequality. Given a collection of estimators (i.e., algorithms) $P_n \rightarrow \hat{Q}_k(P_n)$, the loss-function-based cross-validation selector is

$$k_n = \hat{K}(P_n) = \arg \min_k E_{B_n} P^1_{n,B_n} L(\hat{Q}_k(P^0_{n,B_n})),$$

where $B_n \in \{0, 1\}^n$ is a random variable that splits the data into a training set $\{i : B_n(i) = 0\}$ and validation set $\{i : B_n(i) = 1\}$, P^0_{n,B_n} is the empirical distribution of the training set, P^1_{n,B_n} is the empirical distribution of the validation set, and $L(\cdot)$ our dissimilarity measure: a loss function. The estimator that results is referred to as the discrete super learner: $\hat{Q}(P_n) = \hat{Q}_{\hat{K}(P_n)}(P_n)$.

We consider a loss function that satisfies

$$\sup_Q \frac{\mathrm{var}_{P_0}\{L(Q) - L(Q_0)\}}{P_0\{L(Q) - L(Q_0)\}} \leq M_2$$

and is uniformly bounded:

$$\sup_{O,Q} | L(Q) - L(Q_0) | (O) < M_1 < \infty,$$

where the supremum is over the support of P_0 and over the possible estimators of Q_0 that will be considered.

Under the assumption that our loss function L is uniformly bounded over the support of P_0, the remainder between the dissimilarity of the cross-validation selector and the dissimilarity of the oracle selector at fixed n hold uniformly in all data-generating distributions. This demonstrates that the cross-validation selector approximates the performance of the oracle selector by distance $\log(K(n))/n$. Precisely, for quadratic loss functions, the cross-validation selector satisfies the following oracle inequality:

(continued)

$$E_{B_n}\{P_0 L(\hat{Q}_{k_n}(P^0_{n,B_n})) - L(Q_0)\} \le (1 + 2\delta) E_{B_n} \min_k P_0\{L(\hat{Q}_k(P^0_{n,B_n})) - L(Q_0)\}$$
$$+2C(M_1, M_2, \delta)\frac{1 + \log K(n)}{np},$$

for $\delta > 0$, where the constant $C(M_1, M_2, \delta) = 2(1 + \delta)^2(M_1/3 + M_2/3)$, p is the proportion of subjects in the validation set, and $K(n)$ the number of algorithms in the collection. These results generalize for estimated loss functions $L_n(Q)$ that approximate a fixed loss function $L(Q)$.

3.2 Defining the Estimation Problem

Recall that we have vectors of covariates L_t, for each time t ($t = 0, \ldots, T + 1$). For our data structures in the IVF and HIV studies, covariates at each time point t are $\bar{L} = (L_0, L_1, L_2)$, and $T = 1$. The set of interventions is given by $\bar{A} = (A_0, A_1)$, and we can represent the data on one randomly sampled subject as a time-ordered data structure:

$$O = (L_0, A_0, L_1, A_1, L_2 = Y),$$

where we assume L_t occurs before A_t. We denote the binary final measurement L_2 by Y, which represents the outcome of interest. The sample is composed of n i.i.d. draws $O_1, \ldots, O_n$ of the random variable O. Realizations of these random variables are denoted $o_1, \ldots, o_n$. The probability distribution of O can be factorized according to the time-ordering of the data:

$$p(O) = p(L_0) \times p(L_1 \mid A_0, L_0) \times p(Y \mid A_1, L_1, A_0, L_0)$$
$$\times p(A_0 \mid L_0) \times p(A_1 \mid L_1, A_0, L_0).$$

Recall that $\mathcal{D} = (d_1, \ldots, d_K)$ is a set of dynamic rules and each d_k encodes a time sequence $d_{k,t}$ ($t = 1, \ldots, T$) of rules. Additionally, Y^d is the (counterfactual) value that would have been observed had the subject been set to follow rule d at all time points. In Chap. 1, we discussed the goal of estimating, for each rule $d \in \mathcal{D}$, the population mean of Y had everyone followed rule d, i.e., $E(Y^d)$. (With a binary Y, as we have here, this is equivalent to the probability that $Y^d = 1$.) This parameter can be represented as a function of the observed data-generating distribution P, using the g-computation formula, under a set of strong identifiability assumptions:

$$\Psi^d(P) = \sum_{l_0} \sum_{l_1} P(Y = 1 \mid A_1 = d(\bar{l}_1), \bar{L}_1 = \bar{l}_1, A_0 = d(l_0))$$
$$\times P(L_1 = l_1 \mid A_0 = d(l_0), L_0 = l_0) P(L_0 = l_0).$$

For an estimator of $\Psi^d(P)$ to be consistent, one historically needed to correctly specify all of the conditional density factors in the density representation of p.

Estimation of such a conditional density can be difficult when L_1 is a vector with more than a few continuous variables. The sequential regression approach of Robins (2000) avoids estimation of conditional densities, but instead only requires estimation of conditional means (as described in, e.g., Robins and Ritov 1997). It is based on the following iterative sequence of conditional means $Q^d = (Q_2^d, Q_1^d, Q_0^d)$, where we define, with generality to nonbinary Y, as

$$Q_2^d(L_1, L_0) = E(Y \mid A_1 = d(L_1, L_0), L_1, A_0 = d(L_0), L_0),$$
$$Q_1^d(L_0) = E(Q_2^d(L_1, L_0) \mid A_0 = d(L_0), L_0),$$
$$Q_0^d = E(Q_1^d(L_0)).$$

It follows from the above representation of $\Psi^d(P)$ that Q_0^d is an equivalent representation of $\Psi^d(P)$.

In this chapter, we are interested in prediction. The estimation of conditional means is a component of the estimation of $\Psi^d(P)$, but also an interesting parameter separately. Thus, suppose that we are interested in estimating a conditional mean. This becomes what we refer to as a counterfactual prediction problem.

$$\bar{Q}_{Y^d} = Q_1^d(L_0) = E(Q_2^d(L_1, L_0) \mid A_0 = d(L_0), L_0) = E(Y^d \mid L_0).$$

The first regression $Q_2^d = E(Y \mid A_1 = d(L_1, L_0), L_1, A_0 = d(L_0), L_0)$ can be defined as the minimizer of the expected loss. Since $Y\{0, 1\}$ or $Y \in (0, 1)$ the log-likelihood loss is a reasonable choice. The loss function for the next regression $Q_1^d = E(Q_2^d(L_1, L_0) \mid A_0 = d(L_0), L_0)$ can be the same log-likelihood loss but with Y replaced by the previous $\bar{Q}_2^d$. Thus, this is now a loss function that is indexed by an unknown nuisance parameter Q_2^d. Nonetheless, at each step we can the use super learner with this loss function, treating the loss function as known by plugging in the super learner fit obtained at the previous step. We describe this sequential estimation procedure in the next section.

3.3 Sequential Super (Machine) Learning

We can estimate the parameter $\bar{Q}_{Y^d}$ by nesting a series of regressions, starting at the last time point and moving backwards in time toward L_0, inspired by the sequential regression approach described on the previous page. We call this sequential super learning since we use super learning at each step, treating previous super learner fit as an outcome for the next regression. We will use the notation $L_{t:0} = (L(j) : j = t, \ldots, 0)$.

Algorithm. *Super Learning for Sequential Prediction in Longitudinal Data*

For each rule d:

★ Obtain an estimator $Q^d_{T+1,n}$ of $Q^d_{T+1}(L_{0:T})$ with super learning.

For $t = T + 1$ to $t = 1$

★ Define $Q^d_{t,n}(L_{0:t-1})$ as the outcome in next regression and use super learning to estimate $E(Q^d_t(L_{0:t-1}) \mid A_{t-2:0} = d(L_{t-2:0}), L_{t-2:0})$.

Save the final estimator $Q^d_{t=1,n}(L_0)$ as estimator of $\bar{Q}_{Y^d} = E(Y^d \mid L_0)$.

Specifically, the sequential super learner for $\bar{Q}_{Y^d}$ is constructed as follows:

1. Let $Q^d_{T+2,n} = Y$.
2. Set $t = T + 1$.
3. For time point t, create a data set of n observations where each observation has an outcome $Q^d_{t+1,n}(L_{t+1:0})$, and covariates $A_{0:t}, L_{0:t}$. Fit the K candidate regression algorithms within V-fold cross-validation. Recall that $B_n \in \{0, 1\}^n$ is a random variable that splits the data into a training set $\{i : B_n(i) = 0\}$ and validation set $\{i : B_n(i) = 1\}$. The data set is divided into a training set containing $\frac{V-1}{V}^{\text{ths}}$ of the data and a validation set containing the remaining $\frac{1}{V}^{\text{th}}$ of the data in each of V folds. For each $v = 1, \ldots, V$, for each $k = 1, \ldots, K$, train the k-th algorithm on the training set $T(v)$, while the $V(v)$ validation set is run through the fitted algorithm to obtain cross-validated predicted values. This results in a predicted value $Z^d_{k,t,i}$ for each algorithm k and subject i, $i = 1, \ldots, n$.

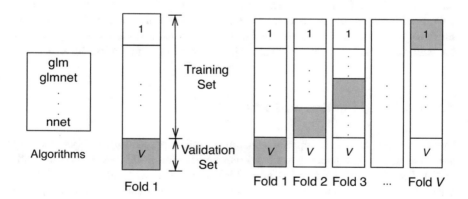

4. Posit a family of weighted combinations of the K algorithms that is a convex combination indexed by α, and select the α_n that minimizes the cross-validated empirical mean of the loss function.

5. Run all K algorithms on the full data set and combine the candidate fits with the α_n vector to build the super learner function and obtain predicted values under the setting that each individual followed rule d. These predicted values represent $Q_{t,n}^d(L_{0:t,i})$, $i = 1, \ldots, n$.

ID	Y_{glm}^d	...	Y_{nnet}^d
1	0.54	...	0.42
.	.	.	.
n	0.09	...	0.12

6. Set $t \to t - 1$ and repeat the above steps 3, 4 and 5.

7. Continue to iterate the sequential prediction algorithm until $t = 1$ and save the resulting object $Q_{t=1,n}^d$, which represents the desired estimator of $\bar{Q}_{Y^d} = E(Y^d \mid L_0)$.

3.4 Computation

With newer advances in parallelization and cloud computing, computational challenges are rapidly being addressed. Software is currently available in R, H2O, and SAS: berkeleybiostats.github.io, including SAS macros and links to R packages on CRAN. The implementation of a sequential super learner can be computational intensive in the context of big data. Thoughtful consideration should be given to programming language, number of algorithms, and number of time points included to maintain applied relevance while remaining computationally feasible.

3.5 Notes and Further Reading

Further details of the asymptotic and finite sample properties of super learning are discussed in key papers (van der Laan and Dudoit 2003; van der Laan et al. 2007). See also van der Laan et al. (2006), van der Vaart et al. (2006), van der Laan et al. (2004), Dudoit and van der Laan (2005), and Keleş et al. (2002). The sequential super learner has been described for a conditional intensity of a counting process in atrial fibrilation (Brooks 2012). This super learner involves defining an overall loss function $L(O, \bar{Q}_{L_t})$ as the sum over all t-specific loss functions $L(O, \bar{Q}_{L_t}) = \sum_t L(O_t, \bar{Q}_{L_t})$. Super learning in longitudinal data with missingness was also described in Díaz et al. (2015). Extensive references on machine learning and ensemble methods can be found in Chap. 3 of *Targeted Learning* (2011).

Chapter 4
LTMLE

Sherri Rose and Mark J. van der Laan

Sequential decision making is a natural part of existence. Humans make a myriad of decisions each day, and many decisions are typically involved when considering a single particular goal amidst an unpredictable and uncertain environment. Any action could impact future states and, importantly, the options available later. What if we had an automated way to understand the impact of decisions? And a means of evaluating differing decision sequences?

As discussed in the three previous chapters, we are often interested in evaluating treatment rules that extend over time, i.e., dynamic rules, and these rules can have complex classifiers. This is common in clinical and public health research, and, in statistics, this work relies on sequential decision theory. There are a number of other important types of problems in longitudinal and complex data structures, enumerated throughout this book, and some also introduced in Chap. 1. In this chapter, we focus on describing the LTMLE in the context of dynamic rules, although the approach is general, as demonstrated in later chapters.

S. Rose (✉)
Department of Health Care Policy, Harvard Medical School, 180 Longwood Ave, Boston, MA 02115, USA
e-mail: rose@hcp.med.harvard.edu

M. J. van der Laan
Division of Biostatistics and Department of Statistics, University of California, Berkeley, 101 Haviland Hall, #7358, Berkeley, CA 94720, USA
e-mail: laan@berkeley.edu

© Springer International Publishing AG 2018
M.J. van der Laan, S. Rose, *Targeted Learning in Data Science*,
Springer Series in Statistics, https://doi.org/10.1007/978-3-319-65304-4_4

This is the second chapter focusing on estimation, and now we turn to questions of effect. We started with carefully defining the research question, including the data, model, and target parameter of the probability distribution of the data. Then, in the previous chapter, we presented estimation of sequential prediction functions with super learning. We discussed that for multiple time point interventions, we could simply average over our estimator of $E(Y^d \mid L_0)$ to estimate $E(Y^d)$. This, however, will not lead to an optimal estimator, and we now describe employing targeting at each step to get an improved estimator for $E(Y^d)$. Thus, we are now ready for the estimation of causal effects using LTMLE. Note that we use the abbreviation *LTMLE* for *longitudinal targeted maximum likelihood estimator* as well as *longitudinal targeted minimum loss-based estimation.*

4.1 LTMLE in Action: When to Start HIV Treatment

Recall from Chap. 1, our discussion of the rule-specific mean for the question of when to start antiretroviral treatment among therapy-naive HIV-infected individuals. For many years, this was an open question, although there is now generally consensus regarding the benefits of early initiation. Randomized and observational studies considered various thresholds for CD4 count, such as 200–500 cells/mm^3 in intervals of 50, where falling below the threshold indicates one should start antiretroviral treatment. (The gap between the first observed CD4 count below the threshold and treatment initiation has also been debated, and commonly used windows include 3 months and 6 months.) Here, suppose we consider only two thresholds: 350 and 500, and an initiation window of 6 months. In the United States, guidelines set by the Department of Health and Human Services for treatment of asymptomatic individuals fluctuated between a cutoff of 500 and one of 350 from 1998 to 2011, before it was changed to "all" in 2012. We carry this example through the chapter as a demonstrative example.

4.2 Defining the Estimation Problem

Recall that we write the data structure on one randomly sampled subject as:

$$O = (L_0, A_0, L_1, A_1, L_2 = Y),$$

with covariates $\bar{L} = (L_0, L_1, L_2)$, and $T = 1$ and a vector of covariates L_t, for each time t ($t = 0, \ldots, T + 1$), as well as the set of interventions $\bar{A} = (A_0, A_1)$. For L_0, the vector contains CD4, viral load, age, sex, intravenous drug use status, and chronic hepatitis C status. At L_1, the vector contains only CD4 count and whether the subject died. The binary final measurement $L_2 = Y$, which represents our outcome of interest, death. If a subject dies by L_1, they are also recorded as having died by L_2.

Our intervention nodes $\bar{A} = (A_0, A_1)$ are defined based on whether treatment had been initiated at or by that time point. Our sample contains n i.i.d. draws $O_1, \ldots, O_n$ of the random variable O, and realizations of these random variables are given as $o_1, \ldots, o_n$. Recall also that the probability distribution of O can be factorized according to the time-ordering of the data:

$$p(O) = p(L_0) \times p(L_1 \mid A_0, L_0) \times p(Y \mid A_1, L_1, A_0, L_0)$$
$$\times p(A_0 \mid L_0) \times p(A_1 \mid L_1, A_0, L_0).$$

In general, we have a set of rules $\mathcal{D} = \{d_1, \ldots, d_K\}$, where each d_k encodes a time sequence $d_{k,t}$ ($t = 1, \ldots, T$) of rules, and Y^d is the counterfactual outcome that would have been observed had the subject been set to follow rule d at all time points. In our simplified example, we consider only the classifier CD4 count, and two levels from the previous guidelines on when to start treatment: 350 and 500 cells/mm^3. Thus, at each intervention node:

$$d_{k,t}(\text{CD4}_t) = \begin{cases} 1 & \text{if CD4}_t < \theta_k \\ 0 & \text{otherwise}, \end{cases}$$

where $\theta = (500, 350)$, 1 indicates that treatment has been initiated, and we have that $\mathcal{D} = \{d_1, d_2\}$. We wish to estimate, for each of our two rules $d \in \mathcal{D}$, the population mean of Y had everyone followed rule d, i.e., $E(Y^d)$. As introduced in Chap. 1, $E(Y^d)$ can be represented as a function of the observed data-generating distribution P with the g-computation formula:

$$\Psi^d(P) = \sum_{l_0} \sum_{l_1} P(Y = 1 \mid A_1 = d(\bar{l}_1), \bar{L}_1 = \bar{l}_1, A_0 = d(l_0))$$
$$\times P(L_1 = l_1 \mid A_0 = d(l_0), L_0 = l_0) P(L_0 = l_0).$$

We also know that Q_0^d is an equivalent representation of $\Psi^d(P)$, where Q_0^d is given in the following iterative sequence of conditional means $Q^d = (Q_2^d, Q_1^d, Q_0^d)$:

$$Q_2^d(L_1, L_0) = E(Y \mid A_1 = d(L_1, L_0), L_1, A_0 = d(L_0), L_0),$$
$$Q_1^d(L_0) = E(Q_2^d(L_1, L_0) \mid A_0 = d(L_0), L_0),$$
$$Q_0^d = E(Q_1^d(L_0)).$$

4.3 What Does It Mean to Follow a Rule?

Before describing the LTMLE algorithm to estimate $\Psi^d(P) = E(Y^d)$, it is essential to be explicit about what it means to "follow a rule." The key to designing our study, data structure, analysis, and appropriately interpreting our estimate, is the experiment we wish we could have conducted, but could not.

- What is the population mean outcome *had all subjects followed the treatment rule to initiate antiretroviral therapy within 6 months of dropping below a CD4 count of 350 cells/mm^3?*
- What is the population mean outcome *had all subjects followed the treatment rule to initiate antiretroviral therapy within 6 months of dropping below a CD4 count of 500 cells/mm^3?*

However, given that we have observational data with time-dependent confounding, we did not in fact force patients to follow either of these rules. Thus, the next step is to be precise about how we encode whether a subject in our hypothetical observational study is following one of our two rules.

Suppose we have a hypothetical study where we follow all therapy-naive HIV-infected individuals for 6 years. (For the moment, additional time points will be useful in this expository subsection material.) The three panels in Fig. 4.1 represent three hypothetical individuals in the study. In the first panel, "Patient 1" drops below 500 cells/mm^3 for the first time in month 6, drops below 350 cells/mm^3 for the first time in month 30, and initiates treatment in month 66. They follow regime d_1, where $\theta = 500$, up to month 12, where we see that they have not initiated treatment within 6 months of first dropping below 500 cells/mm^3. Patient 1 follows d_2, where $\theta = 350$, up to month 36, where, they have not initiated treatment within 6 months of first dropping below 350 cells/mm^3. At month 36 and later, Patient 1 is following neither d_1 or d_2.

Hypothetical Patient 2 appears in the second panel of Fig. 4.1. This patient drops below 500 cells/mm^3 for the first time in month 12, drops below 350 cells/mm^3 for the first time in month 36, and initiates treatment in month 16. Therefore, Patient 2 follows d_1 for the entirety of the study length, all 72 months, because they initiated treatment within 6 months of dropping below 500 cells/mm^3. It does not matter what occurs after this initiation given the manner in which we have defined our rules; the subject is following d_1. Patient 2 is following d_2 up to month 16, when they initiate treatment. Because d_2 is defined by only initiating treatment once the subject falls below 350 cells/mm^3, they are no longer following this rule when they start treatment before ever dropping below 350 cells/mm^3.

Patient 3 is following both d_1 and d_2 for the entire study period of 72 months. They never drop below 500 or 350 cells/mm^3 and they never start treatment. Since the rules are only not being followed when treatment is either (a) not initiated after dropping below the specified θ or (b) initiated before or too far after dropping below the specified θ, Patient 3 is always following both rules.

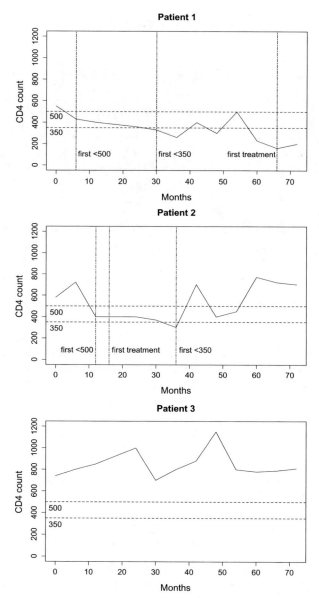

Fig. 4.1 Illustrations of CD4 count trajectory for three hypothetical individuals as well as when they started treatment. (Values and data points do not reflect any real patient data.)

4.4 LTMLE for When to Start Treatment

We have now defined our target parameter as a mapping from an infinite dimensional parameter of the probability distribution of the unit data structure into the parameter space, which included carefully translating our scientific research question into a statistical estimation problem. Our LTMLE requires deriving the efficient influence curve of the pathwise derivative of the target parameter, specifying a loss function, positing a fluctuation working submodel through the initial (or current) estimator so the linear span of the score when the fluctuation is zero includes the efficient influence curve, and an iterative maximization algorithm. This iterative maximization continues until the maximum likelihood estimators of the fluctuation parameters are near zero. The LTMLE will solve the efficient score equation and thereby inherit the double robustness and asymptotic efficiency for the substitution estimator of our target parameter.

4.4.1 Determining the Efficient Influence Curve

Let $P \in \mathcal{M}$ be given. Suppose that we know the treatment mechanism $g = g(P)$. In that case our statistical model is given by the smaller model $\mathcal{M}(g) = \{P_1 \in \mathcal{M} : g(P_1) = g\}$ defined by all possible densities of O in which the factors of the treatment mechanism are defined by g. In such a model we could estimate ψ^d with an inverse probability of treatment weighted estimator $\psi^d_{n,IPTW}$ using the known g:

$$\psi^d_{n,IPTW} = \frac{1}{n} \sum_{i=1}^{n} \frac{I(\bar{A}_i(T) = \bar{d}_T(\bar{L}_i(T)))}{g_{0:T}(\bar{A}_i(T), \bar{L}_i(T))} Y_i.$$

Since $\psi^d_{n,IPTW}$ is a sample mean, it follows that it is an (asymptotically) linear estimator at any $P \in \mathcal{M}(g)$ with influence curve

$$D^d(P)(O) = \frac{I(\bar{A}(T) = \bar{d}_T(\bar{L}(T)))}{g_{0:T}(\bar{A}(T), \bar{L}(T))} Y - \Psi^d(P).$$

An important result from efficiency theory is that the influence curve at P of a regular asymptotically linear estimator of a target parameter is a gradient at P of the pathwise derivative of the target parameter. Therefore, we know that $D^d(P)$ is a gradient at P in the model $\mathcal{M}(g)$. The canonical gradient of $\Psi^d : \mathcal{M}(g) \to \mathbb{R}$ at P is defined by its projection on the tangent space $T_g(P)$ of $\mathcal{M}(g)$:

$$D^{d*}(P) = \Pi(D^d(P) \mid T_g(P)).$$

Because the tangent space of g in model $\mathcal{M}$ is orthogonal to the tangent space of $\mathcal{M}(g)$ it follows that $D^{d*}(P)$ is also the canonical gradient of $\Psi^d : \mathcal{M} \to \mathbb{R}$ for our actual model $\mathcal{M}$. So we can conclude that the task of finding the canonical gradient/efficient influence curve at P of our target parameter $\Psi^d : \mathcal{M} \rightarrow \mathbf{R}$ is reduced to computing the projection of the initial gradient $D^d(P)$ onto the tangent space at P for the model $\mathcal{M}(g)$. The tangent space $T_g(P)$ equals the orthogonal sum of the tangent spaces $T_{g,t}(P)$ of the conditional density of $L(t)$ given $\bar{L}(t-1), \bar{A}(t-1)$, $t = 0, \ldots, T + 1$. This tangent space $T_{g,t}(P)$ is given by all functions in $L_0^2(P)$ of $(\bar{A}(t-1), \bar{L}(t))$ with conditional mean zero, given $\bar{L}(t-1), \bar{A}(t-1)$. From this we learn that the projection $D^{d*}(P)$ of $D^d(P)$ onto $T_g(P)$ equals the sum of the projections of $D^d(P)$ onto $T_{g,t}(P)$, and the latter projection is given by

$$D_t^{d*}(P) \equiv \Pi(D^d(P) \mid T_{g,t}(P)) = E(D^d(P)(O) \mid \bar{L}(t), \bar{A}(t-1))$$
$$- E(D^d(P)(O) \mid \bar{L}(t-1), \bar{A}(t-1)).$$

The latter projection can be rewritten by integrating out $A(t : T)$, which establishes the following formula:

$$D_t^{d*}(P) = \frac{I(\bar{A}(t-1) = \bar{d}_{t-1})}{g_{t-1}(\bar{A}(t-1), \bar{L}(t-1))} \left\{ \bar{Q}_{t+1}^d(\bar{L}(t)) - \bar{Q}_t^d(\bar{L}(t-1)) \right\}.$$

So we can conclude that the efficient influence curve of $\Psi^d : \mathcal{M} \rightarrow \mathbf{R}$ is given by

$$D^{d*}(P)(O) = \sum_{t=0}^{T+1} D_t^{d*}(P)$$
$$= \bar{Q}_1^d(L(0)) - \Psi^d(P)$$
$$+ \sum_{t=1}^{T+1} \frac{I(\bar{A}(t-1) = \bar{d}_{t-1})}{g_{t-1}(\bar{A}(t-1), \bar{L}(t-1))} \left\{ \bar{Q}_{t+1}^d(\bar{L}(t)) - \bar{Q}_t^d(\bar{L}(t-1)) \right\}.$$

The first term $D_0^{d,*}(P)$ represents the score component for the distribution of $L(0)$, while the terms in the sum over t represent the score components for the distribution of $L(t)$, given $\bar{L}(t-1), \bar{A}(t-1)$).

4.4.2 Determining the Loss Function and Fluctuation Submodel

We need to determine a loss function for $\bar{Q}_t^d$ and submodel $\{\bar{Q}_t^d(\epsilon_t, g) : \epsilon\}$ through $\bar{Q}_t^d$ at $\epsilon = 0$ with score $D_t^{d*}(P)$, $t = T + 1, \ldots, 0$. Because the parameters $\bar{Q}_t^d$ are iteratively defined, we will also define this combination of loss function and submodel in an iterative manner, starting at $t = T + 1$ and ending up at $t = 0$. A valid loss function for $\bar{Q}_{T+1}^d$ is given by the Bernoulli log-likelihood loss function

$$L(\bar{Q}_{T+1}^d)(O) = - \left\{ Y \log \bar{Q}_{T+1}^d(\bar{L}(T)) + (1 - Y) \log\{1 - \bar{Q}_{T+1}^d(\bar{L}(T))\} \right\}.$$

Consider the following submodel $\{\bar{Q}^d_{T+1}(\epsilon_{T+1}, g):\epsilon_{T+1}\}$ through $\bar{Q}^d_{T+1}$ at $\epsilon_{T+1}=0$:

$$\text{logit } Q^d_{T+1}(\epsilon_{T+1}, g) = \text{logit } Q^d_{T+1} + \epsilon_{T+1}\frac{I(\bar{A}_T = \bar{d}_T)}{g_{0:T}},$$

where $I(\bar{A}_{t-1} = \bar{d}_{t-1})$ denotes the indicator that rule d was followed from $t = 0$ to $t - 1$ and $g_{0:t-1} = g_0 \times \cdots \times g_{t-1} = P(A_0 \mid L_0) \cdots P(A_{t-1} \mid \bar{L}_{t-1}, \bar{A}_{t-1})$. Notice that indeed the score at $\epsilon = 0$ equals $D^{d*}_{T+1}(P)$:

$$\frac{d}{d\epsilon}L(\bar{Q}^d_{T+1}(\epsilon, g))\bigg|_{\epsilon=0} = D^{d*}_{T+1}(P).$$

Let $t = T$. We now want to determine a loss function and submodel for $\bar{Q}^d_t$, where we can treat $\bar{Q}^d_{t+1}$ as a given outcome. Treating $\bar{Q}^d_{t+1}$ as a given outcome, a valid loss function for $\bar{Q}^d_t$ is again given by the log-likelihood loss function:

$$L_{\bar{Q}^d_{t+1}}(\bar{Q}^d_t)(O) = -\{\bar{Q}^d_{t+1}(\bar{L}(t)) \log \bar{Q}^d_t(\bar{L}(t - 1)) + (1 - \bar{Q}^d_{t+1}) \log\{1 - \bar{Q}^d_t(\bar{L}(t - 1))\}.$$

Consider the following submodel $\{\bar{Q}^d_t(\epsilon_t, g) : \epsilon\}$ through $\bar{Q}^d_t$ at $\epsilon_t = 0$:

$$\text{logit } Q^d_t(\epsilon_t, g) = \text{logit } Q^d_t + \epsilon_t\frac{I(\bar{A}_{t-1} = \bar{d}_{t-1})}{g_{0:t-1}}.$$

Indeed, the score $\frac{d}{d\epsilon_t}L_{\bar{Q}^d_{t+1}}(\bar{Q}^d_t(\epsilon_t, g))$ at $\epsilon_t = 0$ equals $D^{d,*}_t(P)$. In this way, we have defined sequentially a loss function $L_{\bar{Q}^d_{t+1}}(\bar{Q}^d_t)$ and submodel $\{\bar{Q}^d_t(\epsilon_t, g) : \epsilon_t\}$ through $\bar{Q}^d_t$ at $\epsilon_t = 0$ with score $D^{d*}_t(P)$, $t = T + 1, \ldots, 1$.

Finally, given $\bar{Q}^d_1(L(0))$, we define the marginal (Bernoulli) log-likelihood loss function:

$$L_{\bar{Q}^d_1}(\bar{Q}^d_0) = -\{\bar{Q}^d_1(L(0)) \log \bar{Q}^d_0 + (1 - \bar{Q}^d_1(L(0))) \log\{1 - \bar{Q}^d_0\}\}.$$

As submodel through $\bar{Q}^d_0$ we select

$$\text{logit } Q^d_0(\epsilon_0) = \text{logit } Q^d_0 + \epsilon_0.$$

Again, the score $\frac{d}{d\epsilon_0}L_{\bar{Q}^d_1}(\bar{Q}^d_0(\epsilon_0))$ at $\epsilon_0 = 0$ equals the desired component $D^{d*}_0(P)$.

4.4.3 LTMLE Algorithm

Sequentially, updating the initial estimator $\bar{Q}^d_{t,n}$ of $\bar{Q}^d_t$ with an MLE of $\bar{Q}^d_t(\epsilon_{t,n}, g_n)$, starting at $t = T + 1$ until $t = 0$ defines the following TMLE algorithm.

Algorithm. *LTMLE for the Rule-Specific Mean*

For each rule d:
  ⋆ Obtain estimators $Q_{T+1,n}^d$ and $g_{T,n}$ with super learning.
  For $t = T + 1$ to $t = 1$
    ⋆ Hold $Q_{t,n}^d$ fixed and compute maximum likelihood estimate
    $\epsilon_{t,n} = \arg\min_\epsilon P_n L_{\bar{Q}_{t+1}^d}(\bar{Q}_t^d(\epsilon, g))$ of ϵ_t in submodel:
    logit $Q_t^d(\epsilon_t, g) = $ logit $Q_{t,n}^d + \epsilon_t(I(\bar{A}_{t-1} = \bar{d}_{t-1})/g_{0:t-1})$.
    ⋆ Set the updated estimator $Q_{t,n}^{d*} = Q_{t,n}^d(\epsilon_{t,n}, g_n)$.
    ⋆ Define $Q_{t,n}^{d*}$ as the outcome in next regression and use super
    learning to estimate $E(Q_t^d(L_{0:t-1}) \mid A_{t-2:0} = d(L_{t-2:0}), L_{t-2:0})$.
  Save the final estimator $Q_{0,n}^{d*} = 1/n \sum_{i=1}^n Q_{t=1,n}^{d*}(L_{0,i})$ of $\Psi^d(P)$.

Specifically, the LTMLE for $E(Y^d)$ in our HIV treatment example with two intervention time points $\bar{A} = (A_0, A_1)$ and $T = 1$ is constructed as described below. Also, recall from Sect. 4.3 the definition for 'following' the two rules $\mathcal{D} = \{d_1, d_2\}$.

1. Let $Q_{3,n}^d = Y$ and set $t = T + 1 = 2$.
2. Consider the original data set S_n of n observations. Create counterfactual 'rule-specific' versions of the data S_n for each of our two rules $\mathcal{D} = \{d_1, d_2\}$. In these counterfactual data sets S_{d_1} and S_{d_2}, the values of $\bar{A} = (A_0, A_1)$ are set for each observation based on what they would be under rules d_1 and d_2.
3. Obtain estimators $Q_{2,n}^d$, $g_{1,n}$, and $g_{0,n}$ with super learning using S_n.
4. Estimate predicted outcomes under the observed values for $\bar{A}$, as well as counterfactual predicted outcomes $Q_{2,n}^{d_1}$ and $Q_{2,n}^{d_2}$ for each rule using S_{d_1}, S_{d_2}, and the super-learning-based fit for $Q_{2,n}^d$ from the previous step.
5. For each rule d_k: Hold $Q_{2,n}^d$ fixed and compute the maximum likelihood estimate $\epsilon_{2,n} = \arg\min_\epsilon P_n L_{\bar{Q}_3^d}(\bar{Q}_2^d(\epsilon, g))$ of ϵ_2 in the submodel:

$$\text{logit } Q_2^d(\epsilon_2, g) = \text{logit } Q_{2,n}^d + \epsilon_2(I(\bar{A}_{t=1} = \bar{d}_{t=1})/g_{0:1}),$$

where $g_{0:1} = g_0 \times g_1$ and $I(\bar{A}_{t=1} = \bar{d}_{t=1})$ for an individual is an indicator that observation is following rule d_k through $t - 1 = 1$. Set the updated estimator $Q_{2,n}^{d*} = Q_{2,n}^d(\epsilon_{2,n}, g_n)$.
6. Define $Q_{2,n}^{d*}$ as the outcome in the next super-learning-based regression $Q_{1,n}^d$, setting $t = 1$.
7. Repeat steps 4 and 5 for t=1. Then, set $t = 0$.
8. Save the final estimators

$$Q_{0,n}^{d_1*} = 1/n \sum_{i=1}^n Q_{t=1,n}^{d_1*}(L_{0,i}) \text{ and}$$

$$Q_{0,n}^{d_2*} = 1/n \sum_{i=1}^n Q_{t=1,n}^{d_2*}(L_{0,i}).$$

We ignored issues such as patient drop-out in this example for didactic purposes, and refer readers to Chap. 15 for an example of LTMLE with an explicit censoring mechanism. This demonstrative LTMLE also estimated separate submodels for each d_k and t. Other choices in implementation are possible, including the use of joint submodels. Software is currently available in R and SAS: berkeleybiostats.github.io, including the R package ltmle on CRAN (Lendle et al. 2017).

4.5 Analysis of TMLE and Inference

This section contains technical details on the analysis of TMLE and inference. Some readers may wish to skip this material. Let $\bar{Q}^d = (\bar{Q}^d_t : t = T + 1, \ldots, 0)$ denote this sequentially defined parameter. Notice that the efficient influence curve at P only depends on P though $\bar{Q}^d = \bar{Q}^d(P)$ and $g = g(P)$. For notational convenience, we will denote $D^{d*}(P)$ with $D^{d*}(\bar{Q}^d, g)$ as well. Let $\bar{Q}^d_n$ and g_n be the initial estimators of $\bar{Q}^d(P_0)$ and $g_0 = g(P_0)$, respectively. The above TMLE algorithm defines the TMLE $\bar{Q}^{d*}_n$.

4.5.1 TMLE Solves Efficient Influence Curve Equation

The TMLE solves the efficient influence curve equation:

$$P_n D^{d*}(Q^{d*}_n, g_n) = 0.$$

4.5.2 Second-Order Remainder for TMLE

For any pathwise differentiable target parameter $\Psi : \mathcal{M} \to \mathbf{R}$ with canonical gradient $D^*(P)$ at P one can define a second-order remainder:

$$R_2(P, P_0) \equiv \Psi(P) - \Psi(P_0) - (P - P_0)D^*(P),$$

where one can use that $PD^*(P) = 0$. This results in the second-order Taylor expansion of $P \to \Psi(P)$ at P_0:

$$\Psi(P) - \Psi(P_0) = -P_0 D^*(P) + R_2(P, P_0).$$

Applying this general approach to our problem, we define the second-order remainder:

$$R^d_2(P, P_0) = \Psi^d(P) - \Psi^d(P_0) + P_0 D^{d*}(P).$$

One can obtain an explicit expression for this second-order remainder. In order to emphasize its dependence on $\bar{Q}^d$ and g we will also denote this remainder with $R_{20}^d(\bar{Q}^d, g, \bar{Q}_0^d, g_0)$. Inspection of the closed form expression of $R_2^d(P, P_0)$ (not shown here) shows that it consists of a sum of integrals that integrate a product of a difference of parameter of g with its true value (e.g., $(g_{0:t} - g_{0,0:t})$) with a difference of parameter of $\bar{Q}^d$ with its true value (e.g., $\bar{Q}_t^d - \bar{Q}_{0,t}^d$). As a result, the second-order remainder has a so-called double robust structure in the sense that

$$R_2^d(\bar{Q}^d, g, \bar{Q}_0^d, g_0) = 0 \text{ if } \bar{Q}^d = \bar{Q}_0^d \text{ or } g = g_0.$$

More importantly, by using the Cauchy-Schwarz inequality

$$\left(\int fg \, dP_0 \right)^2 \leq \int f^2 \, dP_0 \int g^2 \, dP_0,$$

this product structure of $R_{20}()$ allows one to bound $R_2^d(P, P_0)$ in terms of products of an $L^2(P_0)$-norm of a difference of parameter of $g(d(\bar{L}(T)), \bar{L}(T))$ with its true value and a difference of a parameter of $\bar{Q}^d$ with its true value. Specifically, we can bound $R_2^d(P, P_0)$ with a sum of terms of the type $\| \bar{Q}_{t+1}^d - \bar{Q}_{t+1}^d(P_0) \|_{P_0} \| g_{0:t} - g_{0,0:t} \|_{P_0}$. This bounding relies on the positivity assumption that $g_{0,0:T}(d(\bar{L}(T)), \bar{L}(T)) > \delta > 0$ for some $\delta > 0$. We can apply this second-order expansion at the TMLE $(\bar{Q}_n^{d*}, g_n)$. This results in the following identity:

$$\Psi^d(\bar{Q}_n^{d*}) - \Psi^d(\bar{Q}_0^d) = -P_0 D^{d*}(\bar{Q}_n^{d*}, g_n) + R_{20}(\bar{Q}_n^{d*}, g_n, \bar{Q}_0^d, g_0).$$

Combined with $P_n D^{d*}(Q_n^{d*}, g_n) = 0$, this results in

$$\Psi^d(\bar{Q}_n^{d*}) - \Psi^d(\bar{Q}_0^d) = (P_n - P_0)D^{d*}(\bar{Q}_n^{d*}, g_n) + R_{20}(\bar{Q}_n^{d*}, g_n, \bar{Q}_0^d, g_0). \tag{4.1}$$

4.5.3 Asymptotic Efficiency

This provides a perfect basis for establishing asymptotic efficiency of the TMLE $\Psi^d(\bar{Q}_n^{d*})$ of $\Psi^d(\bar{Q}_0^d)$. Firstly, assume that

$$R_{20}(\bar{Q}_n^{d*}, g_n, \bar{Q}_0^d, g_0) = o_P(n^{-1/2}). \tag{4.2}$$

By the above mentioned Cauchy-Schwarz bound (4.2) holds if $\| g_{n,0:t} - g_{0,0:t} \|_{P_0} \| \bar{Q}_{t+1,n}^{d*} - \bar{Q}_{0,t+1}^{d*} \|_{P_0} = o_P(n^{-1/2})$ for all $t = 1, \ldots, T$. For example, the latter will hold if we estimate each of the nuisance parameters at a rate faster than $n^{-1/4}$ with respect to $\| \cdot \|_{P_0}$-norm. On the other hand, knowledge about g_0 may allow one to estimates g_0 at a significantly faster rate than $n^{-1/4}$, in which case one can estimate $\bar{Q}_0^d$ at a significantly slower rate than $n^{-1/4}$, as long as the product of the rates is of smaller order than $n^{-1/2}$. In observational studies in which little is known about g_0, in order to satisfy this assumption (4.2) we will need to use highly adaptive estimators such as a super learner in which the library includes the highly adaptive lasso estimator (see Chap. 6).

Combining (4.2) with (4.1) results in the following equation:

$$\Psi^d(\bar{Q}_n^{d*}) - \Psi^d(\bar{Q}_0^d) = (P_n - P_0)D^{d*}(\bar{Q}_n^{d*}, g_n) + o_P(n^{-1/2}).$$

In addition, we will assume

$$D^{d*}(\bar{Q}_n^{d*}, g_n) \text{ falls in a } P_0\text{-Donsker class with probability tending to 1.} \qquad (4.3)$$

For example, if O is a J-dimensional vector, we could define the Donsker class as all real-valued multivariate cadlag functions on a multivariate cube $\prod_{j=1}^{J}[0, \tau_j] \subset \mathbf{R}_{\geq 0}^J$ with sectional variation norm smaller than a given constant $M < \infty$ (see Chap. 6). In practice, this corresponds with avoiding estimators that overfit the data. This Donsker class assumption can be avoided by using cross-validated TMLE (see Chap. 7). Given assumption (4.2) one certainly expects that the following consistency assumption holds:

$$P_0\{D^{d*}(\bar{Q}_n^{d*}, g_n) - D^{d*}(\bar{Q}_0^d, g_0)\}^2 \to 0 \text{ in probability, as } n \to \infty. \qquad (4.4)$$

Empirical process theory teaches us that if f_n is a random function of O that falls in a P_0-Donsker class with probability tending to 1 and $P_0 f_n^2 \to 0$ in probability as $n \to \infty$, then $(P_n - P_0)f_n = o_P(n^{-1/2})$. Application of this fundamental empirical process result shows that

$$(P_n - P_0)D^{d*}(\bar{Q}_n^{d*}, g_n) = (P_n - P_0)D^{d*}(\bar{Q}_0^d, g_0) + o_P(n^{-1/2}).$$

We have now shown the desired asymptotic efficiency of the TMLE:

$$\Psi^d(\bar{Q}_n^{d*}) - \Psi^d(\bar{Q}_0^d) = (P_n - P_0)D^{d*}(\bar{Q}_0^d, g_0) + o_P(n^{-1/2}). \qquad (4.5)$$

We can formulate this as a formal theorem.

Theorem 4.1. *Assume (4.2), (4.3) and (4.4). Then, $\Psi^d(\bar{Q}_n^{d*})$ is an asymptotically efficient estimator of $\Psi^d(\bar{Q}_0^d)$. In particular,*

$$\sqrt{n}(\psi_n^{d*} - \psi_0^d) \Rightarrow_d N(0, \sigma_0^{d2}),$$

where $\sigma_0^{d2} = P_0\{D^{d*}(P_0)\}^2$ is the variance of the efficient influence curve.

4.5.4 Inference

An immediate consequence of the above established asymptotic linearity and efficiency is that $\Psi^d(\bar{Q}_n^{d*}) \pm 1.96\sigma_n/n^{1/2}$ is an asymptotic 0.95-confidence interval, where σ_n^{d2} is a consistent estimator of the variance σ_0^{d2} of the efficient influence curve. We can estimate σ_0^{d2} naturally with the empirical sample variance $P_n\{D^{*d}(\bar{Q}_n^{d*}, g_n)\}^2$ of the estimated efficient influence curve $D^{*d}(\bar{Q}_n^{d*}, g_n)$.

Suppose now that one is concerned with estimating $E(Y^d)$ for a collection of rules d varying over a set $\mathcal{D}$. One could now define the vector valued target parameter as $\Psi^{\mathcal{D}}(P) = (\Psi^d(P) : d \in \mathcal{D})$. Let $\bar{Q}^{\mathcal{D}} = (\bar{Q}^d : d \in \mathcal{D})$. In that case, the above analysis shows that, if we assume the above assumptions (4.2), (4.3), and (4.4) for all $d \in \mathcal{D}$, then the TMLE $\psi_n^{\mathcal{D}*} = \Psi^{\mathcal{D}}(\bar{Q}_n^{\mathcal{D}*})$ of $\psi_0^{\mathcal{D}} = \Psi^{\mathcal{D}}(\bar{Q}_0^{\mathcal{D}})$ is asymptotically linear with vector-valued influence curve $D^{\mathcal{D}*}(P_0) = (D^{d*}(P_0) : d \in \mathcal{D})$. In this case, we have that $\sqrt{n}(\psi_n^{\mathcal{D}*} - \psi_0^{\mathcal{D}}) \Rightarrow_d N(0, \Sigma_0)$ as $n \to \infty$, where the asymptotic covariance matrix Σ_0 of the normal limit distribution is given by the covariance matrix of the vector influence curve

$$\Sigma_0 = P_0\{D^{\mathcal{D}*}(P_0)\}\{D^{\mathcal{D}*}(P_0)\}^\top.$$

We can estimate this covariance matrix consistently with the empirical covariance matrix

$$\Sigma_n = P_n\{D_n^{\mathcal{D}*}\}\{D_n^{\mathcal{D}*}\}^\top,$$

where $D_n^{\mathcal{D}*}$ is the plug-in estimator of $D^{\mathcal{D}*}(P_0)$, as above described for the single valued parameter ψ_0^d. This result allows one to carry out simultaneous inference for $(EY^d : d \in \mathcal{D})$. For example,

$$\psi_n^{d*} \pm q_{0.95,n}\sigma_n^d/n^{1/2}$$

is an asymptotic 0.95-simultaneous confidence interval for ψ_0^d, where $q_{0.95,n}$ is the 0.95-quantile of $\max_j | Z(j) |$, $Z \sim N(0, \rho_n)$, and ρ_n is the correlation matrix of Σ_n. By the delta method, the asymptotic linearity of the TMLE of $\Psi^{\mathcal{D}}(P_0)$ also implies the influence curve (and thus inference) of a plug-in TMLE of any differentiable function of $(EY^d : d \in \mathcal{D})$. One class of examples of such a differentiable summary measure of $(EY^d : d \in \mathcal{D})$ is a projection of this dose-response curve $(EY^d : d \in \mathcal{D})$ onto a working marginal structural model $(m_\beta(d) : d \in \mathcal{D})$ (Petersen et al. 2014).

4.6 Notes and Further Reading

Many estimation techniques have been developed for dynamic interventions (Murphy 2003; Robins 2004; Moodie et al. 2007; van der Laan 2006a; van der Laan and Petersen 2007; Robins et al. 2008b; Bembom and van der Laan 2008; Orellana et al. 2010). Notably, Robins (2000) and Bang and Robins (2005) present a sequential regression estimator for the mean outcome under a static rule. Previous work developing LTMLE includes van der Laan and Gruber (2012) and Petersen et al. (2014). This chapter also benefited from conversations and prior collaborations with Susan Gruber, Maya Petersen, Michael Rosenblum, Sharon-Lise Normand, and Mireille Schnitzer.

Part II
Additional Core Topics

Chapter 5
One-Step TMLE

Mark J. van der Laan, Wilson Cai, and Susan Gruber

In this chapter, we will present one-dimensional universal least favorable parametric submodels for the TMLE of univariate and multivariate target parameters. They guarantee that a single TMLE-update of the initial estimator already solves the efficient influence curve equation. We explain why this type of one-step TMLE is more stable than an iterative TMLE. By the fact that the one-step TMLE for high-dimensional or even infinite-dimensional target parameters is a substitution estimator, it follows that it completely respects the structure of the infinite dimensional parameter. The content of this chapter partly relies on van der Laan and Gruber (2016). As an example, we present a one-step TMLE of a complete treatment-specific survival function.

5.1 Local and Universal Least Favorable Submodels

Let's first consider one-dimensional target parameters. A least favorable model at P is a model $\mathcal{S}^* = \{P_{\epsilon,h^*} : \epsilon\}$, dominated by P, for which $P_{\epsilon=0,h^*} = P$, and that maximizes the submodel specific Cramer-Rao lower bound for the asymptotic vari-

M. J. van der Laan (✉)
Division of Biostatistics and Department of Statistics, University of California, Berkeley, 101 Haviland Hall, #7358, Berkeley, CA 94720, USA
e-mail: laan@berkeley.edu

W. Cai
Division of Biostatistics, University of California, Berkeley, 101 Haviland Hall, #7358, Berkeley, CA, USA
e-mail: wcai@berkeley.edu

S. Gruber
Department of Population Medicine, Harvard Medical School, Harvard Pilgrim Health Care Institute, 401 Park Drive, Boston, MA 02215, USA
e-mail: susan_gruber@harvardpilgrim.org

© Springer International Publishing AG 2018
M.J. van der Laan, S. Rose, *Targeted Learning in Data Science*,
Springer Series in Statistics, https://doi.org/10.1007/978-3-319-65304-4_5

ance of a regular asymptotically linear estimator of $\Psi(P_{\epsilon=0})$ for submodel $\{P_{\epsilon,h} : \epsilon\}$ defined by

$$CR(h \mid P) \equiv \frac{\left(\frac{d}{d\epsilon}\Psi(P_{\epsilon,h})\big|_{\epsilon=0}\right)^2}{-P \frac{d^2}{d\epsilon^2} \log \frac{dP_{\epsilon,h}}{dP}\big|_{\epsilon=0}}.$$

It maximizes $CR(h \mid P)$ over all such parametric submodels $\{P_{\epsilon,h} : \epsilon\}$ with h varying over some index set whose closure of the linear span generates the full tangent space $T(P) \subset L_0^2(P)$ of the model at P. Given the pathwise differentiability with canonical gradient $D^*(P)$, denoting the score of $\{P_{\epsilon,h} : \epsilon\}$ at $\epsilon = 0$ with S_h, it follows that this criterion for a submodel can be represented as follows:

$$CR(h \mid P) = \frac{(PD^*(P)S_h)^2}{PS_h^2},$$

By the Cauchy-Schwarz inequality, it follows that this is maximized over all scores in the tangent space $T(P)$ by $S = D^*(P)$. Thus, a least favorable model can also be defined as any parametric model through P that has a score at P equal to $D^*(P)$.

By using a second-order Taylor expansion of $\epsilon \to P \log dP_{\epsilon,h}/dP$ at $\epsilon = 0$ and that this equals the information PS_h^2, it follows that, under some smoothness assumptions on the submodels, the criterion can also be represented as

$$CR(h \mid P) = \lim_{\epsilon \to 0} \frac{(\Psi(P_{\epsilon,h}) - \Psi(P))^2}{-2P \log dP_{\epsilon,h}/dP}.$$

This shows that $CR(h \mid P)$ equals the square change in the target parameter divided by the change in log-likelihood at P at an infinitesimal ϵ. Therefore, we will say that the path $\{P_{\epsilon,h^*} : \epsilon\}$ that maximizes $CR(h \mid P)$ follows at $\epsilon = 0$ (i.e., locally) a path of maximal change in target parameter per unit of information.

To stress that the desired optimality property only applies locally, we will refer to such a submodel as a *locally* (i.e., at $\epsilon = 0$) least favorable submodel.

This latter representation of the criterion is intuitively appealing. A sensible goal of a submodel $\{P_\epsilon : \epsilon\}$ through P is that a small fluctuation of P yields a maximal change in target parameter, making the MLE $\epsilon_n = \arg\max_\epsilon P_n \log dP_\epsilon/dP$ (as used in TMLE) for this parametric model locally all about fitting the target parameter, not wasting data for anything else.

The intuition of TMLE has always been to minimally increase the empirical fit of the initial estimator while achieving the desired bias reduction for the target parameter, measured by solving $P_n D^*(P_n^*)$ with a good estimator P_n^* of P_0 (so not worse than P_n^0). However, if P_n^0 is far away from P_0, the MLE ϵ_n^0 will be far from local. Even though it moves in the right direction at $\epsilon \approx 0$, there is no guarantee that it follows a path of maximal change in target parameter per change in distribution once ϵ moves farther away from zero. In the end, that means that the TMLE might not have followed such a targeted path after all, and it might have taken various iterations to finally end up with a local $\epsilon_n^K \approx 0$ at which point the algorithm stops. The distribution P_n^0 might have changed much more than needed to obtain the bias reduction in the target parameter. That is, the desired bias reduction came at an unnecessary cost of data fitting so that $\Psi(P_n^*)$ will have larger finite sample variance than needed. Based on this insight, we would like to construct a TMLE that is based on a path that at each ϵ (not just at $\epsilon = 0$) follows a path of maximal change in target parameter per unit of information. We will refer to such a path as a *universal* least favorable submodel.

Definition 5.1. Suppose that, given a $P \in \mathcal{M}$, $U\mathrm{lfm}(P) = \{P_\epsilon : \epsilon \in (-a, a)\} \subset \mathcal{M}$ is a parametric submodel dominated by P, such that $P_{\epsilon=0} = P$ and for each $\epsilon \in (-a, a) \subset \mathbf{R}$, we have

$$\frac{d}{d\epsilon} \log \frac{dP_\epsilon}{dP} = D^*(P_\epsilon). \tag{5.1}$$

Then, we say that $U\mathrm{lfm}(P)$ is a universal least favorable submodel through P.

That is, this least favorable model is not only least favorable at $\epsilon = 0$, it is a least favorable model at each $P_\epsilon \in U\mathrm{lfm}(P)$. This chapter proposes such universal least favorable submodels and corresponding targeted maximum likelihood and targeted minimum loss-based estimators.

A very nice by-product of these universal least favorable submodels is that the TMLE always "converges" in one step, as shown in next subsection. This reflects the above intuition of a universal least favorable submodel as a shortest path submodel in the sense that it achieves the desired bias reduction at minimal increase in empirical log-likelihood.

5.2 A Universal Least Favorable Submodel for Targeted *Maximum* Likelihood Estimation

Let P_n^0 be an initial estimator of P_0. Suppose that, given a $P \in \mathcal{M}$, we can construct a universal least favorable parametric model $U\mathrm{lfm}(P) = \{P_\epsilon : \epsilon \in (-a, a)\} \subset \mathcal{M}$. If we use this as parametric submodel in the TMLE, then the TMLE converges in one step. That is, let

$$\epsilon_n^0 = \arg\max_\epsilon P_n \log \frac{dP_{n,\epsilon}^0}{dP_n^0}.$$

One can replace the maximum ϵ_n^0 by the local maximum closest to $\epsilon = 0$, which is what we recommend in case the selected universal least favorable submodel allows for multiple local maxima. Let $P_n^1 = P_{n,\epsilon_n^0}^0$. Since ϵ_n^0 is a local maximum it solves its score equation, given by $P_n D^*(P_n^1) = 0$. That is, it achieves the goal of solving the desired efficient influence curve equation in one step. Further iteration will not yield further updates: the next MLE

$$\epsilon_n^1 = \arg \max_\epsilon P_n \log \frac{dP_{n,\epsilon}^1}{dP_n^1} = 0.$$

Therefore, the TMLE of $\psi_0 = \Psi(P_0)$ is given by the one-step TMLE $\psi_n^* = \Psi(P_n^1)$.

In addition, we strongly suspect that a TMLE using such a universal least favorable model will often perform better in finite samples than an iterative TMLE using a local (nonuniversal) least favorable submodel. In addition, it is philosophically superior by always following a path along ϵ in which the rate of square change in the parameter by unit of information is maximized at each ϵ-value.

5.2.1 Analytic Formula

This motivates us to consider if such a universal least favorable model exists and can be constructed. The answer is, yes, as our constructions below demonstrate. In the following we use p_ϵ for the density of P_ϵ with respect to P, so that $p = 1$, but we will still use p (in case one wants to use the formulas for densities with respect to another dominating measure). For $\epsilon \geq 0$, we recursively define

$$p_\epsilon = p \exp\left(\int_0^\epsilon D^*(P_x)dx \right), \tag{5.2}$$

and, for $\epsilon < 0$, we recursively define

$$p_\epsilon = p \exp\left(-\int_\epsilon^0 D^*(P_x)dx \right).$$

Theorem 5.1. *Consider the definition of* $\{P_\epsilon : \epsilon \in (-a, a)\}$ *above. We have that* $\{P_\epsilon : \epsilon \in (-a, a)\}$ *is a set of probability distributions dominated by* P, $P_{\epsilon=0} = P$, *and, for each* $\epsilon \in (-a, a)$, *we have*

$$\frac{d}{d\epsilon} \log \frac{dP_\epsilon}{dP} = D^*(P_\epsilon).$$

Proof. It follows trivially that for each ϵ, $\frac{d}{d\epsilon} \log p_\epsilon = D^*(P_\epsilon)$. It remains to verify that p_ϵ satisfies $\int p_\epsilon(o)dP(o) = 1$ (obviously, $p_\epsilon \geq 0$). Define $C(\epsilon, P) \equiv \int p_\epsilon dP$. Consider the probability density $p_{\epsilon,1} = C(\epsilon, P)^{-1}p_\epsilon$. We have that its score at ϵ is given by:

$$S(\epsilon, P) = \frac{1}{C(\epsilon, P)} \frac{d}{d\epsilon} C(\epsilon, P) + D^*(P_\epsilon).$$

We know that $P_\epsilon S(\epsilon, P) = 0$. Since $P_\epsilon D^*(P_\epsilon) = 0$, this implies that $\frac{d}{d\epsilon} C(\epsilon, P) = 0$. Thus, $C(\epsilon, P) = C(0, P) = 1$. This completes the proof. $\square$

Note that this recursive relation (5.2) allows one to recursively solve for $p_{\epsilon+d\epsilon}$, given $\{p_x : x \in [0, \epsilon]\}$, in the sense that (e.g.) for $\epsilon > 0$,

$$\frac{p_{\epsilon+d\epsilon}}{p_\epsilon} = \exp(D^*(P_\epsilon)d\epsilon) = (1 + d\epsilon D^*(P_\epsilon)).$$

This differential equation is equivalent to stating that $\frac{d}{d\epsilon} \log p_\epsilon = D^*(P_\epsilon)$. This implies a practical construction that starts with $p_{x_0=0} = p$ and recursively solves for

$$p_{x_j} = p_{x_{j-1}}(1 + (x_j - x_{j-1})D^*(P_{x_{j-1}})), \quad j = 1, \ldots, N,$$

for an arbitrary fine grid $0 = x_0 < x_1 < \ldots < x_N = a$. Similarly, one determines recursively

$$p_{-x_j} = p_{-x_{j-1}}(1 - (x_j - x_{j-1})D^*(P_{-x_{j-1}})), \quad j = 1, \ldots, N.$$

If the model $\mathcal{M}$ is nonparametric, then this practical construction is a submodel of $\mathcal{M}$. But if the model is restricted, the practical construction above might select probability distributions P_{x_j} that are not an element of $\mathcal{M}$, even though it has score at x_j equal to $D^*(P_{x_j})$ in the tangent space at P_{x_j} of the model $\mathcal{M}$. Nonetheless, this practical construction of this least favorable model can be used for any model $\mathcal{M}$, as long as one can extend the target parameter Ψ to be well defined on the probability distributions in this discrete approximation of the theoretical least favorable model. The TMLE will still only require one step and be asymptotically efficient for the actual model $\mathcal{M}$ under regularity conditions. In addition, in the next subsection, Theorem 5.2 proves that under mild regularity conditions, quite surprisingly, the theoretical formula (5.2) for this universal least favorable model, defined as a limit of the above practical construction when the partitioning gets finer and finer, is an actual submodel of $\mathcal{M}$. Another way of viewing this result is that by selecting the partitioning finely enough in the above practical construction $\{p_{x_j}, p_{-x_j} : j = 0, \ldots, N\}$, we obtain a sequence of densities that are arbitrarily close to the model $\mathcal{M}$. Below we will also provide an alternative to the above practical construction that does preserve the submodel property while it still approximates the theoretical formula (5.2).

5.2.2 Universal Least Favorable Submodel in Terms of a Local Least Favorable Submodel

An alternative representation of the above analytic formula (5.2) is given by a product integral representation. Let $d\epsilon > 0$. For $\epsilon \geq 0$, we define

$$p_{\epsilon+d\epsilon} = p \prod_{x \in (0,\epsilon]} (1 + D^*(P_x)dx),$$

and for $\epsilon < 0$, we define

$$p_{\epsilon-d\epsilon} = p \prod_{x \in [\epsilon,0)} (1 - D^*(P_x)dx).$$

In other words, $p_{x+dx} = p_x(1 + D^*(P_x)dx)$, or, another way of thinking about this, is that p_{x+dx} is obtained by constructing a least favorable model through P_x with score $D^*(P_x)$ at P_x, and evaluate it at parameter value dx, slightly away from zero. This suggests the following generalization of the universal least favorable model whose practical analogue will now still be an actual submodel of $\mathcal{M}$.

Let $0 = x_0 < x_1 < \ldots \leq x_N = a$ be an equally spaced fine grid for the interval $[0, a]$. Let $h = x_j - x_{j-1}$ be the width of the partition elements. We will provide a construction for P_{x_j}, $j = 0, \ldots, N$. This construction is expressed in terms of a mapping $P \to \{P_\delta^{\text{lfm}} : \delta \in (-a, a)\} \subset \mathcal{M}$ that maps any $P \in \mathcal{M}$ into a local least favorable submodel of $\mathcal{M}$ through P at $\delta = 0$ and with score $D^*(P)$ at $\delta = 0$, where a is some positive number. For any estimation problem defined by $\mathcal{M}$ and Ψ one is typically able to construct such a local least favorable submodel, so that this is hardly an assumption. Let $P_{x=0} = P$. Let $p_{x_1} = p_{x_0,h}^{\text{lfm}}$, and, in general, let $p_{x_{j+1}} = p_{x_j,h}^{\text{lfm}}$, $j = 1, 2, \ldots, N - 1$. Similarly, let $-a = -x_N < -x_{N-1} < \ldots < -x_1 < x_0 = 0$ be the corresponding grid for $[-a, 0]$, and we define $p_{-x_{j+1}} = p_{-x_j,-h}^{\text{lfm}}$, $j = 1, \ldots, N - 1$. In this manner, we have defined P_{x_j}, P_{-x_j}, $j = 0, \ldots, N$, and, by construction, each of these are probability distributions in the model $\mathcal{M}$. The choice N defines an end value a, but one does not need to *a priori* select N. One only needs to select a small $dx = x_j - x_{j-1}$, and continue until the first local MLE is reached. This construction is all we need when using the universal least favorable submodel in practice, such as in the TMLE.

This practical construction implies a theoretical formulation by letting N converge to infinity (i.e., let the width of the partitioning converge to zero). That is, an analytic way of representing this universal least favorable submodel, given the local least favorable model parameterization $(\epsilon, P) \to p_\epsilon^{\text{lfm}}$, is given by: for $\epsilon > 0$ and $d\epsilon > 0$, we have

$$p_{\epsilon+d\epsilon} = p_{\epsilon,d\epsilon}^{\text{lfm}}.$$

This allows for the recursive solving for p_ϵ starting at $p_{\epsilon=0} = p$, and since $p_{\epsilon,h}^{\text{lfm}} \in \mathcal{M}$, its practical approximation will never leave the model $\mathcal{M}$.

Utilizing that the least favorable model $h \to p_{\epsilon,h}^{\text{lfm}}$ is continuously twice differentiable with a score $D^*(P_\epsilon)$ at $h = 0$, we obtain a second-order Taylor expansion

$$p_{\epsilon,d\epsilon}^{\text{lfm}} = p_\epsilon + \frac{d}{dh}p_{\epsilon,h}^{\text{lfm}}\bigg|_{h=0} d\epsilon + O((d\epsilon)^2) = p_\epsilon(1 + d\epsilon D^*(P_\epsilon)) + O((d\epsilon)^2),$$

so that we obtain

$$p_{\epsilon+d\epsilon} = p_\epsilon(1 + d\epsilon D^*(P_\epsilon)) + O((d\epsilon)^2).$$

This implies:

$$p_\epsilon = p \exp\left(\int_0^\epsilon D^*(P_x)dx\right).$$

Thus, we obtained the exact same representation (5.2) as above. This proves that, under mild regularity conditions, this analytic representation (5.2) is a submodel of $\mathcal{M}$ after all. But, when using its practical implementation and approximation, one should use an actual local least favorable submodel in order to guarantee that one stays in the model. We formalize this result in the following theorem.

Theorem 5.2. *Let O be a maximal support so that the support of a $P \in M$ is a subset of O. Suppose there exists a mapping $P \to \{P_\delta^{\text{lfm}} : \delta \in (-a, a)\} \subset M$ that maps any $P \in M$ into a local least favorable submodel of M through P at $\delta = 0$ and with score $D^*(P)$ at $\delta = 0$, where a is some positive number independent of P. In addition, assume the following type of second-order Taylor expansion:*

$$p_{\epsilon,d\epsilon}^{\text{lfm}} = p_\epsilon + \frac{d}{dh} p_{\epsilon,h}^{\text{lfm}}\Big|_{h=0} d\epsilon + R_2(p_\epsilon, d\epsilon),$$

where

$$\sup_\epsilon \sup_{o \in O} | R_2(p_\epsilon, d\epsilon)(o) | = O((d\epsilon)^2).$$

We also assume that $\sup_\epsilon \sup_{o \in O} | D^*(P_\epsilon) p_\epsilon | (o) < \infty$. Then, the universal least favorable $\{p_\epsilon : \epsilon\}$ defined by (5.2) is an actual submodel of M. Its definition corresponds with $p_{\epsilon+d\epsilon} = p_{\epsilon,d\epsilon}^{\text{lfm}}$ whose corresponding practical approximation will still be a submodel.

5.3 Example: One-Step TMLE for the ATT

The iterative TMLE for estimating the average treatment effect among the treated (ATT) parameter returns to the data several times to make a sequence of local moves that updates the estimate of $\bar{Q}_n(A, W)$ and $\bar{g}_n(A, W)$ at each iteration. In contrast, the one-step TMLE using the universal least favorable submodel fits the data once, where the MLE step requires a series of micro updates within a much smaller local neighborhood defined by a tuning parameter step size, $d\epsilon$. When there is sufficient information in the data for estimating the target parameter these two approaches can be expected to have comparable performance. When there is sparsity in the data theory suggests the one-step TMLE will be more stable, having lower variance than the iterative TMLE.

Let $O = (W, A, Y) \sim P_0$ and let M be a nonparametric statistical model. Let $\Psi : M \to \mathbb{R}$ be defined by $\Psi(P) = E_P(E_P(Y \mid A = 1, W) - E_P(Y \mid A = 0, W) \mid A = 1)$. The efficient influence curve of Ψ at P is given by van der Laan et al. (2013b):

$$D^*(P)(O) = H_1(g, q)(A, W)(Y - \bar{Q}(A, W)) + \frac{A}{q}\{\bar{Q}(1, W) - \bar{Q}(0, W) - \Psi(P)\},$$

where $g(a \mid W) = P(A = a \mid W)$, $\bar{Q}(a, W) = E_P(Y \mid A = a, W)$, $q = P(A = 1)$, and

$$H_1(g, q)(A, W) = \frac{A}{q} - \frac{(1 - A)g(1 \mid W)}{qg(0 \mid W)}.$$

We note that

$$\Psi(P) = \Psi_1(Q_W, \bar{Q}, g, q) = \int \{\bar{Q}(1, w) - \bar{Q}(0, w)\} \frac{g(1 \mid w)}{q} dQ_W(w),$$

where Q_W is the probability distribution of W under P. So, if we define $Q = (Q_W, \bar{Q}, g, q)$, then $\Psi(P) = \Psi_1(Q)$. For notational convenience, we will use $\Psi(P)$ and $\Psi(Q)$ interchangeably. Since we can estimate Q_W and q with their empirical probability distributions, we are only interested in a universal least favorable submodel for $(\bar{Q}, g)$. We can orthogonally decompose $D^*(P) = D_1^*(P) + D_2^*(P) + D_3^*(P)$ in $L_0^2(P)$ into scores of $\bar{Q}, g$, and Q_W, respectively, where

$$D_1^*(P) = H_1(g, q)(A, W)(Y - \bar{Q}(A, W))$$
$$D_2^*(P) = H_2(Q)(W)(A - g(1 \mid W))$$
$$D_3^*(P) = \frac{g(1 \mid W)}{q}\{\bar{Q}(1, W) - \bar{Q}(0, W) - \Psi(Q)\},$$

and

$$H_2(Q)(W) = \frac{\bar{Q}(1, W) - \bar{Q}(0, W) - \Psi(Q)}{q}.$$

Thus the component of the efficient influence curve corresponding with $(\bar{Q}, g)$ is given by $D_1^*(Q) + D_2^*(Q)$.

We consider the following loss-functions and local least favorable submodels for $\bar{Q}$ and g (van der Laan et al. 2013b):

$$L_1(\bar{Q})(O) = -\{Y \log \bar{Q}(A, W) + (1 - Y) \log(1 - \bar{Q}(A, W))\}$$
$$\text{Logit}\bar{Q}_\epsilon^{\text{lfm}} = \text{Logit}\bar{Q} - \epsilon H_1(g, q)$$
$$L_2(g)(O) = -\{A \log g(1 \mid W) + (1 - A) \log g(0 \mid W)\}$$
$$\text{Logit}\bar{g}_\epsilon^{\text{lfm}} = \text{Logit}\bar{g} - \epsilon H_2(Q).$$

We now define the sum loss function $L(\bar{Q}, g) = L_1(\bar{Q}) + L_2(g)$ and local least favorable submodel $\{Q_\epsilon^{\text{lfm}}, g_\epsilon^{\text{lfm}} : \epsilon\}$ through $(\bar{Q}, g)$ at $\epsilon = 0$ satisfying

$$\frac{d}{d\epsilon}L(\bar{Q}_\epsilon^{\text{lfm}}, g_\epsilon^{\text{lfm}})\bigg|_{\epsilon=0} = D_1^*(Q) + D_2^*(Q).$$

Thus, we can conclude that this defines indeed a local least favorable submodel for $(\bar{Q}, g)$.

In our previous work on the TMLE for the ATT, we implemented the TMLE based on the local least favorable submodel $\{\bar{Q}_{\epsilon_1}^{\text{lfm}}, \bar{g}_{\epsilon_2}^{\text{lfm}} : \epsilon_1, \epsilon_2\}$, using a separate ϵ_1 and ϵ_2 for $\bar{Q}$ and $\bar{g}$. This TMLE can also be implemented using a single ϵ by regressing a dependent variable vector (Y, A) on a stacked design matrix consisting of an offset and covariate H, the vector $(H_1(g, q)(A, W), H_2(Q)(W))$. This TMLE require several iterations until convergence, whether it is implemented using a single ϵ or separate (ϵ_1, ϵ_2).

The universal least favorable submodel (5.3) is now defined by the following recursive definition: for $\epsilon \geq 0$ and $d\epsilon > 0$,

$$
\begin{aligned}
\text{Logit}\bar{Q}_{\epsilon+d\epsilon} &= \text{Logit}\bar{Q}^{\text{lfm}}_{\epsilon,d\epsilon} \\
&= \text{Logit}\bar{Q}_{\epsilon} - d\epsilon H_1(g_{\epsilon}, q) \\
\text{Logit}\bar{g}_{\epsilon+d\epsilon} &= \text{Logit}\bar{g}^{\text{lfm}}_{\epsilon,d\epsilon} \\
&= \text{Logit}\bar{g}_{\epsilon} - d\epsilon H_2(Q_W, \bar{Q}_{\epsilon}, q).
\end{aligned}
$$

Similarly, we have a recursive relation for $\epsilon < 0$, but since all these formulas are just symmetric versions of the $\epsilon > 0$ case, we will focus on $\epsilon > 0$. This expresses the next $(Q_{\epsilon+d\epsilon}, g_{\epsilon+d\epsilon})$ in terms of previously calculated $(Q_x, g_x : x \leq \epsilon)$, thereby fully defining this universal least favorable submodel. This recursive definition corresponds with the following integral representation of this universal least favorable submodel:

$$
\begin{aligned}
\text{Logit}\bar{Q}_{\epsilon} &= \text{Logit}\bar{Q} - \int_0^{\epsilon} H_1(g_x, q)dx \\
\text{Logit}\bar{g}_{\epsilon} &= \text{Logit}\bar{g} - \int_0^{\epsilon} H_2(Q_W, \bar{Q}_x, q)dx.
\end{aligned}
$$

Let's now explicitly verify that this indeed satisfies the key property of a universal least favorable submodel. Clearly, it is a submodel and it contains (Q, g) at $\epsilon = 0$. The score of $\bar{Q}_{\epsilon}$ at ϵ is given by $H_1(g_{\epsilon}, q)(Y - \bar{Q}_{\epsilon})$ and the score of g_{ϵ} at ϵ is given by $H_2(Q_W, \bar{Q}_{\epsilon}, q)(A - \bar{g}_{\epsilon}(W))$, so that

$$
\begin{aligned}
\frac{d}{d\epsilon} L(\bar{Q}_{\epsilon}, g_{\epsilon}) &= H_1(g_{\epsilon}, q)(Y - \bar{Q}_{\epsilon}) + H_2(Q_W, \bar{Q}_{\epsilon}, q)(A - \bar{g}_{\epsilon}(W)) \\
&= D_1^*(Q_W, \bar{Q}_{\epsilon}, g_{\epsilon}, q) + D_2^*(Q_W, \bar{Q}_{\epsilon}, g_{\epsilon}, q),
\end{aligned}
$$

explicitly proving that indeed this is a universal least favorable model for $(\bar{Q}, g)$.

The TMLE based on the universal least favorable submodel above is implemented as follows, given an initial estimator $(\bar{Q}, g)$. One first determines the sign of the derivative at $h = 0$ of $P_n L(\bar{Q}_h, g_h)$. Suppose that the derivative is negative so that it decreases for $h > 0$. Then, one keeps iteratively calculating $(\bar{Q}_{\epsilon+d\epsilon}, g_{\epsilon+d\epsilon})$ for small $d\epsilon > 0$, given $(\bar{Q}_x, g_x : x \leq \epsilon)$, until $P_n L(\bar{Q}_{\epsilon+d\epsilon}, g_{\epsilon+d\epsilon}) \geq P_n L(\bar{Q}_{\epsilon}, g_{\epsilon})$, at which point the desired local maximum likelihood ϵ_n is attained. The TMLE of $(\bar{Q}_0, g_0)$ is now given by $\bar{Q}_{\epsilon_n}, g_{\epsilon_n}$, which solves $P_n\{D_1^*(Q_{\epsilon_n}) + D_2^*(Q_{\epsilon_n})\} = 0$, where $Q_{\epsilon_n} = (Q_{W,n}, \bar{Q}_{\epsilon_n}, g_{\epsilon_n}, q_n)$, and $Q_{W,n}, q_n$ are the empirical counterparts of $Q_{W,0}, q_0$. Since, we also have $P_n D_3^*(Q_{\epsilon_n}) = 0$, it follows that $P_n D^*(Q_{\epsilon_n}) = 0$. The (one-step) TMLE of $\Psi(Q_0)$ is given by the corresponding plug-in estimator $\Psi(Q_{\epsilon_n})$.

Simulation. In van der Laan and Gruber (2016), we present two simulation studies demonstrating these properties. Here we report on the first simulation. The iterative TMLE was implemented using a single ϵ, the closest analog to the one-step TMLE. For details, including source code, we refer to van der Laan and Gruber

(2016). For this study, 1000 datasets were generated at two sample sizes, $n = 100$ and $n = 1000$. Two normally distributed covariates and one binary covariate were generated as $W_1 \sim N(0, 1)$, $W_2 \sim N(0, 1)$, $W_3 \sim Bern(0.5)$. All covariates are independent. Treatment assignment probabilities are given by $P(A = 1 \mid W) = expit(-0.4 - 0.2W_1 - 0.4W_2 + 0.3W_3)$. A binary outcome, Y was generated by setting $P(Y = 1 \mid A, W) = expit(-1.2 - 1.2A - 0.1W_1 - 0.2W_2 - 0.1W_3)$. The true value of the ATT parameter is $\psi_0 = -0.1490$. There are no theoretical positivity violations (treatment assignment probabilities were typically between 0.07 and 0.87), but at the smaller sample size there is less information in the data for estimating g_0 within some strata of W. This suggests that some of the generated data sets will prove more challenging to the iterative TMLE than to the one-step TMLE. Estimates were obtained using correct and misspecified logistic regressions for the initial estimates of Q_0 and g_0. Q_{cor} was estimated using a logistic regression of Y on A, W_1, W_2, W_3. Q_{mis} was estimated using a logistic regression of Y on A, W_1. We estimated g_{cor} using a logistic regression of A on W_1, W_2, W_3, and g_{mis} was estimated using a logistic regression of A on W_1. Bias, variance, mean squared error (MSE), and relative efficiency ($RE = MSE_{one-step} / MSE_{iter}$) are shown in Table 5.1. $RE < 1$ indicates the one-step TMLE has better finite sample efficiency than the iterative TMLE.

The one-step and iterative TMLEs exhibit similar performance when $n = 1000$, with RE = 1. When $n = 100$, the iterative TMLE failed to converge for 24 of the 1000 datasets. The performance of the two TMLEs on the remaining 976 datasets was quite similar. However, the fact that the bias, variance, and MSE of the one-step TMLE are larger when evaluated over all 1000 datasets tells us that the 24 omitted datasets where the iterative TMLE failed were among the most challenging. One way to repair the performance of the iterative TMLE is to bound predicted outcome probabilities away from 0 and 1. We re-analyzed the same 1000 datasets enforcing bounds on $\bar{Q}_n$ of $(10^{-9}, 1-10^{-9})$ for both estimators. This minimal bounding prevents the iterative TMLE from failing, and should not introduce truncation bias. Bounding $\bar{Q}_n$ allowed the iterative TMLE to produce a result for all analyses. Enforcing bounds had no effect on estimates produced by the one-step TMLE. This confirms that the strategy of taking many small steps within a local neighborhood whose boundaries shift minutely with each iteration helps avoid extremes. Although the iterative TMLE no longer failed when $\bar{Q}_n$ was bounded, it had higher variance and MSE than the one-step TMLE. Efficiency gains of the one-step TMLE were between 7 and 28%. See Table 5.1.

5.4 Universal Least Favorable Model for Targeted *Minimum* Loss-Based Estimation

Let's now generalize this construction of a universal least favorable with respect to log-likelihood loss to general loss functions so that the resulting universal least favorable submodels can be used in the more general targeted minimum loss-based estimation methodology. We now assume that $\Psi(P) = \Psi_1(Q(P))$ for some parameter $Q : \mathcal{M} \to Q(\mathcal{M})$ defined on the model and real valued function Ψ_1. Here $Q(\mathcal{M}) =$

Table 5.1 Simulation study

	Bias		Variance		MSE		
	One-step	Iterative	One-step	Iterative	One-step	Iterative	RE
n = 1000							
Q correct							
g_{cor}	−0.00042	−0.00042	0.00059	0.00059	0.00059	0.00059	1.00
g_{mis}	−0.00050	−0.00050	0.00057	0.00057	0.00057	0.00057	1.00
Q misspecified							
g_{cor}	−0.00035	−0.00035	0.00059	0.00059	0.00059	0.00059	1.00
g_{mis}	0.01210	0.01210	0.00049	0.00048	0.00063	0.00063	1.00
n = 100, all runs							
Q correct							
g_{cor}	0.00049		0.00694		0.00693		
g_{mis}	−0.00215		0.00635		0.00635		
Q misspecified							
g_{cor}	0.00113		0.00685		0.00684		
g_{mis}	0.01226		0.00528		0.00543		
n = 100, (24 runs omitted)							
Q correct							
g_{cor}	0.00296	0.00295	0.00679	0.00678	0.00679	0.00679	1.00
g_{mis}	0.00023	0.00023	0.00621	0.00621	0.00621	0.00620	1.00
Q misspecified							
g_{cor}	0.00357	0.00363	0.00671	0.00669	0.00671	0.00670	1.00
g_{mis}	0.01474	0.01473	0.00509	0.00509	0.00530	0.00530	1.00
n = 100, Q bounded[a]							
Q correct							
g_{cor}	0.00049	−0.00182	0.00694	0.00781	0.00693	0.00781	0.89
g_{mis}	−0.00215	−0.00168	0.00635	0.01033	0.00635	0.01033	0.62
Q misspecified							
g_{cor}	0.00113	−0.00052	0.00685	0.00738	0.00684	0.00738	0.93
g_{mis}	0.01226	0.01031	0.00528	0.00592	0.00543	0.00602	0.90

[a]Bounding $\bar{Q}_n$ had no effect on estimates produced when $n = 1000$

Bias, variance, MSE and RE of the one-step TMLE and iterative TMLE over 1000 Monte Carlo simulations ($n = 1000$ and $n = 100$)

$\{Q(P) : P \in \mathcal{M}\}$ denotes the parameter space of this parameter Q. Let $L(Q)(O)$ be a loss-function for $Q(P)$ in the sense that $Q(P) = \arg\min_{Q \in Q(\mathcal{M})} PL(Q)$. With slight abuse of notation, let $D^*(P) = D^*(Q(P), G(P))$ be the canonical gradient of Ψ at P, where $G : \mathcal{M} \to G(\mathcal{M})$ is some nuisance parameter. We consider the case that the efficient influence curve is in the tangent space of Q, so that a least favorable submodel does not need to fluctuate G: otherwise, just include G in the definition of Q. Given, (Q, G), let $\{Q_\epsilon^{\text{lfm}} : \epsilon \in (-a, a)\} \subset Q(\mathcal{M})$ be a local least favorable model w.r.t. loss function $L(Q)$ at $\epsilon = 0$ so that

$$\frac{d}{d\epsilon} L(Q_\epsilon^{\text{lfm}})\Big|_{\epsilon=0} = D^*(Q, G).$$

The dependence of this submodel on G is suppressed in this notation.

Let $0 = x_0 < x_1 < \ldots < x_N = a$ be an equally spaced fine grid for the interval $[0, a]$. Let $h = x_j - x_{j-1}$ be the width of the partition elements. We present a construction for Q_{x_j}, $j = 0, \ldots, N$. Let $Q_{x=0} = Q$. Let $Q_{x_1} = Q_{x_0, h}^{\text{lfm}}$, and, in general, let $Q_{x_{j+1}} = Q_{x_j, h}^{\text{lfm}}$, $j = 1, 2, \ldots, N-1$. Similarly, let $-a = -x_N < -x_{N-1} < \ldots < -x_1 < x_0 = 0$ be the corresponding grid for $[-a, 0]$, and we define $Q_{-x_{j+1}} = Q_{-x_j, -h}^{\text{lfm}}$, $j = 1, \ldots, N-1$. In this manner, we have defined Q_{x_j}, Q_{-x_j}, $j = 0, \ldots, N$, and, by construction, each of these are an element of the parameter space $Q(\mathcal{M})$. This construction is all we need when using this submodel in practice, such as in the TMLE.

An analytic way of representing this loss-function specific universal least favorable submodel for $\epsilon \geq 0$ (and similarly for $\epsilon < 0$) is given by: for $\epsilon > 0$, $d\epsilon > 0$,

$$Q_{\epsilon + d\epsilon} = Q_{\epsilon, d\epsilon}^{\text{lfm}}, \tag{5.3}$$

allowing for the recursive solving for Q_ϵ starting at $Q_{\epsilon=0} = Q$, and since $Q_{\epsilon, h}^{\text{lfm}} \in Q(\mathcal{M})$, its practical approximation never leaves the parameter space $Q(\mathcal{M})$ for Q.

Let's now derive a corresponding integral equation. Assume that for some $\dot{L}(Q)(O)$, we have

$$\frac{d}{dh} L(Q_{\epsilon, h}^{\text{lfm}}) \Big|_{h=0} = \dot{L}(Q_\epsilon) \frac{d}{dh} Q_{\epsilon, h}^{\text{lfm}} \Big|_{h=0}.$$

Then, by the local property of a least favorable submodel,

$$\frac{d}{dh} Q_{\epsilon, h}^{\text{lfm}} \Big|_{h=0} = \frac{D^*(Q_\epsilon, G)}{\dot{L}(Q_\epsilon)}.$$

Utilizing that the local least favorable model $h \to Q_{\epsilon, h}^{\text{lfm}}$ is twice continuously differentiable with derivative $D^*(Q_\epsilon, G)/\dot{L}(Q_\epsilon)$ at $h = 0$, we obtain the following second-order Taylor expansion:

$$\begin{aligned}
Q_{\epsilon, d\epsilon}^{\text{lfm}} &= Q_\epsilon + \frac{d}{dh} Q_{\epsilon, h}^{\text{lfm}} \Big|_{h=0} d\epsilon + O((d\epsilon)^2) \\
&= Q_\epsilon + \frac{D^*(Q_\epsilon, G)}{\dot{L}(Q_\epsilon)} d\epsilon + O((d\epsilon)^2).
\end{aligned}$$

Note that Q_ϵ can also be represented as $Q_{\epsilon, 0}^{\text{lfm}}$. This implies the following recursive analytic definition of the universal least favorable model through Q:

$$Q_\epsilon = Q + \int_0^\epsilon \frac{D^*(Q_x, G)}{\dot{L}(Q_x)} dx. \tag{5.4}$$

Similarly, for $\epsilon < 0$, we obtain

$$Q_\epsilon = Q - \int_\epsilon^0 \frac{D^*(Q_x, G)}{\dot{L}(Q_x)} dx.$$

As with the log-likelihood loss (and thus $Q(P) = P$), this shows that, under regularity conditions, this analytic representation for Q_ϵ is an element in $Q(\mathcal{M})$, although using it in a practical construction (in which integrals are replaced by sums) might easily leave the model space $Q(M)$. On the other hand, our above practical construction in terms of the local least favorable model and discrete grid represents the desired practical implementation of this universal least favorable submodel. The following theorem formalizes this result stating that the analytic formulation (5.4) is indeed a universal least favorable submodel.

Theorem 5.3. *Given, any (Q, G) compatible with model M, let $\{Q_\delta^{\text{lfm}} : \delta \in (-a, a)\} \subset Q(M)$ be a local least favorable model w.r.t. loss function $L(Q)$ at $\delta = 0$ so that*

$$\frac{d}{d\delta} L(Q_\delta^{\text{lfm}})\bigg|_{\delta=0} = D^*(Q, G).$$

Assume that for some $\dot{L}(Q)(O)$, we have

$$\frac{d}{d\epsilon} L(Q_\epsilon^{\text{lfm}})\bigg|_{\epsilon=0} = \dot{L}(Q) \frac{d}{d\epsilon} Q_\epsilon^{\text{lfm}}\bigg|_{\epsilon=0}.$$

Consider the corresponding model $\{Q_\epsilon : \epsilon\}$ defined by (5.4). It goes through Q at $\epsilon = 0$, and, it satisfies that for all ϵ

$$\frac{d}{d\epsilon} L(Q_\epsilon) = D^*(Q_\epsilon, G). \tag{5.5}$$

In addition, suppose that the $a > 0$ in the local least favorable submodel above can be chosen to be independent of the choice $(Q, G) \in \{Q_\epsilon, G_\epsilon : \epsilon\}$, and assume the following second-order Taylor expansion:

$$Q_{\epsilon, d\epsilon}^{\text{lfm}} = Q_\epsilon + \frac{d}{dh} Q_{\epsilon, h}^{\text{lfm}}\bigg|_{h=0} d\epsilon + R_2(Q_\epsilon, G, d\epsilon)$$

$$= Q_\epsilon + \frac{D^*(Q_\epsilon, G)}{\dot{L}(Q_\epsilon)} d\epsilon + R_2(Q_\epsilon, G, d\epsilon),$$

where

$$\sup_\epsilon \sup_{o \in O} | R_2(Q_\epsilon, G, d\epsilon)(o) | = O((d\epsilon)^2).$$

We also assume that $\sup_\epsilon \sup_{o \in O} | \frac{D^*(Q_\epsilon, G)}{\dot{L}(Q_\epsilon)}(o) | < \infty$.
Then, we also have $\{Q_\epsilon : \epsilon\} \subset Q(M)$.

Proof. Let $\epsilon > 0$. We have

$$\frac{d}{d\epsilon} L\left(Q + \int_0^\epsilon \frac{D^*(Q_x, G)}{\dot{L}(Q_x)} dx\right) = \dot{L}(Q_\epsilon) \frac{d}{d\epsilon} Q_\epsilon$$

$$= \dot{L}(Q_\epsilon) \frac{D^*(Q_\epsilon, G)}{\dot{L}(Q_\epsilon)}$$

$$= D^*(Q_\epsilon, G).$$

This completes the proof of (5.5). The submodel statement was already shown above, but we now provided formal sufficient conditions. $\square$

5.5 Universal Canonical One-dimensional Submodel for a Multidimensional Target Parameter

Let $\Psi : \mathcal{M} \to H$ be a Hilbert-space valued pathwise differentiable target parameter. Typically, we simply have $H = \mathbf{R}^d$ endowed with the standard inner product $\langle x, y \rangle = \sum_{j=1}^{d} x_j y_j$. However, we also allow that $\Psi(P)$ is a function $t \to \Psi(P)(t)$ from $\tau \subset \mathbf{R}$ to $\mathbf{R}$ in a Hilbert space $L^2(\Lambda)$ endowed with inner product $\langle h_1, h_2 \rangle = \int h_1(t) h_2(t) d\Lambda(t)$, where Λ is a user supplied positive measure with $\int d\Lambda(t) < \infty$. For notational convenience, we will often denote the inner product $\langle h_1, h_2 \rangle$ with $h_1^\top h_2$, analogue to the typical notation for the inner product in $\mathbf{R}^d$. Let $\| h \| = \sqrt{\langle h, h \rangle}$ be the Hilbert space norm, which would be the standard Euclidean norm in the case that $H = \mathbf{R}^d$. Let $D^*(P)$ be the canonical gradient. If $H = \mathbf{R}^d$, then this is a d-dimensional canonical gradient $D^*(P) = (D_j^*(P) : j = 1, \ldots, d)$, but in general $D^*(P) = (D_t^*(P) : t \in \tau)$. Let $L(p) = -\log p$, where $p = dP/d\mu$ is a density of $P \ll \mu$ w.r.t. some dominating measure μ. In this section we will construct a one-dimensional submodel $\{P_\epsilon : \epsilon \geq 0\}$ through P at $\epsilon = 0$ so that, for any $\epsilon \geq 0$,

$$\frac{d}{d\epsilon} P_n L(p_\epsilon) = \| P_n D^*(P_\epsilon) \| . \tag{5.6}$$

The one-step TMLE P_{ϵ_n} with $\epsilon_n = \arg\min_\epsilon P_n L(P_\epsilon)$, or ϵ_n chosen large enough so that the derivative is smaller than (e.g.) $1/n$, now solves $\frac{d}{d\epsilon} P_n L(P_\epsilon)\big|_{\epsilon=0} = 0$ (or $< 1/n$), and thus $\| P_n D^*(P_{\epsilon_n}) \| = 0$ (or $< 1/n$). Note that $\| P_n D^*(P_{\epsilon_n}) \| = 0$ implies that $P_n D_t^*(P_{\epsilon_n}) = 0$ for all $t \in \tau$ so that the one-step TMLE solves all desired estimating equations.

Consider the following submodel: for $\epsilon \geq 0$, we define

$$p_\epsilon = p \Pi_{[0,\epsilon]} \left(1 + \frac{\{P_n D^*(P_x)\}^\top D^*(P_x)}{\| D^*(P_x) \|} dx \right)$$

$$= p \exp \left(\int_0^\epsilon \frac{\{P_n D^*(P_x)\}^\top D^*(P_x)}{\| D^*(P_x) \|} dx \right). \tag{5.7}$$

Theorem 5.4. *We have $\{p_\epsilon : \epsilon \geq 0\}$ is a family of probability densities, its score at ϵ is a linear combination of $D_t^*(P_\epsilon)$ for $t \in \tau$, and is thus in the tangent space at $T(P_\epsilon)$, and*

$$\frac{d}{d\epsilon} P_n L(P_\epsilon) = \| P_n D^*(P_\epsilon) \| .$$

As a consequence, we have $\frac{d}{d\epsilon} P_n L(P_\epsilon) = 0$ implies $\| P_n D^*(P_\epsilon) \| = 0$.

As before, our practical construction below demonstrates that, under regularity conditions, we actually have that $\{p_\epsilon : \epsilon\} \subset \mathcal{M}$ is also a submodel.

The normalization by $\| D^*(P_x) \|$ is motivated by a practical analogue construction below and provides an important intuition behind this analytic construction. However, we can replace this by any other normalization for which the derivative of the log-likelihood at ϵ equals a norm of $P_n D^*(P_\epsilon)$. To illustrate this let's consider the case that $H = \mathbf{R}^d$. For example, we could consider the following submodel. Let $\Sigma_n(P_x) = P_n\{D^*(P_x)D^*(P_x)^\top\}$ be the empirical covariance matrix of $D^*(P_x)$, and let $\Sigma_n^{-1}(P_x)$ be its inverse. We could then define for $\epsilon > 0$,

$$p_\epsilon = p \exp\left(\int_0^\epsilon \{P_nD^*(P_x)\}^\top \Sigma_n^{-1} D^*(P_x)dx\right).$$

In this case, we have

$$\frac{d}{d\epsilon}P_nL(P_\epsilon) = P_nD^*(P_\epsilon)^\top \Sigma_n(P_\epsilon)^{-1}P_nD^*(P_\epsilon).$$

This seems to be an appropriately normalized norm, equal to the Euclidean norm of the orthonormalized version of the original $D^*(P_\epsilon)$, so that the one-step TMLE will still satisfy that $\| P_nD^*(P_{\epsilon_n}) \| = 0$.

It is not clear to us if these choices have a finite sample implication for the resulting one-step TMLE (asymptotics is the same), and if one choice would be better than another. Either way, the resulting one-step TMLE ends up with a P_{ϵ_n} satisfying $P_nD^*(P_{\epsilon_n}) = 0$ (or $o_P(1/\sqrt{n})$), which is the only key ingredient in the proof of the asymptotic efficiency of the TMLE.

5.5.1 Practical Construction

Let's define a local least favorable submodel $\{p_\delta^{\mathrm{lfm}} : \delta\} \subset \mathcal{M}$ by the following local property: for all δ

$$\frac{d}{d\delta}\log p_\delta^{\mathrm{lfm}}\bigg|_{\delta=0}^\top \delta = D^*(P)^\top\delta.$$

For the case that $H = \mathbf{R}^d$, this corresponds with assuming that the score of the submodel at $\delta = 0$ equals the canonical gradient $D^*(P)$, while, for a general Hilbert space, it states that the derivative of $\log p_\epsilon$ in the direction δ (a function in H) equals $\langle D^*(P), \delta\rangle = \int D_t^*(P)\delta(t)d\Lambda(t)$.

Consider the log-likelihood criterion $P_nL(P_\delta^{\mathrm{lfm}})$, and note that its derivative at $\delta = 0$ in the direction δ equals $\langle P_nD^*(P), \delta\rangle = \{P_nD^*(P)\}^\top\delta$. For a small number dx, we want to maximize the log-likelihood over all δ with $\| \delta \| \le dx$, and locally, this corresponds with maximizing its linear gradient approximation:

$$\delta \to \{P_nD^*(P)\}^\top\delta.$$

By the Cauchy-Schwarz inequality, it follows that this is maximized over δ with $\| \delta \| \le dx$ by

$$\delta_n^*(P, dx) = \frac{P_nD^*(P)}{\| P_nD^*(P) \|}dx \equiv \delta_n^*(P)dx,$$

where we defined $\delta_n^*(P) = P_nD^*(P)/ \| P_nD^*(P) \|$. We can now define our update $P_{dx} = P_{\delta_n^*(P,dx)}^{\mathrm{lfm}}$. This process can now be iterated by applying the above with P

replaced by P_{dx}, resulting in an update P_{2dx}, and in general P_{Kdx}. So this updating process is defined by the differential equation:

$$P_{x+dx} = P_{x,\delta_n^*(P_x)dx}^{\text{lfm}},$$

where $P_{x,\delta}^{\text{lfm}}$ is the local least favorable multidimensional submodel above but now through P_x instead of P.

Assuming that the local least favorable model $h \to p_{x,h}^{\text{lfm}}$ is continuously twice differentiable with a score $D^*(P_x)$ at $h = 0$, we obtain a second-order Taylor expansion

$$\begin{aligned}
p_{x,\delta_n^*(P_x)dx}^{\text{lfm}} &= p_x + \left\{ \frac{d}{dh} p_{x,h}^{\text{lfm}} \bigg|_{h=0} \right\}^\top \delta_n^*(P_x)dx + O((dx)^2) \\
&= p_x(1 + \{\delta_n^*(P_x)\}^\top D^*(P_x)dx) + O((dx)^2),
\end{aligned}$$

so that, under mild regularity conditions, we obtain

$$p_{x+dx} = p_x(1 + \{\delta_n^*(P_x)\}^\top D^*(P_x)dx) + O((dx)^2).$$

This implies:

$$p_x = p \exp\left(\int_0^\epsilon \frac{\{P_n D^*(P_x)\}^\top}{\| P_n D^*(P_x) \|} D^*(P_x)dx \right).$$

So we obtained the exact same analytical representation (5.7) as above. Since the above practical construction starts out with $P \in \mathcal{M}$ and never leaves the model $\mathcal{M}$, this proves that, under mild regularity conditions, this analytic representation (5.7) is actually a submodel of $\mathcal{M}$ after all. However, for the purpose of keeping practical implementation and approximation in the model $\mathcal{M}$, one should use the practical construction above based on an actual local least favorable submodel. We can formalize this in a theorem analogue to Theorem 5.2, but instead such a theorem will be presented in Sect. 5.7 for the more general targeted minimum loss-based estimation methodology.

The above practical construction provides us with an intuition for the normalization by $\| P_n D^*(P_x) \|$.

5.5.2 Existence of MLE or Approximate MLE ϵ_n

Since

$$P_n \log p_\epsilon = \int_0^\epsilon \| P_n D^*(P_x) \| \, dx,$$

and its derivative thus equals $\| P_n D^*(P_\epsilon) \|$, we have that the log-likelihood is non-decreasing in ϵ.

If the local least favorable submodel in the practical construction of the one-dimensional universal canonical submodel $\{p_\epsilon : \epsilon \geq 0\}$ (5.7) only contains densities with supremum norm smaller than some $M < \infty$ (e.g., this is assumed by the model $\mathcal{M}$), then we will have that $\sup_{\epsilon \geq 0} \sup_{o \in O} p_\epsilon(o) < M < \infty$. This implies that $P_n \log p_\epsilon$ is bounded from above by $\log M$. Let's first assume that $\lim_{\epsilon \to \infty} P_n \log p_\epsilon < \infty$. Thus, the log-likelihood is a strictly increasing function until it becomes flat, if ever. Suppose that $\limsup_{x \to \infty} \| P_n D^*(P_x) \| > \delta > 0$ for some $\delta > 0$. Then it follows that the log-likelihood converges to infinity when x converges to infinity, which contradicts the assumption that the log-likelihood is bounded from above by $\log M < \infty$. Thus, we know that $\limsup_{x \to \infty} \| P_n D^*(P_x) \| = 0$ so that we can find an ϵ_n so that for $\epsilon > \epsilon_n \| P_n D^*(P_\epsilon) \| < 1/n$, as desired.

Suppose now that we are in a case in which the log-likelihood converges to infinity when $\epsilon \to \infty$, so that our bounded log likelihood assumption is violated. This might correspond with a case in which each p_ϵ is a continuous density, but p_ϵ starts approximating an empirical distribution when $\epsilon \to \infty$. Even in such a case, one would expect that we will have that $\| P_n D^*(P_\epsilon) \| \to 0$, just like an NPMLE of a continuous density of a survival time solves the efficient influence curve equation for its survival function.

The above practical construction of the submodel, as an iterative local maximization of the log-likelihood along its gradient, strongly suggests that even without the above boundedness assumption the derivative $\| P_n D^*(P_\epsilon) \|$ will converge to zero as $\epsilon \to \infty$ so that the desired MLE or approximate MLE exists. Our initial practical implementations of this one-step TMLE of a multivariate target parameter demonstrate that it works well and that finding the desired maximum or approximate maximum is not an issue. We will demonstrate the implementation and practical demonstration of such a one-step TMLE in the next section.

5.5.3 Universal Score-Specific One-Dimensional Submodel

In the above two subsections we could simply replace $D^*(P)$ by a user supplied $D(P)$, giving us a theoretical one-dimensional parametric model $\{P_\epsilon : \epsilon\}$ so that the derivative $\frac{d}{d\epsilon} P_n L(P_\epsilon)$ at ϵ equals $\| P_n D(P_\epsilon) \|$, so that a corresponding one-step TMLE will solve $P_n D(P_{\epsilon_n}) = 0$. Similarly, given a local parametric model whose score at $\epsilon = 0$ equals $D(P)$ will yield a corresponding practical construction of this universal submodel. One can also use such a universal score-specific submodel to construct one-step TMLE of a one-dimensional target parameter with extra properties by making it solve not only the efficient influence curve equation but also other equations of interest (such as the $P_n D_1^*(Q_n^*) = P_n D_2^*(Q_n^*) = 0$ in Sect. 8.4). In the current literature, solving multiple score equations typically required an iterative TMLE based on a local score-specific submodel, so that these estimation problems can be revisited with this new one-step TMLE (see our supplementary material).

5.6 Example: One-Step TMLE, Based on Universal Canonical One-Dimensional Submodel, of an Infinite-Dimensional Target Parameter

An open problem has been the construction of an efficient substitution estimator $\Psi(P_n^*)$ of a pathwise differentiable infinite dimensional target parameter $\Psi(P_0)$ such as a survival function. Current approaches would correspond with incompatible estimators such as using a TMLE for each $\Psi(P_0)(t)$ separately, resulting in a non-substitution estimator such as a nonmonotone estimator of a survival function. In this section we demonstrate, through a causal inference example, that our universal canonical submodel allows us to solve this problem with the one-step TMLE defined in the previous section.

Let $O = (W, A, T) \sim P_0$, where W are baseline covariates, $A \in \{0, 1\}$ is a point-treatment, and T is a survival time. Consider a statistical model $\mathcal{M}$ that only makes assumptions about the conditional distribution $g_0(a \mid W) = P_0(A = a \mid W)$ of A, given W. Let $W \to d(W) \in \{0, 1\}$ be a given dynamic treatment satisfying $g_0(d(W) \mid W) > 0$ a.e. Let $\Psi : \mathcal{M} \to H$ be defined by:

$$\Psi(P)(t) = E_P P(T > t \mid A = d(W), W), \ t \geq 0.$$

Under a causal model and the randomization assumption this equals the counter-factual survival function $P(T_d > t)$ of the counterfactual survival time T_d under intervention d.

Let H be the Hilbert space of real valued functions on $\mathbf{R}_{\geq 0}$ endowed with inner product $h_1^\top h_2 = \langle h_1, h_2 \rangle = \int h_1(t)h_2(t)d\Lambda(t)$ for some user-supplied positive and finite measure Λ. The norm on this Hilbert space is thus given by $\| h \| = \sqrt{hh^\top} = \sqrt{\int h(t)^2 d\Lambda(t)}$. Let $\bar{Q}_t(A, W) = P(T > t \mid A, W)$, $Y(t) = I(T > t)$, Q_W the marginal probability distribution of W, and $Q = (\bar{Q}, Q_W)$. The efficient influence curve $D^*(P) = (D_t^*(P) : t \geq 0)$ is defined by:

$$D_t^*(P)(O) = \frac{I(A = d(W))}{g(A \mid W)}(Y(t) - \bar{Q}_t(A, W)) + \{\bar{Q}_t(d(W), W) - \Psi(P)(t)\}$$
$$\equiv D_{1,t}^*(g, \bar{Q}) + D_{2,t}^*(P),$$

where $D_{1,t}^*(g, \bar{Q})$ is the first component of the efficient influence curve that is a score of the conditional distribution of T, given A, W. Notice that $\Psi(P) = \Psi_1(Q_W, \bar{Q}) = (Q_W \bar{Q}_t : t \geq 0)$. We will estimate $Q_{W,0}$ with the empirical distribution of $W_1, \ldots, W_n$, so that a TMLE will only need to target the estimator of the conditional survival function $\bar{Q}_0$ of T, given A, W. Let $q(t \mid A, W)$ be the density of T, given A, W and let q_n be an initial estimator of this conditional density. For example, one might use machine learning to estimate the conditional hazard $q_0/\bar{Q}_0$, which then implies a corresponding density estimator q_n. We are also given an estimator g_n of g_0.

The universal canonical one-dimensional submodel (5.7) applied to $p = q_n$ is defined by the following recursive relation: for $\epsilon > 0$,

$$q_{n,\epsilon} = q_n \exp\left(\int_0^\epsilon \frac{\{P_n D_1^*(g_n, \bar{Q}_{n,x})\}^\top D_1^*(g_n, \bar{Q}_{n,x})}{\parallel D_1^*(g_n, \bar{Q}_{n,x}) \parallel} dx\right).$$

To obtain some more insight in this expression, we note, for example, that the inner product is given by:

$$\{P_n D_1^*(g_n, \bar{Q}_{n,x})\}^\top D_1^*(g_n, \bar{Q}_{n,x})(o) = \int_t P_n D_{1,t}^*(g_n, \bar{Q}_{n,x}) D_{1,t}^*(g_n, \bar{Q}_{n,x})(o) d\Lambda(t), \quad (5.8)$$

and similarly we have such an integral representation of the norm in the denominator. Our Theorem 5.4, or explicit verification, shows that for all $\epsilon \geq 0$, $q_{n,\epsilon}$ is a conditional density of T, given A, W, and

$$\frac{d}{d\epsilon} P_n \log q_{n,\epsilon} = \parallel P_n D_1^*(g_n, \bar{Q}_{n,\epsilon}) \parallel.$$

Thus, if we move ϵ away from zero, the log-likelihood increases, and, one searches for the first ϵ_n so that this derivative is smaller than (e.g.) $1/n$. Let $q_n^* = q_{n,\epsilon_n}$, and let $\bar{Q}_{n,t}^*(A, W) = \int_t^\infty q_n^*(s \mid A, W) ds$ be its corresponding conditional survival function, $t \geq 0$. Then our one-step TMLE of the d-specific survival function $\Psi(P_0)$ is given by $\psi_n^* = \Psi(Q_{W,n}, \bar{Q}_n^*) = Q_{W,n} \bar{Q}_n^*$:

$$\psi_n^*(t) = \frac{1}{n} \sum_{i=1}^n \bar{Q}_{n,t}^*(d(W_i), W_i).$$

Since q_n^* is an actual conditional density, it follows that ψ_n^* is a survival function. Suppose that the derivative of the log-likelihood at ϵ_n equals zero exactly (instead of being smaller than $1/n$). Then, we have $\parallel P_n D^*(g_n, Q_{W,n}, \bar{Q}_n^*) \parallel = 0$, so that for each $t \geq 0$, $P_n D_t^*(g_n, Q_{W,n}, \bar{Q}_n^*) = 0$, making $\psi_n^*(t)$ a standard TMLE of $\psi_0(t)$, so that its asymptotic linearity for a fixed t can be established accordingly. Let's now consider a proof of weak convergence of $\sqrt{n}(\psi_n^* - \psi_0)$ as a random function. Firstly, for simplicity, let's assume that an exact MLE is obtained so that $P_n D^*(g_n, Q_{W,n}, \bar{Q}_n^*) = 0$. Combined with $\Psi(Q_n^*) - \Psi(Q_0) = -P_0 D^*(g_n, Q_n^*) + R_2((Q_n^*, g_n), (Q_0, g_0))$, where $R_2() = (R_{2t}() : t \in \tau)$ for an explicitly defined $R_{2t}(P, P_0)$, we then obtain

$$\psi_n^* - \psi_0 = (P_n - P_0) D^*(g_n, Q_n^*) + R_2((Q_n^*, g_n), (Q_0, g_0)).$$

We now assume that $\{D_t^*(P) : P \in \mathcal{M} t \in \tau\}$ is a P_0-Donsker class, $\sup_{t \in \tau} P_0\{D_t^*(g_n, Q_n^*) - D_t^*(g_0, Q_0)\}^2 \to 0$ in probability, and $\sup_t |R_{2t}((Q_n^*, g_n), (Q_0, g_0))| = o_P(n^{-1/2})$. Then, it follows that

$$\sqrt{n}(\psi_n^* - \psi_0) = \sqrt{n}(P_n - P_0) D^*(P_0) + o_P(n^{-1/2}) \Rightarrow_d G_0.$$

That is, $\sqrt{n}(\psi_n^* - \psi_0)$ converges weakly as a random element of the cadlag function space endowed with the supremum norm to a Gaussian process G_0 with covariance structure implied by the covariance function $\rho(s, t) = P_0 D_s^*(P_0) D_t^*(P_0)$. In particular, if g_0 is known, then $R_{2t}((Q_n^*, g_0), (Q_0, g_0)) = 0$, so that the second-order term condition $\sup_t | R_{2t}((Q_n^*, g_n), (Q_0, g_0)) |= o_P(n^{-1/2})$ is automatically satisfied with $o_P(n^{-1/2})$ replaced by 0. This also allows the construction of a simultaneous confidence band for ψ_0. Due to the double robustness of the efficient influence curve, under appropriate conditions, one can also obtain asymptotic linearity and weak convergence with an inefficient influence curve under misspecification of either g_n or $\bar{Q}_n$.

If we only have $\| P_n D^*(P_n^*) \|= o_P(n^{-1/2})$ (instead of 0), then the above proof still applies so that we now obtain:

$$\sqrt{n}(\psi_n^* - \psi_0) = (P_n - P_0)D^*(P_0) + r_n,$$

but where now $\| r_n \|= o_P(1/\sqrt{n})$. In this case we obtain asymptotic efficiency and weak convergence in the Hilbert space $L^2(\Lambda)$, beyond the point-wise efficiency of $\psi_n^*(t)$. However, in practice, one can actually track the supremum norm $\| P_n D^*(P_{\epsilon_n}) \|_\infty= \sup_t | P_n D_t^*(P_{\epsilon_n}) |$, and if one observes that for the selected ϵ_n this supremum norm is smaller than $1/n$, then, we still obtain the asymptotic efficiency in supremum norm above.

Regarding the practical construction of $q_{n,\epsilon}$, we could use the following infinite dimensional local least favorable submodel through a conditional density q given by

$$q_\delta^{\text{lfm}} = q(1 + \delta^\top D_1^*(g, \bar{Q})),$$

and follow the practical construction described in the previous section for general local least favorable submodels. Notice that here $\delta^\top D_1^*(g, \bar{Q}) = \int \delta(t) D_{1,t}^*(g, \bar{Q}) d\Lambda(t)$. In order to guarantee that the supremum norm of the density q_δ^{lfm} for local δ with $\| \delta \|< dx$ remains below a universal constant $M < \infty$, one could present such models in the conditional hazard on a logistic scale that bounds the hazard between $[0, M]$. However, we suspect that this will not be an issue in practice, and since it may be necessary for the continuous density $q_{n,\epsilon}$ to approximate an empirical distribution in some sense in order to solve $\| P_n D^*(P_\epsilon) \|= 0$, we do not want to prevent this from happening.

Moore and van der Laan (2009a,b,c) proposed an iterative TMLE of $S_d(t_0)$ for a given t_0, which is defined as follows. Let $q_n(t|A, W)$ and $g_n(A | W)$ be initial estimators of $q_0(t | A, W)$ and $g_0(A | W)$. Let

$$L(q)(O) = - \sum_{t \leq T} \{I(T = t) \log q(t | A, W) + (1 - I(T = t)) \log(1 - q(t | A, W))\}$$

be the log-likelihood loss function for $q(t | A, W)$. We define the local least favorable submodel through q_n as follows:

$$\text{logit}\bar{q}_n(\varepsilon)(t|A, W) = \text{logit}\bar{q}_n^0(t|A, W) + \varepsilon H_{a,n}^*(t, A, W),$$

where the estimated time-dependent clever covariate is given by

$$H^*_{a,n}(t, A, W) = \left(\frac{I(A = a)}{g_n(a|W)}\right)\left(\frac{\bar{Q}_n(t_0|A, W)}{\bar{Q}_n(t|A, W)}\right)I(t \leqslant t_0).$$

We have $\frac{d}{d\epsilon}L(q_{n,\epsilon})$ at $\epsilon = 0$ equals $D^*_{1,t}(\bar{Q}, g)$. Let $Q_{W,n}$ be the empirical probability distribution of W. The first-step TMLE update is defined by $\epsilon_n = \arg\min_\epsilon P_n L(q_{n,\epsilon})$ and $q^1_n = q_{n,\epsilon_n}$. This updating process is iterated until $\epsilon_n \approx 0$. The final update is denoted with $q^*_n(t \mid A, W)$. Let $\bar{Q}^*_n(t \mid A, W)$ be the corresponding survival curve. The iterative TMLE of $\psi_0(t_0) = S_{d,0}(t_0)$ is given by

$$\psi^*_n(t_0) = \frac{1}{n}\sum_{i=1}^{n}\bar{Q}^*_n(t_0 \mid A = 1, W_i).$$

Simulation. Firstly, we have

$$W_1 \sim Bern(0.5); W_2 \sim Bern(0.5); A \sim Bern(0.15 + 0.5W_1);$$

$$T \sim \exp\left(\frac{1 + 0.5W_1 - 0.5A}{100}\right).$$

In this case

$$S_d(t) = 1 - \frac{\Phi(5 \times 10^{-3}t) + \Phi(10^{-2}t)}{2}, t \geqslant 0,$$

where Φ is the cumulative distribution function for exponential distribution with rate equal to 1. The second simulation is identical to the first, except that now $A \sim Bern(0.05 + 0.5W)$. The goal of these two simulations is to compare the one-step TMLE with the iterative TMLE that separately estimates $S_d(t)$ at each point t.

Figure 5.1a provides the iterative TMLE and one-step TMLE for a single data set with $n = 100$ observations from the two data generating distributions. Clearly, it follows that the iterative TMLE is not monotone, while the one-step TMLE is an actual survival curve. The iterative TMLE is particularly erratic for the data set from the second data generating distribution. Figure 5.1b provides the relative efficiencies at each time point from 0 to 400. In order to demonstrate the confounding, we also present the Kaplan-Meier estimator among the observations with $A_i = 1, i = 1, \ldots, n$. We also show the estimate of the treatment-specific survival curve based on the initial estimator. These results show that the iterative and one-step TMLE are both unbiased, but that the iterative TMLE is twice as efficient for $n = 100$. Finally, Fig. 5.1c presents the estimators for $n = 1000$, demonstrating that both the one-step TMLE and the iterative TMLE are asymptotically efficient, and that the above gain in efficiency represents a finite sample gain that disappears asymptotically.

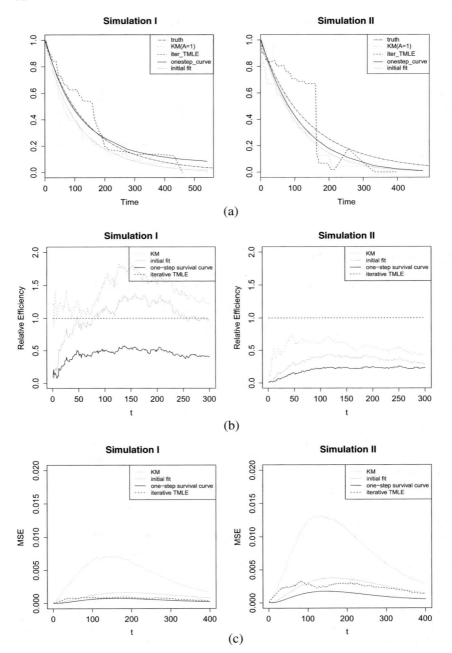

Fig. 5.1 Single data set for the two simulation settings (**a**). Monte Carlo approximation of relative efficiency against iterative TMLE, as a function of t, for sample size 100 (**b**) and 1000 (**c**)

5.7 Universal Canonical One-Dimensional Submodel for Targeted Minimum Loss-Based Estimation of a Multidimensional Target Parameter

For the sake of presentation we will focus on the case that the target parameter is Euclidean valued, i.e. $H = \mathbf{R}^d$, but the presentation immediately generalizes to infinite dimensional target parameters, as in the previous section. Let's now generalize the construction of a universal canonical submodel to the more general targeted minimum loss based estimation methodology. We now assume that $\Psi(P) = \Psi_1(Q(P)) \in \mathbf{R}^d$ for some target parameter $Q : \mathcal{M} \to Q(\mathcal{M})$ defined on the model and real valued function $\Psi_1 : Q(\mathcal{M}) \to \mathbf{R}^d$. Let $L(Q)(O)$ be a loss-function for $Q(P)$ in the sense that $Q(P) = \arg\min_{Q \in Q(\mathcal{M})} PL(Q)$. Let $D^*(P) = D^*(Q(P), G(P))$ be the canonical gradient of Ψ at P, where $G : \mathcal{M} \to G(\mathcal{M})$ is some nuisance parameter. We consider the case that the linear span of the components of the efficient influence curve $D^*(P)$ is in the tangent space of Q, so that a least favorable submodel does not need to fluctuate G: otherwise, one just includes G in the definition of Q. Given, (Q, G), let $\{Q_\delta^{\mathrm{lfm}} : \delta\} \subset Q(\mathcal{M})$ be a local d-dimensional least favorable model w.r.t. loss function $L(Q)$ at $\delta = 0$ so that

$$\frac{d}{d\delta} L(Q_\delta^{\mathrm{lfm}}) \Big|_{\delta=0} = D^*(Q, G).$$

The dependence of this submodel on G is suppressed in this notation.

Consider the empirical risk $P_n L(Q_\delta^{\mathrm{lfm}})$, and note that its gradient at $\delta = 0$ equals $P_n D^*(Q, G)$. For a small number dx, we want to minimize the empirical risk over all δ with $\| \delta \| \le dx$, and locally, this corresponds with maximizing its linear gradient approximation:

$$\delta \to \{P_n D^*(Q, G)\}^\top \delta.$$

By the Cauchy-Schwarz inequality, it follows that this is maximized over δ with $\| \delta \| \le dx$ by

$$\delta_n^*(Q, dx) = \frac{P_n D^*(Q, G)}{\| P_n D^*(Q, G) \|} dx \equiv \delta_n^*(Q) dx,$$

where we defined $\delta_n^*(Q) = P_n D^*(Q, G) / \| P_n D^*(Q, G) \|$. We can now define our update $Q_{dx} = Q_{\delta_n^*(Q, dx)}^{\mathrm{lfm}}$. This process can now be iterated by applying the above with Q replaced by Q_{dx}, resulting in an update Q_{2dx}, and in general Q_{Kdx}. So this updating process is defined by the differential equation:

$$Q_{x+dx} = Q_{x, \delta_n^*(Q_x) dx}^{\mathrm{lfm}},$$

where $Q_{x,\delta}^{\mathrm{lfm}}$ is the local least favorable multidimensional submodel above but now through Q_x instead of Q.

Assume that for some $\dot{L}(Q)(O)$, we have

$$\frac{d}{dh} L(Q_{x,h}^{\mathrm{lfm}}) \Big|_{h=0} = \dot{L}(Q_x) \frac{d}{dh} Q_{x,h}^{\mathrm{lfm}} \Big|_{h=0}. \tag{5.9}$$

Then,

$$\frac{d}{dh} Q_{x,h}^{\mathrm{lfm}} \bigg|_{h=0} = \frac{D^*(Q_x, G)}{\dot{L}(Q_x)}.$$

Utilizing that the local least favorable model $h \to Q_{x,h}^{\mathrm{lfm}}$ is continuously twice differentiable with a score $D^*(Q_x, G)$ at $h = 0$, we obtain a second-order Taylor expansion

$$
\begin{aligned}
Q_{x,\delta_n^*(Q_x)dx}^{\mathrm{lfm}} &= Q_x + \frac{d}{dh} Q_{x,h}^{\mathrm{lfm}} \bigg|_{h=0} \delta_n^*(Q_x)dx + O((dx)^2) \\
&= Q_x + \frac{D^*(Q_x, G)^\top}{\dot{L}(Q_x)} \delta_n^*(Q_x)dx + O((dx)^2).
\end{aligned}
$$

This implies the following recursive analytic definition of the universal canonical submodel through Q:

$$Q_\epsilon = Q + \int_0^\epsilon \frac{D^*(Q_x, G)^\top}{\dot{L}(Q_x)} \delta_n^*(Q_x)dx. \tag{5.10}$$

Let's now explicitly verify that this indeed satisfies the desired condition so that the one-step TMLE solves $P_n D^*(Q_{\epsilon_n}, G) = 0$. Only assuming (5.9) it follows that

$$
\begin{aligned}
\frac{d}{d\epsilon} P_n L(Q_\epsilon) &= P_n \frac{d}{d\epsilon} L(Q_\epsilon) \\
&= P_n \dot{L}(Q_\epsilon) \frac{d}{d\epsilon} Q_\epsilon \\
&= P_n \dot{L}(Q_\epsilon) \frac{D^*(Q_\epsilon, G)^\top}{\dot{L}(Q_\epsilon)} \delta_n^*(Q_\epsilon) \\
&= P_n D^*(Q_\epsilon, G)^\top \delta_n^*(Q_\epsilon) \\
&= \{P_n D^*(Q_\epsilon, G)\}^\top \frac{P_n D^*(Q_\epsilon, G)}{\| P_n D^*(Q_\epsilon, G) \|} \\
&= \frac{\sum_{j=1}^d \{P_n D_j^*(Q_\epsilon, G)\}^2}{\| P_n D^*(Q_\epsilon, G) \|} \\
&= \| P_n D^*(Q_\epsilon, G) \|.
\end{aligned}
$$

In addition, under some regularity conditions, so that the following derivation in terms of the local least favorable submodel applies, it also follows that $Q_\epsilon \in Q(\mathcal{M})$. This proves the following theorem.

Theorem 5.5. *Given any (Q, G) compatible with model $\mathcal{M}$, let $\{Q_\delta^{\mathrm{lfm}} : \delta \in B_a(0)\} \subset Q(\mathcal{M})$ be a local least favorable model w.r.t. loss function $L(Q)$ at $\delta = 0$ so that*

$$\frac{d}{d\delta} L(Q_\delta^{\mathrm{lfm}}) \bigg|_{\delta=0} = D^*(Q, G).$$

Here $B_a(0) = \{x : \| x \| < a\}$ *for some positive number a. Assume that for some* $\dot{L}(Q)(O)$, *we have*

$$\frac{d}{d\epsilon} L(Q_\epsilon^{\text{lfm}})\Big|_{\epsilon=0} = \dot{L}(Q) \frac{d}{d\epsilon} Q_\epsilon^{\text{lfm}}\Big|_{\epsilon=0}.$$

Consider the corresponding univariate model $\{Q_\epsilon : \epsilon\}$ *defined by (5.10). It goes through Q at* $\epsilon = 0$, *and, it satisfies that for all* ϵ

$$P_n \frac{d}{d\epsilon} L(Q_\epsilon) = \| P_n D^*(Q_\epsilon, G) \|, \tag{5.11}$$

where $\| x \| = \sqrt{\sum_{j=1}^d x_j^2}$ *is the Euclidean norm.*

In addition, assume that a in $B_a(0)$ *can be chosen to be independent of the choice* (Q, G) *in* $\{(Q_\epsilon, G) : \epsilon > 0\}$, *and assume the following second-order Taylor expansion: for* $h = (h_1, \ldots, h_d)$,

$$Q_{\epsilon,h}^{\text{lfm}} = Q_\epsilon + \frac{d}{dh} Q_{\epsilon,h}^{\text{lfm}}\Big|_{h=0} h + R_2(Q_\epsilon, G, \| h \|)$$

$$= Q_\epsilon + \frac{D^*(Q_\epsilon, G)}{\dot{L}(Q_\epsilon)} h + R_2(Q_\epsilon, G, \| h \|),$$

where

$$\sup_\epsilon \sup_{o \in O} | R_2(Q_\epsilon, G, \| h \|)(o) | = O((\| h \|^2).$$

We also assume that $\sup_\epsilon \sup_{o \in O} \frac{|D^*(Q_\epsilon, G)}{\dot{L}(Q_\epsilon)}(o) | < \infty$. *Then, we also have* $\{Q_\epsilon : \epsilon \geq 0\} \subset \mathcal{M}.$

Chapter 6
Highly Adaptive Lasso (HAL)

Mark J. van der Laan and David Benkeser

In this chapter, we define a general nonparametric estimator of a d-variate function valued parameter ψ_0. This parameter is defined as a minimizer of an expectation of a loss function $L(\psi)(O)$ that is guaranteed to converge to the true ψ_0 at a rate faster than $n^{-1/4}$, for all dimensions d: $\sqrt{d_0(\psi_n, \psi_0)} = O_P(n^{-1/4-\alpha(d)/8})$, where $d_0(\psi, \psi_0) = P_0 L(\psi) - P_0 L(\psi_0)$ is the loss-based dissimilarity. This is a remarkable result because this rate does not depend on the underlying smoothness of ψ_0. For example, ψ_0 can be a function that is discontinuous at many points or nondifferentiable. The only assumption we need to assume is that ψ_0 is right-continuous with left-hand limits, and has a finite variation norm, so that ψ_0 generates a measure (just as a cumulative distribution function generates a measure on the Euclidean space).

M. J. van der Laan (✉)
Division of Biostatistics and Department of Statistics, University of California, Berkeley, 101 Haviland Hall, #7358, Berkeley, CA 94720, USA
e-mail: laan@berkeley.edu

D. Benkeser
Department of Biostatistics and Boinformatics, Rollins School of Public Health, Emory University, 1518 Clifton Rd. NE, Atlanta, GA 30322, USA
e-mail: benkeser@emory.edu

© Springer International Publishing AG 2018
M.J. van der Laan, S. Rose, *Targeted Learning in Data Science*,
Springer Series in Statistics, https://doi.org/10.1007/978-3-319-65304-4_6

We refer to our general estimator as the highly adaptive lasso (HAL) estimator. This name stems from the fact that it can be represented as a minimizer of the empirical risk of the loss over linear combinations of indicator basis functions under the constraint that the sum of the absolute value of the coefficients is bounded by a data adaptively determined constant. For example, our result proves that our highly data-adaptive lasso estimator (using the squared error loss) of a regression function $E_0(Y \mid W)$, based on observing n i.i.d. observations $O_i = (W_i, Y_i)$, $i = 1, \ldots, n$, converges to the truth at a rate faster than $n^{-1/4}$, for every dimension d of W: $\sqrt{\int (\psi_n(w) - \psi_0)^2(w) dP_0(w)} = o_P(n^{-1/4})$. This rate seems to contradict the well known minimax rates of convergence from the nonparametric density and regression estimation literature. However, these minimax rates are developed for estimation of these true functions at a single point, while our result is in terms of a loss-based dissimilarity, which often corresponds with a square of an L^2-norm.

The HAL estimator appears to be much 'smarter' than local smoothers, even though these local smoothers achieve the minimax rates. For example, consider histogram regression estimator using a partitioning of the covariate space for which each element of the partitioning has a diameter $O(h)$. The bias of such a histogram regression estimator will then be $O(h)$ at any point w, while the variance is $O(1/(nh^d))$. Thus, the optimal rate for h minimizing MSE (i.e., setting that variance equal to the square of the bias) is given by $h = O(n^{-1/(d+2)})$, giving an MSE that is also $O(n^{-1/(d+2)})$. For $d = 1$ this rate is slightly better than the rate of HAL-estimator, but for $d \geq 2$, the rate of this histogram regression estimator is worse than our rate, and will get worse and worse as dimension grows. This phenomena is often referred to as the curse of dimensionality.

For a kernel regression estimator using kernels that are orthogonal to polynomials in W of a certain degree k and bandwidth h, assuming that ψ_0 is k-times continuously differentiable, the bias is $O(h^k)$, the variance is $O(1/(nh^d))$. In this case, the optimal rate for the bandwidth is $h = O(n^{-1/(2k+d)})$, resulting in a rate of convergence $O(n^{-k/(2k+d)})$. Contrary to the HAL estimator, this kernel regression estimator assumes smoothness of ψ_0, but even when the degree k of assumed smoothness is large, for large dimensions d, this rate will typically be much worse than $n^{-1/4}$. This demonstrates that the HAL estimator is asymptotically superior to local smoothers w.r.t. its capability to approximate a true function ψ_0.

A fortunate fact is that the critical rate for estimators of the nuisance parameters in a TMLE is $n^{-1/4}$, so that a TMLE, using super learners for the relevant nuisance parameters that include the HAL estimator in its library, is guaranteed to be asymptotically efficient under essentially no conditions. Before the introduction of this HAL estimator, there was no estimator that was guaranteed to be asymptotically efficient without strong smoothness conditions, and the general wisdom was that such an estimator would simply not exist.

6.1 Statistical Formulation of the Estimation Problem

Suppose that we observe n independent and identically distributed copies $O_1, \ldots, O_n$ of a multidimensional random variable O with probability distribution P_0. Let $\mathcal{M}$ be the statistical model for P_0, and suppose that the target parameter $\Psi : \mathcal{M} \to \Psi = \Psi(\mathcal{M})$ is a function valued parameter: that is, $\Psi(P_0)$ is a function from a subset $\mathbf{R}^d$ to the real line. For example, $\Psi(P_0)$ could be a regression function $w \to E_0(Y \mid W = w)$ or a conditional density function $(w, y) \to p_{Y,0}(y \mid w)$. We will consider the case that the parameter space $\Psi(\mathcal{M})$ consists of all multivariate real valued functions $f : [0, \tau] \subset \mathbf{R}^d \to \mathbf{R}$, up to possibly some smoothness conditions.

Specifically, we will only assume that ψ_0 is right-continuous with left-hand limits (i.e., cadlag), and that its variation norm is finite. These are the assumptions one needs on a function f so that it generates a measure so that an integral $\int h(x) df(x)$ is well defined. Indeed, this will allow us to represent a function f as a sum of integrals of indicator functions with respect to df, providing the basis for our estimation procedure. In addition, we assume that we have a loss function $(O, \psi) \to L(\psi)(O)$ for ψ_0 so that $P_0 L(\psi_0) = \min_{\psi \in \Psi} P_0 L(\psi)$. We will assume that the loss function is uniformly bounded in the sense that $\sup_{\psi \in \Psi, o} \mid L(\psi)(o) \mid < \infty$, where the supremum over o is over a support of P_0. We will also assume that the loss function yields a quadratic dissimilarity $d_0(\psi, \psi_0) = P_0\{L(\psi) - L(\psi_0)\}$, which formally corresponds with the following assumption:

$$\sup_{\psi \in \Psi} \frac{\| L(\psi) - L(\psi_0) \|_{P_0}^2}{d_0(\psi, \psi_0)} < \infty.$$

The latter is a standard property that has been established for standard loss functions such as the log-likelihood loss and squared error loss, as long as the loss-function is uniformly bounded. These two assumptions are the only properties of the loss function needed for establishing a finite sample oracle inequality for the cross-validation selector.

Suppose we estimate ψ_0 with a discrete super learner defined by set of candidate estimators based on V-fold cross-validation. By the oracle inequality for cross-validation, the super learner will perform asymptotically as well as the oracle selector that selects the best estimator among the set of candidate estimators. This itself does not provide any guarantee that the super learner is consistent or converges to the truth ψ_0 at a rate in sample size faster than a certain specified minimal rate. In this chapter we will present an estimator whose rate of convergence is guaranteed to be faster than $n^{-1/4}$, for any dimension d. By including this estimator in the library of the super learner, the (discrete or continuous) super learner is also guaranteed to converge at a rate faster than $n^{-1/4}$. In the next chapter, we will study this estimator in detail in the context of estimating a regression function, and demonstrate its implementation and remarkable practical performance with simulations.

6.2 Representation of a Cadlag Function as a Linear Combination of Basis Functions

For a given vector $x \in \mathbf{R}^d$ and subset $S \subset \{1, \ldots, d\}$ of indices, we define $x(S) = (x_j : j \in S)$ and $x(S^c) = (x_j : j \in S^c)$, where $S^c = \{j : j \notin S\} \subset \{1, \ldots, d\}$ is the complementary set of indices of S. For a given cadlag function $\psi : [0, \tau] \to \mathbf{R}$ and a given subset S, we can define the function $x \to \psi(x(S), 0(S^c))$, which is called the S-specific section ψ_S of ψ. Since ψ_S is cadlag and has finite variation norm, ψ_S generates a finite measure on $[0(S), \tau(S)]$. In other words, ψ_S can be viewed as the analogue of a multivariate cumulative distribution function, without the requirement that it only assigns positive mass. We will also refer to ψ_S as a measure, meaning the measure ψ_S generates. In fact, ψ_S equals a difference of two monotone functions (i.e., cumulative distribution functions without enforcing that they start at 0 and end at 1). In the same way as a cumulative distribution function assigns a measure to a rectangle $(a, b]$ and any measurable set, ψ_S assigns a measure to a such a set. For example, for a univariate function ψ, we have $\psi((a, b]) = \psi(b) - \psi(a)$, and for a bivariate function ψ, we have $\psi((a, b]) = \psi(b_1, b_2) - \psi(a_1, b_2) - \psi(a_2, b_1) + \psi(a_1, a_2)$. As a result, an integral $\int_{(a,b]} \psi(dx)$ is well defined, and represents the measure ψ assigns to the cube $(a, b]$.

A typical definition of the variation norm of ψ is given by $\int_{[0,\tau]} |\psi(dx)|$. In this chapter, we will define the variation norm of ψ as the sum of the variation norms of all its sections ψ_S

$$\| \psi \|_v = \psi(0) + \sum_{S \subset \{1, \ldots, d\}} \int_{(0(S), \tau(S)]} |\psi_S(dx)| .$$

In words, the variation norm of a d-variate real valued cadlag function is defined as the sum over all subsets of $\{1, \ldots, d\}$ of the absolute value integral with respect to ψ over the variables in that subset, while setting the remaining variables equal to zero.

For example, the variation norm of a bivariate real valued function

$$\| \psi \|_v = \psi(0, 0) + \int_0^{\tau_1} |\psi(dx_1, 0)| + \int_0^{\tau_2} |\psi(0, dx_2)| + \int_0^{\tau_1} \int_0^{\tau_2} |\psi(dx_1, dx_2)| .$$

For trivariate real valued functions, we have an integral for each subset of $\{1, 2, 3\}$ over the corresponding variables in that subset, setting the remaining variables equal to zero.

For any d-variate cadlag function ψ with $\| \psi \|_v < \infty$, we have the following representation of ψ:

$$\psi(x) = \psi(0) + \sum_{S \subset \{1, \ldots, d\}} \int_{(0(S), x(S)]} \psi_S(dx') .$$

For example, for a bivariate real valued function ψ we have

$$\psi(x_1, x_2) = \psi(0,0) + \int_0^{x_1} \psi(dx_1', 0) + \int_0^{x_2} \psi(0, dx_2') + \int_0^{x_1} \int_0^{x_2} \psi(dx_1, dx_2).$$

Note that we can also write this as follows:

$$\psi(x) = \psi(0) + \sum_{S \subset \{1,\ldots,d\}} \int_{(0(S),\tau(S)]} I(x' \le x(S)) \psi_S(dx').$$

Suppose we approximate the measure ψ by a discrete measure ψ_m with m support points, or equivalently, we approximate the two cumulative distribution functions ψ_1, ψ_2 in the representation $\psi = \psi_1 - \psi_2$ by two discrete cumulative functions ψ_{1m}, ψ_{2m}: $\psi_m = \psi_{1m} - \psi_{2m}$. We make sure that $\psi_{m,S}$ is a discrete approximation of ψ_S for each subset $S \subset \{1,\ldots,d\}$: that is, ψ_m puts mass on the d-dimensional cube $(0,\tau]$ but also on all the lower dimensional edges $(0(S),\tau(S)]$ of $[0,\tau]$. For each given subset S, let $\{s_j(S) : j\}$ be the support points of $\psi_{m,S}$, and let $d\psi_{m,S,j}$ denote the pointmass that $\psi_{m,S}$ assigns to this point.

For such a discrete approximation ψ_m, we have

$$\psi_m(x) = \psi(0) + \sum_{S \subset \{1,\ldots,d\}} \sum_j I(s_j(S) \le x(S)) d\psi_{m,S,j}.$$

That is, $\psi_m(\cdot)$ is a linear combination of basis functions $x \to \phi_{j,S}(x) = I(x(S) \ge s_j(S))$ with corresponding coefficients $d\psi_{m,S,j}$ across $S \subset \{1,\ldots,d\}$ and support points s_j indexed by j. In addition, note that the variation norm of ψ_m is the sum of the absolute values of its coefficients:

$$\| \psi_m \|_v = \psi(0) + \sum_{S \subset \{1,\ldots,d\}} \sum_j | d\psi_{m,S,j} |.$$

Below we define an estimator $\psi_{n,\lambda}$ of ψ_0 that minimizes the empirical risk $\psi \to P_n L(\psi)$ over all such linear combinations of these indicator basis functions for a specified set support points (i.e., basis functions), and under the constraint that the sum of the absolute value of the coefficients is smaller or equal than λ. Our proposed HAL estimator is then defined by $\psi_n = \psi_{n,\lambda_n}$, where λ_n is the cross-validation selector. This estimator $\psi_{n,\lambda}$ is equivalent with minimizing the empirical risk over all discrete measures ψ_m with variation norm smaller or equal than λ. In order to understand how to select the support points, we want to show that the minimizer of the empirical risk over all measures (continuous and discrete) is equivalent with minimizing the empirical risk over all discrete measures with a particular support defined by the actual n observations $O_1,\ldots,O_n$. As a result, by defining the support points accordingly, this MLE $\psi_{n,\lambda}$ actually equals the minimizer of the empirical risk over all functions with variation norm smaller than λ. The latter estimator is theoretically analyzed below and shown to converge to its true counterpart ψ_0 at a faster rate than $n^{-1/4}$.

6.3 A Minimum Loss-Based Estimator (MLE) Minimizing over all Functions with Variation Norm Smaller than λ

Consider the estimator $\psi_{n,\lambda} = \arg\min_{\psi, \|\psi\|_v \leq \lambda} P_n L(\psi)$ defined as the minimizer of the empirical risk over all functions ψ in the parameter space Ψ that have variation norm smaller than λ. Let $\psi_{0,\lambda} = \arg\min_{\psi, \|\psi\|_v \leq \lambda} P_0 L(\psi)$ the minimizer of the true risk. If $\lambda > \| \psi_0 \|_v$, then we have $\psi_{0,\lambda} = \psi_0$.

Let $d_0(\psi, \psi_0) = P_0 L(\psi) - P_0 L(\psi_0)$ be the loss-based dissimilarity. We can prove that $d_0(\psi_{n,\lambda}, \psi_{0,\lambda}) = O_P(n^{-(0.5+\alpha(d)/4)})$ where $\alpha(d) = 1/(d+1)$.

Theorem 6.1. *Let* $\Psi_\lambda = \{\psi \in \Psi : \| \psi \|_v \leq \lambda\}$. *We assume*

$$\sup_{\psi \in \Psi_\lambda} \frac{\| L(\psi) \|_v}{\| \psi \|_v} < \infty$$

$$\sup_{\psi \in \Psi_\lambda} \frac{\| L(\psi) - L(\psi_{0,\lambda}) \|_{P_0}^2}{d_0(\psi, \psi_{0,\lambda})} < \infty$$

Then, $d_0(\psi_{n,\lambda}, \psi_{0,\lambda}) = O_P(n^{-(0.5+\alpha(d)/4)})$ *where* $\alpha(d) = 1/(d+1)$.
Specifically, if $\lambda > \lambda_0 \equiv \| \psi_0 \|_v$, *then* $d_0(\psi_{n,\lambda}, \psi_0) = O_P(n^{-(0.5+\alpha(d)/4)})$.

Proof. We have

$$0 \leq d_0(\psi_{n,\lambda}, \psi_{0,\lambda}) = P_0\{L(\psi_{n,\lambda}) - L(\psi_{0,\lambda})\}$$
$$= -(P_n - P_0)\{L(\psi_{n,\lambda}) - L(\psi_{0,\lambda})\} + P_n\{L(\psi_{n,\lambda}) - L(\psi_{0,\lambda})\}$$
$$\leq -(P_n - P_0)\{L(\psi_{n,\lambda}) - L(\psi_{0,\lambda})\}.$$

We assumed that $\sup_{\psi \in \Psi} \frac{\|L(\psi)\|_v}{\|\psi\|_v} < \infty$. Since $L(\psi_{n,\lambda}) - L(\psi_{0,\lambda})$ falls in a P_0-Donsker class of all cadlag functions with variation norm smaller than a constant, it follows that the right-hand side is $O_P(1/\sqrt{n})$, and thus $d_0(\psi_{n,\lambda}, \psi_{0,\lambda}) = O_P(n^{-1/2})$. We also assumed that there exists an $M_2 < \infty$ so that $P_0\{L(\psi) - L(\psi_{0,\lambda})\}^2 \leq M_2 P_0\{L(\psi) - L(\psi_{0,\lambda})\}$ for all $\psi \in \Psi$ with $\| \psi \|_v < \lambda$. As a consequence, we have $\| L(\psi_{n,\lambda}) - L(\psi_{0,\lambda}) \|_{P_0}^2 = O_P(1/\sqrt{n})$. By empirical process theory we have that $\sqrt{n}(P_n - P_0)f_n \to_p 0$ if f_n falls in a P_0-Donsker class with probability tending to 1, and $P_0 f_n^2 \to_p 0$ as $n \to \infty$. Applying this to $f_n = L(\psi_{n,\lambda}) - L(\psi_{0,\lambda})$ shows that $(P_n - P_0)(L(\psi_{n,\lambda}) - L(\psi_{0,\lambda})) = o_P(1/\sqrt{n})$, which proves $d_0(\psi_{n,\lambda}, \psi_{0,\lambda}) = o_P(1/\sqrt{n})$.

We now apply Lemma 6.1 below with $\mathcal{F}_n = \{L(\psi) - L(\psi_{0,\lambda}) : \| \psi \|_v \leq \lambda\}$, envelope bound $M_n = \lambda$, $\alpha = \alpha(d)$ (see van der Vaart and Wellner 1996), and $r_0(n) = n^{-1/4}$, which proves that

$$| \sqrt{n}(P_n - P_0)f_n | = O_P(n^{-\alpha(d)/4}).$$

Here we rely on the result in van der Vaart and Wellner (1996) that proves that the class of d-variate cadlag functions with variation norm smaller than a universal constant is a Donsker class with an entropy bounded as in Lemma 6.1 with $\alpha = \alpha(d)$. This proves $d_0(\psi_{n,\lambda}, \psi_{0,\lambda}) = O_P(n^{-(0.5+\alpha(d)/4)})$. $\square$

A theorem in van der Vaart and Wellner (2011) establishes the following result for a Donsker class $\mathcal{F}_n$ with envelope F_n: If $Pf^2 \leq \delta^2 PF_n^2$, then

$$E \parallel G_n \parallel_{\mathcal{F}_n} \leq J(\delta, \mathcal{F}_n, L^2) \left(1 + \frac{J(\delta, \mathcal{F}_n, L_2)}{\delta^2 \sqrt{n} \parallel F_n \parallel_{P_0}} \right) \parallel F_n \parallel_{P_0},$$

where

$$J(\delta, \mathcal{F}_n, L^2) = \sup_\Lambda \int_0^\delta \left(\log(1 + N(\epsilon \parallel F_n \parallel_{P_0}, \mathcal{F}_n, L^2(\Lambda))) \right)^{0.5} d\epsilon$$

is the entropy integral from 0 to δ. Here $\parallel G_n \parallel_{\mathcal{F}_n} = \sup_{f \in \mathcal{F}_n} G_n(f)$ and $G_n(f) = \sqrt{n}(P_n - P_0)f$. A simple corollary of this theorem is the following lemma.

Lemma 6.1. *Consider $\mathcal{F}_n$ with $\parallel F_n \parallel_{P_0} < M_n$ and $\sup_\Lambda \sqrt{\log(1 + N(\epsilon \parallel F_n \parallel_{P_0}, \mathcal{F}_n, L^2(\Lambda)))} < 1/\epsilon^{1-\alpha}$. Then,*

$$E \sup_{f \in \mathcal{F}_n, \parallel f \parallel_{P_0} < r_0(n)} | G_n(f) | \leq \{r_0(n)/M_n\}^\alpha M_n + \{r_0(n)/M_n\}^{2\alpha-2} n^{-0.5}.$$

If $r_0(n) < n^{-1/4}$, one should select $r_0(n) = n^{-1/4}$ in the above right hand side, giving the bound:

$$E \sup_{f \in \mathcal{F}_n, \parallel f \parallel_{P_0} < r_0(n)} | G_n(f) | \leq \{n^{-0.25}/M_n\}^\alpha M_n + \{M_n\}^{2-2\alpha} n^{-\alpha/2}.$$

6.4 The HAL Estimator

Above, we defined candidate estimators $\psi_{n,\lambda} = \hat{\Psi}_\lambda(P_n)$. Let λ vary over a set of K_n values for which the largest value is larger than $\parallel \psi_0 \parallel_v$. Here we select K_n so that $K_n < n^p$ for some finite p. Consider a V-fold cross-validation scheme, and let $P_{n,v}^0$, $P_{n,v}^1$ be the training sample and validation sample corresponding with sample split $v, v = 1, \ldots, V$. The cross-validation selector of λ is then defined as follows:

$$\lambda_n = \arg\min_\lambda \frac{1}{V} \sum_{v=1}^V P_{n,v}^1 L(\hat{\Psi}_\lambda(P_{n,v}^0)).$$

Our proposed estimator of ψ_0 is given by $\psi_n = \psi_{n,\lambda_n} = \hat{\Psi}_{\lambda_n}(P_n)$. By the finite sample oracle inequality for the cross-validation selector we have:

$$d_0(\psi_n, \psi_0) = O_P(n^{-(0.5+\alpha(d)/4)}) + O_P(\log K_n/n) = O_P(n^{-(0.5+\alpha(d)/4)}).$$

One can include this estimator ψ_n in the library of a super learner that includes many other algorithms, thereby guaranteeing that the super learner is not only asymptotically equivalent with the oracle selected estimator, but also has a minimal performance $d_0(\psi_n, \psi_0) = O_P(n^{-(0.5+\alpha(d)/4)})$.

Let's now discuss the implementation of this HAL estimator. Suppose that $L(\psi)(O)$ depends on ψ through $\psi(W)$ where $W = f(O) \in \mathbf{R}^d$ for some specified function f: for example, $O = (Y, W)$ and $L(\psi)(O) = (Y - \psi(W))^2$. We note that $P_n L(\psi)$ only depends on ψ through $(\psi(W_i) : i = 1, \ldots, n)$, suggesting that we should be able to replace the minimization over $\mathbf{\Psi}_\lambda$ by a finite dimensional minimization problem.

For each set $S \subset \{1, \ldots, d\}$, let $W_i(S)$ be the subvector $(W_{ij} : j \in S), i = 1, \ldots, n$. Recall the representation ψ_m of the discrete approximation of ψ, where now the support points of ψ_S are given by $\{W_i(S) : i = 1, \ldots, n\}$:

$$\psi_m(w) = \psi_m(0) + \sum_{S \subset \{1,\ldots,d\}} \sum_{j=1}^{n} I(W_j(S) \le w(S)) d\psi_{m,S,j}.$$

That is, $\psi_m(\cdot)$ is a linear combination of basis functions $x \to \phi_{j,S}(w) = I(w(S) \ge W_j(S))$ with corresponding coefficients $d\psi_{m,S,j}$ across $S \subset \{1, \ldots, d\}$ and $j = 1, \ldots, n$. We claim that the minimizer $\psi_{n,\lambda}$ is attained by such a discrete measure ψ_m.

Let's define

$$\psi_\beta = \beta(0) + \sum_{S \subset \{1,\ldots,d\}} \sum_{j=1}^{n} \beta_j(S) \phi_{j,S},$$

and a corresponding subspace

$$\mathbf{\Psi}_{n,\lambda} = \left\{ \psi_\beta : \beta, \beta(0) + \sum_{S \subset \{1,\ldots,d\}} \sum_{j=1}^{n} |\beta_j| < \lambda \right\}.$$

That is, we claim that $\psi_{n,\lambda} = \psi_{\beta_n}$, where

$$\beta_n = \arg \min_{\beta, \sum_{S \subset \{1,\ldots,d\}} \sum_{j=1}^{n} |\beta_j(S)| \le \lambda} P_n L\left(\psi_\beta\right).$$

Notice that the number of basis functions is given by $m = (2^d - 1)n$, so that computation of $\psi_{n,\lambda}$ requires minimizing over m-dimensional vectors β under the constraint that its L_1-norm is bounded by λ.

6.5 Further Dimension Reduction Considerations

For d reasonable large, the number of basis functions $m = (2^d - 1)n$ cannot be stored in memory, making the computation of the MLE $\psi_{n,\lambda}$ non feasible. Since the empirical risk $P_n L(\psi)$ only depends on ψ through n values $\{\psi(W_i) : i = 1, \ldots, n\}$, one might be able to further reduce the number of basis functions while still attaining the minimum of the empirical risk. Our theorem proves that any $\psi_{n,\lambda}$ attaining the minimum will converge to $\psi_{0,\lambda}$ at the desired rate. In fact, it suffices to achieve the minimum up to an approximation error that is smaller than this rate. This suggest that for finite samples it might suffice to work with a much smaller subset of these basis functions even though all of these types of basis functions will be included as sample size increases so that any function can be arbitrarily well approximated. Developing computationally feasible algorithms that approximate the desired $\psi_{n,\lambda}$ will be an important area of future research.

We propose the following strategy for defining a super learner incorporating the HAL estimator. The key step is to construct a sequence of nested candidate estimators for which the last estimator in this sequence is the full HAL estimator. For example, the first estimator might be the lasso estimator only including the one-way indicator functions in the HAL-representation, while the k-th estimator would incorporate all multiway indicator functions up to the k-th order, $k = 1, \ldots, d$. However, it makes sense to use a much finer sequence of candidate estimators so that the memory storage and computer speed increases gradually along this sequence. For example, one might propose a possibly data-adaptive ordering of all the multi-way indicator basis functions, starting out with one-way, then to two-way, etc. This would require ordering the one-way indicator basis functions, and the two-way indicator basis functions, etc. One might now define a sequence of candidate estimators by defining them as the lasso including the first K_j basis functions in this sequence, $K_1 < K_2 < \ldots < K_M$, $j = 1, \ldots, M$, where K_M is the total number of basis functions in the HAL estimator. Each of these candidate estimators are now included in the library of the super learner. By the oracle inequality, this super learner is at least as good as the full HAL estimator that includes all KM basis functions. Instead of truly computing the super learner, we would compute the candidate estimators along this sequence, each time tracking the cross-validated risk and once the cross-validated risk appears to flatten out or even deteriorates, we define the last estimator as our final estimator. The validity of this proposal relies on the assumption that the more aggressive estimators in the remaining sequence will not achieve a better performance than the selected one. In this manner, for a given sample size n, one expects that the number of selected basis functions will be bounded by $O(n)$, thereby making the estimator computable.

6.6 Applications

We introduced the HAL as a general nonparametric estimator of a d-variate function valued parameter defined as a minimizer of an expectation of a loss function. In this section, we consider applying HAL to the problem of estimating the conditional mean of a real-valued outcome. Specifically, we discuss the case that the observed data consist of n i.i.d. copies of the random variable $O = (W, Y) \sim P_0 \in \mathcal{M}$, where $\mathcal{M}$ is the nonparametric statistical model. The only constraint we will place on this model is that for every $P \in \mathcal{M}$, the conditional mean of Y given W implied by P, say $\bar{Q}_P$, has a finite variation norm. We consider using the highly adaptive lasso to estimate $\bar{Q}_0 = \arg\min_{\bar{Q}} P_0 L(\bar{Q})$, where $L(\bar{Q})(o) = \{y - \bar{Q}(w)\}^2$ is squared error loss.

6.6.1 Constructing the Highly Adaptive Lasso

Recall that HAL can be viewed as the minimizer of the empirical risk over a special linear combinations of indicator basis functions under the constraint that the sum of the absolute value of the coefficients is less than or equal to a data adaptively chosen constant. In this section, we illustrate how these basis functions and the estimator are constructed in simple univariate and bivariate settings.

Consider that the observed data consist of $n = 500$ independent copies of $W \sim \text{Uniform}(-4, 4)$ and $Y = 2\sin(\pi/2|W|) + \epsilon$, where ϵ is drawn independently of W a Normal(0,1) distribution. The basis functionsused by HAL consist of n in-

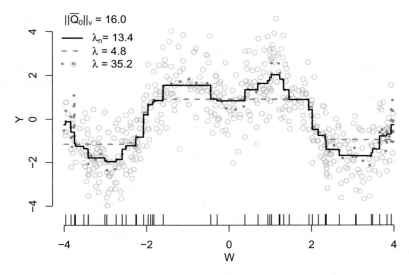

Fig. 6.1 The highly adaptive lasso in the univariate setting

dicators of the observed data values: $\phi_j(w) = I(w \geq w_j)$ for $j = 1, \ldots, n$. To select the bound on the variation norm, we used ten-fold cross validation to select from 100 possible bounds ranging from 0 to about 350. We illustrate the fit from three of these choices in Fig. 6.1. The solid line is the HAL estimator, which uses the cross-validation-selected value $\lambda_n = 13.9$. The dashed and dotted lines represent choices that are smaller and larger respectively than the true variation norm $\|\bar{Q}_0\|_v = 16$. The ticks at the bottom of the figure are placed at the 46 support points of $\bar{Q}_n$ with a nonzero coefficient. The choice of 4.8 as bound on the variation norm (dashed line) visibly over-smooths the data, while the bound of 35.2 appears to provide a reasonable approximation and is similar with the prediction from the HAL estima-tor. However, the larger bound does appear to produce more noise near the edges of the support. Theory dictates that any choice of bound larger than the true norm will yield an estimator with the properties established in the previous chapter. Nev-ertheless, the HAL estimator will exhibit superior performance in finite samples by allowing for selection of a bound smaller than the true norm. The oracle inequality guarantees that so long as at least one bound larger than the true norm is considered as a candidate bound, then we will eventually select a bound that is larger than the true variation norm.

We now illustrate the estimator in the bivariate setting and where W has a discrete component. We drew W_1 from a Uniform$(-4,4)$ distribution and also drew W_2 independently from a Bernoulli(0.5) distribution. We let $Y = -0.5W_1 + W_2W_1^2/2.75 + W_2 + \epsilon$ where ϵ was drawn from a Normal$(0,1)$ distribution. Notice that this data generatingdistribution implies an interaction between W_1 and W_2 in $\bar{Q}_0$, with a lin-

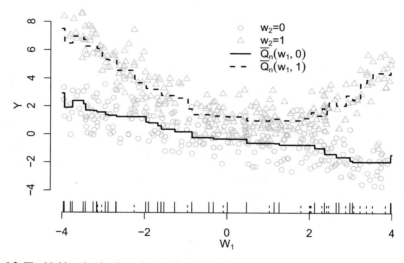

Fig. 6.2 The highly adaptive lasso in the bivariate setting

ear relationship between W_1 and the mean of Y whenever $W_2 = 0$ and a quadratic relationship otherwise. To construct the HAL estimator in this setting, we first created n basis functions corresponding with indicators at the observed values of W_1: $\phi_{1,j}(w) = I(w_1 \geq w_{1,j})$ for $j = 1,\ldots,n$. Next, we added basis functions for the subset consisting only of W_2: $\phi_{2,j}(w) = I(w_2 \geq w_{2,j})$ for $j = 1,\ldots,n$. Note that because W_2 is binary, there was only be a single unique basis function to be added, $\phi_2(w) = I(w_2 \geq 1)$. Finally, we created bivariate basis functions of the form $\phi_{12,j}(w) = I(w_1 \geq w_{1,j}, w_2 \geq w_{2,j})$ for $j = 1,\ldots,n$. These basis functions number fewer than n due to binary W_2. It was unnecessary to add basis functions $\phi_{12,j}(w)$ for any j for which $w_{2,j} = 0$ due to the fact that for any such j we had already placed support on this zero-edge by including $\bar{\phi}_{1,j}$. This illustrates that the number of basis functions in a given sample will be at most $n(2^d - 1)$, while in practice the number may be far fewer depending on the particular data set.

Figure 6.2 illustrates a random draw of size $n = 500$ from this data generating mechanism. Two lines are shown corresponding with the estimate of $\bar{Q}_0$ when $W_2 = 1$ (upper dashed line) and when $W_2 = 0$ (lower solid line). The solid tick marks across the bottom of the figure indicate the univariate basis functions with a non-zero coefficient in $\bar{Q}_n$. Accordingly, these marks corresponding with jumps in both $\bar{Q}_n(\cdot, 0)$ and $\bar{Q}_n(\cdot, 1)$. The dashed tick marks indicate the bivariate basis functions with nonzero coefficients and thus correspond with values of a jump in $\bar{Q}_n(\cdot, 1)$, but not $\bar{Q}_n(\cdot, 0)$. Notice that, as expected these ticks occur most frequently when $W_1 > 2$, corresponding with the values for which $\bar{Q}_0(w_1, 0)$ is decreasing in w_1, while $\bar{Q}_1(w_1, 1)$ is increasing. This example illustrates how the higher-order basis functions used by the HAL estimator act similarly to cross-product interaction terms in standard regression approaches.

6.6.2 Prediction Simulation

We evaluated the finite-sample performance of the HAL estimator relative to other nonparametric algorithms: regression trees (Breiman et al. 1984), random forests (Breiman 2001), gradient boosted machines (GBM) (Friedman 2001), kernel regression (Nadaraya 1964; Watson 1964), support vector machines (SVM) (Hearst et al. 1998), and polynomial multivariate adaptive regressions splines (Polynomial MARS) (Friedman 1991). We considered three types of data generating mechanisms, which we call smooth, jumps, and sinusoidal. For each type of data generating mechanism, we varied the dimension of W and considered $d \in \{1, 3, 5\}$ and sample sizes $n \in \{500, 1000, 2000\}$. Performance was judged based on R^2, which was calculated on an independent test set of size $N = 1e4$, where for a given estimator $\bar{Q}_n$, we define

$$R^2 = 1 - \frac{\sum_{i=1}^{N}\{Y_i - \bar{Q}_n(W_i)\}^2}{\sum_{i=1}^{N}\{Y_i - \bar{Y}_N\}^2} .$$

Each setting was designed so that the optimal R^2 value was $R^2_{opt} = 0.80$, where

$$R^2_{opt} = 1 - \frac{E_{P_0}\{Y - \bar{Q}_0(W)\}^2}{\text{Var}_0(Y)}$$

is the value of R^2 obtained when using the true regression function $\bar{Q}_0$. This value can be viewed as an upper bound on the performance of any estimator.

The distribution of W was as follows: $W_1 \sim \text{Uniform}(-4, 4), W_2 \sim \text{Uniform}(-4,4), W_3 \sim \text{Bernoulli}(0.5), W_4 \sim \text{Normal}(0, 1), W_5 \sim \text{Gamma}(2, 1)$. For dimension d, call the target parameter $\bar{Q}_0^d(W)$ and let $W = (W_j : j = 1, \ldots, d\})$. We define $Y = \bar{Q}_0^d(W) + \epsilon$ where $\epsilon \sim \text{Normal}(0, 1)$.

The "smooth" regression functions for $d = 1, 3, 5$ respectively were defined as

$$\bar{Q}_0^1(w) = 0.05w_1 + 0.42w_1^2 ;$$
$$\bar{Q}_0^3(w) = 0.07w_1 - 0.28w_1^2 + 0.5w_2 + 0.25w_2w_3 ;$$
$$\bar{Q}_0^5(w) = 0.1w_1 - 0.3w_1^2 + 0.25w_2 + 0.5w_3w_2 - 0.5w_4 + 0.04w_5^2 - 0.1w_5 .$$

The "jump" regression functions were defined as

$$\bar{Q}_0^1(w) = -2.7I(w_1 < -3) + 2.5I(w_1 > -2) - 2I(w_1 > 0) + 4I(w_1 > 2) - 3I(w_1 > 3) \,;$$

$$\bar{Q}_0^3(w) = -2I(w_1 < -3)w_3 + 2.5I(w_1 > -2) - 2I(w_1 > 0) + 2.5I(w_1 > 2)w_3$$
$$- 2.5I(w_1 > 3) + I(w_2 > -1) - 4I(w_2 > 1)w_3 + 2I(w_2 > 3) \,;$$

$$\bar{Q}_0^5(w) = -I(w_1 < -3)w_3 + 0.5I(w_1 > -2) - I(w_1 > 0) + 2I(w_1 > 2)w_3 - 3I(w_1 > 3)$$
$$+ 1.5I(w_2 > -1) - 5I(w_2 > 1)w_3 + 2I(w_2 > 3) + 2I(w_4 < 0)$$
$$- I(w_5 > 5) - I(w_4 < 0)I(w_1 < 0) + 2w_3.$$

The "sinusoidal" regression functions were defined as

$$\bar{Q}_0^1(w) = 2\sin(0.5\pi|w_1|) + 2\cos(0.5\pi|w_1|) \,;$$
$$\bar{Q}_0^3(w) = 4w_3 I(w_2 < 0)\sin(0.5\pi|w_1|) + 4.1I(w_2 \geq 0)\cos(0.5\pi|w_1|) \,;$$
$$\bar{Q}_0^5(w) = 3.8w_3 I(w_2 < 0)\sin(0.5\pi|w_1|) + 4I(w_2 > 0)\cos(\pi|w_1|/2) + 0.1w_5\sin(\pi w_4)$$
$$+ w_3\cos(|w_4 - w_5|).$$

Figure 6.3 displays the results of the simulation study with rows representing the different data generating mechanisms and columns representing the different dimensions of W. The margins of the figure show the results aggregated across data generating mechanisms of a particular dimension (bottom margin) and aggregated across different dimensions of a particular data generating mechanism (right margin). In each plot, the algorithms have been sorted by their average R^2 value across the three sample sizes with the highest R^2 at the top of the figure and the lowest R^2 at the bottom.

Beginning with the top row corresponding to the "smooth" data generating mechanisms, we find that all algorithms other than random forests perform well when $d = 1$, with kernel regression performing the best in this case. However, as the dimension increases, the relative performance of kernel regression decreases, while the relative performance of HAL increases. Across all dimensions the SVM had the best overall performance; however, the performance of the GBM and HAL were comparable. In the second row corresponding with the "jumps" scenario, we see that the HAL performs extremely well, nearly achieving the optimal R^2 when $n = 2000$ for all dimensions. In the third row corresponding with the "sinusoidal" scenario, we find that somewhat surprisingly the kernel regression performs the best across all dimensions. This appears to be due in part to superior performance relative to other estimators when $n = 500$. For the larger sample sizes, the R^2 achieved by kernel regression, random forests, and HAL are similarly high. The far bottom right plot shows the results over all simulations with algorithms sorted by average R^2 and we see that HAL had the highest average R^2 followed by kernel regression and random forests. Overall, HAL performed well relative to competitors in all scenarios and particularly well in the jump setting, where local smoothness assumptions fail.

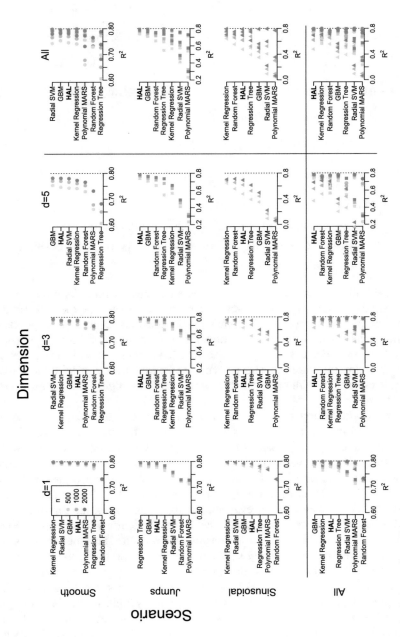

Fig. 6.3 Simulation study results

Though the estimator was not ranked highest for the smooth and sinusoidal data generating mechanisms, its performance was comparable to the best-performing machine learning algorithms, which are generally considered to be state-of-the-art.

6.6.3 Prediction Data Analysis

We separately analyzed five publicly available data sets listed with citation in Table 6.1. Sample sizes for the data sets ranged from 201 to 654 and d ranged from four to eleven. Inaddition to the nonparametric methods evaluated in simulations,

Table 6.1 Data sets analysed using the HAL estimator and competitors

Name	n	d
cpu (Kibler et al. 1989)	209	6
laheart (Afifi and Azen 1979)	201	11
oecdpanel (Liu and Stengos 1999)	616	6
pima (Smith et al. 1988)	392	7
fev (Rosner 1999)	654	4

we considered estimation of $\bar{Q}_0$ with several parametric methods as well. These included a main terms generalized linear model (GLM), a stepwise GLM based on AIC including two-way interactions, and a generalized additive model (GAM) with the degree of splines determined via ten-fold cross-validation. We also included the super learner and discrete super learner using each of these nine algorithms as candidates.

In order to compare the performance of the various methods across different data sets with different outcomes, we studied the ten-fold cross-validated mean squared-error of each method relative to that of the main terms GLM. Values greater than one correspond to better performance of the GLM. The results of each of the data analyses are shown in Fig. 6.4. The gray dots corresponds to the relative MSE in a particular data set, while the black cross corresponds to the geometric mean across all five studies. The super learner and discrete super learner perform best, followed by the HAL estimator. The HAL estimator performed particularly well on the cpu dataset, where its cross-validated MSE was nearly half that of the main terms GLM.

6.6.4 Simulation for Missing Data

Recall that a remarkable feature of HAL is its guaranteed convergence rate of faster than $n^{-1/4}$ regardless of the dimension d. This rate is exactly the critical rate needed for initial estimates of nuisance parameters that guarantees efficiency of the resulting TMLE. Therefore, it is of great interest to determine the extent to which this remarkable asymptotic result yields well-behaved TMLE estimators in finite

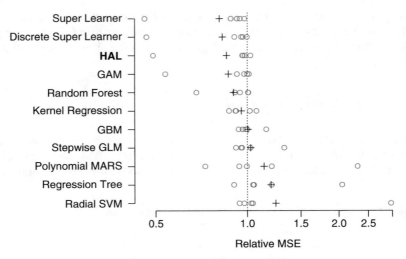

Fig. 6.4 Relative cross-validated mean squared error of methods in five real data sets. *Circle* = result on a single data set, *cross* = geometric mean over five data sets

samples. To study this question, we considered the same nine data generating distributions used in the prediction simulation. However, we additionally introduced missingness to this data structure and let $A = 1$ denote that the outcome Y was observed. We simulated missingness using in the smooth scenario as follows: letting $g_0(w) = P(A = 1|W = w)$, for the "smooth" setting,

$$g_0^1(W) = \text{expit}\{(w_1 + 4)^{1/2} - w_1/2\},$$

$$g_0^3(W) = \text{expit}\{1 + (w_1 + 4)^{1/2} - w_1/2 + w_2 w_1/5 - w_2^2/10\}, \text{ and}$$

$$g_0^5(w) = \text{expit}\{(1 + (w_1 + 4)^{1/2} - w_1/2 + w_2 w_1/5 - w_2^2/10 + w_4 + w_5/5 - w_4 w_3 w_1/5))\};$$

for the "jump" setting,

$$g_0^1(W) = \text{expit}\{-3 + 2I(w_1 < -3) - 1.5I(w_1 \geq -3, w_1 < -1.5)$$
$$+ 0.5I(w_1 \geq -1.5, w_1 < 0.5) - 2I(w_1 \geq 0.5, w_1 < 2) + 2.4I(w_1 \geq 2))\,,$$

$$g_0^3(w) = \text{expit}\{0.1I(w_1 < -3)w_3 + I(w_2 < 0)2.5 + 1.5I(w_1 \geq -3, w_1 < -1.5)$$
$$+ 2I(w_1 \geq -1.5, w_1 < 0.5) - 0.8I(w_1 \geq 0.5, w_1 < 2)$$
$$+ 0.75I(w_1 \geq 2)w_3 + w_3 - 2I(w_1 < 0, w_2 > 0)\}\,, \text{ and}$$

$$g_0^5(w) = \text{expit}\{1 + 0.1I(w_1 < -3)w_3 + I(w_2 < 0)2.5 + 1.5I(w_1 \geq -3, w_1 < -1.5)$$
$$+ 2I(w_1 \geq -1.5, w_1 < 0.5) - 0.8I(w_1 \geq 0.5, w_1 < 2) + 0.75I(w_1 \geq 2)w_3$$
$$+ w_3 - 2I(w_1 < 0, w_2 > 0) + I(w_4 < -1) + 2I(w_4 < -2) - 3I(w_4 > 0)$$
$$+ 2I(w_5 < 3) - 1.7I(w_4 < 0)w_3\}\,;$$

and for the "sinusoidal" setting,

$$g_0^1(W) = \text{expit}\{2 + \sin(w_1)\} ,$$

$$g_0^3(w) = \text{expit}\{(2 + \sin(w_1 w_3) + \cos(w_2 w_3) + \sin(w_1 w_2)\} , \text{ and}$$

$$g_0^5(w) = \text{expit}\{(1.5 + \sin(w_1 w_3) + \cos(w_2 w_3) + \sin(w_1 w_2) + \sin(|w_4|) - w_3 \cos(w_5^{1/2})\} .$$

We generated 500 replications of each of the nine data generating distributions at sample sizes of 500, 2000, and 5000 and estimated $\psi_0 = E_0\{E_0(Y \mid A = 1, W)\}$ using TMLEs based on different nuisance parameter estimators. In particular, we considered estimating $\bar{Q}_0$ and g_0 using the same nonparametric estimators that were used in the predictionsimulation, as well as using a super learner and discrete super

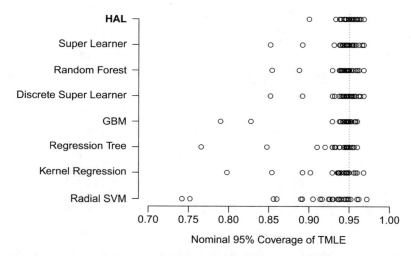

Fig. 6.5 Coverage of Wald style confidence intervals about TMLE estimators based on different nuisance parameter estimators. The results are ordered by the average absolute distance from 95% coverage

learner with those estimators as candidates. We were interested in assessing the extent to which the various TMLEs achieved an approximately normal sampling distribution in finite samples, which we assessed by computing the coverage of 95% Wald-style confidence intervals based on the true asymptotic variance of the TMLE and by visually examining histograms of the sampling distributions.

The coverage of the Wald style confidence intervals across the 27 different simulation settings are illustrated in Fig. 6.5. TMLE estimators using HAL to estimate nuisance parameters performed remarkably well; their coverage was estimated to be only approximately 1.1% off of the nominal 95% coverage on average and was no lower than 90% in any simulation. The TMLE estimators based on super learner also yielded confidence intervals that had remarkably good coverage; however, the performance of HAL-based TMLEs was notably better in the smaller sample sizes

for the univariate "jump" setting. The same is true of GBM-based TMLEs, which
has excellent coverage in all but two of the "jump" settings, where the coverage was
found to be quite poor hurting these estimators' overall performance.

The benefit of the fast convergence rate of HAL is apparent in the histograms
shown in Fig. 6.6, which illustrate the sampling distribution of the TMLE in the 5-
variate "jump" scenario at sample sizes 500 and 5000. The top row shows that the
HAL-based TMLE achieves approximate normality and is minimally biased, even
in small samples. In the larger sample, the HAL-based TMLE has little bias and the
sampling distribution is well-approximated by the Normal distribution shown. In
contrast, the kernel regression-based TMLE exhibits serious bias in small samples
and we clearly see that its bias is not converging to zero faster than $n^{-1/2}$.

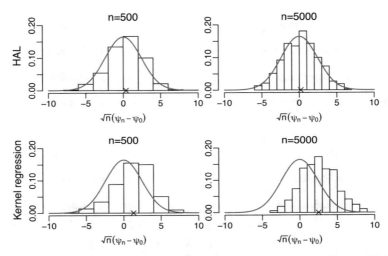

Fig. 6.6 Sampling distribution of standardized TMLE estimators based on HAL and based on
kernel regression. The asymptotic distribution of an efficient estimator is shown in the *solid line*.
The means of the estimators are indicated on each horizontal axis

6.6.5 Conclusion

In this section we examined the practical construction and performance of the HAL
estimator. We found that the estimator performs remarkably well for the purpose
of prediction, as well as for estimating relevant intermediate nuisance parameters
for a TMLE. These results indicate that the HAL makes a valuable contribution
towards building a robust super learner library and there are likely to be real benefits
to its incorporation in practice. Earlier results on the HAL were presented in the
conference paper Benkeser and van der Laan (2016).

Chapter 7
A Generally Efficient HAL-TMLE

Mark J. van der Laan

We will present a TMLE of ψ_0 that is asymptotically efficient at any $P \in \mathcal{M}$. This is a remarkable statement since we only assume strong positivity, some global bounds, and a finite variation norm of $\bar{Q}_0, \bar{G}_0$. This estimation problem for the treatment specific mean will be our key example to demonstrate a general one-step TMLE that is guaranteed to be asymptotically efficient for any model and pathwise differentiable target parameter, essentially only assuming a positivity assumption, also guaranteeing strong identifiability of the target parameter.

The key of our one-step TMLE is that it uses a super learner as initial estimator that includes the highly adaptive lasso estimator as a candidate estimator in the library. Therefore we will refer to such a TMLE with highly adaptive lasso TMLE (HAL-TMLE). By our formal results for the HAL estimator, we know that the super learner will converge at a rate faster than $n^{-1/4}$ with respect to the loss-based dissimilarity, and that is typically sufficient for establishing that the second-order remainder in a TMLE analysis is $o_P(n^{-1/2})$. The latter is the key condition in the efficiency proof for a TMLE.

In this chapter, we focus on demonstrating this general HAL-TMLE for the treatment specific mean, and subsequently demonstrate how our proof is easily generalized to general bounded models and target parameters. We refer to our paper van der Laan (2017) for a presentation of a completely general HAL-TMLE and HAL-CV-TMLE with a general efficiency theorem, even allowing for unbounded models.

M. J. van der Laan (✉)
Division of Biostatistics and Department of Statistics, University of California, Berkeley,
101 Haviland Hall, #7358, Berkeley, CA 94720, USA
e-mail: laan@berkeley.edu

© Springer International Publishing AG 2018
M.J. van der Laan, S. Rose, *Targeted Learning in Data Science*,
Springer Series in Statistics, https://doi.org/10.1007/978-3-319-65304-4_7

7.1 Treatment Specific Mean

Suppose we observe n i.i.d. observations (W_i, A_i, Y_i), $i = 1, \ldots, n$, of a random variable (W, A, Y) with probability distribution P_0, where W is a d-dimensional covariate vector, A is a binary treatment, and Y is a binary outcome. Let's consider an observational study and assume that it is known that A is independent of W, given a d_1-dimensional covariate subvector W_1 of W. Let $\bar{G}(W) = P(A = 1 \mid W) = P(A = 1 \mid W_1)$ and $\bar{Q}(W) = E(Y \mid A = 1, W)$. In addition, suppose that we know that $\bar{G}_0(W) > \delta > 0$ for some $\delta > 0$, and that the functions $\bar{Q}_0, \bar{G}_0$ are cadlag and have finite variation norm smaller than some universal constant C. We also assume that $\delta_1 < \bar{Q}_0(W) < 1 - \delta_1$ for some $\delta_1 > 0$. This δ_1 constraint is not very essential since it can be removed if we replace the log-likelihood loss by the squared error loss function for $\bar{Q}_0$ in our definition of the HAL estimator and the TMLE. Suppose that we have no other knowledge about P_0. This defines a highly nonparametric model $\mathcal{M}$ for P_0, involving known overall bounds C, δ, δ_1 and $\bar{G}(W) = \bar{G}(W_1)$. Our target parameter mapping $\Psi : \mathcal{M} \to \mathbb{R}$ is defined by $\Psi(P) = E_P E_P(Y \mid A = 1, W)$. This target parameter is pathwise differentiable at any $P \in \mathcal{M}$ with canonical gradient $D^*(P)(O) = A/\bar{G}(W)(Y - \bar{Q}(W)) + \bar{Q}(W) - \Psi(P)$. Since $\Psi(P) = Q_W \bar{Q}$, we will also denote the target parameter with $\Psi(Q)$, where $Q = (Q_W, \bar{Q})$.

7.1.1 HAL-TMLE

Let $L_1(\bar{Q})(O) = -A\{Y \log \bar{Q}(W) + (1-Y) \log(1-\bar{Q}(W))\}$ and $L_2(\bar{G}) = -\{A \log \bar{G}(W) + (1 - A) \log(1 - \bar{G}(W))\}$ be the log-likelihood loss functions for $\bar{Q}_0$ and $\bar{G}_0$, respectively. Let $\bar{Q}_n$ and $\bar{G}_n$ be loss-based super learners of $\bar{Q}_0$ and $\bar{G}_0$ that include the logistic HAL estimator as a candidate in its library, using an upper bound for λ equal to C. Note that $\bar{G}_0$ only depends on W though W_1 so that $\bar{G}_n$ will only concern fitting logistic lasso regressions linear in the indicator basis functions of W_1. In this logistic lasso estimator the linear combination of the indicator basis functions is used to approximate $\log \bar{Q}_0/(1 - \bar{Q}_0)$ and $\log \bar{G}_0/(1 - \bar{G}_0)$. The bound λ for the L_1-norm of the coefficient vector of the linear combination of basis functions for the logit of $\bar{Q}_n$ implies that $\bar{Q}_n$ is uniformly bounded away from 0 and 1, and similarly, the bound λ for the L_1-norm of the coefficient vector of the linear combination of the basis functions for the login of $\bar{G}_n$ implies that $\bar{G}_n$ is uniformly bounded away from 0 and 1.

By our result for the HAL estimator, we have

$$d_{01}(\bar{Q}_n, \bar{Q}_0) = O_P(n^{-(0.5+\alpha(d)/4)})$$
$$d_{02}(\bar{G}_n, \bar{G}_0) = O_P(n^{-(0.5+\alpha(d_1)/4)}).$$

We will truncate $\bar{G}_n$ from below by δ to guarantee that it is uniformly bounded away from zero with probability 1. Since $\bar{G}_0 > \delta > 0$, this will not affect its rate of convergence. For notational convenience, we still denote this estimator with $\bar{G}_n$. Similarly, we will truncate $\bar{Q}_n$ from above and below by δ_1, so that $\delta_1 < \bar{Q}_n < 1 - \delta_1$. Again, by the fact that $\delta_1 < \bar{Q}_0 < 1 - \delta_1$, it follows that this truncation of $\bar{Q}_n$ will not affects its rate of convergence to $\bar{Q}_0$. We know that the Kullback-Leibler dissimilarity is equivalent with the $L^2(P_0)$-norm if the densities are uniformly bounded away from zero. Therefore, under our bounds δ_1, δ for $\bar{Q}_0$ and $\bar{G}_0$, it follows that we also have

$$\| \bar{Q}_n - \bar{Q}_0 \|^2_{P_0} = O_P(n^{-(0.5+\alpha(d)/4)})$$
$$\| \bar{G}_n - \bar{G}_0 \|^2_{P_0} = O_P(n^{-(0.5+\alpha(d_1)/4)}),$$

where $\| f \|_{P_0} = \sqrt{P_0 f^2}$ is the $L^2(P_0)$-norm.

Consider the submodel $\{\bar{Q}_{n,\epsilon} : \epsilon\}$ defined by

$$\text{Logit}\bar{Q}_{n,\epsilon} = \text{Logit}\bar{Q}_n + \epsilon H(\bar{G}_n),$$

where $H(\bar{G}_n) = 1/\bar{G}_n(W)$. This submodel combined with the loss function $L_1(\bar{Q})$ generates the desired component of the efficient influence curve $D^*(Q,G)$:

$$\frac{d}{d\epsilon} L_1(\bar{Q}_{n,\epsilon})\bigg|_{\epsilon=0} = D_Y(\bar{Q}_n, \bar{G}_n),$$

where $D_Y(\bar{Q}, \bar{G})(O) = \frac{A}{\bar{G}(W)}(Y - \bar{Q}(W))$. We estimate the probability distribution $Q_{W,0}$ of W with the empirical probability distribution $Q_{W,n}$ of $W_1, \ldots, W_n$. Let $\epsilon_n = \arg\min_\epsilon P_n L(\bar{Q}_{n,\epsilon})$. The TMLE of $\bar{Q}_0$ is given by $\bar{Q}_n^* = \bar{Q}_{n,\epsilon_n}$, and let $Q_n^* = (\bar{Q}_n^*, Q_{W,n})$ be the TMLE of $(\bar{Q}_0, Q_{W,0})$. The TMLE of ψ_0 is given by the substitution estimator $\Psi(Q_n^*) = Q_{W,n}\bar{Q}_n^*$.

7.1.2 Asymptotic Efficiency

We have $P_n D^*(Q_n^*, \bar{G}_n) = 0$, and we also have the identity $\Psi(Q_n^*) - \Psi(Q_0) = -P_0 D^*(Q_n^*, G_n) + R_{20}(P_n^*, P_0)$, where

$$R_{20}(P_n^*, P_0) = P_0 \frac{\bar{G}_n - \bar{G}_0}{\bar{G}_n}(\bar{Q}_n^* - \bar{Q}_0).$$

This yields the starting point:

$$\Psi(Q_n^*) - \Psi(Q_0) = (P_n - P_0)D^*(Q_n^*, G_n) + R_{20}(P_n^*, P_0).$$

Since the variation norm of Q_n^* and G_n is bounded by C, and $\bar{G}_0 > \delta > 0$, it follows that the variation norm of $D^*(Q_n^*, G_n)$ is bounded by C/δ up to a small factor. This shows that $D^*(Q_n^*, G_n)$ falls in a P_0-Donsker class with probability 1. It also follows from our consistency of $\bar{Q}_n$ and $\bar{G}_n$ that $P_0\{D^*(Q_n^*, G_n) - D^*(Q_0, G_0)\}^2 \to 0$ in probability. This proves that $(P_n - P_0)D^*(Q_n^*, G_n) = (P_n - P_0)D^*(Q_0, G_0) + o_P(n^{-1/2})$. Now, we note that by $\bar{G}_n > \delta$, and the Cauchy-Schwarz inequality:

$$R_{20}(P_n^*, P_0) \leq \frac{1}{\delta} \| \bar{G}_n - \bar{G}_0 \|_{P_0} \| \bar{Q}_n - \bar{Q}_0 \|_{P_0},$$

where $\| f \|_{P_0}^2 = P_0 f^2$ is the $L^2(P_0)$-norm. By our convergence results in the $L^2(P_0)$-norm it follows that

$$R_{20}(P_n^*, P_0) = O_P(n^{-(0.5+\alpha(d)/8+\alpha(d_1)/8)}) = o_P(n^{-1/2}).$$

This proves that

$$\Psi(Q_n^*) - \Psi(Q_0) = (P_n - P_0)D^*(Q_0, G_0) + o_P(n^{-1/2}),$$

and thereby that $\Psi(Q_n^*)$ is an asymptotically efficient estimator of ψ_0. This proves the following theorem.

Theorem 7.1. *Consider the statistical model $\mathcal{M}$ on P_0 that assumes $\bar{G}_0(W) > \delta > 0$ for some $\delta > 0$; $\bar{Q}_0, \bar{G}_0$ are cadlag and have finite variation norm smaller than some universal constant C; $\delta_1 < \bar{Q}_0(W) < 1 - \delta_1$ for some $\delta_1 > 0$. Let $\Psi : \mathcal{M} \to \mathbb{R}$ be defined by $\Psi(Q_0) = Q_{W,0}\bar{Q}_0$. Consider the TMLE $\Psi(Q_n^*) = Q_{W,n}\bar{Q}_n^*$ defined above. We have that $\Psi(Q_n^*)$ is an asymptotically efficient estimator of ψ_0.*

An asymptotic 0.95-confidence interval for ψ_0 is given by:

$$\psi_n \pm 1.96\sigma_n / \sqrt{n},$$

where

$$\sigma_n^2 = P_n\{D^*(Q_n^*, G_n)\}^2.$$

The consistency of Q_n^*, G_n in $L^2(P_0)$-norm, $G_n > \delta > 0$, and that the variation norm of $D^*(Q_n^*, G_n^*)$ is bounded by the variation norm of Q_n^*, G_n, implies that

$$\sigma_n^2 \to_p \sigma_0^2 = P_0\{D^*(Q_0, G_0)\}^2 \text{ as } n \to \infty.$$

This proves that this 0.95-confidence interval has indeed asymptotic coverage equal 0.95.

7.2 General HAL-TMLE and Asymptotic Efficiency

Let's now generalize our analysis above to the analysis of a general TMLE for any statistical model and target parameter. Let $O_1, \ldots, O_n$ be n i.i.d. observations on a random variable O with probability distribution P_0 known to be an element of the statistical model $\mathcal{M}$. Let $\Psi : \mathcal{M} \to \mathbb{R}$ be a pathwise differentiable target parameter with canonical gradient $D^*(P)$. Suppose that $\Psi(P) = \Psi_1(Q(P))$ for some parameter $Q : \mathcal{M} \to Q(\mathcal{M})$, and suppose that $D^*(P) = D_1^*(Q(P), G(P))$ for some nuisance parameter $G : \mathcal{M} \to G(\mathcal{M})$. Let $L_1(Q)$ and $L_2(G)$ be loss functions for Q and G, respectively. For example, Q might consist of various variation independent components, each having its own loss function, and $L_1(Q)$ would be defined as the sum-loss function. Similarly, one might use a sum-loss function for a multiple component parameter G. We define the loss-based dissimilarities as $d_{01}(Q, Q_0) =$

$P_0 L_1(Q) - P_0 L_1(Q_0)$ and $d_{02}(G, G_0) = P_0 L_2(G) - P_0 L_2(G_0)$. We assume that $Q(P)$ and $G(P)$ are d_1 and d_2-dimensional cadlag functions for all $P \in \mathcal{M}$. If Q has multiple variation independent components, each having its own loss-function, then this corresponds with assuming that each component is a d_1-variate real valued cadlag function. We also assume that $\sup_{P \in \mathcal{M}} \| Q(P) \|_v < \infty$ and $\sup_{P \in \mathcal{M}} \| G(P) \|_v < \infty$. Let $d = \max(d_1, d_2)$. We also assume that $\sup_{P \in \mathcal{M}} \| D^*(Q(P), G(P)) \|_v < \infty$, but the latter will typically be implied by assuming that the variation norm of $Q(P)$ and $G(P)$ are uniformly bounded, uniformly in $P \in \mathcal{M}$.

Let G_n, Q_n be HAL estimators of G_0, Q_0. That is, one defines

$$Q_{n,\lambda} = \arg \min_{Q \in Q(\mathcal{M}), \|Q\|_v \leq \lambda} P_n L_1(Q),$$

λ_n as the cross-validation selector, and one sets $Q_n = Q_{n,\lambda_n}$. Similarly, one defines

$$G_{n,\lambda} = \arg \min_{G \in G(\mathcal{M}), \|G\|_v \leq \lambda} P_n L_2(G),$$

λ_n as the cross-validation selector, and one sets $G_n = G_{n,\lambda_n}$. By our general result for the HAL estimator, we have

$$d_{01}(Q_n, Q_0) = O_P(n^{-(0.5+\alpha(d_1)/4)})$$
$$d_{02}(G_n, G_0) = O_P(n^{-(0.5+\alpha(d_2)/4)}),$$

where these are the worst-case rates corresponding with models for which the parameter spaces for Q and G contain all cadlag functions with a variation norm smaller than some constant. If the parameter spaces are actual subspaces of this nonparametric parameter space, then the rate will be better, as shown in van der Laan (2017). We can replace G_n, Q_n also be a super learner where these HAL estimators are included in its library.

Let $\{Q_{n,\epsilon} : \epsilon\}$ be a parametric submodel through Q_n at $\epsilon = 0$ so that the linear span of

$$\frac{d}{d\epsilon} L_1(Q_{n,\epsilon}) \Big|_{\epsilon=0}$$

includes $D^*(Q_n, G_n)$. Let ϵ_n be so that $d_{01}(Q_{n,\epsilon_n}, Q_0) = O_P(n^{-(0.5+\alpha(d)/4)})$, and

$$P_n D^*(Q_{n,\epsilon_n}, G_n) = o_P(n^{-1/2}).$$

Let $Q_n^* = Q_{n,\epsilon_n}$.

A natural candidate for ϵ_n is defined as the MLE $\epsilon_n = \arg \min_\epsilon P_n L_1(Q_{n,\epsilon})$. For example, if $\{Q_{n,\epsilon} : \epsilon\}$ is a universal least favorable submodel, so that $\frac{d}{d\epsilon} L_1(Q_{n,\epsilon}) = D^*(Q_{n,\epsilon}, G_n)$ for all ϵ, then $P_n D^*(Q_{n,\epsilon_n}, G_n) = 0$. Under appropriate regularity conditions, even without enforcing the universal least favorable submodel property, one can show that the faster than $n^{-1/4}$-consistency of Q_n, G_n implies that $P_n D^*(Q_{n,\epsilon_n}, G_n) = o_P(n^{-1/2})$. One could also define ϵ_n as the solution of $0 = P_n D^*(Q_{n,\epsilon_n}, G_n) = 0$.

Let $R_{20}((Q, G), (Q_0, G_0)) \equiv \Psi(Q) - \Psi(Q_0) + P_0 D^*(Q, G)$. Then, it follows that

$$\Psi(Q_n^*) - \Psi(Q_0) = (P_n - P_0)D^*(Q_n^*, G_n) + R_{20}((Q_n^*, G_n), (Q_0, G_0)) + o_P(n^{-1/2}).$$

Since $R_{20}()$ is a second-order remainder, it involves integrals over products of a difference between Q_n^* and Q_0 and a difference between G_n and G_0. Since the model $\mathcal{M}$ avoids singularities by having a uniformly bounded efficient influence function, using Cauchy-Schwarz inequality, one should be able to bound $R_{20}((Q, G), (Q_0, G_0))$ by $O(\max(d_{01}(Q, Q_0), d_{02}(G, G_0)))$. As a consequence,

$$R_{20}((Q_n^*, G_n), (Q_0, G_0)) = O_P(n^{-0.5+\alpha(d)/4}) = o_P(n^{-1/2}).$$

Suppose that $(Q, G) \rightarrow D^*(Q, G)$ is continuous at P_0 in the sense that if $d_{01}(Q_n, Q_0)$ and $d_{02}(G_n, G_0)$ converge to zero as $n \rightarrow \infty$, then $P_0\{D^*(Q_n, G_n) - D^*(Q_0, G_0)\}^2 \rightarrow 0$ as $n \rightarrow \infty$. Under this continuity condition, we have $P_0\{D^*(Q_n^*, G_n) - D^*(Q_0, G_0)\}^2 \rightarrow_p 0$ as $n \rightarrow \infty$. Since $D^*(Q_n^*, G_n)$ falls with probability 1 in the Donsker class of multivariate real valued cadlag functions with a variation norm bounded by universal constant, it follows that $(P_n - P_0)D^*(Q_n^*, G_n) = (P_n - P_0)D^*(Q_0, G_0) + o_P(n^{-1/2})$. This proves

$$\Psi(Q_n^*) - \psi_0 = (P_n - P_0)D^*(Q_0, G_0) + o_P(n^{-1/2}),$$

and thus asymptotic efficiency of $\Psi(Q_n^*)$. This proves the following theorem.

Theorem 7.2. *Let $O_1, \ldots, O_n$ be n i.i.d. observations on a random variable O with probability distribution P_0 known to be an element of the statistical model $\mathcal{M}$. Let $\Psi : \mathcal{M} \rightarrow \mathbf{R}$ be a pathwise differentiable target parameter with canonical gradient $D^*(P)$. Suppose that $\Psi(P) = \Psi_1(Q(P))$ for some parameter $Q : \mathcal{M} \rightarrow Q(\mathcal{M})$, and suppose that $D^*(P) = D_1^*(Q(P), G(P))$ for some nuisance parameter $G : \mathcal{M} \rightarrow G(\mathcal{M})$. Let $L_1(Q)$ and $L_2(G)$ be loss functions for Q and G, respectively. We assume that $Q(P)$ and $G(P)$ are d_1 and d_2-dimensional cadlag functions for all $P \in \mathcal{M}$, $\sup_{P \in \mathcal{M}} \| Q(P) \|_v < \infty$ and $\sup_{P \in \mathcal{M}} \| G(P) \|_v < \infty$. Let $d = \max(d_1, d_2)$. We also assume that $\sup_{P \in \mathcal{M}} \| D^*(Q(P), G(P)) \|_v < \infty$. Assume $R_{20}((Q, G), (Q_0, G_0)) = O(\max(d_{01}(Q, Q_0), d_{02}(G, G_0)))$, and that $(Q, G) \rightarrow D^*(Q, G)$ is continuous at P_0 in the sense that if $d_{01}(Q_n, Q_0)$ and $d_{02}(G_n, G_0)$ converge to zero as $n \rightarrow \infty$, then $P_0\{D^*(Q_n, G_n) - D^*(Q_0, G_0)\}^2 \rightarrow 0$ as $n \rightarrow \infty$.*

Let Q_n, G_n be HAL estimators. We have

$$d_{01}(Q_n, Q_0) = O_P(n^{-(0.5+\alpha(d_1)/4)})$$
$$d_{02}(G_n, G_0) = O_P(n^{-(0.5+\alpha(d_2)/4)}),$$

Let $\{Q_{n,\epsilon} : \epsilon\}$ be a parametric submodel through Q_n at $\epsilon = 0$ so that the linear span of

$$\frac{d}{d\epsilon} L_1(Q_{n,\epsilon})\Big|_{\epsilon=0}$$

includes $D^*(Q_n, G_n)$. *Let* ϵ_n *be so that* $d_{01}(Q_{n,\epsilon_n}, Q_0) = O_P(n^{-(0.5+\alpha(d)/4)}))$, *and*

$$P_n D^*(Q_{n,\epsilon_n}, G_n) = o_P(n^{-1/2}).$$

Let $Q_n^* = Q_{n,\epsilon_n}$. *The TMLE* $\Psi(Q_n^*)$ *is asymptotically efficient.*

7.3 Discussion

In this chapter, we established asymptotic efficiency of the one-step TMLE of the treatment specific mean (and thus for the average treatment effect) if one uses a super learner as an initial estimator that includes the HAL estimator in its library.

> The key is that these HAL estimators of the nuisance parameters converge to their truth counterparts at a faster rate than the critical rate $n^{-1/4}$. We were able to prove this only assuming strong positivity, uniformly bounded loss functions, and by assuming that the nuisance parameters have a variation norm bounded by a universal constant.

It is also possible to establish asymptotic efficiency of a TMLE when only assuming that the true nuisance parameters have a finite variation norm, thereby allowing for models that are unbounded with respect to variation norm, still assuming a uniform model bound with respect to supremum norm (so that the loss functions and efficient influence curve are uniformly bounded). In this case, one uses a sieve of bounded models $\mathcal{M}_n$, allowing the universal bounds for $\mathcal{M}_n$ to increase or decrease with n so that for n large enough the true nuisance parameters are captured by the n-specific model $\mathcal{M}_n$: i.e. $P_0 \in \mathcal{M}_n$ for $n > N_0 = N(P_0)$. By using CV-TMLE, one does not even need a sieve for controlling the variation norm bounds. For both the TMLE and CV-TMLE, we can even allow that the supremum norm bounds of a sieve $\mathcal{M}_n$ converge to infinity, as long as it converges slowly enough so that the second-order term in the oracle inequality for the super learner still converges to zero at a faster rate than $n^{-1/2}$. We also demonstrated that these results immediately generalize to general models and target parameters. We refer to van der Laan (2017) for the precise theorems showcasing these general results for general models as well as for the treatment specific mean example.

Due to using the HAL estimators for the nuisance parameters Q and G, there is no need to rely on double robustness of the efficient influence curve defined by $R_{20}((Q, G), (Q_0, G_0)) = 0$ if either $Q = Q_0$ or $G = G_0$. In van der Laan (2014b) we demonstrate that for such double robust estimation problems it is possible to construct TMLE that remain asymptotically linear even when one of the two nuisance parameter estimators is inconsistent. If, for example, Q_n converges to a misspecified Q, then the remainder $R_{20}((Q_n, G_n), (Q_0, G_0))$ is not second order anymore, but, instead, behaves as $R_{20}((Q, G_n), (Q_0, G_0))$. The latter can be written as a function of G_n minus that same function applied to G_0. Therefore, in order to make $R_{20}((Q, G_n), (Q_0, G_0))$ asymptotically linear, G_n will have to be a TMLE targeting the required functional of $G \to R_{20}((Q, G), (Q_0, G_0))$. Indeed, the proposed TMLEs in van der Laan (2014b) involve fluctuation models for both Q and G so that the resulting TMLE (Q_n^*, G_n^*) of (Q_0, G_0) targets not only the target parameter ψ_0 but also these extra nuisance parameters. Even though these complications will not enhance the asymptotic behavior of the HAL-TMLE, it might still enhance the finite sample behavior of the HAL-TMLE.

Chapter 8
HAL Estimator of the Efficient Influence Curve

Mark J. van der Laan

The construction of an efficient estimator of a pathwise differentiable target parameter $\Psi : \mathcal{M} \to \mathbb{R}$ relies on the ability to evaluate its canonical gradient $D^*(P)$ at an initial estimator P of P_0 based on an original i.i.d. sample from P_0. The efficient influence curve $D^*(P)$ is defined as the canonical gradient of the pathwise derivative of the target parameter along parametric submodels through P. It is always possible to represent the pathwise derivative of the target parameter along a parametric submodel as a covariance of a gradient $D(P) \in L_0^2(P)$ with the score of the submodel. The canonical gradient is now defined as the projection of this gradient on the tangent space at P, where the tangent space is defined as the closure of the linear span of all scores one can generate with a parametric submodel through P.

Characterizing the tangent space is not a hard problem, and is often represented as the range of a linear score operator that maps underlying paths into the score for the resulting submodel through P. However, carrying out this projection of $D(P)$ onto the tangent space can be a difficult optimization problem and does not necessarily allow for a closed form solution. General formulas for the efficient influence curve are given by the Hilbert space analogues of the least squares regression formula, $X(X'X)^{-1}X'Y$, involving the inverse of the so called information operator defined by the composition of the score operator and its adjoint. For many problems, the inverse of this infinite dimensional information operator does not exist in closed form and can be very hard to implement algorithmically.

However, the projection formulation of the canonical gradient shows that the canonical gradient is the least squares regression of the gradient on a large regres-

M. J. van der Laan (✉)
Division of Biostatistics and Department of Statistics, University of California, Berkeley,
101 Haviland Hall, #7358, Berkeley, CA 94720, USA
e-mail: laan@berkeley.edu

© Springer International Publishing AG 2018
M.J. van der Laan, S. Rose, *Targeted Learning in Data Science*,
Springer Series in Statistics, https://doi.org/10.1007/978-3-319-65304-4_8

103

sion model represented by the tangent space, under an infinite sample of P. In other words, each score in the tangent space represents a candidate regression curve (as a function of O). The true regression curve, i.e., the regression curve in this model minimizing the distance to the gradient, equals the efficient influence curve. If the tangent space is a range of a linear score operator, each regression curve is identified by an underlying function, which can be viewed as the unknown parameter in this regression model. Moreover, due to the linearity of the score operator, the regression model is linear in this parameter/function. As a result, the efficient influence curve can be formulated as a linear least squares regression problem for an infinite dimensional linear model. In this chapter we present a machine learning method that involves taking a sample from P, and fitting a highly adaptive lasso (HAL) least squares linear regression estimator of the efficient influence curve. The HAL estimator can be replaced by other machine learning algorithms, but our theoretical results for the HAL estimator make the HAL estimator a particularly good choice. This approach avoids having to solve the mathematical optimization problem, but instead we let the machine learning algorithm estimate the regression surface $D^*(P)$.

8.1 Formulation of HAL Least Squares Linear Regression Estimator of the Efficient Influence Curve

Let $\Psi : \mathcal{M} \to \mathbf{R}$ be a statistical target parameter that is pathwise differentiable at a probability distribution P in the statistical model $\mathcal{M}$ with canonical gradient $D^*(P)$. Let $O \sim P$ be a multidimensional random variable in $\mathbf{R}^d$. Given P, our goal is to evaluate the efficient influence curve $O \to D^*(P)(O)$ as a function of O.

One common approach for determining the efficient influence curve is to first find an initial gradient $D(P)$ of the pathwise derivative and then project it on the tangent space $T(P)$ at P. Finding an initial gradient can be achieved by determining an initial estimator of $\Psi(P)$ under sampling n i.i.d. observations from P, and determining the influence curve of this estimator. This influence curve is then the initial gradient $D(P)$. The initial estimator can be selected as simply as possible (there is no reason to prefer one gradient above the other, since all project into the canonical gradient).

One approach for finding an initial gradient is to first extent the parameter $\Psi : \mathcal{M} \to \mathbf{R}$ to a nonparametric model $\Psi^e : \mathcal{M}_{np} \to \mathbf{R}$ so that $\Psi^e(P) = \Psi(P)$ for $P \in \mathcal{M}$. Subsequently, one then finds the gradient of this pathwise derivative of this nonparametric extension Ψ^e. The latter can be computed as the influence curve of any regular asymptotically linear estimator in this nonparametric model, or it can be calculated through the functional delta method as the influence curve of $\Psi^e(P_n)$ where P_n is the empirical probability distribution of $O_1, \ldots, O_n$: here one might first approximate O by a discrete approximation O_m so that $\Psi^e(P_n)$ will indeed be asymptotically linear, and then determine the limit of the resulting influence curve as the approximation error converges to zero (i.e., $m \to \infty$). Different nonparametric extensions will result in different influence curves/gradients, and one might want to select an extension for which this calculation of the influence curve of $\Psi^e(P_n)$ is easy.

Another important observation for determining an initial gradient is the following. One can focus on finding an initial gradient in a submodel of M defined by treating an orthogonal nuisance parameter as known. That is, suppose $\Psi(P) = \Psi_1(Q(P))$ for some parameter Q. Then, an orthogonal nuisance parameter is a parameter for which the scores of parametric submodels only fluctuating the nuisance parameter are orthogonal to scores of parametric submodels only fluctuating Q. For example, suppose that the density $p(O) = q(O)g(O)$ factorizes in two variation independent factors q and g and that $\Psi(P)$ only depends on P through the factor q. In addition, let the model in terms of densities be of the form $M = \{p = qg : q \in Q, g \in G\}$ for parameter spaces Q and G for these two factors q and g, respectively. In that case, one can define the submodel $M(g) = \{p = qg : q \in Q\} \subset M$ by making g known. The efficient influence function of Ψ at P in the submodel $M(g)$ in which g is known is identical to the efficient influence function in the actual model M. Finding a gradient in the model $M(g)$ is often very straightforward: in general, the smaller the model, the easier it is to find a gradient. For example, in censored data models one can define the gradient as the influence curve of an inverse probability of censoring weighed estimator in the model in which the censoring mechanism is known. This is what we will do in each of our examples in this chapter.

The tangent space is often defined as the closure of the range of a linear score operator $A_P : (H, \langle \cdot, \cdot \rangle_H) \to L_0^2(P)$, where $(H, \langle \cdot, \cdot \rangle_H)$ is an underlying Hilbert space H with inner product $\langle h_1, h_2 \rangle_H$ for any pair $h_1, h_2 \in H$. For example, consider a model $\{P_\theta : \theta \in \Theta\}$. A parametric submodel through P_θ is now of the form $\{P_{\theta_{\epsilon,h}} : \epsilon\}$ where h denotes a direction varying over some set $\mathcal{H}$ and $\mathcal{H}$ is embedded in a Hilbert space. Let H be the closure of the linear span of $\mathcal{H}$ within this Hilbert space. The score of this submodel could be represented as a mapping $A_P(h) = \frac{d}{d\epsilon} \log p_{\theta_{\epsilon,h}}\big|_{\epsilon=0}$, and, using a natural parametrization $\theta_{\epsilon,h}$, A_P will be a linear operator. The tangent space is now given by the closure of the range of $A_P : (H, \langle \cdot, \cdot \rangle_H) \to L_0^2(P)$.

The efficient influence curve can then be defined as $D^*(P) = A_P(h^*)$, where

$$h^* = \arg \min_{h \in H} P\{D(P) - A_P(h)\}^2.$$

We treat P as known in $D(P)$ as well as in $A_P(h)$, so that $A_P(h^*(P))$ represents the least squares regression of a known outcome $D(P)(O)$ on the regression model $\{A_P(h) : h\}$ with unknown parameter h, while $h^*(P)$ is the true parameter value. Subsequently, we take a sample $O_1, \ldots, O_n \sim P$, and estimate $h^*(P)$ with a machine learning algorithm based on this data set. Let P_n be the empirical probability distribution.

Suppose that H is a Hilbert space of d-variate real valued functions. Let $H_M \subset H$ be the subset of cadlag functions with variation norm smaller than M, where the variation norm of a function h is defined as $\| h \|_v = \sum_{S \subset \{1,\ldots,d\}} \int | dh_S(u_S) |$. Here $u_S \to h_S(u_S) = h(u_S, 0_{S^c})$ is the section of h that sets the components not in S equal to zero, and the sum is over all subsets of $\{1, \ldots, d\}$. In addition, suppose that $\| h^* \|_v < M$ for some M. We can estimate h^* with the finite sample estimator:

$$h_{n,M} = \arg \min_{h \in H_M} P_n\{D(P) - A_P(h)\}^2.$$

Let $\hat{h}_M$ denote the estimator so that $h_{n,M} = \hat{h}_M(P_n)$. Since we do not know how to select M large enough, we select $M = M_n$ with the cross-validation selector:

$$M_n = \arg\min_{M} E_{B_n} P^1_{n,B_n} \left\{ D(P) - A_P(\hat{h}_M(P^0_{n,B_n})) \right\}^2.$$

Our estimator of h^* is given by

$$h_n \equiv h_{n,M_n} = \hat{h}_{M_n}(P_n),$$

resulting in the estimator $A_P(h_n)$ of $D^*(P)$.

We will now show how we determine $h_{n,M}$ through fitting a high dimensional linear regression model with the lasso algorithm. We can represent any cadlag function h with finite variation norm as $h(x) = \sum_{S \subset \{1,...,d\}} \int \phi_{xS}(u_S) dh_S(u_S)$, where $u_S \to h_S(u_S) = h(u_S, 0_{S^c})$ is the section of h that sets the components not in S equal to zero, and $\phi_{xS}(u_S) = \prod_{j \in S} I(x_j \geq u_j)$ is the product of indicator basis functions. Since A_P is a linear operator we have:

$$A_P(h)(O) = A_P \left(x \to \sum_{S \subset \{1,...,d\}} \int \phi_{xS}(u_S) dh_S(u_S) \right)(O)$$

$$= \sum_{S \subset \{1,...,d\}} \int A_P \left(x \to \phi_{xS}(u_S) \right)(O) dh_S(u_S),$$

Thus,

$$h_{n,M} = \arg\min_{h \in H_M} \frac{1}{n} \sum_{i=1}^{n} \left\{ D(P)(O_i) - \sum_{S \subset \{1,...,d\}} \int A_P \left(x \to \phi_{xS}(u_S) \right)(O_i) dh_S(u_S) \right\}^2.$$

Note that this is an infinite dimensional minimum least squares linear regression estimator, where the outcome $Y_i = D(P)(O_i)$, the main terms are $(A_P(\phi_{\cdot S}(u_S))(O_i) : u_S, S)$ with corresponding coefficients $(dh_S(u_S) : u_S, S)$, where the sum of the absolute values of these coefficients is enforced to be smaller than M. A study of this problem will typically show that this minimum is attained by h for which dh_S only puts positive mass on at most n values $u_{S,j}$, $j = 1, \ldots, n$, for each $S \subset \{1, \ldots, d\}$. In that case, this infinite dimensional minimum least squared linear regression problem becomes a finite dimensional linear regression $Y_i = \sum_{S \subset \{1,...,d\}} \sum_{j=1}^{n} \phi_{S,j}(O_i)\beta_{S,j} + e_i$, where $\sum_S \sum_{j=1}^{n} |\beta_{S,j}| \leq M$ and $\phi_{S,j}(O_i) = A_P \left(\phi_{\cdot S}(u_{S,j}) \right)$.

Thus we can now define the standard lasso linear regression estimator:

$$\beta_{n,M} = \arg\min_{\beta,\|\beta\|_1 \leq M} \frac{1}{n} \sum_{i=1}^{n} \left\{ D(P)(O_i) - \sum_{S \subset \{1,...,d\}} \sum_{j=1}^{n} \phi_{S,j}(O_i)\beta_{S,j} \right\}^2.$$

This defines

$$D^*_{n,M}(P)(O) = \sum_{S \subset \{1,...,d\}} \sum_{j=1}^{n} \phi_{S,j}(O)\beta_{n,M,S,j}.$$

Let M_n be the cross-validation selector, and $\beta_n = \beta_{n,M_n}$. Then, our approximation of the efficient influence curve $D^*(P)$ is given by:

$$D_n^*(P)(O) = D_{n,M_n}^*(P)(O) = \sum_{S \subset \{1,\ldots,d\}} \sum_{j=1}^{n} \phi_{S,j}(O) \beta_{n,S,j}.$$

8.2 Rate of Convergence of the HAL Estimator of the Efficient Influence Curve

We have the following theorem establishing that this estimator $D_n^*(P)$ converges in $L^2(P)$-norm to $D^*(P)$ at a rate faster than $n^{-1/4}$.

Theorem 8.1. *Let $\Psi : \mathcal{M} \to \mathbb{R}$ be pathwise differentiable at P with canonical gradient $D^*(P)$. Let $D(P)$ be a gradient at P which is a uniformly bounded function. Let $A_P : (H, \langle \cdot, \cdot \rangle_H) \to L_0^2(P)$ be a linear score operator from an underlying Hilbert space to $L_0^2(P)$, so that the tangent space $T(P) = \overline{R}(A_P)$ at P is given by the closure of the range of A_P. Suppose that $D^*(P) = A_P(h^*)$ for some $h^* = h^*(P) \in H$. Suppose that H consists of d-variate real valued functions, and that there exists a version of $h^*(P)$ that is cadlag and has a finite variation norm. Let $H_M \subset H$ be the subset of cadlag functions with variation norm smaller than M: we have $h^*(P) \in H_M$ for $M > \| h^* \|_v$. Assume that $A_P(H_M)$ is a P-Donsker class: we note that, if the class of functions $\{A_P (x \to \phi_{x_S}(u_S)) : u_S, S\}$ is a P-Donsker class, then $A_P(H_M)$ is a P-Donsker class.*

Let $O_1, \ldots, O_n$ be a sample of n i.i.d. copies of $O \sim P$. Define the subspace $\Psi_M = A_P(H_M) = \{A_P(h) : h \in H_M\}$, and define

$$\psi_{n,M} = \arg \min_{\psi \in \Psi_M} P_n(D(P) - \psi)^2.$$

Above we showed that this estimator can be defined as a lasso least squares linear regression estimator under the constraint that the sum of the absolute values of the coefficients is bounded by M, where the outcome is $D(P)(O_i)$ and the main terms are a finite subset of $A_P (x \to \phi_{x_S}(u_S)) (O_i) : u_S, S\}$ defined by restricting u_S to a finite set of $O(n)$-values, for each $S \subset \{1, \ldots, d\}$.

Let M_n be the cross-validation selector over a uniformly bounded set:

$$M_n = \arg \min_M E_{B_n} P_{n,B_n}^1 (D(P) - \hat{\Psi}_M(P_{n,B_n}^0))^2.$$

Let $\psi_n = \psi_{n,M_n}$ be our estimator $D_n^*(P)$ of $D^*(P)$. Then,

$$\int \{D_n^*(P) - D^*(P)\}^2(o) dP(o) = o_P(n^{-1/2}).$$

With a little more work, as in van der Laan (2017), utilizing finite sample empirical process results in van der Vaart and Wellner (2011), assuming that the entropy of $A_P(H_M)$ is of same order as entropy of H_M, we can obtain an actual rate $o_P(n^{-(1/2+\alpha(d)/4)})$, where $\alpha(d) = 1/(d+1)$.

Proof. The Donsker class statement is a consequence of the fact that a convex Hull of a Donsker class is also a Donsker class. This also implies that

$$\left\{ \{D(P) - A_P(h)\}^2 : h \in H_M \right\}$$

is a P-Donsker class. It remains to prove the $o_P(n^{-1/2})$-consistency result. Let $L(\psi) = (D(P) - \psi)^2$ be the squared error loss. Note that

$$\psi_M = \Psi_M(P) = \arg \min_{\psi \in \Psi_M} PL(\psi),$$

and

$$\psi_{n,M} = \arg \min_{\psi \in \Psi_M} P_n L(\psi).$$

We want to prove that $\int (\psi_{n,M} - \Psi_M(P))^2 dP = o_P(n^{-1/2})$. Our desired result for the estimator $\psi_n = \psi_{n,M_n}$ now follows from the finite sample oracle inequality for the cross-validation selector.

We have

$$0 \le d_P(\psi_{n,M}, \psi_M) \equiv PL(\psi_{n,M}) - PL(\psi_M)$$
$$= (P - P_n)\{L(\psi_{n,M}) - L(\psi_M)\} + P_n\{L(\psi_{n,M}) - L(\psi_M)\}$$
$$\le -(P_n - P)\{L(\psi_{n,M}) - L(\psi_M)\}.$$

By assumption, $\{L(\psi) - L(\psi_M) : \psi \in \Psi_M\}$ is a P-Donsker class, so that, by empirical process theory, the right-hand side is $O_P(n^{-1/2})$, and thus $d_P(\psi_{n,M}, \psi_M) = O_P(n^{-1/2})$. Since $L(\psi)$ is the squared error loss, we can bound $P\{L(\psi_{n,M}) - L(\psi_M)\}^2$ by a universal constant times $P\{L(\psi_{n,M}) - L(\psi_M)\}$ (see e.g., van der Laan and Dudoit 2003). Thus, this proves that $P\{L(\psi_{n,M}) - L(\psi_M)\}^2 \to_p 0$ as $n \to \infty$. By empirical process theory, this proves that $(P_n - P)\{L(\psi_{n,M}) - L(\psi_M)\} = o_P(n^{-1/2})$. Thus we have shown $d_P(\psi_{n,M}, \psi_M) = o_P(n^{-1/2})$. $\square$

8.2.1 Application to Estimate Projection of Initial Gradient onto Subtangent Spaces

One does not need to apply the HAL estimator to estimate the projection of the initial gradient on the *full* tangent space. For example, suppose that the tangent space $T(P)$ allows for an orthogonal decomposition $\sum_{j=1}^{K} T_j(P)$, where $T_j(P)$ is the tangent space of one of the K nuisance parameters, such as factors that make up the density p. In this common scenario, we have $\Pi(D(P) \mid T(P)) = \sum_{j=1}^{K} \Pi(D(P) \mid T_j(P))$, so that it suffices to compute the projection of $D(P)$ onto $T_j(P)$, for each j separately. For some of the tangent spaces, the projection might be easily analytically determined. For the remaining tangent spaces, we can then apply the above HAL estimator to each j-specific projection separately, where now $T_j(P)$ is the closure of the range of a score operator $A_j : (H_j, \langle \cdot, \cdot \rangle_{H_j}) \to L_0^2(P)$. This approach will be applied to some of our examples.

8.2.2 Using the Actual Data Set from the True Data Distribution

If P represents a consistent estimator $\hat{P}_n$ of P_0 in the sense that $D^*(\hat{P}_n)$ converges to $D^*(P_0)$, then we conjecture that it is fine to use the actual observations $O_1, \ldots, O_n$ from P_0 in the formulation of our estimator, instead of a new sample from $\hat{P}_n$. In this case, we view the HAL estimator $D_n^*(\hat{P}_n)$ as an estimator of $D^*(P_0)$. Of course, in this case the validity of our estimator of $D^*(P_0)$ now depends on $\hat{P}_n$ approximating P_0 as n converges to infinity. The advantage of this approach for estimation of $D^*(P_0)$ is that it does not require resampling from a data distribution P. For example, in many problems the efficient influence curve $D^*(P)$ only depends on P through some parameters (Q, G) say. These parameters might not identify an actual data distribution. An estimator (Q_n, G_n) of (Q_0, G_0) does now not imply a data distribution $\hat{P}_n$ that we can resample from. So we would then have to determine a $\hat{P}_n$ that is compatible with our estimates (Q_n, G_n). This might be easy, but could also be cumbersome in some problems.

The rational for the HAL estimator of $D^*(P_0)$ using the actual i.i.d. sample from P_0 is as follows. Firstly, we can apply our theorem at $P = P_0$, which shows that our lasso estimator converges to $D^*(P_0)$ in $L^2(P_0)$ at a rate faster than $n^{-1/4}$. However, this lasso estimator uses as outcome the unknown $D(P_0)$ and also uses main terms $A_{P_0}(x \to \phi_{x,S}(u_{S,j}))$ in the linear regression model that depend on P_0. If $D(P_0)$ is replaced by a consistent $D(\hat{P}_n)$, then it will be straightforward to show that the lasso estimator is still consistent, and the rate would still be faster than $n^{-1/4}$ if $D(\hat{P}_n)$ converges to $D(P_0)$ at the same or faster rate. Finally, one wants to show that replacing the unknown main terms in the linear regression model by the estimated versions using $\hat{P}_n$ still preserves the consistency. We believe that the latter is not hard to show under a reasonably weak condition. In the remaining sections we consider various examples.

8.3 Truncated Mean Based on Current Status Data

Let $O = (C, \Delta = I(T \leq C))$, where T and C are independent. Let $F(t) = P(T \leq t)$ and $G(t) = P(C \leq t)$ be the two cumulative distribution functions of T and C, respectively. Let g be the Lebesgue density of G, and let $\bar{F} = 1 - F$. Let $P = P_{F,G}$

be the true probability distribution of O. Let F be unspecified, while G might be restricted to a set $\mathcal{G}$, so that $\mathcal{M} = \{P_{F,G} : F, G \in \mathcal{G}\}$ is the statistical model. Let $\Psi : \mathcal{M} \to \mathbf{R}$ be defined by $\Psi(P) = \int (1 - F)(t)r(t)dt$, where r is a given function. By selecting $r(t) = t$, $\Psi(P)$ equals the mean of T, and by selecting r equal to a truncated version of the identity function, it yields a truncated mean. Many other functionals can be generated by selecting an appropriate r. Note that the density $p(c, \delta) = F(c)^\delta (1 - F(c))^{1-\delta} g(c)$ of P factorizes in a factors only depending on F and g, while $\Psi(P) = \Psi(F)$ only depends on F. Therefore, it suffices to determine the efficient influence curve in the model $\mathcal{M}(G)$ in which G is known. In this model $\mathcal{M}(G)$, we can use the following gradient:

$$D(F, G) = \frac{r(C)}{g(C)}(1 - \Delta) - \Psi(F).$$

Note that indeed, $\psi_n = \frac{1}{n}\sum_{i=1}^n \frac{r(C_i)}{g(C_i)}(1 - \Delta_i)$ is an unbiased estimator of $\Psi(F)$ with influence curve $D(F, G)$, so that indeed $D(F, G)$ is a gradient in the model $\mathcal{M}(G)$. Let $dF_{\epsilon,h} = (1 + \epsilon h)dF$ be a submodel through F at $\epsilon = 0$ with score $h \in L_0^2(F)$. The score of $\{P_{F_{\epsilon,h},G} : \epsilon\}$ is given by

$$A_F(h)(O) = E_F(h(T) \mid O) = \frac{\int_0^C h(t)dF(t)}{F(C)}\Delta + \frac{\int_C^\infty h(t)dF(t)}{1 - F(C)}(1 - \Delta).$$

Thus, the score operator is given by this linear operator $A_F : L_0^2(F) \to L_0^2(P_{F,G})$. The efficient influence curve is defined as $D^*(F, G) = A_F(h^* - Fh^*)$, where

$$h^* = \arg \min_{h \in L^2(F)} P\{D(F, G) - A_F(h - Fh)\}^2.$$

We can represent $h(t) = h(0) + \int I(t \geq x)dh(x) = h(0) + \int \phi_x(t)dh(x)$, where $\phi_x(t) = I(t \geq x)$. Using this representation it follows that $Fh = h(0) + \int \bar{F}(x)dh(x)$. Substitution of this representation for h into the above expression yields:

$$h^* = \arg \min_h P\left\{D(F, G) - \int A_F(\phi_x)dh(x) + \int \bar{F}(x)dh(x)\right\}^2$$

$$= \arg \min_h P\left\{D(F, G) - \int \{\bar{F}(x \mid \cdot) - \bar{F}(x)\}dh(x)\right\}^2,$$

where $\bar{F}(x \mid C, \Delta) = P(T > x \mid C, \Delta)$. Let $O_1, \ldots, O_n \sim_{iid} P_{F,G}$ be a sample from $P_{F,G}$. Let M be an upper bound for the variation norm of h^*. Then, we can estimate h^* with

$$h_{n,M}^* = \arg \min_{h, \|h\|_v < M} \frac{1}{n}\sum_{i=1}^n \left\{D(F, G)(C_i, \Delta_i) - \int \{\bar{F}(x \mid C_i, \Delta_i) - \bar{F}(x)\}dh(x)\right\}^2.$$

This minimum is attained at a discrete measure dh which only puts mass on $\{C_1, \ldots, C_n\}$:

$$h^*_{n,M} = \arg\min_{h,\|h\|_v < M} \frac{1}{n} \sum_{i=1}^{n} \left\{ D(F,G)(C_i, \Delta_i) - \sum_{j=1}^{n} \{\bar{F}(C_j \mid C_i, \Delta_i) - \bar{F}(x)\} dh(C_j) \right\}^2.$$

Let $\beta_j = dh(C_j)$, $\| \beta \|_1 = \sum_{j=1}^{n} | \beta_j |$, and define

$$\beta_{n,M} = \arg\min_{\beta,\|\beta\|_1 < M} \frac{1}{n} \sum_{i=1}^{n} \left\{ D(F,G)(C_i, \Delta_i) - \sum_{j=1}^{n} \beta_j \{\bar{F}(C_j \mid C_i, \Delta_i) - \bar{F}(x)\} \right\}^2.$$

This is a standard lasso least squares regression estimator with L_1-constraint set at M. Let M_n be the cross-validation selector, and define $\beta_n = \beta_{n,M_n}$. Then, our estimator of the efficient influence curve $D^*(P)$ is given by:

$$D^*_n(P)(C, \Delta) = \sum_{j=1}^{n} \beta_{n,j} \{\bar{F}(C_j \mid C, \Delta) - \bar{F}(x)\}.$$

By Theorem 8.1, this estimator converges to $D^*(P)$ in $L^2(P)$-norm at a rate faster than $n^{-1/4}$.

8.4 Truncated Mean Based on Interval Censored Data

Let $O = (C_m, \Delta_m = I(T \leq C_m) : m = 1, \ldots M)$ be a general interval censored data structure, where T is a time to event, and $C = (C_1, \ldots, C_M)$ is a vector of continuous valued monitoring times. Suppose that the monitoring times and T are all larger than 0. For simplicity, we consider the case that the number of monitoring time M is fixed, but we suggest that our results below are generalizable to the case that M is random. We assume that T and C are independent. Let $F(t) = P(T \leq t)$ be the cumulative distribution function of T, and let G and g denote the probability distribution and density of C. Note that the probability distribution $P = P_{F,G}$ of O is indexed by the true distributions F and G of T and C, respectively. Let $\mathcal{M} = \{P_{F,G} : F, G \in \mathcal{G}\}$ be the statistical model for P defined by leaving F unspecified, while we might have restrictions/knowledge on the distribution of C defined by the set $\mathcal{G}$. Let $\Psi^f(F) = \int r(t)\bar{F}(t)dt$ be the full data parameter of interest, where $\bar{F} = 1 - F$. Under some support conditions on g, this parameter is identifiable from P, so that $\Psi : \mathcal{M} \to \mathbb{R}$ satisfies $\Psi(P_{F,G}) = \Psi^f(F)$. Let $D^*(F,G)$ be the canonical gradient of Ψ at $P_{F,G}$.

Note that the density of $P_{F,G}$ is given by $p_{F,G}(o) = g(c_1, \ldots, c_M)F(C(o))$, where $C(o)$ is the coarsening for T implied by $O = o$, and for a set A, $F(A) = P(T \in A)$. Thus $C(o) = (L(o), R(o)]$ is an interval, where $L(O)$ is the largest monitoring time C_j for which $\Delta_j = 0$, while $R(O)$ is the smallest monitoring time C_j for which $\Delta_j = 1$. If $\Delta_1 = 1$, then we define $L(O) = 0$, and if $\Delta_M = 0$, then $R(O) = \infty$. Thus, the density of O factorizes in a F and G factor so that the efficient influence curve in our model $\mathcal{M}$ is the same as in the model $\mathcal{M}(G) = \{P_{F,G} : F\} \subset \mathcal{M}$ in which G is known.

Our first task is to determine an initial gradient $D(F, G)$ in the model $\mathcal{M}(G)$. For that purpose, let's define the random variable $\bar{C}$ as the outcome of the following experiment: generate C and then randomly draw one of the M monitoring times, each one receiving probability $1/M$. Let $\bar{\Delta} = I(T \leq \bar{C})$. Let $\bar{g} = \frac{1}{M} \sum_{m=1}^{M} g_m$ be the univariate density of $\bar{C}$, where g_m is the marginal density of C_m. An initial estimator of $\Psi^f(F)$ is given by the IPCW estimator $\frac{1}{n} \sum_{i=1}^{n} (1 - \bar{\Delta}_i) r(\bar{C}_i) / \bar{g}(\bar{C}_i)$ based on the reduced current status data structure $(\bar{C}, \bar{\Delta} = I(T \leq \bar{C})$. The influence curve of this estimator is given by $(1 - \bar{\Delta}) r(\bar{C}) / \bar{g}(\bar{C}) - \Psi(F)$. Strictly speaking this is not an influence curve since it is not only a function of O, but is also random through the random pick involved in selecting $\bar{C}$. So let's take the conditional expectation of this influence curve, given O. This yields $\frac{1}{M} \sum_{m=1}^{M} (1 - \Delta_m) r(C_m) / \bar{g}(C_m) - \Psi(F)$. Let's verify if indeed the expectation of this equals zero, which then shows that this is the influence curve of a linear unbiased estimator in model $\mathcal{M}(G)$. We have:

$$E_{P_{F,G}} \frac{1}{M} \sum_{m=1}^{M} (1 - \Delta_m) r(C_m) / \bar{g}(C_m) = \frac{1}{M} \sum_{m=1}^{M} E_G \bar{F}(C_m) r(C_m) / \bar{g}(C_m)$$

$$= \frac{1}{M} \sum_{m=1}^{M} \int_c \bar{F}(c) r(c) \frac{g_m(c)}{\bar{g}(c)} dc$$

$$= \int r(c) \bar{F}(c) dc = \Psi(F).$$

This proves that indeed we can select the initial gradient:

$$D(F, G)(O) = \frac{1}{M} \sum_{m=1}^{M} (1 - \Delta_m) r(C_m) / \bar{g}(C_m) - \Psi(F).$$

Let $A_F : L_0^2(F) \to L_0^2(P_{F,G})$ be the score operator that maps the score $h \in L_0^2(F)$ of the submodel $\{dF_{\epsilon,h} = (1 + \epsilon h) dF : \epsilon\}$ at $\epsilon = 0$ into the score $A_F(h)$ of $\{P_{F_{\epsilon,h},G} : \epsilon\}$ at $\epsilon = 0$. We have

$$A_F(h)(O) = E_F(h(T) \mid O) = E_F(h(T) \mid T \in (L(O), R(O)]) = \frac{\int_{L(O)}^{R(O)} h(t) dF(t)}{F(R(O)) - F(L(O))}.$$

The efficient influence curve $D^*(F, G)$ can thus be defined as $A_F(h^* - Fh^*)$, where

$$h^* = h^*(F, G) = \arg \min_{h \in L^2(F)} P_{F,G} \{D(F, G) - A_F(h - Fh)\}^2.$$

We can represent $h(t) = h(0) + \int I(t \geq x) dh(x) = h(0) + \int \phi_x(t) dh(x)$, where $\phi_x(t) = I(t \geq x)$. Using this representation it follows that $Fh = h(0) + \int \bar{F}(x) dh(x)$. Substitution of this representation for h into the above definition of h^* yields:

$$h^* = \arg \min_h P \left\{ D(F, G) - \int A_F(\phi_x) dh(x) + \int \bar{F}(x) dh(x) \right\}^2$$

$$= \arg\min_h P\left\{D(F,G) - \int\{\bar{F}(x\mid\cdot) - \bar{F}(x)\}dh(x)\right\}^2,$$

where $\bar{F}(x\mid O) = P(T > x \mid T \in (L(O), R(O)])$.

Let $O_1, \ldots, O_n \sim_{iid} P_{F,G}$ be a sample from $P_{F,G}$. Then, we can estimate h^* with

$$h^*_{n,M} = \arg\min_{h, \|h\|_v < M} \frac{1}{n}\sum_{i=1}^n\left\{D(F,G)(O_i) - \int\{\bar{F}(x\mid O_i) - \bar{F}(x)\}dh(x)\right\}^2.$$

Let $\{x_1, \ldots, x_J\}$ be the set of monitoring times that appear as $L(O_i)$ or $R(O_i)$ across the n coarsenings $(L(O_i), R(O_i)]$, $i = 1, \ldots, n$. Notice that $J = J(n)$ is at most $2n$, and will be a little smaller than $2n$ if there are coarsenings that have as left point 0 or as right point ∞. We suggest that for each M, the minimum $h^*_{n,M}$ is attained at a discrete measure dh which only puts mass on $\{x_1, \ldots, x_J\}$:

$$h^*_{n,M} = \arg\min_{h, \|h\|_v < M} \frac{1}{n}\sum_{i=1}^n\left\{D(F,G)(O_i) - \sum_{j=1}^J\{\bar{F}(x_j\mid O_i) - \bar{F}(x_j)\}dh(x_j)\right\}^2.$$

Let $\beta_j = dh(x_j)$, $\|\beta\|_1 = \sum_{j=1}^n |\beta_j|$, and define

$$\beta_{n,M} = \arg\min_{\beta, \|\beta\|_1 < M} \frac{1}{n}\sum_{i=1}^n\left\{D(F,G)(O_i) - \sum_{j=1}^J\beta_j\{\bar{F}(x_j\mid O_i) - \bar{F}(x_j)\}\right\}^2.$$

This is a standard lasso least squares linear regression estimator with L_1-constraint set at M, outcome $D(F,G)(O_i)$ and J main terms $\bar{F}(x_j\mid O_i) - \bar{F}(x_j)$, $j = 1, \ldots, J$. Let M_n be the cross-validation selector, and define $\beta_n = \beta_{n,M_n}$. Then, our estimator of the efficient influence curve $D^*(F,G)$ is given by:

$$D^*_n(F,G)(O) = \sum_{j=1}^J \beta_{n,j}\{\bar{F}(x_j\mid O) - \bar{F}(x_j)\}.$$

By Theorem 8.1, this estimator converges to $D^*(F,G)$ in $L^2(P_{F,G})$-norm at a rate faster than $n^{-1/4}$.

8.5 Causal Effect of Binary Treatment on Interval Censored Time to Event

Let $O = (W, A, C_m, \Delta_m = I(T \le C_m) : m = 1, \ldots M)$ be a general interval censored data structure, where T is a time to event, and $C = (C_1, \ldots, C_M)$ is a vector of continuous valued monitoring times, W are baseline covariates, and A is a binary treatment. Let $\Delta = (\Delta_1, \ldots, \Delta_M)$. Suppose that the monitoring times and T are all larger than 0. We assume that T and C are independent, given A, W. Let $F(t \mid A, W) = P(T \leq t \mid A, W)$ be the cumulative distribution function of T, given A, W, and let $G_c(\cdot \mid A, W)$ and $g_c(\cdot \mid A, W)$ denote the conditional probability distribution and density of C, given A, W. Let $g(a \mid W) = P(A = a \mid W)$ be the conditional probability of $A = a$, given W. Let Q_W be the probability distribution of W. Let $Q = (Q_W, F)$ and $G = (G_c, g)$. Note that the probability distribution $P = P_{Q,G}$ of O is indexed by Q and G, respectively. Let $\mathcal{M} = \{P_{Q,G} : Q, G \in \mathcal{G}\}$ be the statistical model for P defined by leaving Q unspecified, while we might have restrictions/knowledge on the conditional distribution of C, given A, W, and A, given W, defined by the set $\mathcal{G}$. Let $\Psi_a^f(Q) = \int r(t)\bar{F}_a(t)dt$, where $\bar{F}_a(t) = E_P P(T > t \mid A = a, W)$, be the full data parameter of interest. Under some support conditions on g_c, g, this parameter is identifiable from P, so that $\Psi : \mathcal{M} \to \mathbf{R}$ satisfies $\Psi(P_{Q,G}) = \Psi^f(Q)$. Let $D^*(Q, G)$ be the canonical gradient of Ψ at $P_{Q,G}$.

The density $p_{Q,G}$ of $P_{Q,G}$ is given by

$$p_{Q,G}(o) = q_W(w)g(a \mid w)g_c(c \mid A, W)F(C(o) \mid a, w),$$

where $C(o)$ is the coarsening for T implied by $O = o$, and $F(C(o) \mid a, w) = P(T \in C(o) \mid A = a, W = w)$. Thus $C(o) = (L(o), R(o)]$ is an interval, where $L(O)$ is the largest monitoring time C_j for which $\Delta_j = 0$, while $R(O)$ is the smallest monitoring time C_j for which $\Delta_j = 1$. If $\Delta_1 = 1$, then we define $L(O) = 0$, and if $\Delta_M = 0$, then $R(O) = \infty$. Thus, the density of O factorizes in a Q and G factor so that the efficient influence curve in our model $\mathcal{M}$ is the same as in the model $\mathcal{M}(G) = \{P_{Q,G} : Q\} \subset \mathcal{M}$ in which $G = (G_c, g)$ is known.

Our first task is to determine an initial gradient $D(Q, G)$ in the model $\mathcal{M}(G)$. For that purpose, let's define the random variable $\bar{C}$ as the outcome of the following experiment: generate C and then randomly draw one of the M monitoring times, each one receiving probability $1/M$. Let $\bar{\Delta} = I(T \leq \bar{C})$. Let $\bar{g}_c(\bar{c} \mid a, w) = \frac{1}{M} \sum_{m=1}^{M} g_{c,m}(\bar{c} \mid a, w)$ be the univariate density of $\bar{C}$, given $A = a, W = w$, where $g_{c,m}(\cdot \mid A, W)$ is the conditional density of C_m, given A, W. An initial estimator of $\Psi_a^f(Q)$ is given by the IPCW estimator

$$\frac{1}{n} \sum_{i=1}^{n} (1 - \bar{\Delta}_i)r(\bar{C}_i)\frac{I(A_i = a)}{\bar{g}_c(\bar{C}_i \mid A_i, W_i)g(A_i \mid W_i)},$$

based on the reduced current status data structure $(W, A, \bar{C}, \bar{\Delta} = I(T \leq \bar{C})$. The influence curve of this estimator is given by $(1 - \bar{\Delta})r(\bar{C})I(A = a)/\{\bar{g}_c(\bar{C} \mid A, W)g(A \mid W)\} - \Psi_a(Q)$. Strictly speaking this is not an influence curve since it is not only a function of O, but is also random through the random pick involved in selecting $\bar{C}$. So let's take the conditional expectation of this influence curve, given O. This yields the following initial gradient:

$$D(Q, G)(O) = \frac{1}{M} \sum_{m=1}^{M} (1 - \Delta_m)r(C_m)\frac{I(A = a)}{\bar{g}_c(C_m \mid A, W)g(A \mid W)} - \Psi_a(Q).$$

The tangent space of Q_W is given by $T_{Q_W}(P) = L_0^2(Q_W)$, and

$$\Pi(D(Q,G) \mid T_{Q_W}(P)) = \int r(t)(\bar{F}(t \mid a, W) - \bar{F}_a(t))dt \equiv D_W^*(Q).$$

Thus, it remains to project $D(Q,G)$ on the tangent space generated by the parameter $F(\cdot \mid A, W)$. Let $L_0^2(F)$ denote the Hilbert space of functions of (W, A, T) with conditional mean zero, given W, A. In the model $\mathcal{M}(G)$ the score operator $A_F : L_0^2(F) \to L_0^2(P_{Q,G})$ that maps the score $h \in L_0^2(F)$ of $dF_{\epsilon,h}(t \mid A, W) = (1 + \epsilon h(t \mid A, W))dF(t \mid A, W)$ into the score of $\{P_{Q_W, F_{\epsilon,h}, G} : \epsilon\}$ is given by:

$$
\begin{aligned}
A_F(h)(O) &= E_F(h(T \mid A, W) \mid A, W, T \in (L(O), R(O)]) \\
&= \frac{\int_{L(O)}^{R(O)} h(t \mid A, W)dF(t \mid A, W)}{F(R(O) \mid A, W) - F(L(O) \mid A, W)}.
\end{aligned}
$$

Thus the tangent space $T_F(P)$ generated by F is given by $\overline{R}(A_F)$, the closure of the range of this score operator $A_F : L_0^2(F) \to L_0^2(P_{Q,G})$. Let $Fh = E(h \mid A, W) = \int h(t \mid A, W)dF(t \mid A, W)$ and represent an $h \in L_0^2(F)$ with $h - Fh$ for an $h \in L^2(F)$.

The efficient influence curve $D^*(P) = D^*(Q, G)$ is the projection $D_W^*(P)$ of $D(Q, G)$ onto $T_{Q_W}(P)$ plus the projection $D_T^*(P)$ of $D(P)$ onto $\overline{R}(A_F)$. Thus, the efficient influence curve is the orthogonal sum $D^*(Q, G) = D_W^*(Q) + D_T^*(Q, G)$. Since $\overline{R}(A_F)$ is embedded in the space of functions of O with conditional mean zero, given A, W, we have that $D_T^*(P)$ is also the projection of $D_1(P) \equiv D(P) - E(D(P) \mid A, W)$ onto $\overline{R}(A_F)$. We will also denote $D_1(P)$ with $D_1(Q, G)$. We have $D_T^*(Q, G) = A_F(h^*)$, where

$$h^*(P) = \arg \min_{h \in L^2(F)} P\{D_1(Q, G) - A_F(h) + Fh\}^2,$$

where we recall that for a function $h(O)$, $Fh = E(h \mid A, W)$ depends on A, W. We can obtain this minimum by minimizing the conditional expectation of the squared error loss, given (A, W), over all functions of T, which then defines an optimal $T \to h^*(T \mid A, W)$, and by doing this for all possible values of (A, W), it yields the full solution $(T, A, W) \to h^*(P)(T \mid A, W)$. Let's denote this optimal $T \to h(T)$, for this (A, W)-specific minimization problem, with $h^*(P \mid A, W)$, and note $h^*(P)(T \mid A, W) = h^*(P \mid A, W)$:

$$h^*(P \mid A, W) = \arg \min_{h \in L^2(F_T)} E_P\left(\{D_1(Q, G) - A_F(h) + Fh\}^2 \mid A, W\right),$$

where we define $L^2(F_T)$ as the space of functions $T \to h(T)$ of $T = (T_1, T_2)$ only. Notice that for a given A, W, we now view $A_F : L^2(F_T) \to L^2(P_{Q,G|A,W})$ as a linear operator from $L^2(F_T)$ (functions of T) into the space of functions of $(\tilde{T}, \Delta)$.

We can represent $h(t) = h(0) + \int I(t \geq x)dh(x) = h(0) + \int \phi_x(t)dh(x)$, where $\phi_x(t) = I(t \geq x)$. Using this representation it follows that $Fh = h(0) + \int \bar{F}(x \mid A, W)dh(x)$. Substitution of this representation for $h \in L^2(F_T)$ into the above definition of h^* yields:

$$h^* = \arg\min_h E_P\left(\left\{D_1(Q,G) - \int A_F(\phi_x)dh(x) + \int \bar{F}(x \mid A, W)dh(x)\right\}^2 \mid A, W\right)$$

$$= \arg\min_h E_P\left(\left\{D_1(Q,G) - \int \{\bar{F}(x \mid \cdot) - \bar{F}(x \mid A, W)\}dh(x)\right\}^2 \mid A, W\right),$$

where $\bar{F}(x \mid O) = P(T > x \mid T \in (L(O), R(O)], A, W)$.

Let $(C_1, \Delta_1), \ldots, (C_n, \Delta_n) \sim_{iid} P_{Q,G \mid A, W}$ be a sample from the conditional distribution $P_{Q,G \mid A, W}$ of O, given A, W. Let $O_i = (C_i, \Delta_i, A, W)$, $i = 1, \ldots, n$, denote the resulting n observations. Then, we can estimate $h^*(P \mid A, W)$ with

$$h_{n,M}^* = \arg\min_{h, \|h\|_v < M} \frac{1}{n} \sum_{i=1}^n \left\{D_1(Q,G)(O_i) - \int \{\bar{F}(x \mid O_i) - \bar{F}(x \mid A_i, W_i)\}dh(x)\right\}^2.$$

Let $\{x_1, \ldots, x_J\}$ be the set of monitoring times that appear as $L(O_i)$ or $R(O_i)$ across the n coarsenings $(L(O_i), R(O_i)]$, $i = 1, \ldots, n$. Notice that $J = J(n)$ is at most $2n$, and will be a little smaller than $2n$ if there are coarsenings that have as left point 0 or as right point ∞. We suggest that for all M the minimum $h_{n,M}^* \in L^2(F_T)$ is attained at a discrete measure dh which only puts mass on $\{x_1, \ldots, x_J\}$:

$$h_{n,M}^* = \arg\min_{h, \|h\|_v < M} \frac{1}{n} \sum_{i=1}^n \left\{D_1(Q,G)(O_i) - \sum_{j=1}^J \{\bar{F}(x_j \mid O_i) - \bar{F}(x_j \mid A, W)\}dh(x_j)\right\}^2.$$

Let $\beta_j = dh(x_j)$, $\|\beta\|_1 = \sum_{j=1}^n |\beta_j|$, and define

$$\beta_{n,M}(A, W) = \arg\min_{\beta, \|\beta\|_1 < M} \frac{1}{n} \sum_{i=1}^n \left\{D_1(Q,G)(O_i) - \sum_{j=1}^J \beta_j\{\bar{F}(x_j \mid O_i) - \bar{F}(x_j \mid A, W)\}\right\}^2.$$

This is a standard lasso least squares linear regression estimator with L_1-constraint set at M, outcome $D_1(Q,G)(O_i)$ and J main terms $\bar{F}(x_j \mid O_i) - \bar{F}(x_j)$, $j = 1, \ldots, J$. Let M_n be the cross-validation selector, and define $\beta_n(A, W) = \beta_{n,M_n}(A, W)$. Then, our estimator of the efficient influence curve $(C, \Delta) \rightarrow D_T^*(Q,G)(C, \Delta, A, W)$ is given by:

$$D_{T,n}^*(Q,G)(C, \Delta, A, W) = \sum_{j=1}^J \beta_{n,j}(A, W)\{\bar{F}(x_j \mid C, \Delta, A, W) - \bar{F}(x_j \mid A, W)\}.$$

By Theorem 8.1, for each A, W, this estimator converges to

$$(C, \Delta) \rightarrow D_T^*(Q,G)(C, \Delta, A, W)$$

in $L^2(P_{Q,G \mid A, W})$-norm at a rate faster than $n^{-1/4}$.

To summarize, for a given A, W, the above method allows one to estimate the function $(C, \Delta) \rightarrow D^*(P)(C, \Delta, A, W)$ by fitting a lasso linear least squares regression in approximately n covariates. For the computation of the one-step estimator, the TMLE, or for the influence curve based variance estimator of such an efficient estimator, one typically just needs to know the efficient influence curve at the actual observations $O_i \sim P_0$. Therefore, for each observation i, one will have to run the above procedure to estimate the efficient influence curve $D^*(P)$ at this O_i. For many TMLE one will need to evaluate the whole score $(C, \Delta) \rightarrow D_T^*(P)(C, \Delta, A, W)$ as a function. Fortunately, the above procedure gives this for free without extra work.

Nonconditional Maximization Approach. From above we have $D_T^*(Q, G) = A_F(h^*)$, where

$$h^*(P) = \arg \min_{h \in L^2(F)} P\{D_1(Q, G) - A_F(h) + Fh\}^2.$$

Above we used the approach to minimize the conditional expectation, given A, W, for each A, W separately. Instead, here we go for direct minimization. Let d be the dimension of (W, A, T). Consider the representation $h(x) = \sum_{S \subset \{1,\dots,d\}} \int \phi_{xS}(u_S) dh_S(u_S)$. Let $H(u_S, A, W) = E(\phi_{(W,A,T)_S}(u_S) \mid W, A)$ be the conditional probability that $(W, A, T)_S > u_S$, given W, A. Using this representation it follows that $Fh = h(0) + \sum_S \int H(u_S, A, W) dh(u)$. Substitution of this representation for $h \in L^2(F)$ into the above definition of h^* yields:

$$h^* = \arg \min_h P\left\{D_1(Q, G) - \sum_S \int \{E(\phi_{(T,A,W)_S}(u_S) \mid O) - H(u_S, A, W)\} dh_S(u_S)\right\}^2.$$

Let $O_i = (C_i, \Delta_i, A, W)$, $i = 1, \dots, n$, be an i.i.d. sample from $P_{Q,G}$. Then, we can estimate $h^*(P \mid A, W)$ with $h_{n,M}^*$ defined as the minimizer over h with $\| h \|_v < M$ of

$$\frac{1}{n} \sum_{i=1}^n \left\{D_1(Q, G)(O_i) - \sum_S \int \{E(\phi_{(T_i,A_i,W_i)_S}(u_S) \mid O_i) - H(u_S, A_i, W_i)\} dh_S(u_S)\right\}^2.$$

Suppose that $h_{n,M}^*$ is a discrete measure so that $dh_{n,M,S}^*$ only puts positive mass on $u_{S,j}$, $j = 1, \dots, J_S$, for each $S \subset \{1, \dots, d\}$ (e.g., the analogue of the 2n support points in the previous section). Let

$$X(S, j)(O_i) \equiv E(\phi_{(T_i,A_i,W_i)_S}(u_{S,j}) \mid O_i) - H(u_{S,j}, A_i, W_i),$$

and $\beta_{S,j} = dh_S(u_{S,j})$. As in our general presentation, we then obtain $A_F(h_{n,M_n}) = \sum_S \sum_j \beta_{n,S,j}^* X(S, j)(O)$, where $\beta_n = \beta_{n,M_n}$,

$$\beta_{n,M} = \arg \min_{\beta, \|\beta\|_1 < M} \frac{1}{n} \sum_{i=1}^n \left\{D_1(Q, G)(O_i) - \sum_S \sum_{j=1} X(S, j)(O_i)\beta_{S,j}\right\}^2,$$

and M_n is the cross-validation selector.

The above approach becomes computationally intractable when the dimension of W gets large. As shown in van der Laan et al. (2015), one can often define a dimension reduction of O and corresponding model and target parameter so that the efficient influence curve at P is still the same as it was for the original formulation. In this case the above method is still tractable.

8.6 Bivariate Survival Function Based on Bivariate Right-Censored Data

In this section we demonstrate the HAL estimator of the efficient influence curve of the bivariate survival function based on bivariate right-censored data. This is easily extended to the HAL estimator of the efficient influence curve of the d-variate survival function for d-variate right-censored data, for general dimension $d \geq 2$. Let $O = (\tilde{T}_1 = \min(T_1, C_1), \Delta_1 = I(T_1 \leq C_1), \tilde{T}_2 = \min(T_2, C_2), \Delta_2 = I(T_2 \leq C_2))$, where $T = (T_1, T_2)$ and $C = (C_1, C_2)$ are independent with cumulative distribution functions F and G, respectively. Let $P_{F,G}$ be the probability distribution of O. Let $\Delta = (\Delta_1, \Delta_2)$ and $\tilde{T} = (\tilde{T}_1, \tilde{T}_2)$, so that we can denote $O = (\tilde{T}, \Delta)$. Consider the statistical model $\mathcal{M}$ in which F and G are unspecified. Let $\Psi : \mathcal{M} \to \mathbf{R}$ be the bivariate survival probability $\bar{F}(t_{10}, t_{20}) = \int_{((t_{10},t_{20}),\infty)} dF(x_1, x_2)$. We will also use the notation $t_0 = (t_{10}, t_{20})$. We will denote $\Psi(P_{F,G})$ also with $\Psi(F)$. The density of O factorizes in a F and G factor. As a consequence, the efficient influence curve of Ψ at $P_{F,G}$ in the model $\mathcal{M}$ is the same as the efficient influence curve at $P_{F,G}$ in the model $\mathcal{M}(G) = \{P_{F,G} : F\}$ in which G is known. A gradient of Ψ in the model $\mathcal{M}(G)$ is given by

$$D(F, G)(O) = \kappa_{t_0}(O) \frac{I(\Delta = (1, 1))}{\bar{G}(\tilde{T})} - \Psi(F),$$

where $\kappa_t(O) = I(\tilde{T} > t_0)$. This is the influence curve of the simple IPCW estimator defined as $\frac{1}{n} \sum_{i=1}^{n} \kappa_{t_0}(O_i) I(\Delta_i = (1, 1)) / \bar{G}(\tilde{T}_i)$ and is thus indeed a gradient. In the model $\mathcal{M}(G)$ the score operator $A_F : L_0^2(F) \to L_0^2(P_{F,G})$ that maps the score $h \in L_0^2(F)$ of $dF_{\epsilon,h}(t) = (1 + \epsilon h(t)) dF(t)$ into the score of $\{P_{F_{\epsilon,h},G} : \epsilon\}$ is given by:

$$A_F(h)(O) = E_F(h(T) \mid \tilde{T}, \Delta)$$

$$= h(T) I(\Delta = (1, 1)) + \frac{\int_{\tilde{T}_1}^{\infty} h(t_1, \tilde{T}_2) F^{01}(dt_1, \tilde{T}_2)}{\bar{F}^{01}(\tilde{T}_1, \tilde{T}_2)} I(\Delta = (0, 1))$$

$$+ \frac{\int_{\tilde{T}_2}^{\infty} h(\tilde{T}_1, t_2) F^{10}(\tilde{T}_1, dt_2)}{\bar{F}^{10}(\tilde{T}_1, \tilde{T}_2)} I(\Delta = (0, 1))$$

$$+ \frac{\int_{\tilde{T}_1}^{\infty} \int_{\tilde{T}_2}^{\infty} h(t_1, t_2) dF(t_1, t_2)}{\bar{F}(\tilde{T}_1, \tilde{T}_2)} I(\Delta = (0, 0)),$$

where $F^{01}(t_1, t_2) = \frac{d}{dt_2} F(t_1, t_2)$, $\bar{F}^{01}(t_1, t_2) = \int_{t_1}^{\infty} F^{01}(ds_1, t_2)$, $F^{10}(t_1, t_2) = \frac{d}{dt_1} F(t_1, t_2)$ and $\bar{F}^{10}(t_1, t_2) = \int_{t_2}^{\infty} F^{10}(t_1, ds_2)$. Let $Fh = \int h dF$ and represent an $h \in L_0^2(F)$ with $h - Fh$ for an $h \in L^2(F)$. The efficient influence curve is given by $A_F(h^*)$, where

$$h^*(P) = \arg \min_{h \in L^2(F)} P\{D(F, G) - A_F(h) + Fh\}^2.$$

We can represent a bivariate cadlag function $h \in L^2(F)$ with finite variation norm as follows:

$$h(t_1, t_2) = h(0, 0) + \int_0^{t_1} h(dx_1, 0) + \int_0^{t_2} h(0, dx_2) + \int_0^{t_1} \int_0^{t_2} h(dx_1, dx_2)$$

$$= h(0, 0) + \int \phi_{1,x_1}(t_1) h(dx_1, 0) + \int \phi_{2,x_2}(t_2) h(0, dx_2)$$

$$+ \int \phi_{x_1,x_2}(t_1, t_2) h(dx_1, dx_2),$$

where $\phi_{1,x_1}(t) = I(t_1 \geq x_1)$, $\phi_{2,x_2}(t) = I(t_2 \geq x_2)$, and $\phi_{x_1,x_2}(t) = I(t_1 \geq x_1, t_2 \geq x_2)$. Thus, in this way we have written the function h as a linear infinite combination of indicator functions $\phi_{1,x_1}, \phi_{2,x_2}, \phi_x$ across all x-values. Since A_F is a linear operator, this yields the following representation of $A_F(h)$:

$$A_F(h) = h(0, 0) + \int A_F(\phi_{1,x_1}) h(dx_1, 0) + \int A_F(\phi_{2,x_2}) h(0, dx_2)$$

$$+ \int A_F(\phi_{x_1,x_2}) h(dx_1, dx_2)$$

$$\equiv h(0, 0) + \int \bar{F}_1(x_1 \mid O) h(dx_1, 0) + \int \bar{F}_2(x_2 \mid O) h(0, dx_2)$$

$$+ \int \bar{F}(x_1, x_2 \mid O) h(dx_1, dx_2),$$

where we denoted the conditional probabilities $T_1 > x_1$, given O, $T_2 > x_2$, given O, and $T > (x_1, x_2)$, given O, with $\bar{F}_1(x_1 \mid O)$, $\bar{F}_2(x_2 \mid O)$ and $\bar{F}(x \mid O)$, respectively. Note also that

$$Fh = h(0, 0) + \int \bar{F}_1(x_1) h(dx_1, 0) + \int \bar{F}_2(x_2) h(0, dx_2) + \int \bar{F}(x_1, x_2) h(dx_1, dx_2),$$

where $\bar{F}_1(x_1) = P_F(T_1 > x_1)$, $\bar{F}_2(x_2) = P_F(T_2 > x_2)$ and $\bar{F}(x_1, x_2) = P_F(T_1 > x_1, T_2 > x_2)$. Thus, we have proven the following representation of $A_F(h - Fh)$ for any cadlag function h with finite variation norm:

$$A_F(h - Fh) = \int (\bar{F}_1(x_1 \mid O) - \bar{F}_1(x_1)) h(dx_1, 0) + \int (\bar{F}_2(x_2 \mid O) - \bar{F}_2(x_2)) h(0, dx_2)$$

$$+ \int (\bar{F}(x \mid O) - \bar{F}(x)) h(dx)$$

Let $O_1, \ldots, O_n$ be an i.i.d. sample from $P_{F,G}$. Then we can approximate h^* with

$$h_{n,M}^* = \arg \min_{h, \|h\|_v < M} \frac{1}{n} \sum_{i=1}^n \{D(F, G)(O_i) - A_F(h - Fh)(O_i)\}^2 .$$

We claim that this minimum is attained by a discrete measure h for which $h(dx_1, 0)$ only puts mass on $\{\tilde{T}_{1i} : i\}$, $h(0, dx_2)$ only puts mass on $\{\tilde{T}_{2i} : i\}$, and $h(dx_1, dx_2)$ only puts mass on $\{(\tilde{T}_i : i\}$. Let's denote these three set of support points with $\{m_{1j} :$

$j = 1, \ldots, n\}$, $\{m_{2j} : j = 1, \ldots, n\}$ and $\{m_j : j = 1, \ldots, n\}$, respectively. For such a step-function h we have:

$$A_F(h - Fh)(O_i) = \sum_j (\bar{F}_1(m_{1j} \mid O_i) - \bar{F}_1(m_{1j}))h(dm_{j1}, 0)$$

$$+ \sum_j (\bar{F}_2(m_{2j} \mid O_i) - \bar{F}_2(m_{2j}))h(0, dm_{2j}) + \sum_j (\bar{F}(m_j \mid O_i) - \bar{F}(m_j))h(dm_j).$$

This yields a representation of $h^*_{n,M}$ as a finite dimensional linear regression least squares estimator. Let $\beta_{1j} = h(dm_{1j}, 0)$, $\beta_{2j} = h(0, dm_{2j})$, $\beta_j = h(dm_j)$, and let $\beta = (\beta_{1j}, \beta_{2j}, \beta_j : j = 1, \ldots, n)$ be the vector with all these components. In addition, let $X_i = (X_{1i}(j), X_{2i}(j), X_i(j) : j = 1, \ldots, n)$, where

$$X_{1i}(j) = \bar{F}_1(m_{1j} \mid O_i) - \bar{F}_1(m_{1j})$$
$$X_{2i}(j) = \bar{F}_2(m_{2j} \mid O_i) - \bar{F}_2(m_{2j})$$
$$X_i(j) = \bar{F}(m_j \mid O_i) - \bar{F}(m_j).$$

In addition, let $\| \beta \|_1 = \sum_{j=1}^n |\beta_{1j}| + |\beta_{2j}| + |\beta_j|$ be the L^1-norm of this vector β of coefficients. We can now represent $A_F(h - Fh)$ as a linear regression model:

$$A_F(h - Fh)(O_i) = \sum_j \beta_{1j}(\bar{F}_1(m_{1j} \mid O_i) - \bar{F}_1(m_{1j})) + \sum_j \beta_{2j}(\bar{F}_2(m_{2j} \mid O_i) - \bar{F}_2(m_{2j}))$$

$$+ \sum_j \beta_j(\bar{F}(m_j \mid O_i) - \bar{F}(m_j))$$

$$\equiv \beta^\top X_i,$$

while the variation norm of h is defined by $\| \beta \|_1$.

For a given M, let's define constrained least squares regression estimator:

$$\beta_{n,M} = \arg \min_{\beta, \|\beta\|_1 < M} \frac{1}{n} \sum_{i=1}^n \left\{ D(F, G)(O_i) - \beta^\top X_i \right\}^2.$$

This is a standard lasso least squares regression estimator with L_1-constraint set at M. Let M_n be the cross-validation selector of M, and define $\beta_n = \beta_{n,M_n}$. Then, our estimator of the efficient influence curve $D^*(P)$ is given by:

$$D^*_n(P)(O) = \beta_n^\top X_i$$

$$= \sum_{j=1}^n \beta_{1nj}\bar{F}_1(m_{1j} \mid O_i) - \bar{F}_1(m_{1j})) + \sum_j \beta_{2nj}(\bar{F}_2(m_{2j} \mid O_i) - \bar{F}_2(m_{2j}))$$

$$+ \sum_j \beta_{nj}(\bar{F}(m_j \mid O_i) - \bar{F}(m_j)).$$

By our theoretical result, this estimator $D^*_n(P)$ converges to $D^*(P)$ in $L^2(P)$-norm at a rate faster than $n^{-1/4}$.

8.7 Causal Effect of Binary Treatment on Bivariate Survival Probability Based on Bivariate Right-Censored Data

Consider the extended right-censored data structure for bivariate survival data:

$$O = (W, A, \tilde{T}_1 = \min(T_1, C_1), \Delta_1 = I(T_1 \le C_1), \tilde{T}_2 = \min(T_2, C_2), \Delta_2 = I(T_2 \le C_2)),$$

where W are baseline covariates, $A \in \{0, 1\}$ is a binary treatment. We use the notation of the previous subsection, such as $T = (T_1, T_2)$, $C = (C_1, C_2)$, $\Delta = (\Delta_1, \Delta_2)$ and $\tilde{T} = (\tilde{T}_1, \tilde{T}_2)$. Thus, we can denote this observed data structure with $O = (W, A, \tilde{T}, \Delta)$. Let Q_W be the marginal probability distribution of W, $g(a \mid W) = P(A = a|W)$ be the treatment mechanism, and let $F(\cdot \mid A, W)$ and $G_c(\cdot \mid A, W)$ denote the conditional cumulative distribution functions of T and C, given A, W, respectively. Let $G = (g, G_c)$ and $Q = (Q_W, F)$. Note that the probability distribution $P_{Q,G} = P_{Q_W, F, g, G_c}$ of O is indexed by these four parameters. Consider the statistical model $\mathcal{M} = \{P_{Q,G} : Q, G \in \mathcal{G}\}$ in which Q_W and $F(\cdot \mid A, W)$ are unspecified, while we might have some knowledge on the treatment and censoring mechanism so that $G = (g, G_c)$ might be restricted to a certain set $\mathcal{G}$.

Let $\Psi : \mathcal{M} \to \mathbf{R}$ be the treatment specific bivariate survival probability at $t_0 = (t_{10}, t_{20})$

$$\Psi(P) = \bar{F}_a(t_0) = \int P(T > t_0 \mid A = a, W = w)Q_W(dw)$$

$$= \int \bar{F}(t_0 \mid A = a, W = w)Q_W(dw).$$

Since $\Psi(P)$ only depends on P through $Q = (Q_W, F)$, we also use the notation $\Psi(Q)$. The density of O factorizes in a Q and $G = (g, G_c)$ factor. As a consequence, the efficient influence curve of Ψ at $P_{Q,G}$ in the model $\mathcal{M}$ is the same as the efficient influence curve at $P_{Q,G}$ in the model $\mathcal{M}(G) = \{P_{Q,G} : Q\}$ in which G is known.

A gradient of Ψ in the model $\mathcal{M}(G)$ is given by

$$D(Q, G)(O) = \kappa_{t_0}(O)\frac{I(\Delta = (1, 1), A = a)}{g(A \mid W)\bar{G}_c(\tilde{T} \mid A, W)} - \Psi(Q),$$

where $\kappa_t(O) = I(\tilde{T} > t_0)$. This is the influence curve of the simple IPCW estimator defined as $\frac{1}{n}\sum_{i=1}^n \kappa_{t_0}(O_i)I(\Delta_i = (1, 1), A_i = a)/\{g(A_i \mid W_i)\bar{G}(\tilde{T}_i \mid A_i, W_i)\}$ and is thus indeed a gradient. The tangent space $T_{Q_W}(P)$ generated by fluctuations $dQ_W = (1 + \epsilon S_W(W))dQ_W$ with $S_W \in L_0^2(Q_W)$ is $L_0^2(Q_W)$ itself. Let

$$D_W^*(Q, G) = E_P(D(Q, G)(O) \mid W) = \bar{F}(t_0 \mid A = a, W) - \bar{F}_a(t_0).$$

Note that this represents the projection of $D(Q, G)$ onto the tangent space $T_{Q_W}(P)$, and thus represents a component of the efficient influence curve $D^*(P)$.

Let $L_0^2(F)$ denote the Hilbert space of functions of (W, A, T) with conditional mean zero, given W, A. In the model $\mathcal{M}(G)$ the score operator $A_F : L_0^2(F) \to$

$L_0^2(P_{Q,G})$ that maps the score $h \in L_0^2(F)$ of $dF_{\epsilon,h}(t \mid A, W) = (1 + \epsilon h(t \mid A, W))dF(t \mid A, W)$ into the score of $\{P_{Q_W, F_{\epsilon,h}, G} : \epsilon\}$ is given by:

$$
\begin{aligned}
A_F(h)(O) &= E_F(h(T \mid A, W) \mid W, A, \tilde{T}, \varDelta) \\
&= h(T \mid A, W)I(\varDelta = (1, 1)) \\
&\quad + \frac{\int_{\tilde{T}_1}^{\infty} h(t_1, \tilde{T}_2 \mid A, W)F^{01}(dt_1, \tilde{T}_2 \mid A, W)}{\bar{F}^{01}(\tilde{T}_1, \tilde{T}_2 \mid A, W)}I(\varDelta = (0, 1)) \\
&\quad + \frac{\int_{\tilde{T}_2}^{\infty} h(\tilde{T}_1, t_2 \mid A, W)F^{10}(\tilde{T}_1, dt_2 \mid A, W)}{\bar{F}^{10}(\tilde{T}_1, \tilde{T}_2 \mid A, W)}I(\varDelta = (0, 1)) \\
&\quad + \frac{\int_{\tilde{T}_1}^{\infty} \int_{\tilde{T}_2}^{\infty} h(t_1, t_2 \mid A, W)dF(t_1, t_2 \mid A, W)}{\bar{F}(\tilde{T}_1, \tilde{T}_2 \mid A, W)}I(\varDelta = (0, 0)),
\end{aligned}
$$

where $F^{01}(t_1, t_2 \mid A, W) = \frac{d}{dt_2}F(t_1, t_2 \mid A, W)$, $\bar{F}^{01}(t_1, t_2 \mid A, W) = \int_{t_1}^{\infty} F^{01}(ds_1, t_2 \mid A, W)$, $F^{10}(t_1, t_2 \mid A, W) = \frac{d}{dt_1}F(t_1, t_2 \mid A, W)$ and $\bar{F}^{10}(t_1, t_2 \mid A, W) = \int_{t_2}^{\infty} F^{10}(t_1, ds_2 \mid A, W)$. Thus the tangent space $T_F(P)$ generated by F is given by $\overline{R}(A_F)$, the closure of the range of this score operator $A_F : L_0^2(F) \to L_0^2(P_{Q,G})$. Let $Fh = E(h \mid A, W) = \int h(t \mid A, W)dF(t \mid A, W)$ and represent an $h \in L_0^2(F)$ with $h - Fh$ for an $h \in L^2(F)$.

The efficient influence curve $D^*(P) = D^*(Q, G)$ is the projection $D_W^*(P)$ of $D(Q, G)$ onto $T_{Q_W}(P)$ plus the projection $D_T^*(P)$ of $D(P)$ onto $\overline{R}(A_F)$. Thus, the efficient influence curve is the orthogonal sum $D^*(Q, G) = D_W^*(Q, G) + D_T^*(Q, G)$. Since $\overline{R}(A_F)$ is embedded in the space of functions of O with conditional mean zero, given A, W, we have that $D_T^*(P)$ is also the projection of $D_1(P) \equiv D(P) - E(D(P) \mid A, W)$. We have $D_T^*(Q, G) = A_F(h^*)$, where

$$
h^*(P) = \arg \min_{h \in L^2(F)} P\{D_1(Q, G) - A_F(h) + Fh\}^2,
$$

where we recall that for a function $h(O)$, $Fh = E(h \mid A, W)$ depends on A, W. We can obtain this minimum by minimizing the conditional expectation of the squared error loss, given (A, W), over all functions of T, which then defines an optimal $T \to h^*(T, A, W)$, and by doing this for all possible values of (A, W), it yields the full solution $(T, A, W) \to h^*(P)(T, A, W)$. Let's denote this optimal $T \to h(T)$, for this (A, W)-specific minimization problem, with $h^*(P \mid A, W)$, and note $h^*(P)(T, A, W) = h^*(P \mid A, W)$:

$$
h^*(P \mid A, W) = \arg \min_{h \in L^2(F_T)} E_P\left(\{D_1(Q, G) - A_F(h) + Fh\}^2 \mid A, W\right),
$$

where we define $L^2(F_T)$ as the space of functions $T \to h(T)$ of $T = (T_1, T_2)$ only. Notice that for a given A, W, we now view $A_F : L^2(F_T) \to L^2(P_{\mid A, W})$ as a linear operator from $L^2(F_T)$ (functions of T) into the space of functions of $(\tilde{T}, \varDelta)$.

For a given A, W, we now proceed in the same way as in the previous section, analogue to the example "Causal effect of binary treatment on interval censored time to event". We will not repeat these calculations here.

8.8 Discussion

Our interval censored data and bivariate right-censored data examples represent problems where the efficient influence curve does not exist in closed form, and the construction of efficient estimators has been extremely challenging, accordingly. These examples demonstrate that we can estimate these complex efficient influence curves with HAL least squares linear regression, thereby making the construction of a one-step estimator or TMLE relatively straightforward. This is quite a remarkable result!

In our bivariate survival function example with covariates, we demonstrated that we can even estimate the efficient influence curve of the causal effect of a binary treatment A on a bivariate survival function controlling for a large dimensional covariate vector W, by using such lasso regression estimators for a given (A, W). This makes it now possible to develop a TMLE of this causal effect on a bivariate survival function with bivariate right-censored data. We also demonstrated this same approach for the efficient influence curve of the causal effect of a binary treatment on a truncated mean of the survival time based on an extended interval censored data structure $O = (W, A, \Delta_1, C_1, \ldots, \Delta_M, C_M)$. This allows us estimate a causal effect of treatment on a time to event that is subject to interval censoring. Again, the latter represents another very interesting estimation problem with important practical applications which will need to be pursued in the future. Our HAL estimator of $D^*(P)$ relies on an i.i.d. sample from P. However, we also show (without formal proof) that if the goal is to estimate $D^*(P_0)$ and P represents a consistent estimator of P_0, then we could simply apply the HAL estimator to the original sample from P_0 instead. This makes our proposed HAL estimator particularly convenient.

The approach for estimation of the efficient influence curve presented in this chapter provides an alternative to the methods proposed in Frangakis et al. (2015); Luedtke et al. (2015a); van der Laan et al. (2015). The latter type of research, which started with the inspiring article Frangakis et al. (2015), concerns computerizing the estimation of the efficient influence curve (and thereby efficient estimation) without the need for being trained in efficiency theory. Clearly, the approach presented in this chapter still requires the user to formulate an initial gradient, the score operator and the corresponding regression problem. Nonetheless, importantly, it avoids the need for closed form representations of the canonical gradient and mathematical and numerical computation of the projection of an initial gradient on the tangent space, but instead utilizes the state of the art in machine learning for prediction to approximate this latter projection.

Chapter 9
Data-Adaptive Target Parameters

Alan E. Hubbard, Chris J. Kennedy, and Mark J. van der Laan

What factors are most important in predicting coronary heart disease? Heart disease is the leading cause of death and serious injury in the United States. To address this question we turn to the Framingham Heart Study, which was designed to investigate the health factors associated with coronary heart disease (CHD) at a time when cardiovascular disease was becoming increasingly prevalent. Starting in 1948, the prospective cohort study began monitoring a population of 5209 men and women, ages 30–62, in Framingham, Massachusetts. Those subjects received extensive medical examinations and lifestyle interviews every 2 years that provide longitudinal measurements that can be compared to outcome status. The data has been analyzed in countless observational studies and resulted in risk score equations used widely to assess risk of coronary heart disease. In our case, we conduct a comparison analysis to Wilson et al. (1998) using the data-adaptive variable importance approach described in this chapter.

A. E. Hubbard (✉) · C. J. Kennedy
Division of Biostatistics, University of California, Berkeley, 101 Haviland Hall, #7358, Berkeley, CA 94720, USA
e-mail: hubbard@berkeley.edu; ck37@berkeley.edu

M. J. van der Laan
Division of Biostatistics and Department of Statistics, University of California, Berkeley, 101 Haviland Hall, #7358, Berkeley, CA 94720, USA
e-mail: laan@berkeley.edu

© Springer International Publishing AG 2018
M.J. van der Laan, S. Rose, *Targeted Learning in Data Science*,
Springer Series in Statistics, https://doi.org/10.1007/978-3-319-65304-4_9

This chapter describes using the data to both **define and estimate** a target
parameter, drawing inferences about this parameter.

Consider predictors R (blood pressure, total cholesterol, smoking status, diabetes
status, age, and others) and an outcome, where $Y = 1$ for CHD diagnosis and $Y = 0$
for no CHD diagnosis. We want to know which of these variables is "most important" for explaining (predicting) Y, so this is a variable importance measure (VIM)
estimation problem. We propose a procedure that targets the VIM, *one variable
at a time*, as opposed to deriving variable importance measures as a byproduct of
some parametric model or machine learning procedure (Grömping 2009; Auret and
Aldrich 2011). In this way, we can optimize performance of the VIM estimators as
well as derive robust inference even when deriving such measures data adaptively.
We do so by a combination of using a data-adaptive parameter approach (Hubbard
and van der Laan 2016; Hubbard et al. 2016) and cross-validated targeted maximum
likelihood estimation (CV-TMLE).

Others (van der Laan 2006b) have advocated for estimation of variable importance measures via parameters motivated by causal inference, and we have applied
such techniques to rank variables by importance for acute trauma patient outcomes
(Hubbard et al. 2013), quantitative trait loci (Wang et al. 2011a,b, 2014), biomarker
discovery (Bembom et al. 2009), and health care spending (Rose 2018). We loop
through each variable of $A \in R$, defining W as everything else: $W = R \setminus A$. For one
loop, define the data, for a particular variable of interest, as $O = (W, \Delta, \Delta * (A, Y))$,
where Δ is missingness indicator for either A or Y (=1 if both not missing, 0 otherwise). Assume for now that A is discrete and there is a known "highest risk" level
(a_H) and a lowest risk level (a_L). Then, a candidate parameter for variable importance that would allow comparisons across different candidate predictors (each with
their own (a_L, a_H)):

$$E_W\{E(Y \mid A = a_H, W) - E(Y \mid A = a_L, W)\}, \tag{9.1}$$

or a weighted average of the stratified mean differences comparing, within strata W,
subjects with a_H versus those with a_L. Though we do not emphasize causal interpretations in this chapter, under standard identification assumptions, (9.1) identifies
$E\{Y(a_H) - Y(a_L)\}$, where $Y(a)$ is the counterfactual outcome for patient if, possibly contrary to fact, A was set to a. More generally, (9.1) is a VIM, which can
be compared (and ranked in importance) across the different covariates in R, and
it is also a pathwise-differentiable parameter with the possibility of deriving semi-
parametrically efficient, asymptotically normally distributed estimators. However,
the story becomes more complex if a_H and a_L are not a priori known, but must be
"discovered" data adaptively.

9.1 Example: Defining Treatment or Exposure Levels

Candidates for A are not always discrete and even when they are, there are often no objectively defined high- and low-risk levels for predictors. Thus, consider the situation where one uses the data to define the low-risk level a_L and the high-risk level a_H. Let P_n define the empirical distribution. An algorithm applied to P_n could be used to define these low- and high-risk levels in A, or $(a_L(P_n), a_H(P_n))$. Let $\hat{Q}$: $\mathcal{M}_{NP} \to Q = \{Q(P) : P \in \mathcal{M}\}$ be an estimator of the true regression $Q_0(A, W) = E_{P_0}(Y \mid A, W)$, and $Q_n = \hat{Q}(P_n)$ is its realization when applied to the data. If A were discrete with arbitrary levels $\mathcal{A} = (a_1, a_2, \ldots, a_i, \ldots, a_K)$, we could define $a_L(P_n)$ and $a_H(P_n)$ as

$$a_L(P_n) = \arg\min_{a \in \mathcal{A}} \frac{1}{n} \sum_{i=1}^{n} Q_n(a, W_i), \tag{9.2}$$

$$a_H(P_n) = \arg\max_{a \in \mathcal{A}} \frac{1}{n} \sum_{i=1}^{n} Q_n(a, W_i). \tag{9.3}$$

That is, one "discovers" levels $(a_L(P_n), a_H(P_n))$ that maximize the substitution estimate of (9.1) according to some regression estimate $Q_n(A, W)$. This can be used to define a *data-adaptive target parameter*:

$$\Psi_{a_L(P_n), a_H(P_n)}(P) = E_P\{E_P(Y \mid A = a_H(P_n), W) - E_P(Y \mid A = a_L(P_n), W)\} \tag{9.4}$$

for which the substitution estimator is

$$\frac{1}{n} \sum_{i=1}^{n} \{Q_n(a_H(P_n), W_i) - Q_n(a_L(P_n), W_i)\}. \tag{9.5}$$

Because of the dual use of this data, this substitution estimator (9.5) will suffer from overfitting bias. To illustrate this, consider a data-generating distribution where

$$E_W\{E(Y \mid A = a_H(P_n), W) - E(Y \mid A = a_L(P_n), W)\} = 0,$$

the estimate (9.5) will always be positively biased (it is always ≥ 0). A common concern is that exploratory exercises like this will suffer from erroneous findings (Ioannidis 2008; Broadhurst and Kell 2006). On the other hand, if sample splitting is done such that (1) a training sample was used to define $(a_L(P_{n,tr}), a_H(P_{n,tr}))$, and (2) a separate estimation sample was used to estimate $E_W\{E(Y \mid A = a_H(P_{n,tr}), W) - E(Y \mid A = a_L(P_{n,tr}), W)\}$, then valid statistical inference is possible. Of course, with such a sample splitting method the power is heavily reduced due to the reduction in estimation sample size.

This chapter presents a data-adaptive procedure that uses the data to define the target parameter, estimate it consistently and efficiently, and derive robust measures of uncertainty and confidence intervals. As in Hubbard et al. (2016), we discuss methods that use sample splitting to avoid bias from overfitting, but still use the entire data set to estimate a data-adaptive parameter. These methods apply in circumstances where there is little constraint on how the data is explored to generate potential parameters of interest. Such methods can capitalize on the very large sample sizes and/or very high dimension associated with "Big Data". We first describe the data-adaptive target parameter approach (Hubbard and van der Laan 2016; Hubbard et al. 2016) that uses repeated sample splitting, and subsequently we enhance this approach with CV-TMLE (Chap. 27 in van der Laan and Rose 2011). We will demonstrate the technique to the Framingham Heart Study.

9.2 Methodology for Data-Adaptive Parameters

Let $O_1, \ldots, O_n$ be i.i.d. with probability distribution P_0, and assume that it is known that P_0 is an element of a specified statistical model $\mathcal{M}$. We let P_n represent the empirical distribution of this random sample of n draws from P_0. Cross-validation is a key ingredient of our proposed method. To simplify the presentation, we present the relevant procedures in the context of V-fold cross-validation.

V-fold cross-validation involves the following steps: (1) $\{1, \ldots, n\}$ is divided into V equal size subgroups, (2) for each v, an estimation-sample is defined by the v-th subgroup of size n/V, while the parameter-generating sample is its complement. For split v, let P_{n,v^c} be the empirical distribution of the parameter-generating sample, and $P_{n,v}$ is the empirical distribution of its compliment, which we call the estimation-sample. For an observation O_i, let $Z_i \in \{1, \ldots, V\}$ denote the label of the subgroup that contains O_i. For split v, the parameter-generating sample P_{n,v^c} is used to generate a target parameter mapping $\Psi_{P_{n,v^c}} : \mathcal{M} \to \mathbf{R}$, and let $\hat{\Psi}_{P_{n,v^c}} : \mathcal{M}_{NP} \to \mathbf{R}$ be the estimator mapping of this target parameter.

For the sake of statistical inference, the choice of target parameter mapping and corresponding estimator mapping can be informed by P_{n,v^c}, but not by $P_{n,v}$. We define the sample-split data-adaptive statistical target parameter as $\Psi_n : \mathcal{M} \to \mathbf{R}$ with

$$\Psi_n(P) = Ave\{\Psi_{P_{n,v^c}}(P)\} \equiv \frac{1}{V} \sum_{v=1}^{V} \Psi_{P_{n,v^c}}(P).$$

The statistical estimand of interest is thus

$$\psi_{n,0} = \Psi_n(P_0) = Ave\{\Psi_{P_{n,v^c}}(P_0)\} = \frac{1}{V} \sum_{v=1}^{V} \Psi_{P_{n,v^c}}(P_0). \tag{9.6}$$

This target parameter mapping depends on the data, which is the reason for calling it a *data-adaptive target parameter*. Given an estimator on each estimation sample, $\hat{\Psi}_{P_{n,v^c}}(P_{n,v})$, the corresponding estimator of the data-adaptive estimand $\psi_{n,0}$ is given by:

$$\psi_n = \hat{\Psi}(P_n) = Ave\{\hat{\Psi}_{P_{n,v^c}}(P_{n,v})\} = \frac{1}{V}\sum_{v=1}^{V}\Psi_{P_{n,v^c}}(P_{n,v}). \tag{9.7}$$

In Hubbard et al. (2016) and Hubbard and van der Laan (2016) we showed that $\sqrt{n}(\psi_n - \psi_{n,0})$ converges in distribution to mean zero normal distribution with variance σ^2 under weak regularity conditions, whose variance σ^2 can be consistently estimated, allowing the construction of confidence intervals and hypothesis tests. Note that in this methodology the estimator of $\Psi_{P_{n,v^c}}(P_0)$ is only based on $P_{n,v}$. In a later section, we present a CV-TMLE approach that estimates $\Psi_{P_{n,v^c}}(P_0)$ with a TMLE based on $P_{n,v}$, but where the initial estimators of the nuisance parameters in the TMLE can be based on P_{n,v^c}. Thus, in this CV-TMLE approach only the targeting step in the TMLE (which only involves fitting a low dimensional coefficient ϵ) is based on the separate sample $P_{n,v}$.

9.3 TMLE of v-Specific Data-Adaptive Parameter

Consider the data-adaptive parameter (9.4), where we have data adaptively determined the a_L and a_H on a single "training" sample, P_{n,v^c}, so the parameter of interest is

$$\Psi_{P_{n,v^c}}(P_0) = E_{W,0}\{E_0(Y \mid A = a_H(P_{n,v^c}), W) - E_0(Y \mid A = a_L(P_{n,v^c}), W)\} \tag{9.8}$$

Like above, define $Q(A, W) \equiv E_P(Y \mid A, W)$, with $Q_0(A, W) = E_0(Y \mid A, W)$ being the true regression function. Treating the $(a_L(P_{n,v^c}), a_H(P_{n,v^c}))$ as fixed after being determined (by some algorithm) in the training/parameter generating sample, then the estimator of (9.8) can be just the difference of estimators of two "adjusted" means, a problem well known in causal inference literature (e.g., see Chap. 4 in *Targeted Learning*, van der Laan and Rose 2011). Let $Q_{n,v} = \hat{Q}(P_{n,v})$ be an estimate of Q_0 based on the estimation sample $P_{n,v}$, $v = 1, \ldots, V$, where $\hat{Q}$ denotes a particular estimator. Consider the substitution estimator equivalent to (9.5),

$$\hat{\Psi}_{P_{n,v^c}}(P_{n,v}) = \frac{1}{n_V}\sum_{i:Z_i=v}\{Q_{n,v}(a_H(P_{n,v^c}), W_i) - Q_{n,v}(a_L(P_{n,v^c}), W_i)\}, \tag{9.9}$$

where $Q_{n,v} = \hat{Q}(P_{n,v})$.

In words, this is the difference of averages of the predicted values (based on a fit of Q on the estimation sample, or $Q_{n,v}$) across the observations on the estimation at the observed covariates, W_i, and the variable of interest, A, set at two values determined on the training sample, $a_H(P_{n,v^c})$ versus $a_L(P_{n,v^c})$. If Q is estimated in a very large (semiparametric model) by using, for instance, machine learning methods (such as the super learner; van der Laan et al. 2007), then the bias would be reduced relative to estimation according to a misspecified parametric model. However, such a substitution estimator is overly biased and not asymptotically linear, so that robust statistical inference based on this estimator is highly problematic. However, as is the subject of this book, a targeted maximum likelihood estimator based on this initial estimator reduces bias and under weak assumptions, has an asymptotically normal sampling distribution. Let's define such a TMLE.

The efficient influence curve of $\Psi_{a_L(P_{n,v^c}),a_H(P_{n,v^c})}(P_0) = E_{P_0}\{Q_0(a_H(P_{n,v^c}), W) - Q_0(a_L(P_{n,v^c}), W)\}$ is given by

$$D^*_{P_{n,v^c}}(O) = \left\{ \frac{I(A = a_H(P_{n,v^c}))}{g_0(a_H(P_{n,v^c}) \mid W)} - \frac{I(A = a_L(P_{n,v^c}))}{g_0(a_L(P_{n,v^c}) \mid W)} \right\} (Y - Q_0(A, W))$$
$$+ Q_0(a_H(P_{n,v^c}), W) - Q_0(a_L(P_{n,v^c}), W) - \Psi_{a_L(P_{n,v^c}),a_H(P_{n,v^c})}(P_0).$$

This suggests the following least favorable submodel $\{Q_{n,v,\epsilon} : \epsilon\}$ through $Q_{n,v}$ at $\epsilon = 0$:

$$\text{Logit} Q_{n,v,\epsilon}(A, W) = \text{Logit} Q_{n,v}(A, W) + \epsilon H_{P_{n,v^c}}(A, W; g), \qquad (9.10)$$

where

$$H_{P_{n,v^c}}(A, W; g) = \frac{I(A = a_H(P_{n,v^c}))}{g(a_H(P_{n,v^c}) \mid W)} - \frac{I(A = a_L(P_{n,v^c}))}{g(a_L(P_{n,v^c}) \mid W)}, \qquad (9.11)$$

and $g(a \mid W) \equiv P(A = a \mid W)$. By estimating g on the estimation sample we obtain the so-called clever covariate, $H_{P_{n,v^c}}(A, W; g_{n,v})$, providing the resulting TMLE of $\Psi_{P_{n,v^c}}(P_0)$:

$$\hat{\Psi}^{TMLE}_{P_{n,v^c}}(P_{n,v}) = \frac{1}{n_V} \sum_{i:Z_i=v} \{Q_{n,v,\epsilon_{(n,v)}}(a_H(P_{n,v^c}), W_i) - Q_{n,v,\epsilon_{(n,v)}}(a_L(P_{n,v^c}), W_i)\}, \qquad (9.12)$$

where $\epsilon_{(n,v)}$ is the maximum likelihood estimate of the coefficient ϵ in front of $H_{P_{n,v^c}}(\cdot; g_{n,v})$ based on $P_{n,v}$.

Let's now discuss estimation of g_0. Although A can have many levels, we only need to predict $A = a$ for two values, $A = a_L(P_{n,v^c})$ and $A = a_H(P_{n,v^c})$. On the estimation sample $P_{n,v}$, g_0 could be estimated with a multinomial outcome machine learning algorithm. Alternatively, one can use more commonly implemented machine learning algorithms for binary outcomes by running separate logistic regressions for fitting $g_0(a \mid W)$ for each a separately, and normalizing the estimates so that the resulting estimate of g_0 is a proper conditional probability distribution. Since we only need to know the distribution g_0 at two values, one could use the latter approach to fit a conditional distribution of $A^* \mid W$, where $A^* = A$ if $A \in \{a_L(P_{n,v^c}), a_H(P_{n,v^c})\}$ and it equals a third value otherwise. This will be the approach used below in our data analysis.

Along with the TMLE, comes the estimated influence curve at each observation O_i in the estimation sample $P_{n,v}$, which is given by

$$D^*_{n,v,P_{n,v^c}}(O) = \left[\frac{I(A = a_H(P_{n,v^c}))}{g_{n,v}(a_H(P_{n,v^c})|W)} - \frac{I(A = a_L(P_{n,v^c}))}{g_{n,v}(a_L(P_{n,v^c})|W)} \right] \{Y - Q_{n,v,\epsilon_{(n,v)}}(A, W)\}$$
$$+ \left[Q_{n,v,\epsilon_{(n,v)}}(a_H(P_{n,v^c}), W) - Q_{n,v,\epsilon_{(n,v)}}(a_L(P_{n,v^c}), W) \right] - \hat{\Psi}^{TMLE}_{P_{n,v^c}}(P_{n,v}).$$
$$(9.13)$$

These estimated influence curve values (9.13) provide us with an estimate of the standard error of the TMLE $\hat{\Psi}^{TMLE}_{P_{n,v^c}}(P_{n,v})$ of $\Psi_{P_{n,v^c}}(P_0)$:

$$se(\hat{\Psi}^{TMLE}_{P_{n,v^c}}) = \sqrt{\frac{\widehat{var}(D^*_{n,v,P_{n,v^c}}(O))}{n/V}}$$

where $\widehat{var}(D(O))$ is the sample variance of $D(O)$ w.r.t. estimation sample $P_{n,v}$.

9.4 Combining v-Specific TMLEs Across Estimation Samples

We can define an average split-specific data-adaptive parameter as in (9.6) above:

$$\Psi_n(P_0) = \frac{1}{V} \sum_{v=1}^{V} \Psi_{P_{n,v^c}}(P_0).$$

One can estimate this as an average of the split-specific TMLE estimates, just as in (9.7):

$$\hat{\Psi}(P_n) = \frac{1}{V} \sum_{v=1}^{V} \hat{\Psi}^{TMLE}_{P_{n,v^c}}(P_{n,v}). \qquad (9.14)$$

The asymptotic variance of this estimator can be estimated as

$$\sigma_n^2 = \frac{1}{V} \sum_{v=1}^{V} P_{n,v}(D^*_{n,v,P_{n,v^c}})^2, \qquad (9.15)$$

where (9.15) is the average of the V sample-specific estimates of the variance of the v-specific influence curves. The standard error of the estimator $\hat{\Psi}(P_n)$ can be estimated as

$$se(\hat{\Psi}(P_n)) = \sigma_n / \sqrt{n}.$$

As shown in Hubbard et al. (2016) and Hubbard and van der Laan (2016), under weak conditions, $\hat{\Psi}(P_n)$ is a consistent and asymptotically linear estimator of (9.6) and the above standard error provides valid asymptotic 0.95-confidence intervals $\psi_n \pm 1.96\sigma_n / \sqrt{n}$ for $\Psi_n(P_0)$.

9.5 CV-TMLE

The v-specific TMLE of $\Psi_{P_{n,v^c}}(P_0)$ is only based on the sample $P_{n,v}$ of size n/V. Fortunately, there is a modification of the procedure presented in the previous section that accomplishes (the apparently) conflicting goals for the same statistical assumptions presented in theorem 1 in Hubbard et al. (2016), that is (1) using more of the data for estimating the data-generating distributions used in the estimator of the data-adaptive parameter, and (2) not increasing bias via over-fitting: CV-TMLE (Chap. 27 in van der Laan and Rose 2011) provide the theory and general framework showing, for instance, that CV-TMLE is more robust than standard TMLE (it can guarantee asymptotic sampling distribution results in an even bigger statistical model). In addition, CV-TMLE can also be used for estimating the type of data-adaptive parameters highlighted in this chapter. For instance, it is particularly useful for both using the data to estimate an optimal treatment rule (Luedtke and van der Laan 2016b), as well as to estimate the impact of using such a rule on the mean outcome (van der Laan and Luedtke 2014; Luedtke and van der Laan 2016a). Before discussing the estimation of our particular data-adaptive parameter in our variable importance application, we first provide a general description of CV-TMLE for general data-adaptive parameters as presented in van der Laan and Luedtke (2014).

9.6 CV-TMLE for Data-Adaptive Parameters

Let $\mathcal{D}$ be an index set for a collection of possible definitions of a parameter. For example in our case, this would be the set of all values of (a_L, a_H). In addition, assume that, for each $d \in \mathcal{D}$, we have a statistical target parameter $\Psi_d : \mathcal{M} \to \mathbb{R}$. For example, if d represents a certain treatment rule, then we might define $\Psi_d(P) = E_P Y_d$. Let $\hat{d} : \mathcal{M}_{NP} \to \mathcal{D}$ be an algorithm that maps an empirical distribution into an estimate of a desired index d_0. In our example, $\hat{d}(P_n) = (a_L(P_n), a_H(P_n))$ and the corresponding target parameter (9.8) learned on P_{n,v^c} would be written as:

$$\Psi_{\hat{d}(P_{n,v^c})}(P) = E_P\{E_P(Y \mid A = a_H(P_{n,v^c}), W) - E_P(Y \mid A = a_L(P_{n,v^c}), W)\}.$$

In this chapter we are concerned with presenting a method that provides an estimator and statistical inference for the following data-adaptive target parameter (as in 9.6) indexed by $\hat{d}$:

$$\psi_{0,n} = Ave\{\Psi_{\hat{d}(P_{n,v^c})}(P_0)\} = \frac{1}{V} \sum_{v=1}^{V} \Psi_{\hat{d}(P_{n,v^c})}(P_0).$$

Below we present a modification of the average of TMLE estimators presented above, which is called the CV-TMLE. This CV-TMLE will be denoted with ψ_n^*. Previous results (van der Laan and Luedtke 2014) have shown that this CV-TMLE

ψ_n^* provides robust statistical inference, without relying on the empirical process condition (i.e., Donsker class condition) that restricts the adaptivity of $\hat{d}(P_n)$ and the corresponding estimators of $\Psi_{\hat{d}(P_{n,v^c})}(P_0)$.

For each target parameter Ψ_d, let $D_d^*(P_0)$ be its efficient influence curve at P_0. Assume that $\Psi_d(P_0) = \Psi_d(Q_0^d)$ only depends on P_0 through a parameter Q_0^d, and assume that $D_d^*(P_0) = D_d^*(Q_0^d, g_0^d)$ depends on P_0 through Q_0^d and a nuisance parameter g_0^d; these nuisance parameters are indexed by d because the choice d of target parameter can affect the definition of these parameters.

The canonical gradient $D_d^*(P)$ of the pathwise derivative of $\Psi_d : \mathcal{M} \to \mathbf{R}$ implies a second-order Taylor expansion with second-order term $R_d(Q^d, Q_0^d, g^d, g_0^d)$:

$$\Psi_d(Q^d) - \Psi_d(Q_0^d) = (P - P_0)D_d^*(Q^d, g^d) + R_d(Q^d, Q_0^d, g^d, g_0^d).$$

Let $\hat{Q}^d : \mathcal{M}_{NP} \to \mathcal{Q}^d$ and $\hat{g}^d : \mathcal{M}_{NP} \to \mathcal{G}^d$ be initial estimators of Q_0^d and g_0^d, respectively; $L^d(Q^d)$ is a valid loss function for Q_0^d such that $Q_0^d = \arg\min_{Q^d} P_0 L^d(Q^d)$; $\{Q^d(\epsilon) : \epsilon\}$ is a submodel through Q^d at $\epsilon = 0$ with a univariate or multivariate parameter ϵ so that the linear span of the generalized score includes the efficient influence curve at (Q^d, g^d):

$$D_d^*(Q^d, g^d) \in \langle \frac{d}{d\epsilon} L^d(Q_\epsilon^d) \Big|_{\epsilon=0} \rangle,$$

where $\langle f \rangle = \{\sum_j \beta_j f_j : \beta\}$ denotes the linear space spanned by the components of f. For a sample with empirical distribution P_n, let $\{\hat{Q}^d(P_n) : \epsilon\}$ be this submodel through the estimator $\hat{Q}^d(P_n)$ at $\epsilon = 0$, using $\hat{g}^d(P_n)$. Let's consider the case that the TMLE only requires one-step, which can be formally arranged by using a universal least favorable submodel $\{\hat{Q}_\epsilon^d : \epsilon\}$. We define

$$\epsilon_{n,v} = \arg\min_\epsilon P_{n,v} L^d(Q_{n,v^c,\epsilon}^d),$$

where the submodel $\{Q_{n,v^c,\epsilon}^d : \epsilon\}$ through $Q_{n,v^c}^d = \hat{Q}^d(P_{n,v^c})$ at $\epsilon = 0$ uses $g_{n,v^c}^d = \hat{g}^d(P_{n,v^c})$ as estimator of g_0^d. This defines a first step TMLE update of Q_{n,v^c}^d based on $P_{n,v}$. If we consider the case that the TMLE converges in one-step, as can always be arranged by using a universal least favorable submodel, and is the case in our example, then this implies that this first step TMLE already solves its efficient influence curve score equation

$$P_{n,v} D_d^*(Q_{n,v^c,\epsilon_{n,v}}^d, g_{n,v^c}^d) = 0.$$

Since this holds for each v, we then also have

$$\frac{1}{V} \sum_{v=1}^{V} P_{n,v} D_d^*(Q_{n,v^c,\epsilon_{n,v}}^d, g_{n,v^c}^d) = 0. \tag{9.16}$$

This key equation (9.16) represents the desired efficient score equation for our target parameter. In this case that the TMLE only takes one step, this key score equation can also be established with a single MLE (common in v):

$$\epsilon_n = \arg\min_{\epsilon} \frac{1}{V} \sum_{v=1}^{V} P_{n,v} L^d(Q^d_{n,v^c,\epsilon}).$$

In general, if the TMLE is defined by a multiple step TMLE algorithm, then one uses this multiple step TMLE algorithm applied to $P_{n,v}$ to determine the TMLE update Q^{d*}_{n,v^c} of Q^d_{n,v^c} for each v separately, and, again, one could pool across v at each step of such an multiple step TMLE algorithm. One can view the TMLE-update Q^{d*}_{n,v^c} as a TMLE update based on $P_{n,v}$ in which the initial estimator Q^d_{n,v^c} was based on an external sample P_{n,v^c}.

For notational convenience, we use the notation $Q_{n,v^c} = \hat{Q}^{\hat{d}(P_{n,v^c})}(P_{n,v^c})$, and similarly, we define $g_{n,v^c} = \hat{g}^{\hat{d}(P_{n,v^c})}(P_{n,v^c})$. In the following we assume a one-step TMLE, but the generalization to iterative TMLE is immediate.

The key assumption about ϵ_n and a corresponding update Q_{n,v^c,ϵ_n} is that it solves the cross-validated empirical mean of the efficient influence curve:

$$\frac{1}{V} \sum_{v=1}^{V} P_{n,v} D^*_{\hat{d}(P_{n,v^c})}(Q_{n,v^c,\epsilon_n}, g_{n,v^c}) = o_P(1/\sqrt{n}). \tag{9.17}$$

If one uses a full TMLE update, then, as we showed above, this equation holds with $o_P(1/\sqrt{n})$ replaced by 0, and, if one uses initial estimators of Q^d_0, g^d_0 that converge at a rate faster than $n^{-1/4}$, then it is possible to show that in great generality the first step TMLE will still satisfy (9.17).

The proposed estimator of $\psi_{0,n}$ is given by

$$\psi^*_n \equiv \frac{1}{V} \sum_{v=1}^{V} \Psi_{\hat{d}(P_{n,v^c})}(Q_{n,v^c,\epsilon_n}).$$

In the current literature we have referred to this estimator as the CV-TMLE. The only twist relative to the original CV-TMLE is that we change our target on each training sample into the training sample specific target parameter implied by $\hat{d}(P_{n,v^c})$ on the training sample, while in the original CV-TMLE formulation, the target would still be $\Psi_d(P_0)$. With this minor twist, the (same) CV-TMLE is now used to target the average of training sample specific target parameters averaged across the V training samples. General asymptotic theorems for this CV-TMLE are presented in Sect. A.1.

Suppose g^d_0 is known and that we use its known value so that $\hat{g}^d(P_n) = g^d_0$. Consider the estimator

$$\sigma^2_n = \text{Ave}_v \left[P_{n,v} \left\{ D^*_{\hat{d}(P_{n,v^c})}(Q_{n,v^c,\epsilon_n}, g_{n,v^c}) \right\}^2 \right] \tag{9.18}$$

of the asymptotic variance $\sigma^2_0 = P_0\{D^*_{d_0}(Q, g_0)\}^2$ of the CV-TMLE ψ^*_n. In words, it is the average across v of the v-specific sample variance of the influence curve of the v-specific CV-TMLE $\Psi_{\hat{d}(P_{n,v^c})}(Q_{n,v^c,\epsilon_n})$ as estimator of $\Psi_{\hat{d}(P_{n,v^c})}(P_0)$ based on estimation sample $P_{n,v}$. Since the estimates $(Q_{n,v^c,\epsilon_n}, g_{n,v^c})$ needed for calculating the plug-in estimate of the v-specific efficient influence curve are already needed for the estimator ψ^*_n, this estimate of the variance of $D^*_{d_0}$ is computationally free. Finally, given the results in Sect. A.1, an asymptotic 0.95-confidence interval for $\psi_{0,n}$ is given by $\psi_n^* \pm 1.96\sigma_n/\sqrt{n}$. This confidence interval is also asymptotically valid if g_0^d is unknown and both $\hat{Q}^d$ and $\hat{g}^d$ are consistent estimators of Q_0^d and g_0^d. This same variance estimator and confidence interval can also be used for the case that g_0 is not known and $\hat{g}(P_n)$ is an MLE of g_0 according to some correctly specified model. In that case, the theorem tells us that it is an asymptotically conservative confidence interval if $\hat{Q}^d$ is inconsistent. Either way, we recommend this confidence interval in general as long as one can rely on $\hat{g}^d$ being a consistent estimator.

9.7 CV-TMLE for Variable Importance Measure

The modification to the algorithm discussed above (9.3) involves two small changes: Q_0 and g_0 are estimated on the training sample (the same sample as is used above to determine the a_L, a_H), and estimating the coefficient, ϵ, in front of the clever covariate, is not just done on the corresponding estimation sample, but on the entire sample. Thus, (9.12) is modified to the following:

$$\psi_{n,v}^{CV-TMLE} = \frac{1}{n_V} \sum_{i:Z_i=v} \{Q_{n,v^c,\epsilon_n}(a_H(P_{n,v^c}), W_i) - Q_{n,v^c,\epsilon_n}(a_L(P_{n,v^c}), W_i)\}. \quad (9.19)$$

Here, the differences between (9.12) and (9.19) are that (1) $Q_{n,v} = \hat{Q}(P_{n,v})$ and $g_{n,v} = \hat{g}(P_{n,v})$ changes to $\hat{Q}(P_{n,v^c})$ and $\hat{g}(P_{n,v^c})$, respectively, and (2) epsilon changes from $\epsilon_{n,v}$ (estimated only on the estimation sample $P_{n,v}$) to ϵ_n (estimated on entire sample). This also requires changing the definition of the clever covariate (9.11) to:

$$H_{P_{n,v^c}}(A, W; g_{n,v^c}) = \frac{I(A = a_H(P_{n,v^c}))}{g_{n,v^c}(a_H(P_{n,v^c})|W)} - \frac{I(A = a_L(P_{n,v^c}))}{g_{n,v^c}(a_L(P_{n,v^c})|W)}. \quad (9.20)$$

Thus, ϵ_n is the result of a logistic regression as in (9.10) on Y_i on the covariate $H_{P_{n,v^c}}(A_i, W_i; g_{n,v^c})$ using as offset $\text{Logit}Q_{n,v^c}(A_i, W_i)$, $i = 1, \ldots, n$. This ϵ_n provides the updated Q_{n,v^c,ϵ_n} for each $v = 1, \ldots, V$. Finally, the estimator of the target data-adaptive variable importance measures, $\psi_{0,n}$ is given by

$$\hat{\Psi}(P_n) = \frac{1}{V} \sum_{v=1}^{V} \psi_{n,v}^{CV-TMLE}. \quad (9.21)$$

In addition, for each $v = 1, \ldots, V$, one estimates the influence curve as:

$$D_{P_{n,v^c}}^*(O) = \left[\frac{I(A = a_{H,n,v^c})}{g_{n,v^c}(a_{H,n,v^c}|W)} - \frac{I(A = a_{L,n,v^c})}{g_{n,v^c}(a_{L,n,v^c}|W)}\right]\{Y - Q_{n,v^c,\epsilon_n}(A, W)\}$$
$$+ [Q_{n,v^c,\epsilon_n}(a_{H,n,v^c}, W) - Q_{n,v^c,\epsilon_n}(a_{L,n,v^c}, W)] - \psi_{n,v}^{CV-TMLE}.$$

Finally, once these modifications are made, we can derive inference equivalently as done in (9.15).

This provides an alternative estimator for the same original average data-adaptive parameter, but convenient asymptotics are available in a bigger model, as shown by our theorem in Sect. A.1. Heuristically, one should expect much more robust estimation as the constituent parameters necessary for estimation (Q_0, g_0, ϵ) are now estimated on a larger proportion of the data than the original algorithm described in Sect. 9.3. This CV-TMLE represents a complex algorithm, but fortunately for this application, there is an R package available, described in more detail below.

9.8 Software for Data-Adaptive VIMs: `varImpact`

We provide a software package named varImpact, implemented in the R programming language (R Development Core Team 2016). varImpact is available on the Comprehensive R Archive Network (CRAN) and Github.[1] varImpact implements the variable importance algorithm described in this chapter, along with additional data cleaning, reporting, and related features that facilitate variable importance analysis in real-world datasets. We describe each step of the varImpact algorithm below as well as the parallelization approach.

1. Preprocessing. varImpact begins by preprocessing the datasets, handling factor and numeric variables separately. Variables are removed from the analysis if they exceed a missingness threshold (default of 50%) or have insufficient variation. This serves to protect against overfitting and focus the analysis on variables with reasonable measurement rates. The variation step analyzes the density of each variable and removes those where the 10th and 90th percentiles are equal. When serving as the variable of interest (but not as an adjustment variable) numeric variables are discretized into ten quantiles, provided that they include more than ten distinct values. Missingness indicators are generated for the remaining adjustment variables, for incorporation into the adjustment set, and missing values are imputed by k-nearest neighbors; median and zero-replacement imputation are also supported. The dataset is partitioned into the V folds for CV-TMLE, with $V = 10$ recommended in order to fully utilize the power of CV-TMLE. In the case of a binary outcome variable this splitting is stratified on the outcome in order to maximize power. The same partitioning is used for each variable that is analyzed.

2. Observational Study Per Variable. Now we can construct a data-adaptive observational study of each variable to estimate how the most impactful change in that variable influences the outcome, controlling for all other adjustment variables (covariates and missingness indicators). Each variable in turn is considered to be a multivalued treatment or intervention. We first estimate the mean potential outcome at each level of the treatment using the training data. We then use the held-out test data to estimate our variable importance measure as the mean difference between the level with the highest estimated mean outcome and the level with the lowest. This is the estimated average treatment effect when the current variable is set at its

---

[1] http://github.com/ck37/varImpact/.

"best" level compared to its "worst." varImpact loops over the CV-TMLE folds as it analyzes each variable, treating each fold as the test set and the complementary folds as the training set.

3. Per-fold Analysis. At each fold iteration (minimum 2 CV-TMLE folds) there are several steps in the analysis. The bins of discretized numeric variables are further aggregated by penalized histogram density estimation (Rozenholc et al. 2010; Mildenberger et al. 2009) to avoid small cell sizes. There is the option to hierarchically cluster the adjustment variables and then select the top ten most representative variables (medoids), where we use HOPACH (van der Laan and Pollard 2003) as clustering algorithm. This dimensionality reduction can drastically speed up computation but can easily result in bias or loss in power. For factor variables we also check for the minimum number of observations in each cell when the factor levels are cross-tabulated against a binary outcome variable. Covariates with small cell sizes can be skipped to save computation time and mitigate overfitting.

We then estimate the adjusted mean outcome at each level or bin $a \in A$ for the training set using TMLE.[2] The a-specific adjusted mean outcome is denoted with $\theta_0(a) \equiv E_{0,w}\{Q_0(a, W)\}$ and we denote the estimates for a specific training sample v^c with $\hat{\theta}_{P_{n,v^c}}(a)$. We identify the bin/level associated with the highest and lowest mean outcomes:

$$a_L(P_{n,v^c}) = \arg \min_{a \in A} \hat{\theta}_{P_{n,v^c}}(a)$$
$$a_H(P_{n,v^c}) = \arg \max_{a \in A} \hat{\theta}_{P_{n,v^c}}(a).$$

Observations with the variable's value in the "high" bin (a_H) are effectively the treatment group, and observations in the "low" bin (a_L) are the control group. The associated SuperLearner model fits for the outcome regression (Q) and propensity score (g) on the training set are saved. Then on the corresponding test set we apply the saved Q and g model fits to make the required predictions on the estimation sample.

4. Pooling of Per-fold Results. Once all the nuisance parameters and a_L, a_H have been estimated for each of the V training samples, we can carry out the estimation of ϵ based on the complete data set (i.e., union of V test samples). We actually construct a separate clever covariate for each of the two levels a_L, a_H (Eq. (9.20)) and estimate the bivariate fluctuation coefficient ϵ_n with logistic regression. We then fluctuate the predicted outcomes to target our mean outcome under the two levels, and separately for each test set we calculate the split-specific targeted mean outcome and associated influence curve values. Within each fold the difference of the targeted mean under the high level (a_H) and the targeted mean under the low level (a_L) is our fold-specific ATE (Eq. (9.19)). Similarly, within each fold the difference of the influence curve for a_H and the influence curve for a_L yields the influence curve for the ATE. The estimated sample variance $\sigma_{n,v}^2$ of the fold-specific ATE is the fold-specific estimated sample variance $\sigma_{n,v}^2$ of the influence curve (Eq. (9.18)).

---

[2] The estimated g is truncated to bounds of [0.025, 0.975] as in the TMLE R-package (Gruber and van der Laan 2012a). As in the TMLE R-package, we use nonnegative least squares as the meta-learner for both Q and g.

5. Point Estimation and Inference. To generate the final results we combine the per-fold parameter estimates. We take the mean of the per-fold ATEs as our point estimate of the ATE (Eq. (9.21)): $\psi_n = \text{Ave}\{\hat{\psi}_{P_{n,v^c}}(P_{n,v})\}$, and the mean of the per-fold sample variances as the estimated sample variance: $\sigma_n^2 = \text{Ave}\{\sigma_{n,v}^2\}$. The estimated standard error is $\sqrt{\sigma_n^2/n}$. We report a normal-based confidence interval and a one-sided p-value based on the null hypothesis: $H_0 : \psi_{n,0} \leq 0$. A two-sided test would not be appropriate because the treatment levels a_H and a_L were selected to yield a positive treatment effect. Any negative treatment effect estimate is an indication that the procedure did not find a treatment effect for that particular variable using the identified levels.

6. Reporting. In the final reporting stage we adjust for multiple comparisons, determine final statistical significance, and flag any variables with inconsistent results across the V folds. Our multiple comparison adjustment controls the false discovery rate through the Benjamini-Hochberg procedure (Benjamini and Hochberg 1995). We declare the variables with FDR-adjusted p-value smaller than 0.05 as statistically significant. We state that the results for a particular categorical variable are consistent if one selects the same a_L and a_H levels across the V folds. This criterion could be made more flexible by the analyst, such as requiring only a certain minimum percentage of folds selecting the same a_L and a_H levels. For numeric variables we define consistency as all V folds showing the same directionality for the low and high quantiles. In other words, a consistent result is that for every CV-TMLE fold a_L is a lower quantile than a_H, or alternatively for every CV-TMLE fold the a_L is a higher quantile than a_H. Variables are sorted by ascending p-value and their rank, parameter estimate, naive p-value, FDR-adjusted p-value, and 95% confidence interval are listed.

7. Parallelization. Executing a semiparametric observational study on each variable in a dataset is computationally demanding. This is doubly true for complex SuperLearner libraries that are necessary for accurate outcome and propensity score estimation. CV-TMLE compounds the requisite computation, as it essentially conducts the observational study multiple times per variable. To address this varImpact supports parallelization using the future package, and can seamlessly use multiple cores on a machine or multiple machines in a cluster. This can yield drastic improvements in the total elapsed, or "wall-clock", time for an analysis.

9.9 Data Analysis: Framingham Heart Study

We apply our variable importance estimation method to the Framingham Coronary Heart Disease cohort. Wilson et al. (1998) developed sex-specific risk prediction algorithms for coronary heart disease using discretized blood pressure and cholesterol measurements combined with a few additional variables. The risk equations were developed from multivariate regression stratified by sex to estimate the association of blood pressure,[3] cholesterol, age, smoking, and diabetes with future coronary heart disease (CHD). Cholesterol and blood pressure were binned into categories, allowing nonmonotonic relationships to be modeled within a linear framework. The regression modeling was intentionally simple so that concise risk scoring rules could be implemented by practitioners. The results remain widely used by clinicians for assessing patient risk of coronary heart disease.

We analyze similar data using the same variables and categorical discretization but with the varImpact software implementing the CV-TMLE methodology for data-adaptive target parameters. In particular, we provide comparison analyses for table 5 from the original paper. As in the paper all analyses are stratified by gender. (The analyzed data is 57% women and 43% men.) We analyze the publicly available Framingham Longitudinal Data dataset as provided by the Biologic Specimen and Data Repository Information Coordinating Center at the National Heart, Lung, and Blood Institute. We restrict our analysis to period 3, when LDL and HDL cholesterol measurements were collected, and remove subjects who had experienced CHD in periods 1 or 2. We only evaluate the impact of covariates used in the original paper.

Before we begin we note a few differences in our data as compared with the original study. (1) Wilson et al. use confidential data from the Framingham original cohort and offspring cohort. Their data were collected between 1971 and 1974 whereas ours were collected primarily in 1968. (2) The paper's dataset includes offspring of the original cohort, whereas ours does not. This could feasibly change the data generating processes. (3) The 1998 paper does not specify how missing data was handled. We presume that records with missing data were dropped on a per-table basis. (4) The Framingham Longitudinal Data dataset was anonymized to protect patient confidentiality, which likely has some influence on our resulting analysis. Therefore we recommend viewing these results as suggestive rather than conclusive.

9.9.1 Super Learner Library

We use the following super learner library using the SuperLearner package (Polley and van der Laan 2013; Polley et al. 2017) library for Q and g estimation (with inspiration from the thorough library in Pirracchio et al. 2014), using R version 3.3.2. The library was developed by optimizing the predictive accuracy of the outcome regression, which we recommend as a helpful exercise prior to estimating variable importance.

---

[3] Blood pressure levels are defined by JNC-V (Joint National Committee 1993): optimal (systolic ≤ 120 mm Hg and diastolic ≤ 80 mm Hg), normal blood pressure (systolic 120–129 mm Hg or diastolic 80–84 mm Hg), high normal blood pressure (systolic 130–139 mm Hg or diastolic 85–89 mm Hg), hypertension stage I (systolic 140–159 mm Hg or diastolic 90–99 mm Hg), and hypertension stage II–IV (systolic ≥ 160 or diastolic ≥ 100 mm Hg). "When systolic and diastolic pressures fell into different categories, the higher category was selected for the purposes of classification."

- Logistic regression
- Elastic net (1) with only main terms, and (2) with main terms and two-way interactions; each with six configurations using the *glmnet* package (Friedman et al. 2010, version 2.0-5), ($\alpha \in \{0, 0.2, 0.4, 0.6, 0.8, 1.0\}$)
- Bayesian linear regression using the *arm* package (Gelman et al. 2010, version 1.9-3)
- Multivariate adaptive regression splines with three configurations (degree $\in$ $\{1, 2, 3\}$) using the *earth* package (Milborrow et al. 2014, version 4.4.7)
- Bagging using the *ipred* package (Peters and Hothorn 2009, version 0.9-5)
- Random Forest with four configurations (mtry $\in \{1, 2, 4, 7\}$) using the *randomForest* package (Liaw and Wiener 2002, version 4.6-12)
- Extreme gradient boosting (Chen and Guestrin 2016) with 12 hyperparameter configurations (trees $\in \{100, 1000\} \times$ depth $\in \{1, 2, 3\} \times$ learning rate $\in \{0.1, 0.001\}$) using the *xgboost* package (version 0.6-4)
- Outcome mean, included for performance benchmarking and as a check against overfitting

We used the default hyperparameters provided by the SuperLearner package unless otherwise specified. We refrained from any dimensionality reduction of the adjustment variables in order to maximize (1) statistical power, (2) plausibility of the randomization assumption, and (3) comparability with the original study.

9.9.2 Results

We present the variable importance results stratified by gender below. The initial table reports aggregated results for all variables; a reference line is added at FDR p-value = 0.05. The second table lists the constituent parameter estimates and identified levels (a_L, a_H) for each CV-TMLE fold. We applied the CV-TMLE for $V = 2$.

Female. We see in Table 9.1 that HDL and diabetes are estimated to have significant and consistent impacts on coronary heart disease. The implication is that among women in this dataset, risk for coronary heart disease could be reduced by raising HDL levels to 60+ mg/DL and preventing the occurrence of diabetes. Both results agree with the findings from the original study. Blood pressure is the highest ranked variable, but its high and low levels are inconsistently identified providing a harder to interpret definition of its variable importance. Smoking status shows a low, nonsignificant impact on CHD, distinctly contrary to the high and significant effect estimated in the original study. Table 9.2 shows that the inconsistency of blood pressure is due to a_L being identified as the "high" level of BP in one CV-TMLE fold but as the "normal-optimal" level in the other fold. We also see that the HDL levels are as expected: high HDL is estimated to have the lowest rate of CHD, and low HDL to have the highest rate of CHD. LDL by contrast does show consistently identified a_L and a_H levels across CV-TMLE folds.

Male. In the male results (Table 9.3) we again see diabetes and HDL ranked highly and with comparable point estimates. This time age and LDL show consistent, large, and statistically significant effects. Smoking status once again shows a small effect, although this time it is marginally statistically significant. We do not find a consistent set of high and low levels for blood pressure. The fold-specific results in Table 9.4 show that the same a_L and a_H for HDL were selected in males as in females. The LDL levels are as expected, with low LDL estimated to have a lower expected CHD risk compared to high levels. Similarly, the a_L for age is the lowest age bin as expected.

Table 9.1 Female variable importance results for combined estimates

Rank	Variable	Type	Estimate	CI 95	p-value	Adj. p-value	Consistent
1	BP	Factor	0.1119	(0.0388–0.185)	0.0014	0.0081	No
2	HDL	Factor	0.1102	(0.0257–0.195)	0.0053	0.0125	Yes
3	Diabetes	Ordered	0.1373	(0.0296–0.245)	0.0062	0.0125	Yes
4	Age	Ordered	0.0449	(−0.00552–0.0954)	0.0405	0.0607	Yes
5	Smoking	Ordered	0.0122	(−0.0362–0.0605)	0.3108	0.3730	Yes
6	LDL	Factor	−0.0193	(−0.0771–0.0385)	0.7433	0.7433	No

Table 9.2 Female variable importance results by estimation sample

Variable	Est_v1	Est_v2	Low_v1	High_v1	Low_v2	High_v2	Consistent
BP	0.1780	0.0458	High	Stage2_4	Normal-optimal	Stage2_4	No
HDL	0.1534	0.0670	[60,999)	[0,35)	[60,999)	[0,35)	Yes
Diabetes	0.0910	0.1836	(0.9999999,1]	(1,2]	(0.9999999,1]	(1,2]	Yes
Age	0.0029	0.0870	(0.999999,1]	(1,10]	(1,5]	(5,10]	Yes
Smoking	0.0126	0.0117	(0.9999999,1]	(1,2]	(0.9999999,1]	(1,2]	Yes
LDL	−0.0452	0.0067	[130,160)	[0,130)	[0,130)	[160,999)	No

Table 9.3 Male variable importance results for combined estimates

Rank	Variable	Type	Estimate	CI 95	p-value	Adj. p-value	Consistent
1	Age	Ordered	0.1609	(0.102–0.219)	0.0000	0.0000	Yes
2	LDL	Factor	0.1693	(0.106–0.232)	0.0000	0.0000	Yes
3	HDL	Factor	0.1623	(0.09–0.235)	0.0000	0.0000	Yes
4	Diabetes	Ordered	0.1552	(0.0739–0.236)	0.0001	0.0001	Yes
5	BP	Factor	0.0982	(0.0384–0.158)	0.0006	0.0008	No
6	Smoking	Ordered	0.0356	(−0.0159–0.0871)	0.0879	0.0879	Yes

Table 9.4 Male variable importance results by estimation sample

Variable	Est_v1	Est_v2	Low_v1	High_v1	Low_v2	High_v2	Consistent
Age	0.1339	0.1878	(0.999999,1]	(1,10]	(0.999999,1]	(1,10]	Yes
LDL	0.1139	0.2246	[0,130)	[160,999)	[0,130)	[160,999)	Yes
HDL	0.1542	0.1705	[60,999)	[0,35)	[60,999)	[0,35)	Yes
Diabetes	0.1897	0.1207	(0.9999999,1]	(1,2]	(0.9999999,1]	(1,2]	Yes
BP	0.1295	0.0670	Normal-optimal	Stage1	Normal-optimal	Stage2_4	No
Smoking	0.0333	0.0378	(0.9999999,1]	(1,2]	(0.9999999,1]	(1,2]	Yes

9.10 Discussion

Data-adaptive parameters as a general concept opens up enormous opportunities for estimating relevant scientific parameters when the experiment and current hypotheses do not sufficiently constrain the parameter of interest to apply the more conventional approach based upon prespecified parameters. Given that one will lose power by not prespecifying the parameter of interest, care must be given to fully utilize the amount of information contained in the data to estimate the adaptively-defined parameter, and to develop valid confidence intervals. CV-TMLE achieves these two goals. In addition, the algorithm can be relatively trivially parallelized and the influence-curve based inference avoids time-consuming bootstrap procedures. Thus, the approach can be adapted to exploratory data analysis in high dimensional, big data contexts. Finally, when parameters are pre-specified, CV-TMLE brings the estimator closer to complete automation, as now even issues of adaptivity of machine learning algorithms used in estimation of the data-generating distribution do not affect the asymptotics, so one can derive trustworthy inference with minimal assumptions.

Thus, this is one step closer to statistical algorithms that will require minimal input from users and yield relatively efficient results in very big statistical models. When one adds the data-adaptive component of parameter definition, the potential for automation becomes even greater, as even the parameter of interest no longer needs (precise) pre-specification. As a consequence, this CV-TMLE approach for data-adaptive target parameters represents an important step on the way to bringing relatively unsophisticated users the promise of high-performance exploratory statistical algorithms. In this manner we are moving towards a situation that is analogue to one in which someone with little knowledge of the mechanics of a motor vehicle can still be a safe and effective user of the machinery.

Acknowledgments. This work was partially supported through a Patient-Centered Outcomes Research Institute (PCORI) Pilot Project Program Award (ME-1306-02735). Disclaimer: All statements in this chapter, including its findings and conclusions, are solely those of the authors and do not necessarily represent the views of the Patient Centered Outcomes Research Institute (PCORI), its Board of Governors or Methodology Committee.

Chapter 10
C-TMLE for Continuous Tuning

Mark J. van der Laan, Antoine Chambaz, and Cheng Ju

A TMLE of a causal quantity of interest first constructs an initial estimator of the relevant part of the likelihood of the data and then updates this initial estimator along a least favorable parametric model that uses the initial estimator as an off-set. The least favorable parametric model typically depends on an orthogonal nuisance parameter such as the treatment and censoring mechanism. This nuisance parameter is not needed to evaluate the target parameter, and, in fact, is orthogonal to the target parameter in the sense that a maximum likelihood estimator would completely ignore this nuisance parameter, or, at least, its scores are orthogonal to the scores of the relevant part of the likelihood.

However, the orthogonal nuisance parameter plays a crucial role in determining the best way to update the initial estimator, as directed by the canonical gradient (i.e., efficient influence curve) of the pathwise derivative. In a standard TMLE, one would estimate this nuisance parameter with an estimator that aims for an optimal performance for the nuisance parameter itself. For example, one might estimate it with a super learner based on the log-likelihood loss-function of the nuisance parameter.

M. J. van der Laan (✉)
Division of Biostatistics and Department of Statistics, University of California, Berkeley, 101 Haviland Hall, #7358, Berkeley, CA 94720, USA
e-mail: laan@berkeley.edu

A. Chambaz
MAP5 (UMR CNRS 8145), Université Paris Descartes, 45 rue des Saints-Péres, 75270 Paris cedex 06, France
e-mail: antoine.chambaz@parisdescartes.fr

C. Ju
Division of Biostatistics, University of California, Berkeley, 101 Haviland Hall, #7358, Berkeley, CA 94720, USA
e-mail: cju@berkeley.edu

© Springer International Publishing AG 2018
M.J. van der Laan, S. Rose, *Targeted Learning in Data Science*,
Springer Series in Statistics, https://doi.org/10.1007/978-3-319-65304-4_10

Even though the TMLE is asymptotically efficient, if the initial estimator and the nuisance parameter estimator are well behaved, one might wonder if it would not make more sense to evaluate the fit of the nuisance parameter with respect to how well the resulting TMLE succeeds in reducing the MSE with respect to the target parameter during the targeting step of the TMLE algorithm.

This issue is of enormous practical importance in causal inference in the case that the target parameter is weakly supported by the data (i.e., lack of positivity). In this case the efficient influence curve can take on very large values so that the maximum likelihood estimator along the least favorable submodel (whose score spans the efficient influence curve) can be ill behaved and thereby hurt the initial estimator with respect to the target parameter. For example, if a particular potential confounder that affects treatment decisions happens to be an instrumental variable that has no effect on the outcome, then including it in the fit of the treatment mechanism only harms the TMLE in finite samples. This insight has resulted in a variety of proposals in the literature that prescreens covariates based on their potential effect on the outcome, removes the ones that have weak effects, and then runs one of the available estimators.

However, before we jump into this, we should be aware of the enormous dangers that come with such an approach (an approach that clearly ignores the likelihood principle). Consider a covariate that has an effect on the outcome of interest that is of the order $n^{-1/2}$. Such covariates would correspond with t-statistics (evaluating the effect of the covariate on the outcome) that are of the order 1. That is, their signal is real but are within the noise level so that a prescreening method would easily remove this covariate. However, not including this covariate in the TMLE (or any other estimator) would result in an estimator that has bias of the order $n^{-1/2}$. As a consequence, such a TMLE would not even be asymptotically linear, even in the case that the variance of the efficient influence curve is perfectly well behaved. One might counter this argument by stating that one should simply make these prescreening methods more conservative as sample size increases. However, these screening methods are based on marginal regressions, easily misjudging their effect in the presence of other confounders.

Therefore, the basic message is that an effort to improve an estimator in the context of sparsity (measured by the variance of the efficient influence curve), one can easily destroy the good asymptotic properties of the estimator. Collaborative TMLE (C-TMLE) takes on this enormous challenge by being grounded in theory.

The C-TMLE is tailored such that it does not affect the asymptotic behavior of the TMLE by pushing the selected estimator for the orthogonal nuisance parameter towards the most nonparametric estimator as sample size increases. Simultaneously, it provides potentially dramatic gains in practical performance with the stepwise building of the estimator (from parametric to nonparametric), each time choosing the move for which the maximum likelihood estimator for the corresponding least-favorable submodel results in maximal improvement of the fit of the corresponding TMLE relative to the off-set. Moreover, the latter approach is completely supported by the collaborative double robustness of the efficient influence curve, which shows that the orthogonal nuisance parameter only has to adjust for covariates that are needed to fit the residual bias of the initial estimator with respect to its true counterpart. In this manner, the C-TMLE makes sure that instrumental variables will only be included in the fit of the treatment mechanism at large enough sample sizes for which the parametric maximum likelihood estimator extension in the update step using this covariate results in a statistically significant gain in fit. Indeed, simulations have shown that such C-TMLEs are rarely worse than the standard TMLE, and can be much better when the data are sparse.

The previous literature on C-TMLE (e.g. van der Laan and Gruber 2010; Gruber and van der Laan 2010b; Wang et al. 2011a; Schnitzer et al. 2016; van der Laan and Rose 2011), focused on tuning discrete steps, such as evaluating the addition of a covariate to the treatment mechanism. In this chapter, we focus on C-TMLEs that tune a continuous valued tuning parameter of the fit of the orthogonal nuisance parameter, such as selecting the L_1-penalty in a lasso regression of the treatment mechanism (or a bandwidth of a kernel regression smoother). As we show, this dramatically changes the story when comparing C-TMLE with TMLE. Instead of C-TMLE not affecting the asymptotic linearity of the TMLE, we demonstrate that the C-TMLE can reduce the second-order remainder of the TMLE in its Taylor expansion to the point that the C-TMLE is asymptotically linear while the TMLE is not (e.g., in the case that a strong positivity assumption holds, but the nuisance parameters converge to their true counterparts at rates that are too slow). In addition, in practice we observe dramatic gains of our C-TMLE in nonsparse settings.

10.1 Formal Motivation for Targeted Tuning of Nuisance Parameter Estimator in TMLE

Defining the Estimation Problem. Suppose we observe n i.i.d. copies of a random variable O with probability distribution P_0 known to be an element of a statistical model $\mathcal{M}$: i.e., $P_0 \in \mathcal{M}$. Let $\Psi : \mathcal{M} \to \mathbb{R}^d$ be a target parameter of interest that is pathwise differentiable at any $P \in \mathcal{M}$ with efficient influence curve $D^*(P)$. Suppose that $\Psi(P)$ only depends on P through a parameter $Q = Q(P)$, and, for notational convenience, we will also use the notation $\Psi(Q(P))$ for $\Psi(P)$. The efficient influence curve $D^*(P)$ depends on P through $Q(P)$ and an additional nuisance parameter we will denote with $G(P)$. We will also denote $D^*(P)$ with $D^*(Q(P), G(P))$. Let $R_{20}(P, P_0)$ be defined by the second-order remainder in a first order Taylor expansion of the target parameter as follows:

$$\Psi(P) - \Psi(P_0) = -P_0 D^*(P) + R_2(P, P_0).$$

We will also denote $R_2(P, P_0)$ with $R_{20}(Q, G, Q_0, G_0)$ in order to emphasize that it involves second-order differences between (Q, G) and (Q_0, G_0). Thus, we can write

$$\Psi(Q) - \Psi(Q_0) = -P_0 D^*(Q, G) + R_{20}(Q, G, Q_0, G_0).$$

The estimation problem defined by the statistical model $\mathcal{M}$ and its target parameter $\Psi : \mathcal{M} \to \mathbb{R}^d$ has the so called double robust structure if $R_{20}(Q, G_0, Q_0, G_0) = 0$ for all Q, and $R_{20}(Q_0, G, Q_0, G_0)$ for all G. In essence, this states that the second-order remainder involves a sum of integrals over an integrand that can be represented as a product of a difference of a parameter of Q with its true value and a parameter of G with its true value. For example, $R_{20}(Q, G, Q_0, G_0) = \int (H_1(G) - H_1(G_0))(H_2(Q) - H_2(Q_0)) f(Q, G, Q_0, G_0) dP_0$ for some functionals H_1, H_2 and f.

Example 10.1. Let's consider an example to illustrate these quantities. Let $O = (W, A, Y)$, where W is a vector of baseline covariates, A is a binary treatment and Y a binary outcome. Consider the statistical model that leaves the distribution $Q_{W,0}$ of W, and the conditional distribution of Y, given A, W, unspecified, while we might know that the conditional distribution G_0 of A, given W, falls in a set $\mathcal{G}$. Let's denote this statistical model with $\mathcal{M}$ so that we know that $P_0 \in \mathcal{M}$. Let $\bar{Q}_0(W) = E_0(Y \mid A = 1, W)$, $\bar{G}_0(W) = E_0(A|W)$, and $\Psi(P) = E_P E_P(Y \mid A = 1, W) = Q_{W,0} \bar{Q}_0$ is the target parameter of interest. Let $Q = (Q_W, \bar{Q})$ so that we can also denote $\Psi(P_0)$ with $\Psi(Q_0)$. In this example the efficient influence curve $D^*(P)$ of $\Psi : \mathcal{M} \to \mathbb{R}$ at any $P \in \mathcal{M}$ is given by

$$D^*(P) = \frac{A}{\bar{G}(W)}(Y - \bar{Q}(W)) + \bar{Q}(W) - \Psi(P),$$

and

$$\Psi(P) - \Psi(P_0) = -P_0 D^*(P) + R_{20}(P, P_0),$$

where

$$R_{20}(P, P_0) = P_0 \frac{\bar{G} - \bar{G}_0}{\bar{G}}(\bar{Q} - \bar{Q}_0).$$

We will also denote this remainder with $R_{20}(Q, G, Q_0, G_0)$.

Family of Candidate Nuisance Parameter Estimators Indexed by Continuous Tuning Parameter h. Let $L_1(G)$ be a loss-function for G_0 so that $P_0 L_1(G_0) = \min_{G \in G(\mathcal{M})} P_0 L_1(G)$. Let $\{\hat{G}_h : h \in [0, 1]\}$ be a family of candidate estimators $\hat{G}_h : \mathcal{M}_{NP} \to \mathcal{G}$ of G_0 indexed by a continuous valued index $h \in [0, 1]$, where the estimates $\{G_{n,h} = \hat{G}_h(P_n) : h\}$ are ordered from most nonparametric at $h = 0$ to most parametric at $h = 1$. For the easiest interpretation of our formal results, we suggest to think of the index parameter h so that $h \approx b_n(h)$ for a bias $b_n(h)$ of $G_{n,h}$ such as defined by the loss-based dissimilarity $b_n(h)^2 = P_0 L_1(G_{n,h}) - P_0 L_1(G_0)$. Such an indexing exists and can be constructed in terms of the bias function $b_n(h)$ as long as it is monotone increasing in h: define $\hat{G}_{1,b} = \hat{G}_{b_n^{-1}(b)}$ and index the family of candidate estimators by b as $\{\hat{G}_{1,b} : b\} = \{\hat{G}_h : h\}$. Specifically, we assume that the empirical risk $h \to P_n L(G_{n,h})$ is increasing in h. For example, for some large $M < \infty$, one might define an h-specific MLE

$$G_{n,h} = \arg \min_{G \in \mathcal{G}, \|G\|_v \leq (1-h)M} P_n L_1(G), \qquad (10.1)$$

where $\| G \|_v$ is the variation norm defined in Chap. 6 on the highly adaptive lasso. In Chap. 6 we showed that by representing G with a linear combination of indicator basis functions, $G_{n,h}$ can be implemented as a lasso estimator defined as an MLE over all linear combinations under the constraint that the sum of the absolute value of the coefficients is restricted to be smaller than $(1 - h)M$.

TMLE Depending on Choice of Nuisance Parameter Estimator. Let $L(Q)$ be a loss-function for Q_0 so that $P_0 L(Q_0) = \min_{Q \in \mathcal{Q}(\mathcal{M})} P_0 L(Q)$. Consider a least favorable submodel $\{Q_\epsilon : \epsilon\} \subset \mathcal{Q}(\mathcal{M})$ through Q at $\epsilon = 0$ so that the linear span of the components of the generalized score $\frac{d}{d\epsilon} L(Q_\epsilon)\big|_{\epsilon=0}$ includes $D^*(Q, G)$. Note that this submodel also depends on G so that we will also use the notation $Q_{\epsilon,G}$ in order to emphasize this dependence. Let $\hat{Q} : \mathcal{M}_{NP} \to \mathcal{Q}(\mathcal{M})$ be an initial estimator of Q_0. Given this submodel and initial estimator $(Q_n = \hat{Q}(P_n), G_{n,h})$ of (Q_0, G_0), one can construct a one-step or iterative TMLE $Q^*_{n,h}$. For example, if one uses a universal least favorable submodel, one can use the one-step TMLE $Q^*_{n,h} = Q_{n,\epsilon_n,G_{n,h}}$, where $\epsilon_n = \arg \min_\epsilon P_n L(Q_{n,\epsilon,G_{n,h}})$. Given the sequence $\{G_{n,h} : h\}$ of candidate estimators for G_0, this defines now a sequence of candidate TMLEs $\{Q^*_{n,h} : h\}$ of Q_0, all solving the efficient influence curve equation $0 = P_n D^*(Q^*_{n,h}, G_{n,h}) = 0$.

TMLE Using Cross-Validation Selector for h. A natural approach for selecting the index h is to use L_1-loss based cross-validation. In that case, one defines a random split $B_n \in \{0, 1\}^n$ in a training sample $\{i : B_n(i) = 0\}$ and validation sample $\{i : B_n(i) = 1\}$, with respective empirical probability distributions P^0_{n,B_n}, P^1_{n,B_n}, and one selects h with the cross-validation selector

$$h_{n,CV} = \arg \min_h E_{B_n} P^1_{n,B_n} L_1(\hat{G}_h(P^0_{n,B_n})).$$

In our example (10.1) $G_{n,h_{n,CV}}$ is the highly adaptive lasso estimator (HAL) proposed in Chap. 6, and we have shown that for each fixed h the loss-based dissimilarity $d_{01}(G_{n,h}, G_{0,h}) = P_0 L_1(G_{n,h}) - P_0 L_1(G_{0,h})$ converges at a rate at least as fast as $n^{-(0.5+\alpha(d)/4)}$, where $\alpha(d) = 1/(d + 1)$ and $G_{0,h} = \arg \min_{G \in \mathcal{G}, \|G\|_v < hM} P_0 L_1(G)$. In addition, in that case we know that $G_n = G_{n,h_{n,CV}}$ converges at least as fast as this rate.

Example 10.2. In our example we select $L_1(G)(O) = -\{A \log \bar{G}(W) + (1 - A) \log(1 - \bar{G}(W))\}$ as the log-likelihood loss function. $G_{n,h}$ could be defined as a lasso logistic linear regression with L_1-constraint $(1 - h)M$ as in (10.1), and $G_{n,h_{n,CV}}$ would be the lasso estimator that uses internal cross-validation to select the constraint.

In a regular TMLE framework we would select h with $h_{n,CV}$ and use the TMLE $Q^*_n = Q^*_{n,h_{n,CV}}$ resulting in the TMLE $\Psi(Q^*_n)$ of $\Psi(Q_0)$. If Q_n is consistent at a particular rate, then a TMLE $\Psi(Q^*_{n,h_{n,CV}})$ using a best estimator $G_{n,h_{n,CV}}$ of G_0 might already be asymptotically efficient in which case such a choice $h_{n,CV}$ is appropriate asymptotically. For example, our proposed HAL-TMLE in Chap. 7 relies on an HAL-estimator Q_n that converges to Q_0 at a faster rate than $n^{-1/4}$ so that a TMLE based on an HAL-estimator $G_{n,h_{n,CV}}$ of G_0 is asymptotically efficient under very weak regularity conditions.

Potential Improvement of TMLE with C-TMLE. Nonetheless, the second-order remainder for the HAL-TMLE might be substantial in finite samples (recall that for large dimensional O the second-order remainder multiplied by $n^{1/2}$ converges to zero at a very slow rate) so that the C-TMLE discussed in this chapter is still very relevant for finite sample improvement. In van der Laan (2014b) we proposed a TMLE that maps an initial estimator $(Q_n, G_{n,h})$ into a jointly targeted estimator $(Q^*_{n,h}, G^*_{n,h})$ in such a way that the asymptotic linearity of the TMLE $\Psi(Q^*_{n,h})$ is preserved under misspecification of either Q_n or G_n (but not both), as long as both estimators converge at a rate faster than $n^{-1/4}$ to their (possibly misspecified) limits. This type of TMLE is generally recommended to protect its asymptotic linearity against misspecification of one of the estimators. In that case, an under-smoothed choice h_n is not needed from an asymptotic perspective, but $h_{n,CV}$ could be used. However, as we also suggested above for the HAL-TMLE, we suggest that the C-TMLE algorithm might yield improved finite sample performance (especially when the data is sparse for the parameter of interest). Overall, even when the C-TMLE is not needed for improving asymptotic performance, it will still represent an important practical finite sample advance.

The cross-validation selector $h_{n,CV}$ optimizes the selection of h w.r.t. estimation of G_0, while the real goal should be to minimize the MSE of $h \to \Psi(Q^*_{n,h})$ w.r.t. ψ_0. One should realize that, due to the fact that $\Psi(P_0)$ is a smooth (pathwise differentiable) functional of P_0, typically the variance of a $\Psi(Q^*_{n,h})$ behaves asymptotically as $1/n$ across all h values. Thus, for large sample size n the square bias of $\Psi(Q^*_{n,h})$ will dominate the variance of $\Psi(Q^*_{n,h})$ so that h will have to be selected to minimize the bias of $\Psi(Q^*_{n,h})$ over h. Collaborative targeted maximum likelihood (minimum loss) estimation (C-TMLE) aims to achieve this indirectly by (1) building a sequence of TMLEs $(Q^*_{n,h}, G_{n,h})$ whose empirical fits are increasing as h approximates 0 (in the C-TMLE algorithm $Q^*_{n,h}$ uses one of previous $Q_{n,h'}$ for $h' > h$ as initial estimator so that all these fits are nested) and (2) evaluating the choice of h w.r.t. the L-fit of $Q^*_{n,h}$ w.r.t. Q_0. Since $\Psi(Q^*_{n,h})$ is a targeted estimator of ψ_0 for each h, and these h-specific TMLEs $Q^*_{n,h}$ only differ in the depth h of the C-TMLE -targeting step applied to the same initial estimator Q_n, the L-fit of $Q^*_{n,h}$ is a sensible criterion selecting the maximal amount of targeting (i.e., minimal h) that still represents a statistical significant signal.

Example 10.3. Let $\bar{Q}_n$ be an initial estimator of $\bar{Q}_0$. Consider the least favorable submodel $\text{Logit}\bar{Q}_{n,\epsilon,G_{n,h}} = \text{Logit}\bar{Q}_n + \epsilon C(\bar{G}_{n,h})$ through $\bar{Q}_n$ with one-dimensional fluctuation parameter ϵ, where $C(\bar{G})(A, W) = A/\bar{G}(W)$ is often referred to as the clever covariate. Let $\epsilon_{n,h} = \arg\min_\epsilon P_n L(\bar{Q}_{n,\epsilon,G_{n,h}})$ be the MLE, where $L(\bar{Q}) = -\{Y \log \bar{Q}(A, W) + (1 - Y) \log(1 - \bar{Q}(A, W))\}$ is the log-likelihood loss for $\bar{Q}_0$. Let

$\text{Logit}\bar{Q}^*_{n,h} = \text{Logit}\bar{Q}_n + \epsilon_{n,h}C(\bar{G}_{n,h})$, so that $\bar{Q}^*_{n,h}$ is the TMLE using $\bar{Q}_n$ as initial estimator and $\bar{G}_{n,h}$ in the targeting step.

These TMLEs $Q^*_{n,h} = (Q_{W,n}, \bar{Q}^*_{n,h})$ of Q_0 solve

$$P_n D^*(Q^*_{n,h}, G_{n,h}) = 0,$$

and specifically

$$P_n D^*_1(\bar{Q}^*_{n,h}, G_{n,h}) = 0,$$

where $D^*_1(\bar{Q}, \bar{G}) = A/\bar{G}(Y - \bar{Q}(W))$ is the component of the efficient influence curve that is in the tangent space of the conditional distribution of Y, given A, W. The TMLE $\Psi(Q^*_{n,h})$ of $\Psi(Q_0)$ satisfies:

$$\Psi(Q^*_{n,h}) - \Psi(Q_0) = (P_n - P_0)D^*(Q^*_{n,h}, \bar{G}_{n,h}) + R_{20}((Q^*_{n,h}, G_{n,h}, Q_0, G_0).$$

Suppose that the initial estimator $\bar{Q}_n \to \bar{Q} \neq \bar{Q}_0$ is inconsistent. Then, the asymptotic linearity of $\Psi(Q_{n,h})$ relies upon the second-order remainder term

$$P_0(\bar{G}_{n,h} - \bar{G}_0)/\bar{G}_{n,h}(\bar{Q} - \bar{Q}_0) \tag{10.2}$$

to be asymptotically linear. Consider the likelihood based cross-validation selector $h_{n,CV} = \arg\min_h E_{B_n} P^1_{n,B_n} L_1(\hat{\bar{G}}_h(P^0_{n,B_n}))$. Then this term (10.2) will not be asymptotically linear due to $\bar{G}_{n,h_{n,CV}}$ having a bias larger than $1/\sqrt{n}$. That is, $h_{n,CV}$ will trade off the bias and variance of the actual estimator $\bar{G}_{n,h}$ as an estimator of G_0, while one should want to trade off this bias with the variance of $\Psi(Q^*_{n,h})$. Clearly, the variance of a real valued smooth functional $\Psi(Q^*_{n,h})$ (which behaves as $1/n$) is significantly smaller than the variance of the infinite dimensional object $G_{n,h}$.

Nonetheless, it might be that there exist a rate h_n that undersmooths $G_{n,h}$ enough so that this smooth function (10.2) of $G_{n,h}$ is asymptotically linear. We wonder if in that case, an C-TMLE selector h_n will undersmooth appropriately so that (10.2) is asymptotically linear. In general, we wonder if the rate at which the bias of the C-TMLE $\Psi(Q^*_{n,h_n})$ converges to zero is significantly faster than the rate at which the bias of the TMLE $\Psi(Q_{n,h_{n,CV}})$ converges to zero. In other words, does the C-TMLE choice h_n appropriately minimize MSE for the actual target parameter ψ_0?

10.1.1 Contrasting Discrete and Continuous Tuning Parameters

C-TMLE has been studied for discrete sequences $\{G_{n,k} : k\}$ of candidate estimators (van der Laan and Gruber 2010; Gruber and van der Laan 2010b; Wang et al. 2011a; Schnitzer et al. 2016; van der Laan and Rose 2011), in which case any reasonable selector will asymptotically end up selecting the most nonparametric estimator (i.e., asymptotically best) of G_0. As a result, the asymptotic performance of a C-TMLE is equivalent with the TMLE selecting the most nonparametric estimator of G_0 and the TMLE that selects this estimator of G_0 with L_1-based cross-validation. In this discrete scenario, as shown in a variety of articles, the C-TMLE represents a potentially highly significant *finite sample improvement* that does not affect the asymptotic performance of a standard TMLE.

However, when the tuning parameter is continuous, different data-adaptive selectors of h correspond with different rates at which the bias $b_n(h)$ of the TMLE $\Psi(Q_{n,h}^*)$ converges to zero, which then will affect the rate at which these TMLEs converge to ψ_0. Thus, the study of C-TMLE for continuous tuning parameters h creates an opportunity to potentially develop some asymptotic theory for C-TMLE demonstrating asymptotic superiority of the C-TMLE relative to a TMLE using the L_1-loss based cross-validation selector for h. If Q_n is consistent at a particular rate, then a TMLE $\Psi(Q_{n,h_{n,CV}}^*)$ using a best estimator $G_{n,h_{n,CV}}$ of G_0 might already be asymptotically efficient in which case C-TMLE cannot provide an *asymptotic* improvement relative to the standard TMLE. On the other hand, if Q_n is consistent at a low rate or possibly even inconsistent, then $G_{n,h_{CV}}$ might be overly biased so that $\Psi(Q_{n,h_{n,CV}}^*)$ might not even be root-n consistent. In the latter case, the key question is if the C-TMLE is able to select an undersmoothed choice h_n so that the bias of the C-TMLE is of smaller order and hopefully, if possible, it would select a choice h_n so that the bias $b_n(h_n) = o(n^{-1/2})$.

Let's aim to understand this better. Consider the case that the second-order remainder has a double robust structure. Suppose that Q_n happens to be an inconsistent estimator of Q_0. Due to the double robustness structure of the second-order remainder, $R_{20}(Q, G_0, Q_0, G_0) = 0$ for all Q so that a TMLE Q_{n,G_0}^* using the true G_0 in the targeting step would still result in a consistent and asymptotically linear estimator of ψ_0 under weak conditions. However, if one uses an estimator $G_{n,h_{n,CV}}$ whose bias w.r.t. G_0 converges to zero at a slower rate than $n^{-1/2}$, then the TMLE $\Psi(Q_{n,h_{n,CV}}^*)$ will also have a similar order bias so that this TMLE will not even be root-n consistent, and thus also not be asymptotically linear. On the other hand, a data-adaptive selector h_n that aims to minimize MSE of $h \to \Psi(Q_{n,h}^*)$ w.r.t. ψ_0 would try to select an estimator $G_{n,h}$ that has small bias. So in this scenario, assuming the family $\{G_{n,h} : h\}$ of candidate estimators includes such relatively unbiased estimators, a good selector h_n might still result in an asymptotically linear estimator $\Psi(Q_{n,h_n}^*)$. In our lasso example we would expect that the cross-validation selector $h_{n,CV}$ would result in a lasso fit that includes fewer basis functions than the fit implied by a C-TMLE selector h_n. In this paper we present such C-TMLE type selectors that are theoretically superior to the cross-validation selector $h_{n,CV}$.

10.1.2 Key Theoretical Property and Rational for Proposed C-TMLE That Drives Its Asymptotic Superiority Relative to Standard TMLE

Recall that a TMLE solves $P_n D^*(Q_{n,h}^*, G_{n,h}) = 0$, which is the basis for its asymptotic efficiency when both $Q_{n,h}$ and $G_{n,h}$ converge tot their true counterparts at a fast enough rate. Additional theoretical properties for a TMLE are obtained by making it solve additional key estimating equations that drive certain theoretical properties. The key additional equation solved by our proposed C-TMLE Q^*_{n,h_n} is given by

$$0 = P_n \frac{d}{dh} D^*(Q^*_{n,h_n}, G_{n,h})\Big|_{h=h_n}, \tag{10.3}$$

where we really only need that

$$h_n P_n \frac{d}{dh} D^*(Q^*_{n,h_n}, G_{n,h})\Big|_{h=h_n} = o_P(n^{-1/2}). \tag{10.4}$$

Note that the derivative is only w.r.t. h in $G_{n,h}$, not w.r.t. h in $Q_{n,h}$. Let

$$D^+(Q, G_{n,h}) = \frac{d}{dh} D^*(Q, G_{n,h}).$$

Thus, h_n is chosen so that

$$P_n D^+(Q^*_{n,h_n}, G_{n,h_n}) = 0 \text{ or } h_n P_n D^+(Q^*_{n,h_n}, G_{n,h_n}) = o_P(n^{-1/2}). \tag{10.5}$$

Let's now try to understand the rational of solving this equation. Note that

$$P_0 D^*(Q, G_{h+\delta}) = \Psi(Q_0) - \Psi(Q) + R_{20}(Q, G_{h+\delta}, Q_0, G_0)$$
$$P_0 D^*(Q, G_h) = \Psi(Q_0) - \Psi(Q) + R_{20}(Q, G_h, Q_0, G_0)$$
$$P_0 D^*(Q, G_{h+\delta}) - P_0 D^*(Q, G_h) = R_{20}(Q, G_{h+\delta}, Q_0, G_0) - R_{20}(Q, G_h, Q_0, G_0).$$

Let $h_{0,n}$ be a solution of $P_0 D^+(Q^*_{n,h}, G_{n,h}) = 0$. Then, it follows that this oracle choice h_{0n} solves

$$\lim_{\delta \to 0} \frac{R_{20}(Q^*_n, G_{n,h+\delta}, Q_0, G_0) - R_{20}(Q^*_n, G_{n,h}, Q_0, G_0)}{\delta} = 0$$

at the TMLE $Q^*_n = Q^*_{n,h_n}$ itself. Thus, this oracle choice h_{0n} corresponds with locally minimizing

$$h \to R_{20}(Q^*_n, G_{n,h}, Q_0, G_0).$$

Now note that h_n is the empirical analogue of the oracle choice h_{0n} by simply replacing $P_0 D^+(Q^*_{n,h}, G_{n,h})$ by its empirical counterpart $P_n D^+(Q^*_{n,h}, G_{n,h})$. This demonstrates that our C-TMLE choice h_n is indeed highly targeted by aiming to reduce the second-order remainder of the resulting TMLE of ψ_0. Our formal Theorem 10.1 below actually proves that indeed h_n succeeds in achieving this goal relative to $h_{n,CV}$.

Example 10.4. In our running example Eq. (10.3) reduces to h_n solving

$$0 = \frac{1}{n} \sum_{i=1}^n \frac{A_i \frac{d}{dh_n} \bar{G}_{n,h_n}}{\bar{G}^2_{n,h_n}(W_i)} (Y_i - \bar{Q}^*_{n,h_n}(W_i)). \tag{10.6}$$

10.1.3 Implicitly Defined Tuning Parameter

Suppose that we are given a discrete collection of candidate estimators $\hat{G}_\lambda$, possibly indexed by a multivariate tuning parameter $\lambda \in \mathcal{S}$. How could we apply the C-TMLE approach? We could order these candidate estimators by the value of $P_n L_1(\hat{G}_\lambda)$, which creates an ordered sequence of estimators. Let $\mathcal{H} = \{P_n L_1(\hat{G}_\lambda) : \lambda \in \mathcal{S}\}$ be the set of empirical risk values. Assume that the set of candidate estimators densely spans an interval of empirical risk values, so that for all practical purposes we can treat $\mathcal{H}$ as an interval on the real line. For any given $\lambda \in \mathcal{S}$, we define $h(\lambda) = P_n L_1(\hat{G}_\lambda)$, which defines a 1-1 function $h : \mathcal{S} \rightarrow \mathcal{H}$. For any value of $h \in \mathcal{H}$, we can define $\lambda(h)$ as the inverse of $\lambda \rightarrow h(\lambda)$: i.e., for a given h, we select the λ so that $P_n L_1(\hat{G}_\lambda) = h$. This now defines a collection of candidate estimators $\{\hat{G}_{1h} = \hat{G}_{\lambda(h)} : h \in \mathcal{H}\}$ ordered by its value $P_n L_1(\hat{G}_{1h})$. Finally, we can scale h to be in an interval $[0, 1]$. We could now apply our proposed methodology to this sequence of candidate estimators, resulting in a selector h_n of h, and thereby a selector $\lambda_n = \lambda(h_n)$. The analytic derivative w.r.t. h in $D^+(Q, G_h)$ can be approximated with a numerical derivative, so that there is no need to have an analytic expression for $h \rightarrow G_{n,h}$.

10.2 A General C-TMLE Algorithm

The goal of an C-TMLE algorithm is to construct an ordered sequence of TMLEs $(G_{n,h}, Q^*_{n,h})$ so that both $P_n L_1(G_{n,h})$ and $P_n L(Q^*_{n,h})$ are increasing in h: i.e., we want the empirical fits of both estimators to be increasing as h approximates zero. One then uses $L(Q)$-cross-validation to select h. Given an initial estimator Q_n, the ordered sequence $\{G_{n,h} : h\}$ for which $P_n L_1(G_{n,h}$ is decreasing as $h \rightarrow 0$, just defining $Q^*_{n,h}$ as the TMLE using Q_n as initial estimator and $G_{n,h}$ in its targeting step does not guarantee that $P_n L(Q^*_{n,h})$ is decreasing in h as $h \rightarrow 0$. Therefore, a C-TMLE algorithm also has to build a corresponding sequence of initial estimators $Q_{n,h}$ used by the TMLE $Q^*_{n,h}$ so that the desired increase in empirical fit holds. We refer to van der Laan and Gruber (2010) for a general C-TMLE template that provides a recipe for constructing C-TMLE algorithms. Their general template also includes simultaneously building the sequence of estimators $G_{n,h}$. In our setting this sequence is already given, making our setting a special case of the general template in van der Laan and Gruber (2010). Our algorithm below involves a minor modification by replacing the cross-validation selector $\bar{h}_n$ of h by a choice $\tilde{h}_{\bar{h}_n}(P_n)$ in its neighborhood that corresponds with an locally optimal choice, thereby guaranteeing that our C-TMLE solves a desired score equation (10.8) that provides the basis for its asymptotics.

General Algorithm. Select $h^1_n = \arg\min_h P_n L(Q^*_{n,h})$. For any $h > h^1_n$, we define $Q_{n,h} = Q_n$ and $Q^*_{n,h}$ as the TMLE using Q_n as initial estimator and $G_{n,h}$ in its targeting step. Notice that we expect that $P_n L(Q^*_{n,h})$ is indeed decreasing from $h = 1$

to $h = h_n^1$. For any $h \in [h_n^1, 1]$, we also define $\tilde{h}_h(P_n) = h_n^1$. We now update the initial estimator to $Q_n^1 = Q_{n,h_n^1}^*$, and define, for any $h < h_n^1$, $Q_{n,h}^*$ as the TMLE that uses Q_n^1 as initial estimator and $G_{n,h}$ in its targeting step. We then define $h_n^2 = \arg\min_{h<h_n^1} P_n L(Q_{n,h}^*)$. For any $h \in (h_n^2, h_n^1)$, we define $Q_{n,h} = Q_n^1$, $Q_{n,h}^*$ as this TMLE that uses Q_n^1 as initial estimator and $G_{n,h}$ in its targeting step, and $\tilde{h}_h(P_n) = h_n^2$. We have now defined an ordered sequence of estimators $(G_{n,h}, Q_{n,h}, Q_{n,h}^*)$ for $h \in [h_n^2, 1]$ for which $P_n L(Q_{n,h}^*)$ is mostly increasing as h decreases in value, and we have a corresponding $\tilde{h}_h(P_n)$ that maps any $h \in [h_n^2, 1]$ into the next smaller h that corresponds with a minimizer of the risk of $Q_{n,h}^*$. This process is iterated untill we end up at the last value $h = 0$. This results in a complete sequence $(G_{n,h}, Q_{n,h}, Q_{n,h}^*)$ and corresponding $\tilde{h}_h(P_n)$, $h \in [0, 1]$, for which both $P_n L_1(G_{n,h})$ and $P_n L(Q_{n,h}^*)$ are decreasing as h approximates zero.

We now note that this description of the algorithm defines for each $h \in [0, 1]$ a mapping $(\tilde{h}_h, \hat{G}_h, \hat{Q}_h, \hat{Q}_h^*)$ from data P_n into a tuning parameter $\tilde{h}_h(P_n)$, an estimate $G_{n,h}$ of the nuisance parameter G_0, initial estimate $Q_{n,h}$ of Q_0, and a TMLE $Q_{n,h}^*$ defined by $(Q_{n,h}, G_{n,h})$. In particular, it defines a collection of candidate estimator $\hat{Q}_h$ indexed by h. We select h with the cross-validation selector $\bar{h}_n$:

$$\bar{h}_n = \arg\min_h E_{B_n} P_n L(\hat{Q}_h^*(P_{n,B_n}^0)). \tag{10.7}$$

In the typical C-TMLE algorithm we would select $(G_{n,\bar{h}_n}, Q_{n,\bar{h}_n}, Q_{n,\bar{h}_n}^*)$ and thus use $\Psi(Q_{n,\bar{h}_n}^*)$ as our TMLE. However, we want to guarantee that our selector of h solves the following critical score equation

$$\frac{d}{dh} P_n L(Q_{n,G_{n,h}}^*) = 0, \tag{10.8}$$

where the initial $Q = Q_{n,h}$ is not viewed as a function of h in the derivative. For example, if we use a one-step TMLE, then this writes as

$$\frac{d}{dh} P_n L(Q_{n,\epsilon_n(h),G_{n,h}}) = 0.$$

Note that this evaluates the change in empirical risk of the TMLE at a fixed initial estimator due to a change in $G_{n,h}$ in the targeting step. In order to solve (10.8) we replace $\bar{h}_n$ by the actual minimizer $h_n < \bar{h}_n$ in the C-TMLE algorithm that comes right before $\bar{h}_n$:

$$h_n \equiv \tilde{h}_{\bar{h}_n}(P_n). \tag{10.9}$$

Since h_n minimizes $h \to P_n L(Q_{n,G_{n,h}}^*)$ over an interval of h-values, assuming an interior minimum, this choice h_n indeed solves (10.8). Our proposed C-TMLE is now defined by $(h_n, G_n = G_{n,h_n}, Q_n = Q_{n,h_n}, Q_n^* = Q_{n,h_n}^*)$ resulting in the C-TMLE $\Psi(Q_n^*)$ of ψ_0.

10.3 Verifying That C-TMLE Solves Critical Equation (10.4)

In the next subsection we show that the score equation solved by C-TMLE implies the desired critical equation (10.4) if $\epsilon_n(h_n)h_n = o_P(n^{-1/2})$, where we consider the case that the TMLE is a one-step TMLE. Since the size $\epsilon_n(h)$ behaves as the rate of convergence of the initial estimator $Q_{n,h}$ (and thus of the original initial estimator Q_n in the C-TMLE algorithm), and h_n converges to zero at the same rate as the bias of G_{n,h_n}, this condition corresponds with assuming that the product of the rates of convergence of Q^*_{n,h_n} and the bias of G_{n,h_n} is smaller than $n^{-1/2}$. Note that we expect h_n to undersmooth and thus be of smaller order than $h_{n,CV}$. Suppose that the product of the rates of convergence of $G_{n,h_{n,CV}}$ and Q_n is not smaller than $n^{-1/2}$ so that the second-order remainder $R_{20}(Q^*_n, G_{n,h_{n,CV}}, Q_0, G_0)$ for the standard TMLE based on $(Q_n, G_{n,h_{n,CV}})$ is not asymptotically linear. This implies that the TMLE using $G_{n,h_{n,CV}}$ is not asymptotically linear either. Nonetheless, since h_n is smaller than $h_{n,CV}$ we can still have that $\epsilon_n(h_n)h_n = o_P(n^{-1/2})$. In fact, consider the extreme case that Q_n is inconsistent. In that case, one needs $h_n = o(n^{-1/2})$ in order to guarantee asymptotic linearity of the TMLE based on Q_n and G_{n,h_n}. By our Theorem 10.1, $h_n = o(n^{-1/2})$ if there exists a rate $h_{n,1}$ that undersmooths enough so that the bias of $G_{n,h_{n,1}}$ is $o(n^{-1/2})$. In that case, $\epsilon_n(h_n)h_n = o_P(n^{-1/2})$, even though $\epsilon_n(h_n)$ does not even converge to zero. In the second subsection we show that by modifying the least favorable submodel in the definition of the TMLE, we can arrange that our C-TMLE solves (10.3) exactly. In fact, in our second subsection we also show that the corresponding standard TMLE will also solve (10.3) exactly.

10.3.1 Condition for C-TMLE Solving Critical Equation (10.4)

Consider the case that the TMLE $Q^*_{n,h}$ using $Q_n = Q_{n,h}$ as initial estimator and $G_{n,h}$ in its targeting step is given by the first-step TMLE $Q_{n,\epsilon_n(h),G_{n,h}}$. For notational convenience, in this subsection we denote the initial estimator with Q_n, suppressing its dependence on h, since in the following derivatives w.r.t. h treat $Q_{n,h}$ as fixed. Let h_n be the solution of the Eq. (10.8) solved by our C-TMLE selector. Note that

$$\frac{d}{dh}P_nL(Q_{n,\epsilon_n(h),G_{n,h}}) = \frac{d}{dh}P_nL(Q_{n,\epsilon_n(h),G})\Big|_{G=G_{n,h}} + \frac{d}{dh}P_nL(Q_{n,\epsilon,G_{n,h}})\Big|_{\epsilon=\epsilon_n(h)},$$

where the initial estimator is considered fixed in h. Now notice that the first term on the right-hand side equals

$$\frac{d}{d\epsilon_n(h)}P_nL(Q_{n,\epsilon_n(h),G})\Big|_{G=G_{n,h}}$$

up to a scalar $\frac{d}{dh}\epsilon_n(h)$. But the latter equation is the score equation for the MLE $\epsilon_n(h)$ and thus equals zero. Thus, we can conclude that h_n also solves

$$0 = \frac{d}{dh_n} P_n L(Q_{n,\epsilon,G_{n,h_n}}) \bigg|_{\epsilon=\epsilon_n(h_n)}. \tag{10.10}$$

We will now show that (10.10) implies that $P_n \frac{d}{dh_n} D^*(Q_n, G_{n,h_n}) = O(\epsilon_n(h_n))$, which would establish (10.4) if $\epsilon_n(h_n) h_n = o_P(n^{-1/2})$ (e.g., if the initial estimator Q_n and G_{n,h_n} converge to Q_0 and G_0, respectively, at a rate faster than $n^{-1/4}$).

Example 10.5. In our example, we have that (10.10) reduces to

$$0 = \epsilon_n(h_n) \frac{1}{n} \sum_{i=1}^{n} \frac{d}{dh} C(\bar{G}_{n,h}) \bigg|_{h=h_n} (Y_i - \bar{Q}_{n,\epsilon_n(h_n),G_{n,h_n}}).$$

Thus, this implies that

$$0 = \frac{1}{n} \sum_{i=1}^{n} \frac{d}{dh} C(\bar{G}_{n,h_n}) \bigg|_{h=h_n} (Y_i - \bar{Q}_{n,\epsilon_n(h_n),G_{n,h_n}}),$$

which equals (10.6). So we conclude that in our running example the score equation (10.8) solved by the C-TMLE selector h_n corresponds exactly with solving (10.3). We suggest that this exact equivalence between (10.8) and (10.3) holds more generally for universal least favorable submodels. As we will see below, in general, in our proof below we only obtain that (10.8) implies (10.3) up to an error $O(\epsilon_n(h))$, which provides a basis for (10.4).

Suppose that $L(Q_{n,\epsilon,G_{n,h}}) = f(Q_n, \epsilon C(G_{n,h}, Q_n))$ for some functional $(Q, H) \to f(Q, H)$. In other words, assume that the fluctuation $Q_{n,\epsilon,G_{n,h}}$ of Q_n involves augmenting the off-set Q_n with an ϵ-extension $\epsilon C(G_{n,h}, Q_n)$, thereby linking ϵ and $G_{n,h}$ into one term. Let $d_2 f(Q, H)(r) = \frac{d}{d\delta} f(Q, H + \delta r)\big|_{\delta=0}$ be the directional derivative of $H \to f(Q, H)$ at (Q, H) in the direction r. Then,

$$\frac{d}{d\epsilon} L(Q_{n,\epsilon,G_{n,h}}) \bigg|_{\epsilon=0} = d_2 f(Q_n, \epsilon C(G_{n,h}, Q_n))\big|_{\epsilon=0} (C(G_{n,h}, Q_n))$$

$$\frac{d}{dh} L(Q_{n,\epsilon,G_{n,h}}) = d_2 f(Q_n, \epsilon C(G_{n,h}, Q_n)) \left(\epsilon \frac{d}{dh} C(G_{n,h}, Q_n)\right).$$

The first equation shows that we can represent $D^*(Q, G_{n,h}) = \frac{d}{d\epsilon} L(Q_{n,\epsilon,G_{n,h}})\big|_{\epsilon=0}$ as follows:

$$D^*(Q_n, G_{n,h}) = d_2 f(Q_n, \epsilon C(G_{n,h}, Q_n))\big|_{\epsilon=0} (C(G_{n,h}, Q_n)).$$

This representation shows

$$\frac{d}{dh} D^*(Q_n, G_{n,h}) = d_2 f(Q_n, \epsilon C(G_{n,h}, Q_n))\big|_{\epsilon=0} \left(\frac{d}{dh} C(G_{n,h}, Q_n)\right).$$

The second equation shows that

$$
\frac{\frac{d}{dh}L(Q_{n,\epsilon,G_{n,h}})}{\epsilon} = d_2 f(Q_n, \epsilon C(G_{n,h}, Q_n)) \left(\frac{d}{dh} C(G_{n,h}, Q_n) \right)
$$

$$
= d_2 f(Q_n, \epsilon C(G_{n,h}, Q_n))\big|_{\epsilon=0} \left(\frac{d}{dh} C(G_{n,h}, Q_n) \right) + O(\epsilon)
$$

$$
= \frac{d}{dh} D^*(Q_n, G_{n,h}) + O(\epsilon).
$$

Thus, this shows that

$$
\frac{d}{dh} L(Q_{n,\epsilon,G_{n,h}}) = \epsilon \frac{d}{dh} D^*(Q_n, G_{n,h}) + O(\epsilon^2).
$$

This proves that if $P_n \frac{d}{dh_n} L(Q_{n,\epsilon,G_{n,h}})\big|_{\epsilon=\epsilon_n(h)} = 0$ (i.e., Eq. (10.10) is solved), then

$$
0 = \epsilon_n(h) P_n \frac{d}{dh_n} D^*(Q_n, G_{n,h_n}) + O(\epsilon_n(h)^2),
$$

which implies

$$
P_n \frac{d}{dh_n} D^*(Q_n, G_{n,h_n}) = O(\epsilon_n(h)).
$$

We state this result as a formal lemma.

Lemma 10.1. *Let h_n be the solution of the Eq. (10.8) solved by our C-TMLE selector. Then, h_n solves*

$$
0 = \frac{d}{dh_n} P_n L(Q_{n,\epsilon,G_{n,h_n}}) \bigg|_{\epsilon=\epsilon_n(h_n)}.
$$

Suppose that $L(Q_{n,\epsilon,G_{n,h}}) = f(Q_n, \epsilon C(G_{n,h}, Q_n))$ for some functional $(Q, H) \to f(Q, H)$. Assume that $H \to f(Q, H)$ is differentiable at $H = \epsilon C(G_{n,h}, Q_n)$ with derivative $d_2 f(Q_n, H)$ and that this derivative is continuous at direction $\frac{d}{dh} C(G_{n,h}, Q_n)$ in the following sense:

$$
d_2 f(Q_n, \epsilon C(G_{n,h}, Q_n)) \left(\frac{d}{dh} C(G_{n,h}, Q_n) \right) = d_2 f(Q_n, \epsilon C(G_{n,h}, Q_n))\big|_{\epsilon=0} \left(\frac{d}{dh} C(G_{n,h}, Q_n) \right)
$$

$$
+ O(\epsilon).
$$

Then,

$$
P_n \frac{d}{dh_n} D^*(Q_n, G_{n,h_n}) = O(\epsilon_n(h)).
$$

10.3.2 A TMLE and C-TMLE that Solve Equation (10.3) Exactly

In our typical applications we have that even at a G_1 different from $G = G(P)$ $D^*(Q, G_1)$ is an element of the tangent space $T_Q(P)$ of Q at P. In other words, $D^*(Q, G_1)$ represents a score at $\delta = 0$ of a fluctuation model $\{Q_\delta : \delta\} \subset \mathcal{M}$ through Q at $\delta = 0$. In that case, $D^*(Q, G_{h+\delta}) - D^*(Q, G_h)$ is in the tangent space of Q at P, so that also $\frac{d}{dh} D^*(Q, G_h) \in T_Q(P)$. Thus, $D^*(Q, G_h)$ and $D^+(Q, G_h) = d/dh D^*(Q, G_h)$ are both scores of Q at P so that there exists a local least favorable submodel $Q_{\epsilon, h}$ whose linear span of $\frac{d}{d\epsilon} L(Q_{\epsilon, h})$ at $\epsilon = 0$ includes both $D^*(Q, G_h)$ and $D^+(Q, G_h)$. Using this local least favorable submodel now defines a TMLE $Q^*_{n, h}$ that solves both equations $P_n D^*(Q^*_h, G_h) = P_n D^+(Q^*_h, G_h) = 0$ for all h. We can now apply our general C-TMLE algorithm above with this definition of the TMLE $Q^*_{n, h}$. In this case, the C-TMLE selector h_n (i.e., $(G_{n, h_n}, Q^*_{n, h_n})$) will solve equation (10.3) exactly. In fact, even if do not use the C-TMLE algorithm but just use the standard TMLE $Q^*_{n, h_{n, CV}}$ based on initial estimator $(Q_n, G_{n, h_{n, CV}})$ Eq. (10.3) is solved exactly. As a consequence, our asymptotics theorem below is applicable to both the proposed C-TMLE as well as to the standard TMLE targeting both equations and using $h_{n, CV}$. Nonetheless, we expect that the C-TMLE still has a finite sample advantage.

Example 10.6. Consider the C-TMLE algorithm, but let $\bar{Q}^*_{n, h}$ be the one-step TMLE based on $(Q_{n, h}, G_{n, h})$ that uses a two dimensional clever covariate $(C(\bar{G}_{n, h}), \frac{d}{dh} C(\bar{G}_{n, h}))$. Let

$$D_1^+(Q, G_{n, h}) = \frac{d}{dh} D_1^*(Q, G_{n, h}) = \frac{d}{dh} C(G_{n, h})(A, W)(Y - \bar{Q}(A, W)).$$

In that case, we have that for each h $(Q^*_{n, h}, G_{n, h})$ solves

$$P_n D_1^*(Q^*_{n, h}, G_{n, h}) = 0$$
$$P_n D_1^+(Q^*_{n, h}, G_{n, h}) = 0.$$

It also solves

$$P_n D^*(Q^*_{n, h}, G_{n, h}) = 0$$
$$P_n D^+(Q^*_{n, h}, G_{n, h}) = 0,$$

where $D^+(Q, G_{n, h}) = \frac{d}{dh} D^*(Q, G_{n, h})$. By using this definition of TMLE in our C-TMLE algorithm we guarantee that the critical equation $P_n D^+(Q^*_{n, h}, G_{n, h}) = 0$ for all h, not only for our C-TMLE selector $h = h_n$ defined by the C-TMLE algorithm. As a result, we could now also replace the selector h_n by $h = h_{n, CV}$ in our description of the C-TMLE algorithm. In fact, we can also simply use the TMLE based on initial estimator Q_n and $G_{n, h_{n, CV}}$ (but using this two-dimensional least favorable submodel).

10.4 General Theorem for C-TMLE Asymptotic Linearity

We have the following theorem which proves that if there exists a selector h_{1n} for which the C-TMLE is asymptotically linear, then our proposed C-TMLE using h_n will be asymptotically linear.

Theorem 10.1. *Let $\Psi : \mathcal{M} \to \mathbf{R}^d$ be pathwise differentiable at any $P \in \mathcal{M}$ with efficient influence curve $D^*(P) = D^*(Q(P), G(P))$ and $\Psi(P)$ only depends on P through $Q(P)$: abusing notation, we will also denote $\Psi(P)$ with $\Psi(Q)$. Suppose that*

$$\Psi(Q) - \Psi(Q_0) = -P_0 D^*(Q, G) + R_{20}(Q, G, Q_0, G_0)$$

for a remainder $R_{20}()$ that has a DR-structure so that $R_{20}(Q, G_0, Q_0, G_0) = 0$ for all $Q \in Q(\mathcal{M})$. Let $(G_{n,h} : h)$ be a family of candidate estimators of G_0 indexed by scalar h, and let $(Q^*_{n,h} : h)$ be a family of TMLEs using $G_{n,h}$ in the targeting step, so that $P_n D^*(Q^*_{n,h}, G_{n,h}) = 0$ for all h. Let h_n be a given selector, and let $Q^*_n = Q^*_{n,h_n}$ be the corresponding TMLE. Let (Q^*, G_0) be the limit of (Q_{n,h_n}, G_{n,h_n}) so that $P_0\{D^*(Q^*_{n,h_n}, G_{n,h_n}) - D^*(Q^*, G_0)\}^2 \to_p 0$ as $n \to \infty$.
We make the following assumptions:

Existence of Desired Selector: *We assume that for a fixed $Q^* \in \{Q(P) : P \in \mathcal{M}\}$ there exists a sequence h_{1n} that converges to zero and satisfies*

$$P_0 D^*(Q^*, G_{n,h_{1n}}) - P_0 D^*(Q^*, G_0) = (P_n - P_0)D_1(P_0) + o_P(n^{-1/2}). \quad (10.11)$$

Selector Solves Critical Equation: *For such a selector h_{1n}, we have*

$$\left\{\frac{d}{dh_n} P_n D^*(Q^*_n, G_{n,h_n})\right\} * (h_n - h_{1n}) = o_P(n^{-1/2}), \text{ where } Q^*_n = Q^*_{n,h_n}. \quad (10.12)$$

Negligible Second-Order Remainders:

$$P_0 D^*(Q^*, G_{n,h_{1n}}) - P_0 D^*(Q^*, G_{n,h_n}) - \left\{\frac{d}{dh_n} P_0 D^*(Q^*, G_{n,h_n})\right\} * (h_{1n} - h_n) = o_P(n^{-1/2})$$

$$(P_n - P_0)\{D^*(Q^*_{n,h_n}, G_{n,h_n}) - D^*(Q^*, G_0)\} = o_P(n^{-1/2})$$

$$\left\{(P_n - P_0)\frac{d}{dh_n} D^*(Q^*, G_{n,h_n})\right\} * (h_{1n} - h_n) = o_P(n^{-1/2})$$

$$P_n \left\{\frac{d}{dh_n}\{D^*(Q^*_n, G_{n,h_n}) - D^*(Q^*, G_{n,h_n})\}\right\} * (h_{1n} - h_n) = o_P(n^{-1/2})$$

$$\{P_0 D^*(Q^*_n, G_{n,h_n}) - P_0 D^*(Q^*_n, G_0)\} - \{P_0 D^*(Q^*, G_{n,h_n}) - P_0 D^*(Q^*, G_0)\} = o_P(n^{-1/2})$$

Then, $\Psi(Q^*_n)$ is an asymptotically linear estimator of $\Psi(Q_0)$ at $P_0 \in \mathcal{M}$ with influence curve $D^*(Q, G_0) + D_1(P_0)$:

$$\Psi(Q^*_n) - \Psi(Q_0) = (P_n - P_0)\{D^*(Q^*, G_0) + D_1(P_0)\} + o_P(n^{-1/2}).$$

Discussion of Assumptions of Theorem 10.1. We make the following remarks regarding verification of the assumptions. The first assumption (10.11) assumes that there exists a sequence h_{1n} which undersmoothes enough so that the smooth functional $\Phi_{Q^*}(G_n)$ is an asymptotically linear estimator of $\Phi_{Q^*}(G_0)$, where $\Phi_{Q^*}(G) = P_0 D^*(Q^*, G)$. At misspecified $Q^* \neq Q_0$, this might not be possible, and it will be of interest to understand if we can generalize our proof below to show that, nonetheless, $\Psi(Q_n^*)$ has better second-order term behavior than a standard TMLE not solving the critical equation (10.12). The second assumption (10.12) holds if $P_n D^+(Q_n^*, G_{n,h_n}) = 0$, but as discussed we do not need an exact solution. Our claim is that this assumption comes down to assuming $\left\{\frac{d}{dh_n} P_n D^*(Q_n^*, G_{n,h_n})\right\} * h_n = o_P(n^{-1/2})$, since h_{1n} can be chosen as the fastest rate to zero for which we still have asymptotic linearity (10.11). The first of the "second-order remainder"-assumption corresponds with assuming $(h_{1n} - h_n)^2 = o_P(n^{-1/2})$ if $h \to P_0 D^*(Q^*, G_{n,h})$ is twice continuously differentiable. This will hold if h_{1n}^2 and h_n^2 are both $o_P(n^{-1/2})$. Thus, using our bias-interpretation of h_n, the bias of G_{n,h_n} has to go to zero at a faster rate than $n^{-1/4}$. Suppose that $D^*(Q_n^*, G_{n,h_n})$ falls in a P_0-Donsker class with probability tendon to 1; $D^+(Q_n^*, G_{n,h_n})$ falls in a P_0-Donsker class with probability tending to 1; $P_0\{D^*(Q_n^*, G_{n,h_n}) - D^*(Q^*, G_0)\}^2 \to_p 0$; $P_0\{D^+(Q_n^*, G_{n,h_n}) - D^*(Q^*, G_0)\}^2 \to_p 0$. Then, by empirical process theory, and $\max(h_{1n}, h_n) \to_p 0$, it follows that the second, third and fourth "second-order remainder"-assumption hold. Thus, these three second-order remainder assumption only rely on the consistency of (Q_n^*, G_{n,h_n}) w.r.t (Q^*, G_0) and a Donker-class condition. Finally, the remainder in the fifth "second-order remainder" assumption can generally be represented as $\int (H_1(Q_n^*) - H_1(Q^*))(H_2(G_{n,h_n}) - H_2(G_0)) f(Q_n^*, G_n, Q^*, G_0) dP_0$ for certain functionals H_1, H_2 and f. In that case, assuming away singularities (i.e., assuming strong positivity), this assumption would hold if $\| H_1(Q_n^*) - H_1(Q^*) \|_{P_0} \| H_2 \|_{P_0} = o_P(n^{-1/2})$.

Proof of Theorem 10.1. We have

$$0 = (P_n - P_0)D^*(Q_{n,h_n}^*, G_{n,h_n}) + P_0 D^*(Q_{n,h_n}^*, G_{n,h_n}).$$

By our second-order assumption $(P_n - P_0)D^*(Q_{n,h_n}^*, G_{n,h_n}) = (P_n - P_0)D^*(Q^*, G_0) + o_P(n^{-1/2})$. Let $Q_n^* = Q_{n,h_n}^*$. Using that

$$\Psi(Q_n^*) - \Psi(Q_0) = -P_0 D^*(Q_n^*, G_0) + R_{20}(Q_n^*, G_0, Q_0, G_0)$$

and, by assumption, $R_{20}(Q_n^*, G_0, Q_0, G_0) = 0$, it follows

$$\begin{aligned}
P_0 D^*(Q_n^*, G_{n,h_n}) &= P_0 D^*(Q_n^*, G_0) + \{P_0 D^*(Q_n^*, G_{n,h_n}) - P_0 D^*(Q_n^*, G_0)\} \\
&= \Psi(Q_0) - \Psi(Q_n^*) + P_0 D^*(Q_n^*, G_{n,h_n}) - P_0 D^*(Q_n^*, G_0).
\end{aligned}$$

So we have shown

$$\Psi(Q_n^*) - \Psi(Q_0) = (P_n - P_0)D^*(Q^*, G_0) + P_0 D^*(Q_n^*, G_{n,h_n}) - P_0 D^*(Q_n^*, G_0) + o_P(n^{-1/2}).$$

By our second-order assumption, we have

$$P_0 D^*(Q_n^*, G_{n,h_n}) - P_0 D^*(Q_n^*, G_0) = P_0 D^*(Q^*, G_{n,h_n}) - P_0 D^*(Q^*, G_0) + o_P(n^{-1/2}).$$

We now write

$$P_0 D^*(Q^*, G_{n,h_n}) - P_0 D^*(Q^*, G_0) = P_0 D^*(Q^*, G_{n,h_n}) - P_0 D^*(Q^*, G_{n,h_{1n}})$$
$$+ P_0 D^*(Q^*, G_{n,h_{1n}}) - P_0 D^*(Q^*, G_0)$$

By our "existence of desired selector" assumption, the second term equals $(P_n - P_0)D_1(P_0) + o_P(n^{-1/2})$. By our second-order assumption, we also have the following Taylor expansion at h_n:

$$P_0 D^*(Q^*, G_{n,h_n}) - P_0 D^*(Q^*, G_{n,h_{1n}}) = -\left\{\frac{d}{dh_n} P_0 D^*(Q^*, G_{n,h_n})\right\} * (h_{1n} - h_n) + o_P(n^{-1/2}).$$

Now, we write

$$-\left\{\frac{d}{dh_n} P_0 D^*(Q^*, G_{n,h_n})\right\} * (h_{1n} - h_n) = \left\{\frac{d}{dh_n}(P_n - P_0) D^*(Q^*, G_{n,h_n})\right\} * (h_{1n} - h_n)$$
$$- P_n\left\{\frac{d}{dh_n} D^*(Q^*, G_{n,h_n})\right\} * (h_{1n} - h_n).$$

By our second-order assumption, the first term on the right-hand side is $o_P(n^{-1/2})$. Regarding the second term, we write:

$$P_n\left\{\frac{d}{dh_n} D^*(Q^*, G_{n,h_n})\right\} * (h_{1n} - h_n) = P_n\left\{\frac{d}{dh_n} D^*(Q_n^*, G_{n,h_n})\right\} * (h_{1n} - h_n)$$
$$+ P_n\left\{\frac{d}{dh_n} D^*(Q^*, G_{n,h_n})\right\} * (h_{1n} - h_n) - P_n\left\{\frac{d}{dh_n} D^*(Q_n^*, G_{n,h_n})\right\} * (h_{1n} - h_n).$$

By our second-order assumption, the second term on the right-hand side is $o_P(n^{-1/2})$. By the critical equation assumption on the selector h_n, we have that the first term on the right-hand side is $o_P(n^{-1/2})$ as well. Thus, we have shown

$$\Psi(Q_n^*) - \Psi(Q_0) = (P_n - P_0)\{D^*(Q^*, G_0) + D_1(P_0)\} + o_P(n^{-1/2}),$$

which completes the proof of the theorem. □

10.5 Discussion

In van der Laan and Gruber (2010) we proposed a general template for constructing an iterative algorithm that builds an ordered sequence of TMLEs $(Q_{n,j}^*, G_{n,j}^*)$, $j = 1, \ldots, J$, so that the empirical fit of the relevant part $Q_{n,j}$ and the nuisance parameter $G_{n,j}$ is increasing in j, and using loss-based cross-validation to select a best estimator

Q^*_{n,j_n} of Q_0. In this chapter we assume that such an ordered set of estimators of G_0 is already provided, which represents a special case of this general C-TMLE template: the general template builds a next estimator $G_{n,j+1}$ from $G_{n,j}$ based on evaluating a set of moves, but in this special case, the next $G_{n,j+1}$ is already known (i.e., only one move).

However, we go beyond this past literature on C-TMLE by studying the case that the index parameter h is continuous valued. We focused on double robust estimation problems defined by the product structure of the second-order remainder $R_{20}(Q, G, Q_0, G_0)$, and assume that our family $\{G_{n,h} : h\}$ allows for consistent estimation of G_0. We described a C-TMLE algorithm following the general C-TMLE template with a minor modification to guarantee that the selected estimator G_{n,h_n} solves a score equation for the tuning parameter h. We demonstrates that solving this score equation implies that the TMLE (Q^*_{n,h_n}, G_{n,h_n}) solves a critically important score equation $P_n \frac{d}{dh_n} D^*(Q^*_n, G_{n,h_n}) \approx 0$ where this score is defined as the derivative w.r.t. h of the canonical gradient. We explained why this latter score equation corresponds with locally minimizing (in h) the second-order remainder $R_{20}(Q^*_n, G_{n,h}, Q_0, G_0)$ of he TMLE Taylor expansion.

Moreover, we proved a formal theorem that shows that solving this critical equation beyond the usual efficient influence curve equation $P_n D^*(Q^*_n, G_{n,h_n}) = 0$ guarantees that, if possible, the selector h_n undersmooths enough so that asymptotic linearity of the TMLE $\Psi(Q^*_n)$ is preserved at an inconsistent or slowly converging initial estimator of Q_0. On the other hand, a TMLE only solving $P_n D^*(Q^*_n, G_{n,h_{n,CV}}) = 0$ using a cross-validation selector $h_{n,CV}$ will generally fail to be asymptotically linear when Q^*_n is inconsistent or converges at a slow rate. We also show that by using a least favorable submodel with a two dimensional ϵ that generates both of these scores, we obtain a "special" TMLE that solves both score equations so that the same asymptotics apply to this standard TMLE even when using $h_{n,CV}$. Nonetheless, based on the finite sample rational of the C-TMLE, we expect that the C-TMLE will have a better finite sample performance than this special TMLE. Initial simulations not shown here demonstrate that the C-TMLE and special TMLE can easily outperform the standard TMLE and that the C-TMLE appears to be the best among the three estimators "standard TMLE", "special TMLE", and "C-TMLE".

Part III
Randomized Trials

Chapter 11
Targeted Estimation of Cumulative Vaccine Sieve Effects

David Benkeser, Marco Carone, and Peter Gilbert

Over the last century, effective vaccines have been developed for prevention of disease caused by many pathogens. However, effective vaccines have not yet been developed to prevent infection with the human immunodeficiency virus (HIV). A challenge in developing a vaccine to prevent HIV infection is the substantial heterogeneity in the genetic characteristics of the virus. Preventive HIV vaccines are typically constructed using only several antigens and may protect well against infection caused by virus strains similar to antigens in the vaccine, but fail to protect against disease caused by antigenically dissimilar strains. Therefore, when evaluating preventive HIV vaccines, it is important to study whether and how the efficacy of the vaccine varies with the virus' characteristics—this field of study is called sieve analysis (Gilbert et al. 1998, 2001). The vaccine can be thought of as a sieve, inducing a strain-specific immunity that presents a barrier to infection, while there also may be "holes in the sieve," that is, HIV strains that break through the vaccine barrier.

D. Benkeser (✉)
Department of Biostatistics and Bioinformatics, Rollins School of Public Health,
Emory University, 1518 Clifton Rd. NE, Atlanta, GA 30322, USA
e-mail: benkeser@emory.edu

M. Carone
Department of Biostatistics, University of Washington, F-644 Health Sciences Building,
1705 NE Pacific Street, Seattle, WA 98195, USA
e-mail: mcarone@uw.edu

P. Gilbert
Vaccine and Infectious Disease Division, Fred Hutchinson Cancer Research Center,
1100 Fairview Ave, Seattle, WA 98109, USA
e-mail: pgilbert@scharp.org

© Springer International Publishing AG 2018
M.J. van der Laan, S. Rose, *Targeted Learning in Data Science*,
Springer Series in Statistics, https://doi.org/10.1007/978-3-319-65304-4_11

A sieve effect at a given genetic locus is defined as the difference in vaccine efficacy when comparing viruses matched to the vaccine at this locus to viruses mismatched at this locus. Identification of sieve effects can help guide the selection of antigens that should be included in future, possibly multivalent, vaccines. Such multivalent vaccines may have higher overall efficacy by providing broader protection against genetically diverse viruses.

Statistically, sieve analysis is performed within a competing risks framework. In a setting with competing risks, study trial participants are at risk of experiencing several competing endpoints. In HIV vaccine trials, the various endpoints are defined by the genotype of the virus that causes the infection; each genotype represents a separate type of endpoint. To assess the effect of a treatment on the risk of an endpoint of a given type in competing risks settings, it is common to use either instantaneous or cumulative parameters. The choice of parameter depends on the scientific context (Pintilie 2007) and both have been used to assess vaccine sieve effects (Gilbert 2000). Instantaneous parameters are usually based on the cause-specific hazard function (Prentice et al. 1978; Benichou and Gail 1990; Gaynor et al. 1993; Lunn and McNeil 1995), defined as the instantaneous probability of experiencing an endpoint of a given type among those who have not yet experienced an endpoint. The most common cumulative parameter studied is cumulative incidence, defined as the probability that an event occurs by a given time and is of a particular type. While both parameters are relevant for assessing vaccine sieve effects, the cumulative parameter may be of greater public health relevance when waning vaccine effects are present.

The Aalen-Johansen estimator is commonly used to make inference on cumulative incidence in a sieve analysis (Aalen 1978). This estimator requires few assumptions to achieve several desirable properties. Provided censoring is uninformative, it is consistent. Additionally, if there are no measured prognostic covariates, it is also asymptotically nonparametric efficient. However, informative censoring is a common concern in any longitudinal study and prognostic covariates such as sexual risk behaviors are routinely collected in HIV vaccine trials. By utilizing these covariates, it is possible to weaken assumptions on the censoring mechanism and improve efficiency.

Semiparametric approaches have been devised to incorporate covariates into the analysis of competing risks data and have been applied in sieve analysis. These include proportional hazards regression for cause-specific hazards (Prentice et al. 1978; Lunn and McNeil 1995) or subdistribution hazards (Fine and Gray 1999). These hazard-based approaches can be used to compute estimates of cumulative incidence using known relationships between hazard and incidence, but are more commonly used to estimate hazard-based efficacy parameters. A drawback of these semiparametric approaches, whether for estimating instantaneous or cumulative parameters, is that they require the correct specification of a finite-dimensional regression model.

When this model is incorrect, the target parameter is generally difficult to interpret. For example, the estimand of a misspecified cause-specific Cox model is known to involve the censoring distribution (Struthers and Kalbfleisch 1986), and this is also true of misspecified subdistribution hazard models (Grambauer et al. 2010). The fact that the estimand involves the censoring distribution, typically considered to be a study-specific nuisance rather than a population characteristic of interest, is an undesirable property for assessing treatment efficacy (Stitelman and van der Laan 2011).

Targeted estimators of a marginal survival probability when only one type of endpoint is present were presented in Moore and van der Laan (2009a) and van der Laan and Gruber (2012). These works proposed and evaluated methods for covariate adjustment through machine learning-based estimators, such as the super learner. The estimators were shown to lead to gains in efficiency and robustness to informative censoring. In this chapter, we show how the TMLE developed in van der Laan and Gruber (2012) can be adapted to the setting of competing risks and estimation of cumulative incidence in the context of sieve analysis. We illustrate these methods using data from a recent Phase II preventive HIV vaccine efficacy trial.

11.1 Observed Data

We consider a preventive HIV vaccine efficacy trial that recruits n individuals and measures L_0, a potentially high-dimensional set of baseline characteristics, on each individual. Individuals are assigned, possibly based on L_0, to receive an active vaccine $A_0 = 1$ or control vaccine $A_0 = 0$. Trial participants are asked to attend $K + 1$ regularly scheduled clinic visits to be tested for HIV infection. We consider the case where there is a particular genetic locus of interest and for $k = 1, \ldots, K + 1$, define $L_k = (L_{k,1}, L_{k,2})$ to be a bivariate indicator of infection, where $L_{k,1} = 1$ if a participant is infected with a virus matched to the reference virus in the vaccine at the locus of interest at or before visit k and $L_{k,1} = 0$ otherwise. Similarly, $L_{k,2} = 1$ if a participant is infected with a virus mismatched to the reference virus in the vaccine at or before visit k and $L_{k,2} = 0$ otherwise. Over the course of follow-up some participants may withdraw consent or leave the study for other reasons. For $k = 1, \ldots, K$, we use A_k to denote whether a participant attended clinic visit $k + 1$. If a participant misses a clinic visit, they are considered to be right censored. Thus, the observed data can be represented as n independent copies of $O = (L_0, A_0, L_1, A_1, \ldots, A_K, L_{K+1}) \sim P_0$. We will make no assumptions about P_0, so our statistical model $\mathcal{M}$ is nonparametric.

We note that for $k = 1, \ldots, K$, we could allow L_k to contain time-varying participant characteristics; however, such characteristics are not included in our data analysis. We use $\bar{L}_k = (L_0, \ldots, L_k)$ and $\bar{A}_k = (A_0, \ldots, A_k)$ to denote the history of the time-dependent processes at a given time k. We also use the notation $\bar{0}_m$ and $\bar{1}_m$ to respectively denote zero and one vectors of length m.

11.2 Causal Model and Parameters of Interest

We now define our causal parameter of interest using a structural causal model. We assume that each component of the observed longitudinal data structure is a function of a set of observed parent variables and an unmeasured exogenous error term. The observed parents of L_k are assumed to be $\bar{L}_{k-1}$ and $\bar{A}_k$, while the parents of A_k are assumed to be $\bar{L}_k$ and $\bar{A}_{k-1}$. We can define a post-intervention distribution that represents the distribution the data would have under a specified intervention on $A_0, \ldots, A_K$ that sets these values to $a_0, \ldots, a_K$, respectively. We denote the true post-intervention distribution of a static intervention $\bar{a}$ with $P_0^{\bar{a}}$ and define $L^{\bar{a}} = (L_1^{\bar{a}}, \ldots, L_{K+1}^{\bar{a}})$ to be a counterfactual random variable with this distribution. We are interested in evaluating the mean counterfactual outcome $E_{P_0^{\bar{a}}}(L_{K+1}^{\bar{a}})$ under two interventions: the first assigns $A_0 = 1$ (i.e., active vaccine), the second assigns $A_0 = 0$ (i.e., control vaccine), and both subsequently assign $A_1 = \cdots = A_K = 1$ (i.e., individuals remain under observation for the duration of the study). These interventions may be seen as unnecessarily stringent, since whenever an infection occurs prior to time k, the participant's infection status at time k is known even if the individual was later lost to follow-up. The stochastic intervention on A_k that sets $A_k = 0$ and does not intervene otherwise may be more appropriate. However, it can be shown that both interventions lead to the same observed data parameter. The two static interventions of interest differ only in assignment of A_0, so we will use the shorthand $a = 1$ to refer to the intervention assigning treatment and no censoring and $a = 0$ to refer the intervention assigning control and no censoring.

We are interested in estimating the cumulative incidence of both matched and mismatched infections in the vaccine and placebo arm, which we define as $\psi_{0,j}^a = E_{P_0^a}(L_{K+1,j}^a)$ for $j = 1, 2$ and $a = 0, 1$. These quantities can be used to define a measure of genotype-specific cumulative vaccine efficacy for $j = 1, 2$ as $VE_j = 1 - \psi_{0,j}^1 / \psi_{0,j}^0$, interpreted as the multiplicative reduction in cumulative incidence of type j infections caused by the vaccine. Values of vaccine efficacy near one indicate a highly effective vaccine, small positive values indicate a moderately effective vaccine, and values less than zero indicate a harmful vaccine. We also define a vaccine sieve effect as $VSE = (\psi_{0,1}^0 / \psi_{0,1}^1)/(\psi_{0,2}^0 / \psi_{0,2}^1)$, that is, the ratio (matched vs. mismatched) of the causal cumulative risk ratios (placebo vs. vaccine). Note that the vaccine sieve effect will be greater than one if the efficacy is higher against matched infections and less than one if the efficacy is lower.

For simplicity of exposition, we focus on estimation of $\psi_{0,1}^1$ noting that our labeling of types and treatment arms is arbitrary so that the same methods can be used to estimate each of the four cumulative incidence quantities of interest. These estimates can then be combined to estimate vaccine efficacy and vaccine sieve effects. Furthermore, we also note that our choice of K is arbitrary so that our methods can be applied for pointwise estimation of cumulative incidence, vaccine efficacy, and vaccine sieve effects at any clinic visit.

11.3 Identification

The distribution of the counterfactual variable L^a can be identified using the observed data under the assumptions of sequential randomization and positivity. The identification result we present is based on the general results in Bang and Robins (2005). Beginning at the final time point $K + 1$, we define

$$\bar{Q}_{0,K+1}(\bar{\ell}_K) = \int \ell_{K+1,1} \, dQ_{0,L_{K+1}}(\bar{\ell}_{K+1}) \, ,$$

where $Q_{0,L_{K+1}}$ denotes the conditional distribution of L_{K+1} given $\bar{A}_K = \bar{1}_{K+1}$ and $\bar{L}_K$ implied by P_0. Given $\bar{Q}_{0,K+1}$, we define

$$\bar{Q}_{0,K}(\bar{\ell}_{K-1}) = \int \bar{Q}_{0,K+1}(\bar{\ell}_K) \, dQ_{0,L_K}(\bar{\ell}_K)$$

as the mean of $\bar{Q}_{0,K+1}$ with respect to $Q_{L_K,0}$, the conditional distribution of L_K given $\bar{A}_K = \bar{1}_K$ and $\bar{L}_{K-1}$ implied by P_0. We continue this process, where for $k = 1, \ldots, K - 1$ we define

$$\bar{Q}_{0,k}(\bar{\ell}_{k-1}) = \int \bar{Q}_{0,k+1}(\bar{\ell}_k) \, dQ_{0,L_k}(\bar{\ell}_k) \, .$$

Finally, we define

$$\bar{Q}_{0,0} = \int \bar{Q}_{0,1}(\ell_0) \, dQ_{0,L_0}(\ell_0) \, .$$

We use $\bar{Q} = \bar{Q}(P) = (\bar{Q}_k(P) : k)$ to denote the collection of iterated means and $Q = Q(P) = \{\bar{Q}(P), Q_{L_0}(P)\}$ to denote the set of iterated means along with the distribution of baseline covariates implied by $P \in \mathcal{M}$. Under the causal assumptions previously mentioned, we have that $\bar{Q}_{0,0}$ equals the counterfactual parameter of interest. Thus, we have established that $\psi_{0,1}^1 = \Psi(Q_0)$ under the specified assumptions, where $\Psi : \mathcal{M} \to (0, 1)$ is defined by the iterated mean construction above.

11.4 Efficient Influence Function

For $k = 1, \ldots, K + 1$, we define $G_{A_k} = G_{A_k}(P) = P(A_k = 1 \mid \bar{A}_{k-1} = \bar{1}_k, \bar{L}_k)$ and

$$H_k(G)(o) = \frac{I(\bar{a}_{k-1} = \bar{1}_k)}{\prod_{m=0}^{k-1} G_{A_m}(\bar{\ell}_k)} \, .$$

The efficient influence function of Ψ with respect to our model $\mathcal{M}$ at (Q, g) is

$$D^*(Q, G)(o) = \sum_{k=0}^{K+1} D_k^*(Q, G)(o),$$

where

$$D_{K+1}^*(Q, G)(o) = H_{K+1}(G)(o)\{\ell_{K+1,1} - \bar{Q}_{K+1}(\bar{\ell}_K)\}$$
$$D_k^*(Q, G)(o) = H_k(G)(o)\{\bar{Q}_{k+1}(\bar{\ell}_k) - \bar{Q}_k(\bar{\ell}_{k-1})\} \text{ for } k = 1, \ldots, K, \text{ and}$$
$$D_0^*(Q)(o) = \bar{Q}_1(\ell_0) - \Psi(\bar{Q}).$$

11.5 Initial Estimates

We use the empirical distribution Q_{n,L_0} as initial estimator of the distribution of baseline covariates. To construct initial estimates of $\bar{Q}_0$, we begin at $K + 1$, where the initial estimator of $\bar{Q}_{0,K+1}$ should assign the value one to individuals with $L_{K,1} = 1$ and zero to individuals with $L_{K,2} = 1$. For the remaining individuals, we must estimate the conditional mean of $L_{K+1,1}$ given L_0. In the simplest case, this could be achieved via parametric regression of the outcome $L_{K+1,1}$ on functions of L_0 in the subset of data with $\bar{A}_K = \bar{1}_{K+1}$ and $L_K = (0, 0)$. However, more ideally this estimate would be based on a more flexible technique, such as the super learner including nonparametric tools. Moving to time point K, the estimate of $\bar{Q}_{0,K}$ should assign the value one to individuals with $L_{K-1,1} = 1$ and zero to individuals with $L_{K-1,2} = 1$. For the remaining individuals, $\bar{Q}_{0,K}$ can be estimated using a regression of an estimate of $\bar{Q}_{0,K+1}$ on functions of L_0 in the subset of data with $\bar{A}_{K-1} = \bar{1}_K$ and $L_{K-1} = (0, 0)$. This process continues for each of the remaining time points: previously failed individuals are assigned one or zero depending on the type of failure, while the predicted value from the previous regression serves as the outcome in the next regression in the remaining individuals.

11.6 Submodels and Loss Functions

Given estimators Q_n and G_n, we now define appropriate parametric fluctuation submodels and loss functions to update Q_n. For the distribution of baseline covariates, we use the negative log-likelihood loss, $L(o, Q_{L_0}) = -\log\{dQ_{L_0}(\ell_0)\}$, and submodel $\{(1 + \epsilon D_0^*(\bar{Q}))dQ_{L_0} : \epsilon\}$. For the fluctuation of the estimate of $\bar{Q}_{0,K+1}$, we use the negative log-likelihood loss function

$$L(o, \bar{Q}_{K+1}) = -I(\bar{a}_K = \bar{1}_{K+1})[\ell_{K+1,1}\log\{\bar{Q}_{K+1}(\bar{\ell}_K)\} + (1 - \ell_{K+1,1})\log\{1 - \bar{Q}_{K+1}(\bar{\ell}_K)\}],$$

whose true risk is indeed minimized by $\bar{Q}_{0,K+1}$. An appropriate submodel for $\bar{Q}_{n,K+1}$ given some G is, with a slight abuse of notation, $\{\bar{Q}_{\epsilon,K+1} : \epsilon\}$ where $\bar{Q}_{\epsilon,K+1} = \text{expit}\{\text{logit}(\bar{Q}_{n,K+1}) + \epsilon H_{K+1}(G)\}$. One easily checks that

$$\frac{d}{d\epsilon} L(o, \bar{Q}_{\epsilon,K+1})\bigg|_{\epsilon=0} = D^*_{K+1}(Q,G)(o) .$$

For $k = 1, \ldots, K$ we define a negative log-likelihood loss function for $\bar{Q}_{0,k}$ that is indexed by a given $\bar{Q}_{k+1}$,

$$L_{\bar{Q}_{k+1}}(o, \bar{Q}_k) = -I(\bar{a}_{k-1} = \bar{1}_k)[\bar{Q}_{k+1}(\bar{\ell}_k)\log\{\bar{Q}_k(\bar{\ell}_{k-1})\} + \{1 - \bar{Q}_{k+1}(\bar{\ell}_k)\}\log\{1 - \bar{Q}_k(\bar{\ell}_{k-1})\}] .$$

One can confirm that the true risk of this loss function is minimized by $\bar{Q}_{0,k}$ whenever the index parameter is equal to its true value $\bar{Q}_{0,k+1}$. We again use a logistic submodel for a given G denoted, again with a slight abuse of notation, as $\{\bar{Q}_{\epsilon,k} : \epsilon\}$, where $\bar{Q}_{\epsilon,k} = \text{expit}\{\text{logit}(\bar{Q}_{n,k}) + \epsilon H_k(G)\}$. We can show that

$$\frac{d}{d\epsilon} L_{\bar{Q}_{k+1}}(o, \bar{Q}_{\epsilon,k})\bigg|_{\epsilon=0} = D^*_k(Q,G)(o) .$$

11.7 TMLE Algorithm

The TMLE algorithm follows the initial estimation procedure outlined above, but adds in a fluctuation step at each time point:

1. Generate initial estimates, G_n, of the conditional treatment and censoring mechanisms. These may be obtained through standard methods (e.g., logistic regression for the treatment and Kaplan-Meier for the censoring mechanism), or more ideally using a collection of classical tools along with machine learning techniques combined via the super learner. Use these estimates to compute $H_k(G_n)$ for $k = 1, \ldots, K + 1$.
2. Generate an initial estimate $\bar{Q}_{n,K+1}$ of the first iterated conditional mean as outlined above.
3. Obtain ϵ_n as the coefficient of a logistic regression with $L_{k,1}$ as outcome, the offset $\text{logit}\{\bar{Q}_{n,K+1}(\bar{L}_K)\}$, and $H_{K+1}(G_n)(O)$ as covariate in the subset of data with $\bar{A}_K = \bar{1}_{K+1}$. Set $\bar{Q}^*_{n,K+1} = \bar{Q}_{\epsilon_n,K+1}$. Let $k = K$.
4. Generate an initial estimate $\bar{Q}_{n,k}$ of $\bar{Q}_{0,k}$ as outlined above by first assigning known values when $L_{k-1} \neq (0,0)$ and then estimating unknown values using an appropriate form of regression with $\bar{Q}^*_{n,k+1}$ as the outcome and functions of L_0 as predictors in the subset of data with $\bar{A}_{k-1} = \bar{1}_k$ and $L_{k-1} = (0,0)$.
5. Obtain the next value for ϵ_n by fitting a logistic regression with $\bar{Q}^*_{n,k+1}(\bar{L}_k)$ as outcome, $\text{logit}\{\bar{Q}_{n,k}(\bar{L}_k)\}$ as offset, and $H_k(G_n)(O)$ as covariate in the subset of data with $\bar{A}_k = \bar{1}_{k+1}$. Set $\bar{Q}^*_{n,k} = \bar{Q}_{\epsilon_n,k}$. Set $k = k - 1$.
6. Iterate steps 4–5 until $k = 0$.
7. Obtain estimate $\psi^*_n = \frac{1}{n} \sum_{i=1}^{n} \bar{Q}^*_{n,1}(O_i)$.

11.8 Statistical Properties of TMLE

The TMLE estimator of cumulative incidence is doubly-robust, in that the estimator is consistent if either Q_n consistently estimates Q_0 or G_n consistently estimates G_0. The TMLE estimator will be asymptotically linear under the usual empirical process and rate conditions for TMLE-based estimators. If the treatment and censoring mechanisms are known exactly, the influence function of ψ_n^* is given by $D^*(Q^*, G_0)$, where Q^* is the (possibly misspecified) limit of Q_n^*. In this case, the asymptotic variance of $n^{1/2}(\psi_n^* - \psi_0)$ can be consistently estimated by $\sigma_n^2 = \sum_{i=1}^n D^*(Q_n^*, G_0)(O_i)^2$. However, if the treatment and censoring mechanisms are unknown, as is typical in practice, and Q_n^* is not consistent for Q_0, the asymptotic variance is more complicated. Nevertheless, in such situations, if G_n is an asymptotically efficient estimator within a parametric model G, then we may use σ_n^2 as a conservative estimate of the asymptotic variance.

Confidence intervals for ψ_n^* may be constructed using

$$\left(\psi_n^* - z_{1-\alpha/2}\sigma_n n^{-1/2}, \psi_n^* + z_{1-\alpha/2}\sigma_n n^{-1/2} \right),$$

where z_β is the β-quantile of the standard normal distribution. Similarly, given a fixed $\psi^\circ \in (0, 1)$, a two-sided test of the null hypothesis $\psi_0 = \psi^\circ$ of asymptotic size no larger than α can be constructed by rejecting the null hypothesis whenever $|n^{1/2}(\psi_n^* - \psi^\circ)/\sigma_n| > z_{1-\alpha/2}$. An appealing facet of the influence function-based variance estimation is that the form of the asymptotic variance of a function of multiple estimators is readily available. In sieve analysis, this is quite useful as it allows for the simple construction of confidence intervals and hypothesis tests about genotype-specific vaccine efficacy and vaccine sieve effects.

11.9 HVTN 505 HIV Vaccine Sieve Analysis

We analyzed data from the HVTN 505 study, a recent Phase II preventive HIV vaccine efficacy trial where participants were randomized 1:1 to receive either the candidate vaccine or a placebo (Hammer et al. 2013). Additional information on participants' risk behaviors was collected at recruitment including gender (male, female, transgender male, transgender female), race (white, black, other), BMI, drug use (marijuana, cocaine, poppers, speed, MDMA, other recreational drugs), alcohol use (never, less than once per week, 1–2 days per week, 3–6 days per week, daily), STD status (syphilis, herpes, genital sores, gonorrhea), and sexual risk behaviors (reported number of sexual partners, reported unprotected insertive anal sex with men, reported unprotected recipient sex with men, previously derived behavioral risk score). After receipt of the final dose of vaccine, participants attended visits every 3 months to be tested for HIV. We focus our analysis on the modified intent-to-treat cohort, which included 2504 participants. We refer interested readers to the original publication for more information on the trial's design.

We studied whether the vaccine exhibited sieve effects at amino acid site 169 on the V2 loop of the HIV envelope protein. This locus was chosen because a vaccine with a similar design to the HVTN 505 vaccine exhibited sieve effects at this locus (Rolland et al. 2012). In HVTN 505 there were a total of 17 169-matched and 30 169-mismatched infections. We used TMLE to estimate the cumulative incidence of 169-matched and mismatched infections in the vaccine and placebo arm in each scheduled visit window. These measures were used to estimate the vaccine efficacy against each type of infection, in addition to the vaccine sieve effect at this locus.

We used the super learner to estimate both the iterated conditional means and the conditional censoring distribution. The same library was used for both and consisted

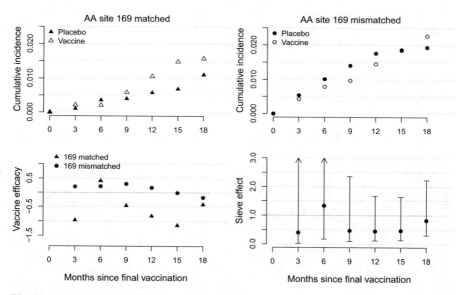

Fig. 11.1 Results from the TMLE analysis of HVTN-505

of 13 algorithms: an unadjusted mean, three main terms-only generalized linear models (using only behavioral risk score, only sexual risk behaviors, and only the five variables most highly correlated with outcome), three main terms-only stepwise regressions based on AIC (using all variables, only sexual risk behaviors, and only the five most highly correlated variables), three random forests (using all variables, only drug/alcohol use, and only sexual risk behaviors), and three gradient boosted machines (using all variables, only drug/alcohol use, and only all sexual risk behaviors).

The results of our analysis are shown in Fig. 11.1. The top row shows the treatment-specific cumulative incidence of HIV infections that were matched (left) or mismatched (right) to the strains in the vaccine insert at the amino acid site of interest. Overall the incidence of mismatched infections was slightly higher than matched infections and in neither case do we see a clear separation between the curves for the two treatment groups. This is reflected in the vaccine efficacy measure shown in the bottom left plot. The vaccine does not appear to confer efficacy against either type of infection. The estimated sieve effect is shown in the bottom right panel, where we see confirmation that there is no evidence of differential vaccine efficacy comparing matched to mismatched infections.

We also performed the analysis using the Aalen-Johansen estimator to examine whether adjusting for covariates had an effect on the results. At 18 months after receipt of the final vaccination the TMLE-estimated sieve effect was 0.83 (95% CI: 0.31–2.23) and the Aalen-Johansen estimated sieve effect was 0.82 (95% CI: 0.23–2.91). Covariate adjustment did not affect the point estimate for the sieve effect, but had a substantial effect on the estimated uncertainty associated with the point estimate. The estimated variance of the TMLE-estimated sieve effect was 60% that of the Aalen-Johansen-estimated sieve effect, which led to a substantially narrower confidence interval about the TMLE estimate.

11.10 Discussion

In this chapter, we illustrated how cumulative incidence, genotype-specific vaccine efficacy, and vaccine sieve effects can be estimated using TMLE.

> There are several benefits to this targeted approach to sieve analysis. Covariate adjustment allows for departures from the assumption of independent censoring by allowing for censoring to depend on baseline covariates. Furthermore, covariate adjustment can lead to gains in efficiency. We can thus obtain estimators of vaccine efficacy and sieve effects that have lower bias and lower variance than the standard, unadjusted approach to sieve analysis.

Our analysis of the HVTN 505 data illustrates that these potential benefits can indeed be realized in real data applications. Adjusting for the large set of risk behaviors collected in the HVTN 505 trial led to a substantially narrower confidence interval about the estimated sieve effect. Targeted learning of vaccine sieve effects appears to be a promising direction for the field.

Chapter 12
The Sample Average Treatment Effect

Laura B. Balzer, Maya L. Petersen, and Mark J. van der Laan

In cluster randomized trials (CRTs), the study units usually are not a simple random sample from some clearly defined target population. Instead, the target population tends to be hypothetical or ill-defined, and the selection of study units tends to be systematic, driven by logistical and practical considerations. As a result, the population average treatment effect (PATE) may be neither well defined nor easily interpretable. In contrast, the sample average treatment effect (SATE) is the mean difference in the counterfactual outcomes for the study units. The sample parameter is easily interpretable and arguably the most relevant when the study units are not sampled from some specific super-population of interest. Furthermore, in most settings the sample parameter will be estimated more efficiently than the population parameter.

L. B. Balzer (✉)
Department of Biostatistics and Epidemiology, School of Public Health and Health Sciences, University of Massachusetts - Amherst, 416 Arnold House, 715 North Pleasant St, Amherst, MA 01003, USA
e-mail: lbalzer@umass.edu

M. L. Petersen
Division of Epidemiology and Division of Biostatistics, University of California, Berkeley, 101 Haviland Hall, #7358, Berkeley, CA 94720, USA
e-mail: mayaliv@berkeley.edu

M. J. van der Laan
Division of Biostatistics and Department of Statistics, University of California, Berkeley, 101 Haviland Hall, #7358, Berkeley, CA 94720, USA
e-mail: laan@berkeley.edu

© Springer International Publishing AG 2018
M.J. van der Laan, S. Rose, *Targeted Learning in Data Science*,
Springer Series in Statistics, https://doi.org/10.1007/978-3-319-65304-4_12

In this chapter, we demonstrate the use of TMLE for estimation and inference of the sample effect in trials with and without pair-matching. We study the asymptotic and finite sample properties of the TMLE for the sample effect and provide a conservative variance estimator. Finite sample simulations illustrate the potential gains in precision and power from selecting the sample effect as the target of inference. This chapter is adapted from Balzer et al. (2016c).

In many studies, the goal is to estimate the impact of an exposure on the outcome of interest. Often the target causal parameter is the PATE: the expected difference in the counterfactual outcomes if all members of some population were exposed and if all members of that population were unexposed. If there are no unmeasured confounders and there is sufficient variability in the exposure assignment (i.e. if the randomization and positivity assumptions hold), then we can identify the PATE as a function of the observed data distribution (Rosenbaum and Rubin 1983b; Robins 1986). The resulting statistical parameter can be estimated with a variety of algorithms, including matching and inverse weighting estimators (e.g., Horvitz and Thompson 1952; Rosenbaum and Rubin 1983b; Shen et al. 2014), simple substitution estimators (e.g., Robins 1986; Snowden et al. 2011), and double robust algorithms (e.g., Robins et al. 1994; van der Laan and Robins 2003; van der Laan and Rubin 2006; van der Laan and Rose 2011).

An alternative causal parameter is the SATE (Neyman 1923; Rubin 1990; Imbens 2004; Imai 2008; Schochet 2013; Imbens and Rubin 2015). The sample effect is the average difference in the counterfactual outcomes for the actual study units. There are several potential advantages to selecting the SATE as the parameter of interest. First, the SATE is readily interpretable as the intervention effect for the sample at hand. Second, the SATE avoids assumptions about randomly sampling from and generalizing to some "vaguely defined super-population of study units" (Schochet 2013). In other words, the sample parameter remains relevant and interpretable if the units were systematically selected for inclusion in the study, as is likely to be common in CRTs. Extensions of the study results to a broader or a different population can be addressed as a distinct research problem, approached with formal tools (e.g., Cole and Stuart 2010; Stuart et al. 2011; Bareinboim and Pearl 2013; Hartman et al. 2015), and do not have to be assumed in the parameter specification. Finally, an estimator of the sample effect is often more precise than the same estimator of the population effect (Neyman 1923; Rubin 1990; Imbens 2004; Imai 2008).

For a randomized trial, Neyman (1923) first proposed estimating the SATE with the unadjusted estimator, which is the difference in the average outcomes among the treated units and the average outcomes among the control units. In this setting, the difference-in-means estimator will be unbiased for the SATE, conditional on the set of counterfactual outcomes for the study units. However, its variance remains unidentifiable as it relies on the correlation of the counterfactual outcomes (Neyman 1923; Rubin 1990; Imbens 2004; Imai 2008). Imbens (2004) later generalized this work for an efficient estimator (i.e. a regular, asymptotically linear estimator, whose influence curve equals the efficient influence curve) in an observational setting. In particular, he showed that an efficient estimator for the population effect was unbiased for the sample effect, conditional on the baseline covariates and the counterfactual outcomes of the study units. He further expressed the variance of an efficient estimator of the SATE in terms of the variance of the same estimator of the PATE minus the variance of the unit-specific treatment effects across the population. This suggested that the standard variance estimator would be biased upwards unless there is no variability in the treatment effect.

In this chapter, we introduce a TMLE for estimation and inference of the sample effect in trials with and without pair-matching. Our results generalize the variance derivations of Imbens (2004) to allow for misspecification of the outcome regression (i.e., the conditional mean outcome, given the exposure and covariates), estimation of the propensity score (i.e. the conditional probability of the receiving the exposure, given the covariates), and adaptive pair-matching (Balzer et al. 2015). Pair-matching is a popular design strategy in CRTs to protect study credibility and to increase power (Klar and Donner 1997; Greevy et al. 2004; Imai et al. 2009; Hayes and Moulton 2009; van der Laan et al. 2013a; Balzer et al. 2015). To the best of our knowledge, TMLE is the first efficient estimator proposed for the sample effect in a pair-matched trial.

As a motivating example, we consider a hypothetical CRT for HIV prevention and treatment. Suppose there are $n = 30$ communities in the trial. In intervention communities, HIV testing is regularly offered and all individuals testing HIV+ are immediately eligible for antiretroviral therapy (ART) with enhanced services for initiation, adherence and retention in care. In control communities, all individuals testing HIV+ are offered ART according to the current standard of care. The primary hypothesis is that the universal "test-and-treat" strategy will reduce the cumulative incidence of HIV over the trial duration. For the purposes of discussion, we focus on the community-level data. Thereby, our results are equally applicable to clustered and nonclustered data structures.

12.1 The Causal Model and Causal Parameters

Consider the following data generating process for a randomized trial with two arms. First, the study units are selected. While some trials obtain a simple random sample from a well-defined target population, in other studies there may not be a clear target population from which units were sampled and about which we wish to make inferences. In the SEARCH trial, for example, 32 communities were selected from Western Uganda (Mbarara region), Eastern Uganda (Tororo region) and the Southern Nyanza Province in Kenya by first performing ethnographic mapping on 54 candidate communities meeting the inclusion criteria (e.g. community size, health care infrastructure and accessibility by a maintained transportation route), and then selecting the 16 pairs best matched on a range of characteristics (e.g. region, occupational mix and migration index) (Balzer et al. 2015). After selection of the study units, additional covariates are often measured. In our running example, additional covariates collected could include male circumcision coverage, measures of HIV prevalence and measures of community-level HIV RNA viral load. Throughout the baseline covariates are denoted W.

Next, the intervention is randomized to the study units. Equal allocation of the intervention can be guaranteed by randomly assigning the intervention to $n/2$ units and the control to remaining units or by randomizing within matched pairs. For ease of exposition, we present the causal model for the simple scenario, where the intervention is completely randomized, but our results are general. (Extensions to pair-matched trials are given in Sect. 12.4.) Let A be a binary variable, reflecting the assigned level of the intervention. For our hypothetical CRT, A equals one if the community was assigned to the treatment (universal test-and-treat) and equals zero if the community was assigned to the control (standard of care). At the end of follow-up, the outcome Y is measured. For our trial, Y is the cumulative incidence of HIV over the relevant time period. The observed data for a given study unit are then

$$O = (W, A, Y).$$

Suppose we observe n i.i.d. copies of O with some distribution P_0. Recall the subscript 0 denotes the true distribution of the observed data. We note that for estimation and inference of the sample and conditional average treatment effects, we can weaken the i.i.d. assumption by conditioning on the vector of baseline covariates $(W_1, W_2, \ldots, W_n)$ (Balzer et al. 2015).

The following structural causal model describes this data generating process (Pearl 1995, 2009a). Each component of the observed data is assumed to be a deterministic function of its parents (variables that may influence its value) and unobservable background factors:

$$W = f_W(U_W) \tag{12.1}$$
$$A = \mathbb{I}(U_A < 0.5)$$
$$Y = f_Y(W, A, U_Y)$$

where the set of background factors $U = (U_W, U_A, U_Y)$ have some joint distribution P_U. By design, the random error determining the intervention assignment U_A is independent from the unmeasured factors contributing the baseline covariates U_W and the outcome U_Y:

$$U_A \perp\!\!\!\perp (U_W, U_Y).$$

Specifically, U_A is independently drawn from a Uniform(0,1). This causal model $\mathcal{M}^F$ implies the statistical model for the set of possible distributions of the observed data O. In a randomized trial, the statistical model $\mathcal{M}$ is semiparametric.

Through interventions on the structural causal model, we can generate the counterfactual outcome Y_a, which is the outcome if possibly contrary-to-fact the unit was assigned $A = a$:

$$W = f_W(U_W)$$
$$A = a$$
$$Y_a = f_Y(W, a, U_Y).$$

In this framework, the counterfactual outcomes Y_a are random variables. For our running example, Y_a is the counterfactual cumulative incidence of HIV if possibly contrary-to-fact the community had been assigned treatment level $A = a$.

The distribution of the counterfactuals can then be used to define the causal parameter of interest. Often, the target of inference is the population average treatment effect:

$$PATE = \mathbb{E}[Y_1 - Y_0].$$

This is the expected difference in the counterfactual outcomes for underlying target population from which the units were sampled. From the structural causal model, we see that the expectation is over the measured factors W and unmeasured factors U_Y, which determine the counterfactual outcomes for the population. In other words, the true value of the PATE does not depend on the sampled values of W or U_Y. For our hypothetical trial, the PATE would be the difference in the expected counterfactual cumulative incidence of HIV if possibly contrary-to-fact all communities in some hypothetical target population implemented the test-and-treat strategy, and expected counterfactual cumulative incidence of HIV if possibly contrary-to-fact all communities in that hypothetical target population continued with the standard of care.

An alternative causal parameter is the sample average treatment effect, which was first proposed in Neyman (1923):

$$SATE = \frac{1}{n} \sum_{i=1}^{n} [Y_{1,i} - Y_{0,i}].$$

This is simply the intervention effect for the n study units. The SATE is a data-adaptive parameter; its value depends on the units included in the study. For recent work on estimation and inference of other data-adaptive parameters, we refer the reader to Chap. 9. The SATE remains interpretable if there is no clear super-population from which the study units were selected. In our running example, the SATE is the average difference in the counterfactual cumulative incidence of HIV under the test-and-treat strategy and under the standard of care for the n study communities.

In a CRT, targeting the sample effect may have several advantages over targeting the population effect. First, there may not be a single real world (as opposed to hypothetical) target population from which the study units were sampled or about which we wish to make inferences. While appropriate analytic approaches can reduce concerns over systematic sampling, the interpretation and policy relevance of the resulting PATE estimate would be unclear. In contrast, targeting the SATE allows us to rigorously estimate the intervention effect in a clearly defined, real world population consisting of all the residents of the study communities. The resulting SATE estimate does not rely on any assumptions about the sampling mechanism, has a clear interpretation, and is generally more precise than an estimate of the PATE. As discussed below, estimators of the sample effect are at least as powerful as those of the population effect and expected to be more powerful when there is effect modification (Rubin 1990; Imbens 2004; Imai 2008).

Clearly, however, it remains of significant policy interest to transport any effect found in a CRT to new populations and settings. However, alternative real world target populations are likely to differ from the current setting in a number of ways that will likely impact the magnitude of the effect. As a result, neither the SATE nor the PATE will apply directly to these new settings. Thus, a desire for generalizability does not constitute an argument for favoring the population parameter over the sample parameter. Instead, the generalization (or transport) of the intervention effect to settings beyond the current sample is best addressed as a distinct research question, making full use of the modern toolbox available (e.g. Cole and Stuart 2010; Stuart et al. 2011; Bareinboim and Pearl 2013; Hartman et al. 2015).

12.2 Identifiability

To identify the above causal effects, we must write them as some function of the observed data distribution P_0 (Imbens 2004; van der Laan and Rose 2011). Under the randomization and positivity assumptions, we can identify the mean counterfactual outcome within strata of covariates (Rosenbaum and Rubin 1983b; Robins 1986):

$$\mathbb{E}[Y_a|W] = \mathbb{E}[Y_a|A = a, W] = \mathbb{E}_0[Y|A = a, W]$$

where the right-most expression is now in terms of the observed data distribution P_0. Briefly, the first equality holds under the randomization assumption, which states that the counterfactual outcome is independent of the exposure, given the measured covariates: $A \perp\!\!\!\perp Y_a|W$. This is equivalent to the no unmeasured confounders assumption (Rosenbaum and Rubin 1983b). The positivity assumption states that the exposure level a occurs with a positive probability within all possible strata of covariates. Both assumptions hold by design in a randomized trial. As a well known result, the PATE is identified as

$$\Psi^{\mathcal{P}}(P_0) = \mathbb{E}_0\Big[\mathbb{E}_0(Y|A = 1, W) - \mathbb{E}_0(Y|A = 0, W)\Big]$$
$$= \mathbb{E}_0[\bar{Q}_0(1, W) - \bar{Q}_0(0, W)]$$

where $\bar{Q}_0(Y|A, W) \equiv \mathbb{E}_0(Y|A, W)$ denotes the conditional mean outcome, given the exposure and covariates. This statistical estimand is also called the g-computation identifiability result (Robins 1986). For our running example, $\Psi^P(P_0)$ is the difference in expected cumulative HIV incidence, given the treatment and measured covariates, and the expected cumulative HIV incidence, given the control and measured covariates, averaged (standardized) with respect to the covariate distribution in the hypothetical target population. As with the causal parameter, there is one true value $\Psi^P(P_0)$ for the population. In a randomized trial, conditioning on the covariates W is not needed for identifiability, but will often provide efficiency gains during estimation (e.g., Fisher 1932; Cochran 1957; Cox and McCullagh 1982; Tsiatis et al. 2008; Moore and van der Laan 2009b; Rosenblum and van der Laan 2010b; European Medicines Agency 2015).

In contrast, the SATE is not identifiable—in finite samples, we cannot strictly write the causal parameter as a function of the observed data distribution P_0. (Asymptotically, the SATE is identifiable, because the empirical mean converges to the expectation and thereby the sample effect converges to the population effect.) To elaborate, we can use the structural causal model (Eq. (12.1)) to rewrite the sample effect as

$$SATE = \frac{1}{n} \sum_{i=1}^{n} [Y_{1,i} - Y_{0,i}]$$

$$= \frac{1}{n} \sum_{i=1}^{n} f_Y(W_i, 1, U_{Y_i}) - f_Y(W_i, 0, U_{Y_i})$$

$$= \frac{1}{n} \sum_{i=1}^{n} \mathbb{E}[Y_{1,i} - Y_{0,i}|W_i, U_{Y_i}].$$

The second equality is from the definition of counterfactuals as interventions on the causal model. The final equality is the conditional average treatment effect (CATE), given the measured baseline covariates as well as the unmeasured factors. The conditional effect was first proposed in Abadie and Imbens (2002) and is the average difference in the expected counterfactual outcomes, treating the measured covariates of the study units as fixed: CATE= $\frac{1}{n} \sum_{i=1}^{n} \mathbb{E}[Y_{1,i} - Y_{0,i}|W_i]$. This representation of the SATE suggests that if we had access to all pre-intervention covariates impacting the outcome (i.e. $\{W, U_Y\}$), then we could apply the results for estimation and inference for the conditional parameter, as detailed in Balzer et al. (2015). In reality, we only measure a subset of these covariates (i.e., W) and only this subset is available for estimation and inference. Therefore, the SATE is not formally identifiable in finite samples. Nonetheless, as detailed below, a TMLE for the population effect will be consistent and asymptotically linear for the sample effect, and the corresponding variance estimator will be asymptotically conservative.

12.3 Estimation and Inference

There are many well-established algorithms for estimation of the population parameter $\Psi^{\mathcal{P}}(P_0)$. Examples include IPW, simple substitution estimators, A-IPW and TMLE (e.g., Horvitz and Thompson 1952; Rosenbaum and Rubin 1983b; Shen et al. 2014; Robins 1986; Snowden et al. 2011; Robins et al. 1994; van der Laan and Robins 2003; van der Laan and Rubin 2006; van der Laan and Rose 2011). In a randomized trial, the unadjusted difference in the average outcomes among the treated units and the average outcome among the control units provides a simple and unbiased estimate of the PATE. Adjusting for measured covariates, however, will generally increase efficiency and study power (e.g., Fisher 1932; Cochran 1957; Cox and McCullagh 1982; Tsiatis et al. 2008; Moore and van der Laan 2009b; Rosenblum and van der Laan 2010b; Shen et al. 2014; European Medicines Agency 2015).

For example, we can obtain a more precise estimator of the PATE by (1) regressing the outcome Y on the exposure A and covariates W, (2) using the estimated coefficients to obtain the predicted outcomes for all units under the exposure and control, and (3) then taking the average difference in the predicted outcomes. For a large class of general linear models, there is no risk of bias if the "working" model for the outcome regression is misspecified (Rosenblum and van der Laan 2010b). This algorithm is called parametric g-computation (Robins 1986) in observational studies and also called analysis of covariance (ANCOVA) (Cochran 1957) in the special case of a continuous outcome and a linear model without interactions. Alternatively, we can obtain a more precise estimator of $\Psi^{\mathcal{P}}(P_0)$ by estimating the known exposure mechanism to capture chance imbalances in the covariate distribution between treatment groups (e.g., van der Laan and Robins 2003; Moore and van der Laan 2009b; Shen et al. 2014). In our running example, the true conditional probability of being assigned to the test-and-treat intervention is $P_0(A = 1|W) = 0.5$. However, with limited numbers of clusters, there is likely to be variation in the baseline covariates across the treatment arms.

We focus our discussion on TMLE, which incorporates estimation of both the outcome regression (the conditional mean outcome given the exposure and covariates) and the propensity score (the conditional probability of receiving the exposure given the covariates, Rosenbaum and Rubin 1983b). In general, TMLE is a double robust estimator; it will be consistent if either outcome regression or the propensity score is consistently estimated. If both functions are consistently estimated at a fast enough rate and there is sufficient variability in the propensity score, the estimator is also asymptotically efficient in that it attains the lowest possible variance among a large class of regular, asymptotically linear estimators. TMLE is also a substitution (plug-in) estimator, which provides stability in the context of sparsity (Gruber and van der Laan 2010b; Balzer et al. 2016a). Finally, TMLE makes use of state-of-the-art machine learning and therefore avoids the parametric assumptions commonly made in other algorithms. In other words, TMLE does not place any unwarranted assumptions on the structure of the data and respects the semiparametric statistical model.

12.3.1 TMLE for the Population Effect

For the population parameter $\Psi^{\mathcal{P}}(P_0)$, a TMLE can be implemented as follows.

- *Step 1. Initial Estimation*: First, we obtain an initial estimator of the outcome regression $\bar{Q}_0(A, W)$. For example, the outcome Y can be regressed on the exposure A and covariates W according to a parametric "working" model (Rosenblum and van der Laan 2010b). Alternatively, we could use an *a priori* specified data-adaptive procedure, such as super learner. The initial estimator is denoted $\bar{Q}_n(A, W)$.

- *Step 2. Targeting*: Second, we update the initial estimator of the outcome regression $\bar{Q}_n(A, W)$ by incorporating information in the propensity score $g_0(A|W) \equiv P_0(A|W)$. Informally, this "targeting" step helps to remove some of the residual imbalance in the baseline covariate distributions across treatment groups. More formally, this targeting step serves to obtain the optimal bias-variance tradeoff for $\Psi^{\mathcal{P}}(P_0)$ and to solve the efficient score equation (Hahn 1998). Briefly, this targeting step is implemented as follows.

  - We calculate the clever covariate based on the known or estimated exposure mechanism $g_n(A|W) \equiv P_n(A|W)$:

$$H_n(A, W) = \left(\frac{\mathbb{I}(A = 1)}{g_n(1|W)} - \frac{\mathbb{I}(A = 0)}{g_n(0|W)} \right).$$

  To estimate the propensity score, we could run logistic regression of the exposure A on the covariates W or use more data-adaptive methods.
  - For a continuous and unbounded outcome, we run linear regression of the outcome Y on the covariate $H_n(A, W)$ with the initial estimator as offset (i.e. we suppress the intercept and set the coefficient on the initial estimator equal to 1). We plug in the estimated coefficient ϵ_n to yield the targeted update: $\bar{Q}_n^*(A, W) = \bar{Q}_n(Y|A, W) + \epsilon_n H_n(A, W)$.
  - For a binary or a bounded continuous outcome (e.g. a proportion),[1] we run logistic regression of the outcome Y on the covariate $H_n(A, W)$ with the $logit(\cdot) = log[\cdot/(1 - \cdot)]$ of the initial estimator as offset. We plug in the estimated coefficient ϵ_n to yield the targeted update:

$$\bar{Q}_n^*(A, W) = logit^{-1}\{logit[\bar{Q}_n(A, W)] + \epsilon_n H_n(A, W)\}.$$

---

[1] Logistic fluctuation can also be used for a continuous outcome that is bounded in $[a, b]$ by first applying the following transformation to the outcome: $Y^* = (Y - a)/(b - a)$. Use of logistic regression over linear regression can provide stability under data sparsity and/or with rare outcomes (e.g., Gruber and van der Laan 2010b).

- *Step 3. Parameter Estimation*: Finally, we obtain a point estimate by substituting the targeted estimates into the parameter mapping:

$$\Psi_n(P_n) = \frac{1}{n} \sum_{i=1}^{n} \left[\bar{Q}_n^*(1, W_i) - \bar{Q}_n^*(0, W_i) \right]$$

where P_n denotes the empirical distribution, placing mass $1/n$ on each observation O_i and $\bar{Q}_n^*(A_i, W_i)$ denotes the targeted estimator of the conditional mean outcome. The sample mean is the nonparametric maximum likelihood estimator of the marginal distribution of the baseline covariates $P_0(W)$.

If the initial estimator for $\bar{Q}_0(A, W)$ is based on a working parametric regression with an intercept and a main term for the exposure and if the exposure mechanism is treated as known (i.e., not estimated), then the updating step can be skipped (Moore and van der Laan 2009b; Rosenblum and van der Laan 2010b). Further precision, however, can be attained by using a data-adaptive algorithm for initial estimation of the outcome regression $\bar{Q}_0(A, W)$ and/or by estimating the exposure mechanism $g_0(A|W)$ (van der Laan and Robins 2003). See Chap. 13 for further details on data-adaptive estimation in CRTs.

Under standard regularity conditions, this TMLE is a consistent and asymptotically linear estimator of the population parameter (van der Laan and Rubin 2006; van der Laan and Rose 2011):

$$\Psi_n(P_n) - \Psi^{\mathcal{P}}(P_0) = \frac{1}{n} \sum_{i=1}^{n} D^{\mathcal{P}}(O_i) + o_P(1/\sqrt{n}).$$

In words, the estimator minus the truth can be written as an empirical mean of an influence curve $D^{\mathcal{P}}(O)$ and a second-order term going to 0 in probability. The influence curve is given by

$$D^{\mathcal{P}}(O) = D_Y(O) + D_W(O)$$
$$D_Y(O) = \left(\frac{\mathbb{I}(A = 1)}{g_0(1|W)} - \frac{\mathbb{I}(A = 0)}{g_0(0|W)} \right)(Y - \bar{Q}_\infty(A, W))$$
$$D_W(O) = \bar{Q}_\infty(1, W) - \bar{Q}_\infty(0, W) - \Psi^{\mathcal{P}}(P_0)$$

where $\bar{Q}_\infty(A, W)$ denotes the limit of the TMLE $\bar{Q}_n^*(A, W)$ and we are assuming the propensity score is known or consistently estimated, as will always be true when the treatment A is randomized. The first term of the influence curve D_Y is the weighted residuals (i.e., the weighted deviations between the observed outcome and the limit of the predicted outcome). The second term D_W is deviation between the limit of the estimated strata-specific association and the marginal association.

The standardized estimator is asymptotically normal with variance given by the variance of its influence curve $D^{\mathcal{P}}(O)$, divided by sample size n (van der Laan and Rubin 2006; van der Laan and Rose 2011). Under consistent estimation of the outcome regression (i.e., when $\bar{Q}_\infty(A, W) = \bar{Q}_0(A, W)$), the TMLE will be asymptotically efficient and achieve the lowest possible variance among a large class of estimators of the population effect. In other words, its influence curve will equal the efficient influence curve, and the TMLE will achieve the efficiency bound of Hahn (1998). Thereby, improved estimation of the outcome regression leads to more precise estimators of the population effect. In finite samples, the variance of the TMLE is well-approximated by the sample variance of the estimated influence curve scaled by sample size:

$$\sigma_n^{2,\mathcal{P}} = \frac{\frac{1}{n} \sum_{i=1}^n \left[D_n^{\mathcal{P}}(O_i) \right]^2}{n}$$

where

$$D_n^{\mathcal{P}}(O) = \left(\frac{\mathbb{I}(A=1)}{g_n(1|W)} - \frac{\mathbb{I}(A=0)}{g_n(0|W)} \right) (Y - \bar{Q}_n^*(A, W)) + \bar{Q}_n^*(1, W) - \bar{Q}_n^*(0, W) - \Psi_n(P_n).$$

12.3.2 TMLE for the Sample Effect

For a randomized trial, Neyman (1923) proposed estimating the SATE with the unadjusted estimator:

$$\Psi_{n,unadj}(P_n) = \frac{\sum_{i=1}^n \mathbb{I}(A_i = 1)Y_i}{\sum_{i=1}^n \mathbb{I}(A_i = 1)} - \frac{\sum_{i=1}^n \mathbb{I}(A_i = 0)Y_i}{\sum_{i=1}^n \mathbb{I}(A_i = 0)}.$$

Conditional on the vector of counterfactual outcomes $\mathbf{Y_a} = \{Y_{a,i} : i = 1, \ldots, n, a = 0, 1\}$, the difference-in-means estimator is unbiased but inefficient. To the best of our knowledge, Imbens (2004) was the first to discuss an efficient estimator (i.e. a regular, asymptotically linear estimator, whose influence curve equals the efficient influence curve) of the sample effect. He proved that an efficient estimator for the PATE was unbiased for the SATE, given the vector of baseline covariates $\mathbf{W} = (W_1, \ldots, W_n)$ and the set of counterfactual outcomes $\mathbf{Y_a}$. We now extend these results to TMLE. Specifically, we allow the estimator of outcome regression $\bar{Q}_0(A, W)$ to converge to a possibly misspecified limit, incorporate estimation of the known propensity score, and suggest an alternate method for variance estimation. In Sect. 12.4, we further extend these results to a pair-matched trial.

The TMLE for the population parameter $\Psi^{\mathcal{P}}(P_0)$, presented in Sect. 12.3.1, also serves as an estimator of the SATE. The implementation is identical. Under typical regularity conditions, the TMLE minus the sample effect behaves as an empirical mean of an influence curve depending on nonidentifiable quantities, and a second-order term going to zero in probability:

$$\Psi_n(P_n) - SATE = \frac{1}{n} \sum_{i=1}^n D^S(U_i, O_i) + o_P(1/\sqrt{n})$$

where

$$D^S(U_i, O_i) = D^C(O_i) - D^{\mathcal{F}}(U_i, O_i)$$
$$D^C(O_i) = D_Y(O_i) - \mathbb{E}_0[D_Y(O_i)|\mathbf{W}] \tag{12.2}$$
$$D^{\mathcal{F}}(U_i, O_i) = Y_{1,i} - Y_{0,i} - [\bar{Q}_0(1, W_i) - \bar{Q}_0(0, W_i)] \tag{12.3}$$

(see Theorem 1 of Balzer et al. 2016c). The first component D^C is the influence curve for the TMLE of the conditional parameter $\Psi^C(P_0) = 1/n \sum_{i=1}^n [\bar{Q}_0(1, W_i) - \bar{Q}_0(0, W_i)]$, which corresponds to the conditional average treatment effect (CATE) under the necessary identifiability assumptions (Balzer et al. 2015). This term depends on the true outcome regression $\bar{Q}_0(A, W)$. Specifically, the conditional expectation of the D_Y component, given the baseline covariates, equals the deviation between the true conditional means and the limits of the estimated conditional means:

$$\mathbb{E}_0[D_Y(O)|\mathbf{W}] = [\bar{Q}_0(1, W) - \bar{Q}_0(0, W)] - [\bar{Q}_\infty(1, W) - \bar{Q}_\infty(0, W)].$$

Under consistent estimation of the outcome regression (i.e. when $\bar{Q}_\infty(A, W) = \bar{Q}_0(A, W)$), this term is zero. The second component $D^{\mathcal{F}}$ is a function of the unobserved factors $U = (U_W, U_A, U_Y)$ and the observed data $O = (W, A, Y)$. This nonidentifiable term captures the deviations between the unit-specific treatment effect and expected effect within covariate strata:

$$\begin{aligned}
D^{\mathcal{F}}(U_i, O_i) &= Y_{1,i} - Y_{0,i} - [\bar{Q}_0(1, W_i) - \bar{Q}_0(0, W_i)] \\
&= Y_{1,i} - Y_{0,i} - [\mathbb{E}(Y_{1,i}|W_i) - \mathbb{E}(Y_{0,i}|W_i)] \\
&= Y_{1,i} - Y_{0,i} - \mathbb{E}[Y_{1,i} - Y_{0,i}|W_i].
\end{aligned}$$

In the last line, the expectation is over the unmeasured factors U_Y that determine the counterfactual outcomes. This term will be zero if there is no variability in the treatment effect across units with the same values of the measured covariates. We also note that there is no contribution to the influence curve D^S from estimation of the covariate distribution, which is considered fixed. In other words, there is no D_W component to the influence curve.

As a result, the standardized estimator of the SATE is consistent and asymptotically normal with mean zero and variance given by the limit of

$$\begin{aligned}
Var[D^S(U, O)] &= Var[D^C(O)] + Var[D^{\mathcal{F}}(U, O)] - 2Cov[D^C(O), D^{\mathcal{F}}(U, O)] \\
&= Var[D^C(O)] - Var[D^{\mathcal{F}}(U, O)]
\end{aligned}$$

(see Theorem 2 of Balzer et al. 2016c). Since the variance of the nonidentifiable $D^{\mathcal{F}}$ component must be greater than or equal to zero, the asymptotic variance of the TMLE as an estimator of the sample effect will always be less than or equal to the asymptotic variance of the same estimator of the conditional effect. They will only have the same precision when there is no variability in the unit-level treatment effect within strata of measured covariates (i.e. when $Var[D^{\mathcal{F}}(U, O)] = 0$). In many settings, however, there will be heterogeneity in the effect, and the TMLE for the SATE will be more precise. Even if the treatment effect is constant within covariate strata, the TMLE for the sample effect (or the conditional effect) will always be at least as precise as the same TMLE for the population effect. They will only have the same efficiency bound when (1) the outcome regression is consistently estimated, (2) there is no variability in the treatment effect *across* strata of measured covariates (i.e. when $Var[D_W(O)] = 0$), and (3) there is no variability in the treatment effect *within* strata of measured covariates. In many settings, there will be effect modification, and focusing on estimation of the SATE will yield the most precision and power.

We can conservatively approximate the influence curve for the TMLE of the sample effect as

$$D_n^S(O_i) = D_{Y,n}(O_i) = \left(\frac{\mathbb{I}(A_i = 1)}{g_n(1|W_i)} - \frac{\mathbb{I}(A_i = 0)}{g_n(= 0|W_i)}\right)(Y_i - \bar{Q}_n^*(A_i, W_i)). \quad (12.4)$$

Thereby, we obtain an asymptotically conservative variance estimator with the sample variance of the weighted residuals scaled by sample size n:

$$\sigma_n^{2,S} = \frac{\frac{1}{n}\sum_{i=1}^n \left[D_n^S(O_i)\right]^2}{n}.$$

As for the PATE, adjusting for predictive baseline covariates can substantially improve power for the SATE by reducing variability in the estimator. Unlike the PATE, however, adjusting for predictive baseline covariates can provide an additional power gain for the SATE by resulting in a less conservative variance estimator. Furthermore, this variance estimator is easy to implement as the relevant pieces are known or already estimated. As a result, this may provide an attractive alternative to the matching estimator of the variance, proposed by Abadie and Imbens (2002) and discussed in Imbens (2004). We note that the bootstrap is inappropriate as the SATE changes with each sample. Fisher's permutation distribution is also not appropriate, because it is testing the strong null hypothesis of no treatment effect ($Y_{1,i} = Y_{0,i}, \forall i$) (Fisher 1935), whereas our interest is in the weak null hypothesis of no average treatment effect.

12.4 Extensions to Pair-Matched Trials

Now consider a pair-matched CRT. In our running example, suppose N candidate communities satisfying the study's inclusion criteria were identified. Of these, the best $n/2$ matched pairs were chosen according to similarity on the baseline covariates of the candidate units. This "adaptive pair-matching" scheme is detailed in Balzer et al. (2015) and also called "nonbipartite matching" and "optimal multivariate matching" in other contexts (Greevy et al. 2004; Zhang and Small 2009; Lu et al. 2011). To the best of our understanding, this pair-matching scheme was been implemented in several CRTS, including the Mwanza trial for HIV prevention (Grosskurth et al. 1995), the PRISM trial for postpartum depression prevention (Watson et al. 2004), the SPACE study for physical activity promotion (Toftager et al. 2011) and the SEARCH trial for HIV prevention and treatment (Balzer et al. 2015). This study design creates a dependence in the data. Specifically, the construction of the matched pairs is a function of the covariates of all candidate sites. As a result, the observed data cannot be treated as n i.i.d. observations nor as $n/2$ i.i.d. paired observations, as current practice sometimes assumes (e.g., Hayes and Moulton 2009; Klar and Donner 1997; Freedman et al. 1997; Campbell et al. 2007). However, once the baseline covariates of the study units are considered to be fixed, we recover $n/2$ conditionally independent units:

$$\bar{O}_j = (O_{j1}, O_{j2}) = ((W_{j1}, A_{j1}, Y_{j1}), (W_{j2}, A_{j2}, Y_{j2}))$$

where the index $j = 1, \ldots, n/2$ denotes the partitioning of the candidate study communities $\{1, \ldots, N\}$ into matched pairs according to their baseline covariates $(W_1, \ldots, W_N)$.

Previously, Imai (2008) generalized Neyman's analysis of the unadjusted estimator for the sample effect in a pair-matched trial. The unadjusted estimator, as the average of the pairwise differences in outcomes, is unbiased but inefficient. For an adaptive pair-matched trial, van der Laan et al. (2013a) detailed the use TMLE for the population effect, and Balzer et al. (2015) for the conditional effect. To the best of our knowledge, Balzer et al. (2016c) were the first to consider using a locally efficient estimator for the sample effect in a pair-matched trial.

The TMLE for the population effect, presented in Sect. 12.3.1, also estimates the sample effect in a pair-matched trial. As before, the TMLE minus the SATE can be written as an empirical mean of a paired influence curve depending on nonidentifiable quantities, and a second-order term going to zero in probability:

$$\Psi_n(P_n) - SATE = \frac{1}{n/2} \sum_{j=1}^{n/2} \bar{D}^S(\bar{U}_j, \bar{O}_j) + o_P(1/\sqrt{n/2})$$

where

$$\bar{D}^S(\bar{U}_j, \bar{O}_j) = \bar{D}^C(\bar{O}_j) - \bar{D}^F(\bar{U}_j, \bar{O}_j)$$

$$\bar{D}^C(\bar{O}_j) = \frac{1}{2}\left[D^C(O_{j1}) + D^C(O_{j2})\right]$$

$$\bar{D}^F(\bar{U}_j, \bar{O}_j) = \frac{1}{2}\left[D^F(U_{j1}, O_{j1}) + D^F(U_{j2}, O_{j2})\right]$$

(Theorem 3 in Balzer et al. 2016c). The first component $\bar{D}^C(\bar{O})$ is the influence curve for the TMLE of the conditional parameter $\Psi^C(P_0) = 1/n \sum_{i=1}^{n} \bar{Q}_0(1, W_i) - \bar{Q}_0(0, W_i)$ in a trial with pair-matching (Balzer et al. 2015). In words, $\bar{D}^C(\bar{O}_j)$ is the average of the pairwise $D^C(O_i)$ components, as defined in Eq. (12.2). The second component $\bar{D}^F(\bar{U}, \bar{O})$ is a nonidentifiable function of the pair's unobserved factors $\bar{U} = (U_{j1}, U_{j2})$ and observed factors $\bar{O}_j = (O_{j1}, O_{j2})$. Specifically, $\bar{D}^F(\bar{U}_j, \bar{O}_j)$ is the average of the pairwise $D^F(U_i, O_i)$ components, as defined in Eq. (12.3). As before, there is no contribution from estimation of the covariate distribution $P_0(W)$, which is considered fixed.

As a consequence, the standardized estimator of the SATE in a pair-matched trial is consistent and asymptotically normal with mean zero and variance given by the limit of

$$Var[\bar{D}^S(\bar{U}_j, \bar{O}_j)] = Var[\bar{D}^C(\bar{O}_j)] - Var[\bar{D}^{\mathcal{F}}(\bar{U}_j, \bar{O}_j)]$$

(Theorem 4 in Balzer et al. 2016c). As before, the variance of the nonidentifiable $\bar{D}^{\mathcal{F}}$ component must be greater than or equal to zero. Therefore, in a pair-matched trial the asymptotic variance of the TMLE as an estimator of the sample effect will always be less than or equal to the asymptotic variance of the same estimator of the conditional effect. Furthermore, by treating the covariate distribution as fixed, the TMLE for the sample (or conditional) effect will always be as or more precise than the TMLE of the population effect in a pair-matched trial. We also briefly note that there is often an additional efficiency gain due to pair-matching. The SATE will be estimated with more precision in a pair-matched trial when the deviations between the true and estimated outcome regressions are positively correlated within matched pairs and/or when the deviations between the treatment effect for a unit and the treatment effect within covariate strata are positively correlated within matched pairs.

We can conservatively approximate the influence curve for the TMLE of the SATE in a pair-matched trial as

$$\bar{D}_n^S(\bar{O}_j) = \tfrac{1}{2}[D_n^S(O_{j1}) + D_n^S(O_{j2})]$$

where $D_n^S(O_i)$ is defined in Eq. (12.4). Thereby, we obtain an asymptotically conservative variance estimator with the sample variance of the estimated paired influence curve, divided by sample size $n/2$:

$$\bar{\sigma}_n^{2,S} = \frac{\frac{1}{n/2}\sum_{j=1}^{n/2}\left[\bar{D}_n^S(\bar{O}_j)\right]^2}{n/2}$$

If we order the observations within matched pairs, such that the first corresponds to the unit randomized to the intervention ($A_{j1} = 1$) and the second to the control ($A_{j2} = 0$) and do not estimate the propensity score $P_0(A) = 0.5$, it follows that

$$\bar{D}_n^S(\bar{O}_j) = (Y_{j1} - \bar{Q}_n^*(1, W_{j1})) - (Y_{j2} - \bar{Q}_n^*(0, W_{j2})).$$

In this case, we can represent the variance estimator as the sample variance of the difference in residuals within matched pairs, divided by $n/2$. This variance estimator will be consistent if there is no heterogeneity in the treatment effect within strata of measured covariates (i.e. if the variance of the $\bar{D}^{\mathcal{F}}$ component is zero) *and* if the outcome regression $\bar{Q}_0(A, W)$ is consistently estimated. Under the same conditions, the TMLE will be efficient (i.e. achieve the lowest possible variance among a large class of regular, asymptotically linear estimators). Otherwise, the TMLE will not be efficient and the variance estimator will be conservative (biased upwards). As before, adjusting for predictive baseline covariates can substantially improve power in two ways: (1) by reducing variability in the estimator, and (2) by resulting in a less conservative variance estimator.

12.5 Simulation

We present the following simulation study to (1) further illustrate the differences between the causal parameters, (2) demonstrate implementation of the TMLE, and (3) understand the impact of the parameter specification on the estimator's precision and attained power. We focus on a randomized trial to illustrate the potential gains in efficiency with pair-matching during the design and adjustment during the analysis. All simulations were carried out in R (R Development Core Team 2016). Full R code is available in Balzer et al. (2016c).

Consider the following data generating process for unit $i = \{1, \ldots, n\}$. First, we generated the background error $U_{Y,i}$ by drawing from a standard normal distribution. Then we generated five baseline covariates from a multivariate normal with means 0 and standard deviation 1. The correlation between the first two covariates $(W1_i, W2_i)$ was 0, and the correlation between the last three $(W3_i, W4_i, W5_i)$ was 0.65. The exposure A_i was randomized such that there were equal numbers of intervention and control units. Recall A_i is a binary indicator, equaling 1 if the unit is randomized to the intervention and 0 if the unit is randomized to the control. For a trial without matching, the intervention was randomly assigned to $n/2$ units and the control to the remaining units. For a trial with matching, we applied the nonbipartite matching algorithm nbpMatch (Beck et al. 2016) to pair units on $\{W1, W4, W5\}$. The outcome Y_i was generated as $Y_i = logit^{-1}[A_i + 0.75W1_i + 0.75W2_i + 1.25W3_i + U_{Y,i} + 0.75A_iW1_i - 0.5A_iW2_i - A_iU_{Y,i}]/5$. We also generated the counterfactual outcomes $Y_{a,i}$ by intervening to set $A_i = a$. For sample sizes of $n = \{30, 50\}$, this data generating process was repeated 5000 times. The true value of the SATE was calculated as the average difference in the counterfactual outcomes for each sample, and the true value of the PATE was calculated by averaging the difference in the counterfactual outcomes over a population of 500,000 units. In this population, the correlations between the observed outcome Y and the baseline covariates were weak to moderate: 0.5 for $W1$, 0.2 for $W2$, 0.6 for $W3$, 0.4 for $W4$ and 0.4 for $W5$.

We compared the performance of the unadjusted estimator to the TMLE with two methods for initial estimation of the outcome regression. Specifically, we estimated $\bar{Q}_0(A, W)$ with logistic regression, including as main terms the exposure A, the covariate $W1$ and an interaction A^*W1. We also estimated $\bar{Q}_0(A, W)$ using super learner with a library that consisted of all possible logistic regressions with terms for the exposure A, a single covariate and their interaction. The unadjusted estimator can be considered as a special case of the TMLE, where $\bar{Q}_n(A, W) = \mathbb{E}_n(Y|A)$. Inference was based on the estimated influence curve and the Student's t-distribution. We constructed Wald-type 95% confidence intervals and tested the null hypothesis of no average effect.

Results. Table 12.1 gives a summary of the parameter values across the 5000 simulated trials. Recall the true value of the SATE depends on the units included in the study, whereas there is one true value of the PATE for the population. The sample effect ranged from 0.17% to 5.94% with a mean of 2.97%. The population effect was constant at 2.98%. As expected, the variability in the SATE decreased with increasing sample size.

Table 12.2 illustrates the performance of the estimators. Specifically, we give the bias as the average deviation between the point estimate and (sample-specific) true value, the standard deviation as the square root of the variance of an estimator for its target, and the MSE. We also show the relative MSE (rMSE) as the MSE of a given estimator divided by the MSE of the unadjusted estimator of the population effect in trial without matching. The attained power, which is the proportion of times the false null hypothesis was rejected, and the 95% confidence interval coverage are also included.

As expected, all estimators were unbiased. In randomized trials, there is no risk of bias due to misspecification of the regression model for $\bar{Q}_0(A, W)$ (e.g., Tsiatis et al. 2008; Moore and van der Laan 2009b; Rosenblum and van der Laan 2010b). Also asexpected, the precision of the estimators improved with increasing sample size and

Table 12.1 Summary of the causal parameters (in %) over 5000 simulations of size $n = \{30, 50\}$

	SATE				PATE			
	min	ave	max	var	min	ave	max	var
$n = 30$	0.17	2.97	5.94	6.5E−3	2.98	2.98	2.98	0
$n = 50$	0.18	2.96	5.14	4.2E−3	2.98	2.98	2.98	0

Table 12.2 Summary of estimator performance over 5000 simulations

Target and design	Estimator	Bias	Std. Dev.	MSE	rMSE	Power	Coverage
		Sample size $n = 30$					
PATE and not matched	Unadj	2.3E−4	2.2E−2	4.8E−4	1.00	0.27	0.95
	TMLE	6.8E−4	1.9E−2	3.6E−4	0.75	0.36	0.94
	TMLE+SL	2.9E−4	1.6E−2	2.6E−4	0.55	0.48	0.93
SATE and not matched	Unadj	3.1E−4	2.0E−2	4.2E−4	0.88	0.27	0.96
	TMLE	7.5E−4	1.7E−2	3.0E−4	0.63	0.39	0.95
	TMLE+SL	3.7E−4	1.4E−2	2.0E−4	0.42	0.52	0.95
SATE and matched	Unadj	5.4E−5	1.5E−2	2.2E−4	0.46	0.37	0.98
	TMLE	3.7E−4	1.4E−2	2.1E−4	0.43	0.44	0.97
	TMLE+SL	1.3E−4	1.1E−2	1.3E−4	0.27	0.58	0.97
		Sample size $n = 50$					
PATE and not matched	Unadj	−1.3E−4	1.7E−2	3.0E−4	1.00	0.41	0.94
	TMLE	1.1E−4	1.5E−2	2.2E−4	0.75	0.53	0.94
	TMLE+SL	−3.1E−6	1.2E−2	1.6E−4	0.53	0.68	0.94
SATE and not matched	Unadj	4.8E−5	1.6E−2	2.5E−4	0.86	0.41	0.96
	TMLE	2.9E−4	1.3E−2	1.8E−4	0.60	0.55	0.96
	TMLE+SL	1.8E−4	1.1E−2	1.1E−4	0.38	0.70	0.97
SATE and matched	Unadj	−1.8E−4	1.1E−2	1.1E−4	0.38	0.59	0.98
	TMLE	−1.6E−4	1.0E−2	1.1E−4	0.36	0.66	0.97
	TMLE+SL	−5.7E−5	8.2E−3	6.7E−5	0.23	0.81	0.98

The rows denote target parameter, the study design, and the estimator: unadjusted, TMLE with logistic regression, and TMLE with super learner ("TMLE+SL")

with adjustment (e.g., Fisher 1932; Cochran 1957; Cox and McCullagh 1982; Tsiatis et al. 2008; Moore and van der Laan 2009b; Rosenblum and van der Laan 2010b). Consider, for example, estimation of the population effect in a trial with $n = 30$ units and without matching. The standard error was 2.2^*10^{-2} for the unadjusted estimator and 1.9^*10^{-2} after adjusting for a single covariate. Incorporating data-adaptive estimation of the conditional mean $\bar{Q}_0(A, W)$ through super learner further reduced the standard error to 1.6^*10^{-2}. Also as expected, precision increased with pair-matching (Imai et al. 2009; van der Laan et al. 2013a; Balzer et al. 2015). For the SATE, the standard error of the unadjusted estimator in the trial without matching was 1.38 times higher with $n = 30$ units and 1.49 times higher with $n = 50$ units than its pair-matched counterpart.

For all estimation algorithms and sample sizes, the impact of the target parameter specification on precision and power was substantial. As predicted by theory, the highest variance was seen with the unadjusted estimator of the PATE. With $n = 50$ units, the MSE of this estimator for the PATE was 2.62 times that of the TMLE with super learner for the SATE in a trial without matching and 4.42 times that of the TMLE with super learner for the SATE in a trial with matching. In the finite sample simulations, the impact of having an asymptotically conservative variance estimator on inference for sample effect was notable. In most settings, the standard deviation of an estimator of the SATE was over-estimated, and the confidence interval coverage was greater than or equal to the nominal rate of 95%. Despite the conservative variance estimator, the TMLE for the sample effect achieved higher power than the same TMLE for the population effect. With $n = 30$ units, the attained power for the TMLE with super learner was 48% for the population effect, 52% for the sample effect without matching and 58% for the sample effect after pair-matching. With $n = 50$ units, the attained power for the TMLE with super learner was 68% for the population effect, 70% for the sample effect without matching and 81% for the sample effect after pair-matching Notably, the power was the same for the unadjusted estimator of the two parameters in the trials without matching. The power of the unadjusted estimator did not vary, because the estimated $D_W(O)$ component of influence curve and thereby its variance were zero:

$$\mathbb{E}_n(Y|A = 1) - \mathbb{E}_n(Y|A = 0) - \Psi_{n,unadj}(P_n) = 0$$

where $\mathbb{E}_n(Y|A)$ denotes the treatment-specific mean. Thus, using the unadjusted estimator sacrificed any potential gains in power by specifying the SATE as the target of inference. In contrast, the TMLE using super learner was able to obtain a better fit of the outcome regression $\bar{Q}_0(A, W)$ and a less conservative variance estimator. As a result, this TMLE was able to achieve the most power. We note that in small trials (e.g. $n \leq 30$) such as early phase clinical trials or cluster randomized trials, obtaining a precise estimate of $\bar{Q}_0(A, W)$ is likely to be challenging. In practice, many baseline covariates are predictive of the outcome, but adjusting for too many covariates can result in over-fitting. Chapter 13 discusses a procedure using cross-validation to data adaptively select from a pre-specified library the optimal adjustment set.

12.6 Discussion

The SATE is an interesting and possibly under-utilized causal parameter. In CRTs, the candidate units are often systematically selected to satisfy the study's inclusion criteria. Often, a matching algorithm is applied to select the best $n/2$ matched pairs (Balzer et al. 2015). As a result, the observed data often do not arise from taking a simple random sample from some hypothetical target population of clusters or matched pairs of clusters. In this setting, the SATE, in contrast to the PATE, remains a readily interpretable quantity that can be rigorously estimated without further assumptions on the sampling mechanism. While generalizability of the study findings and their transport to new settings remains of substantial policy interest, neither the SATE nor the PATE directly addresses this goal; these new settings are likely to differ in important ways from both the current sample and any hypothetical target population from which it was drawn. Instead, generalizability and transportability can be approached as distinct research questions, requiring their own identification results and corresponding optimal estimators (Cole and Stuart 2010; Stuart et al. 2011; Bareinboim and Pearl 2013; Schochet 2013; Hartman et al. 2015). Finally, the sample effect will be estimated with at least as much precision and power as the conditional or population effects.

Acknowledgements Research reported in this chapter was supported by Division of AIDS, NIAID of the National Institutes of Health under award numbers R01-AI074345, R37-AI051164, UM1AI069502 and U01AI099959. The content is solely the responsibility of the authors and does not necessarily represent the official views of the NIH.

Chapter 13
Data-Adaptive Estimation in Cluster Randomized Trials

Laura B. Balzer, Mark J. van der Laan, and Maya L. Petersen

In randomized trials, adjustment for measured covariates during the analysis can reduce variance and increase power. To avoid misleading inference, the analysis plan must be pre-specified. However, it is often unclear *a priori* which baseline covariates (if any) should be included in the analysis. This results in an important challenge: the need to learn from the data to realize precision gains, but to do so in pre-specified and rigorous way to maintain valid statistical inference. This challenge is especially prominent in cluster randomized trials (CRTs), which often have limited numbers of independent units (e.g., communities, clinics or schools) and many potential adjustment variables.

L. B. Balzer (✉)
Department of Biostatistics and Epidemiology, School of Public Health and Health Sciences, University of Massachusetts - Amherst, 416 Arnold House, 715 North Pleasant St, Amherst, MA 01003, USA
e-mail: lbalzer@umass.edu

M. J. van der Laan
Division of Biostatistics and Department of Statistics, University of California, Berkeley, 101 Haviland Hall, #7358, Berkeley, CA 94720, USA
e-mail: laan@berkeley.edu

M. L. Petersen
Division of Epidemiology and Division of Biostatistics, University of California, Berkeley, 101 Haviland Hall, #7358, Berkeley, CA 94720, USA
e-mail: mayaliv@berkeley.edu

© Springer International Publishing AG 2018
M.J. van der Laan, S. Rose, *Targeted Learning in Data Science*,
Springer Series in Statistics, https://doi.org/10.1007/978-3-319-65304-4_13

In this chapter, we discuss a rigorous procedure to data adaptively select the adjustment set, which maximizes the efficiency of the analysis. Specifically, we use cross-validation to select from a pre-specified library the candidate TMLE that minimizes the estimated variance. For further gains in precision, we also propose a collaborative procedure for estimating the known exposure mechanism. Our small sample simulations demonstrate the promise of the methodology to maximize study power, while maintaining nominal confidence interval coverage. We show how our procedure can be tailored to the scientific question (intervention effect for the study sample vs. for the target population) and study design (pair-matched or not). This chapter is adapted from Balzer et al. (2016b).

The objective of a randomized trial is to evaluate the effect of an intervention on the outcome of interest. In this setting, the difference in the average outcomes among the treated units and the average outcomes among the control units provides a simple and unbiased estimator of the intervention effect. Adjusting for measured covariates during the analysis can substantially reduce the estimator's variance and thereby increase study power (e.g., Fisher 1932; Cochran 1957; Cox and McCullagh 1982; Tsiatis et al. 2008; Moore and van der Laan 2009b). Nonetheless, recommendations on how and when to adjust in randomized trials have been conflicting (ICH Harmonised Tripartite Guideline 1998; Pocock et al. 2002; Hayes and Moulton 2009; Austin et al. 2010; Kahn et al. 2014; Campbell 2014; European Medicines Agency 2015). The advice seems to depend on the study design, the unit of randomization, the application and the sample size. As a result, many researchers are left wondering how to adjust for baseline covariates, if at all.

Let n be the number of study units (e.g., communities or schools). Consider a trial where the treatment is randomly allocated to $n/2$ units and the remaining units are assigned to the control. There is a rich literature on locally efficient estimation in this setting (e.g., Tsiatis et al. 2008; Zhang et al. 2008; Rubin and van der Laan 2008; Moore and van der Laan 2009b; Shen et al. 2014). For example, parametric regression can be used to obtain an unbiased and more precise estimate of the intervention effect. Briefly, the outcome is regressed on the exposure and covariates according to a working model. Following Rosenblum and van der Laan (2010b), we use "working" to emphasize that the regression function need not be and often is not correctly specified. This working model can include interaction terms and can be linear or nonlinear. The estimated coefficients are then used to obtain the predicted outcomes for all units under the treatment and the control. The difference or ratio of the average of the predicted outcomes provides an estimate of the intervention effect. For observational studies, this algorithm is sometimes referred to as the parametric g-computation (Robins 1986).

For continuous outcomes and linear working models without interaction terms, this procedure is known as analysis of covariance (ANCOVA) (Cochran 1957), and the coefficient for the exposure is equal to the estimate of the intervention effect. For binary outcomes, Moore and van der Laan (2009b) detailed the potential gains in precision from adjustment via logistic regression for estimating the treatment effect on the absolute or relative scale (i.e., risk difference, risk ratio or odds ratio). Furthermore, the authors showed that parametric maximum likelihood estimation was equivalent to TMLE in this setting. As a result, the asymptotic properties of the TMLE, including double robustness and asymptotic linearity, hold even if the working model for outcome regression is misspecified. Furthermore, this approach is locally efficient in that the TMLE will achieve the lowest possible variance among a large class of estimators if the working model is correctly specified. Rosenblum and van der Laan (2010b) expanded these results for a large class of general linear models. Indeed, the parametric MLE and TMLE can be considered special cases of the double robust estimators of Scharfstein et al. (1999) and semiparametric approaches of Tsiatis et al. (2008) and Zhang et al. (2008). For a recent and detailed review of these estimation approaches, we refer the reader to Colantuoni and Rosenblum (2015).

Now consider a pair-matched trial, where the intervention is randomly allocated within the $n/2$ matched pairs. The proposed estimation strategies have been more limited in this setting. Indeed, the perceived "analytical limitation" of pair-matched trials have led some researchers to shy away from this design (Klar and Donner 1997; Imbens 2011; Campbell 2014). As with a completely randomized trial, the unadjusted difference in treatment-specific means provides an unbiased but inefficient estimate of the intervention effect. To include covariates in the analysis and to potentially increase power, Hayes and Moulton (2009) suggested regressing the outcome on the covariates (but not on the exposure) and then contrasting the observed versus predicted outcomes within matched pairs. Alternatively, TMLE can provide an unbiased and locally efficient approach in pair-matched trials (van der Laan et al. 2013a; Balzer et al. 2015, 2016c). Specifically, the algorithm can be implemented as if the trial were completely randomized: (1) fit a working model for the mean outcome, given the exposure and covariates, (2) obtain predicted outcomes for all units under the treatment and control, and (3) contrast the average of the predicted outcomes on the relevant scale. Inference, however, must respect the pair-matching scheme.

A common challenge to both designs is the selection of covariates for inclusion in the analysis. Many variables are measured prior to implementation of the intervention, and it is difficult to *a priori* specify an appropriate working model. For a completely randomized trial, covariate adjustment will lead to gains in precision if (a) the covariates are predictive of the outcome and (b) the covariates are imbalanced between treatment groups (e.g., Moore et al. 2011). Balance is guaranteed as sample size goes to infinity, but rarely seen in practice. Analogously in a pair-matched trial, covariate adjustment will improve precision if there is an imbalance on predictive covariates after matching.

Limited sample sizes pose an additional challenge to covariate selection. A recent review of randomized clinical trials reported that the median number of participants was 58 with an interquartile range of 27–161 (Califf et al. 2012). Likewise, a recent review of CRTs reported that the median number of clusters was 31 with an interquartile range of 13–60 (Selvaraj and Prasad 2013). In small trials, adjusting for too many covariates can lead to overfitting and inflated Type I error rates (e.g., Moore et al. 2011; Shen et al. 2014; Balzer et al. 2015). Finally, *ad hoc* selection of the adjustment set leads to concerns that researchers will go on a "fishing expedition" to find the covariates resulting in the most power and again risking inflation of Type I error rates (e.g., Pocock et al. 2002; Tsiatis et al. 2008; Olken 2015; Rose 2015).

In summary, covariate adjustment in randomized trials can provide meaningful improvements in precision and thereby statistical power. To avoid misleading statistical inference, the working model, including the adjustment variables, must be specified *a priori*. In practice, sample size often limits the size of the adjustment set, and best set is unclear before the trial's conclusion. In this chapter, we apply the principle of *empirical efficiency maximization* to data adaptively select from a pre-specified library the candidate TMLE, which minimizes variance and thereby maximizes the precision of the analysis (Rubin and van der Laan 2008). We modify this strategy for pair-matched trials. We collaboratively estimate the exposure mechanism for additional gains in precision (van der Laan and Gruber 2010). We also generalize the results for estimation and inference to both the population and sample average treatment effects (Neyman 1923; Balzer et al. 2016c). Finite sample simulations demonstrate the practical performance with limited numbers of independent units, as is common in early phase clinical trials and in CRTs.

13.1 Motivating Example and Causal Parameters

As a motivating example, we consider a community randomized trial to estimate the effect of immediate ART on HIV incidence. Suppose trial is being conducted in 30 communities and extensive covariates were measured at baseline. Further suppose, a subset of these characteristics was used to create the 15 best matched pairs of communities (Balzer et al. 2015). The primary outcome is the cumulative incidence of HIV over the relevant time period. The observed data for a given study community can be denoted $O = (W, A, Y)$ where W represents the vector of baseline covariates, A represents the intervention assignment, and Y denotes the outcome. Specifically, W includes region, HIV prevalence, male circumcision coverage and community-level HIV RNA viral load; A is a binary indicator equalling one if the community was randomized to the treatment and zero if the community was randomized to the control; and Y is the estimated cumulative HIV incidence. We focus on estimation and inference for the population average treatment effect

$$PATE = \mathbb{E}[Y_1 - Y_0] \tag{13.1}$$

and the sample average treatment effect (SATE):

$$SATE = \frac{1}{n} \sum_{i=1}^{n} [Y_{1,i} - Y_{0,i}] \tag{13.2}$$

We refer the reader to Chap. 12 for a detailed discussion of these causal parameters.

13.2 Targeted Estimation in a Randomized Trial Without Matching

In this section, we consider a randomized trial without pair-matching. We assume the observed data consist of n independent, identically distributed (i.i.d.) copies of $O = (W, A, Y)$ with some true, but unknown distribution P_0, which factorizes as

$$p_0(o) = p_0(w)p_0(a|w)p_0(y|a, w).$$

We do not make any assumptions about the common covariate distribution $P_0(W)$ or about the common conditional distribution of the outcome, given the intervention and covariates $P_0(Y|A, W)$. By design, the intervention A is randomized with probability 0.5. Therefore, the exposure mechanism is known: $P_0(A = 1|W) \equiv g_0(1|W) = 0.5$. The statistical model $\mathcal{M}$, describing the set of possible observed data distributions, is semiparametric.

As discussed in the introduction, there are many algorithms available for unbiased and locally efficient estimation of the population effect in a randomized trial (e.g., Tsiatis et al. 2008; Zhang et al. 2008; Rubin and van der Laan 2008; Moore and van der Laan 2009b; Shen et al. 2014). Throughout, our focus is on TMLE. A TMLE for the population effect (Eq. (13.1)) also serves as a consistent and asymptotically linear estimator of the sample effect (Eq. (13.2)). We refer the reader to Chap. 12 for the detailed algorithm. Briefly, the algorithm is implemented as follows.

- Step 1: Initial estimation of the conditional mean outcome, given the exposure and covariates $\bar{Q}_0(A, W)$.
- Step 2: Targeting the initial estimator $\bar{Q}_n(A, W)$ with information in the known or estimated exposure mechanism $g_n(A|W) \equiv P_n(A|W)$. Let $\bar{Q}_n^*(A, W)$ denote the targeted estimator of $\bar{Q}_0(A, W)$.
- Step 3: Parameter estimation by taking the average difference in targeted estimates under intervention and control: $\Psi_n(P_n) = 1/n \sum_{i=1}^n \bar{Q}_n^*(1, W_i) - \bar{Q}_n^*(0, W_i)$.

Under standard regularity conditions, the TMLE is an asymptotically linear estimator of both the population and sample effects (Balzer et al. 2016c). The estimator minus the true effect can be written as an empirical mean of an influence curve and a second-order term going to 0 in probability. As a result, the standardized estimator is asymptotically normal with variance well-approximated by the variance of its influence curve, divided by sample size n.

The influence curve for the TMLE of the population effect (PATE) is given by

$$
\begin{aligned}
D^{\mathcal{P}}(g_0, \bar{Q}_{n,lim}^*)(O) \quad &= \left(\frac{\mathbb{I}(A=1)}{g_0(1|W)} - \frac{\mathbb{I}(A=0)}{g_0(0|W)} \right)(Y - \bar{Q}_{n,lim}^*(A, W)) \\
&+ \bar{Q}_{n,lim}^*(1, W) - \bar{Q}_{n,lim}^*(0, W) - \Psi(P_n)
\end{aligned}
$$

where $\bar{Q}_{n,lim}^*(A, W)$ denotes the limit of the targeted estimator of the conditional mean function $\bar{Q}_0(A, W)$ and where we are assuming the exposure mechanism

$g_0(A|W)$ is known or consistently estimated, as will always be true in a randomized trial (van der Laan and Rose 2011). A plug-in estimator of this influence curve is given by

$$D_n^{\mathcal{P}}(g_n, \bar{Q}_n^*)(O) = \left(\frac{\mathbb{I}(A=1)}{g_n(1|W)} - \frac{\mathbb{I}(A=0)}{g_n(0|W)} \right) (Y - \bar{Q}_n^*(A, W))$$
$$+ \bar{Q}_n^*(1, W) - \bar{Q}_n^*(0, W) - \psi_n^* \qquad (13.3)$$

where ψ_n^* denotes the point estimate. In finite samples, the variance of the TMLE for the PATE is well-approximated by the sample variance of this estimated influence curve, scaled by sample size:

$$\sigma_n^{2,\mathcal{P}} = \frac{\frac{1}{n} \sum_{i=1}^{n} \left[D_n^{\mathcal{P}}(g_n, \bar{Q}_n^*)(O_i) \right]^2}{n}.$$

The influence curve for the TMLE of the sample effect (SATE) relies on nonidentifiable quantities, specifically the counterfactual outcomes $Y_{1,i}$ and $Y_{0,i}$ (Balzer et al. 2016c). Nonetheless, a conservative plug-in estimator of its influence curve is obtained by ignoring these nonidentifiable quantities:

$$D_n^{\mathcal{S}}(g_n, \bar{Q}_n^*)(O) = \left(\frac{\mathbb{I}(A=1)}{g_n(1|W)} - \frac{\mathbb{I}(A=0)}{g_n(0|W)} \right) (Y - \bar{Q}_n^*(A, W)). \qquad (13.4)$$

In finite samples, the variance of the TMLE for the SATE is conservatively approximated by the sample variance of this estimated influence curve, scaled by sample size:

$$\sigma_n^{2,\mathcal{S}} = \frac{\frac{1}{n} \sum_{i=1}^{n} \left[D_n^{\mathcal{S}}(g_n, \bar{Q}_n^*)(O_i) \right]^2}{n}.$$

Comparing Eqs. (13.3) and (13.4), we see that for the SATE there is no variance contribution from the covariate distribution, which is considered fixed. As a result, the sample effect will often be estimated with more precision than the population effect (Neyman 1923; Rubin 1990; Imbens 2004). Indeed, the TMLE for the PATE and the TMLE for the SATE will only have the same efficiency bound if the conditional mean $\bar{Q}_0(A, W)$ is consistently estimated *and* if there is no variability in the intervention effect across units (Balzer et al. 2016c). In many settings, there will be effect heterogeneity, and specifying the SATE as the target of inference can yield more power, especially in large trials. In small trials, however, the gains in precision from targeting the SATE can be attenuated, because this influence curve-based variance estimator is conservative (biased upwards).

Adaptive Pre-specified Approach for Step 1: Initial Estimation. Consider again our hypothetical trial for HIV prevention and treatment. Recall that the outcome Y is cumulative incidence of HIV and bounded between 0 and 1. The first step of the TMLE algorithm is to obtain an initial estimator of the expected outcome, given the exposure and measured covariates $\bar{Q}_0(A, W)$. Suppose that as a working model, we consider running logistic regression[1] of the outcome Y on the treatment A and covariates W. It is unclear *a priori* which covariates should be included in the working model and in what form. For example, baseline HIV prevalence is a known predictor of the outcome and may be imbalanced between the treatment and control groups. Therefore, as initial estimator of $\bar{Q}_0(A, W)$, we could consider a logistic regression working model with an intercept and main terms for the treatment and HIV prevalence. Likewise, there might be substantial heterogeneity in the treatment effect by region and allowing for an interaction between region and the intervention may reduce the variance of the TMLE. Including all the covariates and the relevant interactions in the working model is likely to result in overfitting and misleading inference. To facilitate selection between candidate initial estimators and thereby candidate TMLEs, we propose the following cross-validation selector.

First, we propose a library of candidate working models for initial estimation of the conditional mean outcome $\bar{Q}_0(A, W)$. This library should be pre-specified in the protocol or the analysis plan. A possible library could consist of the following logistic regression working models:

$$\text{logit}[\bar{Q}^{(a)}(A, W)] = \beta_0 + \beta_1 A$$
$$\text{logit}[\bar{Q}^{(b)}(A, W)] = \beta_0 + \beta_1 A + \beta_2 W1$$
$$\text{logit}[\bar{Q}^{(c)}(A, W)] = \beta_0 + \beta_1 A + \beta_2 W2 + \beta_3 A \times W2$$

where, for example, $W1$ denotes baseline prevalence and $W2$ denotes region. Of course, there are many more candidate algorithms, and we are considering this simple set for pedagogic purposes. We also note that the first working model corresponds to the unadjusted estimator.

Second, we need to pre-specify a loss function to measure the performance of the candidate estimators. Following the principle of empirical efficiency maximization (Rubin and van der Laan 2008), we propose using the squared influence curve of the TMLE for the parameter of interest. The expectation of this loss function, called the "risk", is then the asymptotic variance of the TMLE. Thereby, our goal is to select the candidate estimator that maximizes precision. If the target of inference is the population effect, our loss function is

$$\mathcal{L}^{P}(g_0, \bar{Q})(O) = \{D^{P}(g_0, \bar{Q})(O)\}^2 \tag{13.5}$$

where we are not estimating the known exposure mechanism $g_0(A|W) = 0.5$. Since the true influence curve of the TMLE for the sample effect relies on nonidentifiable quantities, our loss function for the SATE is the estimated influence curve-squared:

$$\mathcal{L}^{S}(g_0, \bar{Q})(O) = \{D^{S}_n(g_0, \bar{Q})(O)\}^2 \tag{13.6}$$

---

[1] Logistic regression naturally respects the bounds on this continuous outcome. Prior work has suggested that use of the logistic regression over linear regression can provide stability when there are positivity violations or the outcome is rare (Gruber and van der Laan 2010b).

where again we are not estimating the known exposure mechanism $g_0(A|W) = 0.5$. In this case, the loss function for the SATE corresponds to the L2 squared error loss function: $\mathcal{L}^S(g_0, \bar{Q}) = (Y - \bar{Q}(A, W))^2$.

Next, we need to pre-specify our cross-validation scheme, used to generate an estimate of the risk for each of the candidate estimators. For generality, we present V-fold cross-validation, where the data are randomly split into V partitions, called "folds", of size $\approx n/V$. To respect the limited sample sizes common in early phase clinical trials and in CRTs, leave-one-out cross-validation is often appropriate. Leave-one-out cross-validation corresponds with $V = n$-fold cross-validation, where each fold corresponds to one observation. The cross-validation procedure for initial estimation of the conditional mean $\bar{Q}_0(A, W)$ can be implemented as follows.

(A) For each fold $v = \{1, \ldots, V\}$ in turn,

    (a) Set the observation(s) in fold v to be the validation set and the remaining observations to be the training set.

    (b) Fit each algorithm for estimating $\bar{Q}_0(A, W)$ using only data in the training set. For the above library, we would run logistic regression of the outcome Y on the exposure A and covariates W, according to the working model. Denote the initial regression fits as $\bar{Q}_n^{(a)}(A, W)$, $\bar{Q}_n^{(b)}(A, W)$ and $\bar{Q}_n^{(c)}(A, W)$, respectively.

    (c) For each algorithm, use the estimated fit to predict the outcome(s) for the observation(s) in the validation set under the treatment and the control. For the first algorithm, for example, we would have $\bar{Q}_n^{(a)}(1, W_k)$ and $\bar{Q}_n^{(a)}(0, W_k)$ for observation O_k in the validation set.

    (d) For each algorithm, evaluate the loss function for the observation(s) in the validation set by plugging in the algorithm-specific predictions. For example, if our target of inference were the SATE, we would have for the first algorithm

$$\mathcal{L}^S(g_0, \bar{Q}_n^{(a)})(O_k) = \left[\left(\frac{\mathbb{I}(A_k = 1)}{g_0(1|W_k)} - \frac{\mathbb{I}(A_k = 0)}{g_0(0|W_k)}\right)(Y_k - \bar{Q}_n^{(a)}(A_k, W_k))\right]^2$$

for observation O_k in the validation set. The exposure mechanism is known: $g_0(1|W) = 0.5$.

    (e) For each algorithm, obtain an estimate of the risk by averaging the estimated losses across the observations in validation set v. If our target of inference were the SATE, we would have for the first algorithm

$$Risk_v^{(a)} = \frac{1}{n_v} \sum_{k \in v} \mathcal{L}^S(g_0, \bar{Q}_n^{(a)})(O_k)$$

where n_v denotes the number of observations in validation set v.

(B) For each algorithm, average the estimated risks across the V folds.

(C) Select the algorithm with the smallest cross-validated risk. This is the algorithm yielding the smallest cross-validated variance estimate.

The selected working model is then used for initial estimation of the conditional mean outcome $\bar{Q}_0(A, W)$ in Step 1 of the TMLE algorithm, described above (Sect. 13.2). Specifically, we would re-fit the selected algorithm using all the data $\bar{Q}_n(A, W)$. Since the exposure mechanism was treated as known and our library was limited to simple parametric working models with a main term for the exposure and an intercept, the updating step (Step 2) can be skipped. In other words, the chosen estimator is already targeted $\bar{Q}_n(A, W) = \bar{Q}_n^*(A, W)$ and can be used for Step 3 parameter estimation.

13.3 Targeted Estimation in a Randomized Trial with Matching

Recall the pair-matching scheme for our hypothetical community randomized trial. First, the potential study units were selected. Then a matching algorithm was applied to the baseline covariates of candidate units to create the best 15 matched pairs. The intervention was randomized within the resulting pairs, and the outcome measured with longitudinal follow-up. This pair-matching scheme is considered to be *adaptive*, because the resulting matched pairs are a function of the baseline covariates of all the candidate units (van der Laan et al. 2013a; Balzer et al. 2015, 2016c). This design has also been called "nonbipartite matching" and "optimal multivariate matching" (Greevy et al. 2004; Zhang and Small 2009; Lu et al. 2011).

The adaptive design creates a dependence in the data. Since the construction of the matched pairs is a function of the baseline covariates of all n study units, the observed data do not consist of $n/2$ i.i.d. paired observations, as current practice sometimes assumes (e.g., Klar and Donner 1997; Freedman et al. 1997; Campbell et al. 2007; Hayes and Moulton 2009). Instead, we have n dependent copies of $O = (W, A, Y)$. Nonetheless, there remains substantial conditional independence in the data. Mainly, once we consider the baseline covariates of the study units as fixed, we recover $n/2$ conditionally independent units:

$$\bar{O}_j = (O_{j1}, O_{j2}) = ((W_{j1}, A_{j1}, Y_{j1}), (W_{j2}, A_{j2}, Y_{j2}))$$

where the index $j = 1, \ldots, n/2$ denotes the partitioning of the candidate units $\{1, \ldots n\}$ into matched pairs according to similarity in their baseline covariates $(W_1, \ldots, W_n)$. Throughout subscripts $j1$ and $j2$ index the observations within matched pair j. The conditional distribution of the observed data, given the baseline covariates of the study units, factorizes as

$$\begin{aligned}
P_0(O_1, \ldots, O_n | W_1, \ldots W_n) &= \prod_{j=1}^{n/2} P_0(A_{j1}, A_{j2} | W_1, \ldots, W_n) \times P_0(Y_{j1} | A_{j1}, W_{j1}) \\
&\quad \times P_0(Y_{j2} | A_{j2}, W_{j2}) \\
&= \prod_{j=1}^{n/2} 0.5 \times P_0(Y_{j1} | A_{j1}, W_{j1}) \times P_0(Y_{j2} | A_{j2}, W_{j2}),
\end{aligned}$$

where the second line follows from randomization of the intervention within matched pairs. For estimation and inference of the population effect (PATE), we need to assume that each community's baseline covariates W_i are independently drawn from some common distribution $P_0(W)$. For estimation and inference of the sample effect (SATE), this assumption on the covariate distribution can be weakened (Balzer et al. 2016c).

Despite the dependence in the data, a TMLE for the population or sample effect can be implemented by ignoring the pair-matched design (van der Laan et al. 2013a; Balzer et al. 2016c). In other words, a point estimate is obtained by following the procedure outlined in Sect. 13.2. In Step 1, we obtain an initial estimator of the conditional mean outcome with an *a priori*-specified parametric working model or with a more data-adaptive method (as detailed below). In Step 2, we target the initial estimator by using information in the known or estimated exposure mechanism. Finally in Step 3, we obtain the predicted outcomes for all observations under the treatment and the control, and then take the sample average of the difference in these targeted predictions.

In a trial with adaptive pair-matching, the TMLE is an asymptotically normal estimator of both the population and sample effects (van der Laan et al. 2013a; Balzer et al. 2016c). For the PATE, we could estimate its variance with the sample variance of the estimated influence curve in the nonmatched trial $\frac{1}{n}\sum_{i=1}^{n}[D_n^P(g_n, \bar{Q}_n^*)(O_i)]^2$ divided by n. This variance estimator, however, ignores any gains in precision from pair-matching and will be conservative under reasonable assumptions. A less conservative variance estimator is obtained by accounting for the potential correlations of the residuals within matched pairs:

$$\rho_n(\bar{Q}_n^*)(\bar{O}_j) = \frac{1}{n/2}\sum_{j=1}^{n/2}(Y_{j1} - \bar{Q}_n^*(A_{j1}, W_{j1}))(Y_{j2} - \bar{Q}_n^*(A_{j2}, W_{j2})) \quad (13.7)$$

(van der Laan et al. 2013a). In finite samples, we recommend estimating of the variance of the TMLE for the population effect under pair-matching with

$$\bar{\sigma}_n^{2,P} = \frac{\frac{1}{n}\sum_{i=1}^{n}[D_n^P(g_n, \bar{Q}_n^*)(O_i)]^2 - 2\rho_n(\bar{Q}_n^*)(\bar{O}_j)}{n}.$$

In a pair-matched trial, the TMLE minus the sample effect (SATE) behaves as an empirical mean of an influence curve, depending on nonidentifiable quantities (Balzer et al. 2016c). Nonetheless, a conservative plug-in estimator of its influence curve is given by

$$\bar{D}_n^S(g_n, \bar{Q}_n^*)(\bar{O}_j) = \frac{1}{2}\left[D_n^S(g_n, \bar{Q}_n^*)(O_{j1}) + D_n^S(g_n, \bar{Q}_n^*)(O_{j2})\right]$$

where $D_n^S(g_n, \bar{Q}_n^*)(O)$ is the estimated influence curve for observation O in the nonmatched trial (Eq. (13.4)). In finite samples, we conservatively estimate the variance of the TMLE for the sample effect with the sample variance of the estimated (paired) influence curve divided by $n/2$:

$$\bar{\sigma}_n^{2,S} = \frac{\frac{1}{n/2}\sum_{j=1}^{n/2}[\bar{D}_n^S(g_n, \bar{Q}_n^*)(\bar{O}_j)]^2}{n/2}.$$

Adaptive Pre-specified Approach for Step 1: Initial Estimation. By balancing intervention groups with respect to baseline determinants of the outcome, pair-matching increases the efficiency of the study (e.g., Imai et al. 2009; van der Laan et al. 2013a; Balzer et al. 2015, 2016c). Nonetheless, residual imbalance on the baseline predictors often remains, and adjusting for these covariates during the analysis can further increase efficiency. In our running example, suppose the matched pairs were created before baseline HIV prevalence was measured. As a result, there is likely to be variation across the pairs in baseline prevalence, a known driver of HIV incidence. Adjusting for baseline prevalence during the analysis is likely to increase power via two mechanisms: (1) reducing the variance of the TMLE for the point estimate, and (2) resulting in a less conservative variance estimator. Unfortunately, it is unclear *a priori* whether adjusting for prevalence will yield more power than adjusting for other covariates, such as male circumcision coverage or measures of community-level HIV RNA viral load. With only $n/2 = 15$ (conditionally) independent units, we are limited as to the size of the adjustment set. Adjusting for too many covariates can result in over-fitting. As before, we want to data adaptively select the candidate TMLE (i.e. working regression model), which maximizes the empirical efficiency.

The data-adaptive procedure for initial estimation of the conditional mean outcome $\bar{Q}_0(A, W)$ for a nonmatched trial can be modified for a pair-matched trial. As before, we need to pre-specify our library of candidate estimators, our measure of performance, and the cross-validation scheme. We can use the same library of candidate working models for initial estimation of the conditional mean outcome $\bar{Q}_0(A, W)$. To measure performance, however, we want to use as risk the estimated variance of the TMLE under pair-matching. To elaborate, consider the loss function for the sample effect in a nonmatched trial. Minimizing the sum of squared residuals (Eq. (13.6)) targets the conditional mean outcome $\bar{Q}_0(A, W)$. As a result, the algorithm could select a working model adjusting for a covariate that is highly predictive of the outcome but on which we matched perfectly. In our running example, suppose communities were paired within region, because HIV incidence is expected to be highly heterogeneous across regions. Therefore, minimizing the empirical variance of $D_n^S(g_0, \bar{Q})$ might lead to selection of the candidate TMLE with main terms for the intervention and region. This selection would not improve the precision of the analysis over the unadjusted algorithm. (We already "controlled" for region in the design.) Instead, we want to select the candidate TMLE maximizing precision for the parameter of interest in a pair-matched trial. Thereby, our loss function for the PATE is

$$\bar{\mathcal{L}}^{\mathcal{P}}(g_0, \bar{Q})(\bar{O}_j) = \tfrac{1}{2}\{D_n^{\mathcal{P}}(g_0, \bar{Q})(O_{j1})\}^2 + \tfrac{1}{2}\{D_n^{\mathcal{P}}(g_0, \bar{Q})(O_{j2})\}^2$$
$$-2(Y_{j1} - \bar{Q}(A_{j1}, W_{j1}))(Y_{j2} - \bar{Q}(A_{j2}, W_{j2})), \qquad (13.8)$$

and our loss function for the SATE is

$$\bar{\mathcal{L}}^{\mathcal{S}}(g_0, \bar{Q})(\bar{O}_j) = \{\bar{D}_n^{\mathcal{S}}(g_0, \bar{Q}_n^*)(\bar{O}_j)\}^2. \qquad (13.9)$$

Again, we are treating the exposure mechanism as known: $g_0(A|W) = 0.5$.

Finally, in the cross-validation scheme, the pair should be treated as the unit of (conditional) independence. In other words, when the data are split into V-folds, the pairing should be preserved. In small trials, leave-one-pair-out cross-validation will often be appropriate. With these modifications, we can implement the cross-validation scheme, outlined in Sect. 13.2, to data adaptively select the candidate working model, which minimizes the estimated variance of the TMLE in a pair-matched trial. As before, the selected working model would then be refit using all the data and used to estimate outcomes for all observations under the treatment and control. The average difference in the predicted outcomes would provide an estimate of the intervention effect.

13.4 Collaborative Estimation of the Exposure Mechanism

Even though the intervention A is randomized with balanced allocation, estimating the known exposure mechanism $g_0(A|W) = 0.5$ can increase the precision of the analysis (van der Laan and Robins 2003). As before, we want to respect the study design (i.e., pair-matched or not) as well as adjust for a covariate only if its inclusion improves the empirical efficiency. For example, we will generally not want to adjust for a covariate that is imbalanced between the intervention groups (i.e., predictive of A) but not predictive of the outcome. Likewise, if a given covariate (e.g. $W1$) was included in the working model for conditional mean outcome $\bar{Q}_0(A, W)$, further adjusting for this covariate when estimating the exposure mechanism may not increase precision. To this end, we incorporate C-TMLE approach into our algorithm (see Chap. 10).

Adaptive Pre-specified Approach for Step 2: Targeting. First, we propose a library of candidate estimators of the exposure mechanism $g_0(A|W)$. As before, this library should be pre-specified in the protocol or analysis plan. A possible library could consist of the following logistic regression working models:

$$\text{logit}[g^{(a)}(W)] = \beta_0$$
$$\text{logit}[g^{(b)}(W)] = \beta_0 + \beta_1 W1$$
$$\text{logit}[g^{(c)}(W)] = \beta_0 + \beta_1 W2$$

where, for example, $W1$ is baseline prevalence and $W2$ is male circumcision coverage. Each algorithm would yield a different update to a given initial estimator of the conditional mean outcome $\bar{Q}_n(A, W)$, selected by the data-adaptive procedure for Step 1 for trials without matching and for trials with matching. In other words, each candidate estimator of $g_0(A|W)$ results in a different targeted estimator $\bar{Q}_n^*(A, W)$. We also note that the first working model corresponds to the unadjusted estimator.

To choose between candidate algorithms, we need to pre-specify a measure of performance. As before, we propose using as risk the estimated asymptotic variance of the TMLE, appropriate for the study design (i.e. pair-matched or not) and the scientific question (i.e. population or sample effect). Therefore, our loss functions are

- Without matching and for the PATE: $\mathcal{L}^P(g, \bar{Q}_n)$ as in (13.5)
- Without matching and for the SATE: $\mathcal{L}^S(g, \bar{Q}_n)$ as in (13.6)
- With matching and for the PATE: $\bar{\mathcal{L}}^P(g, \bar{Q}_n)$ as in (13.8)
- With matching and for the SATE: $\bar{\mathcal{L}}^S(g, \bar{Q}_n)$ as in (13.9)

where g denotes a candidate estimator of the exposure mechanism and $\bar{Q}_n$ denotes our selected initial estimator of the outcome regression.

Finally, we need to pre-specify our cross-validation scheme, used to obtain an honest measure of risk and to reduce the potential for over-fitting. As before, we present V-fold cross-validation, where the data are partitioned into V folds of size $\approx n/V$. If matching was used, the partitioning should preserve the pairs. The cross-validation selector for collaborative estimation of the exposure mechanism can be implemented as follows.

(A) For each fold $v = \{1, \ldots, V\}$ in turn,

 (a) Set the observation(s) in fold v to be the validation set and the remaining observations to be the training set.

 (b) Fit the initial estimator of the outcome regression $\bar{Q}_n(A, W)$ using only data in the training set.

 (c) Fit each algorithm for estimating the exposure mechanism using only data in the training set. For the above library, we would run logistic regression of the exposure A on the covariates W, according to the working model. Denote the estimated exposure mechanisms as $g_n^{(a)}(A|W)$, $g_n^{(b)}(A|W)$ and $g_n^{(c)}(A|W)$, respectively.

 (d) For each algorithm, use the estimated fit of the exposure mechanism to target the initial estimator $\bar{Q}_n(A, W)$. Denote the targeted regression fits as $\bar{Q}_n^{(a),*}(A, W)$, $\bar{Q}_n^{(b),*}(A, W)$ and $\bar{Q}_n^{(c),*}(A, W)$, where the superscript corresponds to the algorithm used to estimate the exposure mechanism.

 (e) For each algorithm, obtain targeted predictions of the outcome(s) for the observation(s) in the validation set under the treatment and the control. With the first algorithm for fitting the exposure mechanism, for example, we would have $\bar{Q}_n^{(a),*}(1, W_k)$ and $\bar{Q}_n^{(a),*}(0, W_k)$ for observation O_k in the validation set.

 (f) For each algorithm, evaluate the loss function for the observation(s) in the validation set by plugging in the algorithm-specific predictions. For example, if our target of inference were the SATE in a nonmatched trial, we would have for the first algorithm

$$\mathcal{L}^S(g_n^{(a)}, \bar{Q}_n^{(a),*})(O_k) = \left[\left(\frac{\mathbb{I}(A_k = 1)}{g_n^{(a)}(1|W_k)} - \frac{\mathbb{I}(A_k = 0)}{g_n^{(a)}(0|W_k)} \right) (Y_k - \bar{Q}_n^{(a),*}(A_k, W_k)) \right]^2$$

for observation O_k in the validation set.

(g) For each algorithm, obtain an estimate of the risk by averaging the estimated losses across the observations in validation set v. If our target of inference were the SATE in a nonmatched trial, we would have for the first algorithm for estimating the exposure mechanism

$$Risk_v^{(a)} = \frac{1}{n_v} \sum_{k \in v} \mathcal{L}^S(g_n^{(a)}, \bar{Q}_n^{(a),*})(O_k)$$

where n_v denotes the number of observations in validation set v.

(B) For each algorithm, average the estimated risks across the V folds.
(C) Select the algorithm with the smallest cross-validated risk. This is the algorithm yielding the smallest cross-validated variance estimate.

The chosen estimator is then used for targeting in Step 2 of the TMLE algorithm.

In this scheme, we are treating the initial estimator of the outcome regression $\bar{Q}_n(A, W)$ as fixed and proposing a second round of cross-validation to select the fit of the exposure mechanism. An alternative would be to build a library of candidate TMLEs indexed by choice of initial estimator of outcome regression and estimator of exposure mechanism, and select among this library using the cross-validated variance of the influence curve as the measure of performance. In cluster randomized trials, we recommend the double cross-validation approach to embrace the collaborative principle for estimating the exposure mechanism and to avoid over-fitting. We only want estimate $g_0(A|W)$ if it further improves efficiency beyond adjustment when estimating $\bar{Q}_0(A, W)$. The double cross-validation approach estimates the exposure mechanism $g_0(A|W)$ in response to the fit of the outcome regression $\bar{Q}_0(A, W)$. For example, we could restrict the library for $g_0(A|W)$ in response to the selection for $\bar{Q}_0(A, W)$: if a given covariate was selected for estimation of the outcome regression, then remove the corresponding algorithm from the library for the exposure mechanism. Finally, the double cross-validation approach allows us to consider a large set of possible candidate TMLEs: all possible combinations of estimators for $\bar{Q}_0(A, W)$ and estimators for $g_0(A|W)$. In trials with many (conditionally) independent units, we could consider the single cross-validation approach, which corresponds to the discrete super learner, or a full super learner approach with loss function as the squared influence curve.

13.5 Obtaining Inference

In summary, we have proposed the following data-adaptive C-TMLE to maximize the precision and power of a randomized trial.

- Step 1. Initial estimation of the conditional mean outcome with the working model $\bar{Q}_n(A, W)$, which was data adaptively selected to maximize the empirical efficiency of the analysis for a nonmatched trial and for a matched trial.

- Step 2. Targeting the initial estimator $\bar{Q}_n(A, W)$ using the estimated exposure mechanism $g_n(A|W)$, which was data adaptively selected to further maximize the empirical efficiency of the analysis.
- Step 3. Obtaining a point estimate by averaging the difference in the targeted predictions:

$$\Psi_n(\bar{Q}_n^*) = \frac{1}{n} \sum_{i=1}^{n} [\bar{Q}_n^*(1, W_i) - \bar{Q}_n^*(0, W_i)].$$

We now need a variance estimator that accounts for the selection process. For this, we propose using a cross-validated variance estimator. As before, the data are split into validation and training sets, respecting the unit of (conditional) independence. The selected TMLE is fit using the data in the training set and used to estimate the influence curve[2] for the observation(s) in the validation set. The sample variance of the cross-validated estimate of the influence curve can then be used for hypothesis testing and the construction of Wald-type confidence intervals. We note that a cross-validated estimate of the influence curve was already calculated to evaluate the performance of the candidate estimators. Therefore, this step does not require any extra calculations; we already have an estimate of the cross-validated variance from our selection procedure. We also note that for small libraries (e.g., two candidate TMLEs), simulations support the use of the standard, as opposed to cross-validated, variance estimator for inference (Balzer et al. 2016b).

13.6 Small Sample Simulations

We present the following simulation studies to demonstrate (1) implementation of the proposed methodology, (2) the potential gains in precision and power from data-adaptive estimation of the conditional mean outcome, (3) the additional gains in precision and power from collaborative estimation of the exposure mechanism, and (4) maintenance of nominal confidence interval coverage. All simulations were conducted in R (R Development Core Team 2016). Full R code is provided in Balzer et al. (2016b).

13.6.1 Study 1

For each unit $i = \{1, \ldots, n\}$, we generated the nine baseline covariates by drawing from a multivariate normal with mean 0 and variance 1. The correlation between the first three covariates $\{W1, W2, W3\}$ and between the second three covariates

---

[2] For the TMLE of the population effect in a pair-matched trial, we also need a cross-validated estimate of the correction term ρ_n (Eq. (13.7)). This term is a function of the residuals, which can be estimated for each pair in the validation set based on targeted estimator $\bar{Q}_n^*(A, W)$, fit with the training set.

$\{W4, W5, W6\}$ was 0.5, while the correlation between the remaining covariates $\{W7, W8, W9\}$ was 0. The exposure A was randomized such that the treatment allocation was balanced overall. For the nonmatched trial, we randomly assigned the intervention to $n/2$ units and the control to the remaining $n/2$ units. For the pair-matched trial, we used the nonbipartite matching algorithm nbpMatch to pair units on covariates $\{W1, \ldots, W6\}$ (Beck et al. 2016), and the exposure A was randomized within the resulting matched pairs. Recall A is a binary indicator, equalling 1 if the unit was assigned the treatment and 0 if the unit was assigned the control. For each unit, the outcome Y was then generated as $Y = 0.4A + 0.25(W1 + W2 + W4 + W5 + U_Y) + 0.25A(W1 + U_Y)$, where U_Y was drawn from a standard normal. We also generated the counterfactual outcomes Y_1 and Y_0 by intervening to set $A = a$. To reflect the limited sample sizes common in early phase clinical trials and in CRTs, we selected a sample size of $n = 40$. This resulted in $n/2 = 20$ conditionally independent units in the pair-matched trial.

For each study design (nonmatched or matched), this data generating process was repeated 2500 times. Recall that the sample effect (Eq. (13.2)) is a data-adaptive parameter; its value changes with each new selection of units. Thereby, for each repetition, the SATE was calculated as the sample average of the difference in the counterfactual outcomes. The SATE ranged from 0.22 to 0.59 with a mean of 0.40. In contrast, the population effect (Eq. (13.1)) is constant and was calculated by averaging the difference in the counterfactual outcomes over a population of 900,000 units. The true value of the PATE was 0.40.

We compared the performance of the unadjusted estimator to TMLE with various approaches to covariate adjustment. Specifically, we implemented the TMLE algorithm, where the initial estimation of the conditional mean outcome $\bar{Q}_0(A, W)$ was based on a linear working model with main terms for the intervention A and the irrelevant covariate $W9$ and where the exposure mechanism was treated as known: $g_0(A|W) = 0.5$. This approach was equivalent to standard MLE and represented the unfortunate scenario where the researcher pre-specified adjustment for a covariate that was not predictive of the outcome.

We also implemented a TMLE with the data-adaptive approach for Step 1 initial estimation of the conditional mean outcome. Our library consisted of ten working linear regression models, each with an intercept, a main term for the exposure A and a main term for one baseline covariate: $\{\emptyset, W1, \ldots, W9\}$, where $\emptyset$ corresponds to the unadjusted estimator. Our measure of performance (i.e. our risk function) was the estimated asymptotic variance of the TMLE, appropriate for the target parameter and study design. We chose the candidate working model based on leave-one-out cross-validation for the nonmatched trial and leave-one-pair-out cross-validation for the matched trial. We also implemented C-TMLE which couples the data-adaptive approach for Step 1 initial estimation of the conditional mean outcome with the data-adaptive approach for Step 2 targeting. For the latter, our library of candidates to estimate the exposure mechanism consisted of ten working logistic regression models, each with an intercept and a main term for one baseline covariate: $\{\emptyset, W1, \ldots, W9\}$. The same loss function and cross-validation scheme were used for C-TMLE.

For the unadjusted estimator and the MLE, inference was based on the estimated influence curve. For the data-adaptive TMLEs, inference was based on the cross-validated estimate of the influence curve (Sect. 13.5). We assumed the standardized estimator followed the Student's t-distribution with $n - 2 = 38$ degrees of freedom for the nonmatched trial and with $n/2 - 1 = 19$ degrees of freedom for the matched trial.

Results. Table 13.1 illustrates the performance of the estimators over the 2500 simulated data sets. Specifically, we show the MSE, the relative MSE (rMSE), the average standard error estimate $\hat{\sigma}$, the attained power and the 95% confidence interval coverage. As expected, matching improved efficiency. The MSE of the unadjusted estimator, for example, was over two times larger in the nonmatched trial than in the pair-matched trial. Furthermore, for the pair-matched trial, targeting the sample effect, as opposed to the population effect, resulted in substantial gains in attained power: 36% with the unadjusted estimator for the PATE and 53% with the same estimator for the SATE. For the trial without matching, targeting the sample parameter increased efficiency (smaller MSE), but did not directly translate into increased power due to the conservative variance estimator for the SATE.

Table 13.1 Summary of estimator performance for Simulation 1

	PATE					**SATE**				
	MSE	rMSE	$\hat{\sigma}$	Power	Cover.	MSE	rMSE	$\hat{\sigma}$	Power	Cover.
Non-matched										
Unadj	6.8E−2	1.00	0.25	0.34	0.94	6.4E−2	1.06	0.25	0.34	0.94
MLE	6.9E−2	0.98	0.25	0.35	0.94	6.5E−2	1.04	0.25	0.35	0.94
TMLE	4.5E−2	1.49	0.20	0.48	0.94	4.2E−2	1.62	0.20	0.48	0.95
C-TMLE	4.3E−2	1.57	0.20	0.48	0.95	4.0E−2	1.70	0.20	0.48	0.96
Matched										
Unadj	3.2E−2	2.10	0.22	0.36	0.99	2.9E−2	2.31	0.18	0.53	0.97
MLE	3.4E−2	2.01	0.22	0.37	0.98	3.1E−2	2.19	0.18	0.53	0.96
TMLE	2.6E−2	2.64	0.19	0.51	0.98	2.3E−2	2.93	0.16	0.65	0.96
C-TMLE	2.5E−2	2.71	0.18	0.53	0.98	2.2E−2	3.03	0.15	0.67	0.96

The rows denote the study design and the estimator: unadjusted, MLE adjusting for $W9$, TMLE with data-adaptive selection of the initial estimator, and C-TMLE with data-adaptive selection. Columns denote estimator performance: MSE as the bias^2 plus the variance; rMSE as the MSE of the unadjusted estimator for the PATE in a nonmatched trial divided by the MSE of another estimator; $\hat{\sigma}$ as the average standard error estimate; power; and coverage

In all scenarios, the MSE of the MLE, adjusting for the irrelevant covariate $W9$, was worse than the other estimators. This demonstrates the potential peril of relying on one pre-specified adjustment variable. Indeed, the TMLE with data-adaptive selection of the initial estimator of $\bar{Q}_0(A, W)$ improved precision over the unadjusted estimator and the MLE. Collaborative estimation of the exposure mechanism $g_0(A|W)$ led to further gains in precision. Consider, for example, estimation of the PATE in a trial without matching. The MSE of the unadjusted estimator was 1.49 times larger than the TMLE and 1.57 times larger than the C-TMLE. The attained power was 34%, 48% and 48%, respectively. As a second example, consider the attained power to detect that the SATE was different from zero in the pair-matched trial. We would have 53% power with the unadjusted estimator and with the MLE, adjusting for the irrelevant covariate $W9$. By incorporating the cross-validation selector for initial estimation of $\bar{Q}_0(A, W)$, the TMLE achieved 65% power. By further incorporating collaborative estimation of the exposure mechanism $g_0(A|W)$, the C-TMLE achieved 67% power.

Overall, the greatest efficiency was achieved with C-TMLE for the SATE in the pair-matched trial. Indeed, the MSE of the unadjusted estimator for the population parameter in the trial without matching was three times larger than the MSE of the C-TMLE for the sample effect in the pair-matched trial. Throughout, the confidence interval coverage was maintained near or above the nominal rate of 95%.

13.6.2 Study 2

For the second simulation study, we increased the complexity of the data-generating process and reduced the sample size to $n = 30$. As before, we generated nine baseline covariates from a multivariate normal with mean 0, variance 1 and the same correlation structure. We also generated a binary variable R, equalling 1 with probability 0.5 and equalling -1 with probability 0.5. The final covariate Z was generated as a function of these baseline covariates and random noise U_Z: $Z = R \times \text{logit}^{-1}(W1 + W4 + W7 + 0.5U_Z)$, where U_Z was drawn independently from a standard normal. As before, the intervention A was randomized with balanced allocation. For the pair-matched trial, we used the nonbipartite matching algorithm nbpMatch to explore two matching sets (Beck et al. 2016). In the first, units were matched on R, a baseline covariate strongly impacting Z. In the second, units were matched on $\{R, W2, W5, W8\}$. For each unit, the outcome Y was then generated as

Table 13.2 Simulation 2: covariate/outcome relationships; adaptive pair-matching schemes

		Correlation 0.5			Correlation 0.5			Correlation 0			
	R	W1	W2	W3	W4	W5	W6	W7	W8	W9	Z
Parents of covariate Z	✓	✓			✓			✓			
Parents of the outcome Y			✓			✓			✓		✓
Matching set 1	✓										
Matching set 2	✓		✓			✓			✓		

$Y = \text{logit}^{-1}[0.75A + 0.5(W2 + W5 + W8) + 1.5Z + 0.25U_Y + 0.75A(W2 - W5) + 0.5AZ]/7.5$, where U_Y was drawn from a standard normal. Thereby, the outcome was a continuous variable bounded in $[0, 1]$ (e.g. a proportion). We also generated the counterfactual outcomes Y_1 and Y_0 by intervening to set $A = a$. For each study design, this data generating process was repeated 2500 times. The SATE and PATE were calculated as before. The SATE ranged from 0.2% to 3.3% with a mean of 1.6%. The true value of the PATE was 1.6%. Table 13.2 depicts the relationship between the baseline covariates and the outcome as well as the adaptive pair-matching schemes.

We compared the same algorithms: the unadjusted estimator, the MLE adjusting for the irrelevant covariate $W9$, the TMLE with data-adaptive estimation of the conditional mean outcome, and the C-TMLE pairing data-adaptive estimation of the conditional mean outcome with data-adaptive targeting. Our library for initial estimation of the conditional mean outcome $\bar{Q}_0(A, W)$ consisted of 12 working logistic regression models, each with an intercept and a main term for the exposure A and a main term for one candidate adjustment variable $\{\emptyset, R, W1, \ldots, W9, Z\}$. Our library for collaborative estimation of the exposure mechanism $g_0(A|W)$ included 12 working logistic regression models, each with an intercept and a main term for one candidate adjustment variable: $\{\emptyset, R, W1, \ldots, W9, Z\}$. We used the same measure of performance and cross-validation scheme. As before, inference was based on the estimated influence curve for the unadjusted estimator and the MLE and on the cross-validated estimate of the influence curve for the TMLEs (Sect. 13.5). We assumed the standardized estimator followed the Student's t-distribution with $n - 2 = 28$ degrees of freedom for the nonmatched trial and with $n/2 - 1 = 14$ degrees of freedom for the matched trial.

Results. The results for the second simulation study are given in Table 13.3 and largely echoed the above findings. Pair-matching, even on a single covariate (i.e., match set 1), improved the precision of the analysis. Targeting the sample effect instead of the population effect further improved efficiency. Incorporating data-adaptive selection of the working model for initial estimation of $\bar{Q}_0(A, W)$ yielded even greater precision, and the most efficient analysis was with C-TMLE. Indeed, the MSE of the unadjusted estimator for the PATE in the nonmatched trial was nearly 4.5 times higher than the MSE of the C-TMLE for the SATE when matching on predictive covariates (i.e., match set 2). This resulted in 29% more power to detect the intervention effect.

For these simulations, there was a notable impact of parameter specification on estimator performance. We first focus on the estimation of the PATE and then on estimation of the SATE. When the population effect was the target of inference, the gains in attained power from pair-matching were attenuated despite the gains in MSE. This was likely due to the slight underestimation of the standard error in the nonmatched trial and overestimation in the pair-matched trial. Indeed, the 95% confidence interval coverage in the nonmatched trial was slightly less than nominal (93–94%), while the coverage when matching well (i.e., match set 2) approached 100%. For this set of simulations, the correction factor ρ_n (Eq. (13.7)) used in variance estimation for the pair-matched design was approximately 0. As a result, the variance estimator in the pair-matched trial was quite conservative, and the cross-validation selection scheme was more optimized for the nonmatched trial. The logistic regression model adjusting for R was selected for initial estimation of $\bar{Q}_0(A, W)$ in 10%

Table 13.3 Summary of estimator performance for Simulation 2

	PATE					SATE				
	MSE	rMSE	$\hat{\sigma}$	Power	Cover.	MSE	rMSE	$\hat{\sigma}$	Power	Cover.
Non-matched										
Unadj	1.8E−4	1.00	0.013	0.24	0.94	1.6E−4	1.12	0.013	0.24	0.95
MLE	1.8E−4	0.95	0.012	0.25	0.93	1.7E−4	1.06	0.012	0.25	0.94
TMLE	1.2E−4	1.50	0.010	0.33	0.94	9.8E−5	1.79	0.010	0.33	0.96
C-TMLE	1.1E−4	1.54	0.010	0.34	0.93	9.5E−5	1.85	0.010	0.34	0.96
Match set 1										
Unadj	1.1E−4	1.54	0.012	0.21	0.98	9.2E−5	1.90	0.011	0.28	0.97
MLE	1.2E−4	1.48	0.012	0.23	0.97	9.7E−5	1.81	0.011	0.29	0.97
TMLE	9.2E−5	1.91	0.010	0.31	0.97	6.9E−5	2.52	0.009	0.40	0.96
C-TMLE	9.0E−5	1.95	0.010	0.33	0.96	6.9E−5	2.53	0.008	0.44	0.95
Match set 2										
Unadj	6.5E−5	2.70	0.011	0.17	0.99	4.6E−5	3.79	0.009	0.37	0.98
MLE	7.3E−5	2.41	0.011	0.20	0.99	5.4E−5	3.27	0.009	0.37	0.98
TMLE	5.3E−5	3.30	0.009	0.28	0.99	3.8E−5	4.66	0.008	0.47	0.98
C-TMLE	5.3E−5	3.28	0.009	0.32	0.99	3.9E−5	4.44	0.007	0.53	0.97

The rows denote the study design and the estimator: unadjusted, MLE adjusting for $W9$, TMLE with data-adaptive selection of the initial estimator, and C-TMLE with data-adaptive selection. Columns denote estimator performance: MSE as the bias^2 plus the variance; rMSE as the MSE of the unadjusted estimator for the PATE in a nonmatched trial divided by the MSE of another estimator; $\hat{\sigma}$ as the average standard error estimate; power; and coverage

of the studies without matching and in 7% of the studies when matching well on R (i.e., match set 1). Furthermore, when matching on several covariates (i.e., match set 2), the selection of working models for $\bar{Q}_0(A, W)$ was very similar to the selection in the nonmatched trial.

In contrast, when estimating the SATE, smaller MSE translated to greater attained power, while maintaining nominal, if not conservative, confidence interval coverage. For example, the attained power of the TMLE was 33% in the nonmatched trial, 40% when matching on a single covariate and 47% when matching on several covariates. Likewise, the attained power of the C-TMLE was 34% in the nonmatched trial, 44% in the trial pair-matching on a single covariate and 53% in trial matching on several covariates. The working model adjusting for R was selected for initial estimation of $\bar{Q}_0(A, W)$ in 10% of the studies without matching and only in 2% of the studies when matching well on R (i.e., match set 1). In the latter, more weight was given to other predictive baseline covariates, such as $W2$ and Z.

13.7 Discussion

This chapter builds on the rich history of covariate adjustment in randomized trials (e.g., Fisher 1932; Cochran 1957; Cox and McCullagh 1982; Tsiatis et al. 2008; Zhang et al. 2008; Moore et al. 2011; Yuan et al. 2012; Shen et al. 2014; Colantuoni and Rosenblum 2015). In particular, Rubin and van der Laan (2008) proposed the principle of *empirical efficiency maximization* as a strategy to select the estimator of conditional mean outcome $\bar{Q}_0(A, W)$ that minimized the empirical variance of the estimated efficient influence curve. Their procedure, however, relied on solving a weighted nonlinear least squares problem. Our approach only requires researchers to take the sample variance.

Recent developments in C-TMLE proposed collaborative estimation of the exposure mechanism to achieve the greatest bias reduction in the targeting step of TMLE in a observational study. In randomized trials, there is no risk of bias from regression model misspecification (Rosenblum and van der Laan 2010b). Thereby, the collaborative approach, implemented here, serves only to increase precision by estimating the known exposure mechanism. This chapter generalizes this scheme for estimation and inference of both the population and sample average treatment effects in randomized trials with and without pair-matching. Therefore, our procedure dispels the common concern of "analytical limitation" to pair-matched trials (e.g., Klar and Donner 1997; Imbens 2011; Campbell 2014). Since the step-by-step algorithm (including the library definition) is pre-specified, there is no risk of bias or misleading inference from *ad hoc* analytic decisions. Furthermore, including the unadjusted estimator as a candidate obviates the need for guidelines on whether or not to adjust (e.g., Moore et al. 2011; Colantuoni and Rosenblum 2015). Finally, our procedure is tailored to the scientific question (population vs. sample effect) and study design (with or without pair-matching). Decisions about whether to adjust and how to adjust are made with a rigorous and principled approach, removing some of the "human art" from statistics.

Acknowledgements Research reported in this chapter was supported by Division of AIDS, NIAID of the National Institutes of Health under award numbers R01-AI074345, R37-AI051164, UM1AI069502 and U01AI099959. The content is solely the responsibility of the authors and does not necessarily represent the official views of the NIH.

Part IV
Observational Data

Chapter 14
Stochastic Treatment Regimes

Iván Díaz and Mark J. van der Laan

Standard statistical methods to study causality define a set of treatment-specific counterfactual outcomes as the outcomes observed in a hypothetical world in which a given treatment strategy is applied to all individuals. For example, if treatment has two possible values, one may define the causal effect as a comparison between the expectation of the counterfactual outcomes under regimes that assign each treatment level with probability one. Regimes of this type are often referred to as *static*. Another interesting type of regimes assign an individual's treatment level as a function of the individual's measured history. Regimes like this have been called *dynamic*, since they can vary according to observed pre-treatment characteristics of the individual. Static and dynamic regimes have often been called *deterministic*, because they are completely determined by variables measured before treatment.

Though they are ubiquitous in applied research, deterministic regimes do not provide an appropriate framework to tackle causality questions concerning phenomena that are not subject to direct intervention. For example, in public health research, realistic regimes often fail to put the treatment variable into a deterministic state (e.g., it is unrealistic to set an individuals exercise regime according to a deterministic function), or are the result of implementing policies that target stochastic changes in the behavior of a population (e.g., the use of mass media messages advertising condom use is deterministic at the community level but stochastic at the individual level, because each individual will decide to adopt or not treatment depending upon exogenous factors). In addition, causal effects for deterministic regimes may be unidentifiable because the regime of interest is not supported in the observed

I. Díaz (✉)

Division of Biostatistics and Epidemiology, Department of Healthcare Policy and Research, Weill Cornell Medical College, Cornell University, 402 East 67th Street, New York, NY 10065, USA
e-mail: ild2005@med.cornell.edu

M. J. van der Laan

Division of Biostatistics and Department of Statistics, University of California, Berkeley, 101 Haviland Hall, #7358, Berkeley, CA 94720, USA
e-mail: laan@berkeley.edu

© Springer International Publishing AG 2018
M.J. van der Laan, S. Rose, *Targeted Learning in Data Science*,
Springer Series in Statistics, https://doi.org/10.1007/978-3-319-65304-4_14

data (e.g., health problems are expected to prevent certain portions of the population from higher levels of physical activity). This poses a problem for interpretation of the causal effects based on deterministic regimes, because the estimated effects correspond to regimes that cannot be implemented in practice.

> In this chapter, we consider a generalization of dynamic regimes in which the assigned treatment is also allowed to depend on the natural value of treatment (i.e., the treatment value observed under no intervention). Because the treatment level assigned under this regime cannot be determined until the natural value of treatment is observed, we have called these regimes *stochastic*. Other names found in the literature include stochastic policies, random interventions, randomized dynamic strategies, modified treatment policies, etc. Stochastic regimes are allowed to depend on the natural value of treatment, and can therefore always be defined to be relevant and realistic.

To illustrate this, consider the following two examples:

Example 14.1. Tager et al. (1998) carried out a study with the main goal of assessing the effect of leisure-time physical activity (LTPA) on mortality in the elderly. In principle, one could consider a set of hypothetical worlds corresponding to deterministic regimes on LTPA, for example setting the LTPA level deterministically to each of it possible values. Though conceivable in principle, counterfactual outcomes defined in this way are unsatisfactory because one could not possibly implement a regime that sets an individual's physical activity level deterministically. As a solution, consider a regime that assigns treatment as a function of the natural value of treatment. For example, an individual whose current physical activity level is a may be assigned $a + \delta$ under the regime. More realistically, this regime may be assigned only to individuals for whom it is feasible, where feasibility may be determined according to other covariates, such as health status and the current physical activity level. A regime of this type is more realistic than any deterministic regime and, arguably, may be implemented in the real world. The definition, identification, and estimation of a causal effect defined in this way was first developed by Díaz and van der Laan (2012) and further considered by Haneuse and Rotnitzky (2013).

Example 14.2. Mann et al. (2010) carried out a study analyzing the causal effect of air pollution levels on respiratory symptoms in children with asthma (Fresno Asthmatic Children's Environment Study, FACES). A central aim of the study is to investigate the effect of NO_2 air concentrations on wheezing in asthmatic children. In particular, it may be of interest to assess the effect of a regime that reduces NO_2 air concentrations in the right tail of the NO_2 distribution. An example of such a regime would be to enforce NO_2 levels below a certain threshold. Under this regime, compliant units, those below the selected threshold, may have no incentive to reduce their pollution levels and thus may remain unchanged under the regime. Units above the threshold are likely to reduce their pollution levels only to achieve the threshold and be in compliance with the policy, but may not have any incentive to carry out additional reductions. Causal inference methods to assess the effect of this type of regime were developed by Díaz and van der Laan (2013a).

In addition to aiding in defining more meaningful target causal parameters, stochastic treatment regimes provide a more tractable estimation framework for continuous exposures. Parameters such as causal dose-response curves are hard to estimate in the nonparametric model because they are not pathwise differentiable (e.g., they have an infinite efficiency bound). Other methods, such as those based in categorization of the continuous treatment, fail to use the continuous nature of the treatment and thus are not adequate to answer questions regarding interventions. Stochastic treatment regimes can also be used to tackle common problems, such as identification and estimation of the natural direct effect (NDE), community interventions, individualized treatment regimes, and intention to treat rules.

14.1 Data, Notation, and Parameter of Interest

Let A denote a treatment variable, let Y denote a continuous or binary outcome, and let W denote a vector of observed pre-treatment covariates. Let $O = (W, A, Y)$ represent a random variable with distribution P_0, and let $O_1, \ldots, O_n$ denote a sample of n i.i.d. observations of O. We assume $P_0 \in \mathcal{M}$, where $\mathcal{M}$ is the nonparametric model defined as all continuous densities on O with respect to a dominating measure ν. Let p_0 denote the corresponding probability density function. Then,

$$p_0(o) = p_0(y \mid a, w) p_0(a \mid w) p_0(w).$$

We denote $g_0(a \mid w) = p_0(a \mid w)$, $\bar{Q}_0(a, w) = E_0(Y \mid A = a, W = w)$, and $q_{W,0}(w) = p_0(w)$, as well as $Pf = \int f(o) dP(o)$ for a given function $f(o)$ and a general distribution function $P \in \mathcal{M}$. We use P_n to denote the empirical distribution of $O_1, \ldots, O_n$.

We assume the following nonparametric structural equation model (NPSEM):

$$W = f_W(U_W); \quad A = f_A(W, U_A); \quad Y = f_Y(A, W, U_Y). \tag{14.1}$$

This set of equations represents a mechanistic model that is assumed to generate the observed data O, and it encodes several assumptions. First, there is an implicit temporal ordering: Y is assumed to occur after A and W, and A is assumed to occur after W. Second, each variable is assumed to be generated as deterministic function of the observed variables that precede it, plus an exogenous variable, denoted by U. Each exogenous variable is assumed to contain all unobserved causes of the corresponding observed variable. We assume the following independence condition on the exogenous variables:

$$U_A \perp\!\!\!\perp U_Y. \tag{14.2}$$

This assumption plays a crucial role to achieve the identification result of the causal effect of A on Y from the observed data distribution, described in Sect. 14.1.1. The set of allowed directed acyclic graphs (DAG) implied by this assumption is given in Fig. 14.1.

The causal effect of A on Y is defined as follows. Consider a hypothetical modification to NPSEM (14.1) in which the equation corresponding to A is removed, and A is set equal to a hypothetical regime $d(A, W)$. Regime d depends on the treat-

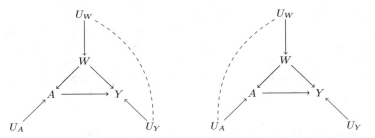

Fig. 14.1 Set of allowed directed acyclic graphs. *Dashed lines* represents correlations, *solid arrows* represent causal relations

ment level that would be assigned in the absence of the regime, A, as well as the covariates W. In our illustrative examples, these regimes may be defined as follows.

Example 14.1 (Continued). Let the distribution of A conditional on $W = w$ be supported in the interval $(l(w), u(w))$. That is, the maximum possible amount of physical activity for an individual with covariates $W = w$ is $u(w)$. Then one could define

$$d(a, w) = \begin{cases} a + \delta & \text{if } a < u(w) - \delta \\ a & \text{if } a \geq u(w) - \delta, \end{cases} \qquad (14.3)$$

where δ some pre-specified amount of physical activity, for example 2 h per week. Under this regime, individuals for whom it is feasible are required to perform δ more units of physical activity. Interesting modifications to this regime may be obtained by allowing δ to be a function of w, therefore allowing the researcher to specify a different increase in physical activity as a function of covariates such as health status, age, etc.

Díaz and van der Laan (2012) interpret this stochastic treatment regimes in terms of a change in the probabilistic mechanism used to assign exposure level. Haneuse and Rotnitzky (2013) point out that such interpretation may be undesirable, arguing as follows. Consider a new distribution for physical activity in which treatment is assigned according to a location-shifted version of the pre-intervention distribution. If stochastic regimes are interpreted as changing the treatment assignment mechanism for a location-shifted distribution, an individual with a physical activity level of 30 min may receive a treatment level of 10 min under the new regime. This may be problematic as such intervention could reduce the physical activity level at the individual level. We note, however, that the population distribution of the exposure is the same under both interventions, and thus they lead to exactly the same counterfactual distributions. As a result, the interpretation adopted is inconsequential for the definition, identification, and estimation of the causal effect.

Example 14.2 (Continued). An interesting regime may be given by $d(a, w) = aI(a \leq \delta) + \delta I(a > \delta)$, where $I(x)$ is the indicator function that equals one if x is true and zero otherwise. Under this regime, all localities are required to have a pollution level of at most δ.

We define casual quantities in terms of the distribution of the outcome variables in a hypothetical world in which the stochastic regime is assigned instead of the natural value of treatment. In NPSEM (14.1), this counterfactual outcome is defined as $Y_{d(A,W)} = f_Y(d(A, W), W, U_Y)$.

14.1.1 Identification

The next step in the causal inference road map is identification of the causal parameter. Identification is necessary because the counterfactual variable $Y_{d(A,W)}$ is generally not observed. Thus, estimation of the expectation $E(Y_{d(A,W)})$ is possible only if it can be expressed as a function of the distribution P_0 of the observed data. This is achieved as follows. Using the law of iterated expectations, we can write

$$E(Y_{d(A,W)}) = \int_{a \in \mathcal{A}} \int_{w \in \mathcal{W}} E(Y_{d(a,w)} \mid A = a, W = w)g_0(a \mid w)q_{W,0}(w)d\nu(a, w),$$

where $\mathcal{A}$ and $\mathcal{W}$ are the support of the distributions of A and W, respectively. NPSEM (14.1) and assumption (14.2) imply

1. $Y_{d(a,w)} \perp\!\!\!\perp A \mid W$, and
2. $Y_{d(a,w)} = Y$ in the event $A = d(a, w)$.

Thus, the expectation $E(Y_{d(A,W)})$ is identified by

$$E(Y_{d(A,W)}) = \int_{a \in \mathcal{A}} \int_{w \in \mathcal{W}} E(Y_{d(a,w)} \mid A = d(a, w), W = w)g_0(a \mid w)q_{W,0}(w)d\nu(a, w)$$

$$= \int_{a \in \mathcal{A}} \int_{w \in \mathcal{W}} E(Y \mid A = d(a, w), W = w)g_0(a \mid w)q_{W,0}(w)d\nu(a, w)$$

$$= E_{P_0}\{\bar{Q}(d(A, W), W)\}. \tag{14.4}$$

We define the parameter of interest as a mapping $\Psi : \mathcal{M} \to \mathbb{R}$ that takes an element P in a statistical model $\mathcal{M}$ and maps it to a real number $\Psi(P)$. The true value of the parameter is given by the mapping evaluated at the true distribution $P_0 \in \mathcal{M}$, and is denoted by $\psi_0 = \Psi(P_0)$. The statistical parameter of interest is then given by

$$\Psi(P) = E_P\{\bar{Q}(d(A, W), W)\}, \tag{14.5}$$

where $\bar{Q}$ denotes the conditional expectation of Y corresponding to the distribution P. Note that this parameter depends only on $Q = (\bar{Q}, g, q_W)$. Therefore, in an abuse of notation, we will use the expressions $\Psi(Q)$ and $\Psi(P)$ interchangeably.

14.1.2 Positivity Assumption

In the identification result derived above, we implicitly assumed that $a \in \mathcal{A}(w)$ implies $d(a, w) \in \mathcal{A}(w)$, for all w in $\mathcal{W}$, where $\mathcal{A}(w)$ denotes the support of A conditional on $W = w$. This assumption is often referred to as the positivity assumption, and it ensures that the regime under consideration is supported in the observed data. Without this assumption, the integrals in Eq. (14.4) could be not well defined, since the conditioning set in $E(Y_{d(a,w)} \mid A = d(a, w), W = w)$ may be empty. Arguably, this assumption is much more easy to attain than the assumption required for static regimes, which states that all treatment levels considered by the regime have a positive probability in the support of W. In particular, a stochastic regime can always be defined such that positivity holds, which is precisely what we have done in (14.3).

A regime that does not satisfy positivity also poses a problem for interpretability of the resulting causal effect, since it does not occur naturally in the population. For illustration, consider Example 27.1. Assume w represents the covariate profile of individuals with coronary heart disease (CHD), and that some individuals with CHD have a natural value of treatment equal to the maximum physical activity level a recommended for their condition. Assume also that all other individuals diagnosed with CHD have LTPA values below the maximum recommended. In this case, the regime $d(a, w) = a + 2$ does not satisfy positivity, and therefore its effect cannot be estimated. This regime would also be of little interest since it would be unrealistic to enforce it on individuals with CHD.

14.2 Optimality Theory for Stochastic Regimes

In the remainder of this chapter we pursue the development of locally efficient, $\sqrt{n}$-consistent estimators for $\Psi(P)$, focusing on Example 14.1. It is not possible to construct $\sqrt{n}$-consistent estimators of $\Psi(P)$ in Example 14.2 if $d(a, w) = aI(a \leq \delta) + \delta I(a > \delta)$. This is because the parameter is not pathwise differentiable, and therefore it is not possible to construct a $\sqrt{n}$-consistent estimator. The reason for this can be explained intuitively by looking at the parameter definition

$$\Psi(P) = E\{\bar{Q}(d(A, W), W)\}$$
$$= E\{\bar{Q}(AI(A \leq \delta) + \delta I(A > \delta), W)\}$$
$$= E\{\bar{Q}(A, W)I(A \leq \delta)\} + E\{\bar{Q}(\delta, W)I(A > \delta)\}.$$

The term $E\{\bar{Q}(\delta, W)I(A > \delta)\}$ in this expression involves estimation of the causal effect of a static intervention setting the continuous exposure to $A = \delta$. Efficient estimation theory is not available for estimation of such parameters in the non-parametric model (Bickel et al. 1997b), since all possible gradients of the pathwise derivative would necessarily need to include a Dirac delta function at δ. An alternative approach to overcome this issue is to redefine the regime $d(a, w)$ so that the parameter becomes pathwise differentiable. Such approach is taken by Díaz and van der Laan (2013a); the interested reader is encouraged to consult the original research article.

In the remainder of this chapter we will assume *piecewise smooth invertibility* of $d(a, w)$. That is, for each $w \in \mathcal{W}$, we assume that the interval $\mathcal{I}(w) = (l(w,), u(w))$ may be partitioned into subintervals $\mathcal{I}_j(w) : j = 1, \ldots, J(w)$ such that $d(a, w)$ is equal to some $d_j(a, w)$ in $\mathcal{I}_j(w)$ and $d_j(\cdot, w)$ has inverse function $h(\cdot, w)$ with derivative $h'(\cdot, w)$. This assumption was first introduced by Haneuse and Rotnitzky (2013), and is necessary to establish the efficient influence function (EIF) given below in Lemma 14.1.

The EIF is a key element in semiparametric efficient estimation, since it defines the linear approximation of any efficient and regular asymptotically linear estimator. As a result, its variance is the asymptotic efficiency bound for all regular asymptotically linear estimators (Bickel et al. 1997b).

Lemma 14.1 (Efficient Influence Function). *The EIF of (14.5) is given by*

$$D(P)(o) = H(a, w)\{y - \bar{Q}(a, w)\} + \bar{Q}(d(a, w), w) - \Psi(P), \qquad (14.6)$$

where

$$H(a, w) = \sum_{j=1}^{J(w)} I\{h_j(a, w) \in I_j(w)\} \frac{g(h_j(a, w) \mid w)}{g(a \mid w)} h'_j(a, w).$$

Lemma 14.1 is a generalization of a result proved by Díaz and van der Laan (2012). We also use the alternative notation $H(g)(a, w)$ to stress the dependence of H on g.

Example 14.1 (Continued). Using the piecewise smooth invertibility of $d(a, w)$ defined in (14.3), the covariate H is found to be equal to

$$H(a, w) = I(a < u(w)) \frac{g_0(a - \delta \mid w)}{g_0(a \mid w)} + I(a \geq u(w) - \delta).$$

Note that, for an individual i such that $A_i \in [u(W_i) - \delta, u(W_i))$, H is equal to

$$H(A_i, W_i) = \frac{g_0(A_i - \delta \mid W_i)}{g_0(A_i \mid W_i)} + 1.$$

The presence of two terms in this covariate indicates that such observation represents two different types of observations under the stochastic regime. The first term appears because the outcome Y_i represents the outcome under the stochastic regime for all observations j such that $W_j = W_i$ and $A_j = A_i - \delta$. The second term appears because the outcome for observation A_i is its own outcome under the stochastic regime.

The following lemma provides a result establishing the double robustness of estimators that solve the EIF estimating equation.

Lemma 14.2 (Unbiased Estimating Equation and Double Robustness). *Let* $D(O \mid \bar{Q}, g, \psi_0)$ *be the estimating function implied by the EIF* $D(P)(O)$:

$$D(O \mid \bar{Q}, g, \psi_0) = H(g)(A, W)\{Y - \bar{Q}(A, W)\} + \bar{Q}(d(A, W), W) - \psi_0,$$

We have that $E_{P_0} D(O \mid \bar{Q}, g, \psi_0) = 0$ *if either g is such that* $H(g) = H(g_0)$, *or* $\bar{Q} = \bar{Q}_0$.

The previous lemma provides some intuition into the double robustness of estimators based on the EIF. If either g_0 or $\bar{Q}_0$ are known, it is possible to construct an unbiased estimating equation. Then, under the conditions outlined in Chap. 5 of van der Vaart (1998), the estimator can be shown to be consistent and asymptotically normal. Because $\bar{Q}_0$ and g_0 are generally unknown, it is not possible to plug in their values in D to obtain an unbiased estimating equation. Instead, estimator of these quantities must be used. This poses additional challenges in the construction of an estimator for ψ_0, in particular regarding its asymptotic distribution. In the following section we develop the theory required to obtain a doubly robust, locally efficient estimator of ψ_0, focusing on a targeted minimum loss based estimators (TMLE). TML estimators, as we will see, can be shown to be doubly robust in the sense that they are consistent if either g_0 or $\bar{Q}_0$ can be estimated consistently. In addition, they are efficient and asymptotically normal if both of these parameters are consistently estimated with certain convergence rates.

14.3 Targeted Minimum Loss-Based Estimation

The EIF D given above plays a central role in the definition of the TMLE. We start by considering its decomposition as $D = D_Y + D_{A,W}$, where

$$D_Y(\bar{Q}, g)(O) = H(g)(A, W)(Y - \bar{Q}(A, W))$$

denotes the projection of the EIF D on the tangent space of the model $\mathcal{M}$ corresponding to $p_0(y \mid a, w)$. Here, $D_{A,W}$ denotes the remainder term, which could be further decomposed into terms D_A and D_W, corresponding to the projections on the tangent spaces of $g_0(a \mid w)$ and $q_{w,0}(w)$, respectively.

A standard TMLE, as originally defined for pathwise differentiable parameters by van der Laan and Rubin (2006), would proceed by computing initial estimators of $\bar{Q}_0$, g_0, and $q_{w,0}$. These estimators would then be updated using D_Y, D_A, and D_W, respectively, in a way such that the EIF estimating function is zero when computed at the updated estimates. Achieving a solution of the EIF estimating equation guarantees, under regularity assumptions, that the estimator enjoys optimality properties such as double robustness and local efficiency. TML estimators defined in this way generally require iteratively optimizing a loss function for the likelihood of the observed data, which may increase programming efforts and require more computational time and power. The reader interested in the construction of standard TML estimators for Example 14.1 is encouraged to consult Díaz and van der Laan (2012).

In this chapter we take a different approach to define a TMLE of the target parameter ψ_0, where we focus exclusively on solving the component of the EIF estimating equation corresponding to D_Y. We will see that this leads to an estimator that does not require iteration, and yet leads to the same asymptotic optimality properties of the standard TMLE of Díaz and van der Laan (2012). Haneuse and Rotnitzky (2013) constructed a similar estimator focusing on parametric models for $\bar{Q}_0$ and g_0. Because parametric models are often misspecified, these estimators are generally inconsistent and can jeopardize the validity of conclusions extracted from an otherwise carefully well planned and executed study.

Assume without loss of generality that Y is supported in $\{0, 1\}$ or $(0, 1)$. TMLE of ψ_0 is performed in the following steps:

1. *Initial estimators.* Obtain initial estimators g_n and $\bar{Q}_n$ of g_0 and $\bar{Q}_0$. In general, the functional form of g_0 and $\bar{Q}_0$ will be unknown to the researcher. Since consistent estimation of these quantities is key to achieve asymptotic efficiency of ψ_n, we advocate for the use of data-adaptive predictive methods that allow flexibility in the specification of these functional forms. We discuss this issue further in Sect. 14.4 below.

2. *Compute auxiliary covariate.* For each subject i, compute the auxiliary covariate

$$H_n(A_i, W_i) = \sum_{j=1}^{J(W_i)} I\{A_i \in I_j(W_i)\} \frac{g_n(h_j(A_i, W_i) \mid W_i)}{g_n(A_i \mid W_i)} h'_j(A_i, W_i).$$

3. *Solve estimating equations.* Estimate the parameter ϵ in the logistic regression model

$$\mathrm{logit}\bar{Q}_{\epsilon,n}(a, w) = \mathrm{logit}\bar{Q}_n(a, w) + \epsilon H_n(a, w), \tag{14.7}$$

by fitting a standard logistic regression model of Y_i on $H_n(A_i, W_i)$, with no intercept and with offset $\mathrm{logit}\bar{Q}_n(A_i, W_i)$. Alternatively, fit the model

$$\mathrm{logit}\bar{Q}_{\epsilon,n}(a, w) = \mathrm{logit}\bar{Q}_n(a, w) + \epsilon$$

with weights $H_n(A_i, W_i)$. In either case, denote the estimate of ϵ by ϵ_n.

4. *Update initial estimator and compute 1-TMLE.* Update the initial estimator as $\bar{Q}_n^\star(a, w) = \bar{Q}_{n,\epsilon_n}(a, w)$, and define the 1-TMLE as

$$\psi_n = \frac{1}{n} \sum_{i=1}^{n} \bar{Q}_n^\star(d(A_i, W_i), W_i).$$

14.3.1 Asymptotic Distribution of TMLE

A key property of the TML estimator defined above is that, by virtue of the logistic regression model (14.7), the TMLE satisfies $P_n D_Y(\bar{Q}_n^\star, g_n) = 0$. To simplify the notation, let us denote $\bar{Q}_d(a, w) = \bar{Q}(d(a, w), w)$. Straightforward algebra shows that, for any $\bar{Q}_d$,

$$P_0 \bar{Q}_d - \psi_0 = -P_0 D_Y(\bar{Q}, g) + R(P, P_0),$$

where

$$R(P, P_0) = -\int \{H(g) - H(g_0)\}\{\bar{Q} - \bar{Q}_0\} dP_0. \tag{14.8}$$

Applying this to $\bar{Q}_n^\star$, and adding and subtracting ψ_n, we obtain

$$\psi_n - \psi_0 = (P_n - P_0) D_Y(\bar{Q}_n^\star, g_n) - (P_n - P_0) \bar{Q}_{n,d}^\star + R(\hat{P}^\star, P_0),$$

where $R(\hat{P}^\star, P_0)$ denotes (14.8) with $\bar{Q}$ replaced by $\bar{Q}_n^\star$ and g replaced by g_n. This now gives

$$\psi_n - \psi_0 = (P_n - P_0) D(\bar{Q}_n^\star, g_n) + R(\hat{P}^\star, P_0).$$

Provided that

1. $D(\bar{Q}_n^\star, g_n)$ converges to $D(P_0)$ in $L_2(P_0)$ norm, and
2. the size of the class of functions considered for estimation of $\bar{Q}_n^\star$, and g_n is bounded (technically, there exists a Donsker class $\mathcal{F}$ of functions of o so that $D(\bar{Q}_n^\star, g_n) \in \mathcal{F}$ with probability tending to one),

results from empirical process theory (e.g., theorem 19.24 of van der Vaart 1998) allow us to conclude that

$$\psi_n - \psi_0 = (P_n - P_0) D(P_0) + R(\hat{P}^\star, P_0).$$

In addition, if

$$R(\hat{P}^\star, P_0) = o_P(1/\sqrt{n}), \tag{14.9}$$

we obtain that $\psi_n - \psi_0 = (P_n - P_0) D(P_0) + o_P(1/\sqrt{n})$. Thus, the central limit theorem can be used to establish

$$\sqrt{n}(\psi_n - \psi) \to N(0, V(D(P_0))).$$

This implies, in particular, that ψ_n is a $\sqrt{n}$-consistent estimator of ψ_0, it is asymptotically normal, and it is locally efficient. Wald-type confidence intervals may be now obtained as $\psi_n \pm z_\alpha \sigma_n / \sqrt{n}$, where

$$\sigma_n^2 = \frac{1}{n} \sum_{i=1}^n D^2(\bar{Q}_n^\star, g_n)(O_i)$$

is an estimator of $V(D(P_0))$. Alternatively, the bootstrap may be used to obtain an estimator σ_n^2.

14.4 Initial Estimators

The condition that $R(\hat{P}, P_0)$ converges to zero in the sense of (14.9) is necessary to obtain the consistency, asymptotic normality, and local efficiency of the TMLE $\hat{\psi}$. Condition (14.9) would be trivially satisfied if g_0 and $\bar{Q}_0$ where known to belong to a parametric family of functions, and were estimated using maximum likelihood or some other type of M-estimator. When working with high-dimensional observational data, it has been long recognized that parametric models can seldom be correctly specified, except in rare and often trivial cases. Model misspecification would then lead to a violation of condition (14.9), which, from the arguments of the previous section, would result in inconsistent estimators of ψ_0. Because they would invalidate the result of a well designed and executed study, we discourage the use of estimators based on parametric models, except in cases in which their correctness can be established from subject-matter scientific knowledge.

As an alternative, methods developed in the field of statistical learning can be used to estimate $\bar{Q}_0$ and g_0. Because statistical learning methods are concerned with finding estimates that resemble the true data generating functions as closely as possible, they are more likely to yield consistent estimators, in contrast to parametric models. We encourage the use of ensemble learners, which are capable of exploiting the advantages of a library of candidate estimation algorithms simultaneously. In particular, super learning, discussed in Chap. 3, is a technique whose optimal properties have been demonstrated theoretically and empirically. Super learning of a conditional expectation such as $\bar{Q}_0$ has been extensively discussed, for example, in the references included in Chap. 3 of this book as well as Chap. 3 of *Targeted Learning* (2011). In the remainder of this section we discuss the problem of estimating the conditional probability density function $g_0(a \mid w)$ for a continuous variable A. This problem has received considerably less attention from the statistical learning research community.

14.4.1 Super Learning for a Conditional Density

If A is continuous, the conditional density g_0 may be defined as the minimizer of the negative log-likelihood loss function. That is $g_0 = \arg\min_{f \in \mathcal{F}} R(f, p_0)$, where $\mathcal{F}$ is the space of all nonnegative functions of (a, w) satisfying $\int f(a, w)da = 1$, and $R(f) = -\int \log f(a, w)dP_0(o)$. An estimator $\hat{g}$ is seen here as an algorithm that takes a training sample $\mathcal{T} \subseteq \{O_i : i = 1, \ldots, n\}$ as an input, and outputs an estimated function $g_n(a, w)$.

We use cross-validation to construct an estimate $R_n(g_{n,k})$ of the risk $R(g_{n,k})$ as follows. Let $\mathcal{V}_1, \ldots, \mathcal{V}_J$ denote a random partition of the index set $\{1, \ldots, n\}$ into J validation sets of approximately the same size. That is, $\mathcal{V}_j \subset \{1, \ldots, n\}$; $\bigcup_{j=1}^{J} \mathcal{V}_j = \{1, \ldots, n\}$; and $\mathcal{V}_j \cap \mathcal{V}_{j'} = \emptyset$. In addition, for each j, the associated training sample is given by $\mathcal{T}_j = \{1, \ldots, n\} \setminus \mathcal{V}_j$. Denote by $\hat{g}_{\mathcal{T}_j}$ the estimated density function obtained by training the algorithm using only data in the sample $\mathcal{T}_j$. The cross-validated risk of an estimated density g_n is defined as

$$R_n(g_n) = -\frac{1}{J} \sum_{j=1}^{J} \frac{1}{|\mathcal{V}_j|} \sum_{i \in \mathcal{V}_j} \log g_{n,\mathcal{T}_j}(A_i, W_i). \qquad (14.10)$$

Consider now a finite collection $\mathcal{L} = \{g_{n,k} : k = 1, \ldots, K_n\}$ of candidate estimators for g_0. We call this collection a *library*. We define the stacked predictor as a convex combination of the predictors in the library:

$$g_{n,\alpha}(a \mid w) = \sum_{k=1}^{K_n} \alpha_k g_{n,k}(a \mid w),$$

and estimate the weights α as the minimizer of the cross-validated risk $\hat{\alpha} = \arg\min R_n(g_{n,\alpha})$, subject to $\sum_{k=1}^{K_n} \alpha_k = 1$. The final estimator is then defined as $g_{n,\hat{\alpha}}$.

14.4.2 Construction of the Library

Consider a partition of the range of A into k bins defined by a sequence of values $\beta_0 < \cdots < \beta_k$. Consider a candidate for estimation of $g_0(a \mid w)$ given by

$$g_{n,\beta}(a \mid w) = \frac{\widehat{Pr}(A \in [\beta_{t-1}, \beta_t) \mid W = w)}{\beta_t - \beta_{t-1}}, \quad \text{for } \beta_{t-1} \le a < \beta_t. \qquad (14.11)$$

Here $\widehat{Pr}$ denotes an estimator of the true probability $Pr_0(A \in [\beta_{t-1}, \beta_t) \mid W = w)$ obtained through a hazard specification and the use of an estimator for the expectation of a binary variable in a repeated measures dataset as follows. Consider the following factorization

$$Pr(A \in [\beta_{t-1}, \beta_t) \mid W = w) = Pr(A \in [\beta_{t-1}, \beta_t) \mid A \ge \beta_{t-1}, W = w) \times$$

$$\prod_{j=1}^{t-1} \{1 - Pr(A \in [\beta_{j-1}, \beta_j) \mid A \ge \beta_{j-1}, W = w)\}.$$

The likelihood for model (14.11) is proportional to

$$\prod_{i=1}^{n} Pr(A_i \in [\beta_{t-1}, \beta_t) \mid W) = \prod_{i=1}^{n} \left[\prod_{j=1}^{t-1} \{1 - Pr(A_i \in [\beta_{j-1}, \beta_j) \mid A_i \ge \beta_{j-1}, W_i)\} \right] \times$$

$$Pr(A_i \in [\beta_{t-1}, \beta_t) \mid A_i \ge \beta_{t-1}, W_i),$$

which corresponds to the likelihood for the expectation of the binary variable $I(A_i \in [\beta_{j-1}, \beta_j))$ in a repeated measures data set in which the observation of subject i is repeated k times, conditional on the event $A_i \geq \beta_{j-1}$.

Thus, each candidate estimator for g_0 is indexed by two choices: the sequence of values $\beta_0 < \cdots < \beta_k$, and the algorithm for estimating the probabilities $Pr_0(A_i \in [\beta_{t-1}, \beta_t) | A_i \geq \beta_{t-1}, W_i)$. The latter is simply a conditional probability, and therefore any standard prediction algorithm may be used as a candidate. In the remainder of this section we focus on the selection of the location and number of bins, implied by the choice of β_j values.

Denby and Mallows (2009) describe the histogram as a graphical descriptive tool in which the location of the bins can be characterized by considering a set of parallel lines cutting the graph of the empirical cumulative distribution function (ECDF). Specifically, given a number of bins k, the equal-area histogram can be regarded as a tool in which the ECDF graph is cut by $k + 1$ equally spaced lines parallel to the x axis. The usual equal-bin-width histogram corresponds to drawing the same lines parallel to the y axis. In both cases, the location of the cutoff points for the bins is defined by the x values of the points in which the lines cut the ECDF. As pointed out by the authors, the equal-area histogram is able to discover spikes in the density, but it oversmooths in the tails and is not able to show individual outliers. On the other hand, the equal-bin-width histogram oversmooths in regions of high density and does not respond well to spikes in the data, but is a very useful tool for identifying outliers and describing the tails of the density.

As an alternative to find a compromise between these two approaches, the authors propose a new histogram in which the ECDF is cut by lines $x + cy = bh$, $b = 1, \ldots, k+1$; where c and h are parameters defining the slope and the distance between lines, respectively. The parameter h identifies the number of bins k. The authors note that $c = 0$ gives the usual histogram, whereas $c \to \infty$ corresponds to the equal-area histogram.

Thus, we can define a library of candidate estimators for the conditional density in terms of (14.11) by defining values of the vector β through different choices of c and k, and considering a library for estimation of conditional probabilities. Specifically, the library is given by the Cartesian product

$$\mathcal{L} = \{c_1, \ldots, c_{m_c}\} \times \{k_1, \ldots, k_{m_k}\} \times \{\widehat{Pr_1}, \ldots, \widehat{Pr_{m_P}}\},$$

where the first is a set of m_c candidate values for c, the second is a set of m_k candidate values for k, and the third is a set of m_P candidates for the probability estimation algorithm. The use of this approach will result in estimators that are able to identify regions of high density as well as provide a good description of the tails and outliers of the density. The inclusion of various probability estimators allows the algorithm to find possible nonlinearities and higher-order interactions in the data. This proposed library may be augmented by considering any other estimator. For example, there may be expert knowledge leading to believe that a normal distribution (or any other distribution) with linear conditional expectation could fit the data. A candidate algorithm that estimates such a density using maximum likelihood may be added to the library. This algorithm was first proposed by Díaz and van der Laan (2011), the reader interested in more details and applications is encouraged to consult the original research article.

14.5 Notes and Further Reading

The contents of this chapter are based on previous work by Díaz and van der Laan (2011, 2012); Díaz and van der Laan (2013a). We have also included here some improvements proposed by Haneuse and Rotnitzky (2013). The reader interested in applications to real data and further discussion is encouraged to consult the original research articles. The reader interested in further discussion of the general theory of stochastic interventions is referred to Robins et al. (2004); Korb et al. (2004); Eberhardt and Scheines (2006); Pearl (2009b) and Dawid and Didelez (2010), among others.

As we briefly mentioned in the introduction of this chapter, stochastic regimes may also be used to tackle standard causal inference problems. For example, van der Laan (2014a) discusses the use of stochastic regimes to define and estimate causal effects in causal networks. Sapp et al. (2014) present an application of stochastic regimes to estimation of variable importance measures with interval-censored outcomes. Applications of stochastic interventions to causal inference under mediation may be found in Naimi et al. (2014) and Zheng and van der Laan (2012a). The latter authors present an important result showing that, in the case of the natural direct effect (NDE), using a stochastic intervention approach may result in weaker identifiability conditions. Therefore, adopting a stochastic regime interpretation of the NDE may be desirable as the estimated parameter represents a causal effect in a larger causal model, in comparison with the standard approach (van der Laan et al. 2014). Further discussions and other applications may be found in Young et al. (2014) and van der Laan et al. (2014).

Chapter 15
LTMLE with Clustering

Mireille E. Schnitzer, Mark J. van der Laan, Erica E. M. Moodie,
and Robert W. Platt

Breastfeeding is considered best practice in early infant feeding, and is recommended by most major health organizations. However, due to the impossibility of directly allocating breastfeeding as a randomized intervention, no direct experimental evidence is available. The PROmotion of Breastfeeding Intervention Trial (PROBIT) was a cluster-randomized trial that sought to evaluate the effect of a hospital program that encouraged and supported breastfeeding, thereby producing indirect evidence of its protective effect on infant infections and hospitalizations.

In this chapter, we use causal inference techniques to estimate the effect of different durations of breastfeeding (a longitudinal exposure) on the number of periods of hospitalization throughout the first year after birth. Because hospitalizations may also affect the continuation of breastfeeding, we consider them a time-varying confounder. We demonstrate two g-computation approaches and an implementation of LTMLE that take into account an outcome that is partially determined by time-varying confounders and the clustering that arises from the nature of the study design.

M. E. Schnitzer (✉)
Faculté de pharmacie, Université de Montréal, #2236, Pavillon Jean-Coutu 2940, chemin de la Polytechnique, Montreal, QC H3C 3J7, Canada
e-mail: mireille.schnitzer@umontreal.ca

M. J. van der Laan
Division of Biostatistics and Department of Statistics, University of California, Berkeley, 101 Haviland Hall, #7358, Berkeley, CA 94720, USA
e-mail: laan@berkeley.edu

E. E. M. Moodie · R. W. Platt
Department of Epidemiology, Biostatistics, and Occupational Health, McGill University, Purvis Hall, 1020 Pine Ave West, Montreal, QC H3A 1A2, Canada
e-mail: erica.moodie@mcgill.ca; robert.platt@mcgill.ca

© Springer International Publishing AG 2018
M.J. van der Laan, S. Rose, *Targeted Learning in Data Science*,
Springer Series in Statistics, https://doi.org/10.1007/978-3-319-65304-4_15

15.1 The PROBIT Study

The PROBIT study was held in the country of Belarus from June 1996 to December 1997 (Kramer et al. 2001, 2002). In this study, maternal hospitals and their corresponding polyclinics were randomized to receive lactation management training, which emphasizes ways to encourage longer durations of exclusive breastfeeding. In order to optimize efficiency, the randomization occurred between 17 pairs of hospitals matched on region, rural versus urban status, number of deliveries per year, and breastfeeding initiation rates upon discharge. However, due to two hospital withdrawals and one case of record falsification, only 31 clusters completed the study and the complete paired structure was lost. Within the hospital clusters, recruitment was limited to pregnant women who intended to breastfeed their child. In particular, the study enrolled healthy, full-term, singleton breastfed infants weighing ≥2500 g. Baseline data included maternal demographic, educational, and smoking information, details about previous pregnancies, and infant information (sex, birth weight, gestational age, and Apgar score; Finster and Wood 2005). Follow-up visits occurred throughout the year post-birth at 1, 2, 3, 6, 9, and 12 months. At these visits, extensive information on infant feeding, growth, illnesses, and hospitalizations was collected. Within the 31 clusters, a total of 17,044 mother-infant pairs participated in the study and had recorded data. Necessary baseline data was missing for eight subjects, bringing the sample size to 17,036 mother-infant pairs. Table 15.1 describes the baseline characteristics adjusted for in the analysis.

The initial analyses (Kramer et al. 2001) found a significant effect of the encouragement trial on gastrointestinal infections, the primary outcome. In subsequent work, we carried out a causal analysis of the effect of breastfeeding duration on the number of gastrointestinal infections throughout the year (Schnitzer et al. 2014). In this chapter, we investigate the effect of breastfeeding duration on infant hospitalizations. In particular, we are interested in knowing whether the number of hospitalizations would decrease with longer durations of breastfeeding. Because the survey only collected information on whether an infant was hospitalized between visits, the outcome of interest at 12 months is the number of intervals recording a hospitalization.

15.1.1 Observed Data

Corresponding with the PROBIT, we consider longitudinal data, taken from each mother-infant pair (defined as the subject), of the form

$$O = (L_0, C_1, L_1, A_1, C_2, L_2, \ldots, L_{K-1}, A_{K-1}, C_K, Y),$$

where subscripts indicate at which time point the measurement was taken. L_0 represents all measured baseline covariates including hospital center, C_t represents whether a visit did not occur, L_t represents time varying covariates, A_t is an indicator for the exposure level (taking value 1 if the infant is breastfed throughout the tth in-

Table 15.1 Characteristics at baseline of the 17,044 mother-infant pairs in the PROBIT dataset

Characteristic	Summary		N. missing
Numeric variables	Median	IQR[a]	
Age of mother (years)	23	(21,27)	
N. previous children	0	(0,1)	
Gestational age (months)	40	(39,40)	
Infant weight (kg)	3.4	(3.2,3.7)	
Infant height (cm)	52	(50,53)	
Apgar score[b]	9	(8,9)	5
Head circumference (cm)	35	(34,36)	3
Binary variables	N.	%	
Smoked during pregnancy	389	2.28	
History of allergy	750	4.40	
Male child	8827	52	
Cesarean	1974	12	
Mother's Education			2
Some high school	663	4	
High school	5497	32	
Some university	8568	50	
University	2316	14	
Geographic region			
East Belarus, urban	5615	33	
East Belarus, rural	2706	16	
West Belarus, urban	4380	26	
West Belarus, rural	4343	25	

[a]IQR: inter-quartile range
[b]The Apgar score is an assessment of newborn health (range 1–10) where 8+ is vigorous, 5–7 is mildly depressed and 4– is severely depressed (Finster and Wood 2005). A range of 5–10 was observed in PROBIT due to entry restrictions on weight and health at baseline. Table and caption reproduced from Schnitzer et al. (2014)

terval), and Y is the outcome of interest measured at the Kth time point. Figure 15.1 represents the order of the measurements collected at time t. We will use $\bar{X}_t$ to denote the history of X up to and including X_t for any time dependent variable.

In the PROBIT, there were $K = 6$ follow-up visits. Y is defined as the number of time periods over the first year in which the child had at least one hospitalization and therefore takes integer values between zero and six. The scientific question of interest involves the effect of breastfeeding measured over time $\{A_t; 1 \leq t \leq 5\}$ on Y. L_t is whether or not an infant was hospitalized in the time period $(t-1, t)$. Therefore, if we take L_6 to be an indicator for at least one hospitalization between times five and six, $Y = \sum_{t=1}^{6} L_t$. At the beginning of the study, all mothers attempted breastfeeding so that we could define $A_0 = 1$ for all subjects. In addition, since we only retained the subset with complete baseline data, we can define $C_0 = 0$ (denoting uncensored at baseline) for all subjects.

Once a subject missed a visit (or the needed information was not collected at a visit), we artificially censored them for all future visits. This did not greatly reduce the available data as item missingness was uncommon. Table 15.2 gives the number

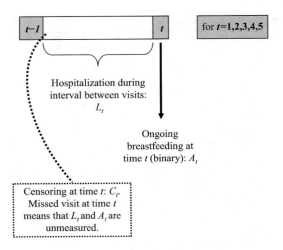

Fig. 15.1 Time ordering of the variables in the PROBIT study. At each follow-up time point, breastfeeding status (A_t) and hospitalization over the past interval (L_t) were noted. Censoring occurring at time t ($C_t = 1$) indicates that later breastfeeding and infection status were not observed

Table 15.2 Censoring, number of hospitalizations and mothers still breastfeeding by time point

Time point	1	2	3	4	5	6
Month	1	2	3	6	9	12
N. censored	156	81	73	148	139	797
Cumulative N.	156	237	310	458	597	1394
Cumulative %	0.9	1.4	1.8	2.7	3.5	8.2
N. hospitalized	626	640	646	1265	1163	887
N. breastfeeding	15,392	13,128	10,765	6893	4717	–

of censored, hospitalized, and breastfeeding subjects at each time point. Note that we did not report the number breastfeeding at time point 6 (month 12) because we do not incorporate this information in the analysis. The total number of intervals with hospitalizations observed for never-censored patients was 4785.

15.1.2 Causal Assumptions

In order to proceed in defining counterfactuals, we require the assumptions of no interference and sequential positivity.

- *No interference:* The potential outcomes of one subject are not dependent on the exposures of others. In our context, this corresponds with the assumption that one infant's breastfeeding does not impact another infant's probability of hospitalization given the second infant's breastfeeding status.

- *Sequential positivity:* For every possible history $\bar{L}_{t-1}$ with $A_{t-1} = 1$, the probability of either continuing or stopping breastfeeding at time t must be greater than zero for all subjects. Because we do not consider regimes where breastfeeding is stopped and restarted, we do not require positivity for A_t when $A_{t-1} = 0$. We also must have that for every possible history $\bar{A}_{t-1}, \bar{L}_{t-1}$ the probability of censoring at time t is less than one.

For the PROBIT, the assumption of no interference requires that the breastfeeding status of one mother does not influence the outcome of another's child. We believe this to be very plausible because mothers spent short periods of time in the hospital which limited their interaction. Regarding positivity, in our estimation of the probabilities of continuing or ceasing breastfeeding conditional on previous continued breastfeeding (see Sect. 15.3.1), we did not observe any values approaching zero. In addition, the probabilities of censoring were quite low for all values of the coefficients, suggesting that positivity is not a concern here.

Let $\bar{a} = (a_1, a_2, \ldots, a_{K-1})$ denote a fixed breastfeeding regimen. For instance, breastfeeding past the first time period, then stopping before the second would be written as $(1, 0, 0, \ldots, 0)$. Because breastfeeding is approximately monotone, we will only consider monotone longitudinal exposures; that is, we compare the relative effects of different stopping times of breastfeeding. Also let $\bar{a}_t = (a_1, \ldots, a_t)$ be the component of the fixed regimen up until time point t.

In order to define the causal parameter of interest, we consider a hospital level intervention that imposes a specific duration of breastfeeding on each subject. We can then define the counterfactual variable $L_t^{\bar{a}}; t > 0$ as the observation L_t that an individual would have had if they had followed the assigned breastfeeding regimen $\bar{a}_{t-1}$ and remained uncensored. Similarly, $Y^{\bar{a}}$ is the counterfactual number of hospitalizations that would have been observed under breastfeeding regimen $\bar{a} = \bar{a}_5$. The individual counterfactual data corresponding to this intervention is $O^a = (L_0, L_1^{\bar{a}}, L_2^{\bar{a}}, \ldots, L_{K-1}^{\bar{a}}, Y^{\bar{a}})$. The target of inference is the marginal mean counterfactual outcome, denoted $\psi_0^{\bar{a}} = E(Y^{\bar{a}})$. Equivalently, we estimate the mean number of periods hospitalized had all infants been exposed to various breastfeeding stopping times.

In order for this causal parameter to be estimable, we also require sequential consistency and sequential exchangeability (Robins 2000).

- *Sequential Consistency:* The consistency assumption in the longitudinal setting is that $\bar{L}_t^{\bar{a}} = \bar{L}_t$ when $\bar{A}_{t-1} = \bar{a}_{t-1}$. Equivalently, we observe the sequence of counterfactual variables defined under the treatment regimen actually observed.

- *Sequential Exchangeability:* This assumption is the independence of the counterfactual intermediate variables and the most recent intervention nodes (exposure and censoring) conditional on the past, $L_t^{\bar{a}} \perp A_{t-1}, C_t \mid \bar{L}_{t-1}, \bar{A}_{t-2}, C_{t-1} = 0$ for $t = 1, \ldots, K + 1$ (where $\bar{A}_{-1}$ is taken to be a null variable and removed).

Sequential consistency assumes that we observe the potential outcome that would have been observed under the intervention of assigning a duration of breastfeeding. This assumes that the specific time within the interval that breastfeeding is ceased does not impact the counterfactual. One might alternatively assume that the assignment leaves the exact stopping time within the interval up to the subject, but this perspective requires additional exchangeability requirements (VanderWeele and Hernán 2013).

For the assumption of sequential exchangeability described above, we must assume that all baseline and time dependent confounders of both breastfeeding and censoring have been adjusted for in $\bar{L}_t$. Specifically, we assume that $\bar{L}_t$ is sufficient to control for confounding of breastfeeding A_t and that censoring C_t is ignorable given $\bar{L}_{t-1}$. In Fig. 15.1 we see that censoring at a visit is not allowed to depend on recent hospitalizations. While this is unrealistic, the low percentage of censoring suggests that the violation may not greatly impact the analysis. Overall, while the exchangeability assumption is not verifiable and generally difficult to fully believe, we argue in Schnitzer et al. (2014) that this assumption is strengthened by controlling for an indicator of cluster.

15.1.3 Model and Parameter

In order to define the model for the observed and counterfactual data, we assume that in a population where patients are clustered into hospital centers, we observe randomly drawn hospitals from some large population. Let O_{mi} denote the observation vector for patient i in hospital cluster m. We define the mth hospital's observed data as $O_m^c = (O_{mi}; i \in Z_m)$ where Z_m represents the set of subjects belonging to hospital m. We suppose that the cluster observations O_m^c are identically and independently drawn with probability distribution P_0^c. Let $O^c \sim P_0^c$ denote this random variable. The probability distribution P_0^c of O^c is a member of some model space $\mathcal{M}^c$. The marginal probability distribution P_0 of a randomly selected patient in a randomly selected hospital cluster corresponds to a mixture probability distribution. This marginal distribution P_0 can be written as a function of the cluster distribution such that $P_0 = P_0(P_0^c)$. We place all model restrictions directly on the marginal model space $\mathcal{M}$, of which the true P_0 is a member. We then restrict the model space for the cluster-specific probability distributions $\mathcal{M}^c = \{P_0^c : P_0 \in \mathcal{M}\}$ to satisfy the constraints placed on the marginal model space $\mathcal{M}$.

At the marginal level, we assume that the true distribution function of O can be factorized according to the time ordering for an individual as

$$P_0 = \underbrace{\prod_{t=1}^{K} Q_{0,L_t}(L_t \mid \bar{C}_t, \bar{A}_{t-1}, \bar{L}_{t-1}) Q_{0,L_0}(L_0)}_{Q_0} \times$$

$$\underbrace{\prod_{t=1}^{K-1} G_{0,A_t}(A_t \mid \bar{L}_t, \bar{C}_t, \bar{A}_{t-1}) \prod_{t=1}^{K} G_{0,C_t}(C_t \mid \bar{A}_{t-1}, \bar{L}_{t-1}, \bar{C}_{t-1})}_{G_0}$$

where Q_0 is the joint conditional distribution of the L_t variables. $Q_{0,L_t}, t = 1, \ldots, K$ are the distributions of each L_t, conditional on the information prior to L_t, and Q_{0,L_0} is the distribution of the baseline covariates. Similarly, G_0 is the conditional distribution of the exposure and censoring variables that can be decomposed into the distributions at each time point, denoted $G_{0,A_t}, t = 1, \ldots, K - 1$ and $G_{0,C_t}, t = 1, \ldots, K$. The model $\mathcal{M}$ is nonparametric up to restrictions on the treat and censoring distribution G_0.

Under the causal assumptions described above, the parameter of interest $\psi_0^{\bar{a}}$ can be identified with the usual g-formula applied to the distribution P_0. Suppose we fix the assigned exposure regimen to $\bar{a}$ for all subjects (so that the A_t are no longer random) and fix that all subjects are fully observed (so that $C_t = 0$ for $t = 1, \ldots, K$). We then define the marginal counterfactual distribution function $Q_0^{\bar{a}}$ corresponding to this static intervention $\bar{a}$. The g-formula for this counterfactual distribution is given by

$$Q_0^{\bar{a}}(\bar{L}_K^{\bar{a}}) = \prod_{t=1}^{K} Q_{0,L_t}(L_t \mid \bar{C}_t = 0, \bar{A}_{t-1} = \bar{a}_{t-1}, \bar{L}_{t-1}) Q_{0,L_0}(L_0). \tag{15.1}$$

The target parameter of interest, specifically the marginal mean under a fixed breast-feeding regimen $\bar{a}$, can then be described as

$$\psi_0^{\bar{a}} = \Psi(P_0(P_0^c)) = \Psi(P_0) = \Psi(Q_0^{\bar{a}}) = E_{Q_0^{\bar{a}}}(Y^{\bar{a}}),$$

where the expectation is taken under the true counterfactual data generating function $Q_0^{\bar{a}}$.

In Sects. 15.2 and 15.3 we proceed with estimation of the target parameter as though the subject level data across hospitals are all independent and identically distributed with probability distribution P_0 in model $\mathcal{M}$ and treating the target parameter as a function of P_0. We then establish in Sect. 15.4 the asymptotic linearity of this i.i.d. LTMLE respecting that only the clusters are i.i.d., and provide formal inference. We conjecture that this i.i.d. type LTMLE is in fact asymptotically efficient for our model $\mathcal{M}^c$, assuming consistent estimation of the nuisance parameters, but this is not formally established in this chapter.

15.2 Two Parametrizations of the g-Formula

The above g-formula (15.1) can be directly used to estimate $\psi_0^{\bar{a}}$ if we treat all subjects as identically and independently distributed. This is done by using the subject level data to estimate each of the quantities in the formula, producing predictions of the potential outcome for each subject, and averaging over all subjects to estimate the expectation. This is called the g-computation approach (Robins 1986). In settings where the time dependent variables are binary, the g-formula can be simplified to

$$\psi_0^{\bar{a}} = \int_{L_0} \sum_{l_1 = \{0,1\}} \cdots \sum_{l_{K-1} = \{0,1\}} E(Y \mid C_K = 0, \bar{A}_{K-1} = \bar{a}, \bar{L}_{K-1} = \bar{l}_{K-1}) \times \quad (15.2)$$
$$Pr(L_{K-1} = l_{K-1} \mid \bar{C}_{K-1} = 0, \bar{A}_{K-2} = \bar{a}_{K-2}, \bar{L}_{K-2} = \bar{l}_{K-2}) \times$$
$$\cdots Pr(L_1 = l_1 \mid C_1 = 0) Q_{0,L_0}(L_0) dL_0.$$

For estimation using g-computation, we must estimate the conditional mean of Y and the conditional probabilities for $L_t = 1, l = 1, \ldots, K$, although no estimation method is prespecified for any of these quantities. We then calculate a prediction of each conditional expectation and probability in Eq. (15.2) for each subject, i. The Q_{0,L_0} can be estimated using the empirical density so that $Q_{n,L_0}(L_{0i}) = 1/n$ for each subject (with baseline variables L_{0i}). Then, the predicted values for the conditional expectation and probabilities are combined according to Eq. (15.2), where the integral is replaced by summation over all subjects, i.

15.2.1 g-Computation for the PROBIT

The g-computation algorithm must be slightly modified when the outcome of interest is a longitudinal count outcome. This is because a component of the outcome is deterministic (not random) when conditioning on the information available from prior time points. Specifically, since $Y = L_6 + \sum_{t=1}^{5} L_t$, if we condition on $\bar{L}_5$ only L_6 is random. Hence, we note that $E(Y \mid \bar{C}_6 = 0, \bar{A}_5 = \bar{a}_5, \bar{L}_5) = E(L_6 \mid \bar{C}_6 = 0, \bar{A}_5 = \bar{a}_5, \bar{L}_5) + \sum_{t=1}^{5} L_t$. Therefore we must only model L_6 to obtain predictions of the conditional expectation of the outcome used in the g-computation algorithm.

Notably, when the dimension of L_t increases or when L_t contains noncategorical variables, the decomposition of the g-formula (15.1) is increasingly complicated. Sampling methods may be required for estimation and the computational burden will increase exponentially in the number of time points. The following section describes an alternative factorization that allows for computational time that is linear in the number of time points and can handle higher dimensional L_t without added complications.

15.2.2 Sequential g-Computation

An alternative decomposition of the counterfactual data distribution, leading to a different g-formula was introduced by Bang and Robins (2005), as also discussed in Chap. 3 and 4 of this book. To understand this decomposition, first note that under the Law of Iterated Expectations, we have that $E(Y^{\bar{a}}) = E(E(Y^{\bar{a}} \mid \bar{L}_{K-1}^{\bar{a}}))$. If we repeatedly apply this principle, we have that

$$E(Y^{\bar{a}}) = E(E(\ldots E(E(Y^{\bar{a}} \mid \bar{L}_{K-1}^{\bar{a}}) \mid \bar{L}_{K-2}^{\bar{a}}) \mid \ldots \mid L_0)).$$

Now, due to the sequential exchangeability of A_K and C_K, we can also write $E(Y^{\bar{a}}) = E\{E(Y^{\bar{a}} \mid C_K = 0, \bar{A}_{K-1} = \bar{a}_{K-1}, \bar{L}^{\bar{a}}_{K-1})\}$. By consistency (we observe $Y = Y^{\bar{a}}$ when $C_K = 0$ and $\bar{A}_{K-1} = \bar{a}_{K-1}$ and that $L_t = L^{\bar{a}}_t$ when $\bar{A}_{t-1} = \bar{a}_{t-1}$), $E(Y^{\bar{a}}) = E\{E(Y \mid C_K = 0, \bar{A}_{K-1} = \bar{a}_{K-1}, \bar{L}_{K-1})\}$, which is estimable from the data. If we similarly apply the sequential exchangeability and consistency at all time points in the nested expectations, we observe that the target parameter can be expressed as a sequence of estimable expectations given by van der Laan and Gruber (2012)

$$\bar{Q}_t(\bar{L}_{t-1}) = E(Y \mid C_t = 0, \bar{A}_{t-1} = \bar{a}_{t-1}, \bar{L}_{t-1}), t = K, \dots, 2 \qquad (15.3)$$

and $\bar{Q}_1(L_0) = E(\bar{Q}_2 \mid L_0)$. The overbar in $\bar{Q}_t(\bar{L}_{t-1})$ denotes a mean. Note that we will generally write $\bar{Q}_t = \bar{Q}_t(\bar{L}_{t-1})$ as shorthand throughout the document. The key observation (and an essential exercise for all interested readers) is that $\bar{Q}_t = E(\bar{Q}_{t+1} \mid C_t = 0, \bar{A}_{t-1} = \bar{a}_{t-1}, \bar{L}_{t-1})$ which applies to all $t = 1, \dots, K$ if we define $\bar{Q}_K = Y$ and $\bar{A}_0 = \bar{a}_0 = 1$ always.

g-Computation can be applied to this g-formula as well. The algorithm is then as follows:
For $t = K, \dots, 1$,

1. Fit a model for $\bar{Q}_t = E(\bar{Q}_{t+1} \mid C_t = 0, \bar{A}_{t-1} = \bar{a}_{t-1}, \bar{L}_{t-1})$ by taking the previously estimated $\bar{Q}_{n,t+1}$ as an outcome in a regression. This model may be fit taking all uncensored subjects ($C_t = 0$) who were treated according to the regimen of interest up to time $t - 1$ ($\bar{A}_{t-1} = \bar{a}_{t-1}$).
2. Use the above model fit to estimate $\bar{Q}_{n,t}$ for *all* subjects with $C_{t-1} = 0$.

Take the empirical mean of $\bar{Q}_{n,1}$ over all subjects. This is the *sequential g-computation* estimate for $E(Y^{\bar{a}})$.

15.2.3 Sequential g-Computation for the PROBIT

The sequential g-computation estimator can also be modified to take into account the longitudinal count outcome of interest, although it is less straight-forward.

- In the first step ($t = 6$), we estimate

$$\bar{Q}_6 = E(Y \mid C_6 = 0, \bar{A}_5 = \bar{a}_5, \bar{L}_5) = E(L_6 \mid C_6 = 0, \bar{A}_5 = \bar{a}_5, \bar{L}_5) + \sum_{t=1}^{5} L_t.$$

Therefore, it is only necessary to fit a model for the random component, defined as $\bar{Q}_{L_6 1} = E(L_6 \mid C_K = 0, \bar{A}_5 = \bar{a}_5, \bar{L}_5)$. We obtain predictions of $\bar{Q}_{L_6 1}$ for every subject with $C_5 = 0$.
- In the next step ($t = 5$), using the predictions of $\bar{Q}_{0,L_6 1}$ from the previous step, we estimate

$$\bar{Q}_5 = E(\bar{Q}_6 \mid C_5 = 0, \bar{A}_4 = \bar{a}_4, \bar{L}_4) =$$

$$E(\bar{Q}_{L_6 1} \mid C_5 = 0, \bar{A}_4 = \bar{a}_4, \bar{L}_4) + E(L_5 \mid C_5 = 0, \bar{A}_4 = \bar{a}_4, \bar{L}_4) + \sum_{t=1}^{4} L_t.$$

Table 15.3 Decomposition of $\bar{Q}_t$ for the g-computation of the PROBIT longitudinal count outcome

Nested expectation of Y		Decomposition (extraction of deterministic counts and separated modeling)	Condition
$\bar{Q}_6$	$=$	$\bar{Q}_{L_61} + L_5 + L_4 + L_3 + L_2 + L_1$	$Pa(Y)$
$\bar{Q}_5$	$=$	$\bar{Q}_{L_62} + \bar{Q}_{L_51} + L_4 + L_3 + L_2 + L_1$	$Pa(L_5)$
$\bar{Q}_4$	$=$	$\bar{Q}_{L_63} + \bar{Q}_{L_52} + \bar{Q}_{L_41} + L_3 + L_2 + L_1$	$Pa(L_4)$
$\bar{Q}_3$	$=$	$\bar{Q}_{L_64} + \bar{Q}_{L_53} + \bar{Q}_{L_42} + \bar{Q}_{L_31} + L_2 + L_1$	$Pa(L_3)$
$\bar{Q}_2$	$=$	$\bar{Q}_{L_65} + \bar{Q}_{L_54} + \bar{Q}_{L_43} + \bar{Q}_{L_32} + \bar{Q}_{L_21} + L_1$	$Pa(L_2)$
$\bar{Q}_1$	$=$	$\bar{Q}_{L_66} + \bar{Q}_{L_55} + \bar{Q}_{L_44} + \bar{Q}_{L_33} + \bar{Q}_{L_22} + \bar{Q}_{L_11}$	$Pa(L_1)$

We denote $\bar{Q}_{L_62} = E(\bar{Q}_{L_61} \mid C_5 = 0, \bar{A}_4 = \bar{a}_4, \bar{L}_4)$ and $\bar{Q}_{L_51} = E(L_5 \mid C_5 = 0, \bar{A}_4 = \bar{a}_4, \bar{L}_4)$. We obtain predictions of $\bar{Q}_{L_62}$ and $\bar{Q}_{L_51}$ for every subject with $C_4 = 0$.

This process is repeated for $t = 4, 3, 2, 1$ with $7 - t$ models to be fit at each step. In general, defining $\bar{Q}_{L,0} = L_t$, we estimate $\bar{Q}_{L_tk} = E(\bar{Q}_{L_tk-1} \mid C_{t-k+1} = 0, \bar{A}_{t-k} = \bar{a}_{t-k}, \bar{L}_{t-k})$ for all $t = 6, \ldots, 1$ and $k = 1, \ldots, t$. This can be proved for any number of time points by induction. At the final step, we calculate $\bar{Q}_{n,1} = \sum_{t=1}^{6} \bar{Q}_{n,L_t}$ for every subject, and take the empirical mean of this quantity over all subjects to obtain the sequential g-computation estimate of $E(Y^{\bar{a}})$. This decomposition is summarized in Table 15.3.

15.2.4 g-Computation Assumptions

In addition to the causal assumptions, all versions of g-computation require consistent estimation of the relevant components of Q_0 conditional on a set of covariates satisfying sequential exchangeability. Inconsistent estimation of these components may produce finite and asymptotic bias in the estimation of $E(Y^{\bar{a}})$. Models for the estimation of Q_0 components are not prespecified in these algorithms and are inevitably left up to the discretion of the user.

In the case of a longitudinal count outcome, decomposing the random and deterministic components of the Q_0 functions can improve estimation. However, this will also increase computational complexity (fitting $K \times (K+1)/2$ versus K models). When compared through the simulation study presented in Schnitzer et al. (2014), the standard sequential g-computation estimator produced 12% estimation bias and 95% confidence intervals with 43% coverage when all confounders were included in the models. The improved sequential g-computation presented in Sect. 15.2.3, produced 0% estimation bias and 93% coverage for the same simulated datasets.

15.3 LTMLE for a Saturated Marginal Structural Model

Although sequential g-computation is computationally efficient and feasible for large numbers of time points and time dependent confounders, it relies on correct parametric specification of a large number of models. The extension to LTMLE allows for semiparametric efficient estimation, good performance using nonparametric methods for the estimation of the necessary model components (Porter et al. 2011), and double robustness. In the longitudinal setting, double robustness implies consistency when either the $\bar{Q}_t; t = K, \ldots, 1$ models are correctly specified or when the models for treatment (see Sect. 15.3.1) are correctly specified.

15.3.1 Construction of Weights

Let $\bar{g}_t(\bar{L}_{t-1}), t = 2, \ldots, K$ be the probability associated with obtaining a given history of breastfeeding $\bar{a}$ up until time $t - 1$, and no censoring up until time point t, conditional on the observed history $\bar{L}_{t-1}$. Specifically, let

$$\bar{g}_t(\bar{L}_{t-1}) = Pr(C_1 = 0 \mid L_0) \times \prod_{k=2}^{t} \{Pr(C_k = 0 \mid \bar{A}_{k-1} = \bar{a}_{k-1}, C_{k-1} = 0, \bar{L}_{k-1}) \times$$
$$Pr(A_{k-1} = a_{k-1} \mid \bar{A}_{k-2} = \bar{a}_{k-2}, C_{k-1} = 0, \bar{L}_{k-1})\}$$

for $t = 2, \ldots, K$. Further, let $\bar{g}_1(L_0) = Pr(C_1 = 0 \mid L_0)$ be the probability of being uncensored at the first time point, conditional on baseline covariates, L_0.

One can directly use these exposure and censoring probabilities as weights in order to estimate $E(Y^{\bar{a}})$ using inverse probability weighting (IPW). Letting $\bar{g}_{n,K}(\bar{L}_{K-1})$ denote the estimated values of $\bar{g}_K(\bar{L}_{K-1})$ for each individual and $I(\cdot)$ be an indicator function for a logical statement, the IPW estimator can be defined as the empirical solution for $\psi_{n,IPW}^{\bar{a}}$ of the estimating equation $P_n D_{IPW} = 0$ where

$$D_{IPW}(O) = (Y - \psi_{n,IPW}^{\bar{a}}) \frac{I(\bar{A}_{K-1} = \bar{a}, C_K = 0)}{\bar{g}_{n,K}(\bar{L}_{K-1})}.$$

15.3.2 Efficient Influence Function

van der Laan and Gruber (2012) demonstrated how the IPW influence function can be projected onto the nonparametric tangent space in order to obtain the efficient influence function (EIF), $D^*(O)$, used in LTMLE. The EIF can be written as the sum of the components

$$D_t^*(O) = \frac{I(\bar{A}_{t-1} = \bar{a}_{t-1}, C_t = 0)}{\bar{g}_t(\bar{L}_{t-1})} (\bar{Q}_{t+1}(\bar{L}_t) - \bar{Q}_t(\bar{L}_{t-1})) \quad \text{for } t = K, \ldots, 2,$$
$$D_1^*(O) = \frac{I(C_1 = 0)}{\bar{g}_1(L_0)} (\bar{Q}_2(\bar{L}_1) - \bar{Q}_1(L_0)), \text{ and}$$
$$D_0^*(O) = (\bar{Q}_1(L_0) - \psi_0^{\bar{a}}).$$

Following an LTMLE procedure that ignores the clustered nature of the data will produce estimates that solve the efficient estimating equation $P_n D_n^* = 0$. In Sect. 15.4 we revisit the EIF and show how clustering alters variance estimation.

15.3.3 LTMLE

In order to produce a LTMLE that has $D^*(O)$ as its influence function, each $\bar{Q}_{n,t}$; $t = K, \ldots, 2$ used in the sequential g-computation is sequentially updated. Chapter 4 and van der Laan and Gruber (2012) give more insight into how these updates are derived. As described in Schnitzer et al. (2014), given a fixed regimen $\bar{a}$, the general procedure is as follows.

- Using models for censoring and exposure, calculate the probabilities of following the regimen $\bar{g}_{n,t}(\bar{L}_{t-1})$ for each subject, as described in Sect. 15.3.1.
- Set $\bar{Q}_{n,7} = Y$. (If Y is not binary, it should be rescaled to [0,1] using the true bounds; Gruber and van der Laan 2010b.)
  Then, for $t = 6, \ldots, 1$,

  - Fit a model for $E(\bar{Q}_{n,t+1} \mid C_t = 0, \bar{A}_{t-1} = \bar{a}_{t-1}, \bar{L}_{t-1})$. Using this model, predict the conditional outcome for all subjects with $C_{t-1} = 0$ and let this vector be denoted $\bar{Q}_{n,t}$.
  - Construct the covariate $H_t(C_t, \bar{A}_{t-1}, \bar{L}_{t-1}) = I(C_t = 0, \bar{A}_{t-1} = \bar{a}_{t-1})/\bar{g}_{n,t}(\bar{L}_{t-1})$.
  - Update the expectation by running a no-intercept logistic regression with outcome $\bar{Q}_{n,t+1}$, the fit $\text{logit}(\bar{Q}_{n,t})$ as an offset, and the covariate H_t as the unique covariate. Let $\hat{\epsilon}_t$ be the estimated coefficient of H_t.
  - Update the fit of $\bar{Q}_t$ by setting

  $$\bar{Q}_{n,t}^1(O) = \text{expit}\left\{ \text{logit}(\bar{Q}_{n,t}) + \frac{\hat{\epsilon}_t}{\bar{g}_{n,t}(\bar{L}_{t-1})} \right\}$$

  to obtain a predicted value of $\bar{Q}_{n,t}^1$ for all subjects with $C_{t-1} = 0$.
  As a result of the update using the logistic regression on outcome $\bar{Q}_{n,t+1}^1$ with the covariate H_t, this step sets $P_n D_{n,t}^* = 0$ where

  $$D_{n,t}^*(O) = \frac{I(\bar{A}_{t-1} = \bar{a}_{t-1}, C_t = 0)}{\bar{g}_{n,t}(\bar{L}_{t-1})}(\bar{Q}_{n,t+1}^1(O) - \bar{Q}_{n,t}^1(O)),$$

  corresponding with the component $D_t^*(O)$ of the EIF.
  At the final step, note that the model for $\bar{Q}_1$ is fit using only subjects with $C_1 = 0$. The resulting fit $\bar{Q}_{n,1}$ is only conditional on L_0 and is estimated for all subjects.

- We take the mean of $\bar{Q}_{n,1}^1$ over all subjects. (If necessary, transform the mean back to the original scale.) This is the LTMLE for the estimation of $\psi_0^{\bar{a}}$.

Overall, this algorithm effectively solves the estimating equation $P_n D_n^* = 0$ where $D_n^*(O)$ is the EIF with the estimated treatment probabilities $\bar{g}_{n,t}(\bar{L}_{t-1})$ and updated outcome expectations $\bar{Q}_{n,t}^1(O)$.

15.3.4 LTMLE for the PROBIT

The particularities of a longitudinal count outcome can be integrated into the LTMLE procedure as well using a reparametrization of the sequential g-formula. In the first step ($t = 6$) for the estimation of $\bar{Q}_6$, we note that

$$Y - \bar{Q}_6 = (Y - \sum_{r \leq 6} L_r) - E(Y - \sum_{r \leq 6} L_r \mid C_6 = 0, \bar{A}_5 = \bar{a}_5, \bar{L}_5)$$

$$= L_6 - E(L_6 \mid C_6 = 0, \bar{A}_5 = \bar{a}_5, \bar{L}_5).$$

Given an initial fit of $E(L_6 \mid C_6 = 0, \bar{A}_5 = \bar{a}_5, \bar{L}_5)$, we update this fit using the covariate H_6 with respect to the outcome L_6. As in the previous algorithm, this successfully sets $P_n D_{n,6}^* = 0$ where

$$D_{n,6}^*(O) = \frac{I(\bar{A}_5 = \bar{a}_5, C_6 = 0)}{\bar{g}_{n,6}(\bar{L}_5)}(Y - \bar{Q}_{n,6}^1(O)).$$

For $t = 5, \ldots, 2$, define $\tilde{Q}_t = E(Y - \sum_{r \leq t-1} L_r \mid C_{t-1} = 0, \bar{A}_{t-2} = \bar{a}_{t-2}, \bar{L}_{t-2})$. For $t = 1$, define $\tilde{Q}_1 = \bar{Q}_1 = E(Y \mid L_0)$. If we take $\tilde{Q}_6 = Y - L_6$, note that $\bar{Q}_{t+1} - \bar{Q}_t = \tilde{Q}_{t+1} - \tilde{Q}_t$ for all $t = 6, \ldots, 1$. Therefore, if we have an initial fit for $\tilde{Q}_t$, we can update it using the covariate H_t with respect to the outcome which is a fit of $\tilde{Q}_{t+1}$ obtained through a previous iteration of the LTMLE algorithm. This procedure will solve the empirical EIF equation, $P_n D_{n,t}^* = 0$.

The full algorithm is as follows.

- For $t = 6$, fit a model for $\tilde{Q}_6 = E(L_6 \mid C_6 = 0, \bar{A}_5 = \bar{a}_5, \bar{L}_5)$. Produce a prediction for all subjects with $C_5 = 0$ and denote these fits as $\tilde{Q}_{n,6}$.
- Update $\tilde{Q}_{n,6}$ using a logistic regression with covariate H_6 against outcome L_6. Denote the updated fit $\tilde{Q}_{n,6}^1$.

  For $t = 5, \ldots, 1$:

  – Fit a model for $E(\tilde{Q}_{t+1} \mid C_t = 0, \bar{A}_{t-1} = \bar{a}_{t-1}, \bar{L}_{t-1})$ using $\tilde{Q}_{n,t+1}^1$ from the previous step as an outcome. Obtain a prediction for all subjects with $C_{t-1} = 0$. Fit a *second* model for $E(L_t \mid C_t = 0, \bar{A}_{t-1} = \bar{a}_{t-1}, \bar{L}_{t-1})$. Obtain a prediction for all subjects with $C_{t-1} = 0$. The sum of these two predictions is denoted $\tilde{Q}_{n,t}$.
  – Scale $\tilde{Q}_{n,t}$ to $(0,1)$ by dividing by $7 - t$, the maximum possible value. Update $\tilde{Q}_{n,t}$ using a logistic regression with covariate H_t. Produce a prediction of the updated fit for all subjects with $C_{t-1} = 0$, multiply by $7 - t$ to rescale, and denote this as $\tilde{Q}_{n,t}^1$.

- The last step produces an estimate of $\tilde{Q}^1_{n,1}$ for all subjects. The mean of $\tilde{Q}^1_{n,1}$ is the LTMLE estimate for $\psi^{\bar{a}}_0$ using the modified sequential decomposition.

15.4 Variance Estimation and Clustering

We did not assume that the individuals are statistically independent but instead we only assumed that the clusters are independent and identically distributed. Therefore the unit is the hospital so that the variance estimator of the TMLE presented above needs to take into account the statistical dependence of individuals within a cluster. If clustering is ignored, true variability will be underestimated as the clustered individuals will be falsely considered independently distributed.

15.4.1 Distinction Between Clustering and Interference

The concept of clustering might be confused with interference, which is often assumed not to exist in causal analysis (Rubin 1980; Hudgens and Halloran 2008). Interference means that the potential outcomes of one subject are not dependent on the exposures of others. In our context, this corresponds with the assumption that one infant's breastfeeding does not impact another infant's probability of hospitalization given the second infant's breastfeeding status.

In contrast, clustering is a violation of the assumption that the individual level data $(O_{mi} : m, i)$ are independently and identically sampled. Within clusters, the observed data (including exposures and outcomes) may be correlated. However, in the absence of interference, this is assumed to arise due to population similarities within clusters that contrast the differences between clusters. In some settings (such as when the outcome is an infectious disease), interference within clusters (such as hospital centers or communities) may be plausible. However, we do not believe it to be so in the PROBIT example.

15.4.2 Estimation with the EIF

With sufficient clusters, we can reasonably use the (efficient) influence function for variance estimation while accounting for a finite set of known clusters. Let $D^*(O_i)$ represent the value of the EIF for subject i. Let the M clusters be described as $Z_m, m = 1, \ldots, M$ where Z_m represents the set of subjects belonging to cluster m. Let the LTML-estimator for parameter $\psi^{\bar{a}}_0$ be denoted $\psi^{\bar{a}}_n$. Even though there is dependence, our i.i.d. LTMLE described above should still behave as an asymptotically linear estimator with influence curve $D^*(O)$ as sample size increases. Therefore, we will still have

$$\psi_n^{\bar{a}} - \psi_0^{\bar{a}} \approx \frac{1}{n} \sum_{i=1}^{n} D^*(O_i)$$

$$= \frac{1}{n} \sum_{m=1}^{M} \sum_{i \in Z_m} D^*(O_i),$$

where we reindexed the units according to cluster membership. Then by multiplying and dividing by M we have

$$\psi_n^{\bar{a}} - \psi_0^{\bar{a}} = \frac{1}{M} \sum_{m=1}^{M} \sum_{i \in Z_m} D^*(O_i) \frac{M}{n}.$$

Due to the independence between clusters, we can consider cluster to be the experimental unit with EIF equal to $\sum_{i \in Z_m} D^*(O_i) \frac{M}{n}$. Therefore, the variance of the estimator can be approximated by the variance of the cluster-specific EIF when M is sufficiently large.

To estimate the variance, we have

$$Var(\psi_n^{\bar{a}}) \approx Var(\frac{1}{n} \sum_{i=1}^{n} D^*(O_i))$$

$$= \frac{1}{n^2} Var\left(\sum_{m=1}^{M} \sum_{i \in Z_m} D^*(O_i) \right).$$

We note that $E(D^*(O_i)) = 0$ by the definition of influence function, so that $Var(D^*(O_i)) = E(D^*(O_i)^2)$. In addition, $Var(D^*(O_i) \times D^*(O_j)) = E(D^*(O_i) \times D^*(O_j))$ for two same-cluster units i and j. Therefore, the above equals

$$\frac{1}{n^2} \sum_{m=1}^{M} \sum_{i,j \in Z_m} E(D^*(O_i) \times D^*(O_j)) I(i \neq j) + E(D(O_i^*)^2) I(i = j)$$

where $I(\cdot)$ is an indicator for the logical statement argument. If we assume that the influence function covariance between subjects within the same cluster is a cluster-specific constant $\rho_m = E(D^*(O_i) \times D^*(O_j)), i \neq j, i$ and $j \in Z_m$, and that the within-cluster variance for each subject $\sigma_m^2 = E(D^*(O_i)^2), i \in Z_m$ is also constant within clusters, the above simplifies to

$$\frac{1}{n^2} \sum_{m=1}^{M} n_m(n_m - 1)\rho_m + n_m \sigma_m^2$$

where n_m is the number of subjects within cluster m. The values of ρ_m and σ_m^2 can be estimated empirically as the covariances and variances of the EIF within each cluster, respectively.

15.4.3 Simulation Study

In order to observe the importance of accounting for clustering, a simulation study was performed. Data were generated as a simplified version of the PROBIT dataset, similar to the simulation study performed in Schnitzer et al. (2014). Five hundred subjects were generated in each of 31 clusters, resulting in $n = 15{,}500$. The baseline covariates $L_0 = \{W, U\}$ were generated as independent Gaussian variables with cluster-specific means drawn from separate Gaussian distributions. The time dependent variables $(C_1, L_1, A_1, C_2, L_2, A_2, C_3, L_3)$ were generated independently for each subject conditional on the subject's recent history and baseline variables. Binary variables $A_t, t = 1, 2$ indicate continued breastfeeding, $C_t, t = 1, 2, 3$ are censoring indicators, and $L_t, t = 1, 2, 3$ indicate a hospitalization during the preceding time interval. The outcome $Y = \sum_{t=1}^{3} L_t$ is a count variable. The baseline variable U is a pure risk factor of hospitalization and did not otherwise affect censoring or exposure.

To correspond with the associations observed in the real PROBIT data, breastfeeding was specifically made to be less likely to continue when hospitalization was indicated at the current time point. Censoring was less likely if breastfeeding continued at the previous time point and more likely if a hospitalization occurred at the previous time point. Hospitalizations were generated conditional on baseline variables and breastfeeding for the past two visits, so that longer duration of breastfeeding decreased the probability of hospitalization.

The parameter $\psi^{\bar{a}} = E(Y^{\bar{a}})$ was estimated for $\bar{a} = (0, 0)$ and $\bar{a} = (1, 1)$. The parameter of interest, corresponding with the first parameter of interest in the PROBIT study, was $\delta = \psi_0^{(1,1)} - \psi_0^{(0,0)}$. We generated 1000 datasets and estimated the parameter of interest using the modified sequential LTMLE with logistic regression models for the probabilities of censoring, exposure and hospitalization. U was considered to be unmeasured in the analysis. Table 15.4 compares the estimation of standard errors using (1) the EIF variance estimator without taking clustering into account, (2) the EIF variance estimation with clustering and (3) the "pairs" clustered nonparametric bootstrap (Cameron et al. 2008) which resamples clusters rather than subjects.

Table 15.4 Simulation study: comparison of LTMLE variance estimation incorporating versus ignoring clustering

Method	Est S.E.	95% C.I. coverage
EIF no clustering	0.007	85
EIF clustering	0.009	92
bootstrap clustering	0.009	98
True effect = -0.030, mean bias=0%		

For standard error estimation, we resampled the 31 clusters with replacement, calculated the estimates from the resampled data, repeated 200 times, and took the standard error of the 200 estimates. Confidence intervals were calculated by taking the 2.5th and 97.5th quantiles of the resampled estimates. From Table 15.4, we see that variance estimation without considering clustering will underestimate standard errors, leading to suboptimal coverage. Both the clustered EIF approach and the clustered bootstrap approach led to near-optimal coverage with the bootstrap producing slight overcoverage and the EIF slightly undercovering. All simulations and modeling were carried out using R (R Development Core Team 2016). Extensive simulations studies were carried out in Schnitzer et al. (2014) to compare the performance of different estimators (LTMLE with parametric models, LTMLE with super learner, IPW, g-computation and sequential g-computation) in this clustered setting.

15.5 PROBIT Results

The PROBIT data were analyzed using the four methods described above: standard (likelihood) and sequential g-computation, IPW, and LTMLE. Both of the g-computation algorithms and the LTMLE algorithm took into account the longitudinal count outcome. The LTMLE algorithm was implemented in two different ways: once using logistic regressions to estimate all of the outcome model components and the probabilities of treatment and censoring; a second time using super learner (Polley et al. 2011; van der Laan et al. 2007) for these same components. To this end, we used the R library `SuperLearner` (Polley and van der Laan 2013). The library we chose used main terms logistic regression, generalized additive modeling (Hastie 2011), the mean estimate, a nearest neighbour algorithm (Peters and Hothorn 2009) (only when modeling a binary outcome), multivariate adaptive regression spline models (Milborrow et al. 2014), and a stepwise AIC procedure (`stepAIC` from Venables and Ripley 2002). The standard errors for all methods except g-computation were calculated using the variance of the EIF taking clustering into account as described in Sect. 15.4.2. The standard errors for the g-computation methods were estimated using pairs cluster bootstrap (Cameron et al. 2008).

We first investigated the difference in the expected number of intervals involving hospitalizations (in the first year of life) when comparing breastfeeding durations of 3–6 months versus 1–2 months, over 9 months versus 3–6 months, and over 9 months versus 1–2 months. The estimates, standard errors and 95% confidence intervals are given in Table 15.5. For the first comparison, the point estimates for all methods suggested a decrease in the expected number of hospitalized intervals by 12 months, although likelihood g-computation and IPW produced confidence intervals that cross zero. g-Computation appeared sensitive to different parametrizations, as the two versions produced very different point estimates (−0.03 and −0.12).

For the second comparison, all estimators except for IPW produced null point estimates, suggesting that no benefit in terms of hospitalizations is obtained by lengthening breastfeeding past 3–6 months. IPW estimated an increase in hospitalizations for increasing breastfeeding durations from 3–6 months to over 9 months. We believe this conclusion to be implausible. For the third contrast, all methods except for

Table 15.5 Marginal expected number of hospitalizations under different breastfeeding durations

Method	Estimate	S.E.	95% C.I.
3–6 months vs 1–2 months			
g-Comp (likelihood)	−0.03	0.02	(−0.06,0.00)
g-Comp (sequential)	−0.12	0.03	(−0.17,−0.06)
IPW	−0.05	0.03	(−0.10,0.01)
Parametric LTMLE	−0.06	0.01	(−0.08,−0.04)
LTMLE with SL	−0.06	0.01	(−0.08,−0.04)
9+ months vs 3–6 months			
g-Comp (likelihood)	−0.00	0.01	(−0.01,0.01)
g-Comp (sequential)	−0.00	0.01	(−0.03,0.02)
IPW	0.05	0.02	(0.01,0.08)
Parametric LTMLE	−0.00	0.01	(−0.02,0.02)
LTMLE with SL	−0.00	0.01	(−0.02,0.02)
9+ months vs 1–2 months			
g-Comp (likelihood)	−0.03	0.02	(−0.07,0.01)
g-Comp (sequential)	−0.12	0.03	(−0.17,−0.07)
IPW	0.00	0.03	(−0.05,0.06)
Parametric LTMLE	-0.06	0.01	(−0.08,−0.04)
LTMLE with SL	−0.06	0.01	(−0.08,−0.04)

IPW produced essentially the same reduction in hospitalizations as in the first contrast, corresponding in no improvement past 3–6 months breastfeeding duration. For all three contrasts, both of the LTMLE implementations gave identical estimates and confidence intervals, suggesting that super learner did not improve estimation in this application. Note, however, that this was not the case when modeling the number of infections (Schnitzer et al. 2014). One can graphically observe the changes in the estimated expected number of hospitalized intervals as the fixed regimen of breastfeeding increases. Figure 15.2 plots the marginal mean estimates and pointwise confidence intervals obtained with LTMLE using super learner for each breastfeeding duration.

15.6 Discussion

In this chapter, we demonstrated how the LTMLE algorithm can be adapted to a real application. In particular, we showed how LTMLE can be modified to better incorporate a longitudinal count structure and how the variance estimation with the EIF can be modified to adapt to clustering in a multicenter study. The LTMLE analysis of the PROBIT data revealed that there is likely an effect of breastfeeding on hospitalizations over the first 12 months of an infant's life. However, we estimated that no additional benefits are obtained by extending breastfeeding past 6 months. In contrast, our previous work suggested that the marginal expected number of gastroin-

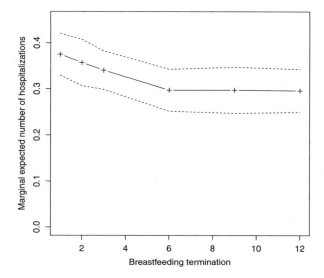

Fig. 15.2 Expected number of periods with hospitalizations including 95% confidence region as estimated by LTMLE and super learning

testinal infections at 12 months continues to decrease as breastfeeding is extended past 9 months (Schnitzer et al. 2014). In terms of the reliability of the statistical estimators, we found that g-computation was sensitive to the choice of parametrization. The IPW estimator concluded that extending breastfeeding past 6 months *increases* the expected number of periods with hospitalizations. This conclusion lacks scientific plausibility and contradicts the results of the other estimators. Given that the maximum value of the unstabilized weights was 225 (with 15,642 uncensored subjects at time six), we expect the anomalous IPW result to be due to treatment model misspecification resulting in bias rather than instability. LTMLE was therefore able to correct this estimate using outcome model components, demonstrating the benefit of using the more sophisticated approach.

Chapter 16
Comparative Effectiveness of Adaptive Treatment Strategies

Romain S. Neugebauer, Julie A. Schmittdiel, Patrick J. O'Connor, and Mark J. van der Laan

In this chapter, we describe secondary analyses of electronic health record (EHR) data from a type 2 diabetes mellitus (T2DM) study of the effect of four adaptive treatment strategies on a time-to-event outcome. More specifically, we describe a TMLE and compare its practical performance to that of three IPW estimators of the same causal estimands defined based on the same non-parametric dynamic marginal structural model. In addition, we evaluate the practical impact of parametric versus data-adaptive estimation of the nuisance parameters on causal inferences from the four estimators considered. Note that the work presented here is a summary of prior results described across several published articles (Neugebauer et al. 2012, 2013, 2014a,b, 2016).

A major goal of clinical care of T2DM is minimization of microvascular and macrovascular complications. Microvascular complications include retinopathy, neuropathy, and nephropathy and can lead to blindness, amputation, severe pain, and kidney failure (requiring dialysis or kidney transplant). Macrovascular complications include heart attack and stroke and can lead to congestive heart failure, brain damage, and death. It has long been hypothesized that aggressive glycemic control through a variety of pharmacological treatments is an effective strategy to reduce

R. S. Neugebauer (✉) · J. A. Schmittdiel
Kaiser Permanente Division of Research, 2000 Broadway, Oakland, CA 94612, USA
e-mail: romain.s.neugebauer@kp.org;julie.a.schmittdiel@kp.org

P. J. O'Connor
HealthPartners Institute, Bloomington, MN 55425, USA
e-mail: Patrick.J.OConnor@HealthPartners.com

M. J. van der Laan
Division of Biostatistics and Department of Statistics, University of California, Berkeley, 101 Haviland Hall, #7358, Berkeley, CA 94720, USA
e-mail: laan@berkeley.edu

© Springer International Publishing AG 2018
M.J. van der Laan, S. Rose, *Targeted Learning in Data Science*,
Springer Series in Statistics, https://doi.org/10.1007/978-3-319-65304-4_16

the occurrence of these devastating complications. Consequently, the management of T2DM involves frequent testing of patients' blood glucose levels, and periodic testing of glycated hemoglobin (A1c) which provides a measure of average glucose levels over a period of about 90 days (10–12 weeks). It is common for T2DM patients to be on multiple glucose-lowering medications due to the progressive nature of this disease. Indeed, glycemic control tends to deteriorate over time, prompting repeated treatment intensification decisions with various glucose-lowering drugs in an on-going effort to achieve recommended levels of glycemic control. Widely accepted guidelines recommend initial treatment of T2DM with metformin, followed by additional medications if glycemic control is not achieved or deteriorates. Addition of insulin is generally considered if adequate glycemic control has not been achieved with two or three noninsulin glucose-lowering medications.

Current T2DM guidelines specify target hemoglobin A1c of <7% for many patients, but also indicate that the benefits and risks of aggressive glucose control vary across patients (Nathan et al. 2006; Skyler et al. 2009; The Diabetes Control and Complications Trial Research Group 1993; Nathan et al. 2005; UK Prospective Diabetes Study (UKPDS) Group 1998; Holman et al. 2008; Action to Control Cardiovascular Risk in Diabetes Study Group 2008; ADVANCE Collaborative Group 2008; Duckworth et al. 2009a; Ray et al. 2009). For this reason, the optimal target levels of A1c for balancing benefits and risks of therapy are not clearly defined. To address this knowledge gap, we aimed to evaluate the impact of progressively less aggressive glucose-lowering strategies on the development or progression of albuminuria, an important biomarker of chronic kidney disease.

Results from the analyses presented in this chapter can be contrasted with results from two randomized experiments. In the ACCORD and ADVANCE clinical trials published from 2008 to 2010 (Gerstein et al. 2008; Patel et al. 2008; Duckworth et al. 2009b), intensive glucose-lowering strategies using multiple classes of glucose-lowering agents succeeded in reducing A1c levels substantially. In the ADVANCE trial, the more intensive therapy arm aimed to reach an A1c level <6.5% and achieved a mean A1c level of 6.5%, compared to a mean level of 7.3% in the control arm. In the ACCORD trial, the more intensive arm aimed for an A1c of <6%, and achieved a mean A1c of 6.4% (vs. 7.5% in controls). There is substantial data from both trials (Ismail-Beigi et al. 2010; O'Connor and Ismail-Beigi 2011) to support the hypothesis (NCEP 2002; Nathan et al. 2009) that, in general, those with T2DM who are treated to lower A1c levels may have lower rates of onset and progression of albuminuria (e.g., HR: 0.79, 0.66–0.93 in ADVANCE).

16.1 The Treatment Intensification Study

Using data from the electronic health records of patients from seven sites of the HMO Research Network (Vogt et al. 2004), a large retrospective cohort study of US adults with T2DM was conducted to inform the following pragmatic clinical question: *Should a T2DM adult currently treated with two or more oral agents or basal insulin intensify treatment the first time an A1c test indicates that the patient's A1c drifts above the recommended level of 7% or should this patient delay TI until a test indicates that the higher A1c thresholds of 7.5%, 8%, or 8.5% are reached?* Below, we briefly summarize the main aspects of the study design, available data, initial analytic approach, and findings from the original analyses.

Enrollment Criteria. As done in clinical trials, inclusion and exclusion criteria were selected to restrict the study cohort to patients for whom the comparative effectiveness research question just described was relevant. Specifically, we searched the entire adult membership of the participating health plans between January 1st 2001 and June 30th 2009 for enrollees who were "failing" current therapy with two or more oral agents or basal insulin. Failing was defined as having an A1c reach or rise above 7% for the first time after being below 7% on the current pharmacotherapy. Cohort entry occurred on the date of this first elevated A1c. Patients whose first elevated A1c was greater than 8.5% were excluded because there is little question that intensifying glucose-lowering therapy is indicated. Patients with limited life expectancy due to certain co-morbid conditions were also excluded from the cohort. These criteria identified 58,671 patients. All patients of this cohort were followed-up from study entry (index date) until the earliest of December 31st 2009 (administrative end of study), plan disenrollment, or death. The median follow-up time for this cohort was about 2.5 years.

EHR Data. To facilitate research, the seven HMORN sites participating in this study developed a standard set of variables in a virtual data warehouse (VDW) based on the various healthcare databases that are maintained by each health plan for operational purposes. These variables capture for each patient: demographic information, all types of clinical encounters with the health plan (ambulatory, emergency department, email, telephone, acute inpatient hospital stay, nonacute institutional stay, laboratory only, radiology only, or other), all diagnoses and procedures from inpatient outpatient and emergency department visits, types of laboratory visits with test results, vital signs, and all prescribed and dispensed medications. These VDW data were assembled on all patients of the study cohort from their index date to their end of follow-up. Subsequently, these data were coarsened into an analytic dataset using a standard format detailed in the next section such that each variable of the dataset could change at most every 90 days only. This choice of analytic unit of time was motivated by the expectation that patients' glycemic control would typically not be monitored with an A1c test more than once every 90 days. We revisit the principle used for coarsening VDW data into an analytic dataset in the next section.

Original Analyses. Ideally, the evaluation of the effect of progressively less aggressive treatment intensification strategies would be conducted with a randomized experiment. For example, the cohort of patients enrolled in this study could have been randomized to several treatment arms—each characterized by a delayed initiation of an intensified treatment (e.g., TI initiation at the index date, at 3 months after the index date, etc.). Contrasts of cumulative risks between any two arms of such a trial would provide a measure of the effect of progressively delayed TI initiation. While such an approach would provide some evidence to address the knowledge gap that motivated this study, inferences from such a trial could not be used directly to inform care because the interventions in this trial are *static*, i.e., determined at randomization only and thus do not reflect how treatment decisions are made in practice. In clinical settings, treatment decisions are personalized over time based on the patient's evolving glycemic control, symptoms and other general considerations of patients' preferences, concurrent conditions, adherence to previously prescribed treatments, and overall health. It is thus more clinically relevant to contrast *dynamic* interventions (a.k.a., adaptive treatment strategies or individualized treatment rules) to inform care than to contrast *static* interventions.

Following this rationale, the original analyses of the data from this study aimed to evaluate the effect of four progressively less aggressive adaptive TI initiation strategies. Each strategy is indexed by an A1c threshold θ =7, 7.5, 8, or 8.5 and requires that a patient initiates TI as soon as (no grace periods allowed) her A1c test reaches or drifts above $\theta\%$ and that she remains on the intensified therapy thereafter. To properly account for time-dependent confounding and informative selection bias, a parametric dynamic marginal structural model (Robins 1998; Murphy et al. 2001; van der Laan and Petersen 2007; Hernan et al. 2006; Robins et al. 2008b; Cain et al. 2010) for counterfactual hazards was fitted using IPW estimation (Robins 1998; Robins et al. 2000; Hernan et al. 2002) for the purpose of contrasting cumulative risks under the four TI strategies just described.

Inferences from these analyses were consistent with those of the ACCORD and ADVANCE randomized experiments and imply that the pattern of results in these trials are applicable to a large population of adults with T2DM treated in routine clinical settings in the US. In particular, findings from this observational study confirmed the benefit of tight glycemic control with respect to the development or progression of albuminuria.

16.2 Data

The observed data on each patient in the cohort consist of measurements on exposure, outcome, and confounding variables updated every 90 days between study entry and end of follow-up. The time (expressed in units of 90 days) when the patient's follow-up ends is denoted by $\tilde{T}$ and is defined as the earliest of the time to failure, i.e., albuminuria development or progression, denoted by T or the time to a right-censoring event denoted by C. When a patient is right-censored, i.e., $\tilde{T} = C$, the type of right-censoring event experienced by the patient is recorded and denoted by Γ with possible values 1, 2, or 3 to represent end of follow-up by administrative end of study, disenrollment from the health plan, or death respectively. For patients with normoalbuminuria at study entry, i.e., microalbumin-to-creatinine ratio (ACR) <30, we defined failure as an ACR measurement indicating either microalbuminuria (ACR 30–300) or macroalbuminuria (ACR>300). For patients with microalbuminuria at study entry, we defined failure as an ACR measurement indicating macroalbuminuria. We thus excluded patients with a baseline ACR measurement missing (5884) or indicating macroalbuminuria (1608), which yielded the sample size $n = 51,179$.

The indicator that the end of follow-up is due to the occurrence of a failure event is denoted by Δ, i.e., $\Delta = 1$ implies that $\tilde{T} = T$ and $\Delta = 0$ implies that $\tilde{T} = C$. Exposure to an intensified treatment was defined as exposure to a glucose-lowering medication that was not used by the patient at study entry. At each time point $t = 0, \ldots, \tilde{T}$, the patient's exposure to an intensified treatment is represented by the binary variable $A_1(t)$, and the indicator of the patient's right-censored status at time t is denoted by $A_2(t)$. We thus have $A_2(t) = 0$ for $t = 0, \ldots, \tilde{T} - 1$ when $\tilde{T} \geq 1$ and $A_2(\tilde{T}) = 1 - \Delta$. The combination $A(t) = (A_1(t), A_2(t))$ is referred to as the action at time t. At each time point $t = 0, \ldots, \tilde{T}$, covariate measurements (e.g., A1c or a particular diagnosis) are denoted by the multidimensional variable $L(t)$.

Table I in Neugebauer et al. (2014b) describes the 48 expert-selected patient attributes represented by these covariates (23 and 10 of which are time-varying and continuous variables, respectively). Each variable of the vector $L(t)$ was constructed such that it can be assumed to occur before the action at time t, $A(t)$, or otherwise assumed not to be affected by the actions at time t and thereafter, $(A(t), A(t+1), \ldots)$. In addition, the covariates at time t include an outcome measurement denoted by $Y(t)$, i.e., $Y(t) \in L(t)$ for $t = 0, \ldots, \tilde{T}$. For each time point $t = 0, \ldots, \tilde{T} + 1$, the outcome is the indicator of past failure, i.e., $Y(t) = I(T \leq t - 1)$. By definition, the outcome is thus 0 for $t = 0, \ldots, \tilde{T}$, missing at $t = \tilde{T} + 1$ if $\Delta = 0$ and, 1 at $t = \tilde{T} + 1$ if $\Delta = 1$. Indeed, when $\Delta = 0$, the patient's end of follow-up is due to occurrence of a right-censoring event during the last follow-up interval $\tilde{T}$ and as a result it is not known to the analyst whether the patient would have experienced failure at that time, i.e., $I(T = \tilde{T})$ and thus $Y(\tilde{T} + 1) = I(T \leq \tilde{T})$ are missing. To simplify notation, we use overbars to denote covariate and exposure histories, e.g., a patient's exposure history through time t is denoted by $\bar{A}(t) = (A(0), \ldots, A(t))$. Following the analytic framework introduced in Robins (1998), we approach the observed data in this study as realizations of n independent and identically distributed copies of $O = (\tilde{T}, \Delta, (1 - \Delta)\Gamma, \bar{L}(\tilde{T}), \bar{A}(\tilde{T}), \Delta Y(\tilde{T} + 1))$ denoted by O_i for $i = 1, \ldots, n$. The longest observed follow-up time is denoted by $K \equiv \max_{i=1,\ldots,n} \tilde{T}_i$ and we have $K = 36$ (9 years) in the treatment intensification study.

More details about the algorithm used for mapping EHR data collected in continuous time into the coarsened exposure, covariate and outcome variables above were described elsewhere (Neugebauer et al. 2012, Appendix E). In short, the guiding principle that informed the construction of the analytic data set was to ensure that all measurements represented by the covariates at time t could not be affected by the exposure at time t and thereafter. The exposure at each time t was defined as 0 (i.e., unexposed to an intensified treatment) for all 90-day intervals t except for the following intervals: (a) the first interval (if any) when a glucose-lowering medication that was not used by the patient at study entry was initiated (i.e., used for at least

1 day of the interval), and (b) all subsequent intervals during which one or more glucose-lowering medication not used by the patient at study entry was used for at least 50% of the days of that interval. Daily use of glucose-lowering drugs were ascribed based on prescription fill dates and drug quantities dispensed extracted from VDW pharmacy data.

For a patient who experiences failure strictly before time interval K (i.e., when $\Delta = 1$ and $\tilde{T} < K$), we extend the definition of her observed data O through $K + 1$ by including the outcome variables $Y(t + 1) = I(T \leq t - 1) = 1$ for $\tilde{T} < t \leq K$. With this extension, the observed data structure becomes:

$$O = (\tilde{T}, \Delta, (1 - \Delta)\Gamma, \bar{L}(\tilde{T}), \bar{A}(\tilde{T}), \Delta\bar{Y}(\tilde{T} + 1, K + 1)), \tag{16.1}$$

where $\Delta\bar{Y}(t, t') = (\Delta Y(t), \ldots, \Delta Y(t'))$ with $t \leq t'$. To simplify expressions below, the outcome $Y(t + 1)$ for $\tilde{T} \leq t \leq K$ when $\Delta = 1$ is also denoted with $L(t + 1)$ and the observed data can thus be expressed as

$$O = (\tilde{T}, \Delta, (1 - \Delta)\Gamma, \bar{L}(\tilde{T}), \bar{A}(\tilde{T}), \Delta\bar{L}(\tilde{T} + 1, K + 1)).$$

Finally, we define $\check{T}(t) = \min(\tilde{T}, t)$ for $t = 0, \ldots, K$.

16.3 Causal Model and Statistical Estimands

We assume the existence of counterfactual covariates whether they are defined from a (nonparametric) structural equation model (Fig. 16.1) that is also assumed to have generated the observed data O (structural modeling framework, Pearl 1995, 2009a) or taken as primitives (Pearl 2010) that are then also linked to the observed data through an identifiability assumption known as the consistency assumption (missing data framework, Neyman 1923; Rubin 1974; Robins 1998).

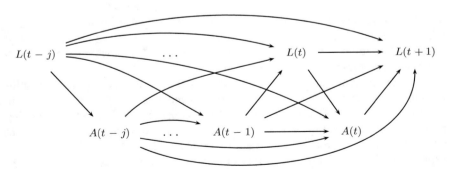

Fig. 16.1 Template of the directed acyclic graph that encodes the time ordering of all variables of the observed data process O. The complete graph can be derived by sequentially drawing the nodes and arcs implied by this template for $t = 0, \ldots, \tilde{T}$ and $1 \leq j \leq t$

In particular, for a given time interval of interest $t_0 \leq K$, we assume the existence of counterfactual outcomes defined by the following individualized action rules (Murphy et al. 2001) $d_\theta = (d_\theta(0), \ldots, d_\theta(t_0))$ where each function, $d_\theta(t)$ for $t = 0, \ldots, t_0$, is a decision rule for determining the action regimen (i.e., a treatment and right-censoring intervention) to be experienced by a patient at time t. More specifically, a decision rule $d_\theta(t)$ maps the action and covariate history measured up to a given time t to an action regimen at time t: $d_\theta(t) : (\bar{L}(t), \bar{A}(t-1)) \mapsto d_\theta(t)(a_1(t), a_2(t))$. In this diabetes study, the four decision rules of interest are defined by $\theta = 7, 7.5, 8, 8.5$ such that $d_\theta(t)((\bar{L}(t), \bar{A}(t-1))$ is equal to:

- $(a_1(t), a_2(t)) = (0, 0)$ (i.e., no use of an intensified treatment and no right-censoring) if and only if the patient was not previously treated with an intensified therapy (i.e., $\bar{A}(t-1) = 0$) and the A1c level at time t (an element of $L(t)$) was lower than or equal to the threshold $\theta\%$.
- $(a_1(t), a_2(t)) = (1, 0)$ (i.e., use of an intensified treatment and no right-censoring) otherwise.

To simplify notation, for any given observed covariate history through time t denoted by $\bar{L}(t)$, the action regimen $(a(0) = d_\theta(0)(L(0)), a(1) = d_\theta(1)(\bar{L}(1), a(0)), \ldots, a(t) = d_\theta(t)(\bar{L}(t), \bar{a}(t-1)))$ through time t is denoted by $d_\theta(\bar{L}(t))$.

The causal estimands of interest denoted by $\psi^{\theta_1, \theta_2}$ are defined as the differences between cumulative risks at a given time t_0 associated with any two distinct dynamic treatment strategies d_{θ_1} and d_{θ_2}:

$$\psi^{\theta_1, \theta_2} = P(Y_{d_{\theta_1}}(t_0 + 1) = 1) - P(Y_{d_{\theta_2}}(t_0 + 1) = 1),$$

where $Y_{d_{\theta_j}}(t_0+1)$ denotes a patient's potential outcome at time t_0+1 if, possibly contrary to fact, the patient experienced the dynamic intervention d_{θ_j} with $\theta_j = 7, 7.5, 8$, or 8.5. Note that unlike the approach taken in the original analyses of the data from this study, we no longer assume a parametric dynamic marginal structural model (MSM) for the counterfactual cumulative risks that define the causal estimands $\psi^{\theta_1, \theta_2}$ nor do we adopt a working MSM (Neugebauer and van der Laan 2007) to approximate the causal estimands $\psi^{\theta_1, \theta_2}$. Motivations for such a nonparametric MSM approach in this study are twofold. First, the large sample size in this study alleviates concerns over the curse of dimensionality which would make the application of such an approach futile otherwise. The curse of dimensionality refers here to the inability to estimate the causal estimands precisely without lowering the dimensionality of the causal estimands of interest through modeling assumptions or the use of a working model. Second, the absence of knowledge about the true functional forms of the four counterfactual survival curves that define the parameter $\psi^{\theta_1, \theta_2}$ means that the specification of a parametric dynamic MSM becomes essentially an arbitrary choice and should thus be expected to lead to biased estimation in practice.

For identifiability of the causal estimands ψ^{θ_1,θ_2}, we make two additional assumptions (Robins 1999) referred to as the sequential randomization assumption (SRA):

$$A(j) \perp Y_{d_\theta}(t_0 + 1) \mid \bar{L}(j), \bar{A}(j-1) \text{ for } j = 0, \ldots, \check{T}(t_0)$$

and the positivity assumption:

$$g_0\Big(A(j) = d_\theta(j)\big(\bar{L}(j), \bar{A}(j-1)\big) \;\Big|\; \bar{Y}(j) = 0, \bar{L}(j), \bar{A}(j-1) = d_\theta\big(\bar{L}(j-1)\big)\Big) > 0$$

$$\text{for } j = 0, \ldots, t_0,$$

$$(16.2)$$

where g_0 is defined by the observed data distribution P_0 and denotes, for each j, the conditional probability of the action at time j given past covariates and actions. The SRA (a.k.a., the assumption of no unmeasured confounders and sources of selection bias) is not testable. In this study, its upholding is motivated by the detailed information available in patient's EHR regarding risk factors for the outcome and the determinants of treatment decisions and rests on the expert-selection of the covariates included in the analytic dataset. Practical violation of the positivity assumption was evaluated by examining the distribution of the estimated probabilities (16.2). Concerns over practical violation of this assumption had also motivated the evaluation of dynamic TI regimens in this study (as opposed to static regimens that may require the absence of exposure to an intensified treatment for a fixed duration) because most T2DM patients in clinical settings with very high A1c tests would be expected to intensify glucose-lowering therapy (i.e., very few such patients would remain on their baseline therapy).

The identifiability assumptions above imply (Bang and Robins 2005) that the cumulative counterfactual risk $P(Y_{d_\theta}(t_0 + 1) = 1)$ and thus the causal estimands ψ^{θ_1,θ_2} can be expressed as a statistical parameter, i.e., a parameter of the observed data distribution (as opposed to the counterfactual data distribution denoted by P below):

$$P(Y_{d_\theta}(t_0 + 1) = 1) =$$

$$E_{P_0}\Big(E_{P_0}\Big[\ldots E_{P_0}\Big(E_{P_0}\Big[E_{P_0}\Big(Y(t_0 + 1) \mid \mathcal{F}(t_0)\Big) \mid \mathcal{F}(t_0 - 1)\Big] \mid \mathcal{F}(t_0 - 2)\Big)\ldots \mid \mathcal{F}(0)\Big]\Big)$$

$$(16.3)$$

with $\mathcal{F}(t) = (\bar{A}(\check{T}(t)) = d_\theta(\bar{L}(\check{T}(t))), \bar{L}(t))$, and where E_{P_0} denotes the expectation under the distribution P_0 of the observed data O. The statistical estimand on the right-hand side of the previous equality is denoted by γ^θ.

16.4 Estimation

Any estimator of the target parameter γ^θ relies on a choice of estimator of the nuisance parameters (action mechanism and the outcome regressions). Therefore, in this section, we detail both a TMLE of the target parameter and different methods for estimation of the nuisance parameters. In Neugebauer et al. (2016), we also described three common alternatives to the TMLE estimator of the target parameter presented here: a standard unbounded IPW estimator (Horvitz-Thompson estimator), the corresponding bounded IPW estimator that guarantees that estimates of the counterfactual survival probabilities are bounded between (0,1), and a hazard-based bounded IPW estimator. Descriptions are not repeated in this section, but the data analyses reported in this chapter include results from these IPW estimators using different estimators of the action mechanism ranging from more parametric to highly nonparametric. We will use the notation $I(\cdot)$ for the indicator that a given event $\cdot$ has occurred.

16.4.1 TMLE

Early evaluation (Neugebauer et al. 2011) of TMLE for CER with longitudinal data from large healthcare databases revealed the practical complexity of the algorithm initially proposed for implementation of targeted learning with time-varying interventions. Subsequently, an alternate algorithm was developed which greatly simplified applications of TMLE in problems with time-varying interventions and a limited number of time-varying covariates (Stitelman et al. 2012). Applications of this alternate algorithm to EHR-based CER studies is however expected to be limited by the fact that these studies such as the one in this chapter often require to control for medium to high-dimensional time-varying covariates. More recently, van der Laan and Gruber (2012) derived an alternate TMLE algorithm for evaluating the effect of time-varying interventions using the general targeted learning estimation road map applied with the key identifiability result (16.3) introduced in Bang and Robins (2005). Compared to previously proposed TMLE algorithms, the new TMLE algorithm further simplifies implementation of targeted learning in CER studies with time-varying interventions and in particular if control for medium to high-dimensional time-varying covariates is needed. The following TMLE algorithm was adapted from the new methodology introduced in van der Laan and Gruber (2012) for application in the treatment intensification study. Each step below is implemented sequentially for a given θ to derive a TMLE of the cumulative risk γ^θ:

1. Estimate the nuisance parameter g_0^θ. The estimator of the probability $g_0(A(t) = d_\theta(t)(\bar{L}(t), \bar{A}(t-1)) \mid \bar{L}(t), \bar{Y}(t) = 0, \bar{A}(t-1) = d_\theta(\bar{L}(t-1)))$ for a given $t = 0, \ldots, t_0$ is denoted by $g_{A(t),n}^\theta$ below.

2. Derive an initial estimator of the nuisance parameter $E_{P_0}(Y(t_0 + 1) \mid \bar{A}(\check{T}(t_0)) = d_\theta(\bar{L}(\check{T}(t_0))), \bar{L}(t_0))$ denoted by $Q_{L(t_0+1)}^\theta(\bar{L}(t_0))$. Note that $\bar{L}(t_0 + 1)$ is always defined in the extended observed data structure (16.1) when $\bar{A}(\check{T}(t_0)) = d_\theta(\bar{L}(\check{T}(t_0)))$ because $\bar{A}(\check{T}(t_0)) = d_\theta(\bar{L}(\check{T}(t_0)))$ implies either (i) $\check{T}(t_0) = t_0$ and $\check{T}(t_0) < \tilde{T}$ or (ii) $\check{T}(t_0) = \tilde{T} = T$ and $\check{T}(t_0) \leq t_0$ (since $\check{T}(t_0) = \tilde{T} = C$

is not possible when $\bar{A}(\check{T}(t_0)) = d_\theta(\bar{L}(\check{T}(t_0))))$. The conditional expectation $Q^\theta_{L(t_0+1)}(\bar{L}(t_0))$ is thus well defined and we have:

$$Q^\theta_{L(t_0+1)}(\bar{L}(t_0)) =$$
$$1 + I(\bar{Y}(t_0) = 0)\Big(E_{P_0}(Y(t_0 + 1) \mid \bar{A}(t_0) = d_\theta(\bar{L}(t_0)), \bar{L}(t_0), \bar{Y}(t_0) = 0) - 1\Big).$$
$$(16.4)$$

This step thus reduces to the estimation of

$$E_{P_0}(Y(t_0 + 1) \mid \bar{A}(t_0) = d_\theta(\bar{L}(t_0)), \bar{L}(t_0), \bar{Y}(t_0) = 0),$$

i.e., the conditional probability that a patient experiences the failure event at time t_0 given (i) that she experienced no such event previously and no censoring event before and at t_0, (ii) that she were continuously treated according to strategy d_θ through t_0, and (iii) her covariates through t_0, $\bar{L}(t_0)$. The initial estimator of the nuisance parameter $Q^\theta_{L(t_0+1)}(\bar{L}(t_0))$ is denoted by $Q^\theta_{L(t_0+1),n}(\bar{L}(t_0))$ and is defined as follows: (i) For a patient who did not experience failure before t_0 and who followed rule d_θ through t_0, $Q^\theta_{L(t_0+1),n}(\bar{L}(t_0))$ is an estimator of the conditional probability just described with possible values restricted to the $[0, 1]$ interval, (ii) For a patient who did experience failure before t_0 and who followed rule d_θ until failure, $Q^\theta_{L(t_0+1),n}(\bar{L}(t_0))$ is set to 1 in accordance with equality (16.4).

3. Update the initial estimator of $Q^\theta_{L(t_0+1)}(\bar{L}(t_0))$. This update is implemented by logistic regression for predicting $Y(t_0 + 1)$ based on an intercept model with an offset variable fitted with weights, and using only data from patients who did not fail before t_0 (i.e., $\bar{Y}(t_0) = 0$) and who followed rule d_θ through t_0 (i.e., $\bar{A}(t_0) = d_\theta(\bar{L}(t_0)))$. The weight and offset associated with the outcome $Y(t_0 + 1)$ from any patient whose data contribute to this logistic regression are defined as $\frac{1}{\prod_{t=0}^{t_0} g^\theta_{A(t),n}}$ and $\mathrm{logit}(Q^\theta_{L(t_0+1),n}(\bar{L}(t_0)))$, respectively. The estimator of the intercept resulting from this weighted logistic regression is denoted by ϵ_n. The updated estimator of $Q^\theta_{L(t_0+1)}(\bar{L}(t_0))$ is denoted by $Q^{\theta,*}_{L(t_0+1),n}(\bar{L}(t_0))$ and is defined as follows: (i) For a patient who did not experience failure before t_0 and who followed rule d_θ through t_0, $Q^{\theta,*}_{L(t_0+1),n}(\bar{L}(t_0))$ is $\mathrm{expit}[\mathrm{logit}(Q^\theta_{L(t_0+1),n}(\bar{L}(t_0))) + \epsilon_n]$ where $\mathrm{expit}(t) = \frac{1}{1+\exp(-t)}$. (ii) For a patient who did experience failure before t_0 and who followed rule d_θ until failure, $Q^{\theta,*}_{L(t_0+1),n}(\bar{L}(t_0))$ is set to 1 in accordance with equality (16.4).

4. Repeat the following two steps for $k = t_0 - 1, \ldots, 0$:

   (a) Derive an initial estimator of $E_{P_0}(Q^\theta_{L(k+2)}(\bar{L}(k + 1)) \mid \bar{A}(\check{T}(k)) = d_\theta(\bar{L}(\check{T}(k))), \bar{L}(k)$ denoted by $Q^\theta_{L(k+1)}(\bar{L}(k))$. Note again that $\bar{L}(k + 1)$ is always defined in the extended observed data structure when $\bar{A}(\check{T}(k)) = d_\theta(\bar{L}(\check{T}(k)))$ because $\bar{A}(\check{T}(k)) = d_\theta(\bar{L}(\check{T}(k)))$ implies either (i) $\check{T}(k) = k$ and $\check{T}(k) < \check{T}$ or (ii) $\check{T}(k) = \check{T} = T$ and $\check{T} \le k$. The conditional expectation $Q^\theta_{L(k+1)}(\bar{L}(k))$ is thus well defined and we have:

$$Q^{\theta}_{\bar{L}(k+1)}(\bar{L}(k)) =$$

$$1 + I(\bar{Y}(k) = 0)\Big(E_{P_0}(Q^{\theta}_{\bar{L}(k+2)}(\bar{L}(k+1)) \mid \bar{A}(k) = d_{\theta}(\bar{L}(k)), \bar{L}(k), \bar{Y}(k) = 0) - 1\Big).$$

(16.5)

This step thus reduces to the estimation of

$$E_{P_0}(Q^{\theta}_{\bar{L}(k+2)}(\bar{L}(k+1)) \mid \bar{A}(k) = d_{\theta}(\bar{L}(k)), \bar{L}(k), \bar{Y}(k) = 0), \qquad (16.6)$$

i.e., the conditional expectation of the continuous measure $Q^{\theta}_{\bar{L}(k+2)}(\bar{L}(k+1))$ (itself a conditional expectation between 0 and 1) characterizing a patient at time $k+1$ given (i) that she did not experience failure before k and no censoring event before and at k, (ii) that she were continuously treated according to strategy d_{θ} through k, and (iii) her baseline and past time-varying covariates through k, $\bar{L}(k)$. Derivation of an estimator for this conditional expectation requires that an estimator $Q^{\theta,*}_{\bar{L}(k+2),n}(\bar{L}(k+1))$ be defined for each patient who did not fail before k and who followed rule d_{θ} through k (i.e., $\bar{Y}(k) = 0$ and $\bar{A}(k) = d_{\theta}(\bar{L}(k))$). Among such patients, some may have followed rule d_{θ} through $k + 1$ and others may only have followed rule d_{θ} through k. For the first group of patients, we already defined an estimator $Q^{\theta,*}_{\bar{L}(k+2),n}(\bar{L}(k+1))$ in the latest "update step". For the second group of patients, the estimator is defined here by extrapolation, i.e., using the same estimator defined in the latest "update step" as if these patients also followed rule d_{θ} at time $k + 1$. The initial estimator of the nuisance parameter $Q^{\theta}_{\bar{L}(k+1)}(\bar{L}(k))$ is denoted by $Q^{\theta}_{\bar{L}(k+1),n}(\bar{L}(k))$ and is defined as follows: (i) For a patient who did not experience failure before k and who followed rule d_{θ} through k, $Q^{\theta}_{\bar{L}(k+1),n}(\bar{L}(k))$ is an estimator of conditional expectation (16.6) above with possible values restricted to the $[0, 1]$ interval. (ii) For a patient who did experience failure before k and who followed rule d_{θ} until failure, $Q^{\theta}_{\bar{L}(k+1),n}(\bar{L}(k))$ is set to 1 in accordance with equality (16.5).

(b) Update the initial estimator of $Q^{\theta}_{\bar{L}(k+1)}(\bar{L}(k))$. This update is implemented by logistic regression for predicting $Q^{\theta}_{\bar{L}(k+2)}(\bar{L}(k + 1))$ based on an intercept model with an offset variable fitted with weights, and using only data from patients who did not fail before k (i.e., $\bar{Y}(k) = 0$) and who followed rule d_{θ} through k (i.e., $\bar{A}(k) = d_{\theta}(\bar{L}(k))$). The weight and offset associated with the outcome $Q^{\theta,*}_{\bar{L}(k+2),n}(\bar{L}(k+1))$ from any patient whose data contribute to this logistic regression are defined as $\frac{1}{\prod_{t=0}^{k} g^{\theta}_{A(t),n}}$ and $\mathrm{logit}(Q^{\theta}_{\bar{L}(k+1),n}(\bar{L}(k)))$, respectively. The estimator of the intercept resulting from this weighted logistic regression is denoted by ϵ_n. The updated estimator of $Q^{\theta}_{\bar{L}(k+1)}(\bar{L}(k))$ is denoted by $Q^{\theta,*}_{\bar{L}(k+1),n}(\bar{L}(k))$ and is defined as follows: (i) For a patient who did not experience failure before k and who followed rule d_{θ} through k, $Q^{\theta,*}_{\bar{L}(k+1),n}(\bar{L}(k))$ is $\mathrm{expit}[\mathrm{logit}(Q^{\theta}_{\bar{L}(k+1),n}(\bar{L}(k))) + \epsilon_n]$. (ii) For a patient who did experience failure before k and who followed rule d_{θ} until failure, $Q^{\theta,*}_{\bar{L}(k+1),n}(\bar{L}(k))$ is set to 1 in accordance with equality (16.5).

5. Derive the estimator of $E_{P_0}(Q^\theta_{L(1)}(L(0)))$ denoted by $Q^\theta_{L(0)}$. For patients who followed rule d_θ at time 0, we already computed an estimator $Q^{\theta,*}_{L(1),n}(L(0))$ in the latest "update step". For all other patients, an estimator $Q^{\theta,*}_{L(1),n}(L(0))$ is defined here by extrapolation, i.e., using the same estimator from the latest "update step" as if these patients also followed rule d_θ at time 0. Thus, an estimator $Q^{\theta,*}_{L(1),n}(L(0))$ is now available for all n patients in the cohort. The average of these estimators is an estimator of $Q^\theta_{L(0)}$ denoted by $Q^{\theta,*}_{L(0),n}$:

$$Q^{\theta,*}_{L(0),n} = \frac{1}{n}\sum_{i=1}^{n} Q^{\theta,*}_{L(1),n}(L_i(0)).$$

This estimator $Q^{\theta,*}_{L(0),n}$ is a TMLE of the counterfactual cumulative risk of interest γ^θ and we also denote it by $\gamma^{\theta,*}_n$.

The vector of nuisance parameters $Q^\theta_{L(t+1)}(\bar{L}(t))$ for $t = 0, \ldots, t_0$ introduced above is denoted by Q^θ_0 and the asymptotic limit of its initial estimator is denoted by $Q^{\theta,\infty}_0 = (Q^{\theta,\infty}_{L(t+1)}(\bar{L}(t)))_{t=0,\ldots,t_0}$. Thus, if the nuisance parameter Q^θ_0 is estimated consistently (e.g., by maximum likelihood using a correctly specified parametric model) then we have $Q^{\theta,\infty}_0 = Q^\theta_0$.

Under regularity conditions, the TMLE estimator $\gamma^{\theta,*}_n$ is asymptotically linear with the following influence curve when the nuisance parameter g^θ_0 is not estimated (i.e., when step 1 above is skipped and all estimates $g^\theta_{A(t),n}$ referenced in all subsequent steps are replaced with their known values):

$$IC^*_\theta(O \mid g^\theta_0, Q^{\theta,\infty}_0) = \sum_{t=0}^{t_0+1} D^*_{\theta,t}(O \mid g^\theta_0, Q^{\theta,\infty}_0) \text{ with}$$

$$D^*_{\theta,t}(O \mid g^\theta_0, Q^{\theta,\infty}_0) = \frac{I(\bar{A}(\check{T}(t-1)) = d_\theta(\bar{L}(\check{T}(t-1))))}{\prod_{j=0}^{\check{T}(t-1)} g_0(A(t) \mid \bar{L}(t), \bar{A}(t-1))}\left(Q^{\theta,\infty}_{L(t+1)}(\bar{L}(t)) - Q^{\theta,\infty}_{L(t)}(\bar{L}(t-1))\right),$$

(16.7)

where $Q^{\theta,\infty}_{L(t_0+2)}(\bar{L}(t_0 + 1)) \equiv Y(t_0 + 1)$ and $\frac{I(\bar{A}(\check{T}(t-1))=d_\theta(\bar{L}(\check{T}(t-1))))}{\prod_{j=0}^{\check{T}(t-1)} g_0(A(t)\mid\bar{L}(t),\bar{A}(t-1))}$ is nil at $t = 0$.

Note that $D^*_{\theta,t}(O \mid g^\theta_0, Q^\theta_0) = 0$ for all t such that either (i) $\tilde{T} = C$ and $C + 1 \le t \le t_0 + 1$ (because we then have $I(\bar{A}(\check{T}(t-1)) = d_\theta(\bar{L}(\check{T}(t-1)))) = 0$), or (ii) $\tilde{T} = T$ and $T + 1 < t \le t_0 + 1$ (because we then have $Q^\theta_{L(t+1)}(\bar{L}(t)) - Q^\theta_{L(t)}(\bar{L}(t-1)) = 0$).

The TMLE estimator $\gamma^{\theta,*}_{n,h}$ is doubly robust in the sense that it is a consistent estimator of γ^θ if either the estimator of the nuisance parameter g^θ_0 is consistent or if the initial estimator of the nuisance parameter Q^θ_0 is consistent. In addition, if both estimators of the nuisance parameters are consistent, $\gamma^{\theta,*}_n$ is then asymptotically linear and efficient in the sense that it attains the semiparametric efficiency bound in the statistical model that may only include constraints on g^θ_0, i.e., the variance of any regular asymptotically linear estimator of γ^θ in such a model (e.g., any of the three

IPW estimators $\gamma_{n,HT}^{\theta}$, $\gamma_{n,bd}^{\theta}$, or $\gamma_{n,h}^{\theta}$) is greater than or equal to the variance of the TMLE $\gamma_n^{\theta,*}$. The IC-based estimator of the variance of $\gamma_n^{\theta,*}$ is conservative only when the vector of probabilities g_0^{θ} are estimated consistently (and asymptotic linearity is preserved). Note that the IC-based estimator of the variance of $\gamma_n^{\theta,*}$ may not be a consistent estimator of the TMLE variance if the estimator of g_0^{θ} is not consistent even when the initial estimator of the nuisance parameter Q_0^{θ} is consistent.

16.4.2 Action Mechanism, g_0^{θ}

In observational studies, the conditional probabilities g_0^{θ} required to implement the TMLE and IPW estimators above are unknown and thus need to be estimated first. Below, we start by describing the decomposition of these nuisance parameters based on five classes of propensity scores (PS). Next, we describe the four approaches considered here for estimating these various PS: two model-based and two data-adaptive approaches.

Decomposition of the Action Mechanism. The conditional probability $g_0(A(t) \mid \bar{L}(t), \bar{Y}(t) = 0, \bar{A}_1(t-1), \bar{A}_2(t-1) = 0)$ for $t = 0 \ldots, t_0$ is referred to as the action mechanism at time t and can be factorized based on the following five PS:

- PS for TI initiation denoted by $\mu_1(t)$

$$g_0(A_1(t) = 1 \mid \bar{L}(t), \bar{Y}(t) = 0, \bar{A}_1(t-1) = 0, \bar{A}_2(t) = 0)$$

- PS for TI continuation denoted by $\mu_2(t)$

$$g_0(A_1(t) = 1 \mid \bar{L}(t), \bar{Y}(t) = 0, \bar{A}_1(t-2), A_1(t-1) = 1, \bar{A}_2(t) = 0)$$

- PS for right-censoring by administrative end of study denoted by $\mu_3(t)$

$$g_0\Big(I(A_2(t) = 1, \Gamma = 1) = 1 \mid \bar{L}(t), \bar{Y}(t) = 0, \bar{A}_1(t-1), \bar{A}_2(t-1) = 0\Big)$$

- PS for right-censoring by disenrollment from the health plan denoted by $\mu_4(t)$

$$g_0\Big(I(A_2(t) = 1, \Gamma = 2) = 1 \mid \bar{L}(t), \bar{Y}(t) = 0,$$
$$\bar{A}_1(t-1), \bar{A}_2(t-1) = 0, I(A_2(t) = 1, \Gamma = 1) = 0\Big)$$

- PS for right-censoring by death denoted by $\mu_5(t)$

$$g_0\Big(I(A_2(t) = 1, \Gamma = 3) = 1 \mid \bar{L}(t), \bar{Y}(t) = 0, \bar{A}_1(t-1), \bar{A}_2(t-1) = 0,$$
$$I(A_2(t) = 1, \Gamma = 1) = 0, I(A_2(t) = 1, \Gamma = 2) = 0\Big).$$

Thus, for patients who did not fail before time t and who followed the decision rule d_θ through time t, an estimate of the nuisance parameter $g_0(A(t) = d_\theta(t)(\bar{L}(t)\bar{A}(t - 1)) \mid \bar{L}(t), \bar{Y}(t) = 0, \bar{A}(t - 1) = d_\theta(\bar{L}(t - 1)))$ can be derived from estimates of the five PS above based on the following expression implied by the factorization of the action mechanism at time t using the product rule:

$$\left(I(\bar{A}_1(t - 1) = 0)\mu_1(t)^{A_1(t)}(1 - \mu_1(t))^{1 - A_1(t)} + I(A_1(t - 1) = 1)\mu_2(t)\right)$$
$$\left(1 - \mu_3(t)\right)\left(1 - \mu_4(t)\right)\left(1 - \mu_5(t)\right).$$

Logistic Models with Pooled Data Over Time. A common approach used in practice by analysts (Hernan et al. 2000; Cole et al. 2003) to estimate each of the five PS above for all time intervals t simultaneously consists in fitting a single model, referred to as a 'pooled model', using data pooled over time t. More specifically here, data were pooled over all follow-up times t to fit a separate main-term logistic model for estimating each of the three PS for right-censoring ($\mu_3(t), \mu_4(t), \mu_5(t)$) and the PS for TI continuation ($\mu_2(t)$). Data were also pooled for all time points $t > 0$ to fit a single main-term logistic model for estimating the PS for TI initiation after $t = 0$ (i.e., $\mu_1(t)$ for $t > 0$). A separate main-term logistic model was fitted for estimating the PS for TI initiation at $t = 0$ (i.e., $\mu_1(0)$). By 'main-term logistic model', we mean a logistic model with only main terms for each explanatory variable considered (i.e., no interaction terms between explanatory variables). The explanatory variables considered were all time-independent covariates and the last measurement of time-varying covariates (Markov assumption). In addition, exposure to TI in the last period was included as an explanatory variable for the three PS for right-censoring and the latest change in A1c was included as an explanatory variable for estimating all PS. All pooled logistic models also included the variable indexing the 90-day follow-up intervals (i.e., t) as an explanatory variable. The resulting estimator of the nuisance parameter g_0^θ is denoted by g_n^θ.

Logistic Models with Data Stratified by Time. In the previous approach, each pooled logistic model encodes the assumption that the associations between the explanatory variables and the PS outcome variable (e.g. death occurrence for PS $\mu_5(t)$) do not change over time. Concern over this assumption (Platt et al. 2009) motivates instead the use of a different logistic model (referred to as a 'stratified model') to estimate each PS at each time point t separately or, at least, the inclusion of interaction terms between the explanatory variables and functions of t in the pooled models. We note that it is such a concern over time-modified confounding that motivated, in the previous section, the specification of two separate models to estimate the PS for TI initiation: one stratified logistic model to estimate $\mu_1(0)$ and one separate pooled logistic model to estimate all $\mu_1(t)$ for $t > 0$ simultaneously. To fully address concerns over time-modified confounding, we also considered a second PS estimation approach in which, for each time point t separately, five main-term logistic models were fitted to estimate each of the five PS. The parameterization of these stratified models are the same as that described in the previous approach with the difference that the time variable t was omitted from all logistic models. The resulting estimator of the nuisance parameter g_0^θ is denoted by $g_{n,t}^\theta$.

Logistic Models with Data-Adaptive Selection of Interaction Terms. To lessen the constraints imposed by main-term logistic models, we considered a third PS estimation approach that is based on extending the previous stratified logistic models by data adaptively including two-way interaction terms between explanatory variables. Due to the large number of explanatory variables in this study and overfitting concerns, we separately implemented the following ad hoc data-adaptive algorithm for selecting the subset of all possible interaction terms to include in each stratified model. For each of the five PS and each time t, we first computed 105 two-way interaction terms based on the 15 explanatory variables that were most significantly associated (smallest p-value) with the PS outcome variable in a univariate logistic regression. Second, for each of these 105 terms, we implemented a logistic regression of the PS outcome variable on the interaction term and the two main terms that define the interaction term. Third, we identified the interaction terms with a p-value lower than 0.05. Finally, if more than 50 interaction terms met this criterion, we selected only the 50 terms with the smallest p-value and added them to the stratified logistic model for the PS. The resulting estimator of the nuisance parameter g_0^θ is denoted by $g_{n,t,\times}^\theta$.

Super Learning. The two model-based estimators g_n^θ and $g_{n,t}^\theta$ described earlier do not reflect real subject-matter knowledge about the adequacy of the expit function chosen to properly represent the true values of the five PS over time. Indeed in practice, PS model specification such as choosing a logistic model is typically rooted in tradition, preference, or convenience. To avoid erroneous inference due to such arbitrary model specifications, data-adaptive estimation of the nuisance parameter g_0^θ has been proposed in practice but consistent estimation then relies on judicious selection of a machine learning algorithm also known as 'learner'. We considered such a learner $g_{n,t,\times}^\theta$ in the previous section but many other learners have been proposed and can be used as potential candidates for estimating the five PS.

Akin to the selection of a parametric model, the selection of a learner does not typically reflect real subject-matter knowledge about the relative suitability of the different learners available, since "in practice it is generally impossible to know a priori which learner will perform best for a given prediction problem and data set" (van der Laan et al. 2007). To address this concern, data-adaptive estimation using ensemble learning methods can be used to hedge against erroneous inference due to arbitrary selection of a learner. Super learning (SL) is one such approach, discussed in detail in Chap. 3, with known theoretical properties that support its application in practice: A super learner performs asymptotically as well (in terms of mean error) or better than any of the candidate learners considered. If the super learner considers a candidate learner defined by a parametric model and if this model happens to be correctly specified then using SL instead of maximum likelihood estimation with correctly specified model only comes at a price of limited increase in prediction variability. Alternate ensemble learning methodology could be substituted for SL but, to our knowledge, the application of any such alternatives could not be theoretically validated by formal finite sample and asymptotic results such that the ones established for the super learning methodology.

In these secondary analyses and for each time point t separately, five super learners were implemented to estimate each of the five PS based on the following ten candidate learners: (a) five learners[1] defined by logistic models with only main terms for the most predictive explanatory variables identified by a significant p-value in univariate regressions with five significance levels ($\alpha=$ 1e−30, 1e−10, 1e−5, 0.1, and 1), and (b) five polychotomous regression learners based on the most predictive explanatory variables identified by a significant p-value in univariate regressions with the same five significance levels. The resulting estimator of the nuisance parameter g_0^θ is denoted by $g_{n,t,SL}^\theta$.

16.4.3 Outcome Regressions, Q_0^θ

The conditional expectations Q_0^θ required to implement the TMLE are unknown and thus need to be estimated first. Below, we describe the two data-adaptive estimators that were considered as *initial* estimators of the nuisance parameters $E_{P_0}(Q_{\bar{L}(k+2)}^\theta(\bar{L}(k+1)) \mid \bar{A}(k) = d_\theta(\bar{L}(k)), \bar{L}(k), \bar{Y}(k) = 0)$. For a given $k = t_0, \ldots, 0$, these estimators are derived using only data from patients who did not fail before time k and who followed rule d_θ through k (i.e., $\bar{Y}(k) = 0$ and $\bar{A}(k) = d_\theta(\bar{L}(k))$). The resulting TMLE are referred to as stratified TMLE by opposition to pooled TMLE in which the initial estimators of Q_0^θ are derived by pooling data from all patients who did not fail before time k (whether or not they followed the dynamic intervention previously) before evaluating these initial estimators at $\bar{A}(k) = d_\theta(\bar{L}(k))$. Note that in studies with small sample sizes. a stratified approach will often not be practical for proper initial estimation of the nuisance parameters Q_0^θ and extrapolation using data from patients who did not experience the relevant treatment history can then improve TMLE performance.

DSA. The Deletion/Substitution/Addition (DSA) algorithm (Sinisi and van der Laan 2004; Neugebauer and Bullard 2010) implements a data-adaptive estimator selection procedure based on cross-validation. It can be used as a machine learning approach for estimating conditional expectations. Here, the DSA was used as an initial estimator of Q_0^θ based on candidate estimators that were restricted to main-term logistic models with the following candidate explanatory variables: all time-independent covariates, the last measurement of time-varying covariates, and the latest change in A1c. To decrease computing time, the DSA was implemented with a single 5-fold cross-validation split, without deletion and substitution moves,

---

[1] Implemented by the `SL.glm` routine available in the `SuperLearner` R package (Polley and van der Laan 2013).

and with a maximum model size (i.e., number of main terms in the logistic models) equal to 10. The resulting estimator of Q_0^θ is denoted by $Q_{n,DSA}^\theta$.

Super Learner. Based on the same rationale that motivated the use of SL to estimate the action mechanism, SL was also considered here to define the initial estimator of the nuisance parameter Q_0^θ. The details of the SL approach that was implemented are described in Neugebauer et al. (2014a). In short, for each time point k a separate super learner was constructed based on the following eight classes of candidate learners (each learner used a different subset of explanatory variables): (a) seven learners defined by main-term logistic models; (b) five learners defined by a stepwise model selection using AIC; (c) five learners defined by neural networks; (d) five learners defined by Bayes regressions; (e) five learners defined by polychotomous regressions; (f) five learners defined by Random Forests; (g) five learners defined by bagging for classification trees; (h) 20 learners defined by generalized additive models with smoothing splines. The set of explanatory variables considered included the variables used in the DSA approach but was also expanded to include two-way interaction terms between these variables and additional summary measures of past covariates (e.g., the average of all past A1c measurement or the number of past A1c measurements above 8%). The addition of summary measures of past covariates was motivated by the fact that the nuisance parameter $Q_{L(k+1)}^\theta(\bar{L}(k))$ is by definition a function of the covariate history $\bar{L}(k)$ in the same way that each PS is potentially a function of past observed covariates. However, while a Markov assumption can often be argued in practice to justify PS estimation using the last treatment and covariate measurements only, it is not clear why a similar approach is reasonable for estimating the nuisance parameters $Q_{L(k+1)}^\theta(\bar{L}(k))$. The resulting estimator of Q_0^θ is denoted by $Q_{n,SL}^\theta$.

16.5 Practical Performance

Results from the TMLE and three IPW estimators are compared to the results from a crude analysis that aims to contrast the survival curves associated with the four treatment intensification strategies of interest d_θ without any adjustment for confounding and selection bias. Such an analysis can be implemented by applying the hazard-based IPW estimator with its stabilized weights set to 1. Table 16.1 provides inferences for the comparison of the four survival curves at 3 years using the crude analysis approach. Inferences in that table can be contrasted to those based on the three IPW estimators implemented with the four PS estimation approaches in Table 16.2 and those based on targeted learning (i.e., TMLE implemented with SL for estimating both nuisance parameters) in Table 16.1. Examination of the results from the crude estimator in Table 16.1 provide some evidence, albeit weak, that is consistent with results from the ACCORD and ADVANCE randomized trials which suggested a beneficial effect of more aggressive therapy initiation rules.

Table 16.1 Comparison of inferences from the crude estimator and untruncated TMLE for the six cumulative RDs ψ^{θ_1,θ_2} at 12 quarters ($t_0 = 11$)

		Crude					TMLE					
θ_1	θ_2	$\psi_n^{\theta_1,\theta_2}$	$\Gamma_n^{\theta_1,\theta_2}$	$\psi_n^{\theta_1,\theta_2,-}$	$\psi_n^{\theta_1,\theta_2,+}$	p	$\psi_n^{\theta_1,\theta_2,*}$	$\Gamma_n^{\theta_1,\theta_2,*}$	$\psi_n^{\theta_1,\theta_2,*,-}$	$\psi_n^{\theta_1,\theta_2,*,+}$	p*	RE
8.5	8.0	0.0013	0.0015	−0.0016	0.0042	0.386	0.0070	0.0038	−0.0004	0.0143	0.064	1.07
8.5	7.5	0.0028	0.0030	−0.0032	0.0088	0.357	0.0221	0.0065	0.0093	0.0349	0.001	1.09
8.5	7.0	0.0335	0.0111	0.0116	0.0553	0.003	0.0386	0.0112	0.0166	0.0606	0.001	1.11
8.0	7.5	0.0015	0.0027	−0.0037	0.0068	0.574	0.0151	0.0059	0.0035	0.0267	0.011	1.09
8.0	7.0	0.0322	0.0111	0.0105	0.0538	0.004	0.0316	0.0112	0.0097	0.0535	0.005	1.11
7.5	7.0	0.0307	0.0108	0.0096	0.0518	0.004	0.0165	0.0103	−0.0038	0.0368	0.110	1.11

The TMLE is derived based on the SL estimators for the nuisance parameters (i.e., $g_{n,t,SL}^{\theta}$ and $Q_{n,SL}^{\theta}$) and is differentiated from the crude estimator by the superscript * notation. The ratio of the standard error of the hazard-based IPW (see Table 16.2) based on $g_{n,t,SL}$ to that of the TMLE is denoted by RE

TMLE inferences in Fig. 16.2 and Tables 16.2 and 16.1 reveal much stronger evidence of a beneficial effect of more aggressive TI strategies and illustrate the estimator's practical performance in adjusting for time-dependent confounding and selection bias. Indeed, whichever the approach adopted for estimating the nuisance parameters, the TMLE estimator indicates an early separation and consistent ordering of the four survival curves suggesting an increasing beneficial effect of more aggressive therapy initiation rules (i.e., of rules indexed by decreasing A1c thresholds). Four to five of the six cumulative risk differences at 3 years are now statistically significant depending on the estimation approach adopted for the nuisance parameter Q_0^{θ}.

Successful performance in bias adjustment with the hazard-based IPW estimator is illustrated in Table 16.2 and reveals similar evidence of a beneficial effect of more aggressive TI strategies. Whichever the approach adopted for estimating the action mechanism, the hazard-based IPW estimator indicates a separation of the four survival curves (results in Neugebauer et al. 2016) that is visually almost identical to that obtained with the TMLE estimator in Fig. 16.2. Three to four of the six cumulative risk differences at 3 years are statistically significant depending on the estimation approach adopted for the action mechanism g_0^{θ}. The performance in bias adjustment with the unbounded IPW estimator largely depends on the approach used to estimate the action mechanism (results in Neugebauer et al. 2016). This IPW estimator reveals the strongest evidence of a beneficial effect of more aggressive TI strategies when it is combined with SL estimation of the action mechanism. The performance in bias adjustment with the bounded IPW estimator follows an intermediate pattern (results in Neugebauer et al. 2016) between that of the unbounded and hazard-based IPW estimators in the sense that while the curves become increasingly and more consistently separated in the expected directions as the PS estimation becomes more nonparametric (similar to results in the unbounded IPW estimator), the distinctions between results from the four PS estimation approaches are less obvious (similar to results in the hazard-based IPW estimator).

We conjecture that the gradual change in the patterns of results across PS estimation approaches as we go from the unbounded, bounded and hazard-based IPW estimators relates to the differences in boundedness and efficiency properties between the three estimators. More specifically, we conjecture that the successive decrease in the three estimators' sensitivity to bias from errors in PS estimation is a consequence of, first, the boundedness property of both the bounded and hazard-based IPW estimators and, second, the improved efficiency of the hazard-based estimator relative to the bounded IPW estimator. This conjecture is further supported by the TMLE results which also show that the estimator is not sensitive to the PS estimation approach adopted (Fig. 16.2). Because the TMLE is a substitution estimator, it also benefits from the boundedness property similar to both the bounded and hazard-based IPW estimators. In the same way that the improved theoretical efficiency properties of the hazard-based IPW estimator compared to the bounded IPW estimator leads to a further decrease in sensitivity to errors in the PS estimates, the improved efficiency property of the TMLE compared to the bounded IPW estimator can explain why the TMLE share its lack of sensitivity to errors in the PS estimates with the hazard-based estimator.

In Table 16.2, we compare the inferences we would derive in practice from various estimation choices that led to results consistent with that of previous randomized experiments. It illustrates the expected theoretical efficiency gains from the hazard-based IPW estimator compared to the bounded and unbounded IPW estimators. It also indicates that the estimated standard error of the hazard-based IPW estimator generally decreases as the PS estimation approach becomes more nonparametric. This apparent gain in estimation efficiency is explained by the concentration of the IPW weights and the decrease in the proportion of large weights from progressively more flexible PS estimation approaches (results in Neugebauer et al. 2016). In fact, is quite remarkable that none of the stabilized IP weights derived from SL are greater than 30. These results thus contradict the common position that model-based PS estimation is preferable in practice because data-adaptive PS estimation leads to larger weights and thus an increase in IPW estimation variability by revealing practical violations of the positivity assumption. Estimation of the IP weights based on arbitrarily specified parametric models can then be viewed as an implicit weight truncation scheme that restricts the proportion of large weights through smoothing with a misspecified model. On the contrary, the results in this study suggest that model-based estimation of the weights can instead lead to artificial practical violation of the positivity assumption (i.e., large weights due to model misspecification) when the positivity assumption is in truth not violated. The practical consequence of such model-based estimation is not only biased inference but also increased uncertainty that both could be avoided with data-adaptive PS estimation.

One motivation for the application of TMLE over (bounded, unbounded, or hazard-based) IPW estimation in practice is the potential for gain in estimation precision that may arise from the efficiency property of TMLE. Table 16.2 illustrates such efficiency gains in this study but indicates however little increase in precision with TMLE compared to the hazard-based IPW estimator: the ratios of the estimated hazard-based IPW standard errors for the six cumulative risk differences at

Table 16.2 Comparison of inferences from six IPW estimators and the DSA-based TMLE of six risk differences (RD) at 3 years (12 quarters, i.e. $t_0 = 11$)

RD estimator	θ_1	θ_2	g_n^θ	$\psi_n^{\theta_1,\theta_2}$	$\Gamma_n^{\theta_1,\theta_2}$	$\psi_n^{\theta_1,\theta_2,-}$	$\psi_n^{\theta_1,\theta_2,+}$	p	RE
Horvitz-Thompson $\psi_{n,HT}^{\theta_1,\theta2}$	8.5	8	$g_{n,t,SL}^\theta$	7.3e−03	4.3e−03	−1.1e−03	0.0156	0.089	1.085
	8.5	7.5	$g_{n,t,SL}^\theta$	0.0231	7.5e−03	8.4e−03	0.0378	2e−03	1.078
	8.5	7	$g_{n,t,SL}^\theta$	0.0506	0.0129	0.0254	0.0758	0	1.082
	8	7.5	$g_{n,t,SL}^\theta$	0.0158	6.8e−03	2.5e−03	0.0292	2e−02	1.078
	8	7	$g_{n,t,SL}^\theta$	0.0433	0.0128	0.0182	0.0685	1e−03	1.083
	7.5	7	$g_{n,t,SL}^\theta$	0.0275	0.0118	4.4e−03	0.0506	0.019	1.079
Bounded IPW $\psi_{n,bd}^{\theta_1,\theta2}$	8.5	8	$g_{n,t,SL}^\theta$	9.8e−03	4.6e−03	8e−04	0.0188	0.032	1.165
	8.5	7.5	$g_{n,t,SL}^\theta$	0.0255	7.9e−03	1e−02	0.041	1e−03	1.138
	8.5	7	$g_{n,t,SL}^\theta$	0.0572	0.0127	0.0322	0.0822	0	1.072
	8	7.5	$g_{n,t,SL}^\theta$	0.0157	7.2e−03	1.5e−03	0.0298	3e−02	1.144
	8	7	$g_{n,t,SL}^\theta$	0.0474	0.0127	0.0224	0.0723	0	1.072
	7.5	7	$g_{n,t,SL}^\theta$	0.0317	0.0117	8.7e−03	0.0547	7e−03	1.075
Hazard-based IPW $\psi_{n,h}^{\theta_1,\theta2}$	8.5	8	g_n^θ	0.0109	5.7e−03	−3e−04	0.022	0.056	
	8.5	7.5	g_n^θ	0.0212	8.7e−03	4.1e−03	0.0382	0.015	
	8.5	7	g_n^θ	0.0387	0.0135	0.0122	0.0652	4e−03	
	8	7.5	g_n^θ	0.0103	7.8e−03	−5e−03	0.0256	0.187	
	8	7	g_n^θ	0.0278	0.0138	8e−04	0.0549	0.044	
	7.5	7	g_n^θ	0.0175	0.0133	−8.5e−03	0.0435	0.187	
Hazard-based IPW $\psi_{n,h}^{\theta_1,\theta2}$	8.5	8	$g_{n,t}^\theta$	9.9e−03	5.5e−03	−9e−04	0.0207	0.071	
	8.5	7.5	$g_{n,t}^\theta$	0.025	7.8e−03	9.6e−03	0.0404	1e−03	
	8.5	7	$g_{n,t}^\theta$	0.0418	0.0126	0.0172	0.0664	1e−03	
	8	7.5	$g_{n,t}^\theta$	0.0151	7.1e−03	1.2e−03	0.029	0.033	
	8	7	$g_{n,t}^\theta$	0.0319	0.0129	6.7e−03	0.0571	0.013	
	7.5	7	$g_{n,t}^\theta$	0.0168	0.012	−6.7e−03	0.0403	0.162	
Hazard-based IPW $\psi_{n,h}^{\theta_1,\theta2}$	8.5	8	$g_{n,t,\times}^\theta$	0.0129	5.5e−03	2.1e−03	0.0237	2e−02	
	8.5	7.5	$g_{n,t,\times}^\theta$	0.0337	7.9e−03	0.0183	0.0491	0	
	8.5	7	$g_{n,t,\times}^\theta$	0.0483	0.0129	0.0229	0.0737	0	
	8	7.5	$g_{n,t,\times}^\theta$	0.0209	6.9e−03	7.3e−03	0.0345	3e−03	
	8	7	$g_{n,t,\times}^\theta$	0.0354	0.0129	1e−02	0.0608	6e−03	
	7.5	7	$g_{n,t,\times}^\theta$	0.0146	0.0121	−9.2e−03	0.0383	0.23	
Hazard-based IPW $\psi_{n,h}^{\theta_1,\theta2}$	8.5	8	$g_{n,t,SL}^\theta$	8e−03	4e−03	1e−04	0.0159	0.046	1.025
	8.5	7.5	$g_{n,t,SL}^\theta$	0.0216	7.1e−03	7.7e−03	0.0356	2e−03	1.024
	8.5	7	$g_{n,t,SL}^\theta$	0.041	0.0125	0.0166	0.0655	1e−03	1.049
	8	7.5	$g_{n,t,SL}^\theta$	0.0136	6.5e−03	1e−03	0.0263	0.035	1.022
	8	7	$g_{n,t,SL}^\theta$	0.033	0.0124	8.6e−03	0.0574	8e−03	1.048
	7.5	7	$g_{n,t,SL}^\theta$	0.0194	0.0114	−3e−03	0.0418	9e−02	1.048
DSA-based TMLE $\psi_n^{\theta_1,\theta2,*}$	8.5	8	$g_{n,t,SL}^\theta$	9.1e−03	3.9e−03	1.4e−03	0.0168	0.021	1
	8.5	7.5	$g_{n,t,SL}^\theta$	0.0238	6.9e−03	0.0102	0.0374	1e−03	1
	8.5	7	$g_{n,t,SL}^\theta$	0.0427	0.0119	0.0194	0.0661	0	1
	8	7.5	$g_{n,t,SL}^\theta$	0.0147	6.3e−03	2.3e−03	0.0271	2e−02	1
	8	7	$g_{n,t,SL}^\theta$	0.0336	0.0119	0.0104	0.0569	5e−03	1
	7.5	7	$g_{n,t,SL}^\theta$	0.0189	0.0109	−2.5e−03	0.0403	0.083	1

For each RD, the estimator for g_0^θ, the point estimate, the estimate of the standard error, its ratio to that of the DSA-based TMLE, the estimate of the lower and upper bound of the 95% confidence interval, and the associated p-value are denoted by g_n^θ, $\psi_n^{\theta_1,\theta_2}$, $\Gamma_n^{\theta_1,\theta_2}$, RE, $\psi_n^{\theta_1,\theta_2,-}$, $\psi_n^{\theta_1,\theta_2,+}$, and p, respectively

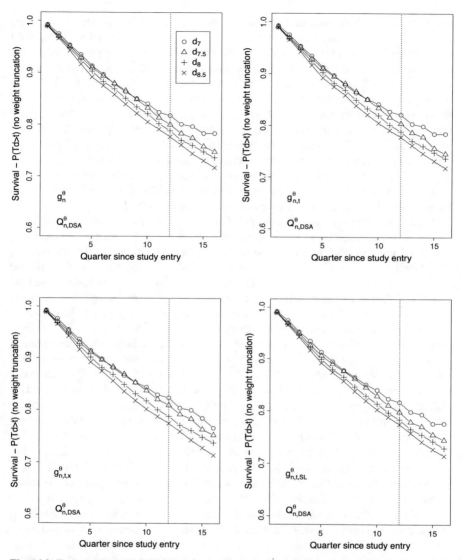

Fig. 16.2 Each plot represents TMLE estimates over 16 quarters of the four counterfactual survival curves corresponding with the four TI initiation strategies d_θ with $\theta = 7, 7.5, 8, 8.5$. The plots located at the *top left, top right, bottom left,* and *bottom right* are obtained based on the estimates g_n^θ, $g_{n,t}^\theta$, $g_{n,t,\times}^\theta$, and $g_{n,t,SL}^\theta$ of the nuisance parameter g^θ, respectively

3 years to the estimated TMLE standard errors range from 2 to 5% only when the action mechanism is estimated with SL. As shown by equality (16.7) defining the influence curve of the TMLE estimator, gain in precision arises from minimization of the prediction errors from the estimate of the nuisance parameter Q_0^θ. Gains in

precision are thus expected to be greatest with TMLE in problems where covariates can predict the outcome well. The small efficiency gains from the TMLE reported in Table 16.2 could thus be the result of either the absence of good predictors of outcomes among the measured covariates or inadequate use of the measured covariates for estimating Q_0^θ (e.g., due to model misspecification such as assuming the Markov assumption). It is this second explanation for poor efficiency gains that motivated the application of SL for flexible estimation of Q_0^θ without relying on the Markov assumption and arbitrary learner choices (such as the main-term logistic models considered by the DSA algorithm). Table 16.1 illustrates the improved efficiency gains that resulted from more nonparametric estimation of the nuisance parameter Q_0^θ with SL: the ratios of the estimated hazard-based IPW standard errors for the six cumulative risk differences at 3 years to the estimated TMLE standard errors range from 7 to 11% when the action mechanism is estimated with SL.

Note again that the validity of TMLE inferences based on the estimator's influence curve defined by (16.7) is only guaranteed when the action mechanism is estimated consistently. To assess the performance of IC-based TMLE inferences, we evaluated the variability of each DSA-based TMLE in Fig. 16.2 based on 10,000 bootstrap samples. For a fair comparison between IC-based and bootstrap-based estimates of TMLE variability, the TMLE point estimates were derived on each bootstrap sample using the same estimated nuisance parameters (i.e., the PS estimator were not reevaluated on each bootstrap sample) because the IC-based TMLE estimator is only consistent for estimating the TMLE standard error when the nuisance parameter g_0^θ is known and it is conservative if the action mechanism is estimated consistently. When the TMLE was based on the estimators g_n^θ, $g_{n,t}^\theta$, or $g_{n,t,x}^\theta$, the ratios of the IC-based estimates of the TMLE standard errors for the six cumulative risk differences at 3 years to the estimates derived from the bootstrap indicated important over-estimation (by up to 61%) of the IC-based estimator of the TMLE standard error (results in Neugebauer et al. 2014a). These ratios clearly converged towards the desired value of 1 as the approach used to estimate the action mechanism became more nonparametric and all ratios became essentially 1 with SL estimation of the action mechanism. Similarly, results indicated that the bootstrap-based estimates of the variance of the bounded IPW estimators based on the more parametric type estimator g_n^θ did not match the IC-based variance estimates when the normalizing constant was ignored from the IC formula (matching estimates were obtained when using the normalizing constant to compute the IC-based estimates). These observations provide further evidence to support the argument that model-based estimation of the action mechanism in this study resulted in biased PS estimates and that SL can be used to hedge against such parametric model misspecification. These results also indicate that unlike what we noted earlier about the TMLE estimator itself, the IC-based estimator of its variance is highly sensitivity to errors in PS estimates. This sensitivity was however not observed for the IC-based estimator of the variance of the hazard-based IPW estimator (results in Neugebauer et al. 2014a).

16.6 Discussion

We demonstrated the feasibility of targeted learning in real-world CER with time-varying exposures based on large EHR databases. We illustrated how this approach can successfully adjust for time-dependent confounding and selection bias and provide gains in efficiency compared to IPW estimation. With both the TMLE and IPW estimators, we demonstrated that SL estimation of the nuisance parameters can result in substantial bias reduction and efficiency gains compared to model-based estimation and that such improved estimation performances come at the cost of increased computing times. In particular, results here contradict the position that SL estimation may not be desirable in practice for fear to reveal violations of the positivity assumption and suggest instead that the estimation approaches for the action mechanism that are commonly used in practice based on arbitrary parametric models can actually lead to artificial violation of the positivity assumption (extreme IP weights) when this assumption is in truth not violated.

These results also suggest that estimation of the nuisance parameter Q_0^θ should generally rely on a richer set of both treatment and covariate histories to realize more substantial efficiency gains with TMLE because simplifying assumptions (e.g., Markov assumption) that may apply to components of the action mechanism are not expected to extend to the nuisance parameter Q_0^θ in general. If these simplifying assumptions are nevertheless made, suboptimal efficiency gains can be expected with TMLE due to inconsistent estimation of Q_0^θ.

In addition, this work points to the importance of the boundedness property in reducing the sensitivity of the TMLE estimator to minor bias in estimation of the action mechanism but also suggests that data-adaptive estimation of the action mechanism is critical for proper IC-based TMLE inference because of the high sensitivity of the IC-based estimator of the TMLE standard error to minor bias in estimation of the action mechanism. This last observation also suggests the use of bootstrapping to evaluate the bias of estimators of the action mechanism and guide the selection of the PS estimator based on which TMLE inference should be derived in practice, e.g., the data-adaptiveness of the estimation approach for g_0^θ may be sequentially increased until there is a match between the IC-based and bootstrap-based estimates of TMLE variance. This PS estimator selection criterion is an alternative to a less targeted approach that consists in increasing the level of data-adaptiveness of the SL estimation approach for the action mechanism until the cross-validated likelihood is stable.

Finally, when using data-adaptive estimation of the nuisance parameters, the IPW and TMLE estimators may no longer be asymptotically linear with the influence curves used to derive the conservative variance estimators presented in this chapter. The application of these variance estimators in practice may thus not be justified theoretically when the nuisance parameters are estimated with SL instead of model-based estimation by maximum likelihood. Recent research (van der Laan 2014b) to derive valid inference when nuisance parameters are estimated with SL led to the development of new IPW and TMLE estimators (Benkeser et al. 2017a) that are asymptotically linear with known influence curves. This approach is currently applicable in point treatment studies only. Additional research is needed to generalize these IPW and TMLE approaches to studies with time-varying exposures and covariates such as the treatment intensification study in this chapter. Until this generalization is available, one may argue that TMLE or IPW estimation in real-world CER with time-varying interventions should rely on model-based nuisance parameter estimation because theoretically valid inferences can then be derived in practice using the conservative variance estimators presented in this chapter. We counterargue that these variance estimators are then only valid when the action mechanism is estimated based on correct models and that the asymptotic coverage of the resulting confidence intervals is 0 if these models are incorrect.

Acknowledgments The authors thank the following investigators from the HMO research network for making data from their sites available for the secondary analyses presented in this chapter: Denise M. Boudreau (Group Health), Connie Trinacty (Kaiser Permanente Hawaii), Gregory A. Nichols (Kaiser Permanente Northwest), Marsha A. Raebel (Kaiser Permanente Colorado), Kristi Reynolds (Kaiser Permanente Southern California). This project was funded under contract no. HHSA290200500161 from the Agency for Healthcare Research and Quality, US Department of Health and Human Services as part of the Developing Evidence to Inform Decisions about Effectiveness (DEcIDE) program. The authors of this report are responsible for its content. Statements in the report should not be construed as endorsement by the Agency for Healthcare Research and Quality or the US Department of Health and Human Services.

Chapter 17
Mediation Analysis with Time-Varying Mediators and Exposures

Wenjing Zheng and Mark J. van der Laan

An exposure often acts on an outcome of interest directly, or indirectly through the mediation of some intermediate variables. Identifying and quantifying these two types of effects contribute to further understanding of the underlying causal mechanism. Modern developments in formal nonparametric causal inference have produced many advances in causal mediation analysis in nonlongitudinal settings. (e.g., Robins and Greenland 1992; Pearl 2001; van der Laan and Petersen 2008; VanderWeele 2009; Hafeman and VanderWeele 2010; Imai et al. 2010b,a; Pearl 2011; Tchetgen Tchetgen and Shpitser 2011a,b; Zheng and van der Laan 2012b; Lendle and van der Laan 2011).

Causal mediation in a longitudinal setting, by contrast, has received relatively little attention. One option is the *controlled direct effect* (e.g., Pearl 2001), which compares the outcomes under different exposure regimens while the mediators are fixed to some common pre-specified values. The analysis is very similar to that of a time-varying exposure in a nonmediation setting; we refer the reader to existing literature on this topic (e.g., Robins and Ritov 1997; Hernan et al. 2000; Stitelman et al. 2011; Petersen et al. 2014). Controlled direct effects are of interest if the exposure effect at a particular mediator value constitutes a meaningful scientific question.

If that is not the case, one may ask a different direct effect question: what would be the effect of exposure on the outcome if the mediator takes its value as if exposure were absent? This question is formalized using the so-called *natural direct*

W. Zheng
Netflix, 100 Winchester Circle, Los Gastos, CA 95032, USA
e-mail: wzheng@netflix.com

M. J. van der Laan (✉)
Division of Biostatistics and Department of Statistics, University of California, Berkeley,
101 Haviland Hall, #7358, Berkeley, CA 94720, USA
e-mail: laan@berkeley.edu

© Springer International Publishing AG 2018
M.J. van der Laan, S. Rose, *Targeted Learning in Data Science*,
Springer Series in Statistics, https://doi.org/10.1007/978-3-319-65304-4_17

effect parameter by Robins and Greenland (1992) and Pearl (2001) in a nonlongitudinal setting. The natural direct effect has a complementary *natural indirect effect*; together they provide a decomposition of the overall effect of the exposure on the outcome. The challenges in extending the above mediation formulation to the longitudinal setting have been studied in Avin et al. (2005), which established that the corresponding natural direct effect and indirect effect parameters would not be identifiable in the presence of confounders of the mediator-outcome relationship that are affected by the exposure. Such confounders, however, are ubiquitous in longitudinal applications.

An alternative, random (stochastic) interventions (RI) based formulation to causal mediation was proposed in Didelez et al. (2006). Under this formulation, a mediator is regarded as an intervention variable, as opposed to a counterfactual variable, onto which a given distribution is enforced. The corresponding natural direct effect and indirect effect parameters have different interpretation than those under the formulations in Robins and Greenland (1992) and Pearl (2001), but their identifiability is at hand even in the presence of exposure-induced mediator-outcome confounder (e.g., VanderWeele et al. 2014a). Zheng and van der Laan (2012a) proposed an RI formulation to causal mediation, through *conditional* mediator distributions, in a survival setting with point exposure and time-varying mediators and confounders. VanderWeele and Tchetgen Tchetgen (2017) proposed an RI formulation, through *marginal* mediator distributions, in a time-varying exposure and mediator setting. In this paper, we extend the work in Zheng and van der Laan (2012a) to formulate causal mediation through conditional mediator distributions in general time-varying exposure and mediator settings with survival or nonsurvival outcomes. The challenges in longitudinal mediation analysis are exemplified in the different effects captured (and corresponding identifiability conditions) under the marginal distribution intervention in VanderWeele and Tchetgen Tchetgen (2017) and the conditional distribution intervention proposed here. We will illustrate these differences with examples and further discussions. This chapter includes content, with permission, from Zheng and van der Laan (2017).

The natural direct and indirect effects proposed here can all be defined in terms of the corresponding version of the mediation functional. We develop a general semiparametric inference framework for this *conditional mediation functional*. More specifically, we derive the efficient influence curves under a locally saturated semiparametric model, establish their robustness properties, present three estimators for the conditional mediation functional, and study the empirical performance of these estimators in a simulation study. The three estimators include: a nested nontargeted substitution estimator, which uses a regression-based representation of the identifying expression, an IPW estimator, and an efficient and multiply robust TMLE.

17.1 The Mediation Formula, Natural Direct, and Natural Indirect Effects

Consider the data structure

$$O = (L_0, A_1, R_1, Z_1, L_1, \ldots, A_t, R_t, Z_t, L_t, \ldots, A_\tau, R_\tau, Z_\tau, L_\tau \supset Y_\tau) \sim P_0,$$

where L_0 encodes baseline covariates, and for $t \geq 1$, A_t encodes the time-varying exposure (and in addition may also include a censoring indicator), Z_t denotes the time-varying mediators. R_t and L_t denote the time-varying covariates. At each time t, R_t encode covariates that are affected by A_t and may in turn affect Z_t and L_t; L_t are covariates that are affected by A_t, R_t, Z_t. L_t may include the outcome process $Y_t \subset L_t$, in particular, L_τ will include the final outcome of interest Y_τ. This data structure allows for confounders of the exposure-outcome relation and exposure-induced confounders of the mediator-outcome relation, both within time and across time. In a survival setting with right censoring, the outcome indicator Y_t indicates whether one has died by time t, i.e. $Y_t = I(\tilde{T} <= t, T \leq C)$ with survival time T, censoring time C and $\tilde{T} = min(T, C)$. We would encode the intervention variables as $A_t = (A_t^C, A_t^E)$, A_t^C is the indicator of remaining uncensored by time t and A_t^E is the exposure of interest at time t. In the case of a survival outcome or if censoring exists, Z_t, R_t, L_t are encoded with a default value after censoring or death. After a linear transformation, one may assume that Y_t is bounded between 0 and 1. The data consists of n i.i.d. copies of O.

We consider an example from diabetic care. Suppose within a large primary healthcare network, all diabetic patients are to receive ongoing education sessions, regular nutrition counseling (meetings with counselor), in addition to their routine care with their family physician. Complications are referred to higher level specialists. Consider a pilot program that integrates simplified referral procedures, enhanced curriculum for education sessions and a more streamlined operations for nutrition counseling (e.g., less wait time, easier scheduling). Each year, patients may opt in and out of this program. Suppose we wish to evaluate how much the effect of the program on long-term control of type 2 diabetes is mediated by changes in attendance of the nutrition counseling meetings. The final outcome of interest is blood glucose levels at 5 years after diagnosis (Y_τ). Observations are taken annually. The exposure of interest A_t is whether patient is in the program at year t; the mediator of interests Z_t is whether the patient has utilized his/her minimum nutrition counseling meetings this year. The time-varying covariates R_t denotes attendance in education sessions and patient knowledge around disease and self-care (as assessed by survey). The time-varying covariates L_t denote health-related covariates such as nutritional status, disease progression, comorbidities, glucose tests results, etc. The covariates R_t are affected by program participation, and may affect utilization of nutrition counseling (better educated patients may have higher utilization of nutrition counseling) as well as subsequent disease progression and nutritional status. The covariates L_t

are affected by current and previous program participation and patient engagement (some captured in R_t and Z_t), and will affect subsequent program participation and patient engagement. Therefore, there are time-varying confounders of the exposure-outcome relationship, as well as time-varying exposure-induced confounders of the mediator-outcome relationship.

From here on, for any $1 \leq t \leq \tau$ and a time-dependent variable V, we will use the boldface $\mathbf{V}_t$ to denote the vector $(V_1, \ldots, V_t)$, use $\mathbf{V}_{s,t}$ to denote the vector $(V_s, \ldots, V_t)$. When referring to the entire vector $\mathbf{V}_\tau$, we will also use the short-hand $\mathbf{V}$. Degenerate indices such as $\mathbf{V}_{-1}$ signify the empty set. We encode the time-ordering of the variables using the following SCM (Pearl 2009a):

$$
\begin{aligned}
L_0 &= f(U_{L_0}), \\
A_t &= f_{A_t}(\mathbf{A}_{t-1}, \mathbf{R}_{t-1}, \mathbf{Z}_{t-1}, \mathbf{L}_{t-1}, U_{A_t}), \\
R_t &= f_{R_t}(\mathbf{A}_t, \mathbf{R}_{t-1}, \mathbf{Z}_{t-1}, \mathbf{L}_{t-1}, U_{R_t}), \\
Z_t &= f_{Z_t}(\mathbf{A}_t, \mathbf{R}_t, \mathbf{Z}_{t-1}, \mathbf{L}_{t-1}, U_{Z_t}), \\
L_t &= f_{L_t}(\mathbf{A}_t, \mathbf{R}_t, \mathbf{Z}_t, \mathbf{L}_{t-1}, U_{L_t}).
\end{aligned}
$$

In words, within each time point, we assume the temporal relation between the measured variables are exposure, then covariates R_t, then mediator, then covariates L_t, then outcome. For each variable V, the model posits that V is an unknown deterministic function of all variables preceding it, and some unmeasured exogenous factors. It is important to note that our formulation, identifying expression, and proposed estimators can be adapted to other choices of temporal ordering.

The observed data structure is generated from the above structural equations model without any intervention, and the likelihood of $O \sim P_0$ can be factored into the following conditional probabilities according to that time-ordering:

$$
\begin{aligned}
p_0(O) = {}& p_{0,L_0}(L_0) \\
& \times \prod_{t=1}^{\tau} \Big(p_{0,A}(A_t \mid \mathbf{A}_{t-1}, \mathbf{R}_{t-1}, \mathbf{Z}_{t-1}, \mathbf{L}_{t-1}) p_{0,R}(R_t \mid \mathbf{A}_t, \mathbf{R}_{t-1}, \mathbf{Z}_{t-1}, \mathbf{L}_{t-1}) \\
& \times p_{0,Z}(Z_t \mid \mathbf{A}_t, \mathbf{R}_t, \mathbf{Z}_{t-1}, \mathbf{L}_{t-1}) p_{0,L}(L_t \mid \mathbf{A}_t, \mathbf{R}_t, \mathbf{Z}_t, \mathbf{L}_{t-1}) \Big). \quad (17.1)
\end{aligned}
$$

The conditional densities of A_t, R_t, Z_t and L_t can depend on t, although we suppressed it in the above notation. In the case of a survival outcome or if censoring exists, subsequent A_t, Z_t, R_t, L_t are assigned a default value with probability 1 after censoring or death, and therefore do not contribute to the likelihood.

17.1.1 Counterfactual Outcome Under Conditional Mediator Distribution

Let $\mathbf{a}$ and $\mathbf{a}'$ be two possible exposure regimens. Let

$$
\Gamma_t^{\mathbf{a}'}(z_t \mid \mathbf{r}_t, \mathbf{z}_{t-1}, \mathbf{l}_{t-1}) \equiv p\left(Z_t(\mathbf{a}') = z_t \mid \mathbf{R}_t(\mathbf{a}') = \mathbf{r}_t, \mathbf{Z}_{t-1}(\mathbf{a}') = \mathbf{z}_{t-1}, \mathbf{L}_{t-1}(\mathbf{a}') = \mathbf{l}_{t-1}\right)
$$

denote the conditional probabilities of the mediators at $t \geq 1$, if the exposure had been set to $\mathbf{A} = \mathbf{a}'$ in the population. At each time t, within each stratum $(\mathbf{r}_t, \mathbf{z}_{t-1}, \mathbf{l}_{t-1})$, this provides a random draw of $Z_t \sim \Gamma_t^{\mathbf{a}'}$. For convenience, we will denote $\bar{\Gamma}^{\mathbf{a}'} = \left(\Gamma_1^{\mathbf{a}'}, \ldots, \Gamma_\tau^{\mathbf{a}'} \right)$.

Consider an intervention on the structural equations model to statically set $A_t = a_t$ and randomly draw $Z_t \sim \Gamma_t^{\mathbf{a}'}$. For $X_t \in \{R_t, L_t\}, t \geq 1$, we will write $X_t(\mathbf{a}, \bar{\Gamma}^{\mathbf{a}'})$ to denote the corresponding covariates resulting from this intervention. In terms of an ideal experiment, the data would be generated as follows. At baseline, we measure the covariates L_0 (say $L_0 = l_0$). At $t = 1$, intervene to set $A_1 = a_1$, and measure the resulting covariates $R_1(\mathbf{a}, \bar{\Gamma}^{\mathbf{a}'})$ (say it's measured to be r_1). Then, intervene to draw Z_1 according $\Gamma_1^{\mathbf{a}'} (\cdot \mid l_0, r_1)$. Suppose we have drawn value z_1, then we measure the resulting covariates $L_1(\mathbf{a}, \bar{\Gamma}^{\mathbf{a}'})$ (say it's measured to be l_1). At time $t = 2$, intervene to set $A_2 = a_2$; measure the resulting covariates $R_2(\mathbf{a}, \bar{\Gamma}^{\mathbf{a}'})$ (say it's measured to be r_2); intervene to draw Z_2 according to $\Gamma_2^{\mathbf{a}'} (\cdot \mid l_0, r_1, z_1, l_1, r_2)$; measure the resulting covariates $L_2(\mathbf{a}, \bar{\Gamma}^{\mathbf{a}'})$. So on. At the end of the experiment, we denote the final outcome as $Y_\tau(\mathbf{a}, \bar{\Gamma}^{\mathbf{a}'})$—this is the outcome if exposures were set to fixed values $\mathbf{a}$ and the mediators were set to have conditional distribution $\Gamma_t^{\mathbf{a}'}$. In contrast to the traditional nonrandom intervention formulation, $\Gamma_t^{\mathbf{a}'}$ is not the would-be mediator on the same person had she been under a different exposure, but simply a random variable whose distribution is specified by a given conditional probability function $\Gamma_t^{\mathbf{a}'}$ and the person's accruing history.

Let's illustrate this ideal experiment with the motivating example. Suppose $\mathbf{a} = 1$ denotes program participation throughout the study, and $\mathbf{a}' = 0$ denotes non-participation. Under the intervention described in the above experiment, at time $t = 1$, a person $L_0 = l_0$ is assigned to the program at year 1 $A_1 = 1$, his education session attendance and knowledge attainment $R_1(\mathbf{a}, \bar{\Gamma}^{\mathbf{a}'})$ is a consequence of this participation (say R_1 is likely to be high attendance and high attainment as a result of program). His nutrition counseling utilization $Z_1 \sim \Gamma_1^0$ will be that of a person with his same baseline characteristics l_0 and his high diabetes education, but who did not participate in the program. His disease progress, nutritional status and comorbidities $L_1(\mathbf{a}, \bar{\Gamma}^{\mathbf{a}'})$ would be a consequence of his baseline l_0, his program participation $A_1 = 1$, his high diabetic education, and of this counseling utilization pattern (Γ_1^0). Hence, the final counterfactual $Y_\tau(1, \bar{\Gamma}^0)$ is the outcome of a person participating in the program, but has the nutrition counseling utilization pattern of an individual sharing his baseline characteristics, disease progression and comorbidities development, and diabetic education, but who otherwise did not participate in the program.

17.1.2 Causal Parameters and Identifiability

From the above formulation, we define as the *conditional mediation formula*:

$$E\left[Y_\tau(\mathbf{a}, \bar{\Gamma}^{\mathbf{a}'}) \right]. \tag{17.2}$$

To contrast the effects of two exposure regimens on an outcome at time τ, the corresponding *natural indirect effect* is defined as

$$E\left[Y_\tau(1,\bar{\Gamma}^1)\right] - E\left[Y_\tau(1,\bar{\Gamma}^0)\right], \tag{17.3}$$

and the *natural direct effect* is

$$E\left[Y_\tau(1,\bar{\Gamma}^0)\right] - E\left[Y_\tau(0,\bar{\Gamma}^0)\right]. \tag{17.4}$$

These two effects provide a decomposition of the total effect

$$E\left[Y_\tau(1,\bar{\Gamma}^1)\right] - E\left[Y_\tau(0,\bar{\Gamma}^0)\right].$$

As we will see in the next section, this total effect is the same mathematical quantity as the traditional total effect measure $E\left[Y_\tau(1)\right] - E\left[Y_\tau(0)\right]$, where $Y_\tau(\mathbf{a})$ is the would-be outcome under exposure $\mathbf{A} = \mathbf{a}$ and no intervention on $\mathbf{Z}$.

In our illustrative example, the pilot program can impact disease progression as a result of increased diabetic education among patients, increased utilization of nutrition counseling, either due to more streamlined operations or due to better educated patients. Suppose compared to other factors, program participation has large impact on increased R_1 (better diabetic-educated patients). Then the variable $R_1(1,\bar{\Gamma}^0) = r_1$ would be relatively high, and $Z_2 \sim \Gamma_2^0(l_0, r_1)$ would be the nutrition counseling utilization of an individual who did not participate in the program, but shares same baseline characteristics $L_0 = l_0$ and has the same high diabetes education. Suppose the program's more streamlined nutrition counseling operations has large impact on increased utilization of this service, then $\Gamma_2^0(l_0, r_1)$ would be heavily distributed around lower utilization and $\Gamma_2^1(l_0, r_1)$ would be distributed around higher utilization for the same individual characterized by (l_0, r_1). Subsequently, $L_1(1,\bar{\Gamma}^0)$ vs $L_1(1,\bar{\Gamma}^1)$ would be the nutritional status and disease progression of individuals with the same baseline l_0, diabetes education level r_1, but with lower vs higher nutrition counseling utilization patterns as a result of programmatic improvements. Then, the indirect effect $E\left[Y_\tau(1,\bar{\Gamma}^0)\right] - E\left[Y_\tau(1,\bar{\Gamma}^1)\right]$ would capture the indirect effect (on disease progression) of differential nutrition counseling utilization due to program's streamlined operation of this service, but not due to program's effect on differential demand for nutrition counseling as a result of better diabetes-educated patients. So this indirect effect compares only the paths from exposures (program) into mediators (nutrition counseling utilization), but not from exposures into covariates (patient education) into mediators. Dual to this, the direct effect $E\left[Y_\tau(1,\bar{\Gamma}^0)\right] - E\left[Y_\tau(0,\bar{\Gamma}^0)\right]$ capture the paths from exposure into final outcome (diabetes control), from exposure into covariates into final outcome, as well as the paths from exposure into covariates into mediator into final outcome. In our example, this last set of paths would be the differential effect on diabetes progression due to increased demand for nutrition counseling as a result of better educated patients.

To proceed with the identifiability result, let us denote $L_t(\mathbf{a}, \mathbf{z})$ the counterfactual covariate at time t under an intervention to set $\mathbf{A} = \mathbf{a}$ and $\mathbf{Z} = \mathbf{z}$. The identifiability of the corresponding causal parameters only rely on the Sequential Randomization Assumptions of Robins (1986) and positivity assumptions.

Lemma 17.1. *Suppose the following assumptions hold*

A1. $(\mathbf{R}_{s\geq t}(\mathbf{a}'), \mathbf{Z}_{s\geq t}(\mathbf{a}'), \mathbf{L}_{s\geq t}(\mathbf{a}')) \perp A_t \mid \mathbf{A}_{t-1}, \mathbf{R}_{t-1}, \mathbf{Z}_{t-1}, \mathbf{L}_{t-1}.$
A2. $(\mathbf{R}_{s\geq t}(\mathbf{a},\mathbf{z}), \mathbf{L}_{s\geq t}(\mathbf{a},\mathbf{z})) \perp A_t \mid \mathbf{A}_{t-1}, \mathbf{R}_{t-1}, \mathbf{Z}_{t-1}, \mathbf{L}_{t-1}.$

In words, A1 and A2 require that conditional on observed history, there are no unmeasured confounders of the relationship between each exposure A_t and all its subsequent covariates and mediators, i.e. A_t is randomized conditional on observed history.

A3. $(\mathbf{R}_{s>t}(\mathbf{a},\mathbf{z}), \mathbf{L}_{s\geq t}(\mathbf{a},\mathbf{z})) \perp Z_t \mid \mathbf{A}_t, \mathbf{R}_t, \mathbf{Z}_{t-1}, \mathbf{L}_{t-1}.$ *Conditional on observed history, there are no unmeasured confounders of the relationship between each mediator Z_t and all its subsequent covariates, i.e. Z_t is randomized conditional on observed past.*

A4. Positivity: for all $t \geq 1$ and all $\mathbf{r}, \mathbf{l}, \mathbf{z}$, (i) if $p_0(\mathbf{a}'_{t-1}, \mathbf{r}_{t-1}, \mathbf{z}_{t-1}, \mathbf{l}_{t-1}) > 0$, then $p_0(a'_t \mid \mathbf{a}'_{t-1}, \mathbf{r}_{t-1}, \mathbf{z}_{t-1}, \mathbf{l}_{t-1}) > 0$; (ii) if $p_0(\mathbf{a}_{t-1}, \mathbf{r}_{t-1}, \mathbf{z}_{t-1}, \mathbf{l}_{t-1}) > 0$, then $p_0(a_t \mid \mathbf{a}_{t-1}, \mathbf{r}_{t-1}, \mathbf{z}_{t-1}, \mathbf{l}_{t-1}) > 0$; (iii) If $p_0(r_t \mid \mathbf{a}_t, \mathbf{r}_{t-1}, \mathbf{z}_{t-1}, \mathbf{l}_{t-1}) > 0$, then $p_0(r_t \mid \mathbf{a}'_t, \mathbf{r}_{t-1}, \mathbf{z}_{t-1}, \mathbf{l}_{t-1}) > 0$; (iv) If $p_0(l_t \mid \mathbf{a}_t, \mathbf{r}_t, \mathbf{z}_t, \mathbf{l}_{t-1}) > 0$, then $p_0(l_t \mid \mathbf{a}'_t, \mathbf{r}_t, \mathbf{z}_t, \mathbf{l}_{t-1}) > 0$; (v) if $p_0(\mathbf{a}_t, \mathbf{r}_t, \mathbf{z}_{t-1}, \mathbf{l}_{t-1}) > 0$ and $p_0(z_t \mid \mathbf{a}'_t, \mathbf{r}_t, \mathbf{z}_{t-1}, \mathbf{l}_{t-1}) > 0$, then $p_0(z_t \mid \mathbf{a}_t, \mathbf{r}_t, \mathbf{z}_{t-1}, \mathbf{l}_{t-1}) > 0.$

Conditions (i) and (ii) require that the exposures of interest are observed within each supported covariate and mediator stratum; (iii) and (iv) require that covariate values supported under $\mathbf{A} = \mathbf{a}$ are also supported under $\mathbf{A} = \mathbf{a}'$, and (v) requires that the mediator values supported under $\mathbf{A} = \mathbf{a}'$ are also supported under $\mathbf{A} = \mathbf{a}$.

Then the conditional mediation formula in (17.2) identifies to

$$E\left[Y_\tau(\mathbf{a}, \bar{\Gamma}^{\mathbf{a}'})\right] = \Psi^{\mathbf{a},\mathbf{a}'}(P_0)$$
$$\equiv \sum_{\mathbf{r},\mathbf{l},\mathbf{z}} y_\tau \, p_{0,L_0}(l_0)$$
$$\times \prod_{t=1}^{\tau} p_{0,R}(r_t \mid \mathbf{a}_t, \mathbf{r}_{t-1}, \mathbf{z}_{t-1}, \mathbf{l}_{t-1}) p_{0,Z}(z_t \mid \mathbf{a}'_t, \mathbf{r}_t, \mathbf{z}_{t-1}, \mathbf{l}_{t-1}) p_{0,L}(l_t \mid \mathbf{a}_t, \mathbf{r}_t, \mathbf{z}_t, \mathbf{l}_{t-1}).$$

Consequently, the natural indirect and direct effects are respectively identified to

$$E\left[Y_\tau(1, \bar{\Gamma}^1)\right] - E\left[Y_\tau(1, \bar{\Gamma}^0)\right] = \Psi^{1,1}(P_0) - \Psi^{1,0}(P_0) \tag{17.5}$$
$$E\left[Y_\tau(1, \bar{\Gamma}^0)\right] - E\left[Y_\tau(0, \bar{\Gamma}^0)\right] = \Psi^{1,0}(P_0) - \Psi^{0,0}(P_0). \tag{17.6}$$

Proof. In Sect. A.2.

As we alluded to in our earlier discussion, when $\mathbf{a} = \mathbf{a}'$, $E\left[Y_\tau(\mathbf{a}, \bar{\Gamma}^{\mathbf{a}})\right] = \Psi^{\mathbf{a},\mathbf{a}}(P_0) = E\left[Y_\tau(\mathbf{a})\right]$. Therefore, the proposed natural direct and indirect effects provide a decomposition of the total effect

$$E[Y_\tau(1)] - E[Y_\tau(0)]$$
$$= \left(E\left[Y_\tau(1, \bar{\Gamma}^1)\right] - E\left[Y_\tau(1, \bar{\Gamma}^0)\right]\right) + \left(E\left[Y_\tau(1, \bar{\Gamma}^0)\right] - E\left[Y_\tau(0, \bar{\Gamma}^0)\right]\right).$$

The consistency assumption is typical of causal inference under a counterfactual framework. The positivity assumptions (i) and (ii) are typical in the study of total exposure effects, whereas (iii)–(v) are unique to the proposed conditional random intervention. Assumptions A1–A3 are the so-called strong sequential randomization assumptions.

In our illustrative example, at a period t, assumptions A1 and A2 requires that the history of program participation, nutrition counseling and covariates (diabetes education, disease progression, comorbidities, nutritional status, lifestyle factors, etc.) account for all the confounders of the relationship between current program participation and current and future covariates and nutrition counseling utilization. Assumption A3 requires that current program participation, and previous history of program participation, nutritional service utilization and other covariates, account for all the confounders of the relationship between current nutrition counseling utilization and current and future covariates. These assumptions would be violated if, for example, there was an unrecorded event that would affect one's current program participation as well as current and future nutrition counseling utilization and covariates.

17.1.3 Longitudinal Mediation Analysis with Marginal vs Conditional Random Interventions

An alternative formulation of longitudinal mediation analysis has been proposed in VanderWeele and Tchetgen Tchetgen (2017) using random interventions with marginal mediator distributions (conditioning only on baseline covariates). Specifically, let $G_t^{a'}$ be the marginal distribution of Z_t (conditioning on baseline L_0) under an ideal experiment setting $\mathbf{A} = \mathbf{a}'$, i.e. $G_t^{a'}(z_t \mid L_0) = p_{Z(a')}(z_t \mid L_0)$. Let $\mathbf{G}^{a'} \equiv (G_1^{a'}, \ldots, G_\tau^{a'})$. In our illustrative example, to generate $Y_\tau(1, \mathbf{G}^0)$, at time $t = 1$, a person with baseline characteristic $L_0 = l_0$ is set to participate in the program, and his diabetes education level is high as a result. But his nutritional service utilization $Z_1 \sim G_1^0$ will be that of a person with same the baseline $L_0 = l_0$ but who does not participate in the program and has a resulting low diabetes education level. His disease progression $L_1(1, \mathbf{G}^0) = l_1$ would be a consequence of his L_0, his program participation, his resulting high diabetes education level, and his counseling service utilization pattern G_1^0. So on. The difference in the formulation of $Y_\tau(\mathbf{a}, \mathbf{G}^{a'})$ and the proposed $Y_\tau(\mathbf{a}, \bar{\Gamma}^{a'})$ lies in that at each time t, $G_t^{a'}$ is drawn as that of a random person with $\mathbf{A} = \mathbf{a}'$ and sharing the same baseline (i.e., marginalizing over all time-varying covariate histories under the influence of $\mathbf{A} = \mathbf{a}'$), whereas $\Gamma_t^{a'}$ is drawn as that of a random person with $\mathbf{A} = \mathbf{a}'$, and sharing the same baseline and same time-varying covariate history (which are under the influence of $\mathbf{A} = \mathbf{a}$). This difference in formulation has implications for applicability, interpretation and identifiability. We discuss each in turn.

In a survival setting, the marginal-intervention counterfactual $Y_\tau(\mathbf{a}, \mathbf{G}^{a'})$ is not well defined since a person who is still alive under $\mathbf{A} = \mathbf{a}$ would be allowed to draw the mediator value of someone under $\mathbf{A} = \mathbf{a}'$ that has died. On the other hand, by conditioning on the person's own time-varying history, the conditional-intervention counterfactual $Y_\tau(\mathbf{a}, \bar{\Gamma}^{\mathbf{a}'})$ circumvents this problem and thus is well defined in the survival setting. Beyond formal definition, such time-varying covariate histories still need to be well-supported under both exposure regimes, as is apparent in the identifiability conditions for $Y_\tau(\mathbf{a}, \bar{\Gamma}^{\mathbf{a}'})$.

Consider now a nonsurvival setting. In our illustrative example, suppose program participation increases patients' diabetic education through its enhanced curriculum, and increases use of nutrition counseling utilization through better patient education and more streamlined counseling services, and both education and nutrition counseling have equal contribution to changing disease progression. Then, at time $t = 1$, under both interventions, the patient education $R_1(1, \bar{\Gamma}^0)$ and $R_1(1, \mathbf{G}^0)$ would be high due to program participation $A_1 = 1$. But the nutritional service utilization G_1^0 would be that of a nonparticipant who has a low diabetic education due to his nonparticipation in the program (e.g., less demand for nutrition counseling due to limited patient knowledge), whereas the nutritional service utilization Γ_1^0 would be that of a nonparticipant who has a high diabetic education but who otherwise did not participate in the program. Therefore, the indirect effect $E\left[Y_\tau(1, \mathbf{G}^0)\right] - E\left[Y_\tau(1, \mathbf{G}^1)\right]$ would capture the effect due to differential nutritional service utilization as a result of both the program's streamlined nutrition counseling services and the increased demand for these services due to program's impact in increasing patient knowledge. It would capture effect due to the paths from program into nutrition service utilization as well as the paths from program into patient education into nutritional service utilization. Contrast this with the indirect effect $E\left[Y_\tau(1, \bar{\Gamma}^0)\right] - E\left[Y_\tau(1, \bar{\Gamma}^1)\right]$, which captures the effect due to differential nutritional service utilization as a result of the program's streamlined nutrition counseling services (but not of increased demand due to better education). Dual to the indirect effect, the direct effect $E\left[Y_\tau(1, \mathbf{G}^0)\right] - E\left[Y_\tau(0, \mathbf{G}^0)\right]$ capture the effects due to direct paths from program into disease progression and the paths from program into patient education into disease. Contrasting these with the definitions and motivating example in Sect. 17.1.2, we see that the different mediation formulations in this longitudinal setting allows one to ask different mediation questions, as the increased complexity of the data structure also offers more options for potential questions of interest.

We saw in (17.5) that $E\left(Y_\tau(\mathbf{a}, \bar{\Gamma}^{\mathbf{a}})\right) = E(Y_\tau(\mathbf{a}))$ as mathematical quantities, though under different ideal experiment formulation. On the other hand, as can be derived from VanderWeele and Tchetgen Tchetgen (2017),

$$E\left(Y_\tau(\mathbf{a}, \mathbf{G}^{\mathbf{a}'})\right) = \sum_{\mathbf{r},\mathbf{l},\mathbf{z}}\Big\{y_\tau p_{0,L_0}(l_0)\prod_{t=1}^{\tau}p_{0,R}(r_t \mid \mathbf{a}_t, \mathbf{r}_{t-1}, \mathbf{z}_{t-1}, \mathbf{l}_{t-1})p_{0,L}(l_t \mid \mathbf{a}_t, \mathbf{r}_t, \mathbf{z}_t, \mathbf{l}_{t-1})$$

$$\times \Big[\sum_{\mathbf{r}',\mathbf{l}'_{\tau-1}}\prod_{t=1}^{\tau}p_{0,R}(r'_t \mid \mathbf{a}'_t, \mathbf{r}'_{t-1}, \mathbf{z}_{t-1}, \mathbf{l}'_{t-1})p_{0,Z}(z_t \mid \mathbf{a}'_t, \mathbf{r}'_t, \mathbf{z}_{t-1}, \mathbf{l}'_{t-1})\prod_{t=1}^{\tau-1}p_{0,L}(l'_t \mid \mathbf{a}'_t, \mathbf{r}'_t, \mathbf{z}_t, \mathbf{l}'_{t-1})\Big]\Big\}.$$

Therefore, $E(Y_\tau(\mathbf{a}, \mathbf{G}^{\mathbf{a}}))$ does not equal $E(Y_\tau(\mathbf{a}))$, in the presence of time-varying confounding. Hence, the total effect measure $E\left(Y_\tau(1, \mathbf{G}^1)\right) - E\left(Y_\tau(0, \mathbf{G}^0)\right)$ is an alternative quantification of the total effect than the traditional $E(Y_\tau(1)) - E(Y_\tau(0))$. Hence, the choice of formulation also depends on which total effect decomposition one wishes to study.

While the proposed conditional distribution intervention provides more flexibility by allowing application in survival setting and decomposition of the standard total effect, this, however, comes at the expense of stronger identifiability conditions. To identify the marginal intervention parameter $E\left(Y_\tau(\mathbf{a}, \mathbf{G}^{\mathbf{a}'})\right)$, one can use the weaker versions of assumptions A2–A3 which only require no unmeasured confounding with respect to the final outcome of interest (as opposed to all subsequent covariates). For instance, in our example, if there was an unrecorded short-term event (e.g., short-term unemployment) that would affect one's current program participation as well as current disease status, lifestyle, nutritional service utilization, but not the final outcome, then $E\left(Y_\tau(\mathbf{a}, \mathbf{G}^{\mathbf{a}'})\right)$ would still be identified, whereas $E\left(Y_\tau(\mathbf{a}, \bar{\Gamma}^{\mathbf{a}'})\right)$ would not. Similarly, if there is unmeasured confounder of program participation (or nutritional counseling utilization) and patient diabetic education R_t, but not of the final outcome, then $E\left(Y_\tau(\mathbf{a}, \mathbf{G}^{\mathbf{a}'})\right)$ would still be identified, whereas $E\left(Y_\tau(\mathbf{a}, \bar{\Gamma}^{\mathbf{a}'})\right)$ would not. Therefore, in nonsurvival settings, the tradeoff between identifiability and effect interpretation and total effect decomposition would need to be carefully weighted. The rest of this chapter is devoted to the statistical inference of $\Psi^{\mathbf{a}, \mathbf{a}'}(P_0)$.

17.2 Efficient Influence Curve

In this section, we establish a general semiparametric inference framework for these parameters. In particular, we derive the Efficient Influence Curves (EIC) of (17.5), (17.6) and (17.5) under a (locally saturated) semiparametric model, and establish their robustness properties. For a given pathwise-differentiable parameter Ψ, under certain regularity conditions, the variance of the EIC of Ψ is a generalized Cramer-Rao lower bound for the variances of the influence curves of asymptotically linear estimators of Ψ. Therefore, the variance of the EIC provides an efficiency bound for the regular and asymptotically linear (RAL) estimators of Ψ. Moreover, under a locally saturated model, the influence curve of any RAL estimator is in fact the EIC. We refer the reader to Bickel et al. (1997b) for general theory of efficient semiparametric inference.

Nested Expectation Representation of the Conditional Mediation Formula. Let $\mathcal{M}$ denote a locally saturated semiparametric model containing the true data generating distribution P_0. Following an important observation by Bang and Robins (2005), we define recursively the following functionals for $t = \tau, \ldots, 1$, at $P \in \mathcal{M}$.

$\bar{Q}_{R_{\tau+1}}^{\mathbf{a},\mathbf{a}'} \equiv Y_\tau.$ Then, for each $t = \tau, \ldots, 1$:

$$\bar{Q}_{L_t}^{\mathbf{a},\mathbf{a}'}(\mathbf{R}_t, \mathbf{Z}_t, \mathbf{L}_{t-1}) \equiv \sum_{l_t} \bar{Q}_{R_{t+1}}^{\mathbf{a},\mathbf{a}'}(\mathbf{R}_t, \mathbf{Z}_t, \mathbf{L}_{t-1}, l_t)\, p\,(l_t \mid \mathbf{A}_t = \mathbf{a}_t, \mathbf{R}_t, \mathbf{Z}_t, \mathbf{L}_{t-1})$$

$$\bar{Q}_{Z_t}^{\mathbf{a},\mathbf{a}'}(\mathbf{R}_t, \mathbf{Z}_{t-1}, \mathbf{L}_{t-1}) \equiv \sum_{z_t} \bar{Q}_{L_t}^{\mathbf{a},\mathbf{a}'}(\mathbf{R}_t, \mathbf{Z}_{t-1}, z_t, \mathbf{L}_{t-1})\, p\,(z_t \mid \mathbf{A}_t = \mathbf{a}_t', \mathbf{R}_t, \mathbf{Z}_{t-1}, \mathbf{L}_{t-1})$$

$$\bar{Q}_{R_t}^{\mathbf{a},\mathbf{a}'}(\mathbf{R}_{t-1}, \mathbf{Z}_{t-1}, \mathbf{L}_{t-1}) \equiv \sum_{r_t} \bar{Q}_{Z_t}^{\mathbf{a},\mathbf{a}'}(\mathbf{R}_{t-1}, r_t, \mathbf{Z}_{t-1}, \mathbf{L}_{t-1})$$

$$\times\, p\,(r_t \mid \mathbf{A}_{t-1} = \mathbf{a}_{t-1}, \mathbf{R}_{t-1}, \mathbf{Z}_{t-1}, \mathbf{L}_{t-1}). \tag{17.7}$$

Evaluating these functionals at the data generating P_0, we obtain an nested expectation-based representation of the identifying expression (17.5):

$$\Psi^{\mathbf{a},\mathbf{a}'}(P_0) = E_{P_0}\left[\bar{Q}_{R_1}^{\mathbf{a},\mathbf{a}'}(P_0)(L_0)\right]. \tag{17.8}$$

We will use $\bar{Q}^{\mathbf{a},\mathbf{a}'}$ to denote the nested expectations $\left(\bar{Q}_{L_t}^{\mathbf{a},\mathbf{a}'}, \bar{Q}_{Z_t}^{\mathbf{a},\mathbf{a}'}, \bar{Q}_{R_t}^{\mathbf{a},\mathbf{a}'} : t \geq 1\right)$. We use P_n to denote the empirical distribution of n i.i.d. copies of $O \sim P_0$. Given a function $O \mapsto f(O)$, $P_n f$ denotes the empirical mean $P_n f \equiv \frac{1}{n}\sum_{i=1}^{n} f(O_i)$.

Efficient Influence Curves for The Mediation Formula and Direct and Indirect Effects. The mediation formula in (17.5) can be considered as the value at P_0 of the map $P \mapsto \Psi^{\mathbf{a},\mathbf{a}'}(P) \equiv E_P\left[\bar{Q}_{R_1}^{\mathbf{a},\mathbf{a}'}(P)(L_0)\right]$ on $\mathcal{M}$. In particular, this map depends on P through $\bar{Q}^{\mathbf{a},\mathbf{a}'}$, i.e. $\Psi^{\mathbf{a},\mathbf{a}'}(P) = \Psi^{\mathbf{a},\mathbf{a}'}(\bar{Q}^{\mathbf{a},\mathbf{a}'})$. Similarly, the natural direct effect in (17.6) and the natural indirect effect in (17.5) are, respectively, the values at P_0 of the maps $P \mapsto \Psi^{NDE}(P) = \Psi^{1,0}(P) - \Psi^{0,0}(P)$ and $P \mapsto \Psi^{NIE}(P) = \Psi^{1,1}(P) - \Psi^{1,0}(P)$.

Theorem 17.1 (Efficient Influence Curve). *Let $\Psi^{\mathbf{a},\mathbf{a}'} : \mathcal{M} \rightarrow \mathbf{R}$ be defined as above. Suppose at $P \in \mathcal{M}$ the conditional probabilities of A_t, R_t, Z_t, L_t, under the likelihood decomposition (17.1), are all bounded away from 0 and 1. The Efficient influence curve of $\Psi^{\mathbf{a},\mathbf{a}'}$ at P is given by $D^{*,\mathbf{a},\mathbf{a}'}(P) \equiv D^{*,\mathbf{a},\mathbf{a}'}(P, \Psi^{\mathbf{a},\mathbf{a}'}(\bar{Q}^{\mathbf{a},\mathbf{a}'}))$, with*

$$D^{*,\mathbf{a},\mathbf{a}'}(P, \psi) \equiv \sum_{t=1}^{\tau}\left(D_{L_t}^{\mathbf{a},\mathbf{a}'}(P) + D_{Z_t}^{\mathbf{a},\mathbf{a}'}(P) + D_{R_t}^{\mathbf{a},\mathbf{a}'}(P)\right) + D_{L_0}^{\mathbf{a},\mathbf{a}'}(P, \psi), \tag{17.9}$$

where

$$D_{L_t}^{\mathbf{a},\mathbf{a}'}(P) \equiv H_{L_t}^{\mathbf{a},\mathbf{a}'}\left\{\bar{Q}_{R_{t+1}}^{\mathbf{a},\mathbf{a}'}(\mathbf{R}_t, \mathbf{Z}_t, \mathbf{L}_t) - \bar{Q}_{L_t}^{\mathbf{a},\mathbf{a}'}(\mathbf{R}_t, \mathbf{Z}_t, \mathbf{L}_{t-1})\right\}$$

$$D_{Z_t}^{\mathbf{a},\mathbf{a}'}(P) \equiv H_{Z_t}^{\mathbf{a},\mathbf{a}'}\left\{\bar{Q}_{L_t}^{\mathbf{a},\mathbf{a}'}(\mathbf{R}_t, \mathbf{Z}_t, \mathbf{L}_{t-1}) - \bar{Q}_{Z_t}^{\mathbf{a},\mathbf{a}'}(\mathbf{R}_t, \mathbf{Z}_{t-1}, \mathbf{L}_{t-1})\right\}$$

$$D_{R_t}^{\mathbf{a},\mathbf{a}'}(P) \equiv H_{R_t}^{\mathbf{a},\mathbf{a}'}\left\{\bar{Q}_{Z_t}^{\mathbf{a},\mathbf{a}'}(\mathbf{R}_t, \mathbf{Z}_{t-1}, \mathbf{L}_{t-1}) - \bar{Q}_{R_t}^{\mathbf{a},\mathbf{a}'}(\mathbf{R}_{t-1}, \mathbf{Z}_{t-1}, \mathbf{L}_{t-1})\right\}$$

$$D_{L_0}^{\mathbf{a},\mathbf{a}'}(P, \psi) \equiv \bar{Q}_{t=1}^{R,\mathbf{a},\mathbf{a}'}(\mathbf{L}_0) - \psi,$$

with

$$H_{L_t}^{\mathbf{a},\mathbf{a}'} \equiv \frac{I(\mathbf{A}_t \equiv \mathbf{a}_t)}{\prod_{j=1}^{t} p_A\left(\mathbf{a}_j \mid \mathbf{a}_{j-1}, \mathbf{R}_{j-1}, \mathbf{Z}_{j-1}, \mathbf{L}_{j-1}\right)} \prod_{j=1}^{t} \frac{p_Z(Z_j \mid \mathbf{a}_j', \mathbf{R}_j, \mathbf{Z}_{j-1}, \mathbf{L}_{j-1})}{p_Z(Z_j \mid \mathbf{a}_j, \mathbf{R}_j, \mathbf{Z}_{j-1}, \mathbf{L}_{j-1})},$$

$$(17.10)$$

$$H_{Z_t}^{\mathbf{a},\mathbf{a}'} \equiv \frac{I(\mathbf{A}_t \equiv \mathbf{a}_t')}{\prod_{j=1}^{t} p_A(a_j' \mid \mathbf{a}_{j-1}', \mathbf{R}_{j-1}, \mathbf{Z}_{j-1}, \mathbf{L}_{j-1})}$$

$$\times \prod_{j=1}^{t-1} \frac{p_L(L_j \mid \mathbf{a}_j, \mathbf{R}_j, \mathbf{Z}_j, \mathbf{L}_{j-1})}{p_L(L_j \mid \mathbf{a}_j', \mathbf{R}_j, \mathbf{Z}_j, \mathbf{L}_{j-1})} \prod_{j=1}^{t} \frac{p_R(R_j \mid \mathbf{a}_j, \mathbf{R}_{j-1}, \mathbf{Z}_{j-1}, \mathbf{L}_{j-1})}{p_R(R_j \mid \mathbf{a}_j', \mathbf{R}_{j-1}, \mathbf{Z}_{j-1}, \mathbf{L}_{j-1})} \quad (17.11)$$

$$H_{R_t}^{\mathbf{a},\mathbf{a}'} \equiv \frac{I(\mathbf{A}_t \equiv \mathbf{a}_t)}{\prod_{j=1}^{t} p_A\left(\mathbf{a}_j \mid \mathbf{a}_{j-1}, \mathbf{R}_{j-1}, \mathbf{Z}_{j-1}, \mathbf{L}_{j-1}\right)} \prod_{j=1}^{t-1} \frac{p_Z(Z_j \mid \mathbf{a}_j', \mathbf{R}_j, \mathbf{Z}_{j-1}, \mathbf{L}_{j-1})}{p_Z(Z_j \mid \mathbf{a}_j, \mathbf{R}_j, \mathbf{Z}_{j-1}, \mathbf{L}_{j-1})},$$

$$(17.12)$$

Moreover, $D^{*,\mathbf{a},\mathbf{a}'}(P)$ is a multiply robust estimating function of $\Psi^{\mathbf{a},\mathbf{a}'}(P)$ in the sense that if one of the following holds:

R1. The conditional probabilities p_R, p_L and p_Z correctly specified;
R2. The conditional probabilities p_A , p_R and p_L are correctly specified.
R3. The conditional probabilities p_A and p_Z are correctly specified.

then $E_{P_0} D^{*,\mathbf{a},\mathbf{a}'}(P) = 0$ implies $\Psi^{\mathbf{a},\mathbf{a}'}(P) = \Psi^{\mathbf{a},\mathbf{a}'}(P_0)$.

Proof. See Sect. A.2.

It is easy to note that if $\mathbf{a} = \mathbf{a}'$, then (17.9) equals the efficient influence curve for the overall treatment effect of a time varying exposure (see e.g., van der Laan and Gruber 2012). The EICs of both the NDE and NIE can be derived from (17.9) by a simple application of the delta method. We state them in a corollary without proof.

Corollary 1 *Suppose the conditions in Theorem 17.1 hold for $a, a' \in \{0, 1\}$. The efficient influence curve of the natural direct effect is given by*

$$D^{*,NDE}(P)(O) = D^{*,1,0}(P) - D^{*,0,0}(P),$$

and the efficient influence curve of the natural indirect effect is given by

$$D^{*,NIE}(P)(O) = D^{*,1,1}(P) - D^{*,1,0}(P).$$

Moreover, $D^{*,NDE}$ and D^*NIE satisfy the same robustness condition in Theorem 17.1 for $\mathbf{a} = 0, 1$ and $\mathbf{a}' = 0, 1$.

The variances $Var_{P_0}(D^{*,\mathbf{a},\mathbf{a}'}(P_0))$, $Var_{P_0}(D^{*,NDE}(P_0))$, and $Var_{P_0}(D^{*,NIE}(P_0))$ are generalized Cramer-Rao lower bounds for the asymptotic variances of the RAL estimators of $\Psi^{\mathbf{a},\mathbf{a}'}(P_0)$, $\Psi^{NDE}(P_0)$, and $\Psi^{NIE}(P_0)$, respectively. Estimators which satisfy the EIC equations will also inherit their robustness properties. We will present four estimators in the next section, two of which are robust and locally efficient.

Notes on Estimating Components of the Efficient Influence Curve. The parameter of interest (17.5) and the corresponding EIC (17.9) are represented in terms of conditional probabilities p_R, p_L, p_Z, p_A. In applications where the covariates or the mediator may be high-dimensional, estimating these conditional densities may be difficult. To proceed with the estimation in these situations, firstly we note that due to the law of iterated expectations

$$
\begin{aligned}
\bar{Q}_{L_t}^{\mathbf{a},\mathbf{a}'}(\mathbf{R}_t, \mathbf{Z}_t, \mathbf{L}_{t-1}) &= E_P\left[\bar{Q}_{R_{t+1}}^{\mathbf{a},\mathbf{a}'}(\mathbf{R}_t, \mathbf{Z}_t, \mathbf{L}_t) \mid \mathbf{R}_t, \mathbf{Z}_t, \mathbf{L}_{t-1}\right] \\
\bar{Q}_{Z_t}^{\mathbf{a},\mathbf{a}'}(\mathbf{R}_t, \mathbf{Z}_{t-1}, \mathbf{L}_{t-1}) &= E_P\left[\bar{Q}_{L_t}^{\mathbf{a},\mathbf{a}'}(\mathbf{R}_t, \mathbf{Z}_t, \mathbf{L}_{t-1}) \mid \mathbf{R}_t, \mathbf{Z}_{t-1}, \mathbf{L}_{t-1}\right] \\
\bar{Q}_{R_t}^{\mathbf{a},\mathbf{a}'}(\mathbf{R}_{t-1}, \mathbf{Z}_{t-1}, \mathbf{L}_{t-1}) &= E_P\left[\bar{Q}_{Z_t}^{\mathbf{a},\mathbf{a}'}(\mathbf{R}_t, \mathbf{Z}_{t-1}, \mathbf{L}_{t-1}) \mid \mathbf{R}_{t-1}, \mathbf{Z}_{t-1}.\mathbf{L}_{t-1}\right]
\end{aligned} \tag{17.13}
$$

Therefore, one may directly estimate the expectation $\bar{Q}_{L_t}^{\mathbf{a},\mathbf{a}'}(\mathbf{R}_t, \mathbf{Z}_t, \mathbf{L}_{t-1})$ by regressing $\bar{Q}_{R_{t+1}}^{\mathbf{a},\mathbf{a}'}(\mathbf{R}_t, \mathbf{Z}_t, \mathbf{L}_t)$ on the covariates $(\mathbf{R}_t, \mathbf{Z}_t, \mathbf{L}_{t-1})$ using a parametric or data-adaptive algorithm, without estimating the conditional probabilities of L_t. Similarly for the expectations corresponding to Z_t and R_t. Secondly, we define

$$
\gamma^{1,s,j}(A_s \mid \mathbf{A}_{s-1}, \mathbf{R}_j, \mathbf{Z}_j, \mathbf{L}_{j-1}) \equiv p(A_s \mid \mathbf{A}_{s-1}, \mathbf{R}_j\mathbf{Z}_j, \mathbf{L}_{j-1})
$$

and

$$
\gamma^{2,s,j}(A_s \mid \mathbf{A}_{s-1}, \mathbf{R}_j, \mathbf{Z}_{j-1}, \mathbf{L}_{j-1}) \equiv p(A_s \mid \mathbf{A}_{s-1}, \mathbf{R}_j, \mathbf{Z}_{j-1}, \mathbf{L}_{j-1}).
$$

Then, we can rewrite the expressions in the EIC as

$$
\begin{aligned}
H_{L_t}^{\mathbf{a},\mathbf{a}'} &= \frac{I(\mathbf{A}_t \equiv \mathbf{a}_t)}{\prod_{j=1}^t p_A\left(a_j \mid \mathbf{a}_{j-1}, \mathbf{R}_{j-1}, \mathbf{Z}_{j-1}, \mathbf{L}_{j-1}\right)} \\
&\times \prod_{j=1}^t \prod_{s=1}^j \frac{\gamma^{1,s,j}(a'_s \mid \mathbf{a}'_{s-1}, \mathbf{R}_j\mathbf{Z}_j, \mathbf{L}_{j-1})}{\gamma^{1,s,j}(a_s \mid \mathbf{a}_{s-1}, \mathbf{R}_j\mathbf{Z}_j, \mathbf{L}_{j-1})} \frac{\gamma^{2,s,j}(a_s \mid \mathbf{a}_{s-1}, \mathbf{R}_j, \mathbf{Z}_{j-1}, \mathbf{L}_{j-1}}{\gamma^{2,s,j}(a'_s \mid \mathbf{a}'_{s-1}, \mathbf{R}_j, \mathbf{Z}_{j-1}, \mathbf{L}_{j-1})} \\
H_{R_t}^{\mathbf{a},\mathbf{a}'} &= \frac{I(\mathbf{A}_t \equiv \mathbf{a}_t)}{\prod_{j=1}^t p_A\left(a_j \mid \mathbf{a}_{j-1}, \mathbf{R}_{j-1}, \mathbf{Z}_{j-1}, \mathbf{L}_{j-1}\right)} \\
&\times \prod_{j=1}^{t-1} \prod_{s=1}^j \frac{\gamma^{1,s,j}(a'_s \mid \mathbf{a}'_{s-1}, \mathbf{R}_j\mathbf{Z}_j, \mathbf{L}_{j-1})}{\gamma^{1,s,j}(a_s \mid \mathbf{a}_{s-1}, \mathbf{R}_j\mathbf{Z}_j, \mathbf{L}_{j-1})} \frac{\gamma^{2,s,j}(a_s \mid \mathbf{a}_{s-1}, \mathbf{R}_j, \mathbf{Z}_{j-1}, \mathbf{L}_{j-1}}{\gamma^{2,s,j}(a'_s \mid \mathbf{a}'_{s-1}, \mathbf{R}_j, \mathbf{Z}_{j-1}, \mathbf{L}_{j-1})}
\end{aligned}
$$

and

$$
\begin{aligned}
H_{Z_t}^{\mathbf{a},\mathbf{a}'} &= \frac{I(\mathbf{A}_t \equiv \mathbf{a}'_t)}{\prod_{j=1}^t p_A(a'_j \mid \mathbf{a}'_{j-1}, \mathbf{R}_{j-1}, \mathbf{Z}_{j-1}, \mathbf{L}_{j-1})} \\
&\times \prod_{j=1}^{t-1} \prod_{s=1}^j \frac{\gamma^{1,s,j}(a'_s \mid \mathbf{a}'_{s-1}, \mathbf{R}_j\mathbf{Z}_j, \mathbf{L}_{j-1})}{\gamma^{1,s,j}(a_s \mid \mathbf{a}_{s-1}, \mathbf{R}_j\mathbf{Z}_j, \mathbf{L}_{j-1})} \\
&\times \prod_{j=1}^t \frac{p_A(a'_j \mid \mathbf{a}'_{j-1}, \mathbf{R}_{j-1}, \mathbf{Z}_{j-1}, \mathbf{L}_{j-1})}{p_A(a_j \mid \mathbf{a}_{j-1}, \mathbf{R}_{j-1}, \mathbf{Z}_{j-1}, \mathbf{L}_{j-1})} \prod_{s=1}^j \frac{\gamma^{2,s,j}(a_s \mid \mathbf{a}_{s-1}, \mathbf{R}_j, \mathbf{Z}_{j-1}, \mathbf{L}_{j-1}}{\gamma^{2,s,j}(a'_s \mid \mathbf{a}'_{s-1}, \mathbf{R}_j, \mathbf{Z}_{j-1}, \mathbf{L}_{j-1})}
\end{aligned}
$$

Note that these conditional probabilities of A_s differ from the conditional (and possibly censoring) probabilities encoded by p_A in that the γs are not conditioning on parents of A_s. However, as we shall see in the following lemma, they offer an alternative to obtain robust estimators that are more suitable to real life settings where L_t and/or Z_t may be high dimensional. We write $\gamma = \left(\gamma^{1,s,t}, \gamma^{2,s,t} : 1 \le t \le \tau, s \le t \right)$. Based on this representation, the robustness conditions in Theorem 17.1 can be generalized.

Corollary 2 *Let γ be defined as above. If one of the following holds,*

R1. The nested regressions $\bar{Q}^{\mathbf{a},\mathbf{a}'}$, as represented in (17.13), are correctly specified;
R2. p_A, $\bar{Q}_{R_t}^{\mathbf{a},\mathbf{a}'}$, $\bar{Q}_{L_t}^{\mathbf{a},\mathbf{a}'}$, and either (p_L, p_R) or γ are correctly specified;
R3. p_A, $\bar{Q}_{Z_t}^{\mathbf{a},\mathbf{a}'}$, and either p_Z or γ are correctly specified,

then $E_{P_0} D^{*,\mathbf{a},\mathbf{a}'}(P) = 0$ implies $\Psi^{\mathbf{a},\mathbf{a}'}(P) = \Psi^{\mathbf{a},\mathbf{a}'}(P_0)$.

17.3 Estimators

In this section, we develop an nested nontargeted substitution estimator, an IPW estimator and a TMLE for the mediation functional (17.5); the estimators for the natural direct and indirect effects can be obtained by taking the corresponding differences. The first two estimators are consistent if the estimates of all the relevant components of P_0 are consistent. On the other hand, the TMLE satisfies the efficient influence curve equation, and hence remains unbiased under the model mis-specifications described in Theorem 17.1. Under appropriate regularity conditions, if all the nuisance parameters are consistently estimated, then TMLE will be asymptotically efficient (e.g., Bickel et al. 1997b; van der Laan and Robins 2003; van der Laan and Rose 2011). Let $p_{n,A}$, $p_{n,L}$, $p_{n,R}$ and $p_{n,Z}$ denote the estimators of the conditional probabilities. We will use shorthand $\bar{p}_n$ to denote these estimators. Let $\bar{Q}_n^{\mathbf{a},\mathbf{a}'} \equiv \left(\bar{Q}_{n,L_t}^{\mathbf{a},\mathbf{a}'}, \bar{Q}_{n,Z_t}^{\mathbf{a},\mathbf{a}'}, \bar{Q}_{n,R_t}^{\mathbf{a},\mathbf{a}'} : t \right)$ denote the estimators of the nested expectations. These may be density-based estimators that are obtained by plugging in the density estimates $p_{n,L}$, $p_{n,R}$ and $p_{n,Z}$ into the definition of the expectations in (17.7), or they may be regression-based estimators that are obtained using the relations in (17.13).

17.3.1 Nontargeted Substitution Estimator

The identification formula in (17.5) which defines that statistical estimand is generally known as the g-computation formula (Robins 1986). Readily, it delivers a nontargeted substitution estimator, which is generally known as the g-computation estimator. To avoid estimation of densities, one can recast it in terms of $\bar{Q}^{\mathbf{a},\mathbf{a}'}$, as they are represented (17.8) and (17.13), and obtain a nontargeted substitution estimator $\Psi^{\mathbf{a},\mathbf{a}'}(\bar{Q}_n^{\mathbf{a},\mathbf{a}'})$ of $\Psi^{\mathbf{a},\mathbf{a}'}(P_0)$, through nontargeted estimates of the regressions $\bar{Q}_n^{\mathbf{a},\mathbf{a}'}$.

To estimate the series of nested conditional expectations $\bar{Q}^{\mathbf{a},\mathbf{a}'}(P_0)$, we can use the following algorithm, which exploits the relations in (17.13) to make use of available regression techniques in the literature.

1. Initiate $\bar{Q}_{R_{\tau+1}}\mathbf{a},\mathbf{a}' \equiv Y_\tau$. In our example this is the final glucose level, one may have dummy values for those that were lost to follow up.

2. At each $t = \tau, \ldots 1$, in decreasing order, we would have obtained estimators $\bar{Q}^{\mathbf{a},\mathbf{a}'}_{n,R_{t+1}}$ from the previous step. We now obtain $\bar{Q}^{\mathbf{a},\mathbf{a}'}_{n,L_t}$, $\bar{Q}^{\mathbf{a},\mathbf{a}'}_{n,Z_t}$, and $\bar{Q}^{\mathbf{a},\mathbf{a}'}_{n,R_t}$, in that order, as follows:

    (a) Regress $\bar{Q}^{\mathbf{a},\mathbf{a}'}_{n,R_{t+1}}(\mathbf{R}_t, \mathbf{Z}_t, \mathbf{L}_t)$ on observed values $(\mathbf{A}_t, \mathbf{R}_t, \mathbf{Z}_t, \mathbf{L}_{t-1})$ among observations that remained uncensored at time t. In our example, the independent variables in this regression would be histories of program participation, diabetic education, and nutrition counseling attendance up to time t and disease progression and other health-related variables up to $t-1$. We then evaluate the fitted function at the observed mediator and covariates histories $\mathbf{R}_t, \mathbf{Z}_t, \mathbf{L}_{t-1}$ and the intervened exposure $\mathbf{A}_t = \mathbf{a}_t$ for these uncensored observations. This can be accomplished by creating a new dataset for these uncensored units with the observed covariates and mediators, but with $\mathbf{A}_t$ set to $\mathbf{a}_t$. This results in the estimates $\bar{Q}^{\mathbf{a},\mathbf{a}'}_{n,L_t}(\mathbf{R}_t, \mathbf{Z}_t, \mathbf{L}_{t-1}) = E\left[\bar{Q}^{\mathbf{a},\mathbf{a}'}_{n,R_{t+1}}(\mathbf{R}_t, \mathbf{Z}_t, \mathbf{L}_t) \mid \mathbf{A}_t = \mathbf{a}_t, \mathbf{R}_t, \mathbf{Z}_t, \mathbf{L}_{t-1}\right]$ for those uncensored individuals.

    (b) Regress the newly minted $\bar{Q}^{\mathbf{a},\mathbf{a}'}_{n,L_t}(\mathbf{R}_t, \mathbf{Z}_t, \mathbf{L}_{t-1})$ on $\mathbf{A}_t, \mathbf{R}_t, \mathbf{Z}_{t-1}, \mathbf{L}_{t-1}$ among observations that remained uncensored at time t. In our example, the independent variables in this regression would be histories of program participation and diabetic education up to time t, and nutrition counseling attendance disease progression and other health-related variables up to $t-1$. We then evaluate the fitted function at the observed mediator and covariate histories $\mathbf{R}_t, \mathbf{Z}_{t-1}, \mathbf{L}_{t-1}$ and the intervened income levels $\mathbf{A}_t = \mathbf{a}'_t$ for these uncensored observations. This results in the estimates $\bar{Q}^{\mathbf{a},\mathbf{a}'}_{n,Z_t}(\mathbf{R}_t, \mathbf{L}_{t-1}, \mathbf{Z}_{t-1})$.

    (c) Regress the newly obtained $\bar{Q}^{\mathbf{a},\mathbf{a}'}_{n,Z_t}(\mathbf{R}_t, \mathbf{L}_{t-1}, \mathbf{Z}_{t-1})$ on $\mathbf{A}_t, \mathbf{R}_{t-1}, \mathbf{Z}_{t-1}, \mathbf{L}_{t-1}$ among observations that remained uncensored at time t. In our example, the independent variables in this regression would be histories of program participation up to time t, and diabetic education, nutrition counseling attendance disease progression and other health-related variables up to $t-1$. We then evaluate the fitted function at the observed mediator and covariate histories $\mathbf{R}_{t-1}, \mathbf{Z}_{t-1}, \mathbf{L}_{t-1}$ and the intervened income levels $\mathbf{A}_t = \mathbf{a}'_t$ for these uncensored observations. This results in the estimates $\bar{Q}^{\mathbf{a},\mathbf{a}'}_{n,R_t}(\mathbf{R}_{t-1}, \mathbf{L}_{t-1}, \mathbf{Z}_{t-1})$.

3. After running the algorithm in step (2) sequentially from $t = \tau$ down to $t = 1$, we now have $\bar{Q}^{\mathbf{a},\mathbf{a}'}_{n,R_1}(L_0)$ for each of the n observations.

The nontargeted substitution estimator is given by

$$\psi_n^{NT\,sub} \equiv \Psi^{\mathbf{a},\mathbf{a}'}(\bar{Q}_n^{\mathbf{a},\mathbf{a}'}) = \frac{1}{n}\sum_{i=1}^{n} \bar{Q}^{\mathbf{a},\mathbf{a}'}_{n,R_1}(L_{0,i}) \tag{17.14}$$

Consistency of $\psi_{n,NT\,sub}^{\mathbf{a},\mathbf{a}'}$ relies on consistency of $\bar{Q}_n^{\mathbf{a},\mathbf{a}'}$. Correct specification of $\bar{Q}^{\mathbf{a},\mathbf{a}'}(P_0)$ under a finite dimensional parametric model is possible only in limited applications. Alternatively, we may use machine learning algorithms, such as super learner. This option is more enticing, especially when used with the regression-based approach, since there are more data-adaptive techniques available to estimate the conditional mean of a binary variable via regression. Variance estimates of the estimator based on a nonparametric bootstrap are not supported by theory and can be expected to be inconsistent. Theoretical results on the asymptotic behavior, such as a central limit theorem, of the resulting estimator $\Psi^{\mathbf{a},\mathbf{a}'}(\bar{Q}_n^{\mathbf{a},\mathbf{a}'})$ are not available, and, there is no reason to expect that such a data-adaptive g-computation estimator has a limit distribution. Moreover, a nontargeted estimator $\bar{Q}_n^{\mathbf{a},\mathbf{a}'}$ of $\bar{Q}^{\mathbf{a},\mathbf{a}'}(P_0)$ is obtained by minimizing a global loss function for $\bar{Q}^{\mathbf{a},\mathbf{a}'}(P_0)$, not for $\Psi^{\mathbf{a},\mathbf{a}'}(P_0)$. This means, in particular, that the bias-variance tradeoff in $\bar{Q}_n^{\mathbf{a},\mathbf{a}'}$ is optimized for the high-dimensional nuisance parameter $\bar{Q}^{\mathbf{a},\mathbf{a}'}(P_0)$, instead of a much lower-dimensional parameter of interest $\Psi^{\mathbf{a},\mathbf{a}'}(P_0)$. As a consequence, the mean squared error of the non-targeted substitution estimator is too high: e.g., such a data-adaptive g-computation estimator does converge at a lower rate than $n^{-1/2}$ under the same conditions under which the proposed targeted estimator will be asymptotically linear and efficient (and thus converge at the parametric rate $n^{-1/2}$). The proposed targeted estimator in Sect. 17.3.3 aims to address these two issues by providing a substitution estimator that is asymptotically linear (under appropriate regularity conditions), and optimizes the bias-variance tradeoff of $\bar{Q}_n^{\mathbf{a},\mathbf{a}'}$ towards $\Psi^{\mathbf{a},\mathbf{a}'}(P_0)$ via an updating step.

17.3.2 IPW Estimator

Instead of estimating the conditional expectations $\bar{Q}^{\mathbf{a},\mathbf{a}'}(P_0)$, one may wish to employ the researcher's knowledge about the treatment assignment and mediator densities. To this end, consider the following function:

$$D^{IPW,\mathbf{a},\mathbf{a}'}(p_A, p_Z)(O)$$

$$\equiv Y_\tau \frac{I(\mathbf{A}_\tau = \mathbf{a}_\tau)}{\prod_{j=1}^\tau p_A(a_j \mid \mathbf{a}_{j-1}, \mathbf{R}_{j-1}, \mathbf{Z}_{j-1}, \mathbf{L}_{j-1})} \prod_{j=1}^\tau \frac{p_Z(Z_j \mid \mathbf{a}'_j, \mathbf{R}_j, \mathbf{Z}_{j-1}, \mathbf{L}_{j-1})}{p_Z(Z_j \mid \mathbf{a}_j, \mathbf{R}_j, \mathbf{Z}_{j-1}, \mathbf{L}_{j-1})}. \quad (17.15)$$

Note that

$$E_{P_0} D^{IPW,\mathbf{a},\mathbf{a}'}(P_0)$$

$$= \sum_{\mathbf{r},\mathbf{z},\mathbf{l}} \Bigg\{ y_\tau \frac{1}{\prod_{j=1}^\tau p_{0,A}(a_j \mid \mathbf{a}_{j-1}, \mathbf{r}_{j-1}, \mathbf{z}_{j-1}, \mathbf{l}_{j-1})} \prod_{j=1}^\tau \frac{p_{0,Z}(z_j \mid \mathbf{a}'_j, \mathbf{r}_j, \mathbf{z}_{j-1}, \mathbf{l}_{j-1})}{p_{0,Z}(z_j \mid \mathbf{a}_j, \mathbf{r}_j, \mathbf{z}_{j-1}, \mathbf{l}_{j-1})}$$

$$\times p_{0,L_0}(l_0) \prod_{t=1}^\tau \big[p_{0,A}(a_t \mid \mathbf{a}_{t-1}, \mathbf{r}_{t-1}, \mathbf{z}_{t-1}, \mathbf{l}_{t-1}) p_{0,R}(r_t \mid \mathbf{a}_t, \mathbf{r}_{t-1}, \mathbf{z}_{t-1}, \mathbf{l}_{t-1})$$

$$\times \, p_{0,Z}(z_t \mid \mathbf{a}_t, \mathbf{r}_t, \mathbf{z}_{t-1}, \mathbf{l}_{t-1}) p_{0,L}(l_t \mid \mathbf{a}_t, \mathbf{r}_t, \mathbf{z}_t, \mathbf{l}_{t-1})\Big]\Big\}$$
$$= \Psi^{\mathbf{a},\mathbf{a}'}(P_0) = 0.$$

Therefore, given estimators $p_{n,A}$ and $p_{n,Z}$, the IPW estimator of $\Psi^{\mathbf{a},\mathbf{a}'}(P_0)$ is given by

$$\psi_n^{IPW} \equiv \frac{1}{n} \sum_{i=1}^{n} D^{IPW,\mathbf{a},\mathbf{a}'}(p_{n,A}, p_{n,Z})(O_i). \tag{17.16}$$

In our example, this estimate can be obtained by taking the weighted average of the final outcome of all those with observed exposure level $\mathbf{A}_\tau = \mathbf{a}_\tau$, using weights $\frac{1}{\prod_{j=1}^{\tau} p_A(a_j \mid \mathbf{a}_{j-1}, \mathbf{R}_{j-1}, \mathbf{Z}_{j-1}, \mathbf{L}_{j-1})} \prod_{j=1}^{\tau} \frac{p_Z(Z_j \mid \mathbf{a}'_j, \mathbf{R}_j, \mathbf{Z}_{j-1}, \mathbf{L}_{j-1})}{p_Z(Z_j \mid \mathbf{a}_j, \mathbf{R}_j, \mathbf{Z}_{j-1}, \mathbf{L}_{j-1})}$. The factors p_A would be the probabilities of having program participation $A_j = a_j$ at each time, under the individual's covariate (diabetes education and health status) and mediator history, and the factors p_Z are the conditional probabilities of nutritional service utilization at each time, given the individual's observed covariates and utilization history, and under the two program exposures considered.

Consistency of ψ_n^{IPW} relies on consistency of p_A and p_Z. As noted in Sect. 17.2, if Z is high dimensional, we may replace estimation of the densities p_Z with estimation of the conditional probabilities $\gamma^{\cdot,s,t}$. These can be estimated by regressing A_s onto the corresponding independent variables, for every pair (s,t). This way, using (17.10), we can rewrite

$$D^{IPW,\mathbf{a},\mathbf{a}'}(p_A, p_Z) = D^{IPW,\mathbf{a},\mathbf{a}'}(p_A, \gamma)$$

$$= Y_\tau \frac{I(\mathbf{A}_\tau \equiv \mathbf{a}_\tau)}{\prod_{j\equiv 1}^{\tau} p_A\left(a_j \mid \mathbf{a}_{j-1}, \mathbf{R}_{j-1}, \mathbf{Z}_{j-1}, \mathbf{L}_{j-1}\right)}$$

$$\times \prod_{j\equiv 1}^{\tau} \prod_{s\equiv 1}^{j} \frac{\gamma^{1,s,j}(a'_s \mid \mathbf{a}'_{s-1}, \mathbf{R}_j \mathbf{Z}_j, \mathbf{L}_{j-1})}{\gamma^{1,s,j}(a_s \mid \mathbf{a}_{s-1}, \mathbf{R}_j \mathbf{Z}_j, \mathbf{L}_{j-1})} \frac{\gamma^{2,s,j}(a_s \mid \mathbf{a}_{s-1}, \mathbf{R}_j, \mathbf{Z}_{j-1}, \mathbf{L}_{j-1}}{\gamma^{2,s,j}(a'_s \mid \mathbf{a}'_{s-1}, \mathbf{R}_j, \mathbf{Z}_{j-1}, \mathbf{L}_{j-1})}. \tag{17.17}$$

If a correct parametric model is specified for p_A and p_Z, then the IPW estimator is an asymptotically linear estimator; its influence curve involves $D^{IPW,\mathbf{a},\mathbf{a}'}$ plus a first-order residue due to estimation of p_A and p_Z, which is part of the parameter definition. In this case, the influence curve for this residual term can be derived using the Delta method. In the more reasonable setting that one uses a data-adaptive estimator of p_A and p_Z, the IPW-estimator is not expected to be asymptotically linear. As shown in van der Laan (2014b), it would be necessary to target the estimators of p_A and p_Z (as in TMLE), while important, this is beyond the scope of this chapter. We will approximate $\sqrt{n}\left(\psi_n^{IPW} - \Psi^{\mathbf{a},\mathbf{a}'}(P_0)\right)$ by the sample variance $\hat{\text{Var}}\, D^{IPW,\mathbf{a},\mathbf{a}'}(p_{n,A}, p_{n,Z})$.

Due to its inverse weighting by treatment and censoring probabilities, this estimator is particularly sensitive to near positivity violations. In particular, if the outcome of interest has a bounded range, the IPW estimator is not guaranteed to stay within this range when the inverse weights become large. Substitution estimators like the nontargeted estimator in (17.14) and the next estimator, TMLE, can mitigate this problem partly by incorporating global information in the parameter map, however, but the effect of near positivity violations still takes form of poor smoothing in these estimators.

17.3.3 TMLE

To maximize finite sample gain and provide more stable estimates in the presence of near positivity violations, one can make use of the substitution principle. TMLE provides a substitution-based estimator which also satisfies the EIC equation, thereby remaining unbiased under model mis-specifications. In a glimpse, our strategy consists of targeted update the initial estimators $\bar{Q}_n^{\mathbf{a},\mathbf{a}'}$ of $\bar{Q}^{\mathbf{a},\mathbf{a}'}(P_0)$ by minimizing a pre-specified loss along a least favorable (with respect to $\Psi^{\mathbf{a},\mathbf{a}'}(P_0)$) submodel through $\bar{Q}_n^{\mathbf{a},\mathbf{a}'}$, then we obtain a substitution estimator of the parameter by evaluating $\Psi^{\mathbf{a},\mathbf{a}'}$ at the updated estimator $\bar{Q}_n^{*,\mathbf{a},\mathbf{a}'}$. A byproduct of this updating procedure is that the $\bar{p}_n$ and $\bar{Q}_n^{*,\mathbf{a},\mathbf{a}'}$ satisfy $P_n D^{*,\mathbf{a},\mathbf{a}'}\left(\bar{Q}_n^{*,\mathbf{a},\mathbf{a}'}, \bar{p}_n\right) = 0$, and hence the estimator $\Psi^{\mathbf{a},\mathbf{a}'}(\bar{Q}_n^{*,\mathbf{a},\mathbf{a}'})$ is multiply robust, as specified in Theorem 17.1 and Corollary 17.2.

From the nested relationships noted (17.13), to update estimators of $\bar{Q}_{L_t}^{\mathbf{a},\mathbf{a}'}$, $\bar{Q}_{Z_t}^{\mathbf{a},\mathbf{a}'}$ and $\bar{Q}_{R_t}^{\mathbf{a},\mathbf{a}'}$, we will use for the loss functions

$$L(\bar{Q}_{L_t}^{\mathbf{a},\mathbf{a}'}) \equiv -\left\{\bar{Q}_{R_{t+1}}^{\mathbf{a},\mathbf{a}'} \log\left(\bar{Q}_{L_t}^{\mathbf{a},\mathbf{a}'}\right) + (1 - \bar{Q}_{R_{t+1}}^{\mathbf{a},\mathbf{a}'})\log\left(1 - \bar{Q}_{L_t}^{\mathbf{a},\mathbf{a}'}\right)\right\},$$

$$L(\bar{Q}_{Z_t}^{\mathbf{a},\mathbf{a}'}) \equiv -\left\{\bar{Q}_{L_t}^{\mathbf{a},\mathbf{a}'} \log\left(\bar{Q}_{Z_t}^{\mathbf{a},\mathbf{a}'}\right) + (1 - \bar{Q}_{L_t}^{\mathbf{a},\mathbf{a}'})\log\left(1 - \bar{Q}_{Z_t}^{\mathbf{a},\mathbf{a}'}\right)\right\},$$

$$L(\bar{Q}_{R_t}^{\mathbf{a},\mathbf{a}'}) \equiv -\left\{\bar{Q}_{Z_t}^{\mathbf{a},\mathbf{a}'} \log\left(\bar{Q}_{R_t}^{\mathbf{a},\mathbf{a}'}\right) + (1 - \bar{Q}_{Z_t}^{\mathbf{a},\mathbf{a}'})\log\left(1 - \bar{Q}_{R_t}^{\mathbf{a},\mathbf{a}'}\right)\right\}. \tag{17.18}$$

Recall that upon linear transformation, our outcome is bounded between 0 and 1, and hence these loss functions are well defined. We define the corresponding least favorable submodels through $\bar{Q}_{L_t}^{\mathbf{a},\mathbf{a}'}$, $\bar{Q}_{Z_t}^{\mathbf{a},\mathbf{a}'}$ and $\bar{Q}_{R_t}^{\mathbf{a},\mathbf{a}'}$, respectively, to be

$$\bar{Q}_{L_t}^{\mathbf{a},\mathbf{a}'}(\epsilon) = \text{expit}\left(\text{logit}\bar{Q}_{L_t}^{\mathbf{a},\mathbf{a}'} + \epsilon\right),$$

$$\bar{Q}_{Z_t}^{\mathbf{a},\mathbf{a}'}(\epsilon) = \text{expit}\left(\text{logit}\bar{Q}_{Z_t}^{\mathbf{a},\mathbf{a}'} + \epsilon\right),$$

$$\bar{Q}_{R_t}^{\mathbf{a},\mathbf{a}'}(\epsilon) = \text{expit}\left(\text{logit}\bar{Q}_{R_t}^{\mathbf{a},\mathbf{a}'} + \epsilon\right), \tag{17.19}$$

and note that

$$\frac{\partial}{\partial \epsilon} H_{L_t}^{\mathbf{a},\mathbf{a}'}(p_A, p_Z) L\left(\bar{Q}_{L_t}^{\mathbf{a},\mathbf{a}'}(\epsilon)\right)\bigg|_{\epsilon=0} = D_{L_t}^{\mathbf{a},\mathbf{a}'}(\bar{Q}^{\mathbf{a},\mathbf{a}'}, p_A, p_Z),$$

$$\frac{\partial}{\partial \epsilon} H_{Z_t}^{\mathbf{a},\mathbf{a}'}(p_A, p_L, p_R) L\left(\bar{Q}_{Z_t}^{\mathbf{a},\mathbf{a}'}(\epsilon)\right)\bigg|_{\epsilon=0} = D_{Z_t}^{\mathbf{a},\mathbf{a}'}(\bar{Q}^{\mathbf{a},\mathbf{a}'}, p_A, p_L, p_R),$$

$$\frac{\partial}{\partial \epsilon} H_{R_t}^{\mathbf{a},\mathbf{a}'}(p_A, p_Z) L\left(\bar{Q}_{R_t}^{\mathbf{a},\mathbf{a}'}(\epsilon)\right)\bigg|_{\epsilon=0} = D_{R_t}^{\mathbf{a},\mathbf{a}'}(\bar{Q}^{\mathbf{a},\mathbf{a}'}, p_A, p_Z).$$

We are now ready to describe the TMLE algorithm, which will targets the estimation of $\bar{Q}_{L_t}^{\mathbf{a},\mathbf{a}'}$, $\bar{Q}_{Z_t}^{\mathbf{a},\mathbf{a}'}$ and $\bar{Q}_{R_t}^{\mathbf{a},\mathbf{a}'}$ sequentially in order of decreasing t.

1. Obtain estimators $p_{n,A}, p_{n,Z}, p_{n,L}, p_{n,R}$ (if high dimensional settings, estimation of $p_{n,Z}, p_{n,L}, p_{n,R}$ can be replaced with the estimators γ_n for γ defined in Sect. 17.2). These estimators will be used to obtain estimates $H_{n,L_t}^{\mathbf{a},\mathbf{a}'}$, $H_{n,Z_t}^{\mathbf{a},\mathbf{a}'}$ and $H_{n,R_t}^{\mathbf{a},\mathbf{a}'}$, see (17.10), (17.11) and (17.12).
2. Initiate $\bar{Q}_{R_{\tau+1}}\mathbf{a},\mathbf{a}' \equiv Y_\tau$.
3. At each $t = \tau, \ldots 1$, in decreasing order, we have obtained targeted estimator $\bar{Q}_{n,R_{t+1}}^{*,\mathbf{a},\mathbf{a}'}$ from a previous step. We now obtain targeted estimator $\bar{Q}_{n,L_t}^{*,\mathbf{a},\mathbf{a}'}$, $\bar{Q}_{n,Z_t}^{*,\mathbf{a},\mathbf{a}'}$, and $\bar{Q}_{n,R_t}^{*,\mathbf{a},\mathbf{a}'}$, in that order, as follows:

(a) Regress $\bar{Q}_{n,R_{t+1}}^{*,\mathbf{a},\mathbf{a}'}(\mathbf{R}_t, \mathbf{Z}_t, \mathbf{L}_t)$ on observed values $(\mathbf{A}_t, \mathbf{R}_t, \mathbf{Z}_t, \mathbf{L}_{t-1})$ among observations that remained uncensored at time t. We then evaluate the fitted function at the observed mediator and covariates histories $\mathbf{R}_t, \mathbf{Z}_t, \mathbf{L}_{t-1}$ and the intervened exposure $\mathbf{A}_t = \mathbf{a}_t$ for these uncensored observations. This results in the estimates $\bar{Q}_{n,L_t}^{\mathbf{a},\mathbf{a}'}(\mathbf{R}_t, \mathbf{Z}_t, \mathbf{L}_{t-1})$ for those uncensored individuals. Update this estimate using $\bar{Q}_{n,L_t}^{*,\mathbf{a},\mathbf{a}'} \equiv \bar{Q}_{n,L_t}^{\mathbf{a},\mathbf{a}'}(\epsilon_{n,L_t})$, where

$$\epsilon_{n,L_t} \equiv \arg\min_\epsilon P_n H_{L_t}^{\mathbf{a},\mathbf{a}'}(\bar{p}_n) L\left(\bar{Q}_{n,L_t}^{\mathbf{a},\mathbf{a}'}(\epsilon)\right).$$

This ϵ_{n,L_t} is the coefficient of a weighted logistic regression of the expectant $\bar{Q}_{n,R_{t+1}}^{*,\mathbf{a},\mathbf{a}'}(\mathbf{R}_t, \mathbf{Z}_t, \mathbf{L}_t)$ on the intercept model with an offset $\mathrm{logit}\big(\bar{Q}_{n,L_t}^{\mathbf{a},\mathbf{a}'}(\mathbf{R}_t, \mathbf{Z}_t, \mathbf{L}_{t-1})\big)$, and weights $H_{L_t}^{\mathbf{a},\mathbf{a}'}(\bar{p}_n)(\mathbf{A}_t, \mathbf{R}_t, \mathbf{Z}_t, \mathbf{L}_{t-1})$.

(b) Next, to obtain an initial estimator $\bar{Q}_{n,Z_t}^{\mathbf{a},\mathbf{a}'}$, we regress the targeted estimate $\bar{Q}_{n,L_t}^{*,\mathbf{a},\mathbf{a}'}(\mathbf{R}_t, \mathbf{Z}_t, \mathbf{L}_{t-1})$ constructed above on $\mathbf{A}_t, \mathbf{R}_t, \mathbf{Z}_{t-1}, \mathbf{L}_{t-1}$ among observations that remained uncensored at time t. We then evaluate the fitted function at the observed mediator and covariate histories $\mathbf{R}_t, \mathbf{Z}_{t-1}, \mathbf{L}_{t-1}$ and the intervened income levels $\mathbf{A}_t = \mathbf{a}'_t$ for these uncensored observations. This results in the initial estimates $\bar{Q}_{n,Z_t}^{\mathbf{a},\mathbf{a}'}(\mathbf{R}_t, \mathbf{L}_{t-1}, \mathbf{Z}_{t-1})$. Update this estimate using $\bar{Q}_{n,Z_t}^{*,\mathbf{a},\mathbf{a}'} \equiv \bar{Q}_{n,Z_t}^{\mathbf{a},\mathbf{a}'}(\epsilon_{n,Z_t})$, where

$$\epsilon_{n,Z_t} \equiv \arg\min_\epsilon P_n H_{Z_t}^{\mathbf{a},\mathbf{a}'}(\bar{p}_n) L\left(\bar{Q}_{n,Z_t}^{\mathbf{a},\mathbf{a}'}(\epsilon)\right).$$

(c) Finally, to obtain an initial estimator $\bar{Q}_{n,R_t}^{\mathbf{a},\mathbf{a}'}$, we regress the targeted estimate $\bar{Q}_{n,Z_t}^{*,\mathbf{a},\mathbf{a}'}(\mathbf{R}_t, \mathbf{Z}_{t-1}, \mathbf{L}_{t-1})$ constructed above on $\mathbf{A}_t, \mathbf{R}_{t-1}, \mathbf{Z}_{t-1}, \mathbf{L}_{t-1}$ among observations that remained uncensored at time t. We then evaluate the fitted function at the observed mediator and covariate histories $\mathbf{R}_{t-1}, \mathbf{Z}_{t-1}, \mathbf{L}_{t-1}$ and the intervened income levels $\mathbf{A}_t = \mathbf{a}_t$ for these uncensored observations. This results in the initial estimates $\bar{Q}_{n,R_t}^{\mathbf{a},\mathbf{a}'}(\mathbf{R}_{t-1}, \mathbf{L}_{t-1}, \mathbf{Z}_{t-1})$. Update this estimate using $\bar{Q}_{n,R_t}^{*,\mathbf{a},\mathbf{a}'} \equiv \bar{Q}_{n,R_t}^{\mathbf{a},\mathbf{a}'}(\epsilon_{n,R_t})$, where

$$\epsilon_{n,R_t} \equiv \arg\min_\epsilon P_n H_{R_t}^{\mathbf{a},\mathbf{a}'}(\bar{p}_n) L\left(\bar{Q}_{n,R_t}^{\mathbf{a},\mathbf{a}'}(\epsilon)\right).$$

4. After running the algorithm in step (3) sequentially from $t = \tau$ down to $t = 1$, we have targeted estimates $\bar{Q}_{n,R_1}^{*,\mathbf{a},\mathbf{a}'}(L_0)$ for each of the n observations.

The TMLE for $\Psi^{\mathbf{a},\mathbf{a}'}$ is given by

$$\psi_n^{TMLE} \equiv \Psi^{\mathbf{a},\mathbf{a}'}(\bar{Q}_n^{*,\mathbf{a},\mathbf{a}'}) = \frac{1}{n}\sum_{i=1}^{n}\bar{Q}_{n,R_1}^{*,\mathbf{a},\mathbf{a}'}(L_{0,i}) \qquad (17.20)$$

By construction, this estimator satisfies $P_n D^{*,\mathbf{a},\mathbf{a}'}(\bar{Q}_n^{*,\mathbf{a},\mathbf{a}'}, \bar{p}_n) = 0$. Consequently, it inherits robustness of EIC described in Sect. 17.2. Under the usual empirical process condition on the estimated efficient influence curve and that the second-order remainder is $o_P(n^{-1/2})$ (i.e., all nuisance parameters are consistently estimated), it is asymptotically linear and efficient at the data-generating P_0, with influence curve $D^{*,\mathbf{a},\mathbf{a}'}(P_0)$. We could estimate the asymptotic variance of $\sqrt{n}\left(\psi_n^{TMLE} - \Psi^{\mathbf{a},\mathbf{a}'}(P_0)\right)$ with the sample variance $\widehat{\mathrm{Var}}D^{*,\mathbf{a},\mathbf{a}'}\left(\bar{Q}_n^{*,\mathbf{a},\mathbf{a}'}, \bar{p}_n\right)$. It is double robust w.r.t. estimation, but its asymptotic linearity (as with the IPW estimator) will be affected by inconsistency of one of the estimators (van der Laan 2014b).

17.4 Simulation

We conduct a simulation to evaluate the comparative performance of these three estimators in estimating the mediation formula (17.5) for survival outcome Y_τ. Consider the data structure $O = (L_0, A_1, R_1, Z_1, L_1, \ldots, A_\tau, R_\tau, Z_\tau, L_\tau)$, with $\tau = 2$. L_0 encodes two baseline covariates L_{01} and L_{02}, A_t encodes a censoring indicator A_t^C of whether patient remained in the study by time t and a binary exposure A_t^E, R_t encodes covariates at time t that are directly affected by A_t and may influence Z_t and L_t, Z_t is a binary mediator of interest, L_t includes a time varying covariate L_t^1 and a death indicator Y_t of whether patient had died by time t. These variables are distributed according to the following data generating distribution

$$L_{01} \sim Bern(0.4); L_{02} \sim Bern(0.6);$$

$$A_t^C \sim Bern\Big(\text{expit}\big(1.5 - 0.8L_{02} - 0.4\big(I(t > 1) \times L_{t-1}^1 + I(t = 1)L_{01}\big)$$
$$+ 0.5I(t > 1)A_{t-1}^E\big)\Big)$$

$$A_t^E \sim Bern\Big(\text{expit}\big(-0.1 + 1.2L_{02} + 0.7\big(I(t > 1) \times L_{t-1}^1 + I(t = 1)L_{01}\big)$$
$$- 0.1I(t > 1)A_{t-1}^E\big)\Big)$$

$$R_t \sim Bern\Big(\text{expit}\big(-0.8 + A_t^E + 0.1\big(I(t > 1) \times L_{t-1}^1 + I(t = 1)L_{01}\big)$$
$$+ 0.3\big(I(t > 1) \times R_{t-1} + I(t = 1)L_{02}\big)\big)\Big)$$

$$Z_t \sim Bern\Big(\text{expit}\big(-0.5 + 0.8L_{02} + 0.8A_t^E + R_t\big)\Big)$$

$$L_t^1 \sim Bern\left(\text{expit}(-1 + 0.3L_{02} + A_t^E + 0.7Z_t - 0.2I(t > 1)L_{t-1}^1)\right)$$

$$Y_t \sim Bern\left(\text{expit}(0.2 + 1.5L_{02} + R_t + 0.2L_t^1 - 0.3A_t^E - 0.3Z_t - 0.2A_t^E \times Z_t\right.$$
$$\left. - 0.1I(t > 1)R_{t-1})\right).$$

Table 17.1 Bias, variance and MSE over 1000 simulations

		Bias		Var	
	n	500	5000	500	5000
All correct	TMLE	6.57e−04	1.87e−05	1.32e−03	1.01e−04
	IPW	1.12e−03	4.63e−05	1.24e−03	1.05e−04
	NTsub	8.42e−04	7.53e−04	9.81e−04	7.93e−05
$\bar{Q}^{a,a'}$ correct	TMLE	1.02e−03	2.61e−04	1.35e−03	1.15e−04
	IPW	4.20e−03	6.33e−03	1.86e−03	1.78e−04
L misspec.	TMLE	4.63e−04	3.11e−05	1.22e−03	1.02e−04
	IPW	1.12e−03	4.63e−05	1.24e−03	1.05e−04
	NTsub	5.28e−03	5.85e−03	6.64e−04	6.30e−05
Z misspec.	TMLE	1.41e−02	5.61e−03	3.02e−03	2.51e−04
	IPW	1.22e−02	1.46e−02	2.37e−03	2.12e−04
	NTsub	7.66e−02	7.73e−02	3.53e−03	2.93e−04
$\bar{Q}^{a,a'}$ misspec.	TMLE	1.90e−04	7.73e−05	1.25e−03	1.02e−04
	NTsub	6.86e−03	6.95e−03	7.33e−04	7.05e−05

After either censoring or death, all subsequent variables take a default value. The target parameter of interest is $\Psi^{1,0}(P_0) \approx 0.912$. To obtain this approximate, we first generated a large sample (1,000,000 observations) using the above distributions by setting $A_t^E = 0$ in the equation for Z_t and $A_t^E = 1$ elsewhere, as well as assigning the indicator $A_t^C = 1$ (i.e. all remain in study), and then take the sample mean outcome Y_τ in this large sample.

Correctly specified conditional probabilities $\bar{p} = (p_A, p_Z, p_R, p_L)$ are obtained using logistic regressions as specified in the data-generating distributions. We estimate $\bar{Q}_{L_t}^{a,a'}$, $\bar{Q}_{Z_t}^{a,a'}$ and $\bar{Q}_{R_t}^{a,a'}$ using the regression-based approach: the so-called correctly specified estimators are obtained using super learner to regress the expectant on all the parent exposure, mediator and covariate history up to point t, described in steps in Sects. 17.3.1 and 17.3.3. The misspecified counterparts of $\bar{p}$ and $\bar{Q}^{a,a'}$ only regress on A_t, in the case of nuisance parameters related to Z_t and R_t, or only regress on A_t and Z_t, in the case of nuisance parameters related to L_t, or uses in marginal distribution, in the case of p_A. We note here that due to the nature of nested regressions, the so-called 'correct' $\bar{Q}^{a,a'}$ do not actually specify the functional form of these expectations, they only adjust for all the relevant terms; this abuse of terminology is meant to contrast with the estimators which omit important covariates. The super learner was implemented using a very limited default library of candidate algorithms, which include glm, stepAIC, bayesglm, each coupled with a correlation-based variable screening method, as well as a no variable screening version. These algorithms rely heavily on main term functional forms and are thus not flexible enough to do a good job in estimation the true functional forms We implement the nontargeted substitution estimator (NTsub), IPW estimator, and TMLE using these nuisance parameter specifications. We considered sample sizes $n = 400$ and $n = 4000$. Bias, variance and mean squared error (MSE) for each sample size were estimated over the 1000 datasets and presented in Table 17.1.

As predicted by general robustness conditions in Corollary 17.2, when the nested regressions $\bar{Q}^{\mathbf{a},\mathbf{a}'}$ are correct and the conditional probabilities are misspecified ("$\bar{Q}^{\mathbf{a},\mathbf{a}'}$ correct"), TMLE provides bias reduction over a misspecified IPW. Similarly, when only the covariate-related nuisance parameters (p_R, p_L and $\bar{Q}_{R_t}^{\mathbf{a},\mathbf{a}'}$, $\bar{Q}_{L_t}^{\mathbf{a},\mathbf{a}'}$) are misspecified ("$L$ misspec."), TMLE also provides substantial bias reduction over the misspecified IPW or NTsub Estimators. When only the Z_t-related nuisance parameters (p_Z and $\bar{Q}_{Z_t}^{\mathbf{a},\mathbf{a}'}$) ("Z misspec."), TMLE provides bias reduction over the misspecified NTsub estimator across sample sizes, but its bias reduction over IPW is only apparent after large sample sizes. When $\bar{Q}^{\mathbf{a},\mathbf{a}'}$ are all misspecified, but correct conditional probabilities p_A, p_Z, p_L, p_R are used in the weights $H^{\mathbf{a},\mathbf{a}'}$ of the TMLE updating step, we still observe bias reduction of TMLE over NTsub in this example.

We recall that the correctly specified $\bar{Q}^{\mathbf{a},\mathbf{a}'}$ in this implementation are only correct up to specification of key main terms, but not functional form, as that is difficult to implement with the sequential regression approach (we could implement it by estimating the actual conditional densities in the likelihood) or would require a much more computer intensive super learner relying on highly data-adaptive estimators (something that one should do for a single data analysis). This probably resulted in certain loss of finite sample gain for TMLE under 'all correct' specifications, as it was expected to be more efficient than the IPW, and this efficiency gain was not apparent until larger sample sizes. By the same argument, this may also have contributed to speed of convergence of the corresponding NTsub estimator.

17.5 Discussion

In this chapter, we proposed a random interventions approach to formulate parameters of interest in longitudinal mediation analysis with time varying mediator and exposures. Specifically, we defined the random interventions based on conditional distributions for the mediator. In comparison to an alternative random interventions formulation based on marginal distributions of the mediator (VanderWeele and Tchetgen Tchetgen 2017), the proposed formulation capture different pathways of mediated effect and allows for application in survival settings, but it also trades-off stronger sequential randomization assumptions.

Under the RI formulation, the treatment of interest as well as the mediator variables are regarded as intervention variables. Under the proposed formulation, one can obtain a total effect decomposition and the subsequent definition of natural direct and indirect effects that are analogous to those in Pearl (2001). The natural direct effect under this formulation has an intrinsic interpretation as a weighted average of controlled direct effects, because controlled direct effects can be considered as a deterministic intervention on the treatment and mediator variables. By regarding the mediator variables as intervention variables, the RI formulation requires external specification of a counterfactual mediator distribution. It is important to note that causal mediation, under either RI or non-RI approaches, presupposes that the mediator of interest is amenable to external manipulation. In applications where such manipulations are not conceivable, we should be cautious that causal mediation can only offer answers to purely mechanistic questions defined under hypothetical experiments.

The second contribution of this paper is a general semiparametric inference framework for the resulting effect parameters. More specifically, efficient influence curves under a locally saturated semiparametric model are derived, and their robustness properties are established. In many applications where the mediator densities are difficult to estimate, regression-based estimators of these iterated expectations are viable alternatives to substitution-based estimators that rely on consistent estimation of the mediator densities. We also developed the nontargeted substitution estimator, IPW estimator and TMLE for the mediational functional.

Part V
Online Learning

Chapter 18
Online Super Learning

Mark J. van der Laan and David Benkeser

We consider the case that the data $O^n = (O(1), \ldots, O(n)) \sim P_0^n$ is generated sequentially by a conditional probability distribution $P_{\theta,t}$ of $O(t)$, given certain summary measures of the past $\bar{O}(t-1) = (O(1), \ldots, O(t-1))$, and where this t-specific conditional probability distribution $P_{\theta,t}$ is identified by a common parameter $\theta \in \Theta$. For example, the experiment at time t generates a new observation $O(t)$ from a probability distribution $\bar{P}_0(\cdot \mid z)$ determined by a fixed dimensional summary measure Z of the past $O(1), \ldots, O(t-1)$, and one would assume that this conditional distribution $\bar{P}_0$ is an element of some semiparametric model. An important special case is that the sample can be viewed as independent and identically distributed observations from a fixed data generating distribution that is known to belong to some semiparametric statistical model, such as the nonparametric model: in this case $\bar{P}_0(\cdot \mid z) = \bar{P}_0(\cdot)$ does not depend on the past. Another important case is that the data is generated by a group sequential adaptive design in which the randomization and or censoring mechanism is a function of summary measures of the observed data on previously sampled groups (van der Laan 2008b; Chambaz and van der Laan 2011a,b,c). More generally, this covers a whole range of time series models.

M. J. van der Laan (✉)
Division of Biostatistics and Department of Statistics, University of California, Berkeley, 101 Haviland Hall, #7358, Berkeley, CA 94720, USA
e-mail: laan@berkeley.edu

D. Benkeser
Department of Biostatistics and Bioinformatics, Rollins School of Public Health, Emory University, 1518 Clifton Rd. NE, Atlanta, GA 30322, USA
e-mail: benkeser@emory.edu

© Springer International Publishing AG 2018
M.J. van der Laan, S. Rose, *Targeted Learning in Data Science*,
Springer Series in Statistics, https://doi.org/10.1007/978-3-319-65304-4_18

The goal is to develop an online estimator of a particular infinite dimensional target parameter of this common mechanism θ, such as a conditional density or regression function. An online estimator is an estimator that updates a current estimator with a new incoming batch of data without having to revisit the past data. Such estimators are scalable to big data and can provide fast updates over time.

The current literature provides such online estimators for parametric models and independent and identically distributed observations, where the online estimators are based on variations of the stochastic gradient descent algorithm (e.g., Bottou 2012, and see Appendix A.3 for a succinct review of these methods). However, we wish to allow for large infinite dimensional models for $\bar{P}_0$ so that we need to develop a highly data-adaptive online estimator. For i.i.d. data the literature provides a super-learner based on cross-validation. Due to the oracle inequality, it follows that such an estimator is asymptotically at least as good as the best estimator in the library of candidate estimators (van der Laan and Dudoit 2003; van der Vaart et al. 2006; van der Laan et al. 2006, 2007; Polley et al. 2011). Due to this oracle property of this ensemble learner we referred to such an estimator as a super learner. In this article we extend this standard cross-validation approach to an online cross-validation selector for dependent time series data as described above.

In this chapter we will propose a class of online cross-validation selectors that select among a library of candidate online estimators such as parametric online estimators indexed by different parametric models. We establish an oracle inequality for the resulting cross-validation selector and show that it proves that the cross-validation selector is asymptotically optimal by being asymptotically equivalent with the oracle selector, under specified conditions. We also extend this cross-validation selector to an infinite family of candidate estimators such as all convex combinations of a set of candidate estimators. We refer to the resulting online estimator as the online super learner since it is guaranteed to do asymptotically as well or better as the best among the class of candidate online estimators and is itself an online estimator.

These online estimators and online super learner can also be applied to a large i.i.d. data set by partitioning the data set in many small subsets and enforcing an ordering of these subsets. However, for i.i.d. the choice of ordering is random. Therefore, for i.i.d. data we extend the online cross-validation selector and its oracle inequality to an average across different orderings of the data set, and propose V specific orderings to make it comparable in performance with a classical V-fold cross-validation selector.

18.1 Statistical Formulation of Estimation Problem

In this section we first formulate the statistical model for the time-series data, and the target parameter of interest for which we wish to develop an online super learner. We also present an example.

18.1.1 Statistical Model

Suppose we observe at each t, a random variable $O(t)$, $t = 1, \ldots, n$. For example, $O(t) = (W(t), Y(t))$ where $W(t)$ is vector of baseline covariates and $Y(t)$ is an outcome measured on a unit. We do not assume that the random variables $O(t)$ are independent across t. Let P_0^n be the true probability distribution of $O(1), \ldots, O(n)$, and let p_0^n be its density w.r.t. a dominating measure μ^n. We factorize the density of $O(1), \ldots, O(n)$ according to the ordering, so that $O(t)$ is drawn from a conditional distribution of $O(t)$, given $O(1), \ldots, O(t-1)$, $t = 1, \ldots, n$, and we make the following conditional independence assumption on these conditional distributions:

$$O(t) \text{ is independent of } \bar{O}(t-1), \text{ given } Z(t), \tag{18.1}$$

where $Z(t) = f_t(\bar{O}(t-1)) \in \mathbf{R}^d$ is a specified d-dimensional extraction from $O(1), \ldots, O(t-1)$ and d is common to all t, $t = 1, \ldots, n$. Let $\mathcal{Z}$ and O be sets so that $P_0^n(O(t) \in O, Z(t) \in \mathcal{Z}) = 1$ for all t.

Let $(t, \theta, z) \to P_{\theta,t}(\cdot \mid z)$ be a specified parameterization that maps a (t, θ, z) into a conditional probability distribution on O. We assume that for each $t = 1, \ldots, n$, we have

$$P_0^n(O(t) \in S \mid Z(t) = z) = P_{\theta_0,t}(S \mid z) \text{ for a } \theta_0 \in \Theta. \tag{18.2}$$

Let $p_{\theta,t}(\cdot \mid z)$ denote the conditional density of $P_{\theta,t}(\cdot \mid z)$ w.r.t. some dominating measure μ. For notational convenience, we will also use the notation $P_{\theta,t,z}$ for the probability distribution $P_{\theta,t}(\cdot \mid z)$ at a particular $z \in \mathcal{Z}$. Let P_θ^n be the probability distribution of O^n implied by $\theta \in \Theta$. This defines now our statistical model for the probability distribution P_0^n of O^n:

$$\mathcal{M}^n = \{P_\theta^n : \theta \in \Theta\}.$$

18.1.2 Statistical Target Parameter and Loss Function

Our statistical target parameter $\Psi^n : \mathcal{M}^n \to \mathbf{\Psi}$ is defined by $\Psi^n(P_\theta^n) = \Psi(\theta)$ for a $\Psi : \Theta \to \mathbf{\Psi}$ for some parameter space $\mathbf{\Psi}$.

Let $(Z, O, \psi) \to L(\psi)(Z, O)$ be a loss function for ψ_0 so that for all $z \in \mathcal{Z}$ and t,

$$P_{\theta_0,t,z} L(\psi_0) = \arg \min_{\psi \in \mathbf{\Psi}} P_{\theta_0,t,z} L(\psi),$$

where we used the notation $P_{\theta_0,t,z} L(\psi) = \int L(\psi)(z, o) dP_{\theta_0,t,z}(o)$.

18.1.3 Regression Example

For the purpose of this chapter, let's have in mind the following example. Suppose that $O(t) = (W(t), Y(t))$. Let's assume that for each t, $P_{o(t)}(\cdot \mid \bar{O}(t - 1)) = \bar{P}(\cdot \mid Z(t))$ for some common conditional probability distribution $\bar{P} \in \mathcal{M}$ and summary measure $Z(t)$. Let $\mathcal{M}$ be a nonparametric model. This model implies $E_0(Y(t) \mid W(t), \bar{O}(t - 1)) = E_0(Y(t) \mid W(t), Z(t))$ and that $E_0(Y(t) \mid W(t) = w, Z(t) = z) = \psi_0(w, z)$ for some common function ψ_0. Let $\bar{P}_{0,y|w,z}$ be the conditional distribution of Y, given $W = w, Z = z$ under $\bar{P}_0$, and let $\bar{P}_{0,z}(\cdot) = \bar{P}_0(\cdot \mid z)$ be the conditional probability distribution of (W, Y), given $Z = z$. Suppose that our target parameter is given by ψ_0. A possible loss function is the squared error loss:

$$L(\psi)(Z, W, Y) = (Y - \psi(Z, W))^2.$$

For a given w, z,

$$E_0((Y(t) - \psi(Z(t), W(t)))^2 \mid W(t) = w, Z(t) = z) = \bar{P}_{0,y|w,z}L(\psi)$$

is minimized by $\psi_0(z, w)$. In particular, this implies that for all $z \in \mathcal{Z}$, $\bar{P}_{0,z}L(\psi)$ is minimized by ψ_0. Alternatively, if Y is binary, we could have selected the log-likelihood loss:

$$L(\psi)(Z, W, Y) = - \{Y \log \psi(Z, W) + (1 - Y) \log(1 - \psi(Z, W))\} .$$

18.2 Cross-Validation for Ordered Sequence of Dependent Experiments

In this section we define an online cross-validation selector for a discrete set of candidate estimators, analogue to a typical cross-validation method used for time-series data. We also define the corresponding oracle selector. Finally, we define an online cross-validation selector for a continuous family of candidate estimators.

18.2.1 Online Cross-Validation Selector

Let $\hat{\Psi}_k$ be candidate estimators that can be applied to data sets $(Z(i), O(i))$ for $i = 1, \ldots, m$ for $m \leq n$, $k = 1, \ldots, K$. For a given point $t_0 \in \{n_l + 1, \ldots, n\}$, we define $\mathcal{T}(t_0) = \{i : i < t_0\}$ as the training sample, and the singleton $O(t_0)$ as the validation sample. Note that the t_0-specific training sample is defined as the past before time t_0, and the t_0-specific validation sample is defined as the next observation. Here n_l is a minimal sample size one requires for the training sample. We consider such a split for each time point t_0 that is larger than some minimal required training sample size n_l: the total number of splits is given by $n - n_l + 1$. The cross-validation scheme measures how well an algorithm trained on the past is able to predict an outcome at the next time point. For each t_0, let P_{t_0-1} be the empirical distribution of the training sample $\{(Z(i), O(i)) : i = 1, \ldots, t_0 - 1\}$.

Given a candidate estimator $\hat{\Psi}_k$ we define its online cross-validated risk as follows:

$$R_{CV,n}(\hat{\Psi}_k) \equiv \frac{1}{n - n_l + 1} \sum_{t_0=n_l+1}^{n} L(\hat{\Psi}_k(P_{t_0-1}))(Z(t_0), O(t_0)).$$

Notice that if $\hat{\Psi}_k$ is an online estimator, then the online cross-validated risk is itself an online estimator. That is, when a new observation $O(n + 1)$ comes in, we create the new observation $(Z(n + 1), O(n + 1))$, update the online estimator $\hat{\Psi}_k(P_n)$ into $\hat{\Psi}_k(P_{n+1})$, evaluate its loss $L(\hat{\Psi}_k(P_{n+1}))(Z(n + 1), O(n + 1))$, and add it to the current online cross-validated risk $R_{CV,n}(\hat{\Psi}_k)$ to obtain the $R_{CV,n+1}(\hat{\Psi}_k)$.

The corresponding online cross-validation selector is defined as

$$k_n = \arg\min_k R_{CV,n}(\hat{\Psi}_k)$$

$$= \arg\min_k \frac{1}{n - n_l + 1} \sum_{t_0=n_l+1}^{n} L(\hat{\Psi}_k(P_{t_0-1}))(Z(t_0), O(t_0)).$$

In our regression example, if we use the squared error loss, then we have

$$k_n = \arg\min_k \frac{1}{n - n_l + 1} \sum_{t_0=n_l+1}^{n} (Y_{t_0} - \hat{\Psi}_k(P_{t_0-1})(Z(t_0), W_{t_0}))^2,$$

and similarly we can define the cross-validation selector for the log-likelihood loss.

The online super learner is defined as follows:

$$\hat{\Psi}(P_t) = \hat{\Psi}_{k_t}(P_t), \quad t = 1, \ldots,.$$

Thus, at t observations, it uses the estimator $\hat{\Psi}_k$ with index $k = k_t$. Such an online super learner could switch from one estimator to another estimator over time. If all the candidate estimators are online estimators, then also the selector k_t is an online selector, and as a consequence, this super learner is as scalable as any of the candidate estimators. Of course, a super learner is trivially parallelized, by running each online learner on its own computer/core.

18.2.2 Online Oracle Selector

Note that the online cross-validated risk estimates the following online cross-validated true risk:

$$\tilde{R}_{CV,n}(\hat{\Psi}_k) = \frac{1}{n - n_l + 1} \sum_{t_0=n_l+1}^{n} P_{\theta_0,t_0,Z(t_0)} L(\hat{\Psi}_k(P_{t_0-1})),$$

which is minimized by ψ_0. The difference between $\tilde{R}_{CV,n}(\hat{\Psi}_k)$ and $\tilde{R}_{CV,n}(\psi_0)$ defines a loss-based dissimilarity between a candidate estimator $\hat{\Psi}_k$ and the true target ψ_0:

$$d_{0n}(\hat{\Psi}_k, \psi_0) = \frac{1}{n - n_l + 1} \sum_{t_0=n_l+1}^{n} P_{\theta_0,t_0,Z(t_0)}\{L(\hat{\Psi}_k(P_{t_0-1})) - L(\psi_0)\}.$$

The online oracle selector is defined as the minimizer of this loss-based dissimilarity:

$$\tilde{k}_n = \arg\min_{k} \frac{1}{n - n_l + 1} \sum_{t_0=n_l+1}^{n} P_{\theta_0,t_0,Z(t_0)}\{L(\hat{\Psi}_k(P_{t_0-1})) - L(\psi_0)\}.$$

In our regression example, when using the squared error loss, this loss-based dissimilarity $d_{0n}(\hat{\Psi}_k, \psi_0)$ equals

$$d_{0n}(\hat{\Psi}_k, \psi_0) = \frac{1}{n - n_l + 1} \sum_{t_0=n_l+1}^{n} \int_w (\hat{\Psi}_k(P_{t_0-1}) - \psi_0)^2(w, Z(t_0)) d\bar{P}_{0,Z(t_0)}(w \mid Z(t_0)).$$

Consider the loss-based dissimilarity for the squared error loss under sampling from the conditional probability distribution $\bar{P}_{0,z}$ of $(W(t_0), Y(t_0))$, given $Z(t_0) = z$:

$$d_{L2,z}(\psi, \psi_0) = \int \{(y - \psi(w, z))^2 - (y - \psi_0(w, z))^2\} d\bar{P}_{0,z}(w, y).$$

Then, we can represent the above loss-based dissimilarity as

$$d_{0n}(\hat{\Psi}_k, \psi_0) = \frac{1}{n - n_l + 1} \sum_{t_0=n_l+1}^{n} d_{L2,Z(t_0)}(\hat{\Psi}_k(P_{t_0-1}), \psi_0).$$

Similarly, for binary Y and the log-likelihood loss, we have

$$d_{0n}(\hat{\Psi}_k, \psi_0) = \frac{1}{n - n_l + 1} \sum_{t_0=n_l+1}^{n} d_{KL,Z(t_0)}(\hat{\Psi}_k(P_{t_0-1}), \psi_0),$$

where

$$d_{KL,z}(\psi, \psi_0) = \bar{P}_{0,z} \log f_{\psi_0,z}/f_{\psi,z},$$

and

$$f_{\psi,z}(y \mid w) = \psi(w, z)^y (1 - \psi(w, z))^{1-y}$$

is the conditional probability distribution of Y, given $W, Z = z$, implied by ψ. In other words, $d_{KL,z}(\psi, \psi_0)$ is the Kullback-Leibler divergence of $\hat{\Psi}_k(\cdot, z)$ and $\psi_0(\cdot, z)$ under the z-specific experiment with probability distribution $\bar{P}_{0,z}$, and $d_{0n}(\hat{\Psi}_k, \psi_0)$ is an average of $d_{KL,z}(\hat{\Psi}_k, \psi_0)$ across a set of z-values $\{Z(t_0) : t_0 = n_l + 1, \ldots, n\}$.

18.2.3 The Online Super Learner for a Continuous Finite Dimensional Family of Candidate Estimators

Consider a given set of K candidate estimators $\hat{\Psi}_k, k = 1, \ldots, K$. Let $\hat{\Psi}_\alpha$ be a combination of these K estimators indexed by a finite dimensional vector of coefficients α. For example, if the parameter space is convex, one could define convex linear combinations $\hat{\Psi}_\alpha = \sum_{k=1}^K \alpha(k) \hat{\Psi}_k$, where $\alpha \in \{x \in \mathbf{R}^K : x(k) \geq 0, \sum_k x(k) = 1\}$. Let $R_{CV,n}(\hat{\Psi}_\alpha)$ be the online cross-validated risk defined above:

$$R_{CV,n}(\hat{\Psi}_\alpha) \equiv \frac{1}{n - n_l + 1} \sum_{t_0=n_l+1}^{n} L(\hat{\Psi}_\alpha(P_{t_0-1}))(Z(t_0), O(t_0)).$$

Let α_n be the cross-validation selector:

$$\alpha_n = \arg\min_\alpha R_{CV,n}(\hat{\Psi}_\alpha).$$

Tracking each online estimator $\hat{\Psi}_\alpha$ for all α only involves tracking the K online estimators $\hat{\Psi}_k$, but α_n is itself not an online estimator since it involves recomputing the minimum for each n. Therefore, we propose to approximate the minimum α_n with a (e.g.,) second-order stochastic gradient descent algorithm, just like we can approximate the MLE with the online stochastic gradient descent algorithm (SGD). For that purpose, let's assume that $L(\hat{\Psi}_\alpha)$ is twice differentiable. Let

$$S_{n,\alpha} = \frac{d}{d\alpha} L(\hat{\Psi}_\alpha(P_{n-1}))$$

be the vector score, and let $S_{n,\alpha}^1 = \frac{d}{d\alpha} S_{n,\alpha}$ the matrix of second derivatives. Given an online estimator α_t, $t = n_l + 1, \ldots,$ let

$$c_n = -\frac{1}{n - n_l + 1} \sum_{t_0=n_l+1}^{n} S_{n,\alpha_{t_0-1}}^1 (Z(t_0), O(t_0)).$$

Let $S_{n,\alpha}^* = c_n^{-1} S_{n,\alpha}$. The second-order SGD online estimator approximating α_n is defined by:

$$\alpha_{n+1}^* = \alpha_n^* + \frac{1}{n+1} S_{n,\alpha_n^*}^* (Z(n+1), O(n+1)).$$

One could refine this step by checking if $R_{CV,n+1}(\alpha_{n+1}^*) \geq R_{CV,n+1}(\alpha_n^*)$, and if not, replacing α_{n+1}^* by a convex linear combination of α_n^* and this candidate α_{n+1}^* for which there is an actual reduction in the online cross-validated risk. By the theoretical results in the above referenced literature on second-order SGD, this online estimator α_n^* can be expected to approximate α_n up to a term $o_P(n^{-1/2})$. Of course, if α is high dimensional, one might want to replace the second-order SGD by a first order SGD.

18.3 An Oracle Inequality for Online Cross-Validation Selector

In this section we present the oracle inequality for two types of loss functions, namely so called quadratic loss-functions for which the loss-based dissimilarity behaves as a square difference between the candidate and true parameter, and other loss functions. Most loss functions are quadratic. One expects such quadratic behavior since the loss-based dissimilarity at a ψ close to ψ_0 corresponds with a local increase at the minimum value ψ_0 at which the first derivative should be zero. For quadratic loss functions we need a remainder that approximately behaves as $1/n$, while for nonquadratic loss functions a remainder that behaves as $1/\sqrt{n}$ suffices. Therefore, just as for the i.i.d. case, we need a separate more involved proof for quadratic loss functions than needed for nonquadratic loss functions (van der Laan and Dudoit 2003; van der Vaart et al. 2006; van der Laan et al. 2006).

18.3.1 Quadratic Loss Functions

We have the following formal theorem comparing the online cross-validation selector with the corresponding oracle selector.

Theorem 18.1. *Consider the above model M^n for the distribution P_0^n of $O^n = (O(1), \ldots, O(n))$, the definition of the target parameter $\Psi^n : M^n \to \Psi$ defined by $\Psi^n(P_\theta^n) = \Psi(\theta)$ where $\Psi : \Theta \to \Psi$, the loss function $L(\psi)(Z, O)$ for $\psi_0 = \Psi(\theta_0)$, and the loss-based dissimilarity $d_{0n}(\hat{\Psi}_k, \psi_0)$. Consider also the above defined online cross-validation selector k_n and online oracle selector $\tilde{k}_n$.*

Assumptions.

A1. *There exist an $M_1 < \infty$ so that*

$$\sup_{\psi \in \Psi} \sup_{i, O(i), Z(i)} \mid L(\psi)(Z(i), O(i)) - L(\psi_0)(Z(i), O(i)) \mid \leq M_1,$$

where the supremum over $Z(i), O(i)$ is taken over a support of the distribution $Z(i), O(i)$.
A2. *There exist an $M_2 < \infty$ so that with probability 1*

$$\sup_{\psi \in \Psi} \frac{P_{\theta_0, i, Z(i)}\{L(\psi) - L(\psi_0)\}^2}{P_{\theta_0, i, Z(i)}\{L(\psi) - L(\psi_0)\}} \leq M_2 < \infty. \tag{18.3}$$

A3. *Assume that there exists a slowly increasing sequence $M_{3n} < \infty$ (e.g., $M_{3n} = \log n$) so that with probability tending to 1, for both $\bar{k}_n = k_n$ and $\bar{k}_n = \tilde{k}_n$, we have*

$$\frac{1}{M_{3n}} < \frac{d_{0n}(\hat{\Psi}_{\bar{k}_n}, \psi_0)}{E_0 d_{0n}(\hat{\Psi}_{\bar{k}_n}, \psi_0)} < M_{3n}.$$

A4.

$$nM_{3n}^{-3} \min_k E_0 d_{0n}(\hat{\Psi}_k, \psi_0) \to \infty \text{ as } n \to \infty.$$

Finite Sample Result. *For any $\delta > 0$, there exists a universal $C(\delta, M_1, M_2) < \infty$ (i.e., universal in n and choice of candidate estimators) so that*

$$d_{0n}(\hat{\Psi}_{k_n}, \psi_0) \leq (1 + 2\delta) d_{0n}(\hat{\Psi}_{\tilde{k}_n}, \psi_0) + Z_n,$$

where $Z_n = Z_{n1} + Z_{n2}$, $P_0^n(Z_{n2} = 0) \to 1$ as $n \to \infty$, and for $n > n_1$ for some $n_1 < \infty$,

$$E_0 Z_{n1} \leq C(\delta, M_1, M_2) \frac{M_{3n}^2 (1 + \log(K(n)))}{n}.$$

*If Assumption **A4** does not hold, then we have*

$$d_{0n}(\hat{\Psi}_{k_n}, \psi_0) = o_P(n^{-1} M_{3n}^3) + o_P(n^{-1} M_{3n}^2 (1 + \log K(n))).$$

Discussion of Assumptions. Assumption A1 states that the loss function has to be uniformly bounded by some constant M_1, uniformly in all possible realizations of $(O(t), Z(t))$ and the candidate estimators of ψ_0. Assumption A2 is precisely the assumption one expects to hold for quadratic uniformly bounded loss functions, as shown in van der Laan and Dudoit (2003) and related articles. Assumption A3 states that the mean 1 centered random variable $d_{0n}(\hat{\Psi}_{k_n}, \psi_0)/E_0 d_{0n}(\hat{\Psi}_{k_n}, \psi_0)$ (and similarly for $\tilde{k}_n$) falls with probability tending to 1 in an interval slowly growing towards its full support $(0, \infty)$. We anticipate that this assumption will hold for any sequence M_{3n} that converges to infinity such as $M_{3n} = \log n$. Assumption A3 is essentially equivalent with assuming that the mean zero centered $\log d_{0n}(\hat{\Psi}_{k_n}, \psi_0) - E_0 \log d_{0n}(\hat{\Psi}_{k_n}, \psi_0)$ falls with probability tending to 1 in an interval $[-\log n, \log n]$. Assumption A4 is not a real assumption since it only affects the precise statement of the result. Given that M_{3n} is a sequence that grows arbitrarily slow to infinity, assumption A4 states that the oracle selected estimator converges to ψ_0 at a rate slower than the rate $1/n$ of an MLE for a correctly specified parametric model. Therefore assumption A4 will typically hold, but either way if somehow one of the candidate estimators converges to the truth at the parametric rate $1/n$, then the online super learner converges at an almost equally fast rate M_{3n}^3/n.

18.3.2 Nonquadratic Loss Functions

For nonquadratic loss functions, the following straightforward theorem can be applied.

Theorem 18.2. *Consider the above model M^n for the distribution P_0^n of $O^n = (O(1), \ldots, O(n))$, the definition of the target parameter $\Psi^n : M^n \to \Psi$ defined by $\Psi^n(P_\theta^n) = \Psi(\theta)$ where $\Psi : \Theta \to \Psi$, the loss function $L(\psi)(Z, O)$ for $\psi_0 = \Psi(\theta_0)$, and* *loss-based dissimilarity $d_{0n}(\hat{\Psi}_k, \psi_0)$. Consider also the above defined online cross-validation selector k_n and online oracle selector $\tilde{k}_n$.*

Assumption.

A1. *There exist an $M_1 < \infty$ so that*

$$\sup_{\psi \in \Psi} \sup_{i, O(i), Z(i)} \mid L(\psi)(Z(i), O(i)) - L(\psi_0)(Z(i), O(i)) \mid \leq M_1,$$

where the supremum over $Z(i), O(i)$ is taken over a support of the distribution $Z(i), O(i)$.

Finite Sample Result. *There exists a universal $C(M_1) < \infty$ (i.e., universal in n and choice of candidate estimators) so that*

$$E_0 d_{0n}(\hat{\Psi}_{k_n}, \psi_0) \leq E_0 d_{0n}(\hat{\Psi}_{\tilde{k}_n}, \psi_0) + C(M_1) \frac{\log^{0.5}(1 + K(n))}{n^{0.5}}.$$

18.4 Special Online-Cross-Validation Selector for Independent Identically Distributed Observations

In this section we consider the case that the observations $O(1), \ldots, O(n)$ are i.i.d. so that $P_{\theta_0, t, z} = \bar{P}_0$ is a common probability distribution in time t that does not depend on summary measures of an observed past, and is an element of some model $\mathcal{M}$. In this case, we could define a cross-validated risk that averages across different orderings, thereby enhancing the precision of the corresponding cross-validation selector. This minor extension and the corresponding oracle inequality is presented in this section.

18.4.1 Online Cross-Validation Selector

Consider an initial ordering $O(1), \ldots, O(n)$. A new ordering $O(\pi(1)), \ldots, O(\pi(n))$ is defined by a permutation $\pi : \{1, \ldots, n\} \to \{1, \ldots, n\}$ that is 1-1 and onto. Consider V such permutations $\pi_1, \ldots, \pi_V$. Let $\hat{\Psi}_k$ be candidate estimators that can be applied to data sets $O(i)$ for i ranging over a subset of $\{1, \ldots, n\}$, $k = 1, \ldots, K(n)$. Let P_{v, t_0} be the empirical distribution based on $O(\pi_v(1)), \ldots, O(\pi_v(t_0))$. Given a candidate estimator $\hat{\Psi}_k$ we define its online cross-validated risk as follows:

$$R_{CV, n}(\hat{\Psi}_k) \equiv \frac{1}{V} \sum_{v=1}^{V} \frac{1}{n - n_l + 1} \sum_{t_0 = n_l + 1}^{n} L(\hat{\Psi}_k(P_{v, t_0 - 1}))(O(\pi_v(t_0))).$$

The corresponding online cross-validation selector is defined as

$$k_n = \arg\min_k R_{CV,n}(\hat{\Psi}_k).$$

The online super learner is defined as follows:

$$\hat{\Psi}(P_t) = \hat{\Psi}_{k_t}(P_t), \ t = n_l, \ldots,.$$

Thus, at t observations, it uses the estimator $\hat{\Psi}_k$ with index $k = k_t$.

18.4.2 Imitating V-Fold Cross-Validation

Suppose that we partition the n observations in V subgroups of observations and let the permutation π_v be defined by an ordering of the n observations for which the last n/V observations belong to the v-th subgroup, $v = 1, \ldots, V$. In addition, we could define $n_l = n(1 - p)$ for $p = 1/V$. In this case, the online cross-validated risk corresponds with evaluating the performance of candidate estimators trained on v-specific training samples of size at least $n(1 - p)$ when applied to each observation in the corresponding v-specific validation sample across each of the V orderings. Thus, in this case the online cross-validated risk is very similar to the cross-validated risk for V-fold cross-validation.

18.4.3 Online Oracle Selector

Note that the online cross-validated risk estimates the following online cross-validated true risk:

$$\tilde{R}_{CV,n}(\hat{\Psi}_k) = \frac{1}{V} \sum_{v=1}^{V} \frac{1}{n - n_l + 1} \sum_{t_0 = n_l + 1}^{n} \bar{P}_0 L(\hat{\Psi}_k(P_{v,t_0-1})),$$

which is minimized by ψ_0. The difference between $\tilde{R}_{CV,n}(\hat{\Psi}_k)$ and $\tilde{R}_{CV,n}(\psi_0)$ defines a loss-based dissimilarity between a candidate estimator $\hat{\Psi}_k$ and the true target ψ_0:

$$d_{0n}(\hat{\Psi}_k, \psi_0) = \frac{1}{V} \sum_{v=1}^{V} \frac{1}{n - n_l + 1} \sum_{t_0 = n_l + 1}^{n} \bar{P}_0 \{ L(\hat{\Psi}_k(P_{v,t_0-1})) - L(\psi_0) \}.$$

The online oracle selector is defined as the minimizer of this loss-based dissimilarity:

$$\tilde{k}_n = \arg\min_k d_{0n}(\hat{\Psi}_k, \psi_0).$$

We can literally copy the above two theorems 18.1 and 18.2 by using the above definition of $d_{0n}(\hat{\Psi}_k, \psi_0)$ and noting that $Z(i)$ is now empty. Just for completeness we state here the analogue of Theorem 18.1.

Theorem 18.3. *Consider the above model $\mathcal{M}^n$ for the distribution P_0^n of $O^n = (O(1), \ldots, O(n))$ in which $O(i) \sim_{iid} \bar{P}_0$, the definition of the target parameter $\Psi^n : \mathcal{M}^n \to \Psi$ defined by $\Psi^n(P^n) = \Psi(\bar{P})$ where $\Psi : \mathcal{M} \to \Psi$, and the loss function $L(\psi)(O)$ for $\psi_0 = \Psi(\bar{P}_0)$. Consider also the above defined online cross-validation selector k_n, online oracle selector $\tilde{k}_n$, and $d_{0n}(\hat{\Psi}_k, \psi_0)$ for i.i.d. data defined in terms of an average over V permutations of O^n.*

Assumptions.

A1. *There exist an $M_1 < \infty$ so that*

$$\sup_{\psi \in \Psi} \sup_{i, O(i)} \mid L(\psi)(O(i)) - L(\psi_0)(O(i)) \mid \le M_1,$$

where the supremum over $O(i)$ is taken over a support of the distribution $\bar{P}_0$.
A2. *There exist an $M_2 < \infty$ so that with probability 1*

$$\sup_{\psi \in \Psi} \frac{\bar{P}_0\{L(\psi) - L(\psi_0)\}^2}{\bar{P}_0\{L(\psi) - L(\psi_0)\}} \le M_2 < \infty.$$

A3. *Assume that there exists a possibly increasing sequence $M_{3n} < \infty$ (e.g., $M_{3n} = \log n$) so that with probability tending to 1,*

$$\frac{1}{M_{3n}} < \frac{d_{0n}(\hat{\Psi}_k, \psi_0)}{E_0 d_{0n}(\hat{\Psi}_k, \psi_0)} < M_{3n} \text{ for all } k = 1, \ldots, K(n).$$

A4.

$$nM_{3n}^{-3} \min_k E_0 d_{0n}(\hat{\Psi}_k, \psi_0) \to \infty \text{ as } n \to \infty.$$

Finite Sample Result. *For any $\delta > 0$, there exists a universal $C(\delta, M_1, M_2) < \infty$ (i.e., universal in n and choice of candidate estimators) so that*

$$d_{0n}(\hat{\Psi}_{k_n}, \psi_0) \le (1 + 2\delta) d_{0n}(\hat{\Psi}_{\tilde{k}_n}, \psi_0) + Z_n,$$

where $Z_n = Z_{n1} + Z_{n2}$ with $P_0^n(Z_{n2} = 0) \to 1$ as $n \to \infty$, and for $n > n_1$ for some $n_1 < \infty$, we have

$$E_0 Z_{n1} \le C(\delta, M_1, M_2) \frac{M_{3n}^2 (1 + \log(K(n)))}{n}.$$

If Assumption **A4** *does not hold, then we have*

$$d_{0n}(\hat{\Psi}_{k_n}, \psi_0) = o_P(n^{-1} M_{3n}^3) + o_P(n^{-1} M_{3n}^2 (1 + \log K(n))).$$

Proof. The proof of this theorem is completely analogue to the proof of Theorem 18.1 and is therefore omitted. The only new observation is that the terms $R_{n,k}, T_{n,k}$ in this proof involve now an average over V terms $R_{n,k,v}, T_{n,k,v}$, $v = 1, \ldots, V$, where we can apply the same proof to $R_{n,k,v}$ for each v. □

18.5 Discussion

Our oracle inequality demonstrates that we can apply loss-based cross-validation based ensemble learning for time-series data $(O(t) : t = 1, \ldots)$, and a general class of time series models, where $O(t)$ can be a high dimensional data structure, and $O(t)$ only depends on the past through a fixed dimensional summary measure $Z(t)$. The online super learner can be used for estimation of any common parameter of the conditional probability distribution of $O(t)$, given $\bar{O}(t-1)$ which is minimizes by a conditional expectation of a loss function. The oracle inequality demonstrates that under weak assumptions, this super learner will be asymptotically equivalent with the oracle selected estimator.

Another important feature of this online super learner is that it is an online estimator and therefore scalable to large data sets. We have demonstrated its application to i.i.d. data to obtain a scalable super learner for i.i.d. data which will be computationally much more tractable than the regular super learner and by averaging across orderings of the data the cross-validation scheme imitates V-fold cross-validation. We refer to Benkeser et al. (2017b) for a data analysis and simulation implementing and practically evaluating this online super learner. When the target parameter is a pathwise differentiable (typically, low dimensional) parameter of P_0^n, then we could develop an online TMLE that is asymptotically normally distributed and efficient (van der Laan and Rubin 2006; van der Laan 2008b; van der Laan and Rose 2011). Such a TMLE, relies on a good initial estimator. By using the online super learner as initial estimator, we obtain a powerful online TMLE. This opens up the construction of online TMLE for pathwise differentiable parameters of flexible/nonparametric time series models of the type defined in this chapter. This is addressed in the next chapter.

Chapter 19
Online Targeted Learning for Time Series

Mark J. van der Laan, Antoine Chambaz, and Sam Lendle

We consider the case that we observe a time series where at each time we observe in chronological order a covariate vector, a treatment, and an outcome. We assume that the conditional probability distribution of this time specific data structure, given the past, depends on the past through a fixed (in time) dimensional summary measure, and that this conditional distribution is described by a fixed (in time) mechanism that is known to be an element of some model space (e.g., unspecified). We propose a causal model that is compatible with this statistical model and define a family of causal effects in terms of stochastic interventions on a subset of the treatment nodes on a future outcome, and establish identifiability of these causal effects from the observed data distribution.

This general formulation of the statistical estimation problem includes many important estimation problems. For example, by selecting empty summary measures of the past, it includes targeted estimation of causal effects based on independent and identically distributed data (Bickel et al. 1997b; Robins and Rotnitzky 1992; van der Laan and Rubin 2006; van der Laan 2008a; van der Laan and Rose 2011). By selecting parametric models for the conditional density it includes classical time series models. It also includes group sequential adaptive designs in which

M. J. van der Laan (✉)
Division of Biostatistics and Department of Statistics, University of California, Berkeley, 101 Haviland Hall, #7358, Berkeley, CA 94720, USA
e-mail: laan@berkeley.edu

A. Chambaz
MAP5 (UMR CNRS 8145), Université Paris Descartes, 45 rue des Saints-Pères, 75270 Paris cedex 06, France
e-mail: antoine.chambaz@parisdescartes.fr

S. Lendle
Pandora Media Inc, 2100 Franklin St, Suite 600, Oakland, CA 94612, USA
e-mail: sam.lendle@gmail.com

© Springer International Publishing AG 2018
M.J. van der Laan, S. Rose, *Targeted Learning in Data Science*,
Springer Series in Statistics, https://doi.org/10.1007/978-3-319-65304-4_19

the treatment allocation for the next subject is based on what has been observed on the previously recruited subjects (van der Laan 2008a; Chambaz and van der Laan 2011a,b). In the latter case, t indexes the t-th subject that enrolls in the trial. It also includes completely new problems not addressed in the literature, such as the estimation of causal effects based on observing a single time series on a single unit or on a specified group of units, under stationarity and Markov type assumptions, but leaving functional forms unspecified.

A key feature of the estimation problem addressed in this article is that the data is ordered and that statistical inference is based on asymptotics in time. Thus, a main feature of our proposed estimators is that they are online estimators in the sense that they can be updated continuously in time and still be computationally feasible, analogue to stochastic gradient descent algorithms for fitting parametric models in the computer science literature (e.g., Bottou 2012). This chapter is a generalization and augmentation of our earlier technical report on online targeted learning for independent and identically distributed data (van der Laan and Lendle 2014).

19.1 Statistical Formulation of the Estimation Problem

Suppose that we observe a time-ordered data structure $O^N = (W(t), A(t), Y(t) : t = 1, \ldots, N)$, where $W(t)$ occurs before $A(t)$ and $A(t)$ occurs before $Y(t)$. Let $A(t)$ denote an exposure or treatment, while $Y(t)$ denotes an outcome of interest at time t. Denote the true probability distribution of O^N with P_0^N. Let $O(t) = (W(t), A(t), Y(t))$ be the t-specific longitudinal data structure so that $O^N = (O(t) : t = 1, \ldots, N)$. We can factorize the probability density of the data O^N w.r.t. an appropriate dominating measure according to the time ordering as follows:

$$p_0^N(o) = \prod_{t=1}^{N} q_{w(t)}(w(t) \mid \bar{o}(t-1)) \prod_{t=1}^{N} g_{a(t)}(a(t) \mid \bar{o}(t-1), w(t))$$

$$\prod_{t=1}^{N} q_{y(t)}(y(t) \mid \bar{o}(t-1), w(t), a(t))$$

$$= \prod_{t=1}^{N} q_{w(t)}(w(t) \mid w^-(t)) \prod_{t=1}^{N} g_{a(t)}(a(t) \mid a^-(t)) \prod_{t=1}^{N} q_{y(t)}(y(t) \mid y^-(t)),$$

where we define $W^-(t) = \bar{O}(t-1)$, $A^-(t) = (\bar{O}(t-1), W(t))$ and $Y^-(t) = (\bar{O}(t-1), W(t), A(t))$ as the histories for $W(t)$, $A(t)$, and $Y(t)$, respectively. Here $q_{w(t)}, g_{a(t)}, q_{y(t)}$ denote the conditional probability densities of $W(t)$, given $W^-(t)$, $A(t)$, given $A^-(t)$, and $Y(t)$, given $Y^-(t)$, respectively. Let μ_w, μ_a and μ_y be the corresponding dominating measures.

19.1.1 Statistical Model: Stationarity and Markov Assumptions

Stationarity Assumption. We assume that each of the conditional densities $q_{w(t)}$, $q_{y(t)}$, $g_{a(t)}$ depends on the past through a fixed dimensional summary measure and are identified by a common (in time t) parameter $\bar{q}_w, \bar{q}_y, \bar{g}$, respectively:

$$q_{w(t)}(w(t) \mid w^-(t)) = q_{t,\bar{q}_w}(w(t) \mid c_{w,t}(w^-(t))),$$
$$g_{a(t)}(a(t) \mid a^-(t)) = g_{t,\bar{g}}(a(t) \mid c_{a,t}(a^-(t))),$$
$$q_{y(t)}(y(t) \mid y^-(t)) = q_{t,\bar{q}_y}(y(t) \mid c_{y,t}(y^-(t))),$$

for functions $(w, c) \to \bar{q}_w(w, c)$ with $(w, c) \in \mathbf{R}^{k_1}$, $(a, c) \to \bar{g}(a \mid c)$ with $(a, c) \in \mathbf{R}^{k_2}$, and $(y, c) \to \bar{q}_y(y \mid c)$ with $(y, c) \in \mathbf{R}^{k_2}$. Here $c_{w,t}()$, $c_{a,t}()$ and $c_{y,t}()$ are functions from the histories $w^-(t)$, $a^-(t)$ and $y^-(t)$ into a vector of fixed (not depending on t) dimensions k_1, k_2 and k_3, respectively.

Markov-Type Assumption. A particular type of example of summary measures are extractions of a recent history of fixed dimension from the complete history:

$$c_{w,t}(w^-(t)) = (o(t - l_w : t - 1)),$$
$$c_{a,t}(a^-(t)) = (o(t - l_a : t - 1), w(t)),$$
$$c_{y,t}(y^-(t)) = (o(t - l_y : t - 1), w(t), a(t)),$$

where we used the notation $O(s : t) = (O(s), \ldots, O(t))$. As we will see later, for the purpose of establishing asymptotics for our estimators, it will be important that our summary measures only cover a finite history, so that our time series has a universally (in time) finite memory.

These model assumptions allow us to deal with the curse of dimensionality and the fact that we only observe a single time series. We will also use the notation $C_w(t) = c_{w,t}(W^-(t))$, $C_a(t) = c_{a,t}(A^-(t))$ and $C_y(t) = c_{y,t}(Y^-(t))$ for the random variables implied by these summary measures of the history for $W(t)$, $A(t)$ and $Y(t)$, respectively.

Nesting Assumption on Summary Measures. Suppose that $c_w(t)$ is a function of $c_y(t - 1)$ and $y(t - 1)$; $c_a(t)$ is a function of $c_w(t)$ and $w(t)$ and $c_y(t)$ is a function of $c_a(t)$ and $a(t)$, across all t. That is, for any variable, its parents are a function of the parents of the previous variable in the ordered sequence of variables and the previous variable itself. This assumption dramatically simplifies the computation of our estimator as will be explained later, but is not essential for the presentation of the estimator and for establishing its asymptotics. This assumption assumes that the summary measures at time t contain all the relevant information from the past for the future of the time series after time t.

Starting Values of Time-Series. We assume that $W^-(0)$, $A^-(0)$ and $Y^-(0)$ are given and are already of dimension k_1, k_2 and k_3, so that the first conditional distributions are already conditioning on enough history to be defined by the common mechanisms $\bar{q}_w, \bar{g}$ and $\bar{q}_y$.

Statistical Model. Even though the conditional densities $q_{w(t)}, q_{y(t)}$ and $g_{a(t)}$ are determined by $\bar{q}_w, \bar{q}_y$ and $\bar{g}$, for the sake of notational convenience, we will often suppress this dependence in the notation. The statistical model for the true probability density p_0^N of O^N is defined as follows:

$$\mathcal{M}^N = \left\{ p^N = \prod_t q_{w(t)} g_{a(t)} q_{y(t)} : \bar{q}_w \in \mathbf{Q}_w, \bar{q}_y \in \mathbf{Q}_y, \bar{g} \in \mathcal{G} \right\},$$

where $\mathbf{Q}_w, \mathbf{Q}_y, \mathcal{G}$ are parameter spaces for $\bar{q}_w, \bar{q}_y, \bar{g}$, respectively. We will consider the case that $\mathbf{Q}_w$ and $\mathbf{Q}_y$ are nonparametric, but the parameter space $\mathcal{G}$ for $\bar{g}_a$ might be restricted. In order to provide concrete results for the definition of the efficient influence curve and the proposed online estimators, we will consider the case that $q_{w(t)} = \bar{q}_w$, $q_{y(t)} = \bar{q}_y$ and $g_{a(t)} = \bar{g}$ and that all parameter spaces are nonparametric. We remark that much of the work is easily generalized to other parameterizations of the time-specific mechanisms $q_{w(t)}, q_{y(t)}$ and $g_{a(t)}$ in terms of $\bar{q}_w, \bar{q}_y$ and $\bar{g}$, respectively. The statistical model $\mathcal{M}$ depends on the starting values $W^-(0), A^-(0), Y^-(0)$ for the time series since the first conditional densities at $t = 1$ depend on these values, but this dependence is suppressed in our notation.

19.1.2 Underlying Causal Model and Target Quantity

Suppose we assume a structural causal model

$$W(t) = f_{w(t)}(W^-(t), U_{w(t)}),$$
$$A(t) = f_{a(t)}(A^-(t), U_{a(t)}),$$
$$Y(t) = f_{y(t)}(Y^-(t), U_{y(t)}),$$
$$t = 1, \ldots, N,$$

for certain deterministic functions $(f_{w(t)}, f_{a(t)}, f_{y(t)} : t = 1, \ldots, N)$ and random exogenous errors $U = (U_{w(t)}, U_{a(t)}, U_{y(t)} : t = 1, \ldots, N)$ with some probability distribution $P_{u,0}$. Let $\mathcal{I} \subset \{1, \ldots, N\}$ be a finite set of K time points, and let $(A(t) : t \in \mathcal{I})$ be the corresponding vector of intervention nodes. Let $g^* = (g_t^* : t \in \mathcal{I})$ denote a collection of conditional distributions for $A(i)$, given $A^-(i)$ across these intervention nodes. Let $\tau_{g^*} \geq \max\{i : i \in \mathcal{I}\}$ be a time point that is larger than or equal to the time point of the last intervention node. We will refer to g^* as our stochastic intervention of interest, and we are concerned with estimation of a causal effect of such a stochastic intervention on the outcome $Y(\tau_{g^*})$.

For that purpose we define a counterfactual random variable $O_{g^*}^N$ defined by the modified system of equations in which the $A(t)$-equations with $t \in \mathcal{I}$ are replaced by the desired stochastic intervention g^*:

$$W_{g^*}(t) = f_{w(t)}(W_{g^*}^-(t), U_{w(t)}),$$
$$A_{g^*}(t) = f_{a(t)}(A_{g^*}^-(t), U_{a(t)}), t \notin \mathcal{I},$$

$$A_{g^*}(t) \sim g_t^*(\cdot \mid A_{g^*}^-(t)), t \in I,$$
$$Y_{g^*}(t) = f_{y(t)}(Y_{g^*}^-(t), U_{y(t)}),$$
$$t = 1, \ldots, N,$$

where $W_{g^*}^-(t)$ and $Y_{g^*}^-(t)$ are the same functions of the past as $W^-(t)$ and $Y^-(t)$, but now applied to the new random variable $O_{g^*}^N = (W_{g^*}(t), A_{g^*}(t), Y_{g^*}(t) : t = 1, \ldots, N)$. The probability distribution $P_{g^*}^N$ of the counterfactual $O_{g^*}^N$ is called the post-intervention probability distribution we would have observed if in fact $(A(t) : t \in I)$ would have been assigned according to our stochastic intervention g^* instead of the actual distribution $g_{a(t)}$ of the observed data.

19.1.3 g-Computation Formula for Post-intervention Distribution

Under the sequential randomization assumption that for all $t \in I_{g^*}$, $A(t)$ is independent of $Y_{g^*}(\tau_{g^*})$, conditional on $A^-(t)$, and the positivity assumption (simply defined by the requirement that the conditional probability distributions in the G-computation formula below are well defined) this probability distribution of $O_{g^*}^N$ equals the probability distribution $P_{g^*}^N$ whose density is given by

$$p_{g^*}^N(o) = \prod_{t=1}^N q_{w(t)}(w(t) \mid w^-(t)) \prod_{t \notin I} g_{a(t)}(a(t) \mid a^-(t)) \prod_{t \in I} g_t^*(a(t) \mid a^-(t-1))$$
$$\times \prod_{t=1}^N q_{y(t)}(y(t) \mid y^-(t)).$$

The latter density is called the g-computation formula in causal inference (Robins 1986), and we note that $p_{g^*}^N$ is identified by the probability distribution P^N of the data O^N.

19.1.4 Statistical Estimand: Intervention-Specific Counterfactual Mean

The statistical target parameter identifying the counterfactual mean $EY_{g^*}(\tau_{g^*})$ is defined as the expectation of $Y_{g^*}(\tau_{g^*})$ under this latter distribution $P_{g^*}^N$. Formally, we define it by the mapping $\Psi_{g^*}^N : \mathcal{M}^N \to \mathbf{R}^d$

$$\Psi_{g^*}^N(P^N) = E_{P_{g^*}^N} Y_{g^*}(\tau_{g^*}).$$

Note that we allow that $Y(t)$ is a d-dimensional vector. We can represent this statistical target parameter as follows:

$$\Psi_{g^*}^N(P^N) = \int_{o(1),\dots,o(\tau)} y(\tau)dP_{g^*}^N(o(1),\dots,o(\tau)) \equiv \Psi_{g^*}(\bar{q}_w, \bar{q}_y, \bar{g}),$$

where the latter notation emphasizes that $\Psi_{g^*}^N$, by our model assumptions, only depends on P^N through $(\bar{q}_w, \bar{q}_y, \bar{g})$.

19.1.5 Sequential Regression Representation of Counterfactual Mean

Let $Q_{y(t)}f = \int_{y(t)} f(y(t), y^c(t))dQ_{y(t)}(y(t) \mid c_y(t))$ denote the conditional expectation of $f(O^N)$ over $Y(t)$ w.r.t. conditional density $q_{y(t)}$ of $Y(t)$, given $C_y(t)$, where all other variables $y^c(t)$ f depends upon are kept fixed. Similarly, we define $G_{a(t)}f = \int_{a(t)} f(a(t), a^c(t))dG_{a(t)}(a(t) \mid c_a(t))$, and $Q_{w(t)}f = \int_{w(t)} f(w(t), w^c(t))dQ_{w(t)}(w(t) \mid c_w(t))$ as conditional expectations of $f(O^N)$ with respect to conditional densities $g_{a(t)}$ and $q_{w(t)}$, respectively. Moreover, we define this for our stochastic intervention g^*: $G_{a(t)}^* f = \int_{a(t)} f(a(t), a^c(t))dG_{a(t)}^*(a(t) \mid C_a^*(t))$. We view $Y(\tau_{g^*}) = Y(\tau_{g^*})(O^N)$ as a function that maps O^N into $Y(\tau_{g^*})$. We note that

$$\Psi_{g^*}^N(\bar{q}_w, \bar{q}_y, \bar{g}) = \left(\prod_{t=1}^{\tau} Q_{w(t)} G_{a(t)}^{*I(t \in \mathcal{I})} G_{a(t)}^{I(t \notin \mathcal{I})} Q_{y(t)} \right) Y(\tau_{g^*}).$$

That is, $E_{P_{g^*}^N} Y_{g^*}(\tau)$ can be represented as a sequential regression, starting with taking the conditional expectation over $Y(\tau)$, given $C_y(\tau)$ (i.e., given the observed past), taking a conditional expectation over $A^*(\tau)$ or $A(\tau)$ depending on if τ is an intervention node, given the observed past, taking a conditional expectation over $W(\tau)$, given $C_w(\tau)$, and going backwards in time t from τ to time $t = 1$, when we end up with a conditional expectation, given $W(1)$, and finally we take the expectation over $W(1)$. In the last conditional expectation at time $t = 1$, we condition on the starting values $W^-(0), A^-(0), Y^-(0)$. Since each conditional expectation over $Y(t)$, $W(t)$ and $A(t)$ is w.r.t. a distribution identified by a common $\bar{q}_y$, $\bar{q}_w$ and $\bar{g}$, this sequential regression formula indeed only depends on these common parameters $(\bar{q}_y, \bar{q}_w, \bar{g})$.

19.1.6 General Class of Target Parameters

For notational convenience, in the sequel, we will often suppress the dependence on N of quantities such as P^N and $P_{g^*}^N$, $\Psi_{g^*}^N$. Before we move on to the estimation problem, let's further generalize this class of target parameters. Let $(\mathcal{I}_1, \tau_1), \dots, (\mathcal{I}_J, \tau_J)$ be a collection of J intervention node sets with corresponding outcome time points. Let g_j^* be a stochastic intervention on $(A(t) : t \in \mathcal{I}_j)$, $j = 1, \dots, J$. We could now define a more general class of parameters as follows:

$$\Psi_J(P) = \frac{1}{J} \sum_{j=1}^{J} E_{P_{g_j^*}^N} Y(\tau_j) = \frac{1}{J} \sum_{j=1}^{J} \Psi_{g_j^*}(P).$$

For example, we could have that $\mathcal{I}_j = \{j\}$ and $\tau_j = j$, so that this parameter equals an average of J counterfactual means of the counterfactual outcome $Y_{g_j^*}(j)$ at time j under a single time point intervention g_j^* on the previous treatment node $A(j)$, $j = 1, \ldots, J$. It is possible to set $J = J(N) = N$, so that the target parameter averages over such single time point interventions over all treatment nodes, ending with the counterfactual mean outcome of $Y(N)$ under an intervention on $A(N)$. Similarly, we could have that $\mathcal{I}_j = \{j, j+1\}$ and $\tau_j = j+1$, so that this parameter $\Psi_J(P)$ equals an average of J counterfactual means of the counterfactual outcome $Y_{g_j^*}(j+1)$ under a two time point intervention g_j^* on the previous two treatment nodes $A(j), A(j+1)$, $j = 1, \ldots, J$. The above sequential regression representation of $\Psi_{g_j^*}(P)$ shows that $\Psi_J(P)$ is an average over j of j-specific sequential regression representations, $j = 1, \ldots, J$.

One is often interested in contrasts that represent a causal effect relative to a baseline treatment regimen. For that purpose, we could define

$$\Psi_J^c(P) = \frac{1}{J} \sum_{j=1}^{J} (\Psi_{g_j^*} - \Psi_{g_{0j}^*})(P),$$

where g_{0j}^* is a baseline treatment regimen on the intervention nodes $(A(t) : t \in \mathcal{I}_j)$. We will focus on the target parameter $\Psi_J : \mathcal{M}^N \to \mathbf{R}$, since the contrast parameters are defined as a difference of two of such target parameters. Our results for Ψ_{g^*} for a single g^* will naturally imply corresponding results for Ψ_J.

19.1.7 Statistical Estimation Problem

Our goal is to construct an estimator $\bar{q}_N^* = (\bar{q}_{w,N}^*, \bar{q}_{y,N}^*)$ of $\bar{q}_0 = (\bar{q}_{w,0}, \bar{q}_{y,0})$ and $\bar{g}_N^*$ of $\bar{g}_0$ so that $\Psi_J(\bar{q}_N^*, \bar{g}_N^*)$ is consistent and satisfies $\sqrt{N}(\Psi_J(\bar{q}_N^*, \bar{g}_N^*) - \Psi_J(\bar{q}_0, \bar{g}_0)) \Rightarrow_d N(0, \Sigma_0)$, as $N \to \infty$. We want this normal limit distribution to correspond with the normal limit distribution of a maximum likelihood estimator in the special case that all data is discrete, thereby guaranteeing that the proposed estimator of $\Psi_J(\bar{q}_0, \bar{g}_0)$ is asymptotically efficient. The normal limit distribution also provides us with statistical inference in terms of confidence intervals and p-values for testing null hypotheses. In particular, for the sake of scalability, we want these estimators $\bar{q}_N^*, \bar{g}_N^*$ to be online-estimators.

To characterize the efficient normal limit distribution, we need to compute the canonical gradient/efficient influence curve of the target parameter mapping $\Psi^J : \mathcal{M}^N \to \mathbf{R}^d$. In order to obtain a concrete efficient influence curve and TMLE, we consider the important case that $q_{w(t)} = \bar{q}_w$, $q_{y(t)} = \bar{q}_y$ and $g_{a(t)} = \bar{g}$, and that these functions are completely unspecified. For any specific parameterizations of $q_{w(t)}, q_{y(t)}$ and $g_{a(t)}$ in terms of $\bar{q}_w, \bar{q}_y$ and $\bar{g}$, and model spaces Q_w, Q_y, G, the next sections can also be viewed a roadmap for computing the efficient influence curve and defining the corresponding one-step and targeted maximum likelihood estimators in terms of the efficient influence curve.

19.2 Efficient Influence Curve of the Target Parameter

The efficient influence curve depends on a series of conditional expectations and ratios of densities. Our first job is to determine the efficient influence curve of this target parameter Ψ_J. One method for deriving the efficient influence curve is to determine the influence curve of the MLE when all random variables are discrete. This yields the efficient influence curve for the case that the data is discrete and its natural counterpart for continuous data is then the conjectured efficient influence curve. The conjecture can then be verified by checking that (1) it is a gradient of the pathwise derivative of the target parameter mapping and (2) an element of the tangent space.

Consider the case that all random variables are discrete. In that case the MLE is well defined for large enough N, and the MLE of $\bar{q}_w, \bar{q}_y, \bar{g}$ equals the empirical counterparts $\bar{q}^e_{W,N}, \bar{q}^e_{Y,N}, \bar{g}^e_N$ based on the N relevant observations $(C_w(t), W(t))$, $t = 1, \ldots, N$, $(C_y(t), Y(t))$, $t = 1, \ldots, N$ and $(C_a(t), A(t))$, $t = 1, \ldots, N$, respectively, treating them as i.i.d. That is, these MLEs are defined as follows:

$$\bar{q}^e_{Y,N}(y(t) \mid c_y(t)) = \frac{\sum_{i=1}^N I(Y(i) = y(t), C_y(i) = c_y(t))}{\sum_{i=1}^N I(C_y(i) = c_y(t))},$$

$$\bar{q}^e_{W,N}(w(t) \mid c_w(t)) = \frac{\sum_{i=1}^N I(W(i) = w(t), C_w(i) = c_w(t))}{\sum_{i=1}^N I(C_w(i) = c_w(t))},$$

$$\bar{g}^e_N(a(t) \mid c_a(t)) = \frac{\sum_{i=1}^N I(A(i) = a(t), C_a(i) = c_a(t))}{\sum_{i=1}^N I(C_a(i) = c_a(t))}.$$

Linearizing yields

$$\Psi_J(\bar{q}^e_{Y,N}, \bar{q}^e_{W,N}, \bar{g}^e_N) - \Psi_J(\bar{q}_{Y,0}, \bar{q}_{W,0}, \bar{g}_0) \approx \frac{1}{N} \sum_{i=1}^N \bar{D}(\bar{q}_0, \bar{g}_0)(\bar{O}(i)),$$

from which we deduce the following efficient influence curve.

Theorem 19.1. *The canonical gradient of* $\Psi_J : \mathcal{M}^N \to \mathbb{R}^d$ *at* $P^N \in \mathcal{M}^N$ *is given by:*

$$D^N(\bar{q}, \bar{g})(O^N) = \frac{1}{N} \sum_{i=1}^N \bar{D}(\bar{q}, \bar{g})(\bar{O}(i))$$

$$\equiv \frac{1}{N} \sum_{i=1}^N \sum_{s=1}^N \bar{D}_s(\bar{q}, \bar{g})(\bar{O}(i))$$

$$\equiv \frac{1}{N} \sum_{i=1}^{N} \sum_{s=1}^{N} \{\bar{D}_{q_{y(s)}}(Y(i), C_y(i)) + \bar{D}_{q_{w(s)}}(W(i), C_w(i))\}$$

$$+ \frac{1}{N} \sum_{i=1}^{N} \sum_{s=1}^{N} \bar{D}_{g_s}(A(i), C_a(i)),$$

where

$$\bar{D}_{q_{y(s)}}(Y(i), C_y(i)) = \frac{1}{J} \sum_{j=1}^{J} I(s \le \tau_j) \frac{h^*_{c_y(s),j}(C_y(i))}{\bar{h}_{c_y}(C_y(i))}$$
$$\{E(Y_{g_j^*}(\tau_j) \mid Y(s) = Y(i), C_y(s) = C_y(i)) - E(Y_{g_j^*}(\tau_j) \mid C_y(s) = C_y(i))\},$$

$$\bar{D}_{q_{w(s)}}(W(i), C_w(i)) = \frac{1}{J} \sum_{j=1}^{J} I(s \le \tau_j) \frac{h^*_{c_w(s),j}(C_w(i))}{\bar{h}_{c_w}(C_w(i))}$$
$$\{E(Y_{g_j^*}(\tau_j) \mid W(s) = W(i), C_w(s) = C_w(i)) - E(Y_{g_j^*}(\tau_j) \mid C_w(s) = c_w(i))\},$$

$$\bar{D}_{\bar{g}_s}(A(i), C_a(i)) = \frac{1}{J} \sum_{j=1}^{J} I(s \le \tau_j, s \notin I_j) \frac{h^*_{c_a(s),j}(C_a(i))}{\bar{h}_{c_a}(C_a(i))}$$
$$\{E(Y_{g_j^*}(\tau_j) \mid A(s) = A(i), C_a(s) = C_a(i)) - E(Y_{g_j^*}(\tau_j) \mid C_a(s) = c_a(i))\},$$

and

$$h_{c_y(i)}(c_y) \equiv p_{c_y(i)}(c_y),$$

$$\bar{h}_{c_y}(c_y) \equiv \frac{1}{N} \sum_{i=1}^{N} h_{c_y(i)}(c_y),$$

$$h^*_{c_y(s),j}(c_y) \equiv p_{g_j^*,c_y(s)}(c_y),$$

$$h_{c_w(i)}(c_w) \equiv p_{c_w(i)}(c_w),$$

$$\bar{h}_{c_w}(c_w) \equiv \frac{1}{N} \sum_{i=1}^{N} h_{c_w(i)}(c_w),$$

$$h^*_{c_w(s),j}(c_w) \equiv p_{g_j^*,c_w(s)}(c_w),$$

$$h_{c_a(i)}(c_a) \equiv p_{c_a(i)}(c_a),$$

$$\bar{h}_{c_a}(c_a) \equiv \frac{1}{N} \sum_{i=1}^{N} h_{c_a(i)}(c_a),$$

$$h^*_{c_a(s),j}(c_a) \equiv p_{g_j^*,c_a(s)}(c_a).$$

Here $p_{c_y(i)}$ denotes the marginal density of $C_Y(i)$ under P^N, and $p_{c_w(i)}$, $p_{c_a(i)}$ are defined in the same manner. In addition, $p_{g^*,c_y(s)}$ denotes the marginal density of $C_Y(s)$ under $P^N_{g^*}$, and $p_{g^*,c_w(s)}$, $p_{g^*,c_a(s)}$ are defined in the same manner.

Efficient Score Components of $\bar{q}_w, \bar{q}_y, \bar{g}$. This efficient influence curve is an orthogonal sum of the three efficient score components generated by $\bar{q}_y$, $\bar{q}_w$ and $\bar{g}$, respectively:

$$D^N(O^N) = D^N_{\bar{q}_y}(O^N) + D^N_{\bar{q}_w}(O^N) + D^N_{\bar{g}}(O^N),$$

$$D^N_{\bar{q}_y}(O^N) = \frac{1}{N} \sum_{i=1}^{N} \bar{D}_{\bar{q}_y}(Y(i), C_y(i)),$$

$$\bar{D}_{\bar{q}_y}(Y(i), C_y(i)) = \sum_{s=1}^{N} \bar{D}_{q_{y(s)}}(Y(i), C_y(i)),$$

$$D_{\bar{q}_w}^N(O^N) = \frac{1}{N} \sum_{i=1}^{N} \bar{D}_{\bar{q}_w}(W(i), C_w(i)),$$

$$\bar{D}_{\bar{q}_w}(W(i), C_w(i)) = \sum_{s=1}^{N} \bar{D}_{q_{w(s)}}(W(i), C_w(i)),$$

$$D_{\bar{g}}^N(O^N) = \frac{1}{N} \sum_{i=1}^{N} \bar{D}_{\bar{g}}(A(i), C_a(i)),$$

$$\bar{D}_{\bar{g}}(A(i), C_a(i)) = \sum_{s=1}^{N} \bar{D}_{g_s}(A(i), C_a(i)).$$

Since each $\bar{D}_{q_{y(s)}}(Y(i), C_y(i))$ has conditional mean zero w.r.t. the distribution of $Y(i)$, given $C_y(i)$, it follows that $D_{\bar{q}_y}^N(O^N)$ is of the form $\frac{1}{N} \sum_i h(Y(i), C_y(i))$ for a common (in i) function h satisfying $\int h(y, c)\bar{q}_y(y \mid c)d\mu_y(y \mid c) = 0$. This proves that $D_{\bar{q}_y}^N(O^N)$ is an element of the tangent space of $\bar{q}_y$. Similarly, it follows that $D_{\bar{q}_w}^N(O^N)$, $D_{\bar{g}}^N(O^N)$ are in the tangent space at P^N of the model $\mathcal{M}^N$.

19.2.1 Monte-Carlo Approximation of the Efficient Influence Curve using the Nesting Assumption

The main ingredient of the TMLE and the other estimators will be the evaluation of the efficient influence curve at O^N, under an initial estimator $(\bar{q}_N, \bar{g}_N)$. In this subsection, we discuss a Monte-Carlo method for approximating this evaluation $D^N(\bar{q}_N, \bar{g}_N)(O^N)$. We will use the following lemma regarding the conditional expectations in the definition of the efficient influence curve.

Lemma 19.1. *Under the nesting assumption, the conditional distribution of $Y_{g^*}(\tau)$, given $Y(s), C_y(s)$, is the same as the conditional distribution of $Y_{g^*}(\tau)$, given $O(1)$, $\ldots, O(s)$; the conditional distribution of $Y_{g^*}(\tau)$, given $W(s), C_w(s)$, is the same as the conditional distribution of $Y_{g^*}(\tau)$, given $\bar{O}(s-1), W(s), C_w(s)$; the conditional distribution of $Y_{g^*}(\tau)$, given $A(s), C_a(s)$, is the same as the conditional distribution of $Y_{g^*}(\tau)$, given $\bar{O}(s-1), W(s), A(s), C_a(s)$.*

Evaluation of the Conditional Expectations in the Efficient Influence Curve. For the sake of presentation, let j be fixed and denote g_j^* with g^*. The nesting assumption implies that we can evaluate the conditional mean $E(Y_{g^*}(\tau) \mid Y(s) = y(s), C_y(s) = c_y(s)) = E(Y_{g^*}(\tau) \mid \bar{O}(s))$ with a straightforward Monte-Carlo simulation approximation based on sequentially drawing from the factors in the intervention specific density p_{g^*}, starting with the node after $O(s)$. Recall the ordered sequence of nodes in O^N. We start with drawing the next node $W(s+1)$ from $q_{w(s+1)}(\cdot \mid c_w(s+1))$, where by this nesting assumption $c_w(s + 1)$ is a function of $(y(s), c_y(s))$, and is thus available. Subsequently, we draw $A(s + 1)$ from either $g_{a(s+1)}$ (if $s + 1 \notin \mathcal{I}_{g^*}$) or g^*_{s+1} (otherwise), where, by the nesting assumption, the parents of $A(s + 1)$ are known as a function of $(c_y(s), y(s), w(s + 1))$. Then, we draw $Y(s + 1)$ from $q_{y(s+1)}$. In this manner we draw the whole future sequence of nodes making up O^N under p_{g^*} until we draw $Y_{g^*}(\tau)$, where, again, by the nesting assumption, for each node we need to draw we can calculate the required parent nodes from the previous realizations. We can repeat this procedure many times, resulting in a large sample of realizations of $Y_{g^*}(\tau)$ from this conditional probability distribution of $Y_{g^*}(\tau)$, given $Y(s), C_y(s)$. By averaging these realizations of $Y_{g^*}(\tau)$ we obtain the Monte-Carlo approximation of $E(Y_{g^*}(\tau) \mid Y(s) = y(s), C_y(s) = c_y(s))$. Similarly, we can draw $Y_{g^*}(\tau)$ from a conditional probability distribution of $Y_{g^*}(\tau)$, given $C_y(s) = c_y(s)$, by first drawing $Y(s)$ and subsequently the remaining future sequence as above.

Similarly, we obtain Monte-Carlo approximations of the other conditional expectations $E(Y_{g^*}(\tau) \mid W(s), C_w(s))$ and $E(Y_{g^*}(\tau) \mid A(s), C_a(s))$ for given values $(w(s), c_w(s))$ and $(a(s), c_a(s))$, respectively. In this manner, the nesting assumption allows us to compute the conditional expectations in the efficient influence curve $D^N(\bar{q}, \bar{g})$ at a given $(\bar{q}, \bar{g})$ in a computationally reasonable fast manner.

Machine Learning Approximation of the h-Density Ratios. Let's now discuss the computation or approximation of the density ratio $h^*_{c_y(s),j}/\bar{h}_{c_y}$ in the efficient influence curve, which implies the same method for approximation of the ratios $h^*_{c_w(s),j}/\bar{h}_{c_w}$ and $h^*_{c_a(s),j}/\bar{h}_{c_a}$. For the sake of presentation, let j be fixed and denote g^*_j with g^*. We are given $P = P_{\bar{q},\bar{g}}$ and $P^* = P_{\bar{q},g^*}$ and our goal is to approximate this ratio $h^*_{c_y(s)}/\bar{h}_{c_y}$ of densities under these two distributions. We sample B observations $O^N_b \sim P_{\bar{q},\bar{g}}$ and we sample B observations $O^{N*}_b \sim P_{\bar{q},g^*}$, $b = 1, \ldots, B$, resulting in $2B$ observations. For each given s, this yields B observations $C^*_{y,b}(s)$ from P^*, $b = 1, \ldots, B$. For each of the draws O^N_b, we randomly sample an $i_b \in \{1, \ldots, N\}$ (i.e. uniform distribution with probability $1/N$), and map this O^N_b into $C_{y,b}(i_b)$. This results in B observations $C_{y,b}(i_b)$ from P. We represent this data set of $2B$ observations $\{C^*_{y,b}(s), C_{y,b}(i_b) : b = 1, \ldots, B\}$ as (C_j, Δ_j), $j = 1, \ldots, 2B$, where $\Delta_j = 1$ indicates that it is one of the B observations $C_{y,b}(i_b)$ from P, while $\Delta_j = 0$ indicates that it is one of the B observations $C^*_{y,b}(s)$ from P^*. We view these $2B$ observations as an i.i.d. sample on a random variable (C, Δ), where $P(\Delta = 1) = 0.5$, while the conditional density $p(c \mid \Delta = 1)$ of C, given $\Delta = 1$, at c equals the density $\bar{h}_{c_y}(c)$, and the conditional density $p(c \mid \Delta = 0)$ of C, given $\Delta = 0$, at c, equals $h^*_{c_y(s)}(c)$. We have

$$\frac{h^*_{c_y(s)}(c)}{\bar{h}_{c_y}(c)} = \frac{p(c \mid \Delta = 1)}{p(c \mid \Delta = 0)} = \frac{P(\Delta = 1 \mid C = c)}{1 - P(\Delta = 1 \mid C = c)}.$$

As a consequence, we can approximate the ratio-function $h^*_{c_y(s)}/\bar{h}_{c_y}$ with the logit of a fit of a logistic regression of $P(\Delta = 1 \mid C)$ based on this sample of $2B$ observations (C_j, Δ_j) using a flexible machine learning algorithm. If C is discrete, we simply use the nonparametric empirical distribution. In this manner, we can use

Monte-Carlo simulation combined with a flexible machine learning algorithm for logistic regression to approximate these ratios $h^*_{c_y(s)}/\bar{h}_{c_y}$. By making B large enough and using a flexible machine learning algorithm, we can control the approximation error. The above can be refined by using each O^N_b to map into N observations $C_{y,b}(1), \ldots, C_{y,b}(N)$, and drawing NB observations from P^* while only B from P. The resulting data set of $2NB$ observations can now be represented as $(C_j, \varDelta_j)$, $j = 1, \ldots, 2NB$ on $(C, \varDelta)$ with $P(\varDelta = 1) = 0.5$ and $P(c \mid \varDelta = 1) = \bar{h}_{c_y}(c)$ and $P(c \mid \varDelta = 0) = h_{y,s}(c)$, as before.

19.2.2 A Special Representation of the Efficient Influence Curve for Binary Variables

Here we provide alternative representations of the components of the efficient influence curve for the nodes that are binary, which immediately imply logistic regression fluctuation models for the TMLE that will be presented later. If $Y(s) \in \{0, 1\}$ is binary, then we have

$$E(Y_{g^*_j}(\tau_j) \mid Y(s), C_y(s)) - E(Y_{g^*_j}(\tau_j) \mid C_y(s)) = H_{y(s),j}(C_y(s))(Y(s) - \bar{q}_Y(1 \mid C_y(s))),$$

where

$$H_{y(s),j}(C_y(s)) \equiv E(Y_{g^*_j}(\tau_j) \mid Y(s) = 1, C_y(s)) - E(Y_{g^*_j}(\tau_j) \mid Y(s) = 0, C_y(s)).$$

Let

$$H_{y(s)}(C_y(s)) = \frac{1}{J} \sum_{j=1}^{J} I(s \le \tau_j) H_{y(s),j}(C_y(s)) \frac{h^*_{c_y(s),j}(C_y(s))}{\bar{h}_{c_y}(C_y(s))},$$

so that

$$\bar{D}_{q_{y(s)}}(\bar{q}, \bar{g})(y(s), c_y(s)) = H_{y(s)}(c_y(s))(y(s) - \bar{Q}_y(1 \mid c_y(s))).$$

Finally, we define

$$H_y(c) \equiv \sum_{s=1}^{N} H_{y(s)}(c), \tag{19.1}$$

so that the efficient influence curve component generated by $\bar{q}_y$ is given by:

$$D^N_{\bar{q}_y}(O^N) = \frac{1}{N} \sum_{i=1}^{N} \bar{D}_{\bar{q}_y}(Y(i), C_y(i)) = \frac{1}{N} \sum_{i=1}^{N} H_y(C_y(i))(Y(i) - \bar{Q}_y(1 \mid C_y(i))).$$

Similarly, if $A(s) \in \{0, 1\}$ is binary, then

$$E(Y_{g^*_j}(\tau_j) \mid A(s), C_a(s)) - E(Y_{g^*_j}(\tau_j) \mid C_a(s)) = H_{a(s),j}(C_a(s))(A(s) - \bar{g}(1 \mid C_y(s))),$$

where

$$H_{a(s),j}(C_a(s)) = E(Y_{g_j^*}(\tau_j) \mid A(s) = 1, C_a(s)) - E(Y_{g*_j}(\tau_j) \mid A(s) = 0, C_a(s)).$$

Let

$$H_{a(s)}(C_a(s)) = \frac{1}{J} \sum_{j=1}^{J} I(s \leq \tau_j, s \notin \mathcal{I}_j) H_{a(s),j}(C_a(s)) \frac{h^*_{c_a(s),j}(C_a(s))}{\bar{h}_{c_a}(C_a(s))},$$

so that

$$\bar{D}_{\bar{g}_s}(\bar{q}, \bar{g})(a(s), c_a(s)) = H_{a(s)}(c_a(s))(a(s) - \bar{g}(1 \mid c_a(s))).$$

Finally, we define

$$H_a(c) \equiv \sum_{s=1}^{N} H_{a(s)}(c), \tag{19.2}$$

so that the efficient influence curve component generated by $\bar{g}$ is given by:

$$D_{\bar{g}}^N(O^N) = \frac{1}{N} \sum_{i=1}^{N} \bar{D}_{\bar{g}}(A(i), C_a(i)) = \frac{1}{N} \sum_{i=1}^{N} H_a(C_a(i))(A(i) - \bar{g}(1 \mid C_a(i))).$$

Suppose $W(s) = (W_k(s) : k = 1, \ldots, K)$ is a vector of binary variables $W_k(s) \in \{0, 1\}$, and let $\bar{q}_{w,k}$ be the conditional density of $W_k(s)$, given $W_{1:k-1}(s), C_W(s))$, implied by $\bar{q}_w$. Let $C_{w_k}(s) = (C_w(s), W_{1:k-1}(s))$ be the parent set for $W_k(s)$, so that we can write $\bar{q}_{w,k}(\cdot \mid C_{w_k}(s))$. We have

$$E(Y_{g_j^*}(\tau_j) \mid W(s), C_w(s)) - E(Y_{g_j^*}(\tau_j) \mid C_w(s))$$
$$= \sum_{k=1}^{K_s} \{ E(Y_{g_j^*}(\tau_j) \mid W_k(s), C_{w_k}(s)) - E(Y_{g_j^*}(\tau_j) \mid C_{w_k}(s)) \}$$
$$= \sum_{k=1}^{K_s} H_{w_k(s),j}(C_{w_k}(s))(W_k(s) - \bar{q}_{w_k}(1 \mid C_{w_k}(s))),$$

where

$$H_{w(k),s,j}(C_{w_k}(s)) \equiv E(Y_{g*_j}(\tau_j) \mid W_k(s) = 1, C_{w_k}(s)) - E(Y_{g_j^*}(\tau_j) \mid W_k(s) = 0, C_{w_k}(s)).$$

Let

$$H_{w_k(s)}(C_{w_k}(s)) = \frac{1}{J} \sum_{j=1}^{J} I(s \leq \tau_j) H_{w_k(s),j}(C_{w_k}(s)) \frac{h^*_{w_k,s,j}(C_{w_k}(s))}{\bar{h}_{w_k}(C_{w_k}(s))},$$

so that

$$\bar{D}_{\bar{q}_{w_k(s)}}(\bar{q}, \bar{g})(w_k(s), c_{w_k}(s)) = H_{w_k(s)}(c_{w_k}(s))(w_k(s) - \bar{q}_{w_k}(1 \mid c_{w_k}(s))).$$

Finally, we define

$$H_{w_k}(c) \equiv \sum_{s=1}^{N} H_{w_k(s)}(c), \tag{19.3}$$

so that the efficient influence curve component generated by $\bar{q}_{w_k}$ is given by:

$$D_{\bar{q}_{w_k}}^N(O^N) = \frac{1}{N}\sum_{i=1}^N \bar{D}_{\bar{q}_{w_k}}(W_k(i), C_{w_k}(i)) = \frac{1}{N}\sum_{i=1}^N H_{w_k}(C_{w_k}(i))(W_k(i) - \bar{q}_{w_k}(1 \mid W_k(i)).$$

Of course, all these component-specific representations provide a corresponding representation of the efficient influence curve by using that $D^N(O^N) = D_{\bar{q}_y}^N(O^N) + D_{\bar{g}}^N(O^N) + \sum_{k=1}^K D_{\bar{q}_{w_k}}^N$.

19.3 First Order Expansions for the Target Parameter in Terms of Efficient Influence Curve

The analysis of any of the estimators relies on a first order Taylor expansion of the target parameter at the plugged-in estimator $\theta_N = (\bar{q}_N, \bar{g}_N)$ of $\theta_0 = (\bar{q}_0, \bar{g}_0)$. In this section we present two of such expansions, one for a standard TMLE (that might use an online super learner as initial estimator, but whose targeting step is not online) and one for the fully online one-step estimator and online TMLE. The two theorems in this section are proved in Sect. A.4.

19.3.1 Expansion for Standard TMLE

Let $\theta = (\bar{q}, \bar{g})$. The next Theorem 19.2 presents a first order expansion of the type

$$\Psi_J(P^{N*}) - \Psi_J(P_0^N) = -\frac{1}{N}\sum_{i=1}^N P_{0,o(i)}\bar{D}(P^{N*}) + R_{2,N}(\theta_N^*, \theta_0),$$

where $P_{0,o(i)}$ denotes the conditional expectation w.r.t. $O(i)$, given $\bar{O}(i-1)$, and $R_{2,N}(\theta, \theta_0)$ is a second-order remainder. In order to demonstrate the relevance of this identity for the analysis of the TMLE, we note the following. This identity combined with the TMLE P^{N*} solving $\frac{1}{N}\sum_{i=1}^N \bar{D}(P^{N*})(\bar{O}(i)) = o_P(N^{-1/2})$ results in the first order expansion:

$$\Psi_J(P^{N*}) - \Psi_J(P_0^N) = \frac{1}{N}\sum_{i=1}^N \{\bar{D}(P^{N*})(\bar{O}(i)) - P_{0,o(i)}\bar{D}(P^{N*})\} + R_{2,N}(\theta_N^*, \theta_0) + o_P(N^{-1/2}).$$

The leading term on the right hand side is a martingale process, allowing us to establish that $\sqrt{N}(\Psi_J(P^{N*}) - \Psi_J(P_0^N))$ converges to a normal limit distribution under the condition that $R_{2,N}(\theta_N^*, \theta_0) = o_P(N^{-1/2})$.

Theorem 19.2. *Let* $\Phi_{x(s)}(x, c_x) = E(Y_{g^*}(\tau) \mid X(s) = x, C_x(s) = c_x), x \in \{w, a, y\}$. *Recall the representation* $D_{g^*}^N(\theta) = \frac{1}{N}\sum_{i=1}^N \bar{D}_{g^*}(\theta)(\bar{O}(i))$ *of the efficient influence curve for* $\Psi_{g^*} : \mathcal{M}^N \to \mathbf{R}$. *We have*

$$-\frac{1}{N}\sum_{i=1}^{N}P_{0,o(i)}\bar{D}_{g^*}(\theta) = \Psi_{g^*}(\theta) - \Psi_{g^*}(\theta_0) + R_{2,g^*,N}(\theta,\theta_0),$$

where $R_{2,g^*,N}(\theta,\theta_0) = R_{21,g^*}(\theta,\theta_0) + R_{22,g^*,N}(\theta,\theta_0)$, $R_{21,g^*}(\theta,\theta_0)$ is defined in Theorem A.3 (A.12), and

$$R_{22,g^*,N}(\theta,\theta_0) \equiv \frac{1}{N}\sum_{i=1}^{N}(P_{0,o(i)} - P_0^N)\left(\bar{D}_{g^*}(\theta) - \bar{D}_{g^*}(\theta_0)\right).$$

Recall $\Psi_J(P) = \frac{1}{J}\sum_{j=1}^{J}\Psi_{g_j^*}(P)$ and its canonical gradient $D^N(P) = \frac{1}{J}\sum_{j=1}^{J}D_{g_j^*}^N(P)$, where $D^N(P) = \frac{1}{N}\sum_{i=1}^{N}\bar{D}(P)(\bar{O}(i))$. Therefore, it follows that

$$-\frac{1}{N}\sum_{i=1}^{N}P_{0,o(i)}\bar{D}(\theta) = \Psi_J(\theta) - \Psi_J(\theta_0) + R_{2N}(\theta,\theta_0),$$

where $R_{2N}(\theta,\theta_0) = R_{21}(\theta,\theta_0) + R_{22,N}(\theta,\theta_0)$, $R_{21}(\theta,\theta_0) = \frac{1}{J}\sum_{j=1}^{J}R_{21,g_j^*}(\theta,\theta_0)$ and $R_{22,N}(\theta,\theta_0) = \frac{1}{J}\sum_{j=1}^{J}R_{22,g_j^*,N}(\theta,\theta_0)$.

Finite Memory Requirement for Control of Second-Order Remainder. The remainder $R_{21,g^*}(\theta,\theta_0)$ is a clear second-order term in differences between θ and θ_0. We note that $R_{22,g^*,N}(\theta,\theta_0)$ is an empirical mean of zero mean random variables, where each random variable converges to zero as θ approximates θ_0. Specifically, we have

$$R_{22,g^*,N}(\theta,\theta_0) = \sum_{x\in\{y,a,w\}}\sum_{s=1}^{\tau}R_{22,x(s),g^*,N}(\theta,\theta_0) \text{ with}$$

$$R_{22,x(s),g^*,N}(\theta,\theta_0) = \frac{1}{N}\sum_{i=1}^{N}(P_{0,o(i)} - P_0^N)(\bar{D}_{g^*,x(s)}(\theta) - \bar{D}_{g^*,x(s)}(\theta_0))$$

$$= \frac{1}{N}\sum_{i=1}^{N}f_{x(s),g^*}^0(\theta,\theta_0)(C_x(i)) \text{ where}$$

$$f_{x(s),g^*}^0(\theta,\theta_0)(C_x(i)) = \{f_{x(s),g^*}(\theta,\theta_0)(C_x(i)) - P_0^N f_{x(s),g^*}(\theta,\theta_0)\},$$

$$f_{x(s),g^*}(\theta,\theta_0)(C_x(i)) \equiv \frac{h_{c_x(s)}^*(C_x(i))}{\bar{h}_{c_x}(C_x(i))}\int_x \Phi_{x(s)}(x,C_x(i))d(P_{x(s)}^* - P_{0,x(s)}^*)(x\mid C_x(i)).$$

Thus, $R_{22,x(s),g^*,N}(\theta,\theta_0)$ is an average $X_N = \frac{1}{N}\sum_{i=1}^{N}f_i$ of N mean zero random variables $f_i = f_{x(s),g^*}^0(\theta,\theta_0)(C_x(i))$ that are functions of $C_x(i)$, where each random variable f_i goes to zero as θ approximates θ_0. From this representation, it follows that the remainder $R_{22,g^*,N}(\theta,\theta_0)$ can only be a second-order term if the memory of the time series is bounded or fast waning. This is discussed in detail in Sect. A.4.

19.3.2 Expansion for Online One-Step Estimator and Online TMLE

The following second-order expansion of the target parameter is needed for the online one-step estimator and online TMLE.

Theorem 19.3. *Let* $(\theta_n : n = 1, \ldots, N)$ *be a sequence of estimators of* $\theta_0 = (\bar{q}_0, \bar{g}_0)$. *We have*

$$-\frac{1}{N} \sum_{i=1}^{N} P_{0,o(i)} \bar{D}_{g^*}(\theta_{i-1}) = \frac{1}{N} \sum_{i=1}^{N} \Psi_{g^*}(\theta_{i-1}) - \Psi_{g^*}(\theta_0) + \bar{R}_{2,g^*,N},$$

where $\bar{R}_{2,g^*,N} = \bar{R}_{21,g^*,N} + \bar{R}_{22,g^*,N}$, $\bar{R}_{21,g^*,N}$ *is defined in Theorem A.5 (A.14), and*

$$\bar{R}_{22,g^*,N} \equiv \frac{1}{N} \sum_{i=1}^{N} (P_{0,o(i)} - P_0^N) \left(\bar{D}_{g^*}(\theta_{i-1}) - \bar{D}_{g^*}(\theta_0) \right).$$

As in the previous theorem, this also implies a corresponding second-order expansion for $\Psi_J(P)$.

In Sect. A.4 we discuss this remainder $\bar{R}_{21,g^*,N}$ and suggest that its form is not as nice as that of R_{21,g^*} and that controlling this remainder might require an additional stronger stationarity assumption.

19.4 TMLE

In this section we define a TMLE based on a least favorable fluctuation model $\{p_{\bar{q}_\epsilon,\bar{g}_\epsilon}^N : \epsilon\}$ through the whole density $p_{\bar{q},\bar{g}}^N$ at $\epsilon = 0$ that satisfies

$$\frac{d}{d\epsilon} \log p_{\bar{q}_\epsilon,\bar{g}_\epsilon}^N \bigg|_{\epsilon=0} (O^N) = D^N(\bar{q}, \bar{g})(O^N). \tag{19.4}$$

19.4.1 Local Least Favorable Fluctuation Model

Firstly, consider the case that $A(t)$, $Y(t)$ are binary and $W(t) = (W_k(t) : k = 1, \ldots, K)$ is a vector of K binaries, so that we can use the binary representation of the efficient influence curve in terms of sums of residuals multiplied by "clever" covariates H_y (19.1), H_a (19.2), H_{w_k} (19.3), $k = 1, \ldots, K$, as defined above. This suggest the following logistic regression fluctuation model through $(\bar{q}, \bar{g})$:

$$\text{Logit}\bar{q}_{y,\epsilon}(1 \mid c_y) = \text{Logit}\bar{q}_y(1 \mid c_y) + \epsilon H_y(c_y),$$
$$\text{Logit}\bar{q}_{w_k,\epsilon}(1 \mid c_{w_k}) = \text{Logit}\bar{q}_{w_k}(1 \mid c_{w_k}) + \epsilon H_{w_k}(c_{w_k}), k = 1, \ldots, K,$$
$$\text{Logit}\bar{g}_{a,\epsilon}(1 \mid c_a) = \text{Logit}\bar{g}(1 \mid c_a) + \epsilon H_a(c_a).$$

This defines now a one-dimensional fluctuation model $\{p_{\bar{q}_\epsilon,\bar{g}_\epsilon} : \epsilon\}$ though the whole density $p_{\bar{q},\bar{g}}$ at $\epsilon = 0$, and it satisfies

$$\frac{d}{d\epsilon} \log p_{\bar{q}_\epsilon,\bar{g}_\epsilon}\bigg|_{\epsilon=0} (O^N) = D^N(\bar{q}, \bar{g})(O^N). \tag{19.5}$$

We could also use a different $\epsilon_1, \epsilon_2, \epsilon_3$ for the fluctuations of $\bar{q}_y, \bar{q}_w, \bar{g}$, respectively, resulting in a 3 dimensional least favorable submodel.

In general, we can use any parametric model $\{p_{\bar{q}_\epsilon,\bar{g}_\epsilon} : \epsilon\}$ with a score at $\epsilon = 0$ that spans $D^N(\bar{q}, \bar{g})$. For example, for a local neighborhood for ϵ around zero, one could use

$$\bar{q}_{y,\epsilon}(y \mid c_y) = \{1 + \epsilon \bar{D}_{\bar{q}_y}(\bar{q}, \bar{g})(y \mid c_y)\}\bar{q}_y(y \mid c_y),$$
$$\bar{q}_{w,\epsilon}(w \mid c_w) = \{1 + \epsilon \bar{D}_{\bar{q}_w}(\bar{q}, \bar{g})(w \mid c_w)\}\bar{q}_w(w \mid c_w),$$
$$\bar{g}_\epsilon(a \mid c_a) = \{1 + \epsilon \bar{D}_{\bar{g}}(\bar{q}, \bar{g})(a \mid c_a)\}\bar{g}(a \mid c_a).$$

To avoid careful bounds on ϵ, one could use the usual exponential fluctuation $c(\epsilon) \exp(\epsilon D)p$ as a model through p at $\epsilon = 0$ with score D, using a normalizing constant $c(\epsilon)$.

19.4.2 One-Step TMLE

Let $(\bar{q}_N^0, \bar{g}_N^0)$ be an initial estimator of $(\bar{q}_0, \bar{g}_0)$, and let $\{p_{\bar{q}_{N,\epsilon}^0, \bar{g}_{N,\epsilon}^0} : \epsilon\}$ be the above defined local least favorable parametric submodel through $p_{\bar{q}_N^0, \bar{g}_N^0}$ at $\epsilon = 0$. Define ϵ_N as a solution of

$$D^N(\bar{q}_{N,\epsilon}, \bar{g}_{N,\epsilon})(O^N) = \frac{1}{N} \sum_{i=1}^N \bar{D}(\bar{q}_{N,\epsilon}, \bar{g}_{N,\epsilon})(\bar{O}(i)) = o_P(N^{-1.2}).$$

One could use the first step of the Newton-Raphson algorithm for solving this equation using an initial estimator ϵ^0:

$$\epsilon_N = \epsilon^0 + \frac{1}{N} \sum_{i=1}^N \gamma_N^{-1} \bar{D}(\bar{q}_{N,\epsilon^0}, \bar{g}_{N,\epsilon^0})(\bar{O}(N)),$$

where

$$\gamma_N = -\frac{d}{d\epsilon^0} \frac{1}{N} \sum_{i=1}^N \bar{D}(\bar{q}_{N,\epsilon^0}, \bar{g}_{N,\epsilon^0}).$$

Then, one only has to evaluate the $\bar{D}$ at a single fit $(\bar{q}, \bar{g})$, which is important given the fact that the evaluation of $\bar{D}(\bar{q}, \bar{g})$ can be quite computer intensive.

19.4.3 Iterative TMLE

Let $k = 0$, and define the MLE

$$\epsilon_N^k = \arg\max_{\epsilon} \log p_{\bar{q}_{N,\epsilon}^k, \bar{g}_{N,\epsilon}^k}(O^N).$$

This defines now the first step TMLE $\bar{q}_N^{k+1} = \bar{q}_{N,\epsilon_N^k}^k$ and $\bar{g}_n^{k+1} = \bar{g}_{N,\epsilon_N^k}^k$. Setting $k = k + 1$ and iterating this updating process defines a k-th step TMLE $(\bar{q}_N^k, \bar{g}_N^k)$ for $k = 1, 2, \ldots$. The corresponding k-th step TMLE of the target parameter $\Psi(\bar{q}_0, \bar{g}_0)$ is now given by the plug-in estimator $\Psi(\bar{q}_N^k, \bar{g}_N^k)$, $k = 1, \ldots$. If one iterates this updating process until a step K at which $\epsilon_N^K \approx 0$, then it follows by (19.5) that

$$D^N(\bar{q}_N^*, \bar{g}_N^*)(O^N) = \frac{1}{N} \sum_{i=1}^{N} \bar{D}(\bar{q}_N^*, \bar{g}_N^*)(O(i)) \approx 0,$$

where $\bar{q}_N^* = \bar{q}_N^K$ and $\bar{g}_N^* = \bar{g}_N^K$.

One-Step TMLE. If $(\bar{q}_N^1, \bar{g}_N^1)$ converge in N at a fast enough rate (e.g., faster than $N^{-1/4}$) to $(\bar{q}_0, \bar{g}_0)$, then one can show that the first step TMLE already solves this efficient influence curve equation up to an asymptotically negligible error:

$$D^N(\bar{q}_N^1, \bar{g}_N^1)(O^N) = o_P(N^{-1/2}).$$

This one-step TMLE is computationally and practically feasible, while a HAL super learner estimator will indeed converge at a rate faster than $N^{-1/4}$ (Chap. 15).

In general, statistically, we recommend to iterate the TMLE update algorithm until the efficient influence curve is significantly smaller than σ_N (e.g. $\sigma_N/10$ or $\sigma_N/\sqrt{N}$), where σ_N is an estimate of the standard error of D^N presented below. In this manner it is guaranteed that a remaining bias due to not precisely solving the efficient influence curve equation is not affecting coverage of the confidence interval for ψ_0.

19.4.4 Analysis of the TMLE

An analysis of this TMLE relies on $M_N(\theta) = \frac{1}{\sqrt{N}} \sum_{i=1}^{N} \{\bar{D}_{g^*}(\theta)(\bar{O}(i)) - P_{0,o(i)}\bar{D}_{g^*}(\theta)\}$ be a martingale process in θ. Under entropy conditions on the parameter space Θ for θ, we can establish asymptotic equicontinuity of this martingale process, so that

$M_N(\theta_N^*) - M_N(\theta_0) = o_P(1)$ if θ_N^* is a consistent estimator of θ_0, and thereby carry out a standard TMLE analysis. This will be presented formally in future work. Our online estimators in the next sections will not rely on such entropy conditions, making their analysis more straightforward.

19.5 Online One-Step Estimator

Consider an online estimator $(\theta_i : i)$ of $\theta_0 = (\bar{q}_0, \bar{g}_0)$. Recall (Theorem 19.3)

$$-\frac{1}{N}\sum_{i=1}^{N} P_{0,o(i)} \bar{D}_{g^*}(\theta_{i-1}) = \frac{1}{N}\sum_{i=1}^{N} \Psi_{g^*}(\theta_{i-1}) - \Psi_{g^*}(\theta_0) + \bar{R}_{2,g^*,N}.$$

Define the following online one-step estimator:

$$\psi_{g^*,N} = \frac{1}{N}\sum_{i=1}^{N} \Psi_{g^*}(\theta_{i-1}) + \frac{1}{N}\sum_{i=1}^{N} \bar{D}_{g^*}(\theta_{i-1})(\bar{O}(i)).$$

We have

$$\psi_{g^*,N} - \psi_{g^*}(\theta_0) = \frac{1}{N}\sum_{i=1}^{N} \Psi_{g^*}(\theta_{i-1}) - \Psi_{g^*}(\theta_0) + \frac{1}{N}\sum_{i=1}^{N} \bar{D}_{g^*}(\theta_{i-1})(\bar{O}(i))$$

$$= \frac{1}{N}\sum_{i=1}^{N} \{\bar{D}_{g^*}(\theta_{i-1}) - P_{0,o(i)}\bar{D}_{g^*}(\theta_{i-1})\} - \bar{R}_{2,g^*,N}$$

$$\equiv \frac{M_N}{\sqrt{N}} - \bar{R}_{2,g^*,N},$$

The martingale central limit theorem now proves the following theorem.

Theorem 19.4. *Assume $\bar{R}_{2,g^*,N} = o_P(N^{-1/2})$; $\max_i \mid \bar{D}_{g^*,i}(\theta_{i-1}) \mid < M$ with probability 1 for some $M < \infty$;*

$$\frac{1}{N}\sum_{i=1}^{N}(P_0^N - P_{0,o(i)})\bar{D}_{g^*,i}(\theta_{i-1})\bar{D}_{g^*,i}(\theta_{i-1})^\top \to 0,$$

as $N \to \infty$. Let $\Sigma(N) = \frac{1}{N}\sum_{i=1}^{N} P_0^N \bar{D}_{g^*,i}(\theta_{i-1})\bar{D}_{g^*,i}(\theta_{i-1})^\top$. Then,

$$\sqrt{N}(\psi_{g^*,N} - \psi_{g^*}(\theta_0)) = M_N + o_P(1),$$

$\Sigma(N)^{-1/2} M_N \Rightarrow_d N(0, I)$, $\sqrt{N}\Sigma(N)^{-1/2}(\psi_{g^*,N} - \psi_{g^*}(\theta_0)) \Rightarrow_d N(0, I)$, as $N \to \infty$. If $\Sigma(N) \to \Sigma_0$ for a fixed Σ_0, then we have

$$\sqrt{N}(\psi_N - \psi_0) \Rightarrow_d N(0, \Sigma_0).$$

Similarly, we can define the online one-step estimator of $\Psi_J(P_0)$ and present the complete analogue theorem for the online one-step estimator of $\Psi_J(P_0)$.

Second-Order Remainder of Online One-Step Estimator Relative to Regular One-Step Estimator. We have $\bar{R}_{2,g^*,N} = \frac{1}{N}\sum_{i=1}^{N} R_{2,g^*,i}(\theta_{i-1},\theta_0)$ for i-specific second-order terms $R_{2,g^*,i}(\theta_{i-1},\theta_0)$. Suppose $R_{2,g^*,i}(\theta_{i-1},\theta_0) = O(i^{-\delta})$ for some $0 < \delta < 1$. In that case, $\bar{R}_{2,g^*,N} = \frac{1}{N}\sum_{i=1}^{N} R_{2,g^*,i}(\theta_{i-1},\theta_0) = O(N^{-\delta})$. In other words, the rate of convergence to zero for the second-order remainder $\bar{R}_{2,g^*,N}$ is the same as the remainder $R_{2,g^*,N}(\theta_N,\theta_0)$ for a regular one-step estimator. However, note that one expects that $\bar{R}_{2,g^*,N}$ is larger than $R_{2,g^*,N}(\theta_N,\theta_0)$ by a constant $1/(-\delta+1)$. In the special case that $R_{2,g^*,i}(\theta_{i-1},\theta_0) = O(i^{-1})$ (i.e., the rate of an MLE for a correctly specified parametric model), then we have $\bar{R}_{2,g^*,N} = O(\log N/N)$, so that $\bar{R}_{2,g^*,N}$ is a $\log N$ factor larger than $R_{2,g^*,N}(\theta_N,\theta_0)$.

19.6 Online TMLE

Even though $(\theta_i : i)$ might be an online estimator, the actual TMLE θ_{N,ϵ_N} of θ_0 presented in the previous section is not an online estimator since ϵ_N is not an online estimator. In this section we develop an online one-step estimator ϵ_N^* for which $\Psi_{g^*}(\theta_{N,\epsilon_N^*})$ is still an asymptotically efficient estimator of $\Psi_{g^*}(\theta_0)$. We refer to this estimator as the online TMLE. The online one-step estimator $(\epsilon_N^* : N)$ takes as input the online estimators $(\theta_N : N)$ of θ_0 and an initial online estimator $(\epsilon_N^0 : N)$. As with the regular TMLE, $\{\theta_{N,\epsilon} : \epsilon\}$ is the least favorable submodel through θ_N at $\epsilon = 0$ so that the linear span of its score $\frac{d}{d\epsilon}\log p_{\theta_{N,\epsilon}}^N(O^N)\big|_{\epsilon=0}$ contains $D^N(\theta_N)(O^N)$. In addition, we assume that ϵ has the same dimension as $\psi_{g^*,0}$.

Let ϵ_N^0 be an initial estimator of ϵ_{0N}, where $\epsilon_{0N} = f_N^{-1}(\psi_{g^*,0})$ is defined as the solution of

$$f_N(\epsilon_{0N}) \equiv \Psi_{g^*}(\theta_{N,\epsilon_{0N}}) = \Psi_{g^*}(\theta_0).$$

If θ_N is a consistent estimator of θ_0, a requirement for our efficiency theorem, then we could set $\epsilon_N^0 = 0$. We use this initial estimator ϵ_N^0 of ϵ_{0N} to construct a first order Taylor expansion $\tilde{f}_N$ of $\epsilon \rightarrow f_N(\epsilon)$ in a neighborhood of ϵ_{0N}:

$$\tilde{f}_N(\epsilon) = f_N(\epsilon_N^0) + \frac{d}{d\epsilon_N^0}f_N(\epsilon_N^0)(\epsilon - \epsilon_N^0).$$

Let f_N^{-1} be the inverse function of f_N, and similarly let $\tilde{f}_N^{-1}$ be the inverse function of $\tilde{f}_N$:

$$\tilde{f}_N^{-1}(\psi) = \epsilon_N^0 + c_N^{-1}(\psi - f_N(\epsilon_N^0)),$$

where we used the short-hand notation

$$c_N \equiv \frac{d}{d\epsilon_N^0} f_N(\epsilon_N^0).$$

We have the following lemma showing that the first order Taylor expansion $\tilde{f}_N^{-1}(\psi_0)$ approximates $\epsilon_{0N} = f_N^{-1}(\psi_0)$ up to a negligible $o_P(N^{-1/2})$ remainder term. In the following lemma, ψ_0 represents $\psi_{g^*,0}$.

Lemma 19.2. *Suppose that for some $M < \infty$, for $k \in \{1, 2\}$, $\| \left(\frac{d}{d\epsilon_N^0} \right)^k f_N(\epsilon_N^0) \| < M$ with probability tending to 1; $\| c_N^{-1} \| < M$ with probability tending to 1; $\| \epsilon_{0N} - \epsilon_N^0 \|^2 = o_P(N^{-1/2})$.*
  Then,

$$R_{0N}(\epsilon_N^0, \epsilon_{0N}) \equiv \tilde{f}_N^{-1}(\psi_0) - f_N^{-1}(\psi_0) = o_P(N^{-1/2}). \tag{19.6}$$

Proof. If f_N is twice differentiable with bounded continuous second derivative, we have

$$f_N(\epsilon) = \tilde{f}_N(\epsilon) + O_P(\| \epsilon - \epsilon_N^0 \|^2).$$

We have

$$\begin{aligned}
\tilde{f}_N^{-1}(\psi_0) - f_N^{-1}(\psi_0) &= -\left\{ \tilde{f}_N^{-1} \tilde{f}_N f_N^{-1}(\psi_0) - \tilde{f}_N^{-1} f_N f_N^{-1}(\psi_0) \right\} \\
&= -\left\{ \epsilon_N^0 + c_N^{-1} \{ \tilde{f}_N(\epsilon_{0N}) - f_N(\epsilon_N^0) \} - \epsilon_N^0 - c_N^{-1} \{ \psi_0 - f_N(\epsilon_N^0) \} \right\} \\
&= -c_N^{-1} \left\{ \tilde{f}_N(\epsilon_{0N}) - f_N(\epsilon_{0N}) \right\} \\
&= -c_N^{-1} O_P(\| \epsilon_{0N} - \epsilon_N^0 \|^2) \\
&= O_P(\| \epsilon_{0N} - \epsilon_N^0 \|^2),
\end{aligned}$$

where we use that c_N^{-1} has a uniformly in N bounded norm. By assumption, $\| \epsilon_{0N} - \epsilon_N^0 \|^2 = o_P(N^{-1/2})$. $\square$

The assumption $\| \epsilon_{0N} - \epsilon_N^0 \|^2 = o_P(N^{-1/2})$ typically corresponds with assuming that θ_N converges to θ_0 at a rate faster than $N^{-1/4}$ with respect to a loss-based dissimilarity. The condition (19.6) is the crucial condition our proposed online estimator ϵ_N^* below relies upon, which, by this lemma, holds under weak additional regularity conditions. We now define the following online one-step estimator ϵ_N^* of ϵ_{0N} in terms of $\tilde{f}_N$:

$$\epsilon_N^* = \frac{1}{N} \sum_{i=1}^N c_N^{-1} \Psi_{g^*}(\theta_{i-1}) + \frac{1}{N} \sum_{i=1}^N \tilde{f}_N^{-1} \bar{D}_{g^*}(\theta_{i-1})(\bar{O}(i)).$$

If we select $\epsilon_N^0 = 0$, then this reduces to

$$\epsilon_N^* = \frac{1}{N} \sum_{i=1}^N c_N^{-1} \bar{D}_{g^*}(\theta_{i-1})(\bar{O}(i)).$$

This defines our online TMLE $\psi_N^* = \Psi_{g^*}(\theta_{N,\epsilon_N^*})$ of $\psi_{g^*,0}$. In order to analyze this online TMLE, we first present the following theorem for this online one-step estimator ϵ_N^*:

Theorem 19.5. Let $M_N = \frac{1}{\sqrt{N}} \sum_{i=1}^{N} \{\bar{D}_{g^*}(\theta_{i-1})(\bar{O}(i)) - P_{0,o(i)}\bar{D}_{g^*,i}\}$ and the second-order remainder

$$\bar{R}(N) = -\frac{1}{N} \sum_{i=1}^{N} P_{0,o(i)}\bar{D}_{g^*}(\theta_{i-1}) - \left\{\frac{1}{N} \sum_{i=1}^{N} \Psi_{g^*}(\theta_{i-1}) - \Psi_{g^*}(\theta_0)\right\},$$

explicitly defined in Theorem 19.3. Recall the above definitions of f_N, $\tilde{f}_N$, c_N, $R_{0N}(\epsilon_N^0, \epsilon_{0N}) \equiv \tilde{f}_N^{-1}(\psi_{g^*,0}) - f_N^{-1}(\psi_{g^*,0})$ and ϵ_{0N} defined by $\Psi_{g^*}(\theta_{N,\epsilon}) = \psi_{g^*,0}$. Let $c_{0N} \equiv \frac{d}{d\epsilon_{0N}} \Psi_{g^*}(\theta_{N,\epsilon_{0N}})$. We have the following expansion:

$$\epsilon_N^* - \epsilon_{0N} = c_N^{-1} M_N / \sqrt{N} + c_N^{-1}\bar{R}(N) + R_{0N}(\epsilon_N^0, \epsilon_{0N}). \tag{19.7}$$

Assumptions. Assume $\bar{R}(N) = o_P(1/\sqrt{N})$, and the martingale consistency conditions on θ_N so that M_N converges in distribution to $Z \sim N(0, \Sigma_0)$. In addition, assume $\epsilon_N^0, \epsilon_{0N}$ converge to zero in probability and $\|\epsilon_N^0 - \epsilon_{0N}\|^2 = o_P(N^{-1/2})$; the first and second derivative of $f_N(\epsilon)$ at 0 are continuous and bounded uniformly in N with probability tending to 1; the inverse of the first derivative $c_N(\epsilon) = \frac{d}{d\epsilon}f_N(\epsilon)$ is continuous in ϵ at $\epsilon = 0$;

Conclusion. Then,

$$\epsilon_N^* - \epsilon_{0N} = c_{0N}^{-1}\frac{M_N}{\sqrt{N}} + o_P(1/\sqrt{N}),$$

and $\sqrt{N}(\epsilon_N^* - \epsilon_{0N}) - c_{0N}^{-1}Z \to 0$ in probability.

Using a simple δ-method argument provided in the proof of Theorem 19.6 below, Theorem 19.5 establishes the asymptotic efficiency of the online TMLE $\Psi_{g^*}(\theta_{N,\epsilon_N^*})$ of $\psi_{g^*,0}$.

Theorem 19.6. Under the same conditions as in Theorem 19.5, we have

$$\Psi_{g^*}(\theta_{N,\epsilon_N^*}) - \Psi_{g^*}(\theta_0) = \frac{M_N}{\sqrt{N}} + o_P(1/\sqrt{N}),$$

and thereby $\sqrt{N}(\Psi_{g^*}(\theta_N(\epsilon_N^*)) - \Psi_{g^*}(\theta_0)) \Rightarrow_d Z$, where $Z \sim N(0, \Sigma_0)$. Thus $\Psi_{g^*}(\theta_N(\epsilon_N^*))$ is an asymptotically efficient estimator of $\psi_{g^*,0}$.

Proof of Theorem 19.6

In this proof and the proof of Theorem 19.5, we will suppress the dependence of the quantities on g^* in our notation. Consider the online TMLE $\psi_N^* = \Psi(\theta_{N,\epsilon_N^*})$ as estimator of ψ_0, and the conclusion of Theorem 19.5 stating $\epsilon_N^* - \epsilon_{0N} = c_{0N}^{-1} M_N / \sqrt{N} + o_P(N^{-1/2})$. We have for a $\tilde{\epsilon}_{0N}$ between ϵ_N^* and ϵ_{0N}

$$
\begin{aligned}
\Psi(\theta_{N,\epsilon_N^*}) - \Psi(\theta_0) &= \Psi(\theta_{N,\epsilon_N^*}) - \Psi(\theta_{N,\epsilon_{0N}}) \\
&= \frac{d}{d\tilde{\epsilon}_{0N}} \Psi(\theta_{N,\tilde{\epsilon}_{0N}})(\epsilon_N^* - \epsilon_{0N})) \\
&= \frac{d}{d\tilde{\epsilon}_{0N}} \Psi(\theta_{N,\tilde{\epsilon}_{0N}}) c_{0N}^{-1} \frac{M_N}{\sqrt{N}} + o_P(1/\sqrt{N}) \\
&= \frac{M_N}{\sqrt{N}} + o_P(1/\sqrt{N}),
\end{aligned}
$$

where we use that $\tilde{\epsilon}_{0N} - \epsilon_{0N}$ converges to zero as $N \to \infty$, so that, by assumption

$$
\frac{d}{d\tilde{\epsilon}_{0N}} \Psi(\theta_{N,\tilde{\epsilon}_{0N}}) c_{0N}^{-1} \to 1 \text{ in probability as } N \to \infty.
$$

Proof of Theorem 19.5

We first prove (19.7). We have

$$
\begin{aligned}
\epsilon_N^* &= \frac{1}{N} \sum_{i=1}^N c_N^{-1} \Psi(\theta_{i-1}) \\
&+ \frac{1}{N} \sum_{i=1}^N \left\{ \tilde{f}_N^{-1} \bar{D}_{g^*}(\theta_{i-1})(\bar{O}(i)) - \tilde{f}_N^{-1} P_{0,o(i)} \bar{D}_{g^*}(\theta_{i-1}) \right\} \\
&+ \frac{1}{N} \sum_{i=1}^N \tilde{f}_N^{-1} P_{0,o(i)} \bar{D}_{g^*}(\theta_{i-1}).
\end{aligned}
$$

By the identity

$$
-P_{0,o(i)} \bar{D}_{g^*}(\theta_{i-1}) = \Psi(\theta_{i-1}) - \Psi(\theta_0) + R_{2,i}(\theta_{i-1}, \theta_0),
$$

the last term equals

$$
\begin{aligned}
&\frac{1}{N} \sum_{i=1}^N \tilde{f}_N^{-1} \{ \Psi(\theta_0) - \Psi(\theta_{i-1}) + R_{2,i}(\theta_{i-1}, \theta_0) \} \\
&= \frac{1}{N} \sum_{i=1}^N \tilde{f}_N^{-1} \{ \Psi(\theta_0) - \Psi(\theta_{i-1}) + \frac{1}{N} \sum_{i=1}^N R_{2,i}(\theta_{i-1}, \theta_0).
\end{aligned}
$$

Regarding the second term, recall $\tilde{f}_N^{-1}(\psi) = \epsilon_N^0 - c_N^{-1} \Psi(\theta_N(\epsilon_N^0)) + c_N^{-1} \psi$ equals a constants plus a linear transformation c_N^{-1}. Thus, $\tilde{f}_N^{-1} a - \tilde{f}_N^{-1} b = c_N^{-1}(a - b)$ for any two numbers a, b. Thus, we have

$$
\begin{aligned}
&\frac{1}{N} \sum_{i=1}^N \left\{ \tilde{f}_N^{-1} \bar{D}_{g^*}(\theta_{i-1})(\bar{O}(i)) - \tilde{f}_N^{-1} P_{0,o(i)} \bar{D}_{g^*}(\theta_{i-1}) \right\} \\
&= \frac{1}{N} \sum_{i=1}^N c_N^{-1} \left\{ \bar{D}_{g^*,i}(\theta_{i-1})(\bar{O}(i)) - P_{0,o(i)} \bar{D}_{g^*}(\theta_{i-1}) \right\}.
\end{aligned}
$$

So we conclude that

$$
\begin{aligned}
\epsilon_N^* &= \frac{1}{N} \sum_{i=1}^N c_N^{-1} \Psi(\theta_{i-1}) + c_N^{-1} \frac{M_N}{\sqrt{N}} \\
&+ \frac{1}{N} \sum_{i=1}^N \tilde{f}_N^{-1} \{ \Psi(\theta_0) - \Psi(\theta_{i-1}) \} + c_N^{-1} \bar{R}_{2,N},
\end{aligned}
$$

where

$$\bar{R}_{2,N} = \frac{1}{N} \sum_{i=1}^{N} R_{2,i}(\theta_{i-1}, \theta_0),$$

and

$$M_N = \frac{1}{\sqrt{N}} \sum_{i=1}^{N} \left\{ \bar{D}_{g^*}(\theta_{i-1})(\bar{O}(i)) - P_{0,o(i)}\bar{D}_{g^*}(\theta_{i-1}) \right\}.$$

Let's now focus on the last term $\frac{1}{N} \sum_{i=1}^{N} \tilde{f}_N^{-1}\{\Psi(\theta_0) - \Psi(\theta_{i-1})\}$, which equals

$$\begin{aligned}
&\frac{1}{N} \sum_{i=1}^{N} \left\{ \epsilon_N^0 - c_N^{-1} f_N(\epsilon_N^0) + c_N^{-1}\psi_0 - c_N^{-1}\Psi(\theta_{i-1}) \right\} \\
&= \{\epsilon_N^0 - c_N^{-1} f_N(\epsilon_N^0) + c_N^{-1}\psi_0\} - \frac{1}{N} \sum_{i=1}^{N} c_N^{-1}(\Psi(\theta_{i-1}) \\
&= \tilde{f}_N^{-1}(\psi_0) - \frac{1}{N} \sum_{i=1}^{N} c_N^{-1}\Psi(\theta_{i-1}).
\end{aligned}$$

Plug this expression back into our expression above for ϵ_N^* and notice that

$$\frac{1}{N} \sum_{i} c_N^{-1}\Psi(\theta_{i-1})$$

cancels out. So we obtain:

$$\epsilon_N^* - \tilde{f}_N^{-1}(\psi_0) = c_N^{-1} \frac{M_N}{\sqrt{N}} + c_N^{-1}\bar{R}_{2,N}.$$

Finally, we use that $\tilde{f}_N^{-1}(\psi_0) = f_N^{-1}(\psi_0) + R_{0N}(\epsilon_N^0, \epsilon_{0N})$ and $f_N^{-1}(\psi_0) = \epsilon_{0N}$ to obtain:

$$\epsilon_N^* - \epsilon_{0N} = c_N^{-1} \frac{M_N}{\sqrt{N}} + c_N^{-1}\bar{R}_{2,N} + R_{0N}(\epsilon_N^0, \epsilon_{0N}).$$

By assumption and Lemma 19.2, we have $R_{0N}(\epsilon_0^N, \epsilon_{0N}) = o_P(N^{-1/2})$. We assumed that $\bar{R}_{2,N} = o_P(1/\sqrt{N})$. Under these assumptions, we have

$$\epsilon_N^* - \epsilon_{0N} = c_N^{-1} \frac{M_N}{\sqrt{N}} + o_P(1/\sqrt{N}).$$

By our continuity and boundedness assumption on the inverse of the first derivative $c_N(\epsilon)$ at $\epsilon = 0$, by $Z_N = M_N/\sqrt{N}$ converging to Z, we have

$$\begin{aligned}
(c_N^{-1} - c_{0N}^{-1})(Z_N) &= -\left\{ c_{0N}^{-1} c_N(c_N^{-1}(Z_N)) - c_{0N}^{-1} c_{0N}(c_N^{-1}(Z_N)) \right\} \\
&= -c_{0N}^{-1}(c_N - c_{0N})(c_N^{-1}(Z_N)) \\
&\to_p 0.
\end{aligned}$$

This completes the proof of Theorem 19.5.

19.7 Online Targeted Learning with Independent Identically Distributed Data

In the next subsection we define the estimation problem for estimating an average causal effect based on independent and identically distributed $(W(t), A(t), Y(t))$, $t = 1, \ldots, N$, and we define the relevant quantities. In Sect. 19.7.2 we present the online one-step estimator and a corresponding theorem establishing asymptotic efficiency. In Sect. 19.7.3 we define the online TMLE.

19.7.1 Online Targeted Learning of the Average Causal Effect

We observe N i.i.d. observations $O(t) = (W(t), A(t), Y(t))$, $t = 1, \ldots, N$, and let P_0 be the common probability distribution of $O(t)$. Suppose that $Y(t)$ is binary or that it is continuous with values in $(0, 1)$. Let $\bar{q}_w, \bar{g}, \bar{q}_y$ be the common marginal density of $W(t)$, the conditional density of $A(t)$, given $W(t)$, and the conditional density of $Y(t)$, given $W(t), A(t)$, under P, respectively. Let $\theta(P) = (\bar{q}_w, \bar{g}, \bar{q}_y)$. Let $\bar{Q}_w, \bar{Q}_y, \bar{G}$ be the corresponding conditional probability distributions. Let $\mathcal{M}$ be a statistical model for P_0 that only makes assumptions about $\bar{g}_0$, and consider the target parameter $\Psi : \mathcal{M} \to \mathbb{R}$ defined by $\Psi(P) = E_P E_P(Y \mid A = 1, W)$. Let $\bar{Q}(W) = E_P(Y \mid A = 1, W)$ and note that $\Psi(P) = \bar{Q}_w \bar{Q} = \int \bar{Q}(w) d\bar{Q}_w(w)$. We will also use the notation $\Psi(\bar{Q}_w, \bar{Q})$ or $\Psi(\theta)$. The efficient influence curve of $\Psi : \mathcal{M} \to \mathbb{R}$ is given by

$$\bar{D}(\theta)(O(t)) = \frac{A(t)}{\bar{g}(A(t) \mid W(t))}(Y(t) - \bar{Q}(W(t))) + \bar{Q}(W(t)) - \Psi(\bar{Q}_w, \bar{Q}).$$

This satisfies the identity:

$$-P_0\bar{D}(\theta) = \Psi(\theta) - \Psi(\theta_0) + R_{20}(\theta, \theta_0),$$

where

$$R_{20}(\theta, \theta_0) = E_{P_0}(\bar{g} - \bar{g}_0)/\bar{g}(1 \mid W)(\bar{Q} - \bar{Q}_0)(W).$$

19.7.2 Online One-Step Estimator

Let $(\bar{Q}_i : i)$ and $(\bar{g}_i : i)$ be online estimators of $\bar{Q}_0$ and $\bar{g}_0$, respectively. This also defines an online estimator $(\psi_i : i)$ of ψ_0:

$$\psi_i = \frac{1}{i} \sum_{j=1}^{i} \bar{Q}_{j-1}(W_j).$$

Notice that we can write $\bar{D}(\theta) = \bar{D}_1(\theta) - \Psi(\theta)$, where $\bar{D}_1(\theta) = A(t)/\bar{g}(A(t) \mid W(t))(Y(t) - \bar{Q}(W(t))) + \bar{Q}(W(t))$. The online one-step estimator is now defined as:

$$\psi_N = \frac{1}{N} \sum_{i=1}^{N} \psi_{i-1} + \frac{1}{N} \sum_{i=1}^{N} \bar{D}(\theta_{i-1}, \psi_{i-1})(O(i))$$

$$= \frac{1}{N} \sum_{i=1}^{N} \bar{D}_1(\theta_{i-1})(O(i)).$$

It follows that

$$\psi_N - \psi_0 = \frac{1}{N} \sum_{i=1}^{N} \left\{ \bar{D}(\theta_{i-1})(O(i)) - P_0 \bar{D}(\theta_{i-1}) \right\}$$

$$+ \frac{1}{N} \sum_{i=1}^{N} R_{20}(\theta_{i-1}, \theta_0). \tag{19.8}$$

The following theorem is an application of the general Theorem 19.4.

Theorem 19.7. *Assume*

$$\bar{R}_N = \frac{1}{N} \sum_{i=1}^{N} R_{20}(\theta_{i-1}, \theta_0) = o_P(N^{-1/2});$$

$\bar{g}(1 \mid W) > \delta > 0$ *a.e. for some* $\delta > 0$,

$$\frac{1}{N} \sum_{i=1}^{N} P_0 \{\bar{D}(\theta_{i-1}) - \bar{D}(\theta_0)\}^2 = o_P(1);$$

$\frac{1}{N} \sum_{i=1}^{N} P_0 \{\bar{D}(\theta_{i-1})\}^2 \to_p \sigma_0^2$ *as* $N \to \infty$, *where* $\sigma_0^2 = P_0 \bar{D}(\theta_0)^2$.
Then,

$$\sqrt{N} (\psi_N - \Psi(\theta_0)) = \frac{1}{\sqrt{N}} \sum_{i=1}^{N} \bar{D}(\theta_0)(O(i)) + o_P(1),$$

and the right-hand side converges in distribution to $N(0, \sigma_0^2)$.

19.7.3 Online TMLE

Consider the local least favorable fluctuation model

$$\text{Logit}\bar{Q}_{N,\epsilon} = \text{Logit}\bar{Q}_N + \epsilon H(\bar{g}_N),$$

where $H(\bar{g})(W) = 1/\bar{g}(1 \mid W)$. The online estimator $(\bar{Q}_i : i)$ and its fluctuation, implies an ϵ-specific online estimator $(\bar{Q}_{i,\epsilon} : i)$ for any given ϵ. We consider the following two online estimators of its mean with respect to W:

$$\psi_{N,\epsilon}^1 \equiv \frac{1}{N} \sum_{i=1}^{N} \bar{Q}_{i-1,\epsilon}(W_i),$$

$$\psi_{N,\epsilon}^2 \equiv \frac{1}{N} \sum_{i=1}^{N} \bar{Q}_{N,\epsilon}(W_i).$$

In the second estimator, we use the empirical mean of $\bar{Q}_{N,\epsilon}$ with respect to distribution of W which is not an online type estimator, while in the first estimator we use a full online estimator to estimate this empirical mean. Let $\theta = (\bar{Q}, \psi)$ represent the outcome regression and corresponding target parameter value also depending on the distribution of W. Then $\theta_{N,\epsilon}^j = (\bar{Q}_{N,\epsilon}, \psi_{N,\epsilon}^j)$, $j = 1, 2$, is its ϵ-specific online estimator. Below, we will present two online one-step estimators ϵ_N^{*j}, $j = 1, 2$, which results in corresponding online TMLE $\psi_N^{*j} = \psi_{N,\epsilon_N^{*j}}^j$, $j = 1, 2$.

Let ϵ_N^{0j} be an initial estimator of ϵ_{0N}^j, where ϵ_{0N}^j is defined as the solution of

$$f_N^j(\epsilon) \equiv \psi_{N,\epsilon}^j = \Psi(\theta_0),$$

or equivalently,

$$\frac{1}{N} \sum_{i=1}^{N} \bar{Q}_{i-1,\epsilon_{0N}^1}(W_i) = \psi_0,$$

$$\frac{1}{N} \sum_{i=1}^{N} \bar{Q}_{N,\epsilon_{0N}^2}(W_i) = \psi_0.$$

If θ_N is a consistent estimator of θ_0, a requirement for our efficiency theorem below, one can select $\epsilon_N^{0j} = 0$.

We use this initial estimator ϵ_N^{0j} of ϵ_{0N}^j to construct a first order Taylor expansion $\tilde{f}_N^j(\epsilon)$ of $f_N^j(\epsilon)$ in a neighborhood of ϵ_{0N}^j:

$$\tilde{f}_N^j(\epsilon) = f_N^j(\epsilon_N^{0j}) + \frac{d}{d\epsilon_N^{0j}} f_N^j(\epsilon_N^{0j})(\epsilon - \epsilon_N^{0j}).$$

We will use the short-hand notation $c_N^j \equiv \frac{d}{d\epsilon_N^{0j}} f_N^j(\epsilon_N^{0j})$, $j = 1, 2$. In other words,

$$\frac{1}{N} \sum_{i=1}^{N} \bar{Q}_{i-1,\tilde{f}_N^1(\epsilon)}(W(i)) = \frac{1}{N} \sum_{i=1}^{N} \bar{Q}_{i-1,\epsilon_N^{01}}(W(i)) + \frac{1}{N} \sum_{i=1}^{N} \frac{d}{d\epsilon_N^{01}} \bar{Q}_{i-1,\epsilon_N^{01}}(W(i))(\epsilon - \epsilon_N^{01})$$

$$= \frac{1}{N} \sum_{i=1}^{N} \bar{Q}_{i-1,\epsilon_N^{01}}(W(i)) + c_N^1(\epsilon - \epsilon_N^{01}),$$

$$\frac{1}{N} \sum_{i=1}^{N} \bar{Q}_{N,\tilde{f}_N^2(\epsilon)}(W(i)) = \frac{1}{N} \sum_{i=1}^{N} \bar{Q}_{N,\epsilon_N^{02}}(W(i)) + \frac{1}{N} \sum_{i=1}^{N} \frac{d}{d\epsilon_N^{02}} \bar{Q}_{N,\epsilon_N^{02}}(W(i))(\epsilon - \epsilon_N^{02})$$

$$= \frac{1}{N} \sum_{i=1}^{N} \bar{Q}_{N,\epsilon}(W(i)) + c_N^2(\epsilon - \epsilon_N^{02}).$$

Since $\bar{Q}_{i,\epsilon}$ is differentiable in ϵ with a bounded derivative (since $\bar{g}_0(W) > \delta > 0$ a..e), we have

$$f_N^j(\epsilon) - \tilde{f}_N^j(\epsilon) = O_P(\| \epsilon - \epsilon_N^{0j} \|^2).$$

Let $f_N^{j,-1}$ be the inverse function of f_N^j, and similarly let $\tilde{f}_N^{j,-1}$ be the inverse function of $\tilde{f}_N^j$:

$$\tilde{f}_N^{j,-1}(\psi) = \epsilon_N^{0j} + c_N^{j,-1}(\psi - f_N^j(\epsilon_N^{0j})).$$

As in our general representation, we have

$$\tilde{f}_N^{j,-1}(\psi_0) - f_N^{j,-1}(\psi_0) = -\left\{ \tilde{f}_N^{j,-1} \tilde{f}_N^j f_N^{j,-1}(\psi_0) - \tilde{f}_N^{j,-1} f_N^j f_N^{j,-1}(\psi_0) \right\}$$

$$= -\left\{ \epsilon_N^{0j} + c_N^{j,-1}\{\tilde{f}_N^j(\epsilon_{0N}^j) - f_N^j(\epsilon_N^{0j})\} - \epsilon_N^{0j} - c_N^{j,-1}\{\psi_0 - f_N^j(\epsilon_N^{0j})\} \right\}$$

$$= -c_N^{j,-1} \left\{ \tilde{f}_N^j(\epsilon_{0N}^j) - f_N^j(\epsilon_N^{0j})\} - \psi_0 + f_N^j(\epsilon_N^{0j}) \right\}$$

$$= -c_N^{j,-1} \left\{ \tilde{f}_N^j(\epsilon_{0N}^j) - f_N^j(\epsilon_{0N}^j) \right\}$$

$$= -c_N^{j,-1} O_P(\| \epsilon_{0N}^j - \epsilon_N^{0j} \|^2)$$

$$= O_P(\| \epsilon_{0N}^j - \epsilon_N^{0j} \|^2),$$

where we use that $c_N^{j,-1}$ is bounded. We assume that $\| \epsilon_{0N}^j - \epsilon_N^{0j} \|^2 = o_P(N^{-1/2})$. This corresponds with assuming that θ_N^j converges at a rate faster than $N^{-1/4}$ w.r.t. loss-based dissimilarity. So we can conclude that for both definitions of the online TMLE we have

$$R_{0N}(\epsilon_N^{0j}, \epsilon_{0N}^j) \equiv \tilde{f}_N^{j,-1}(\psi_0) - f_N^{j,-1}(\psi_0) = o_P(N^{-1/2}).$$

We now define the following online one-step estimator ϵ_N^{*j} of ϵ_{0N}^j:

$$\epsilon_N^{*j} = \frac{1}{N} \sum_{i=1}^{N} c_N^{j,-1}\psi_{i-1} + \frac{1}{N} \sum_{i=1}^{N} \tilde{f}_N^{j,-1} \bar{D}(\theta_{i-1})(\bar{O}(i)).$$

Note we can use here the online definition ψ_{i-1} as in the estimator $j = 1$ above. This defines our online TMLE $\psi_N^{*j} = \psi_{N,\epsilon_N^{*j}}^j$ of ψ_0.

We first present the following theorem for this online estimator ϵ_N^{*j}:

Theorem 19.8. *Let* $M_N = \frac{1}{\sqrt{N}} \sum_{i=1}^{N} \{\bar{D}(\theta_{i-1})(O(i)) - P_0 \bar{D}(\theta_{i-1})\}$ *and the second-order remainder*

$$\bar{R}(N) = \frac{1}{N} \sum_{i=1}^{N} R_{20}(\theta_{i-1}, \theta_0).$$

Let $j \in \{1, 2\}$ *be given. Recall the above definitions of* f_N^j, $\tilde{f}_N^j$, c_N^j, $R_{0N}(\epsilon_N^{0j}, \epsilon_{0N}^j) \equiv \tilde{f}_N^{j,-1}(\psi_0) - f_N^{j,-1}(\psi_0)$ *and* ϵ_{0N}^j *defined by* $\psi_{N,\epsilon}^j = \psi_0$. *Let* $c_{0N}^j \equiv \frac{d}{d\epsilon_{0N}^j} \psi_{N,\epsilon_{0N}^j}$. *We have the following expansion:*

$$\epsilon_N^{*j} - \epsilon_{0N}^j = c_N^{j,-1} M_N / \sqrt{N} + c_N^{j,-1} \bar{R}(N) + R_{0N}(\epsilon_N^{0j}, \epsilon_{0N}^j).$$

Assumptions. *Assume* $\bar{R}(N) = o_P(N^{-1/2})$; $R_{0N}(\epsilon_N^{0j}, \epsilon_{0N}^j) = o_P(N^{-1/2})$; $\bar{g}_0 > \delta > 0$ *a.e. for some* $\delta > 0$; $\frac{1}{N} \sum_{i=1}^{N} P_0 \bar{D}(\theta_{i-1})^2 - P_0 \bar{D}(\theta_0)^2 \to 0$ *in probability, as* $N \to \infty$.

Conclusion. *Then,* $c_{0N}^j \sqrt{N}(\epsilon_N^{*j} - \epsilon_{0N}) \Rightarrow_d Z$ *in probability, where* $Z \sim N(0, \sigma_0^2)$ *and* $\sigma_0^2 = P_0 \bar{D}(\theta_0)^2$.

Using a simple δ-method argument, Theorem 19.5 establishes the asymptotic efficiency of the online TMLE $\Psi(\theta_{N,\epsilon_N^*})$ of ψ_0.

Theorem 19.9. *Let* $j \in \{1, 2\}$ *be given. Under the same conditions as in Theorem 19.5, we have*

$$\psi_N^{*j} - \Psi(\theta_0) = \frac{M_N}{\sqrt{N}} + o_P(1/\sqrt{N}),$$

and thereby $\sqrt{N}(\psi_N^{j*} - \Psi(\theta_0)) \Rightarrow_d Z$, *where* $Z \sim N(0, \sigma_0^2)$. *Thus* ψ_N^{j*} *is an asymptotically efficient estimator of* ψ_0.

Practical Computation of the First Online TMLE. We note that for each ϵ, $\psi_{N+1,\epsilon}^1$ is an online estimator that only requires updating the initial estimator $\bar{Q}_N$ and $\bar{g}_N$ and subsequently $\psi_{N,\epsilon}^1$ with the $(N+1)$-th observation, and thus completely controls computation speed and memory. This suggest the following approach for highly scalable computation of the online TMLE $\psi_N^{1*} = \psi_{N+1,\epsilon_N^{*1}}^1$. A priori select a fine grid of ϵ-values, and for each ϵ-value in this finite set, track the online estimators $\psi_{N+1,\epsilon}$, while also running the online estimator ϵ_N^{*1}. In this manner, one has also available $\psi_{N+1,\epsilon_N^{*1}}$ at each step and there is no need to recompute $\bar{Q}_{i,\epsilon}$ for $i \leq N$ for this new ϵ-value. In addition, one could decide to also track a number of derivatives $\frac{d^k}{d\epsilon^k} \psi_{N,\epsilon}^1$ for each of the ϵ-values in the grid, $k = 1, \dots, K$. In this manner, one can use a K-th order Taylor expansion of $\psi_{N,\epsilon}^1$ to obtain the values at ϵ that are not in the grid. This then allows to use a smaller size grid while still preserving the desired approximation of $\psi_{N,\epsilon}^1$ for all ϵ that could occur as realization of ϵ_N^{1*}.

19.8 Discussion

We have proposed a general class of statistical models for dependent data that is ordered in time. We proposed a particular class of causal quantities and developed online one-step and online TMLEs of their corresponding estimands. We proved theorems that provide sufficient conditions under which these online estimators are asymptotically efficient. The stochastic behavior of the initial online estimator is crucial for the second-order remainder to be negligible, by far the most fundamental condition for asymptotic efficiency.

Analogue to the HAL estimator for i.i.d. data, we can define an HAL estimator that converges to the truth at a faster rate than the critical $N^{-1/4}$, where we now rely on fundamental exponential tail probabilities for martingale processes: we will present the formal results on this HAL estimator and its HAL super learner in future work. This then proves that asymptotic efficient estimation is possible in this highly nonparametric class of statistical models for time series data.

We also highlighted a fundamental condition that controls the amount of dependence that our results allow. We demonstrated our proposed online estimators for estimation of the average causal effect based on i.i.d. data. A very important set of applications is the analysis of a single time series on a single unit, which, in particular, has applications for precision medicine involving tracking data on patients. Our results demonstrate that it is possible to assess causal effects for a single patient based on observing a time series involving experimentation with a particular treatment or exposure variable. It will be interesting to extend the work on optimal dynamic treatments for i.i.d. data to learning the optimal treatment rule for a single time series. In future research, we will implement our proposed estimators, evaluate them in simulations, obtain more insight into the computational speed of these estimators, and their statistical performance. We also aim to apply these estimators in real data.

Part VI
Networks

Chapter 20
Causal Inference in Longitudinal Network-Dependent Data

Oleg Sofrygin and Mark J. van der Laan

Much of the existing causal inference literature focuses on the effect of a single or multiple time-point intervention on an outcome based on observing longitudinal data on n independent units that are not causally connected. The causal effect is then defined as an expectation of the effect of the intervention assigned to the unit on their outcome, and causal effects of the intervention on other units on the unit's outcome are assumed nonexistent. As a consequence, causal models only have to be concerned with causal relations between the components of the unit-specific data structure. Statistical inference is based on the assumption that the n data structures are n independent realizations of a random variable. However, in many CRTs or observational studies of few communities, the number of independent units is not large enough to allow statistical inference based on limit distributions.

In an extreme but not uncommon setting, one may observe a single community of causally connected individuals. Can one still statistically evaluate the causal effect of an intervention assigned at the community level on a community level outcome, such as the average of individual outcomes? This is the question we address in this chapter. Clearly, causal models incorporating all units are needed in order to define the desired causal quantity, and identifiability of these causal quantities under (minimal) assumptions should be established without relying on asymptotics in a number of *independent* units.

O. Sofrygin (✉)
Division of Biostatistics, University of California, Berkeley, 101 Haviland Hall, #7358, Berkeley, CA 94720, USA
e-mail: sofrygin@berkeley.edu

M. J. van der Laan
Division of Biostatistics and Department of Statistics, University of California, Berkeley, 101 Haviland Hall, #7358, Berkeley, CA 94720, USA
e-mail: laan@berkeley.edu

© Springer International Publishing AG 2018
M.J. van der Laan, S. Rose, *Targeted Learning in Data Science*,
Springer Series in Statistics, https://doi.org/10.1007/978-3-319-65304-4_20

As previously noted, the frequently made assumption of independence among units is generally violated when data is collected on a population of connected units, since the network interactions will often cause the exposure of one unit to have an effect on the outcomes of other connected units. Consider a setting with single time point exposure where we collect data on baseline covariates, exposures, and outcomes on N connected units. We might expect that the interactions between any two connected units can cause the exposure of one unit to have an effect on the outcome of the other—an occurrence often referred to as interference or spillover (Sobel 2006; Hudgens and Halloran 2008).

While many past studies have wrongfully assumed no interference for the sake of simplicity,[1] the past decade has also seen a growing body of literature devoted to estimation of causal effects in network-dependent data. Many of these studies have sought to answer questions about the role of social networks on various aspects of public health. For example, Christakis et al. used observational data on subjects connected by a social network in a series of studies that estimated the causal effects of contagion for obesity, smoking, and a variety of other outcomes, finding that many of these conditions are subject to *social contagion* (Christakis and Fowler 2007, 2013). In one study, authors found that an individual's risk of becoming obese increases with each additional obese friend, even after controlling for all *measured* confounding factors (Christakis and Fowler 2007). However, the statistical methods employed by these studies have come under scrutiny due to possibly anti-conservative standard error estimates that did not account for network-dependence among the observed units (Lyons 2010), and possibly biased effect estimates that could have resulted from: model misspecification (Lyons 2010; VanderWeele and An 2013), network-induced homophily (Shalizi and Thomas 2011), and unmeasured confounding by environmental factors (Shalizi and Thomas 2011).

20.1 Modeling Approach

An important ingredient of our modeling approach is the incorporation of network information that describes for each unit i (in a finite population of N units) a set of other units $F_i \subset \{1, \ldots, N\}$ from which this unit may receive input. This allows us to pose a structural equation model for this group of units in which the observed data node at time t of a unit i is only causally affected by the observed data on the units in F_i, beyond exogenous errors. The set F_i includes only the immediate friends of unit i that directly affect unit i's data at each time t. The structural equation model could be visualized through a so-called causal graph involving all N units, which one might call a network.

Our assumptions on the exogenous errors in the structural equation model will correspond with assuming sequential conditional independence of the unit-specific data nodes at time t, conditional on the past of all units. That is, conditional on the most recent past of all units, including the network information, the data on the units

---

[1] See references in Sect. 21.7 for implications of incorrectly ignoring interference.

at the next time point are independent across units. Smaller sets of friends of i, F_i, will result in fewer incoming edges for each node in the causal graph, providing a larger effective sample size for targeting the desired quantity. While these causal graphs allows the units to depend on each other in complex ways, if the size of F_i is bounded universally in N, in conjunction with our assumption of the independence of the exogenous errors, the likelihood of the data on all N units allows statistical inference driven by the number of units, N, rather than the number of communities (e.g., 1). Furthermore, our most recent work generalizes these formal asymptotic results to allow the size of F_i to grow with N (Ogburn et al. 2017).

We will apply the roadmap for targeted learning of a causal effect to define and solve the estimation problem. We start by defining a structural causal model (Pearl 2009a) that determines how each node is generated as a function of parent data nodes and exogenous variables. We then define the causal quantity of interest in terms of stochastic interventions on the unit-specific treatment nodes. The structural causal model could be visualized as a causal graph describing the causal links between the N units. In this presentation, we will assume that the network is measured at baseline, and that it remains static throughout the duration of the follow-up. We also note that these results generalize naturally to situations where the set of friend connections over time (i.e., time-varying node $F_i(t)$) are themselves random (e.g., see van der Laan 2014a).

As mentioned above, our structural equation model also makes strong independence assumptions on the exogenous errors, which imply that the unit specific data-nodes at time t are independent across the N units, *conditionally* on the past of all N units. We refer to this assumption as a sequential conditional independence assumption. Thus, it is assumed that any dependence of the unit-specific data nodes at time t can be fully explained by the observed past on all N units. As a next step in the roadmap, we then establish the identifiability of the causal quantity from the data distribution under transparent additional (often nontestable) assumptions. This identifiability result allows us to define and commit to a statistical model that contains the true probability distribution of the data, and an estimand (i.e., a target parameter mapping applied to true data distribution) that reduces to this causal quantity if the required causal assumptions hold. The statistical model must contain the true data distribution, so that the statistical estimand can be interpreted as a pure statistical target parameter, while under the stated additional causal conditions that were needed to identify the causal effect, it can be interpreted as the causal quantity of interest. The statistical model, and the target parameter mapping that maps data distributions in the statistical model into the parameter values, defines the pure statistical estimation problem.

Because the statistical model does not assume that the data generating experiment involves the repetition of independent experiments, the development of targeted estimators and inference represents novel statistical challenges. Targeted minimum loss-based estimation was developed for estimation in semiparametric models for i.i.d. data (van der Laan and Rubin 2006; van der Laan 2008a; van der Laan and Rose 2011), and extended to a particular form of dependent treatment/censoring allocation as present in group sequential adaptive designs (van der Laan 2008a; Chambaz and van der Laan 2011a,b) and community randomized trials (van der

Laan et al. 2013a). In this chapter, we describe the application of the targeted minimum loss based estimation to new types of complex semiparametric statistical models that allow for network-based interference between observational units.

20.2 Data Structure

Consider a study in which we observe a sample of N units where any two units might be connected via a social network. For a unit i, let $O_i = (W_i, A_i, \bar{Y}_i = (Y_i(t) : t = 1, \ldots, \tau))$ be a time-ordered longitudinal observed data structure, where W_i are baseline-covariates, A_i denotes an action/treatment/exposure intervention node, $Y_i(t)$ denotes the time-dependent outcome process on unit i, and $Y_i(\tau)$ denotes the final outcome on unit i. For example, we may assume that A_i records a particular trait on subject i, such as 'high level of physical activity' and $Y_i(\tau)$ is the indicator of subject i being overweight at time τ. Let F_i be a component of W_i and we assume that F_i denotes the set of friends from which individual i may receive input at any time during the follow-up. Thus, $F_i \subset \{1, \ldots, N\}$. We assume that W_i occurs before A_i, in the sense that A_i is affected by all the past, including W_i.

We define $\mathbf{W} = (W_i : i = 1, \ldots, N)$, $\mathbf{A} = (A_i : i = 1, \ldots, N)$ and similarly we define $\mathbf{Y}(t) = (Y_i(t) : i = 1, \ldots, N)$, for each $t = 1, \ldots, \tau$. Lastly, we define the observed data on N units as $\mathbf{O} = (O_1, \ldots, O_N) \sim P_0^N$, and we note that this observed data can be represented as a single time-ordered data structure:

$$\mathbf{O} = (\mathbf{W}, \mathbf{A}, \bar{\mathbf{Y}} = (\mathbf{Y}(t) : t = 1, \ldots, \tau)).$$

The latter ordering is the only causally relevant ordering. The ordering of units within a time-point is user supplied but inconsequential. We define $Pa(\mathbf{A}) = \mathbf{W}$, $Pa(\mathbf{Y}(t)) = (\mathbf{W}, \mathbf{A}, \bar{\mathbf{Y}}(t-1))$, as the parents of $\mathbf{A}$ and each $\mathbf{Y}(t)$, respectively. The parents of A_i, denoted with $Pa(A_i)$, are defined to be equal to $Pa(\mathbf{A})$, and, the parents of $Y_i(t)$, denoted with $Pa(Y_i(t))$, are also defined to be equal to $Pa(\mathbf{Y}(t))$, for $t = 1, \ldots, \tau$ and $i = 1, \ldots, N$.

20.3 Example

In order to provide the reader with some context for the type of real-world applications of our modeling approach, we present a few examples. Consider a study in which we wish to evaluate the effectiveness of pre-exposure prophylaxis (PrEP) on the rate of new HIV infections in the target population of interest (e.g., among individuals who are at high risk for HIV). A cohort of individuals is followed for several years, and for each individual i the researchers collect data on i's baseline characteristics W_i, that include the set $F_i \subset W_i$ of i's sexual partners. In addition, researchers measure i's treatment status A_i, which is the indicator of receiving PrEP, as assigned at the beginning of the study. We assume that regular tests for new HIV-infections are carried out at certain time intervals, where the HIV status for unit i at time t is indicated by $Y_i(t)$. Let $t = 1, \ldots, \tau$, where the τ-th point represents the end of follow-up. Suppose one is interested in the effect of PrEP at baseline, A_i, on the proportion $1/N \sum_i Y_i(\tau)$ of HIV-infections at the last time point τ. In this setting, it may be of interest to estimate the mean outcome $1/N \sum_i E_0 Y_{i,\mathbf{g}^*}(\tau)$ under some stochastic intervention $\mathbf{g}^*$ on $\mathbf{A} = (A_1, \ldots, A_N)$, which, for example, might assign every individual in the study to receive PrEP. This is an example of a deterministic intervention. Finally, in our model we may assume that the conditional distributions of $Y_i(t)$ and A_i, given the past on all individuals, only depend on the individual pasts of the sexual partners of subject i, beyond the past of subject i itself. In other words, we assumed that the risk of HIV-infection at time t for individual i, depends on the treatment status of his or her sexual partners.

As an example of a stochastic intervention, we may choose to target a random subset of the total set of intervention nodes, $(A_i : i = 1, \ldots, N)$, by focusing on a specific subset of individuals. That is, we could define a stochastic intervention that allows some subjects to have their observed exposures, while setting the exposures of others to pre-specified intervened values. For example, when limited resources allow treatment allocation for only a small number of subjects, the researchers may wish to identify which treatment allocation strategies will maximize the reduction in infection rates. In addition, researcher might be interested in measuring the effectiveness of an intervention that, as a prophylactic measure, treats a random proportion of high-risk individuals. The latter is an example of a truly stochastic intervention. Similarly, the treatment node could be defined as the indicator of condom use, so that the counterfactual mean outcomes evaluate the effect of condom use on the spread of the HIV-epidemic. One could also consider intervening on F_i itself, for example, by decreasing the number $|F_i|$ of sexual partners.

As another potential application, consider a study where we wish to evaluate the effect of physical activity at a single time-point on the future risk of being overweight, where overweight status of individual i is measured at time points $t = 1, \ldots, \tau$ and is denoted as $Y_i(t)$. In this case, the final $Y_i(\tau)$ denotes the outcome of interest for individual i. The encouraging effects of high levels of physical activity among the friends of individual i may affect the future weight of subject i (interference). Similarly, the level of physical activity of individual i is affected by the weight of i's friends at baseline—a notion known as the adaptive treatment allocation. This is an application in which both the treatment nodes and the outcome nodes of an individual are affected by the observed past of i's friends.

20.4 Estimation Problem

We present our structural causal model that models the data generating process for a population of interconnected units. As previously mentioned, we assume that the network is measured at baseline, and that it remains static throughout the duration of the follow-up. We will present a model for the distribution of $(\mathbf{O}, \mathbf{U}) = (O_i, U_i : i = 1, \ldots, N)$, where each O_i denotes the observed data on unit i, and U_i represents a vector of exogenous errors for the structural equations for unit i. This structural causal model allows us to define stochastic interventions denoted with $\mathbf{g}^*$ on the collection of unit-specific treatment nodes (contained in O_i), and corresponding counterfactual outcomes. The causal quantity, denoted with $E\left(1/N \sum_{i=1}^{N} Y_{i,\mathbf{g}^*}(\tau)\right)$, is defined in terms of the (possibly conditional) expectation of the intervention specific counterfactual-outcomes $Y_{i,\mathbf{g}^*}(\tau)$, and it represents a parameter of the distribution of $(\mathbf{O}, \mathbf{U})$. Subsequently, we establish identifiability of the causal quantity from the data distribution P_0 of data $\mathbf{O} = (O_1, \ldots, O_N)$ on the N units, commit to a statistical model $\mathcal{M}$ for the probability distribution P_0 of $\mathbf{O}$, define the statistical target parameter mapping $\Psi : \mathcal{M} \to \mathbb{R}$ that defines the estimand $\Psi(P_0)$, where the latter reduces to the causal quantity under the additional assumptions that are needed to establish the identifiability. The statistical estimation problem is now defined by the data $\mathbf{O} \sim P_0 \in \mathcal{M}$, the statistical model $\mathcal{M}$ and target parameter $\Psi : \mathcal{M} \to \mathbb{R}$.

In order to define causal quantities, we assume that $\mathbf{O}$ is generated by a structural equation model of the following type: first generate a collection of exogenous errors $\mathbf{U}_N = (U_i : i = 1, \ldots, N)$ across the N units, where the exogenous errors for unit i are given by

$$U_i = (U_{W_i}, U_{A_i}, U_{Y_i(1)}, \ldots, U_{Y_i(\tau)}), \quad i = 1, \ldots, N,$$

and then generate $\mathbf{O}$ deterministically by evaluating functions as follows:

$$W_i = f_{W_i}(U_{W_i}), \text{ for } i = 1, \ldots, N \tag{20.1}$$
$$A_i = f_{A_i}(Pa(A_i), U_{A_i}), \text{ for } i = 1, \ldots, N$$
$$Y_i(t) = f_{Y_i(t)}(Pa(Y_i(t)), U_{Y_i(t)}), \text{ for } i = 1, \ldots, N \text{ and } t = 1, \ldots, \tau.$$

These functions f_{W_i}, f_{A_i} and $(f_{Y_i(t)} : t = 1, \ldots, \tau)$ are unspecified at this point, but will be subjected to modeling below. Since $Pa(A_i) = \mathbf{W}$ and $Pa(Y_i(t)) = (\mathbf{W}, \mathbf{A}, \bar{\mathbf{Y}}(t-1))$, an alternative succinct way to represent the above structural equation model is:

$$\mathbf{W} = \mathbf{f}_{\mathbf{W}}(\mathbf{U}_{\mathbf{W}}) \tag{20.2}$$
$$\mathbf{A} = \mathbf{f}_{\mathbf{A}}(Pa(\mathbf{A}), \mathbf{U}_{\mathbf{A}})$$
$$\mathbf{Y}(t) = \mathbf{f}_{\mathbf{Y}(t)}(Pa(\mathbf{Y}(t)), \mathbf{U}_{\mathbf{Y}(t)}), \text{ for } t = 1, \ldots, \tau.$$

Recall that the set of friends, F_i, is a component of W_i and is thus also a random variable defined by this structural equation model. However, one may also decide to condition on F_i, as we do for the case of single time-point outcome (i.e., data structure $\mathbf{O} = (\mathbf{W}, \mathbf{A}, \mathbf{Y})$), which is presented in next chapter.

20.4.1 Counterfactuals and Stochastic Interventions

This structural equation model for

$$\mathbf{O} = (\mathbf{W}, \mathbf{A}, \mathbf{Y}(1), \ldots, \mathbf{Y}(\tau)),$$

allows us to define counterfactuals $Y_d(\tau)$ corresponding with an dynamic intervention d on $\mathbf{A}$ (Gill and Robins 2001). For example, one could define A_i as a particular deterministic function d_i of the parents $Pa(A_i)$ of subject $i = 1, \ldots, N$. Using our previously described study of the effects of physical activity on weight, one could define such an intervention by adding a new physically active friend to all subjects who weren't physically active at baseline. Such an intervention corresponds with replacing the equation for $\mathbf{A}$ with the deterministic equation $d_i(Pa(\mathbf{A}))$. More generally, we can replace the equation for $\mathbf{A}$ that describe a degenerate distribution for drawing $\mathbf{A}$, given $\mathbf{U} = \mathbf{u}$, and $Pa(\mathbf{A})$, by a user-supplied conditional distribution of a new variable $\mathbf{A}_*$, given $Pa(\mathbf{A}_*)$. Such a conditional distribution defines a so-called stochastic intervention (Didelez et al. 2006).

Let $\mathbf{g}^*$ denote our selection of a stochastic intervention identified by the conditional distribution of $\mathbf{A}_*$, given $Pa(\mathbf{A}_*(t))$. For convenience, we represent the stochastic intervention with equation $\mathbf{A}_* = f_{\mathbf{A}_*}(Pa(\mathbf{A}_*), \mathbf{U}_{\mathbf{A}_*})$ in terms of random errors $\mathbf{U}_{\mathbf{A}_*}$. This implies the following modified system of structural equations:

$$\mathbf{W} = \mathbf{f}_{\mathbf{W}}(\mathbf{U}_{\mathbf{W}}) \tag{20.3}$$
$$\mathbf{A}_* = \mathbf{f}_{\mathbf{A}_*}(Pa(\mathbf{A}_*), \mathbf{U}_{\mathbf{A}_*})$$
$$\mathbf{Y}_*(t) = \mathbf{f}_{\mathbf{Y}(t)}(Pa(\mathbf{Y}_*(t)), \mathbf{U}_{\mathbf{Y}(t)}), \text{ for } t = 1, \ldots, \tau,$$

where $Pa(\mathbf{Y}_*(t))$ is the same set of variables as $Pa(\mathbf{Y}(t))$, but where the variables $(\mathbf{A}, \bar{\mathbf{Y}}(t-1))$ are replaced by $(\mathbf{A}_*, \bar{\mathbf{Y}}_*(t-1))$. Let $Y_{i,\mathbf{g}^*}(\tau)$, or short-hand $Y_{i,*}(\tau)$, denote the corresponding final counterfactual outcome for unit i. A causal effect at the unit level could now be defined as a contrast, such as $Y_{i,\mathbf{g}_1^*}(\tau) - Y_{i,\mathbf{g}_2^*}(\tau)$, for two interventions $\mathbf{g}_1^*$ and $\mathbf{g}_2^*$. For example, $\mathbf{g}_1^*$ may correspond with the actual data generating distribution as defined by the above structural equation model for the observed $\mathbf{A}$, while $\mathbf{g}_2^*$ corresponds with adding one overweight friend to each unit i at baseline.

20.4.2 Post-Intervention Distribution and Sequential Randomization Assumption

We assume the following sequential randomization assumption on $\mathbf{U}$,

$$\mathbf{A} \perp \bar{\mathbf{Y}}_{\mathbf{g}^*}, \text{ conditional on } Pa(\mathbf{A}), \tag{20.4}$$

and $\mathbf{U}_{\mathbf{A}_*} \perp \mathbf{U}$, where $\bar{\mathbf{Y}}_{\mathbf{g}^*} = (\mathbf{Y}_{\mathbf{g}^*}(1), \ldots, \mathbf{Y}_{\mathbf{g}^*}(\tau))$. Then, the probability distribution $P_{\mathbf{g}^*}$ of $(\mathbf{W}, \mathbf{A}_*, \bar{\mathbf{Y}}_{\mathbf{g}^*})$ is given by the so-called g-computation formula (Gill and Robins 2001):

$$P_{\mathbf{g}^*}(\mathbf{W}, \mathbf{A}_*, \bar{\mathbf{Y}}_*) = P_{\mathbf{W}}(\mathbf{W}) \prod_{i=1}^{N} \bar{g}_i^*(A_{i,*}(t) \mid Pa(A_{i,*}(t))) \prod_{t=1}^{\tau} \prod_{i=1}^{N} P_{Y_i(t)}(Y_{i,*}(t) \mid Pa(Y_{i,*}(t))),$$

where $P_{Y_i(t)}$ is the conditional distribution of $Y_i(t)$, given $Pa(Y_i(t))$, $Pa(Y_{i,*}(t)) = (\mathbf{W}, \mathbf{A}_*, \bar{\mathbf{Y}}_*(t-1),)$ and $\bar{g}_i^*$ is the i-specific stochastic intervention on A_i. Note that $\bar{g}_i^*$ is defined by the i-specific marginal of the joint density $\mathbf{g}^*$ and we are implicitly assuming that $(A_{1,*}, \ldots, A_{N,*})$ are conditionally independent given $\mathbf{W}$. Thus, under this sequential randomization assumption, the post-intervention distribution $P_{\mathbf{g}^*}$ is identified from the observed data distribution of $\mathbf{O}$ that was generated from this structural equation model.

20.4.3 Target Parameter as the Average Causal Effect (ACE)

One can define the average causal effect as the following target parameter of the distribution of $P_{\mathbf{g}^*}$:

$$E_{P_{\mathbf{g}_1^*}} \left\{ \frac{1}{N} \sum_{i=1}^{N} Y_{i,\mathbf{g}_1^*}(\tau) \right\} - E_{P_{\mathbf{g}_2^*}} \left\{ \frac{1}{N} \sum_{i=1}^{N} Y_{i,\mathbf{g}_2^*}(\tau) \right\}.$$

By setting $\bar{Y}_{\mathbf{g}^*}(\tau) = \frac{1}{N} \sum_{i=1}^{N} Y_{i,\mathbf{g}^*}(\tau)$ we can also write this causal effect as $E(\bar{Y}_{\mathbf{g}_1^*}(\tau) - \bar{Y}_{\mathbf{g}_2^*}(\tau))$. Since the distribution $P_{\mathbf{g}^*}$ is indexed by N, this parameter must also depend on N. In particular, the effect of stochastic intervention on a population of N interconnected units will naturally depend on the size N of that population, and the network information $\mathbf{F}$: i.e., adding a unit will change the dynamics. As we will do in our description of the single time-point outcome case in the following chapter, one might decide to replace these marginal expectations by conditional expectations, where one conditions on $(F_i : i = 1, \ldots, N)$, or even conditions on $(W_i : i = 1, \ldots, N)$. We will now focus on the causal quantity $\psi^F = E_{P_{\mathbf{g}^*}} \bar{Y}_{\mathbf{g}^*}(\tau)$ under user-supplied stochastic intervention, and our results generalize naturally to causal quantities defined by some Euclidean-valued function of a collection of such intervention specific means.

   The parameter $E\bar{Y}_{\mathbf{g}^*}(\tau)$ can be evaluated by taking the expectation of $\bar{Y}_{\mathbf{g}^*}(\tau)$ with respect to this post-intervention distribution $P_{\mathbf{g}^*}$. In practice, however, this integral will be often intractable, in which case one can perform Monte-Carlo integration to obtain a reasonable approximation of $E\bar{Y}_{\mathbf{g}^*}(\tau)$. This process is carried out by sequentially sampling N random variables at a time from each factor of $P_{\mathbf{g}^*}$, until one obtains a single realization of $\bar{Y}_{\mathbf{g}^*}(\tau)$. That is, we first sample a realization of N values $\mathbf{W} = (W_1, \ldots, W_N)$ from $P_{\mathbf{W}}$, followed by a realization of $\mathbf{A}_*$ from $\mathbf{g}^*$, sampled conditionally on $\mathbf{W}$. This is followed by a sample of $\mathbf{Y}_{\mathbf{g}^*}(1) = (Y_{1,\mathbf{g}^*}(1), \ldots, Y_{N,\mathbf{g}^*})$, conditional on $(\mathbf{W}, \mathbf{A}_*)$, and so on, until we obtain a sample of N final outcomes $\mathbf{Y}_{\mathbf{g}^*}(\tau)$, conditional on $(\mathbf{W}, \mathbf{A}_*, \mathbf{Y}_{\mathbf{g}^*}(\tau - 1))$. One Monte-Carlo estimate of the target parameter $E\bar{Y}_{\mathbf{g}^*}(\tau)$ is then obtained by taking the empirical mean $\bar{Y}_{\mathbf{g}^*}(\tau)$ of $\mathbf{Y}_{\mathbf{g}^*}(\tau)$. Finally, by iterating this procedure enough times and averaging over such Monte-Carlo estimates, we can obtain a good enough approximation of $E\bar{Y}_{\mathbf{g}^*}(\tau)$.

20.4.4 Dimension Reduction and Exchangeability Assumptions

The above stated identifiability of ψ^F is not of interest, since we cannot estimate the distribution of $\mathbf{O}$ based on a single observation. Therefore, we will need to make much more stringent assumptions that will allow us to learn the distribution of $\mathbf{O}$ based on a single draw. One could make such assumptions directly on the distribution of $\mathbf{O}$, but below we present these assumptions in terms of assumptions on the structural equations and the exogenous errors.

Beyond the assumptions described above, we also assume that for each node A_i and Y_i, we can define known functions, $Pa(A_i) \to c_{A,i}(Pa(A_i))$ and $Pa(Y_i(t)) \to c_{Y(t),i}(Pa(Y_i(t)))$, that map into a Euclidean set with a dimension that does not depend on N, and corresponding common (in i) functions f_A and $f_{Y(t)}$, so that

$$W_i = f_{W_i}(U_{W_i}), \text{ for } i = 1, \ldots, N \tag{20.5}$$
$$A_i = f_A(c_{A,i}(Pa(A_i)), U_{A_i}), \text{ for } i = 1, \ldots, N$$
$$Y_i(t) = f_{Y(t)}(c_{Y(t),i}(Pa(Y_i(t))), U_{Y_i(t)}), \text{ for } i = 1, \ldots, N, \ t = 1, \ldots, \tau.$$

(As mentioned above, an interesting variation of this structural causal model treats $\mathbf{W}$ as given and thus removes that data generating equation.) Examples of such dimension reductions are $c_{Y(t),i}(Pa(Y_i(t))) = ((\bar{Y}_i(t-1), A_i), (\bar{Y}_j(t-1), A_j : j \in F_i))$, i.e., the observed past of unit i itself and the observed past of its friends, and, similarly, we can define $c_{A,i}(Pa(A_i)) = (W_i, (W_j : j \in F_i))$. By augmenting these reductions to data on maximally K friends, filling up the empty cells for units with fewer than K friends with a missing value, these dimension reductions have a fixed dimension, and include the information on the number of friends. This structural equation model assumes that, across all units i, the data on unit i at the next time point t is a common function of its own past and past of its friends.

20.4.5 Independence Assumptions on Exogenous Errors

Beyond the sequential randomization assumption (20.4), we make the following (conditional) independence assumptions on the exogenous errors. First, we assume that each U_{W_i} is independent of all U_{W_j} such that $F_i \cap F_j = \emptyset$, for $i = 1, \ldots, N$. It then follows that each W_i depends on at most K other W_j, for some universal constant K (weak dependence of $\mathbf{W}$). We also note that this assumption of weak dependence of $\mathbf{W}$ provides us with a statistical model for which we can still prove the asymptotic normality of our resulting estimator. Hence, we will assume a model on the distribution of $\mathbf{W}$ which at minimal makes this assumption. However, additional restrictions on $\mathbf{W}$ are also possible (e.g., a more restricted statistical model that assumes independence of $\mathbf{W}$).

In addition, we assume that conditional on $\mathbf{W}$, all $(U_{A_i} : i = 1, \ldots, N)$ are independent and identically distributed (i.i.d.). Similarly, we assume that for $t = 1, \ldots, \tau$, conditional on $(\mathbf{A}, \bar{\mathbf{Y}}(t-1))$, all $(U_{Y_i(t)} : i = 1, \ldots, N)$ are i.i.d. The important implication of the latter assumptions is that, given the observed past $Pa(\mathbf{Y}(t))$, for any two units i and j that have the same value for their summaries $c_{Y(t),i} = c_{Y(t),j}$ as functions of $Pa(\mathbf{Y}(t))$, we have that $Y_i(t)$ and $Y_j(t)$ are independent and identically distributed, and similarly, we have this statement for the treatment nodes. This allows us to factorize the likelihood of the observed data as done below, parameterized by common conditional distributions $\bar{q}_{0,t}$ and $\bar{g}_0$ that can actually be learned from a single (but growing) $\mathbf{O}$ when $N \to \infty$.

20.4.6 Identifiability: g-Computation Formula for Stochastic Intervention

For notational convenience, let $C_{Y(t),i} = c_{Y(t),i}(Pa(Y_i(t)))$, and let $C^*_{Y(t),i}$ be defined accordingly by replacing $(\mathbf{A}, \bar{\mathbf{Y}})$ with $(\mathbf{A}_*, \bar{\mathbf{Y}}_*)$. Due to the exchangeability and dimension reduction assumptions, the probability distribution $P_{\mathbf{g}^*}$ of $\bar{\mathbf{Y}}_* = (\bar{Y}_{i,*} : i = 1, \ldots, N)$ now simplifies:

$$p_{\mathbf{g}^*}(\mathbf{W}, \mathbf{A}_*, \bar{\mathbf{Y}}_*) = q_{\mathbf{W}}(\mathbf{W}) \prod_{i=1}^{N} \bar{g}^*_i(A_{i,*} \mid Pa(A_{i,*})) \prod_{t=1}^{\tau} \prod_{i=1}^{N} \bar{q}_t(Y_{i,*}(t) \mid C^*_{Y(t),i}))$$

$$\equiv p^{\mathbf{g}^*}(\mathbf{W}, \mathbf{A}_*, \bar{\mathbf{Y}}_*), \tag{20.6}$$

where $q_{\mathbf{W}}$ is the joint marginal probability density of $\mathbf{W}$ and $\bar{q}_t$ is the common conditional density of $Y_i(t)$, given $Pa(Y_i(t))$, where, by our assumptions, these conditional densities are constant in $i = 1, \ldots, N$, as functions of $C_{Y(t),i}$, $t = 1, \ldots, \tau$. We also introduce the notation $q_{\mathbf{Y}(t)}$ for the joint conditional density of $\mathbf{Y}(t)$, given $Pa(\mathbf{Y}(t))$, which is thus parameterized in terms of $\bar{q}_t$. Similarly, we introduce the notation $\mathbf{g}$ to denote the joint conditional density of $\mathbf{A}$, given $Pa(\mathbf{A})$, which is thus parameterized in terms of $\bar{g}$. We introduced the notation $p^{\mathbf{g}^*}$ for the right-hand side in (20.6), which thus represents an expression in terms of the distribution of the data under the assumption that the conditional densities of $Y_i(t)$, given $Pa(Y_i(t))$, are constant in i as functions of $C_{Y(t),i}$, indexed by the choice of stochastic intervention $\mathbf{g}^*$. We also note that one needs the causal model and randomization assumption in order to establish that the right-hand side actually models the counterfactual post-intervention distribution $P_{\mathbf{g}^*}$. This shows that $\psi_0^F = \Psi(P_0)$, where Ψ is a mapping that takes the distribution P_0 of $\mathbf{O}$ and maps it into the real line.

Strictly speaking, the above result does not establish a *desired* identifiability result yet, since we cannot learn P_0 based on only a single draw $\mathbf{O}$. To start with, we note that P_0, $\psi_0^{F,N}$, ψ_0^N are all indexed by N, and we only observed one draw of N units from P_0. Therefore, we still need to show that we can construct an estimator based on a single draw $\mathbf{O}$ that is consistent for ψ_0^N, as $N \to \infty$. For that purpose, we note that the distribution $P^{\mathbf{g}^*}$ is identified by the common conditional densities $\bar{q}_{Y(t)}$, $t = 1, \ldots, \tau$, and $q_{\mathbf{W}}$, where $\mathbf{W} = (W_i : i = 1, \ldots, N)$. We can construct consistent estimators of the common conditional distributions $\bar{q}_{0,t}$ based on MLE that are consistent as $N \to \infty$, which follows from our presentation of estimators and theory. This demonstrates the identifiability of $\bar{q}_{0,t}$ as $N \to \infty$, $t = 1, \ldots, \tau$. In addition, our target parameter involves an average $E_{\mathbf{W}}(\bar{Q}_{\mathbf{g}^*})$ with respect to the joint distribution of $\mathbf{W}$, where $\bar{Q}_{\mathbf{g}^*}(\mathbf{W})$ defines the conditional expectation of the sample-average counterfactual outcomes, i.e., $\bar{Q}_{\mathbf{g}^*}(\mathbf{W}) := E_{\mathbf{g}^*}(\bar{Y}_{\mathbf{g}^*}(\tau) \mid \mathbf{W})$. We note that $E_{\mathbf{W}}(\bar{Q}_{\mathbf{g}^*})$ can be consistently estimated by its empirical counterpart under our weak dependence assumption, as discussed above. This finally demonstrates the desired identifiability of $\psi_0^{F,N}$ from the observed data as $N \to \infty$.

20.4.7 Likelihood and Statistical Model

By our assumptions, we can factorize the probability density of the data $\mathbf{O} = (\mathbf{W}, \mathbf{A}, \bar{\mathbf{Y}} = (\mathbf{Y}(1), \ldots, \mathbf{Y}(\tau)))$ as follows:

$$p(\mathbf{w}, \mathbf{a}, \bar{\mathbf{y}}) = q_{\mathbf{W}}(\mathbf{w}) \prod_{i=1}^{N} \bar{g}(a_i \mid c_{A,i}(\mathbf{w})) \prod_{t=1}^{\tau} \prod_{i=1}^{N} \bar{q}_t(y_i(t) \mid c_{Y(t),i}(\mathbf{w}, \mathbf{a}, \bar{\mathbf{y}}(t-1))). \quad (20.7)$$

In above, we denoted the factor representing the joint marginal density of $\mathbf{W}$ with $q_{\mathbf{W}}$. We denoted the conditional density of each $Y_i(t)$ with $\bar{q}_t$, where these conditional densities at $Y_i(t)$, given $Pa(Y_i(t))$, are constant in i, as functions of $Y_i(t)$ and $C_{Y(t),i}$. Similarly, we modeled the $\mathbf{g}$-factor in terms of common conditional density $\bar{g}$. We also let $\mu_{\mathbf{W}}$, $(\mu_{Y(t)} : t = 1, \ldots, \tau)$ and μ_A denote the corresponding dominating measures of these densities. Let $\bar{q} := (\bar{q}_t : t = 1, \ldots, \tau)$ represent the collection of these conditional densities, so that the distribution of $\mathbf{O}$ is parameterized by $(q_{\mathbf{W}}, \bar{q}, \bar{g})$. The conditional distributions $\bar{q}_t(Y(t) \mid C_{Y(t)})$ are unspecified functions of $Y(t)$ and $C_{Y(t)}$, beyond that for each value of $C_{Y(t)}$, each is a well-defined conditional density, and $q_{\mathbf{W}}$ satisfies our particular independence model, i.e., that $\mathbf{W}$ are at least weakly dependent. Similarly, $\bar{g}$ is an unspecified conditional density. This defines now a statistical parameterization of the distribution of $\mathbf{O}$ in terms of $(q_{\mathbf{W}}, \bar{q}, \bar{g})$, resulting in the following formulation of the statistical model for P_0:

$$\mathcal{M} = \{P_{q_{\mathbf{W}}, \bar{g}, \bar{q}} : q_{\mathbf{W}} \in Q_W, \bar{g} \in \mathcal{G}, \bar{q} \in Q_{\bar{Y}}\}, \quad (20.8)$$

where Q_W, $\mathcal{G}$ and $Q_{\bar{Y}}$ denote the parameter spaces for $q_{\mathbf{W}}$, $\bar{g}$ and $\bar{q} = (\bar{q}_t : t = 1, \ldots, \tau)$, respectively.

20.4.8 Statistical Target Parameter

Let $\mathbf{Y}^{\mathbf{g}^*}(\tau)$ denote the final N outcomes sampled under intervention-specific distribution $P^{\mathbf{g}^*}$ (20.6). Note that $P^{\mathbf{g}^*}$ was defined as a function of the data distribution P of $\mathbf{O}$. Namely, $P^{\mathbf{g}^*}$ was obtained by replacing the $\mathbf{g}$-factor in P with the corresponding joint stochastic intervention $\mathbf{g}^*$. We define our statistical target parameter as $\frac{1}{N} \sum_{i=1}^{N} E\left[Y_i^{\mathbf{g}^*}(\tau)\right]$, which is a function of $P^{\mathbf{g}^*}$. We also note that this parameter is equal to the target causal quantity $\frac{1}{N} \sum_{i=1}^{N} E\left[Y_{i,\mathbf{g}^*}(\tau)\right]$, under the above stated causal assumptions. Thus,

$$\frac{1}{N} \sum_{i=1}^{N} E\left[Y_i^{\mathbf{g}^*}(\tau)\right] = \Psi(P_{q_{\mathbf{W}},\bar{g},\bar{q}}) = \Psi(q_{\mathbf{W}}, \bar{q}) \tag{20.9}$$

depends on the data generating distribution P only through the marginal distribution of $\mathbf{W}$ and the conditional densities $\bar{q} = (\bar{q}_t : t = 1, \ldots, \tau)$.

In addition, this statistical parameter can be represented as $E_{\mathbf{W}}(\bar{Q}(\mathbf{W}))$, where $\bar{Q}(\mathbf{W})$ defines the expectation of $\bar{Y}^{\mathbf{g}^*}$ conditional on $\mathbf{W}$, i.e., $\bar{Q}(\mathbf{W}) := E_{\bar{q},\mathbf{g}^*}(\bar{Y}^{\mathbf{g}^*}(\tau) \mid \mathbf{W})$. This suggests that our parameter $\Psi(P_{q_{\mathbf{W}},\bar{g},\bar{q}})$ can be also represented as a mapping $\Psi(q_{\mathbf{W}}, \bar{Q}^{\mathbf{g}^*})$. Evaluation of $\bar{Q}(\mathbf{W})$ involves taking an expectation over all vectors $\mathbf{Y}(t)$ and $\mathbf{A}^*$, with respect to the product measure of common conditional densities $\bar{q}_t$, for $t = 1, \ldots, \tau$ and g_i^*, for $i = 1, \ldots, N$. In what follows, we focus exclusively on estimation of our target parameter based on the mapping $\Psi(q_{\mathbf{W}}, \bar{q})$. As described previously, we can evaluate this statistical target parameter by performing Monte-Carlo simulations that samples $\bar{Y}^{\mathbf{g}^*}$ from the post-intervention distribution $P^{\mathbf{g}^*}$ by going forward in time and sampling N random variables from each successive factor of $p^{\mathbf{g}^*}$, starting from $q_{\mathbf{W}}$, and all the way up to $\bar{q}_\tau$.

Finally, one can define another statistical parameter by conditioning on all the observed $\mathbf{W}$, namely a parameter given by $E_{\bar{q},\mathbf{g}^*}(\bar{Y}^{\mathbf{g}^*} \mid \mathbf{W})$. This parameter can be still effectively evaluated with the same Monte-Carlo sampling procedure, except that we skip the sampling from the first factor $q_{\mathbf{W}}$ and instead use the observed $\mathbf{W}$ values.

20.4.9 Statistical Estimation Problem

We have now defined a statistical model $\mathcal{M}$ (20.8) for the distribution (20.7) of $\mathbf{O}$, and a statistical target parameter mapping $\Psi : \mathcal{M} \to \mathbf{R}$ (20.9) for which $\Psi(P_{q_{\mathbf{W}},\bar{g},\bar{q}})$ only depends on $(q_{\mathbf{W},0}, \bar{q})$, where $\bar{q} = (\bar{q}_t : t = 1, \ldots, \tau)$. We will also denote this target parameter with $\Psi(q_{\mathbf{W}}, \bar{q})$, with some abuse of notation by letting Ψ represent these two mappings. Given a single draw $\mathbf{O} \sim P_{q_{\mathbf{W},0},\bar{g}_0,\bar{q}_0}$, we want to estimate $\Psi(q_{\mathbf{W},0}, \bar{q}_0)$. In addition, we want to construct an asymptotically valid confidence interval. Recall that our notation suppressed the dependence on N and F of the data distribution $P_{q_{\mathbf{W}},\bar{g},\bar{q}}$, statistical model $\mathcal{M}$, and target parameter Ψ.

20.4.10 Summary

In summary, we defined a structural causal model (20.5), including the stated independence (and i.i.d.) assumptions on the exogenous errors, the dimension reduction assumptions, and the sequential randomization assumption (20.4). This resulted in the likelihood (20.7) and corresponding statistical model $\mathcal{M}$ (20.8) for the distribution P_0 of $\mathbf{O}$. In addition, these assumptions allowed us to write the causal quantity ψ_0^F as a statistical estimand $\Psi(q_{\mathbf{W}}, \bar{q}_{\bar{Y}})$ (20.9): $\psi_0^F = \Psi(q_{\mathbf{W}}, \bar{q})$, where the densities $\bar{q} = (\bar{q}_t : t = 1, \ldots, \tau)$ can be learned from a single draw $\mathbf{O}$ as $N \to \infty$. The pure statistical estimation problem is now defined: $\mathbf{O} \sim P_0 \in \mathcal{M}$, and we want to learn $\psi_0 = \Psi(P_0)$ where $\Psi : \mathcal{M} \to \mathbb{R}$. Under the nontestable causal assumptions, beyond the statistical assumption for $P_0 \in \mathcal{M}$, we can interpret ψ_0 as ψ_0^F. However, even without these nontestable assumptions, one might interpret ψ_0 (and its contrasts) as a purely statistical parameter for the effect measure of interest controlling for the *observed* confounders.

20.5 Efficient Influence Curve

Due to our sequential conditional independence assumption, the log-likelihood of O, i.e., the log of the data-density (20.7) of O, can be represented as a double sum over time-points t and units i. For each t, the sum over i consists of independent random variables, conditional on the past. As a consequence, under appropriate regularity conditions, one can show that the log-likelihood is asymptotically normally distributed. Therefore, we conjecture that we can establish the so-called local asymptotic normality of our statistical model, which involves establishing the asymptotic normality of log-likelihood under sampling from fluctuations (submodels) $P_{\epsilon=1/\sqrt{N}} \subset \mathcal{M}$ of a fixed data distribution P across all possible fluctuations. As shown in van der Vaart (1998), for models satisfying the local asymptotic normality condition, the normal limit distribution of an MLE is an optimal limit distribution based on the convolution theorem (Bickel et al. 1997b).

While it is well known that a regular estimator based on sample of n i.i.d. observations is efficient if and only if it is asymptotically linear with influence curve equal to the efficient influence curve, here we are not interested in asymptotics when we observe n of our data structures that are indexed by this parameter N, as is the case when observing an i.i.d. sample $\mathbf{O}_1, \ldots, \mathbf{O}_n$, where each $\mathbf{O}_i$ describes the data on N causally connected units. Instead, we are interested in the asymptotics in N based on a single draw of $\mathbf{O}$. Nonetheless, the asymptotic behavior of the MLE based on such a single draw $\mathbf{O}^N$ when $N \to \infty$ was analyzed in Section 5 of van der Laan (2014a), which showed that the asymptotic variance of such MLE is still characterized by the efficient influence curve. In more detail, it showed that, under appropriate regularity conditions required for an MLE to be valid (i.e., all observables are discrete, so that MLE is asymptotically well defined), the asymptotic variance of a standardized MLE $\sqrt{N}(\psi_N - \psi_0)$ of the target parameter equals the limit in N of $N P_0 \{D^*(\bar{q}_0, \bar{g}_0)\}^2$, where $P_0 \{D^*(\bar{q}_0, \bar{g}_0)\}^2$ is the variance of the efficient influence curve $D^*(\bar{q}_0, \bar{g}_0)$ (as defined below). As a consequence, our goal should still be to construct an estimator that is asymptotically normally distributed with variance equal to the variance of the efficient influence curve, appropriately normalized. Our proposed TMLE achieves this goal by using the least favorable submodels, whose score span the efficient influence curve.

Theorem 3 of the technical report van der Laan (2012) established a general representation of the efficient influence curve of $E\bar{Y}_{\mathbf{g}^*}(\tau)$ for the longitudinal data structure, assuming the baseline covariates $(W_1, \ldots, W_N)$ are independent. Let

$$D^*_{W_i}(P)(W_i) = E(\bar{Y}_{\mathbf{g}^*}(\tau) \mid W_i) - E(\bar{Y}_{\mathbf{g}^*}(\tau))$$

$$= \frac{1}{N} \sum_{j=1}^{N} \left[E(Y_{j,\mathbf{g}^*}(\tau) \mid W_i) - E(Y_{j,\mathbf{g}^*}(\tau)) \right].$$

Then the EIC is given by

$$D^*(P) = \sum_{i=1}^{N} D^*_{W_i}(P)(W_i) + \sum_{t=1}^{\tau} \sum_{i=1}^{N} D^*_t(P)(Y_i(t), C_{Y(t),i}), \qquad (20.10)$$

where we define D^*_t as:

$$D^*_t(P)(Y_i(t), C_{Y(t),i}) = \frac{1}{N} \sum_{m=1}^{N} \left[\frac{h^*_{t,m}}{\bar{h}_t}(C_{Y(t),i}) \frac{1}{N} \sum_{l=1}^{N} D_{l,t,m}(Y_i(t), C_{Y(t),i}) \right].$$

In above, we also defined $h^*_{t,m}(c) = P_{\bar{q}_0, \mathbf{g}^*}(C_{Y(t),m} = c)$, $h_{t,m}(c) = P_{\bar{q}_0, \bar{g}_0}(C_{Y(t),m} = c)$, and $\bar{h}_t = \frac{1}{N} \sum_m h_{t,m}$, and we define $D_{l,t,m}$ as:

$$D_{l,t,m}(Y_i(t), C_{Y(t),i}) = E(Y_{l,\mathbf{g}^*}(\tau) \mid Y_m(t) = Y_i(t), C_{Y(t),m} = C_{Y(t),i}) - E(Y_{l,\mathbf{g}^*}(\tau) \mid C_{Y(t),m} = C_{Y(t),i}).$$

We will also denote a collection of all $\bar{h}_t$ as $\bar{h}$, namely, $\bar{h}(\bar{q}, \bar{g}) := (\bar{h}_t(\bar{q}, \bar{g}) : t = 1, \ldots, \tau)$.

Note that if we were to assume a different model for the distribution of the baseline covariates $\mathbf{W}$, we would only change the first component of this efficient influence curve, namely, the above defined terms $D^*_{W_i}$. For example, the following chapter presents the efficient influence curve when assuming a fully nonparametric model for the distribution of $\mathbf{W}$. We also note that $D^*_t(\bar{q}_0, \bar{g}_0)$ has conditional mean zero, given $C_{Y(t),i}$. In order for $D^*_t(\bar{q}_0, \bar{g}_0)$ to have a finite variance, the summation over l must reduce to a finite sum, since $Y_m(t)$ is conditionally independent of $Y_l(\tau)$, given $C_{Y(t),m}$, for most m. We now refer to Section 6 of the technical report van der Laan (2012) for the derivation of the above efficient influence curve and a study of its robustness.

We now describe an alternative representation of the above EIC for the case when all $(Y_i(t), A_i)$ are binary and W_i are categorical, for all $i = 1, \ldots, N$ and $t = 1, \ldots, \tau$. In particular, for each $t = 1, \ldots, \tau$, the corresponding t-specific component of the above EIC can be also expressed in terms of sums of residuals $(Y_i(t) - \bar{q}_t(1|C_{Y(t),i}))$, where each residual is multiplied by the "clever covariate" $\bar{H}_t(C_{Y(t),i})$:

$$\sum_{i=1}^{N} D^*_t(Y_i(t), C_{Y(t),i}) = \sum_{i=1}^{N} \bar{H}_t(C_{Y(t),i})(Y_i(t) - \bar{q}_t(1 \mid C_{Y(t),i})), \qquad (20.11)$$

and the clever covariate is defined as:

$$\bar{H}_t(c_{y(t)}) = \frac{1}{N} \sum_{m=1}^{N} \frac{h_{t,m}^*}{\bar{h}_t}(c_{y(t)})\bar{D}_{t,m}(c_{y(t)}),$$

where

$$\bar{D}_{t,m}(c_{y(t)}) = \left[E(\bar{Y}_{\mathbf{g}^*}(\tau) \mid Y_m(t) = 1, C_{Y(t),m} = c_{y(t)}) - E(\bar{Y}_{\mathbf{g}^*}(\tau) \mid Y_m(t) = 0, C_{Y(t),m} = c_{y(t)}) \right].$$

Note that the binary EIC representation in (20.11) will allow us to define the least favorable submodel of our TMLE, as described in Sect. 20.7.

20.6 Maximum Likelihood Estimation, Cross-Validation, and Super Learning

We could estimate the distribution of $\mathbf{W}$ with the empirical distribution that puts mass 1 on $(W_i : i = 1, \ldots, N)$. This choice also corresponds with a TMLE of the intervention specific mean outcome $E(\bar{Y}^{g^*} \mid \mathbf{W})$ that conditions on $\mathbf{W}$, as we formally show in our later sections for the single time-point data structure. If it is assumed that $(W_i : i = 1, \ldots, N)$ are independent, then we estimate the distribution of $\mathbf{W}$ with the NPMLE that maximizes the log-likelihood $\sum_i \log q_{W_i}(W_i)$ over all possible distributions of $\mathbf{W}$ that the statistical model $\mathcal{M}$ allows. In particular, if it is known that W_i are i.i.d., then we would estimate the common distribution of W_i with the empirical distribution that puts mass $1/N$ on W_i, $i = 1, \ldots, N$.

Regarding estimation of $(\bar{q}_{t,0} : t = 1, \ldots, \tau)$, we consider the log-likelihood loss function for each $\bar{q}_t$:

$$L_t(\bar{q}_t) \equiv - \sum_{i=1}^{N} \log \bar{q}_t(Y_i(t) \mid C_{Y(t),i}).$$

Note that $E_0 L_t(\bar{q}_t)$ is minimized in $\bar{q}_t$ by the true $\bar{q}_{t,0}$, since, conditional on $(\mathbf{W}, \mathbf{A}, \bar{\mathbf{Y}}(t-1))$, the true density of $Y_i(t)$ is given by $\bar{q}_{t,0}(\cdot \mid C_{Y(t),i})$, $i = 1, \ldots, N$. In addition, this expectation $E_0 L_t(\bar{q}_t)$ is well approximated by $\frac{1}{N} \sum_{i=1}^{N} \log \bar{q}_t(Y_i(t) \mid C_{Y(t),i})$, since, conditional on $(\mathbf{W}, \mathbf{A}, \bar{\mathbf{Y}}(t-1))$, this is a sum of independent random variables $Y_i(t)$, $i = 1, \ldots, N$. The latter allows us to prove the convergence of the empirical mean process to the true mean process uniformly in large parameter spaces for $\bar{q}_t$, using similar techniques as we use in the Appendix of van der Laan (2014a), based on weak convergence theory in van der Vaart and Wellner (1996). As a consequence, one could pose a parametric model for $\bar{q}_{0,t}$, say $\{\bar{q}_{t,\theta} : \theta\}$, and use standard maximum likelihood estimation

$$\theta_N = \arg \min_{\theta} L_t(q_{t,\theta}),$$

as if the observations $(Y_i(t), C_{Y(t),i})$, $i = 1, \ldots, N$, are independent and identically distributed and we are targeting this common conditional density of $Y_i(t)$ given

$C_{Y(t),i}$. More importantly, we can use loss-based cross-validation and super-learning to fit this function $\bar{q}_{t,0}$ of $(y(t), c_{Y(t)})$, thereby allowing for adaptive estimation of $\bar{q}_t$. Specifically, consider a collection of candidate estimators $\hat{\bar{q}}_{t,k}$ that maps a data set $\{(Y_i(t), C_{Y(t),i}) : i\}$ into an estimate for $k = 1, \ldots, K$. Let P_N^t denote the empirical distribution that puts mass $1/N$ onto each $(Y_i(t), C_{Y(t),i})$. Given a random split vector $B_N \in \{0, 1\}^N$, define $P_{N,B_N}^{t,1}$ and $P_{N,B_N}^{t,0}$ as the empirical distributions of the validation sample $\{i : B_N(i) = 1\}$ and training sample $\{i : B_N(i) = 0\}$, respectively. We define the cross-validation selector k_n of k as

$$k_n = \arg\min_k E_{B_N} P_{N,B_N}^{t,1} L_t(\hat{\bar{q}}_{t,k}(P_{N,B_N}^{t,0}))$$
$$= \arg\min_k E_{B_N} \sum_{i:B_N(i)=1} \log \hat{\bar{q}}_{t,k}(P_{N,B_N}^{t,0})(Y_i(t) \mid C_{Y(t),i}).$$

If $Y_i(t)$ is continuous, one could encode $Y_i(t)$ in terms of binary variables $I(Y_i(t) = l)$ across the different levels l of $Y_i(t)$, and model the conditional distribution/hazard of $I(Y_i(t) = l)$, given $Y_i(t) \geq l$ and $(\mathbf{W}, \mathbf{A}, \bar{\mathbf{Y}}(t-1))$, as a function of $C_{Y(t),i}$ and l, as in van der Laan (2010a,b). Candidate estimators of this conditional hazard can be constructed, possibly smoothing in the level l, by utilizing estimators of predictors of binary variables in the machine learning literature, including standard logistic regression software for fitting parametric models. Similarly, this can be extended to multivariate $Y_i(t)$ by first factorizing the conditional distribution of $Y_i(t)$ in univariate conditional distributions. In this manner, candidate estimators of $\bar{q}_{t,0}$ can be obtained based on a large variety of algorithms from the literature.

We could fit each $\bar{q}_{t,0}$ separately for $t = 1, \ldots, \tau$, but it is also possible to pool across t by constructing estimators and using cross-validation based on the sum loss function

$$L(\bar{q}) = \sum_t L_t(\bar{q}_t).$$

Similarly, we can use the log-likelihood loss-function for $\bar{g}$:

$$L(\bar{g}) = -\sum_{i=1}^N \log \bar{g}(A_i \mid C_{A,i}),$$

and use loss-based cross-validation and super-learning to fit $\bar{g}$.

Given the resulting estimators $q_{\mathbf{W},N}$ of $q_{\mathbf{W},0}$ and $\bar{q}_N = (\bar{q}_t : t = 1, \ldots, \tau)$ of $\bar{q}_0 = (\bar{q}_{t,0} : t = 1, \ldots, \tau)$, one can evaluate $\Psi(q_{\mathbf{W},N}, \bar{q}_N)$ as estimator of $\psi_0 = \Psi(q_{\mathbf{W},0}, \bar{q}_0)$, by applying the previously described Monte-Carlo integration procedure. Since $\bar{q}_N$ is optimized with respect to the bias-variance trade-off to fit $\bar{q}_0$ and not ψ_0, even though this data-adaptive plug-in estimator inherits the rate of convergence at which $\bar{q}_N$ converges to $\bar{q}_0$, it is overly biased for $\Psi(q_{\mathbf{W},0}, \bar{q}_0)$. That is, $\Psi(q_{\mathbf{W},N}, \bar{q}_N)$ will generally not converge to $\Psi(q_{\mathbf{W},0}, \bar{q}_0)$ at rate $1/\sqrt{N}$. The resulting smoothed/regularized maximum likelihood substitution estimators $\Psi(q_{\mathbf{W},N}, \bar{q}_N)$ are not targeted, and are overly biased with respect to the target parameter $\Psi(q_{\mathbf{W},0}, \bar{q}_0)$. As a consequence, estimators of the statistical target parameter will not be asymptotically normally distributed . Thus there is a need for targeted learning (targeting the fit towards ψ_0) instead of MLE.

20.7 TMLE

TMLE will involve modifying an initial estimator $\bar{q}_{t,N}$ into a targeted version $\bar{q}_{t,N}^*$, for $t = 1, \ldots, \tau$, through utilization of an estimator $\bar{g}_N$ of $\bar{g}_0$, a least favorable submodel (with respect to the target parameter ψ_0) $\{\bar{q}_{t,N}^k(\epsilon, \bar{g}_N) : \epsilon\}$ through a current fit $\bar{q}_{t,N}^k$ at $\epsilon = 0$, fitting ϵ for each t and each step k with standard MLE $\epsilon_{t,N}^k$, iterative updating $\bar{q}_{t,N}^{k+1} = \bar{q}_{t,N}^k(\epsilon_{t,N}^k)$, $t = 1, \ldots, \tau$, until convergence in $k = 1, 2, \ldots$. The resulting TMLE of ψ_0 is defined accordingly as the substitution estimator $\Psi(q_{\mathbf{W},N}, \bar{q}_N^*)$. Thus, this TMLE will also involve estimation of the intervention mechanism $\bar{g}_0$. To define such a TMLE we need to use the efficient influence curve of the statistical target parameter, given by 21.12, which implies these least favorable submodels. This least favorable fluctuation model solves the efficient influence curve equation 21.12. That is, we will define a fluctuation model $\{p_{\bar{q}_\epsilon} : \epsilon\}$ through the $\bar{q}_t$ factors of the density p so that it satisfies

$$\frac{d}{d\epsilon} \log p_{\bar{q}_\epsilon} \bigg|_{\epsilon=0} (\mathbf{O}) = D^*(q_{\mathbf{W}}, \bar{q}, \bar{h}(\bar{q}, \bar{g}))(\mathbf{O}). \tag{20.12}$$

20.7.1 Local Least Favorable Fluctuation Model

We now consider the case when all $Y_i(t)$ are binary and all W_i are categorical, for $i = 1, \ldots, N$ and $t = 1, \ldots, \tau$. In this case we can use the binary representation of the EIC in 20.11. In particular, 20.11 shows that each t-specific component of the EIC can be expressed in terms of sums of residuals $(Y_i(t) - \bar{q}_t(1|C_{Y(t),i}))$, where each residual is multiplied by the "clever covariate" $\bar{H}_t(C_{Y(t),i})$. As a result, we can use the following logistic regression fluctuation model through each $\bar{q}_t$, indexed by parameter ϵ_t:

$$\text{Logit}\bar{q}_{t,\epsilon_t}(1 \mid c_{y(t)}) = \text{Logit}\bar{q}_t(1|c_{y(t)}) + \epsilon_t \bar{H}_t(c_{y(t)}),$$

The above model now defines a τ-dimensional fluctuation submodel $\{\bar{q}_t(\epsilon_t, \bar{g}) : \epsilon_t, t = 1, \ldots, \tau\}$ indexed by parameter $\{\epsilon_t : t = 1, \ldots, \tau)$ through the $\bar{q}_t$ components of the density p. Note that this choice of the fluctuation submodel clearly satisfies

$$\frac{d}{d\epsilon} \log p_{\bar{q}_\epsilon} \bigg|_{\epsilon=0} (\mathbf{O}) = D^*(q_{\mathbf{W}}, \bar{q}, \bar{h}(\bar{q}, \bar{g}))(\mathbf{O}).$$

We define the estimator $q_{\mathbf{W},N}$ of $q_{\mathbf{W},0}$ as a NPMLE that puts mass one on the observed $\mathbf{W}$, assuming it is well defined. This implies that $q_{\mathbf{W},N}$ already solves the efficient score equation defined by the EIC component $\sum_{i=1}^{N} D^*_{W_i}(P)(W_i)$. Define the estimator $\bar{g}_N$ of $\bar{g}_0$ and the initial estimators $\bar{q}_N := (\bar{q}_{t,N} : t = 1,\ldots,\tau)$ of $\bar{q}_0 := (\bar{q}_{t,0} : t = 1,\ldots,\tau)$ that were obtained by applying one of the methods outlined in the previous section, either by separately fitting for each t or by pooling across all t. Let $k = 0$, and define the MLE

$$\epsilon^k_{t,N} = \arg\max_{\epsilon_t} \log \bar{q}^k_{t,N,\epsilon_t}(\mathbf{O}), \text{ for } t = 1,\ldots,\tau.$$

Note that this model can be obtained by running $t = 1,\ldots,\tau$ separate regressions. That is, for each t one fits a separate logistic regression that pools the observed outcomes $(Y_i(t) : i = 1,\ldots,N)$, using as offset the initial super learner fit $\bar{q}_{t,N}(1|C_{Y(t),i})$, for $i = 1,\ldots,N$ and the plug-in estimate of the "clever covariate" $\bar{H}_t(\bar{g}_N, \mathbf{g}^*, \bar{q}_N)(C_{Y(t),i})$, evaluated at observations $C_{Y(t),i}$, for $i = 1,\ldots,N$. This now constitutes the first TMLE step, in which we define $\bar{q}^{k+1}_{t,N} := \bar{q}^k_{t,N,\epsilon^k_{t,N}}$, for $t = 1,\ldots,\tau$. Setting $k = k + 1$ and iterating this updating process defines a k-th step TMLE $\bar{q}^k_{t,N}$, for $k = 1,2,\ldots$. The corresponding k-th step TMLE of the target parameter $\Psi(q_{\mathbf{W},0}, \bar{q}_0)$ is now given by the plug-in estimator $\Psi(q_{\mathbf{W},N}, \bar{q}^k_N)$, for $k = 1,2,\ldots$. If one iterates this updating process until a step K at which $\epsilon^K_{N,t} \approx 0$ for all t, then it follows by 20.12 that

$$D^*(q_{\mathbf{W},N}, \bar{q}^*_N, \bar{h}(\bar{q}^*_N, \bar{g}_N))(\mathbf{O}) \approx 0,$$

where $\bar{q}^*_N = \bar{q}^K_N$. If the initial estimator of $\bar{q}_t$ is close enough to the truth, for example, if the rate of convergence is faster than $n^{-1/4}$, then our recent results prove that the single iteration of the TMLE will be enough, which would significantly simplify the computational burden of this algorithm (see Chap. 5).

20.7.2 Estimation of the Efficient Influence Curve

The efficient influence curve (EIC) depends on the data generating distribution P only through $(\bar{q}, \bar{h}(\bar{q}, \bar{g}))$. Given estimators of $\bar{q}_N, \bar{g}_N$, one obtains a plug-in estimator $\bar{h}(\bar{q}_N, \bar{g}_N)$ of $\bar{h}_0 = \bar{h}(\bar{q}_0, \bar{g}_0)$. This suggests that the only goal in estimation of $\bar{g}_0$ is to construct a good estimator of $\bar{h}_0$. The important advantage of this plug-in estimator $\bar{h}(\bar{q}_N, \bar{g}_N)$ of $\bar{h}_0$ is that it fully utilizes the knowledge coded by the statistical model. Nonetheless, when the computational burden of evaluating the plug-in $\bar{h}(\bar{q}_N, \bar{g}_N)$ is too high, it may be of interest to construct a direct estimator of $\bar{h}_0$ separate from an estimator $\bar{q}_N$ of $\bar{q}_0$. For that purpose we note that each $\bar{h}_{t,0}$ is a density of $C_{Y(t),j}$. We can estimate $\bar{h}_{t,0}$ by using a density estimator treating $(C_{Y(t),j} : j = 1,\ldots,N)$ as if these observations are i.i.d. This corresponds with using as loss-function for $\bar{h}_{t,0}$,

$$L(\bar{h}_t)(O) = -\sum_j \log \bar{h}_t(C_{Y(t),j}).$$

However, the evaluation of the plug-in estimate of the clever covariate $\bar{H}_t(\bar{h}_N, \bar{q}_N, \mathbf{g}^*)$ can become computationally very challenging even when using this direct approach to estimate $\bar{h}_{t,0}$. In particular, it is necessary to evaluate

$$\bar{D}_{t,m,N}(c_{y(t)}) = E_{\bar{q}_N,\mathbf{g}^*}(\bar{Y}_{\mathbf{g}^*}(\tau) \mid Y_m(t) = 1, C_{Y(t),m} = c_{y(t)}) - E_{\bar{q}_N,\mathbf{g}^*}(\bar{Y}_{\mathbf{g}^*}(\tau) \mid Y_m(t) = 0, C_{Y(t),m} = c_{y(t)}),$$

and $h^*_{t,m,N}(c_{y(t)}) = P_{\bar{q}_N,\mathbf{g}^*}(C_{Y(t),m} = c_{y(t)})$, for each $m = 1, \ldots, N$ and $t = 1, \ldots, \tau$. Letting

$$\bar{Q}^{\mathbf{g}^*}_{t,m}(y(t), c_{y(t)}) = E(\bar{Y}_{\mathbf{g}^*}(\tau) \mid Y_m(t) = y(t), C_{Y(t),m} = c_{y(t)}),$$

the above can be written as

$$\bar{D}_{t,m}(c_{y(t)}) = \bar{Q}^{\mathbf{g}^*}_{t,m}(1, c_{y(t)}) - \bar{Q}^{\mathbf{g}^*}_{t,m}(0, c_{y(t)}).$$

Note that the following Monte-Carlo simulation approach could be used for evaluating $\bar{Q}^{\mathbf{g}^*}_{t,m,N}(y(t), c_{y(t)})$, when all $C_{Y(t),i}$ are discrete:

1. We start by sampling realizations $\mathbf{O}^{\mathbf{g}^*,k}$ from the estimate of the post-intervention distribution $P_{\mathbf{g}^*,N}$, for $k = 1, \ldots, K$, where each sample $\mathbf{O}^{\mathbf{g}^*,k}$ consists of N observations and K is large.
2. As a next step, we evaluate $\bar{Y}^k_{\mathbf{g}^*}(\tau)$ for each sample k. Thus, all N observations in a single k sample receive the same outcome $\bar{Y}^k_{\mathbf{g}^*}(\tau)$.
3. Next, for a fixed $m = 1, \ldots, N$, we identify all samples among $k = 1, \ldots, K$ which match the specified values $(y(t), c_{y(t)})$.
4. By taking the average of the outcomes $\bar{Y}^k(\tau)$ among these matched samples, we can obtain an estimate of $\bar{Q}^{\mathbf{g}^*}_{t,m,N}(y(t), c_{y(t)})$.

Note that the above algorithm can be parallelized in a trivial manner. Moreover, the approximation of $\bar{Q}^{\mathbf{g}^*}_{t,m,N}(y(t), c_{y(t)})$ can proceed concurrently as the realizations $\mathbf{O}^{\mathbf{g}^*,k}$ are being simulated. This estimate of $\bar{Q}^{\mathbf{g}^*}_{t,m,N}(y(t), c_{y(t)})$ will continue to improve as we generate more and more such samples (as K grows). Furthermore, a substitution estimator for $h^*_{t,m}(c_{y(t)})$ can also be obtained in a similar manner, based on the empirical distribution of the same realizations $\mathbf{O}^{\mathbf{g}^*,k}$, for $k = 1, \ldots, K$.

While this approach provides an algorithmically more tractable way to evaluate $\bar{D}_{t,m}$, it can be computationally very intensive. Furthermore, it cannot handle the case of continuous $C_{Y(t),i}$. An alternative approach is to use loss-based estimation to obtain a good approximation of $\bar{Q}^{\mathbf{g}^*}_{t,m,N}$, based on the same K realizations $\mathbf{O}^{\mathbf{g}^*,k}$ sampled from $P_{\mathbf{g}^*,N}$. Given that $\bar{Y}_{\mathbf{g}^*}$ is known to be continuous with values in $(0, 1)$, one could use the following log-likelihood loss function

$$L_{t,m}(\bar{Q}^{\mathbf{g}^*}_{t,m})(\mathbf{O}^{\mathbf{g}^*,k} : k = 1, \ldots, K) =$$
$$- \sum_{k=1}^{K} \log \left\{ \bar{Q}^{\mathbf{g}^*}_{t,m}(Y^k_m(t), C^k_{Y(t),m})^{\bar{Y}^k_{\mathbf{g}^*}(\tau)} (1 - \bar{Q}^{\mathbf{g}^*}_{t,m}(Y^k_m(t), C^k_{Y(t),m}))^{1 - \bar{Y}^k_{\mathbf{g}^*}(\tau)} \right\}.$$

We could also consider the following squared error loss function for each $\bar{Q}^{\mathbf{g}^*}_{t,m}$:

$$L_{t,m}(\bar{Q}^{\mathbf{g}^*}_{t,m})(\mathbf{O}^{\mathbf{g}^*,k} : k = 1, \ldots, K) \equiv - \sum_{k=1}^{K} (\bar{Y}^k_{\mathbf{g}^*}(\tau) - \bar{Q}^{\mathbf{g}^*}_{t,m}(Y^k_m(t), C^k_{Y(t),m}))^2.$$

Furthermore, since we have K independent realizations from $P_{\mathbf{g}^*,N}$, one could pose a parametric model for $\bar{Q}_{t,m}^{\mathbf{g}^*}$, say $\{\bar{Q}_{t,m,\theta}^{\mathbf{g}^*} : \theta\}$, and use standard maximum likelihood estimation

$$\theta_K = \arg\min_{\theta} L_{t,m}(\bar{Q}_{t,m,\theta}^{\mathbf{g}^*}).$$

This parametric model may also allow pooling over individual observations $m = 1,\ldots,N$, and individual time-points $t = 1,\ldots,\tau$ by assuming that $\bar{Q}_{t,m}^{\mathbf{g}^*}$ can be parametrized by a common parameter $\theta(m,t)$, say $\{\bar{Q}_{\theta(t,m)}^{\mathbf{g}^*} : \theta(m,t)\}$, using the following sum loss function over t and m:

$$L(\bar{Q}^{\mathbf{g}^*}) \equiv \sum_{t=1}^{\tau}\sum_{m=1}^{N} L_{t,m}$$

and fitting this common parameter with standard maximum likelihood estimation

$$\theta_K(m,t) = \arg\min_{\theta} L(\bar{Q}_{\theta(m,t)}^{\mathbf{g}^*}).$$

More importantly, we can apply the above approaches to use loss-based cross-validation and super-learning to fit the above functions $\bar{Q}_{t,m}^{\mathbf{g}^*}$, either separately or by pooling across t or m or both, thereby allowing for adaptive estimation of $\bar{Q}_{t,m}^{\mathbf{g}^*}$. By pooling our modeling of $\bar{Q}_{t,m}^*$ over t and m, this becomes a computationally tractable problem for any number of time-points (τ), since all components of the EIC can be estimated with a single modeling step, for all $t = 1,\ldots,\tau$ and $m = 1,\ldots,N$.

20.8 Summary

In this chapter, we formulated a general causal model for the longitudinal data structure generated by a finite population of causally connected units. We then defined counterfactuals indexed by interventions on the treatment nodes of the units, and corresponding causal contrasts. We established identifiability of the causal quantities from the data observed on the units when observing all units, or a random sample of the units, under appropriate assumptions. Our causal assumptions implied conditional independence across units at time t, conditional on the past of all units. This resulted in a factorized likelihood of the observed data even though the observed data was generated by a single experiment, not by a repetition of independent experiments. To deal with the curse of dimensionality we assumed that a unit's dependence on the past of other units could be summarized by a finite dimensional measure, and that this dependence was described by a common function across the units. We then described the statistical model for the data distribution, the statistical target parameter, and the resulting statistical estimation problem. We demonstrated that cross-validation and super-learning can be used to estimate the different factors of the likelihood. Given the statistical model and statistical target parameter that identifies the counterfactual mean under an intervention, we derived the efficient influence curve of the target parameter. We showed that the EIC characterizes the normal limit distribution of a maximum likelihood estimator, and thus represents an optimal asymptotic variance among estimators of the target parameter. However, due to high dimensionality, maximum likelihood estimators will be ill-defined for finite samples, and smoothing will be required.

Smoothed/regularized maximum likelihood estimators are not targeted and will thereby be overly biased with respect to the target parameter. As a consequence, they generally do not result in asymptotically normally distributed estimators of the statistical target parameter. Therefore, we formulated targeted maximum likelihood estimators of this estimand. The bias of the proposed TMLE is a second-order term involving squared differences $\bar{h}_N - \bar{h}_0$ and $q_N - q_0$ for two nuisance parameters $\bar{h}_0 = \bar{h}(\bar{g}_0, \bar{q}_0)$ and the relevant factor of likelihood $\bar{q}_0$.

Overall, we believe that the statistical study of these causal models for dynamic networks of units provides a fascinating and important area of future research, relying on deep advances in empirical process and statistical estimation theory. While raising new challenges, these advances will be needed to move forward statistical practice.

20.9 Notes and Further Reading

For literature reviews on causal inference with independent subject-level data we refer to a number of books on this topic: Rubin (2006), Pearl (2009a), van der Laan and Robins (2003), Tsiatis (2006), and van der Laan and Rose (2011). We refer to Halloran and Struchiner (1995); Hudgens and Halloran (2008); VanderWeele et al. (2012b); Tchetgen Tchetgen and VanderWeele (2012) for defining different types of causal effects in the presence of causal interference between units. Lacking a general methodological framework, many past studies have assumed away interference for the sake of simplicity. The risk of this assumption is practically demonstrated in Sobel (2006), who shows that ignoring interference can lead to wrong conclusions about the effectiveness of the intervention. We also refer to Donner and Klar (2000), Hayes and Moulton (2009), and Campbell et al. (2007) for reviews on cluster randomized trials and cluster level observational studies.

We refer to the accompanying technical report (van der Laan 2012) for various additional results such as weakening of the sequential conditional independence assumption (still heavily restricting the amount of dependence, but allowing that, even conditional on the observed past, a subject can be dependent on maximally K other subjects), and only observing a random sample of the complete population of causally connected units, among others. More general extensions of this framework will be the target of our future research. Another possible approach to weaken the independence assumption for the baseline covariates $\mathbf{W}$ is to simply define the target parameter conditional on $\mathbf{W}$. This estimator was analyzed in detail in van der Laan (2014a).

Our semiparametric model for possibly network-dependent units generalizes the models in the causal inference literature for independent units. Even though in this chapter our causal model pertains to a single group of units, it subsumes settings in which the units can be partitioned in multiple causally independent subgroups of units. In addition, our models incorporate group sequential adaptive designs in which the treatment allocation for one individual is based on what has been observed on previously recruited individuals in the trial (Hu and Rosenberger 2006; van der Laan 2008a; Chambaz and van der Laan 2011a,b). Our models also permit the outcome of an individual to be a function of the treatments other individuals received. The latter is referred to as interference in the causal inference literature. Thus the causal models proposed in this article do not only generalize the existing causal models for independent units, but they also generalize causal models that incorporate previously studied causal dependencies between units. Finally, we note that our models and corresponding methodology can be used to establish an approach for assessing causal effects of *interventions on the network* on the average of the unit specific outcomes. For example, one might want to know how the community level outcome changes if we change the network structure of the community through some intervention, such as increasing the connectivity between certain units in the community. In this case, our treatment nodes must be defined as properties of the sets F_i, so that a change in treatment corresponds with a change in the network structure. Finally, one may also be interested in causal effects of interventions that both change the network, and also assign an exposure to some of the units.

We refer to Aronow and Samii (2013) for an IPTW approach for estimation of an average causal effect under general interference, relying on the experimental design to generate the required generalized propensity scores. Authors additionally provide finite sample positively biased estimators of the true (nonidentifiable) conditional variance of this IPTW estimator, conditioning on the underlying counterfactuals, relying on knowing the generalized propensity score. Lastly, authors consider asymptotics when multiple independent samples from subpopulations are observed, the number of subpopulations converging to infinity, each sample allowing for their general type of interference. Their innovative approach relies on defining an exposure model that maps the treatment nodes of the N units, and specifying the characteristics of unit i into a generalized exposure of unit i. For example, the user might define this generalized exposure as the vector of exposures of the friends of unit i, beyond the exposure of unit i itself. It defines for each unit i the counterfactual outcome corresponding with the static intervention that sets this generalized exposure to a certain value, same for each unit i, and then defines the counterfactual mean outcome as the expectation of the average of these unit-specific counterfactuals. It weights by the inverse conditional probability of the generalized exposure to obtain an unbiased estimator of the expectation of the average of these counterfactual outcomes.

Our model includes the case of observing many independent clusters of units as a special case. However, by assuming more general conditional independence assumptions we also allow for asymptotic statistical inference when only one population of interconnected units is observed. Additionally, we define causal quantities in terms of stochastic interventions on the N unit-specific exposures, we allow for more general dependencies than interference, and we develop highly efficient estimators that are very different from the above mentioned IPTW-type estimator, overall making our approach distinct from Aronow and Samii (2013).

Acknowledgment This research was supported by NIH grant R01 AI074345-05.

Chapter 21
Single Time Point Interventions in Network-Dependent Data

Oleg Sofrygin, Elizabeth L. Ogburn, and Mark J. van der Laan

Consider a study in which we collect data on N units connected by a social network. For each unit $i = 1, \ldots, N$, we record baseline covariate (W_i), exposure (A_i), and outcome of interest (Y_i) information. We also observe the set F_i that consists of the units in $\{1, \ldots, N\} \backslash \{i\}$ that are connected to and could influence the unit i. The set F_i constitutes "*i's friends*". We allow $|F_i|$, the total number of friends of i, to vary in i. In addition, we assume that $|F_i|$ goes to zero when scaled by $1/N$. For example, F_i could represent the set of all the friends of i in a social network, or the set of all of i's sexual partners in a study of the effects of early HIV treatment initiation.

In this chapter, we allow for the following types of between-unit dependencies: (a) the exposure of each unit can depend on its own baseline covariates and on those of other units in F_i, and (b) the outcome of each unit can depend on its own baseline and exposure covariates of and on those of other units in F_i. An important ingredient of our modeling approach is to assume that any dependence among units is fully described by the known network. Specifically, we assume that the dependence of i's exposure and outcome on other units is limited to the set of i's friends.

O. Sofrygin (✉)
Division of Biostatistics, University of California, Berkeley, 101 Haviland Hall, #7358, Berkeley, CA 94720, USA
e-mail: sofrygin@berkeley.edu

E. L. Ogburn
Department of Biostatistics, Johns Hopkins Bloomberg School of Public Health, 615 N. Wolfe Street, Baltimore, MD 21230, USA
e-mail: eogburn@jhu.edu

M. J. van der Laan
Division of Biostatistics and Department of Statistics, University of California, Berkeley, 101 Haviland Hall, #7358, Berkeley, CA 94720, USA
e-mail: laan@berkeley.edu

© Springer International Publishing AG 2018
M.J. van der Laan, S. Rose, *Targeted Learning in Data Science*,
Springer Series in Statistics, https://doi.org/10.1007/978-3-319-65304-4_21

In addition, in this chapter we focus on the estimation of the sample-average treatment effects under a single time point stochastic intervention among units connected by a social network. We start by proposing a semiparametric statistical model. Next, we define a general single time-point intervention of interest, which may include many possible static, dynamic, or stochastic interventions on N exposures. We then describe how the TMLE framework can be used to estimate the sample-average treatment effects of such interventions. Lastly, we describe a simulation study that showcases potential real-world applications of our estimation framework.

21.1 Modeling Network Data

Suppose P_0^N is the true data generating distribution for N observed and connected units. $\mathbf{O} = (\mathbf{W}, \mathbf{A}, \mathbf{Y}) \sim P_0^N$ denotes the random vector for the N units and $O_i = (W_i, A_i, Y_i)$ for $i = 1, \ldots, N$. The network profile $\mathbf{F} = (F_1, \ldots, F_N)$ is assumed to have been recorded at baseline (i.e., $\mathbf{F} \in \mathbf{W}$). We additionally assume that all Y_i are bounded random variables. The potential for dependence among all N units implies that we generally observe a single draw $\mathbf{O}$ from P_0^N. As a result, additional assumptions are required in order to estimate P_0^N.

21.1.1 Statistical Model

Let $\mathcal{M}$ denote a statistical model containing P_0^N. A series of statistical assumptions are needed in order to learn the true distribution of $\mathbf{O}$ based on a single draw. These assumptions impose restrictions on the set of possible distributions that belong $\mathcal{M}$. Our statistical quantity of interest (target parameter) can then be defined as a mapping Ψ from $\mathcal{M}$ into the real line $\mathbb{R}$. In particular, for any $P_0^N \in \mathcal{M}$ we assume:

A1[1]. Conditional on $\mathbf{F}$, each W_i depends on at most K other observations in $\mathbf{W} = (W_1, \ldots, W_N)$. Formally, if $(W_j : j \in F_i)$ is the set of all observations dependent with W_i then $\max_i |F_i| \le K$ and K must not depend on N;

A2. $\mathbf{A} = (A_1, \ldots, A_N)$ are independent, conditional on $\mathbf{W}$;

A3. $(Y_1, \ldots, Y_N)$ are independent, conditional on $(\mathbf{A}, \mathbf{W})$.

A4. We denote $F_i^* = F_i \cup \{i\}$. For each $i = 1, \ldots, N$, the conditional distribution $P(Y_i \mid \cdot)$ depends on $(\mathbf{W}, \mathbf{A})$ only via $(A_j, W_j : j \in F_i^*)$. Similarly, for each $i = 1, \ldots, N$, the conditional distribution $P(A_i \mid \cdot)$ depends on $\mathbf{W}$ only via $(W_j : j \in F_i^*)$.

---

[1] This assumption will be also referred to as the *weak dependence of* $\mathbf{W}$.

To simplify our notation and to allow additional control on the dimensionality of each $P(Y_i \mid \cdot)$ and $P(A_i \mid \cdot)$, we will also assume the following restrictions:

B1. Each $P(Y_i \mid \cdot)$ is a function of some known summary measures $a_i^s((A_j, W_j) : j \in F_i^*)$ and $w_i^s(W_j : j \in F_i^*)$. Each $P(A_i \mid \cdot)$ is a function of the summary measure $w_i^s(W_j : j \in F_i^*)$. We assume that $w_i^s(\cdot)$ and $a_i^s(\cdot)$ are some known functions that map into a Euclidean set of constant (in i) dimension that does not depend on N, where a_i^s map into some common space $\mathcal{A}^s$, and w_i^s map into some common space $\mathcal{W}^s$.

We will use the following shorthand notation for the above summary measures:

$$
\begin{aligned}
W_i^s &= w_i^s(\mathbf{W}) &= w_i^s(W_j : j \in F_i^*) \in \mathcal{W}^s, \\
A_i^s &= a_i^s(\mathbf{A}, \mathbf{W}) = a_i^s((A_j, W_j) : j \in F_i^*) \in \mathcal{A}^s,
\end{aligned}
\tag{21.1}
$$

Note that A_i^s and W_i^s are allowed to be arbitrary functions of the units' friends, as long as their dimension is fixed, common-in-i, and doesn't depend on N. We will provide some examples of such summaries in our discussion of the simulation and analysis of network data in Sects. 21.5 and 21.4.

The following likelihood is obtained by applying the summary measures to the observed data $\mathbf{O}$:

$$
p^N(\mathbf{O}) = \left[\prod_{i=1}^N p(Y_i \mid A_i^s, W_i^s) \right] \left[\prod_{i=1}^N p(A_i \mid W_i^s) \right] p(\mathbf{W}).
\tag{21.2}
$$

We now make a final restriction on $\mathcal{M}$:

C1. We assume that all Y_i are sampled from the same distribution Q_Y with density given by $q_Y(Y_i \mid a^s, w^s)$, conditional on fixed values of the summaries (A_i^s, W_i^s), for $i = 1, \ldots, N$. Similarly, we assume that all A_i are sampled from the same distribution with density $g(A_i \mid w^s)$, conditional on some fixed value of the summaries $W_i^s = w^s$, for $i = 1, \ldots, N$.

The above assumption implies that, if two units i and j have the same values of baseline summaries W_i^s and W_j^s, and the same values of exposure summaries A_i^s and A_j^s, then i and j will be subject to the same conditional distributions for drawing their exposures and outcomes.

We denote the joint density of conditional network exposures $\mathbf{A}$ given $\mathbf{W}$ by $\mathbf{g}(\mathbf{A} \mid \mathbf{W}) = \prod_{i=1}^N g(A_i \mid W_i^s)$. We also denote the joint distribution of $\mathbf{W}$ by $\mathbf{Q_W}(\mathbf{W})$, making no additional assumptions of independence between $\mathbf{W} = (W_1, \ldots, W_N)$, and we assume $\mathbf{q_W}$ is a well-defined density for $\mathbf{Q_W}$ with respect to some dominating measure. Finally, we introduce the notation $P = P_{\mathbf{Q},\mathbf{G}}$, for $\mathbf{Q} \equiv (\mathbf{Q_W}, Q_Y)$ and we assume the distributions $\mathbf{Q_W}$ and Q_Y are unspecified beyond the above modeling restrictions. We also note that the observed exposure model for $\mathbf{G}$ may be restricted to incorporate the real-world knowledge about the true conditional treatment assignment, for example when the common-in-i density $g(A_i \mid W_i^s)$ is known in a randomized clinical trial.

The modeling restrictions **A1–A4**, **B1**, and **C1** thus define our statistical model $\mathcal{M}$. In particular, the statistical parametrization for the data-generating distribution of **O** is now defined in terms of the distributions **Q** and **G**, and the corresponding statistical model is defined as $\mathcal{M} = \{P_{\mathbf{Q},\mathbf{G}} : \mathbf{Q} \in \mathcal{Q}, \mathbf{G} \in \mathcal{G}\}$, where $\mathcal{Q}$ and $\mathcal{G}$ denote the parameter spaces for **Q** and **G**, respectively. For example, we let $\mathbf{Q}_0$ denote **Q** evaluated at P_0^N.

Finally, using the newly introduced notation we can rewrite the likelihood in 21.2 as:

$$p^N(\mathbf{O}) = \left[\prod_{i=1}^{N} q_Y(Y_i \mid A_i^s, W_i^s)\right]\left[\prod_{i=1}^{N} g(A_i \mid W_i^s)\right]\mathbf{q}_{\mathbf{W}}(\mathbf{W}). \tag{21.3}$$

With the above modeling restrictions in place, we now have a well-defined estimation problem. It is now possible to learn the common-in-i densities q_Y and g from a single (but growing) draw **O**, as $N \to \infty$.

21.1.2 Types of Interventions

The intervention of interest is defined by replacing the conditional distribution **G** with a new user-supplied intervention $\mathbf{G}^*$ that has a density $\mathbf{g}^*$, which we assume is well defined. Namely, $\mathbf{G}^*$ is a multivariate conditional distribution that encodes how each intervened exposure, denoted as A_i^*, is generated conditional on **W**. We note that static or dynamic interventions on **A** correspond to degenerate choices of $\mathbf{g}^*$ while a nondegenerate choice of $\mathbf{g}^*$ is often referred to as a stochastic intervention (e.g., Dawid and Didelez 2010). We assume that **A** and $\mathbf{A}^*$ belong to the same common space $\mathcal{A}$ and we make no further restrictions on $\mathbf{G}^*$. We also define $A_i^{*s} := a_i^s(\mathbf{A}^*)$, where A_i^{*s} denotes the random variable implied by the summary measure mapping from an intervened exposure vector $\mathbf{A}^*$, for $i = 1, \ldots, N$. Finally, we define the post-intervention distribution $P_{\mathbf{Q},\mathbf{G}^*}$ by replacing **G** in $P_{\mathbf{Q},\mathbf{G}}$ with a new user-supplied distribution $\mathbf{G}^*$. We use $\mathbf{O}^* = (W_i, A_i^*, Y_i^*)_{i=1}^N$ to denote the random variable generated under $P_{\mathbf{Q},\mathbf{G}^*}$, with its likelihood given by

$$p_{\mathbf{Q},\mathbf{G}^*}^N(\mathbf{O}^*) = \left[\prod_{i=1}^{N} q_Y(Y_i^* \mid A_i^{*s}, W_i^s)\right]\mathbf{g}^*(\mathbf{A}^* \mid \mathbf{W})\mathbf{q}_{\mathbf{W}}(\mathbf{W}). \tag{21.4}$$

The latter distribution $P_{\mathbf{Q},\mathbf{G}^*}$ is referred to as the g-computation formula for the post-intervention distribution of **O**, under the intervention $\mathbf{G}^*$ (Robins 1987).

21.1.3 Target Parameter: Sample-Average of Expected Outcomes

Our target statistical quantity ψ_0 is defined as a function of the post-intervention distribution in 21.4. Specifically, it is given by

$$\psi_0 = \Psi(P_0^N) = \frac{1}{N} \sum_{i=1}^{N} E_{\mathbf{q}_0, \mathbf{g}^*}[Y_i^*], \qquad (21.5)$$

the expectation of the sample-average of N outcomes among dependent units $i = 1, \ldots, N$, evaluated with respect to the post-intervention distribution $P_{\mathbf{Q}, \mathbf{G}^*}$. We view $\Psi(P_0^N)$ as a mapping from the statistical model $\mathcal{M}$ into $\mathbb{R}$. ψ_0 is defined conditionally on the observed network structure, $\mathbf{F}$, and is also indexed by N. We also define $\bar{Q}(A_i^s, W_i^s) = \int_y y q_Y(y \mid A_i^s, W_i^s) d\mu(y)$ as the conditional mean evaluated under common-in-i distribution Q_Y, and $\bar{Q}_0$ as $\bar{Q}$ evaluated at P_0^N. Our dimension reduction assumptions imply that $E_{P_0^N}[Y_i \mid \mathbf{A}, \mathbf{W}] = \bar{Q}_0(A_i^s, W_i^s)$. Since the target parameter ψ_0 only depends on P_0^N through $\bar{Q}_0$ and $\mathbf{Q}_{\mathbf{W},0}$, with a slight abuse of notation we will interchangeably use $\Psi(P_0^N)$ and $\Psi(\bar{Q}_0, \mathbf{Q}_{\mathbf{W},0})$. Thus, the parameter ψ_0 is indexed by N, $\mathbf{F}$ and $\mathbf{G}^*$ and can be written as

$$\psi_0 = \frac{1}{N} \sum_{i=1}^{N} \int_{\mathbf{a}, \mathbf{w}} \bar{Q}_0(a_i^s(\mathbf{a}, \mathbf{w}), w_i^s(\mathbf{w})) \mathbf{g}^*(\mathbf{a} \mid \mathbf{w}) \mathbf{q}_{\mathbf{W},0}(\mathbf{w}) d\mu(\mathbf{a}, \mathbf{w}), \qquad (21.6)$$

with respect to some dominating measure $\mu(\mathbf{a}, \mathbf{w})$.

To many practitioners it may be of value to define the target quantity as a contrast of two stochastic interventions. For example, one may define $\Psi^{\mathbf{G}_1^*}(P_0^N)$ and $\Psi^{\mathbf{G}_2^*}(P_0^N)$ as the above target parameter evaluated under stochastic interventions $\mathbf{G}_1^*$ and $\mathbf{G}_2^*$, respectively. A target contrast could then be defined as $\Psi^{\mathbf{G}_1^*, \mathbf{G}_2^*}(P_0^N) = \Psi^{\mathbf{G}_1^*}(P_0^N) - \Psi^{\mathbf{G}_2^*}(P_0^N)$. The average treatment effect over N connected units is a special case of $\Psi^{\mathbf{G}_1^*, \mathbf{G}_2^*}(P_0^N)$ for degenerate choices of $\mathbf{G}_1^*, \mathbf{G}_2^*$ defined by $\mathbf{g}_1^*(1^N \mid \mathbf{w}) = 1$ and $\mathbf{g}_2^*(0^N \mid \mathbf{w}) = 1$, for all $\mathbf{w} \in \mathcal{W}$. That is, $\mathbf{G}_1^*$ assigns treatment 1 to all subjects and $\mathbf{G}_2^*$ assigns treatment 0 to all subjects. In what follows, we will only focus on the estimation of the statistical parameter ψ_0 defined for one particular $\mathbf{G}^*$, noting that all of our results generalize naturally to such contrasts.

Additional causal assumptions beyond **A1–A4**, **B1**, and **C1** are required in order for ψ_0 to be interpreted as a causal quantity that measures the sample average of the expected counterfactual outcomes in a network of N units under intervention $\mathbf{G}^*$ (see van der Laan 2014a). These causal assumptions put no further restrictions on the probability distribution P_0^N described above, and our statistical model $\mathcal{M}$ remains the same. Since $\mathcal{M}$ contains the true data distribution P_0^N, it follows that ψ_0 will always have a pure statistical interpretation as the feature $\Psi(P_0^N)$ of the data distribution P_0^N. The causal assumptions play no role in the estimation problem at hand: even when one does not believe any of the untestable causal assumptions, the statistical parameter ψ_0 still represents an effect measure of interest controlling for all *measured* confounders. Finally, the assumption **A1** can be avoided entirely by simply choosing a target parameter ψ_0 that is conditional on the observed baseline covariates $\mathbf{W}$. We refer to van der Laan (2014a) for an in-depth description of the resulting estimation problem.

21.1.4 Sample Average Mean Direct Effect Under Interference

We now illustrate how our definition of the target parameter can be easily extended to define the sample average of expected outcomes indexed by a collection of i-specific stochastic interventions $\{g^*_{F^*_i} : i = 1, \dots, N\}$. In this case, each $g^*_{F^*_i}$ represents an intervention on i's exposure (A_i) and on the exposures of i's friends $(A_j : j \in F_i)$.[2] For example, consider the problem of estimating the direct effect of a binary exposure under interference for a network of N connected individuals. In the context of interference, a direct effect isolates the effect of unit i's exposure on i's own outcome from the effect that the exposures of i's friends may have on i's outcome. In order to capture a direct effect we can define each $g^*_{F^*_i}$ by setting the unit-specific exposure, A_i, to zero or one, while assigning all A_j for $j \in F_i$ to their observed values $(a_j : j \in F_i)$. Alternatively, we could stochastically intervene on the exposures $(A_j : j \in F_i)$ according to their observed conditional distribution $G_0(A_j|W^s_j)$.

Each outcome $Y_i^{g^*_{F^*_i}}$ is generated from its own i-specific post-intervention distri-bution that replaces the observed treatment allocation for i and $j \in F_i$ with $g^*_{F^*_i}$. The sample average of the expected outcomes $Y_i^{g^*_{F^*_i}}$ is then given by

$$\psi_0 = \frac{1}{N} \sum_{i=1}^{N} E_{\mathbf{q}_0, g^*_{F^*_i}} \left[Y_i^{g^*_{F^*_i}} \right]. \tag{21.7}$$

The parameter ψ_0 can be alternatively expressed as:

$$\psi_0 = \frac{1}{N} \sum_{i=1}^{N} \int_{\mathbf{a},\mathbf{w}} \bar{Q}_0(a^s_i(\mathbf{a}, \mathbf{w}), w^s_i(\mathbf{w})) g^*_{F^*_i}(\mathbf{a}|\mathbf{w}) \mathbf{q}_{\mathbf{w},0}(\mathbf{w}) d\mu(\mathbf{a}, \mathbf{w}). \tag{21.8}$$

The direct effect under interference can be thus represented as the contrast of two target parameters. The first parameter is indexed by a collection of interventions that assign each A_i to one, but keeps each set of exposures $(A_j : j \in F_i)$ unchanged. The second parameter is indexed by a collection of interventions that assign each A_i to zero while also keeping each set $(A_j : j \in F_i)$ unchanged. We note that these parameters can be estimated by adapting the framework presented in the following sections and we refer to Section 8 of Sofrygin and van der Laan (2017) for additional details.

21.2 Estimation

In this section we describe estimation of the target parameter using TMLE. We refer to references in Sect. 21.7 for a general overview of TMLE in i.i.d. data. To enhance the finite sample performance of TMLE of the target parameter it is generally of

---

[2] Note that we have previously defined F^*_i as $F_i \cup \{i\}$.

interest to select the smallest possible relevant part of the data generating distribution in the definition of TMLE, which is then estimated and updated by TMLE. In addition, by focusing on what really needs to be estimated, the resulting TMLE can be also simpler to implement. Thus, having defined our target parameter as a mapping $\Psi(P_0^N)$, our next step is to find the most economical representation for this mapping, namely, a new mapping that only depends on the minimal summaries of the joint distribution of N observed units, P_0^N. More precisely, the following lemma shows that our target parameter $\Psi(P_0^N)$ can be represented as an equivalent mapping $\bar{\Psi}(\bar{P}_0)$, where $\bar{P}_0$ is a mixture of N unit-specific components of the joint data-generating distribution P_0^N. We also show that this new mapping $\bar{\Psi}$ depends on $\bar{P}_0$ only through its components $\bar{Q}_0$ and $\bar{q}_{w,0}$, which are defined below.

Lemma 21.1. *Let $P^N \in \mathcal{M}$ and let P_i^s denote the i-specific summary data distribution of $O_i^s = (W_i^s, A_i^s, Y_i)$, assuming it has a well-defined density p_i^s, for each $i = 1, \ldots, N$. Note that the likelihood of each O_i^s can be factorized as follows:*

$$p_i^s(Y_i, A_i^s, W_i^s) = q_Y(Y_i|A_i^s, W_i^s)h_i(A_i^s, W_i^s) = q_Y(Y_i|A_i^s, W_i^s)g_i^s(A_i^s|W_i^s)q_{W_i^s}(W_i^s)$$

$$(21.9)$$

We can now define the mixture distribution $\bar{P}$ as a finite mixture of N unit-specific summary distributions P_i^s, where each P_i^s receives the same weight $1/N$. We assume that $\bar{P}$ has a well-defined density $\bar{p}$, and we let $\bar{O}^s = (\bar{W}^s, \bar{A}^s, \bar{Y}) \sim \bar{P}$ denote one sample drawn from $\bar{P}$. Then it follows that the likelihood of $\bar{O}^s$ can be written as:

$$\bar{p}(\bar{Y}, \bar{A}^s, \bar{W}^s) = \bar{q}_Y(\bar{Y}|\bar{A}^s, \bar{W}^s)\bar{h}(\bar{A}^s, \bar{W}^s) = \bar{q}_Y(\bar{Y}|\bar{A}^s, \bar{W}^s)\bar{g}(\bar{A}^s|\bar{W}^s)\bar{q}_W(\bar{W}^s), \quad (21.10)$$

where $\bar{q}_Y = q_Y$ (q_Y is the density of $Q_Y \in \mathcal{M}$ previously defined in Sect. 21.1, i.e., q_Y is the common-in-i conditional density of Y_i given (A_i^s, W_i^s)), $\bar{h}(a^s, w^s) := 1/N \sum_{i=1}^N h_i(a^s, w^s)$, $\bar{q}_W(w^s) := 1/N \sum_{i=1}^N q_{W_i^s}(w^s)$ and $\bar{g}(a^s|w^s) = \bar{h}(a^s, w^s)/\bar{q}_W(w^s)$. Note that we can perform an analogous exercise on the post-intervention likelihood P_{Q,G^*}^N in 21.4 for $\mathbf{O}^*$, which allows us to write-down the likelihood of the following post-intervention mixture distribution:

$$\bar{p}(\bar{Y}^*, \bar{A}^{*,s}, \bar{W}^s) = q_Y(\bar{Y}^* | \bar{A}^{*,s}, \bar{W}^s)\bar{g}^*(\bar{A}^{*,s} | \bar{W}^s)\bar{q}_W(\bar{W}^s). \quad (21.11)$$

where $\bar{q}_W$ is the factor previously defined for $\bar{p}$, $\bar{g}^*(a^s|w^s) := \bar{h}^*(a^s, w^s)/\bar{q}_W(w^s)$, $\bar{h}^*(a^s, w^s) := 1/N \sum_{i=1}^N h_i^*(a^s, w^s)$ and h_i^* is the density of the joint distribution of $(A_i^{*,s}, W_i^s)$. It now follows that $\Psi(P^N) \equiv \bar{\Psi}(\bar{P})$, where the new mapping $\bar{\Psi}(\bar{P})$ is given by:

$$\bar{\Psi}(\bar{Q}, \bar{q}_W) = E_{\bar{Q}_W}\left[E_{\bar{g}^*}[\bar{Q}(\bar{A}^{*s}, \bar{W}^s) | \bar{W}^s]\right]$$

$$= \int_{w^s \in \mathcal{W}^s, a^s \in \mathcal{A}^s} \bar{Q}(a^s, w^s)\bar{g}^*(a^s | w^s)d\bar{Q}_W(w^s),$$

where $\bar{Q}(a^s, w^s)$ is as previously defined. With the slight abuse of notation we will interchangeably write $\bar{\Psi}(\bar{Q}, \bar{q}_W)$ and $\bar{\Psi}(\bar{P})$, to emphasize the fact that $\bar{\Psi}$ depends on $\bar{P}$ only through $(\bar{Q}, \bar{q}_W)$.

Thus, the above lemma showed that our parameter of interest, $\Psi(P_0^N)$, can be represented as an expectation of $\bar{Y}^*$, namely, it is equal to the parameter of the g-computation formula for the mean outcome $E\bar{Y}_{\bar{g}_0^*}$, under stochastic intervention $\bar{g}_0^*$ (as defined above), where one uses the observed summary data $(\bar{W}^s, \bar{A}^s, \bar{Y}) \sim \bar{P}_0$. Note that the above lemma implies that the estimation of ψ_0 should only be concerned with estimating the relevant components of the mixture $\bar{P}_0$, namely, $\bar{Q}_0$ and $\bar{q}_{W,0}$, which can be done by following the standard TMLE template. Finally, expressing the dependent-data parameter as some function of the mixture $\bar{P}_0$ also implies that the estimation of the factors of $\bar{P}_0$ can be carried out by simply treating the observed dependent units as if they are i.i.d., as shown in Lemma 4.1 of Sofrygin and van der Laan (2017). To summarize, whenever we are concerned with estimating any parameter of $\bar{P}_0$, such as ψ_0 given above, we can ignore the dependence among units O_i^s, $i = 1, \ldots, N$, leading to an i.i.d.-analogue estimator for ψ_0. Of course, while the estimation can be carried out this way, performing valid inference still requires accounting for the existing dependence among units.

Before we can proceed with describing TMLE for estimating ψ_0, we have to define the efficient influence curve (EIC) for our parameter. That is, TMLE for a given parameter is derived from the parameter-specific efficient influence curve (EIC), and in our case this EIC is given by:

$$\bar{D}(P^N)(\mathbf{O}) = \frac{1}{N} \sum_{i=1}^{N} \left\{ \left(\frac{\bar{g}^*}{\bar{g}} (A_i^s \mid W_i^s) \left(Y_i - \bar{Q}(A_i^s, W_i^s) \right) \right) + \left(E_{\mathbf{q}_0, \mathbf{g}^*} \left[Y_i^* \mid \mathbf{W} \right] - E_{\mathbf{q}_0, \mathbf{g}^*} \left[Y_i^* \right] \right) \right\}.$$

$$(21.12)$$

The above EIC will now allow us to define the relevant loss function and the least favorable submodel for estimation of ψ_0. Note that TMLE for ψ_0 will be described in terms of the components of the above EIC, namely, the estimators $\bar{Q}_N$, $\bar{g}_N$, $\bar{g}_N^*$ and $\bar{Q}_{W,N}$ of $\bar{Q}_0$, $\bar{g}_0$, $\bar{g}_0^*$ and $\bar{Q}_{W,0}$, respectively. This will be followed by creating a targeted estimator $\bar{Q}_N^*$ of $\bar{Q}_0$ by updating the initial estimator $\bar{Q}_N$, defining the TMLE ψ_N^* as the corresponding plug-in estimator for the mixture mapping $\bar{\Psi}(\bar{Q}_N^*, \bar{Q}_{W,N})$. We define a targeted update $\bar{Q}_N^*$ based on the loss function for $\bar{Q}_0$ and the least favorable fluctuation submodel through $\bar{Q}_0$ in terms of $\bar{g}_0$ and $\bar{g}_0^*$. The model update $\bar{Q}_N^*$ is such that its score represents the efficient influence curve $\bar{D}$ presented in 21.12. That is, the targeted estimator $\bar{Q}_N^*$ updates $\bar{Q}_N$ by: (1) using the estimated weights $\bar{g}_N^*/\bar{g}_N$, (2) using a parametric submodel $\{\bar{Q}_N(\varepsilon, \bar{g}_N^*/\bar{g}_N)\}$ through the initial estimator $\bar{Q}_N = \bar{Q}_N(0, \bar{g}_N^*/\bar{g}_N)$ at $\varepsilon = 0$, where $\{\bar{Q}_N(\varepsilon, \bar{g}_N^*/\bar{g}_N)\}$ is referred to as the least-favorable submodel, (3) fitting ε with the standard parametric MLE, with ε^N denoting this fit, and finally, (4) defining the targeted (updated) estimator as $\bar{Q}_N^* := \bar{Q}_N(\varepsilon^N, \bar{g}_N^*/\bar{g}_N)$. Finally, since this TMLE ψ_N^* solves the empirical score equation given by the efficient influence curve $\bar{D}$ in 21.12, it follows that ψ_N^* also inherits the double robustness property of this efficient influence curve.

21.2.1 The Estimator $\bar{Q}_{W,N}$ for $\bar{Q}_{W,0}$

We define an estimator $\bar{Q}_{W,N}$ of $\bar{Q}_{W,0}$ by first defining the empirical counterpart $\mathbf{Q}_{W,N}$ of $\mathbf{Q}_{W,0}$ that puts mass one on the observed $\mathbf{W} = (W_1, \ldots, W_N)$, which then implies that the empirical distribution $Q_{W_i^s,N}$ of $Q_{W_i^s,0}$ will put mass one on its corresponding observed $W_i^s = w_i^s(\mathbf{W})$, for $i = 1, \ldots, N$. Hence, for each $w^s \in \mathcal{W}^s$, the empirical counterpart $\bar{Q}_{W,N}(w^s)$ of $\bar{Q}_{W,0}(w^s)$ may be defined as follows:

$$\bar{Q}_{W,N}(w^s) := \frac{1}{N} \sum_{i=1}^{N} I(W_i^s = w^s).$$

21.2.2 The Initial (Nontargeted) Estimator $\bar{Q}_N$ of $\bar{Q}_0$

We assumed there is a common model $\bar{Q}_0$ across all i, and Y_i are conditionally independent given (A_i^s, W_i^s) for all i. Consequently, the estimation of a common $\bar{Q}_N$ can proceed by using the pooled summary data (W_i^s, A_i^s, Y_i), $i = 1, \ldots, N$, as if the sample is i.i.d. across i. Additionally, one can rely on the usual parametric maximum likelihood estimator or loss-based cross-validation for estimating $\bar{Q}_N$. Given that Y_i can be continuous or discrete for some known range $Y_i \in [a, b]$, for $i = 1, \ldots, N$, the estimation of $\bar{Q}_0$ can be based on the following log-likelihood loss,

$$L(\bar{Q})(Y \mid A^s, W^s) = -\sum_{i=1}^{N} \log \left\{ \bar{Q}(A_i^s, W_i^s)^{Y_i} (1 - \bar{Q}(A_i^s, W_i^s))^{1-Y_i} \right\},$$

or the squared error loss

$$L(\bar{Q})(O^s) = -\sum_{i=1}^{N} \left(Y_i - \bar{Q}(A_i^s, W_i^s) \right)^2.$$

Thus, fitting $\bar{Q}_N$ for common $\bar{Q}_0 = E[Y_i \mid A_i^s, W_i^s]$ amounts to using the summary data structure (W_i^s, A_i^s, Y_i), for $i = 1, \ldots, N$. In other words, we use the entire sample of N observations for predicting Y_i. For example, for binary Y_i, $\bar{Q}_N$ can be estimated by fitting a single logistic regression model to all N observations, with Y_i as the outcome, (W_i^s, A_i^s) as predictors, and possibly adding the number of friends, $|F_i|$, as an additional covariate. A vector of unit-specific prediction values is generated by fitting $\bar{Q}_N$, , $(\bar{Q}_N(A_i^s, W_i^s))_{i=1}^{N}$. The predicted values are then used to build an updated version $\bar{Q}_N^*$ of $\bar{Q}_N$.

21.2.3 Estimating Mixture Densities $\bar{g}_0^*$ and $\bar{g}_0$

We now describe a direct approach to estimation of $\bar{g}_0$ that relies on Lemma 4.1 from Sofrygin and van der Laan (2017). This lemma states that a consistent estimator $\bar{g}_N$ of $\bar{g}_0$ can be obtained by taking a pooled sample (A_i^s, W_i^s), for $i = 1, \ldots, N$, and using the usual i.i.d. maximum likelihood-based estimation, as if we were fitting a common-in-i conditional density for A_i^s given W_i^s and treating (A_i^s, W_i^s) as independent observations. For example, if each component of A_i^s is binary, and $|A_i^s| = k$ for all i, the conditional distribution for $\bar{g}_0$ could be factorized in terms of the product of k binary conditional distributions. Each of these binary conditional distributions can be estimated with the usual logistic regression methods. Suppose now that $\mathbf{g}_0$ is known, as will be the case in a randomized clinical trial (RCT). We note that this aforementioned approach to estimating $\bar{g}_0$ can be easily adopted to incorporate the knowledge of true $\mathbf{g}_0$. That is, one could proceed by first simulating a very large number of observations $(A_j^s, W_j^s)_{j=1}^M$ from $(\mathbf{g}_0, \mathbf{Q}_{\mathbf{W},N})$, with $\mathbf{Q}_{\mathbf{W},N}$ that puts mass one on the observed $\mathbf{W}$, and then fitting the maximum likelihood-based estimator for $\bar{g}_0$, as if we were fitting a common model for A_i^s given W_i^s, based on this very large sample.

Note that $\bar{g}_0^* := \bar{h}^*(\mathbf{G}^*, \mathbf{Q}_{\mathbf{W},0})/\bar{q}_W(\mathbf{Q}_{\mathbf{W},0})$ will generally be unknown and hence will also need to be estimated from the data, in particular, since $\bar{g}_0^*$ depends on the true distribution of the data via $\mathbf{Q}_{\mathbf{W},0}$. Therefore, we propose adopting the above approach for estimation of $\bar{g}_0$ towards estimation of $\bar{g}_0^*$ by simply replacing the known $\mathbf{g}_0$ with known $\mathbf{g}^*$. The resulting model fits $\bar{g}_N^*$ and $\bar{g}_N$ are used to obtain N predictions $(\bar{g}_N^*(A_i^s \mid W_i^s)/\bar{g}_N(A_i^s \mid W_i^s))$, for $i = 1, \ldots, N$. These predictions will be used as the unit-level weights for the TMLE update of the estimator $\bar{Q}_N$, as described next.

21.2.4 The TMLE Algorithm

Having defined the estimators $\bar{Q}_N$, $\bar{g}_N$, $\bar{g}_N^*$ and $\bar{Q}_{\mathbf{W},N}$, the TMLE ψ_N^* is obtained by first constructing the model update $\bar{Q}_N^*$ for $\bar{Q}_N$, as described in step 1. below, and then evaluating ψ_N^* as a substitution estimator for the mapping $\bar{\Psi}$, as described in step 2. below.

1. Define the following parametric submodel for $\bar{Q}_N$: $\text{Logit}\bar{Q}_N(\varepsilon) = \varepsilon + \text{Logit}\bar{Q}_N$ and define the following weighted log-likelihood loss function:

$$L^w(\bar{Q}_N(\varepsilon))(\mathbf{O}^s) = -\sum_{i=1}^{N} \log\left\{\bar{Q}_N(\varepsilon)(A_i^s, W_i^s)^{Y_i}(1 - \bar{Q}_N(\varepsilon)(A_i^s, W_i^s))^{1-Y_i}\right\} \frac{\bar{g}_N^*}{\bar{g}_N}(A_i^s \mid W_i^s).$$

(21.13)

The model update $\bar{Q}_N^*$ is defined as $\bar{Q}_N(\varepsilon^N) = \text{Expit}\left(\text{Logit}\bar{Q}_N + \varepsilon^N\right)$, where ε^N minimizes the above loss, i.e., $\varepsilon^N = \arg\min_\varepsilon L^w(\bar{Q}_N(\varepsilon))(\mathbf{O}^s)$. That is, one can fit ε^N by simply running the intercept-only weighted logistic regression using the pooled sample of N observations (W_i^s, A_i^s, Y_i), for $i = 1, \ldots, N$, with outcome Y_i, intercept ε, using offsets $\text{Logit}\bar{Q}_N(A_i^s, W_i^s)$, predicted weights $\bar{g}_N^*(A_i^s \mid W_i^s)/\bar{g}_N(A_i^s \mid W_i^s)$ and no covariates. The fitted intercept is the maximum likelihood fit ε^N for ε, yielding the model update $\bar{Q}_N^*$, which can be evaluated for any fixed (a^s, w^s), by first computing the initial model prediction $\bar{Q}_N(a^s, w^s)$ and then evaluating the update $\bar{Q}_N(\varepsilon^N)$.

2. The TMLE $\psi_N^* = \bar{\Psi}_N(\bar{Q}_N^*, \bar{Q}_{W,N})$ is defined as the following substitution estimator:

$$\psi_N^* = \frac{1}{N} \sum_{i=1}^{N} \int_{\mathbf{a}} \bar{Q}_N^*(a_i^s(\mathbf{a}, \mathbf{W}), w_i^s(\mathbf{W}))\mathbf{g}^*(\mathbf{a} \mid \mathbf{W})d\mu(\mathbf{a}). \qquad (21.14)$$

For nondegenerate $\mathbf{g}^*$, the latter expression for ψ_N^* can be closely approximated by sampling from $\mathbf{g}^*$ and performing Monte Carlo integration. That is, we propose evaluating ψ_N^* by iterating the following procedure over $j = 1, \ldots, M$: (1) Sample N observations $\mathbf{A}_j^* = (A_{j,1}^*, \ldots, A_{j,N}^*)$ from $\mathbf{g}^*(\mathbf{a}|\mathbf{W})$, conditionally on observed $\mathbf{W} = (W_1, \ldots, W_N)$; (2) Apply the summary measure mappings, constructing the following summary dataset $(A_{j,i}^{*s}, W_i^s)$, for $i = 1, \ldots, N$, where each $A_{j,i}^{*s} := a_i^s(\mathbf{A}_j^*, \mathbf{W})$; and (3) Evaluate the Monte Carlo approximation to ψ_N^* for iteration j as:

$$\psi_{j,N}^* = \frac{1}{N} \sum_{i=1}^{N} \bar{Q}_N^*(A_{j,i}^{s*}, W_i^s). \qquad (21.15)$$

The Monte Carlo estimate $\bar{\psi}_N^*$ of ψ_N^* is then obtained by averaging $\psi_{j,N}^*$ across $j = 1, \ldots, M$, where M is chosen large enough to guarantee a small approximation error to ψ_N^*. Finally, we note that one could substantially reduce the computation time of this algorithm by simply re-using the summary datasets $(A_{j,i}^{*s}, W_i^s)$ that were already constructed while estimating $\bar{g}_0^*$ in 21.2.3.

21.3 Inference

We refer to Sofrygin and van der Laan (2017) and van der Laan (2014a) for results that demonstrate that the above defined TMLE is an asymptotically linear estimator for our assumed statistical model $\mathcal{M}$ from Sect. 21.1. This result relies in part on the assumption **A1** (*weak dependence of* $\mathbf{W}$). In particular, this assumption states that each W_i can be dependent on at most K other observations in $\mathbf{W} = (W_1, \ldots, W_N)$, where K is fixed. Let σ_0^2 denote the true asymptotic variance of this TMLE and we refer to Sofrygin and van der Laan (2017) for the actual expression of σ_0^2. Our next goal is to conduct statistical inference by estimating σ_0^2.

We note, however, that while our assumed statistical model $\mathcal{M}$ allowed us to construct a consistent and asymptotically linear TMLE, the same model $\mathcal{M}$ will not allow us to estimate its variance, that is, $\mathcal{M}$ results in nonidentifiable σ_0^2. To clarify, our model assumed no additional knowledge about the joint distribution of $\mathbf{W}$, beyond their weak dependence. However, the weak dependence assumption alone provides us with no means of obtaining a valid joint likelihood for $(W_1, \ldots, W_N)$. Thus, we are unable to perform any type of re-sampling from the joint distribution of $\mathbf{W}$. As a consequence, valid estimation of σ_0^2 will generally require making additional modeling restrictions for $\mathbf{Q_W}$, beyond those that were already assumed.

Conversely, when it is possible to assume a more restricted statistical model that results in a valid likelihood of $\mathbf{W}$ and allows us to fit its joint distribution, then we can also estimate the variance σ_0^2. In that case, one could either apply the empirical analogue estimator of σ_0^2 or employ the parametric bootstrap method, and both of these approaches are described below. Finally, if we are unwilling to make additional modeling restrictions for $\mathbf{Q_W}$ then only the upper bound estimates of σ_0^2 may be obtained. While we propose a possible ad-hoc upper bound estimate of σ_0^2 below, we leave its theoretical validation and the more detailed analysis of this topic for the future research.

In what follows, we provide some examples of specific modeling restrictions on $\mathbf{W}$ (e.g., when assuming $\mathbf{W}$ are i.i.d.) that allow us to consistently estimate the variance of the TMLE. We also describe an ad-hoc approach which may provide an estimate of the upper bound of such variance when not making any additional modeling restrictions for $\mathbf{Q_W}$. Finally, we describe the inference for the parameter that conditions on all observed $\mathbf{W}$ and thus doesn't require making any modeling assumptions about their joint distribution.

21.3.1 Inference in a Restricted Model for Baseline Covariates

Consider a special case where $\mathbf{W}$ are i.i.d. A valid approach to conducting inference is to use the empirical analogue estimator of σ_0^2 based on the plug-in estimates $D_{N,i}$ of $D_{0,i}$. Evaluating each $E[D_{N,i}(O_i^s)D_{N,j}(O_j^s)]$ with respect to its empirical counterpart then results in the following estimator of σ_0^2:

$$\sigma_N^2 = \frac{1}{N} \sum_{i,j} R(i,j) \left[D_{N,i}(O_i^s) D_{N,j}(\bar{O}_j) \right],$$

where

$$
\begin{aligned}
D_{N,i}(O_i^s) &= D_i(\bar{Q}_N, \bar{g}_N, \bar{g}_N^*, \psi_N^*)(O_i^s) \\
&= \frac{\bar{g}_N^*}{\bar{g}_N}(A_i^s \mid W_i^s)\left[Y_i - \bar{Q}_N(A_i^s, W_i^s) \right] \\
&\quad + \left[\int_a \bar{Q}_N^*(a_i^s(\mathbf{a}, \mathbf{W}), w_i^s(\mathbf{W}))\mathbf{g}^*(\mathbf{a} \mid \mathbf{W}) - \Psi_i(\bar{Q}_N^*, \mathbf{Q}_{\mathbf{W},N}) \right]
\end{aligned}
$$

and

$$\Psi_i(\bar{Q}_N^*, \mathbf{Q}_{\mathbf{W},N}) = \int_{\mathbf{w}} \int_a \bar{Q}_N^*(a_i^s(\mathbf{a}, \mathbf{w}), w_i^s(\mathbf{w}))\mathbf{g}^*(\mathbf{a} \mid \mathbf{w}) d\mathbf{Q}_{\mathbf{W},N}(\mathbf{w}).$$

Note that each $\Psi_i(\bar{Q}_N^*, \mathbf{Q}_{\mathbf{W},N})$, for $i = 1, \ldots, N$, can be approximated with Monte-Carlo integration by sampling with replacement from N i.i.d. baseline covariates and then sampling N exposures according to $\mathbf{g}^*$. We also note that for the case of i.i.d. baseline covariates one can derive the actual EIC for this specific model, as it was presented in Section 6.2 of van der Laan (2014a). However, the implementation of its corresponding plug-in asymptotic variance estimator is computationally overwhelming and we hence propose using the above estimator σ_N^2 of σ_0^2. One can then construct a 95% confidence interval (CI) $\psi_N^* \pm 1.96\sigma_N/\sqrt{N}$, which will provide correct asymptotic coverage.

An alternative approach that avoids assuming i.i.d. $\mathbf{W}$ is to assume a statistical model that specifies a particular ordering of observations $i = 1, \ldots, N$. This ordering, combined with the weak dependence of $\mathbf{W}$, allows us to assume a particular statistical graph for the dependence among $(W_1, \ldots, W_N)$, thus defining a unique factorization of the joint likelihood of $\mathbf{W}$. By putting additional modeling restriction on these individual likelihood factors we can obtain a single fit of the joint distribution $\mathbf{Q}_{\mathbf{W}}$. This will in turn allow us to re-sample $\mathbf{W}$. Thus, we would be able to obtain an estimate of each $E[D_{0,i}(O_i^s)D_{0,j}(O_j^s)]$ by utilizing the same plug-in approach as in the above described i.i.d. case.

Parametric Bootstrap. Alternatively, one can employ the parametric bootstrap for the estimation of σ_0^2. Note that this method still requires putting additional restrictions on the joint distribution of $\mathbf{W}$ (e.g., the i.i.d. assumption). Briefly, the parametric bootstrap is based on iterative resampling of N observations at a time from the existing joint fit of the likelihood of the observed data, which is followed by re-fitting the univariate least favorable submodel parameter ϵ and re-evaluating the resulting TMLE. Iterating this procedure enough times allows us to obtain a consistent estimate of the TMLE variance. In more detail, for each of M bootstrap iterations indexed by $b = 1, \ldots, M$, first N covariates $\mathbf{W}^b = (W_1^b, \ldots, W_N^b)$ are sampled with replacement, then the existing model fit $\hat{g}$ is applied to sampling of N exposures $\mathbf{A}^b = (A_1^b, \ldots, A_N^b)$, followed by a sample of N outcomes $\mathbf{Y}^b = (Y_1^b, \ldots, Y_N^b)$ based on the existing outcome model fit $\bar{Q}_N$. Note that we are also assuming that the corresponding bootstrapped random summaries $(W_i^{s,b}, A_i^{s,b} : i = 1, \ldots, N)$ were constructed by applying the summary functions $w_i^s(\cdot)$ and $a_i^s(\cdot)$ to $\mathbf{W}^b$ and $\mathbf{A}^b$, respectively. This bootstrap sample is then applied to obtain the predicted weights based on the existing fits $\bar{g}_N^*/\bar{g}_N$, for $i = 1, \ldots, N$. These bootstrapped predictions are then used as the unit-level weights for performing the TMLE update resulting in a bootstrap-based update of the initial estimator $\bar{Q}_N$. Note that the TMLE model update is the only model fitting step needed at each iteration of the bootstrap, which significantly lowers the computational burden of this procedure. The variance estimate is then obtained by taking the empirical variance of these bootstrapped TMLE estimates. Finally, we evaluate the finite sample performance of this approach in a simulation study in Sect. 21.6, also contrasting it to the performance of the above described empirical analogue variance estimator σ_N^2.

21.3.2 Ad-Hoc Upper Bound on Variance

An alternative approach that avoids putting additional modeling restriction on the joint distribution of $\mathbf{W}$ is to consider various ad-hoc approaches of obtaining conservative estimates of the variance. As one of the examples, consider a plug-in estimate for σ_0^2 that is based on the following estimate of $D_{0,i}$:

$$
D_{N,i}^c(O_i^s) = \frac{\bar{g}_N^*}{\bar{g}_N}(A_i^s \mid W_i^s)\left[Y_i - \bar{Q}_N(A_i^s, W_i^s)\right]
$$
$$
+ \left[\int_{\mathbf{a}} \bar{Q}_N^*(a_i^s(\mathbf{a}, \mathbf{W}), w_i^s(\mathbf{W}))\mathbf{g}^*(\mathbf{a} \mid \mathbf{W}) - \bar{\Psi}_N^*\right],
$$

where

$$
\bar{\Psi}_N^* = \frac{1}{N}\sum_{i=1}^{N}\Psi_i(\bar{Q}_N^*, \mathbf{Q}_{\mathbf{W},N}).
$$

Note that the above i-specific estimates $D_{N,i}$ of $D_{0,i}$ are no longer guaranteed to have mean zero (i.e., they are not necessary properly centered), resulting in the following conservative estimate of the variance:

$$
\frac{1}{N}\sum_{i,j} R(i, j)\left[D_{N,i}^c(O_i^s)D_{N,j}^c(\bar{O}_j)\right].
$$

Our simulations suggest that this estimator can be quite accurate when the number of friends $|F_i|$ is relatively uniform, for $i = 1, \ldots, N$, and that it becomes conservative for networks with skewed friend distributions (simulation results not shown). Finally, for more extreme cases, such as when the number of friends follows a power law distribution, the above plug-in variance estimator becomes overly conservative to the point of being noninformative. It is our conjecture that as the variability of individual $\Psi_i, 0$ increases, this estimator should become more and more conservative. We now leave a more detailed analysis of this estimator as a topic of future research.

21.3.3 Inference for Conditional Target Parameter

Admittedly, the assumption of i.i.d. baseline covariates ($\mathbf{W}$) might be too restrictive for many realistic network data generating scenarios. Similarly, the other suggested approaches will also require making specific modeling assumptions that are not necessarily supported by the data. We now propose an alternative approach for conducting inference by simply giving up on the marginal parameter of interest and performing inference conditionally on the observed baseline covariates $\mathbf{W}$. This approach has been previously described in van der Laan (2014a, Section 8) and it results in a TMLE which is identical to the one we present in this paper. Moreover, this TMLE achieves asymptotic normality with an identifiable asymptotic variance.

The asymptotic variance estimator is given by

$$\sigma_N^2(\mathbf{W}) = \frac{1}{N} \sum_{i=1}^{N} \left[D_{N,Y_i}^2(O_i^s) \right],$$

where

$$D_{N,Y_i}(O_i^s) = \frac{\bar{g}_N^*}{\bar{g}_N}(A_i^s \mid W_i^s)\left(Y_i - \bar{Q}_N(A_i^s, W_i^s)\right).$$

Note that this conditional TMLE doesn't require modeling the distribution of the baseline covariates and thus achieves the asymptotic normality under much weaker set of conditions for **W** (see van der Laan 2014a, Section 8). Thus, conducting conditional inference in a model with weakly dependent baseline covariates is a powerful alternative, especially when one is willing to accept the conditional interpretation of the resulting inference.

21.4 Simulating Network-Dependent Data in R

In this section, we describe a simulation study to evaluate the finite sample and asymptotic behavior of the above-described TMLE. The network-dependent data was simulated using the simcausal R package (Sofrygin et al. 2017). This package facilitates the simulation of complex longitudinal data, including complex network-dependent data, specifically for the types of data-generating distributions described in Sect. 21.1 (for more details see the forthcoming simcausal package vignette on conducting network simulations).

The simulations presented in this chapter were previously described in Ogburn et al. (2017). Briefly, this simulation study intended to mimic a hypothetical study designed to increase the level of physical activity in a highly-connected target community. For each community member indexed by $i = 1, \ldots, n$, the study collected data on i's baseline covariates, denoted W_i, which included the indicator of being physically active, denoted PA_i and the network of friends on each subject, F_i. The exposure or treatment, A_i, was assigned randomly to 25% of the community. Treated individuals received various economic incentives to attend a local gym.

21.4.1 Defining the Data-Generating Distribution for Observed Network Data

First, we describe the network of connections between these units (e.g., social or geographical network) by specifying either a network graph or a probabilistic network graph model for N nodes. Next, one specifies the distribution of the unit-level covariates (node attributes) by parameterizing a structural equation model (SEM) for connected units (van der Laan 2014a). This SEM allows the covariates of one unit to be dependent on the covariates of other connected units via some known functional form which is considered unspecified.

In this simulation, we assumed that the social network was modeled according to the preferential attachment model (Barabási and Albert 1999), where the node degree (number of friends) distribution followed a power law. We start by defining the distribution of the observed network graph with the sampling function provided below.

```
require("igraph")
require("simcausal")
generate.igraph.prefattach <- function(n, power,
  zero.appeal, m, ...) {
  g <- sample_pa(n, power = power, zero.appeal =
    zero.appeal, m = m)
  g <- as.directed(as.undirected(g, mode =
  "collapse"), mode = "mutual")
  sparse_AdjMat <- simcausal::igraph.to.sparse
AdjMat(g)
  NetInd_out <-
    simcausal::sparseAdjMat.to.NetInd(sparse_AdjMat)
  return(NetInd_out$NetInd_k)
}
```

The above network distribution is then added to a simcausal DAG object, which will define the observed data-generating distribution, as shown below.

```
D <- DAG.empty()
Net.prefattach <- network("Net", netfun =
  "generate.igraph.prefattach", power = 0.5,
  zero.appeal = 5, m = 5)
```

In the following example, we define the distributions of the baseline covariates. Note that we define the baseline indicator HUB, which indicates if a person has more or equal to 25 friends. We also define the baseline covariate PA, which indicates if a person is physically active at baseline. Lastly, we define the network baseline summary nF.PA, which calculates for each individual the total number of friends who are physically active (note that some of the simulated baseline covariates are not shown).

```
D <- D + Net.prefattach +
  node("HUB", distr = "rconst", const =
    ifelse(nF >= 25, 1, 0)) +
  node("W1", distr = "rcat.b1",
    probs = c(0.0494, 0.1823, 0.2806, 0.2680, 0.1651,
      0.0546)) +
  node("W2", distr = "rbern", prob = plogis(-0.2)) +
  node("PA", distr = "rbern", prob = W2*0.05 +
```

```
   (1-W2)*0.15) +
node("nF.PA", distr = "rconst", const =
   sum(PA[[1:Kmax]]), replaceNAw0 = TRUE)
```

As a next step, we randomly assign the binary exposure A, to 25% of the population. This exposure corresponds with an informational campaign about the benefits of physical exercise and is intended to promote and sustain attendance of the local gym by community members.

```
D <- D + node("A", distr = "rbern", prob = 0.25)
```

We then define a network summary measure, sum.netA, which depends on the exposures of the individuals' friends, as well as his or her friends' baseline covariate values.

```
D <- D + node("sum.netA", distr = "rconst",
   const = (sum(A[[1:Kmax]])*(HUB==0) +
      sum((W1[[1:Kmax]] > 4)*A[[1:Kmax]])*(HUB==1)),
   replaceNAw0 = TRUE)
```

The outcome Y_i is a binary indicator of maintaining gym membership for a predetermined follow-up period, following the exposure **A**. The outcome is defined by the variable 'Y' below. Note that we assumed that each Y depends on the individual exposure and baseline covariates. It also depends on the network summary, sum.netA, as well as the number of friends who were physically active at baseline, nF.PA

The probability of success for each Y_i is defined a logit-linear function of i's exposure A_i (indicator of receiving the economic incentive), the baseline covariates W_i and the three summary measures of i's friends' exposures and baseline covariates. In particular, we also assumed that the probability of maintaining gym membership increased on a logit-linear scale as a function of the following network summaries: the total number of i's friends who were exposed ($\sum_{j\in F_i} A_j$), the total number of i's friends who were physically active at baseline ($\sum_{j\in F_i} PA_j$) and the product of the two summaries ($\sum_{j\in F_i} A_j \times \sum_{j\in F_i} PA_j$). The economic incentive to attend local gym had a small direct effect on each individual who was not physically active at baseline and no direct effect on those who were already physically active. However, physically active individuals were more likely to maintain gym membership over the follow-up period if they had at least one physically active friend at baseline.

```
D <- D +
   node("Y", distr = "rbern",
      prob = plogis(ifelse(PA == 1,
         +5 - 15*(nF.PA < 1),
         -8.0 + 0.25*A) +
      +0.5*sum.netA + 0.25*nF.PA*sum.netA + 0.5*nF.PA +
      +0.5*(W1-1) - 0.58*W2),
   replaceNAw0 = TRUE)
```

Finally, we define the data-generating distribution based on the preferential attachment network model, as shown below.

```
D.prefattach <- set.DAG(D, n.test = 200)
```

We now call the function `sim`, which simulates a single network of 5000 observations using the above defined data-generating distribution, as shown below. We plot the distribution of the number of friends for this network in Fig. 21.1. We also plot the network graph based on a random sample of 100 units in Fig. 21.2.

```
datO_5K <- sim(D.prefattach, n = 5000)
```

21.4.2 Defining Intervention, Simulating Counterfactual Data and Evaluating the Target Causal Quantity

Next, we define several stochastic and dynamic interventions on the exposure A, as well as the total number of physically active friends nF.PA. We also calculate the corresponding causal effects of these interventions, using the preferential attachment network model. Note that one can easily evaluate the true values of the above causal parameters by simulating intervention-specific counterfactual data and then evaluating the estimated mean of the counterfactual outcomes, as shown in all of the following examples.

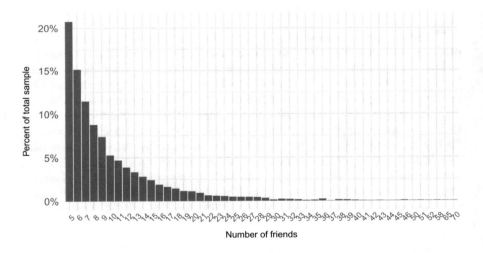

Fig. 21.1 Distribution of the number of friends for a preferential attachment network with 5000 observations

The following code shows an example of evaluating the true counterfactual mean outcome when the exposure is assigned to a random 35% of the population (stochastic intervention):

```
D.prefattach <- D.prefattach +
  action("gstar", nodes = node("A", distr = "rbern",
  prob = aset), aset = 0.35)
datFull <- sim(D.prefattach, actions="gstar",
  n = 50000, rndseed = 54321)
print(psi0_a0.4 <- mean(datFull[["gstar"]]$Y))
[1] 0.15186
```

The following code illustrates the evaluation of the true counterfactual mean outcome under a dynamic intervention that covers only around 10% of the population by intervening (stochastically) only on the most connected individuals, i.e., only on individuals that have more than 15 friends:

```
D.prefattach <- D.prefattach +
action("gHubs",
  nodes = c(node("A", distr = "rbern", prob =
  ifelse(nF >= 20, 0.9, ifelse(nF >= 15, 0.40, 0)))))
datFull <- sim(D.prefattach, actions="gHubs",
  n = 50000, rndseed = 54321)
print(psi0_g.dynamic <- mean(datFull[["gHubs"]]$Y))
[1] 0.1204
```

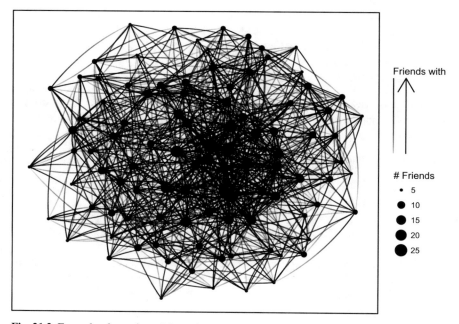

Fig. 21.2 Example of a preferential attachment network for a sample of 100 observations

21.5 Causal Effects with Network-Dependent Data in R

Having defined the simulated network data, as well as the true value of the target causal quantity (the gold standard), we now discuss the estimation of such causal parameters in R. In particular, we will utilize the `tmlenet` R package (Sofrygin and van der Laan 2015) which implements the TMLE for network-dependent data.

Defining Network-Based Summaries. We start by employing the function `def_sW` to define some of the baseline network summaries (W_i^s), as shown in the example below. Note that these summary measures will be automatically evaluated based on the input network data.

```
require("tmlenet")
sW <-  def_sW(W1, W2, WNoise, corrW.F1, corrW.F2,
  corrW.F3, corrW.F4, corrW.F5, HUB =
  ifelse(nF >= 25, 1, 0))
```

Similarly, we employ the function `def_sA` to define some of the "*effective exposure*" or the exposure network summaries (A_i^s), as shown in the example below.

```
sA <-  def_sA(A, nF.PA = sum(PA[[1:Kmax]]),
  replaceNAw0 = TRUE) +
  def_sA(A.PAeq0 = A * (PA == 0)) +
  def_sA(nFPAeq0.PAeq1 = (nF.PA < 1) * (PA == 1)) +
  def_sA(sum.netA = (sum(A[[1:Kmax]])*(HUB==0) +
    sum((W1[[1:Kmax]] > 4)*A[[1:Kmax]])*(HUB==1)),
  sum.netA.sum.netPA = sum.netA*nF.PA,
  replaceNAw0 = TRUE)
```

Regression Models. Next, we specify the regression formulas which will be used to fit the conditional *effective exposure* and the conditional outcome models, as shown below.

```
hform <- "A + sum.netA ~ HUB + PA + nF.PA +
          nFPAeq0.PAeq1"
Qforms <- "Y ~ nF.PA + A.PAeq0 + nFPAeq0.PAeq1 +
          sum.netA + sum.netA.sum.netPA + PA +
          W1 + W2 + corrW.F1 + corrW.F2 +
          corrW.F3 + corrW.F4 + corrW.F5"
```

Interventions. As a next step, we provide some examples for specifying the intervention of interest using the `tmlenet` R package syntax. We start with a an example of a stochastic intervention on each A_i that assigns exposure to a random 35% of the sample.

```
new.sA1.stoch.2 <-  def_new_sA(A =
  rbinom(n = length(A), size = 1, prob = 0.35))
```

Next, we provide an example of a dynamic intervention on each A_i, conditional on the number of friends (nF). This intervention assigns exposure to approximately 10% of the most connected individuals in the network.

```
new.sA1.dyn.4 <- def_new_sA(A = rbinom(n = length(A),
    size = 1, prob = ifelse(nF >= 20, 0.9,
    ifelse(nF >= 15, 0.40, 0)))))
```

Estimation. The following example demonstrates how to run TMLE to estimate the sample-average of expected outcomes for the above defined intervention, using the observed input dataset along with the observed network data.

```
res <- tmlenet(data = datO, sW = sW, sA = sA,
    Ynode = "Y", Kmax = K,
    NETIDmat = attributes(datO)$netind_cl$NetInd,
    intervene1.sA = new.sA1.stoch.2, Qform = Qform,
    hform.g0 = hform, hform.gstar = hform)
```

21.6 Simulation Results

We based our simulation study on the above described data-generating distribution and as before, assumed that the outcome Y_i was a binary indicator of maintaining gym membership for a predetermined follow-up period.

We assumed that it was of interest to examine and estimate the average of the mean counterfactual outcomes $E\left[\bar{Y}^{g^*}\right]$ under various hypothetical interventions g^* on such a community. First, we considered a stochastic intervention g_1^*, which assigned each individual to treatment with a constant probability of 0.35; this differs from the observed allocation of treatment to 25% of the community members. We also considered a scenario in which the aforementioned economic incentive was resource constrained and could only be allocated to up to 10% of community members and estimated the effects of various targeted approaches to allocating the exposure. For example, we considered an intervention g_2^* that targeted only the top 10% most connected members of the community, as such a targeted intervention would be expected to have a higher impact on the overall average probability of maintaining gym membership among the community, when compared to purely random assignment of exposure to 10% of the community. Hypothetical intervention g_3^* assigned an additional physically active friend to individuals with fewer than ten friends. Notably, this intervention aimed at directly intervening on the social network of some members of the community. Finally, we estimated the combined effect of simultaneously implementing both the dynamic intervention g_2^* and the network-based intervention g_3^* on the same community. We report the expected outcome under each of these interventions here; causal effects defined as contrasts of such interventions can be estimated using our methods and these results are available in the supplementary material.

The simulations were performed for sample sizes of $n = 500$, $n = 1000$ and $n = 10{,}000$, where each dataset was sampled according the above described data-generating distribution. The estimation was repeated by sampling 1000 such datasets, conditional on the same network (sampled only once for each sample size). The baseline covariates for each unit i were sampled as i.i.d. We used the TMLE to estimate the sample-average of expected outcomes for the above described interventions g_1^* through g_4^* and evaluated the true finite sample bias and variance of the corresponding estimates. We also used our simulations to compare three different estimators of the asymptotic variance provided by the `tmlenet` R package. In particular, we compared the coverage of the their corresponding asymptotic confidence intervals (CIs). First, we used the plug-in i.i.d. estimator ("*IID Var*") for the variance of the efficient influence curve which treated observations as if they were i.i.d. Second, we used the plug-in variance estimator which correctly adjusted for correlated observations ("*dependent IC Var*"). Finally, we used the previously described parametric bootstrap variance estimator ("*bootstrap Var*"). The simulation results in Fig. 21.3 display the mean length and the coverage of the 95% CIs, stratified by the intervention type, CI type and the observed sample size.

We evaluated the performance of the three approaches to variance estimation described above, as measured by the coverage of the 95% CIs. Our results show that conducting inference while ignoring the nature of the dependence in such datasets

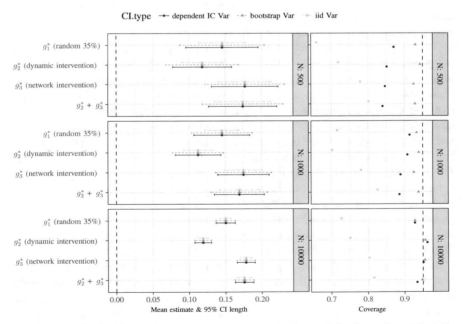

Fig. 21.3 Mean 95% CI length (*left panel*) and coverage (*right panel*) for the preferential attachment network, by sample size, intervention and CI type. Results shown for average expected outcomes only

generally results in anti-conservative variance estimates and under-coverage of CIs, which can be as low as 50% even for very large sample sizes (*"IID Var"* in the right panel of Fig. 21.3. The CIs based on the dependent variance estimates (*"dependent IC Var"* in the right panel of the same figures) obtain nearly nominal coverage of 95% for large enough sample sizes, but can suffer in smaller sample sizes due to lack of asymptotic normality and near-positivity violations. Notably, the CIs based on the parametric bootstrap variance estimates provide the most robust coverage for smaller sample sizes, while attaining the nominal 95% coverage in large sample sizes for nearly all of the simulation scenarios (*"bootstrap Var"* in the right panel of the same figures). One of the surprising finding of this study was the apparent robustness of the parametric bootstrap method for inference in small sample sizes. That is, while it was expected that the highly connected network types used in these simulations can lead to anti-conservative coverage of the asymptotic CIs derived from the dependent-data influence curve variance estimates (e.g., due to near positivity violations and lack of convergence to normality), it was surprising to see that the analogous CIs based on the parametric bootstrap variance estimates resulted in nearly nominal 95% coverage for sample sizes as low as 500. Future work will explore the assumptions under which this parametric bootstrap works and its sensitivity towards violations of those assumptions. Furthermore, our simulations illustrate that by targeting the exposure assignment to highly connected and physically active individuals, one may leveraging the structure of the network to increases the mean probability of sustaining gym membership compared to the similar level of un-targeted coverage of the exposure. We demonstrated the feasibility of estimating effects of interventions on the observed network structure itself, such as intervention g_3^*, which can be also combined with economic incentives, such as the intervention $g_2^* + g_3^*$. Such combined interventions may lead to a larger average expected effect on the community as a whole, especially for resource constrained interventions.

21.7 Notes and Further Reading

The literature on networks and causal inference in network data is rapidly evolving. However, the existing statistical methods for performing estimation and inference for causal effects in networks are limited and the literature on this subject has only recently started to develop (van der Laan 2014a; VanderWeele et al. 2014b; Ogburn and VanderWeele 2014; VanderWeele and An 2013; Tchetgen Tchetgen and VanderWeele 2012). Most of the recently proposed approaches can be categorized as relying on either the assumption of randomized exposures across units (Rosenbaum 2007; Aronow and Samii 2013; Bowers et al. 2013; Walker and Muchnik 2014; Aral and Walker 2011, 2014; Toulis and Kao 2013; Liu and Hudgens 2014; Choi 2014; Basse and Airoldi 2015), or on parametric modeling of the outcome as a particular function of the unit's network.

Some of the parametric approaches applied in the network settings include generalized linear models (GLMs) and generalized estimating equations (GEEs)

(Christakis and Fowler 2013, 2007), methods which have important limitations (Lyons 2010; VanderWeele 2011, 2013; VanderWeele et al. 2012c; Ogburn and VanderWeele 2014). For one, GLMs and similar modeling techniques require making strong, simplifying modeling assumptions about the underlying data generating process. Hence, model misspecification for GEEs and GLMs in the network data settings is a major cause of concern. Perhaps more importantly, performing valid statistical inference with GLMs and other similar statistical techniques generally requires independence of the observational units, an assumption that is unlikely to hold due to the very nature of the network data. It has also been previously described that application of such standard statistical procedures to dependent data will result in invalid and generally anti-conservative statistical inference (Lyons 2010; Ogburn and VanderWeele 2014).

In addition, a few promising methodological approaches to estimation in network data have begun to emerge in recent years. For example, Aronow and Samii (2013) proposed a Horvitz-Thompson estimator in a randomized study settings, defined the so-called "network exposure model" and derived the finite sample estimator of the variance. However, such methods are of limited utility in observational settings. Other proposed approaches for identification and estimation of treatment effects in networks include stochastic actor-oriented models (Steglich et al. 2010), and a linear Bayesian modeling approach that can accommodate for network uncertainty (Toulis and Kao 2013). Another recently proposed approach applied the semiparametric framework of targeted maximum likelihood estimation to the observation network data settings (van der Laan 2014a), yielding valid asymptotic inference, while allowing for a much larger and realistic class of data-generative models.

Part VII
Optimal Dynamic Rules

Chapter 22
Optimal Dynamic Treatment Rules

Alexander R. Luedtke and Mark J. van der Laan

Suppose we observe n independent and identically distributed observations of a time-dependent random variable consisting of baseline covariates, initial treatment and censoring indicator, intermediate covariates, subsequent treatment and censoring indicator, and a final outcome. For example, this could be data generated by a sequential RCT in which one follows up a group of subjects, and treatment assignment at two time points is sequentially randomized, where the probability of receiving treatment might be determined by a baseline covariate for the first-line treatment, and time-dependent intermediate covariate (such as a biomarker of interest) for the second-line treatment. Such trials are often called sequential multiple assignment randomized trials (SMART). A dynamic treatment rule deterministically assigns treatment as a function of the available history. If treatment is assigned at two time points, then this dynamic treatment rule consists of two rules, one for each time point. The mean outcome under a dynamic treatment is a counterfactual quantity of interest representing what the mean outcome would have been if everybody would have received treatment according to the dynamic treatment rule. The optimal dynamic treatment rule is defined as the dynamic treatment rule that maximizes the mean outcome.

Previous approaches, described at the end of this chapter, rely on semiparametric models that make strong assumptions on the data generating process. We instead define the statistical model for the data distribution as nonparametric, beyond possible knowledge about the treatment mechanism (e.g., known in a RCT) and censoring mechanism. In order to not only consider the most ambitious fully optimal rule, we

A. R. Luedtke (✉)
Vaccine and Infectious Disease Division, Fred Hutchinson Cancer Research Center,
1100 Fairview Ave, Seattle, WA 98109, USA
e-mail: aluedtke@fredhutch.org

M. J. van der Laan
Division of Biostatistics and Department of Statistics, University of California, Berkeley,
101 Haviland Hall, #7358, Berkeley, CA 94720, USA
e-mail: laan@berkeley.edu

© Springer International Publishing AG 2018
M.J. van der Laan, S. Rose, *Targeted Learning in Data Science*,
Springer Series in Statistics, https://doi.org/10.1007/978-3-319-65304-4_22

define the V-optimal rules as the optimal rule that only uses a user-supplied subset V of the available covariates. This allows us to consider suboptimal rules that are easier to estimate and thereby allow for statistical inference for the counterfactual mean outcome under the suboptimal rule.

In this chapter, we describe how to obtain semiparametric inference about the mean outcome under the two time point V-optimal rule. We will show that the mean outcome under the optimal rule is a pathwise differentiable parameter of the data distribution, indicating that it is possible to develop asymptotically linear estimators of this target parameter under conditions. In fact, we obtain the surprising result that the pathwise derivative of this target parameter equals the pathwise derivative of the mean counterfactual outcome under a given dynamic treatment rule set at the optimal rule, treating the latter as known. By a reference to the earlier for double robust and efficient estimation of the mean outcome under a given rule (see Chap. 4), we then obtain a CV-TMLE for the mean outcome under the optimal rule. Subsequently, we prove asymptotic linearity and efficiency of this CV-TMLE, allowing us to construct confidence intervals for the mean outcome under the optimal dynamic treatment or its contrast with respect to a standard treatment.

In a SMART the statistical inference would only rely upon a second-order difference between the estimator of the optimal dynamic treatment and the optimal dynamic treatment itself to be asymptotically negligible. This is a reasonable condition if we restrict ourselves to rules only responding to a one-dimensional time-dependent covariate, or if we are willing to make smoothness assumptions. While this condition appears to be necessary when estimating the optimal mean outcome, it is not necessary if the parameter of interest is redefined as the average mean outcome under our cross-validated estimates of the optimal dynamic treatment. This parameter relies on the data through our estimates of the optimal dynamic treatment, and we thus refer to it as a data-adaptive parameter.

22.1 Optimal Dynamic Treatment Estimation Problem

For the sake of presentation, we focus on two time point treatments in this chapter. Suppose we observe n i.i.d. copies $O_1, \ldots, O_n \in O$ of

$$O = (L(0), A(0), L(1), A(1), Y) \sim P_0,$$

where $A(j) = (A_1(j), A_2(j))$, $A_1(j)$ is a binary treatment and $A_2(j)$ is an indicator of not being right censored at "time" j, $j = 0, 1$. That is, $A_2(0) = 0$ implies that $(L(1), A_1(1), Y)$ is not observed, and $A_2(1) = 0$ implies that Y is not observed. Each time point j has covariates $L(j)$ that precede treatment, $j = 0, 1$, and the outcome of interest is given by Y and occurs after time point 1. For a time-dependent process

$X(\cdot)$, we will use the notation $\bar{X}(t) = (X(s) : s \leq t)$, where $\bar{X}(-1) = \emptyset$. Let $\mathcal{M}$ be a statistical model that makes no assumptions on the marginal distribution $Q_{0,L(0)}$ of $L(0)$ and the conditional distribution $Q_{0,L(1)}$ of $L(1)$, given $A(0), L(0)$, but might make assumptions on the conditional distributions $g_{0A(j)}$ of $A(j)$, given $\bar{A}(j-1), \bar{L}(j)$, $j = 0, 1$. We will refer to g_0 as the intervention mechanism, which can be factorized in a treatment mechanism g_{01} and censoring mechanism g_{02} as follows:

$$g_0(O) = \prod_{j=1}^{2} g_{01}(A_1(j) \mid \bar{A}(j-1), \bar{L}(j)) g_{02}(A_2(j) \mid A_1(j), \bar{A}(j-1), \bar{L}(j)).$$

In particular, the data might have been generated by a SMART, in which case g_{01} is known.

Let $V(1)$ be a function of $(L(0), A(0), L(1))$, and let $V(0)$ be a function of $L(0)$. Let $V = (V(0), V(1))$. Consider dynamic treatment rules $V(0) \to d_{A(0)}(V(0)) \in \{0, 1\} \times \{1\}$ and $(A(0), V(1)) \to d_{A(1)}(A(0), V(1)) \in \{0, 1\} \times \{1\}$ for assigning treatment $A(0)$ and $A(1)$, respectively, where the rule for $A(0)$ is only a function of $V(0)$, and the rule for $A(1)$ is only a function of $(A(0), V(1))$. Note that these rules are restricted to set the censoring indicators $A_2(j) = 1$, $j = 0, 1$. Let $\mathcal{D}$ be the set of all such rules. We assume that $V(0)$ is a function of $V(1)$ (i.e., observing $V(1)$ includes observing $V(0)$), but in the theorem below we indicate an alternative assumption. For $d \in \mathcal{D}$, we let:

$$d(a(0), v) \equiv (d_{A(0)}(v(0)), d_{A(1)}(a(0), v(1))).$$

If we assume a structural equation model (Pearl 2009a) for variables stating that

$$L(0) = f_{L(0)}(U_{L(0)})$$
$$A(0) = f_{A(0)}(L(0), U_{A(0)})$$
$$L(1) = f_{L(1)}(L(0), A(0), U_{L(1)})$$
$$A(1) = f_{A(1)}(\bar{L}(1), A(0), U_{A(1)})$$
$$Y = f_Y(\bar{L}(1), \bar{A}(1), U_Y),$$

where the collection of functions $f = (f_{L(0)}, f_{A(0)}, f_{L(1)}, f_{A(1)})$ are unspecified or partially specified, we can define counterfactuals Y_d defined by the modified system in which the equations for $A(0), A(1)$ are replaced by $A(0) = d_{A(0)}(V(0))$ and $A(1) = d_{A(1)}(A(0), V(1))$. Denote the distribution of these counterfactual quantities as $P_{0,d}$, where we note that $P_{0,d}$ is implied by the collection of functions f and the joint distribution of exogenous variables $(U_{L(0)}, U_{A(0)}, U_{L(1)}, U_{A(1)}, U_Y)$. We can now define the causally optimal rule under $P_{0,d}$ as $d_0^* = \arg\max_{d \in \mathcal{D}} E_{P_{0,d}} Y_d$. If we assume a sequential randomization assumption stating that $A(0)$ is independent of $U_{L(1)}, U_Y$, given $L(0)$, and $A(1)$ is independent of U_Y, given $\bar{L}(1), A(0)$, then we can identify $P_{0,d}$ with observed data under the distribution P_0 using the g-computation formula:

$$p_{0,d}(L(0), A(0), L(1), A(1), Y)$$
$$\equiv I(A = d(A(0), V)) q_{0,L(0)}(L(0)) q_{0,L(1)}(L(1) \mid L(0), A(0)) q_{0,Y}(Y \mid \bar{L}(1), \bar{A}(1)),$$
$$(22.1)$$

where $p_{0,d}$ is the density of $P_{0,d}$ and $q_{0,L(0)}$, $q_{0,L(1)}$, and $q_{0,Y}$ are the densities for $Q_{0,L(0)}$, $Q_{0,L(1)}$, and $Q_{0,Y}$, where $Q_{0,Y}$ represents the distribution of Y given $\bar{L}(1), \bar{A}(1)$. We assume that all densities above are absolutely continuous with respect to some dominating measure μ. We have a similar identifiability result/g-computation formula under the Neyman-Rubin causal model (Robins 1987). More generally, for a distribution $P \in \mathcal{M}$ we can define the g-computation distribution P_d as the distribution with density

$$p_d(L(0), A(0), L(1), A(1), Y)$$
$$\equiv I(A = d(A(0), V))q_{L(0)}(L(0))q_{L(1)}(L(1) \mid L(0), A(0))q_Y(Y \mid \bar{L}(1), \bar{A}(1)),$$

where $q_{L(0)}$, $q_{L(1)}$, and q_Y are the counterparts to $q_{0,L(0)}$, $q_{0,L(1)}$, and $q_{0,Y}$ under P.

For the remainder of this chapter, if for a static or dynamic intervention d, we use notation L_d (or Y_d, O_d) we mean the random variable with the probability distribution P_d in (22.1) so that of all our quantities are statistical parameters. For example, the quantity $E_0(Y_{a(0)a(1)} \mid V_{a(0)}(1))$ defined in the next theorem denotes the conditional expectation of $Y_{a(0)a(1)}$, given $V_{a(0)}(1)$, under the probability distribution $P_{0,a(0)a(1)}$ (i.e., g-computation formula presented above for the static intervention $(a(0), a(1))$). In addition, if we write down these parameters for some P_d, we will automatically assume the positivity assumption at P required for the g-computation formula to be well defined. For that it will suffice to assume the following positivity assumption at P:

$$Pr_P\left(0 < \min_{a_1 \in \{0,1\}} g_{0A(0)}(a_1, 1|L(0))\right) = 1$$
$$Pr_P\left(0 < \min_{a_1 \in \{0,1\}} g_{0A(1)}(a_1, 1 \mid \bar{L}(1), A(0))\right) = 1. \tag{22.2}$$

The strong positivity assumption will be defined as the above assumption, but where the 0 is replaced by a $\delta > 0$.

We now define a statistical parameter representing the mean outcome Y_d under P_d. For any rule $d \in \mathcal{D}$, let

$$\Psi_d(P) \equiv E_{P_d} Y_d.$$

For a distribution P, define the V-optimal rule as

$$d_P = \arg\max_{d \in \mathcal{D}} E_{P_d} Y_d.$$

For simplicity, we will write d_0 instead of d_{P_0} for the V-optimal rule under P_0. Define the parameter mapping $\Psi : \mathcal{M} \to \mathbb{R}$ as $\Psi(P) = E_{P_{d_P}} Y_{d_P}$. The first part of this chapter is concerned with inference for the parameter

$$\psi_0 \equiv \Psi(P_0) = E_{P_{0,d_0}} Y_{d_0}.$$

Under our identifiability assumptions, d_0 is equal to the causally optimal rule d_0^*. Even if the sequential randomization assumption does not hold, the statistical parameter ψ_0 represents a statistical parameter of interest in its own right. We will not concern ourselves with the sequential randomization assumption for the remainder of this paper.

The next theorem presents an explicit form of the V-optimal individualized treatment rule d_0 as a function of P_0.

Theorem 22.1. *Suppose $V(0)$ is a function of $V(1)$. The V-optimal rule d_0 can be represented as the following explicit parameter of P_0:*

$$\bar{Q}_{b,20}(a(0), v(1)) = E_0(Y_{a(0),A(1)=(1,1)} \mid V_{a(0)}(1) = v(1))$$
$$-E_0(Y_{a(0),A(1)=(0,1)} \mid V_{a(0)}(1) = v(1)),$$
$$d_{0,A(1)}(A(0), V(1)) = (I(\bar{Q}_{b,20}(A(0), V(1)) > 0), 1),$$
$$\bar{Q}_{b,10}(v(0)) = E_0(Y_{(1,1),d_{0,A(1)}} \mid V(0)) - E_0(Y_{(0,1),d_{0,A(1)}} \mid V(0)),$$
$$d_{0,A(0)}(V(0)) = (I(\bar{Q}_{b,10}(V(0)) > 0), 1),$$

where $a(0) \in \{0, 1\} \times \{1\}$. If $V(1)$ does not include $V(0)$, but, for all $(a(0), a(1)) \in \{\{0, 1\} \times \{1\}\}^2$,

$$E_0(Y_{a(0),a(1)} \mid V(0), V_{a(0)}(1)) = E_0(Y_{a(0),a(1)} \mid V_{a(0)}(1)), \tag{22.3}$$

then the above expression for the V-optimal rule d_0 is still true.

Following Robins (2004), we refer to $\bar{Q}_{b,10}$ and $\bar{Q}_{b,20}$ as the (first and second time point) blip functions.

22.2 Efficient Influence Curve of the Mean Outcome Under V-Optimal Rule

In this section, we establish the pathwise differentiability of Ψ and give an explicit expression for the efficient influence curve. Before presenting this result, we give the efficient influence curve for the parameter $\Psi : \mathcal{M} \to \mathbb{R}$ where $\Psi_d(P) \equiv E_P Y_d$ and the rule $d = (d_{A(0)}, d_{A(1)}) \in \mathcal{D}$ is treated as known. This influence curve was presented in Chap. 4. The parameter mapping Ψ_d has efficient influence curve

$$D^*(d, P) = \sum_{k=0}^{2} D_k^*(d, P),$$

where

$$D_0^*(d, P) = E_P[Y_d \mid L(0), A(0) = d_{A(0)}(V(0))] - E_P Y_d,$$

$$D_1^*(d, P) = \frac{I(A(0) = d_{A(0)}(V(0)))}{g_{A(0)}(O)} (E_P[Y \mid \bar{A}(1) = d(A(0), V), \bar{L}(1)]$$
$$- E_P[Y_d \mid L(0), A(0) = d_{A(0)}(V(0))]),$$

$$D_2^*(d, P) = \frac{I(\bar{A}(1) = d(A(0), V))}{\prod_{j=0}^{1} g_{A(j)}(O)} \left(Y - E_P\left[Y \mid \bar{A}(1) = d(A(0), V), \bar{L}(1)\right]\right). \quad (22.4)$$

Above $(g_{A(0)}, g_{A(1)})$ is the intervention mechanism under the distribution P. We remind the reader that Y_d has the g-computation distribution from (22.1) so that:

$$E_P[Y_d \mid L(0), A(0) = d_{A(0)}(V(0))]$$
$$= E_P\left[E_P\left[Y \mid \bar{A}(1) = d(A(0), V), \bar{L}(1))\right] \mid L(0), A(0) = d_{A(0)}(V(0))\right]$$

At times it will be convenient to write $D_k^*(d, Q^d, g)$ instead of $D_k^*(d, P)$, where Q^d represents both of the conditional expectations in the definitions of D_1^* and the marginal distribution of $L(0)$ under P and g represents the intervention mechanism under P. We will denote these conditional expectations under P_0 for a given rule d by Q_0^d. We will similarly at times denote $D^*(d, P)$ by $D^*(d, Q^d, g)$.

Whenever $D^*(P)$ does not contain an argument for a rule d, this $D^*(P)$ refers to the efficient influence curve of the parameter mapping Ψ for which $\Psi(P) = E_P Y_{d_P}$, where the optimal rule d_P under P is not treated as known. Not treating d_P as known means that d_P depends on the input distribution P in the mapping $\Psi(P)$. The following theorem presents the efficient influence curve of Ψ at a distribution P. The main condition on this distribution P is that it satisfies the nonexceptional law condition that

$$\max_{a_0(0) \in \{0,1\}} Pr_P\left(\bar{Q}_{b,2}((a_0(0), 1), V_{a(0)=(a_0(0),1)}) = 0\right) = 0,$$
$$Pr_P\left(\bar{Q}_{b,1}(V(0)) = 0\right) = 0, \quad (22.5)$$

where $\bar{Q}_{b,2}$ and $\bar{Q}_{b,1}$ are defined analogously to $\bar{Q}_{b,20}$ and $\bar{Q}_{b,10}$ in Theorem 22.1 with the expectations under P_0 replaced by expectations under P. That is, we assume that each of the blip functions under P is nowhere zero with probability 1. Distributions that do not satisfy this assumption have been referred to as "exceptional laws" (Robins 2004). These laws are indeed exceptional when one expects that treatment will have a beneficial or harmful effect in all V-strata of individuals. When one only expects that treatment will have an effect on outcome in some but not all strata of individuals then this assumption may be violated. We will make this assumption about P_0 for all subsequent asymptotic linearity results about $E_0 Y_{d_0}$, and we will assume a weaker but still not completely trivial assumption about the consistency of the optimal rule estimate to some fixed limit for the data-adaptive target parameters in Sect. 22.3.

Theorem 22.2. *Suppose $P \in \mathcal{M}$ is such that $Pr_P(\mid Y \mid < M) = 1$ for some $M < \infty$, P satisfies the positivity assumption (22.2), and P satisfies the nonexceptional law condition (22.5). Then the parameter $\Psi : \mathcal{M} \to \mathbb{R}$ is pathwise differentiable at P with canonical gradient given by*

$$D^*(P) \equiv D^*(d_P, P) = \sum_{k=0}^{2} D_k^*(d_P, P),$$

That is, $D^*(P)$ equals the efficient influence curve $D^*(d_P, P)$ for the parameter $\Psi_d(P) \equiv E_P Y_d$ at the V-optimal rule $d = d_P$, where Ψ_d treats d as given.

The above theorem is proved as Theorem 8 in van der Laan and Luedtke (2014) so the proof is omitted here.

We will at times denote $D^*(P)$ by $D^*(Q, g)$, where Q represents Q^{d_P}, along with portions of the likelihood that suffice to compute the V-optimal rule d_P. We denote d_P by d_Q when convenient. We explore which parts of the likelihood suffice to compute the V-optimal rule in our companion paper, though Theorem 22.1 shows that $\bar{Q}_{b,20}$ and $\bar{Q}_{b,10}$ suffice for d_0 (and analogous functions suffice for a more general d_P). We have the following property of the efficient influence curve, which will provide a fundamental ingredient in the analysis of the CV-TMLE presented in the next section.

Theorem 22.3. *Let d_Q be the V-optimal rule corresponding with Q. For any Q, g, we have*

$$P_0 D^*(Q, g) = \Psi(Q_0) - \Psi(Q) + R_{1d_Q}(Q^{d_Q}, Q_0^{d_Q}, g, g_0) + R_2(Q, Q_0)$$

where, for all $d \in \mathcal{D}$,

$$R_{1d}(Q^d, Q_0^d, g, g_0) \equiv P_0 D^*(d, Q^d, g) - (\Psi_d(Q_0^d) - \Psi_d(Q^d)),$$
$$R_2(Q, Q_0) \equiv \Psi_{d_Q}(Q_0^{d_Q}) - \Psi_{d_0}(Q_0^{d_0}),$$

$\Psi_d(P) = E_P Y_d$ is the statistical target parameter that treats d as known, and $D^*(d, Q_0^d, g_0)$ is the efficient influence curve of Ψ_d at P_0 as given in Theorem 22.2.

From the study of the statistical target parameter Ψ_d in Chap. 4, we know that $P_0 D^*(d, Q^d, g) = \Psi_d(Q_0^d) - \Psi_d(Q^d) + R_{1d}(Q^d, Q_0^d, g, g_0)$, where R_{1d} is a closed form second-order term involving integrals of differences $Q^d - Q_0^d$ times differences $g - g_0$.

22.3 Statistical Inference for the Average of Sample-Split Specific Mean Counterfactual Outcomes Under Data Adaptively Determined Dynamic Treatments

Let $\hat{d} : \mathcal{M} \to \mathcal{D}$ be an estimator that maps an empirical distribution into an individualized treatment rule. Let $B_n \in \{0, 1\}^n$ denote a random vector for a cross-validation split, and for a split B_n, let P_{n,B_n}^0 be the empirical distribution of the training sample $\{i : B_n(i) = 0\}$ and P_{n,B_n}^1 is the empirical distribution of the validation sample $\{i : B_n(i) = 1\}$. Consider a J-fold cross-validation scheme. In J-fold cross-validation, the data is split into J mutually exclusive and exhaustive sets of size approximately n/J uniformly at random. Each set is then used as the validation set once, with the union of all other sets serving as the training set. With probability $1/J$, B_n has value 1 in all indices in validation set $j \in \{1, \ldots, J\}$ and 0 for all indices not corresponding to training set j.

In this section, we first present a method that provides an estimator and statistical inference for the data-adaptive target parameter

$$\tilde{\psi}_{0n} = E_{B_n} \Psi_{\hat{d}(P^0_{n,B_n})}(P_0).$$

Note that this target parameter is defined as the average of data-adaptive parameters, where the data-adaptive parameters are learned from the training samples of size approximately $n(J - 1)/J$. One applies the estimator $\hat{d}$ to each of the J training samples, giving a target parameter value $\Psi_{\hat{d}(P^0_{n,B_n})}(P_0)$, and our target parameter $\tilde{\psi}_{0n}$ is defined as the average across these J target parameters.

22.3.1 General Description of CV-TMLE

Here we give a general overview of the CV-TMLE procedure. In Sect. 22.6 we present a particular CV-TMLE that satisfies all of the properties described in this section. Denote the realizations of B_n with $j = 1, .., J$, and let $d_{nj} = \hat{d}(P^0_{n,j})$ for some estimator of the optimal rule $\hat{d}$. Let

$$(a(0), \bar{l}(1)) \mapsto E_{nj}[Y|\bar{A}(1) = d_{nj}(a(0), v), \bar{L}(1) = \bar{l}(1)]$$

represent an initial estimate of $E_0[Y \mid \bar{A}(1) = d_{nj}(A(0), V), \bar{L}(1)]$ based on the training sample j. Similarly, let $l(0) \mapsto E_{nj}[Y_{d_{nj}}|L(0) = l(0)]$ represent an initial estimate of $E_0[Y_{d_{nj}}|L(0)]$ based on the training sample j. Finally, let $Q_{L(0),nj}$ represent the empirical distribution of $L(0)$ in validation smaple j. We then fluctuate these three regression functions using the following submodels:

$$\left\{ E^{(\epsilon_2)}_{nj}[Y|\bar{A}(1) = d_{nj}(a(0), v), \bar{L}(1) = \bar{l}(1)] : \epsilon_2 \in \mathbb{R} \right\}$$
$$\left\{ E^{(\epsilon_1)}_{nj}[Y_{d_{nj}}|L(0) = l(0)] : \epsilon_1 \in \mathbb{R} \right\}$$
$$\left\{ Q^{(\epsilon_0)}_{L(0),nj} : \epsilon_0 \in \mathbb{R} \right\},$$

where these submodels rely on an estimate g_{nj} of g_0 based on training sample j and are such that:

$$E^{(0)}_{nj}[Y|\bar{A}(1) = d_{nj}(a(0), v), \bar{L}(1)] = E_{nj}[Y|\bar{A}(1) = d_{nj}(a(0), v), \bar{L}(1)]$$
$$E^{(0)}_{nj}[Y_{d_{nj}}|L(0)] = E_{nj}[Y_{d_{nj}}|L(0)]$$
$$Q^{(0)}_{L(0),nj} = Q_{L(0),nj}.$$

Let $Q_{nj}^{d_{nj}}(\epsilon)$ represent the parameter mapping that gives the three regression functions above fluctuated by $\epsilon \equiv (\epsilon_0, \epsilon_1, \epsilon_2)$. For a fixed ϵ, $Q_{nj}^{d_{nj}}(\epsilon)$ only relies on P_{nj}^1 through the empirical distribution of $L(0)$ in validation sample j. Let ϕ be a valid loss function for Q_0^d so that $Q_0^d = \arg\min_{Q^d} P_0\phi(Q^d)$, and let ϕ and the submodels above satisfy

$$D^*(d, Q^d, g) \in \left\langle \left. \frac{d}{d\epsilon}\phi(Q^d(\epsilon)) \right|_{\epsilon=0} \right\rangle,$$

where $\langle f \rangle = \{\sum_j \beta_j f_j : \beta\}$ denotes the linear space spanned by the components of f. We choose ϵ_n to minimize $P_n^1\phi(Q_{nj}^{d_{nj}}(\epsilon))$ over $\epsilon \in \mathbb{R}^3$. We then define the targeted estimate $Q_{nj}^{d_{nj}*} \equiv Q_{nj}^{d_{nj}}(\epsilon_n)$ of $Q_0^{d_{nj}}$. We note that $Q_{nj}^{d_{nj}*}$ maintains the rate of convergence of Q_{nj} under mild conditions that are standard to M-estimator analysis. The key property that we need from the ϵ_n and the corresponding update $Q_{nj}^{d_{nj}*}$ is that it (approximately) solves the cross-validated empirical mean of the efficient influence curve:

$$E_{B_n}P_{n,B_n}^1 D^*(d_{nj}, Q_{nj}^{d_{nj}*}, g_{nj}) = o_{P_0}(1/\sqrt{n}). \tag{22.6}$$

The CV-TMLE implementation presented in the appendix satisfies this equation with $o_{P_0}(1/\sqrt{n})$ replaced by 0. The proposed estimator of $\tilde{\psi}_{0n}$ is given by

$$\tilde{\psi}_n^* \equiv E_{B_n} \Psi_{d_{nj}}(Q_{nj}^{d_{nj}*}).$$

We give a concrete CV-TMLE algorithm for $\tilde{\psi}_n^*$ in Sect. 22.6, but note that other CV-TMLE algorithms can be derived using the approach in this section for different choices of loss function ϕ and submodels.

22.3.2 Statistical Inference for the Data-Adaptive Parameter $\tilde{\psi}_{0n}$

We now proceed with the analysis of this CV-TMLE $\tilde{\psi}_n^*$ of $\tilde{\psi}_{0n}$. We first give a representation theorem for the CV-TMLE.

Theorem 22.4. *Let g_{nj} and d_{nj} represent estimates of g_0 and d_0 based on training sample j. Let $Q_{nj}^{d_{nj}*}$ represent a targeted estimate of $Q_0^{d_{nj}}$ as presented in Sect. 22.3.1 so that $Q_{nj}^{d_{nj}*}$ satisfies (22.6). Let R_{1d} be as in Theorem 22.3. Further suppose that the supremum norm of $\max_j D^*(d_{nj}, Q_{nj}^{d_{nj}*}, g_{nj})$ is bounded by some $M < \infty$ with probability tending to 1, and that*

$$\max_{j\in\{1,\dots,J\}} P_0\{D^*(d_{nj}, Q_{nj}^{d_{nj}*}, g_{nj}) - D^*(d_1, Q^{d_1}, g)\}^2 \to 0 \text{ in probability}$$

for some $d_1 \in \mathcal{D}$ and possibly misspecified Q^{d_1} and g. Finally, suppose that

$$\max_{j \in \{1,\dots,J\}} \left| R_{1d_{nj}}(Q_{nj}^{d_{nj}*}, Q_0^{d_{nj}}, g_{nj}, g_0) \right| = o_{P_0}(n^{-1/2}).$$

Then,

$$\tilde{\psi}_n^* - \tilde{\psi}_{0n} = (P_n - P_0)D^*(d_1, Q^{d_1}, g^{d_1}) + o_{P_0}(n^{-1/2}).$$

Note that d_1 in the above theorem need not be the same as the optimal rule d_0, though later we will discuss the desirable special case where $d_1 = d_0$. The above theorem also does not require that g_0 is known, or even that the limit of our intervention mechanisms g is equal to g_0.

Note in the above theorem that the condition that, if g_0 is known so that all g_{nj} can be correctly specified, it immediately follows that

$$\max_{j \in \{1,\dots,J\}} \left| R_{1d_{nj}}(Q_{nj}^{d_{nj}*}, Q_0^{d_{nj}}, g_{nj}, g_0) \right| = 0.$$

In practice we would recommend estimating g_0 according to a correctly specified model even when g_0 is known, because this can improve efficiency (see Section 2.3.7 of van der Laan and Robins 2003).

If the conditions of the above theorem hold, the asymptotic linearity result implies that

$$\sqrt{n}\left[\tilde{\psi}_n^* - \tilde{\psi}_{0n}\right] \to \text{Normal}(0, \sigma_0^2),$$

where $\sigma_0^2 = P_0 D^*(d_1, Q^{d_1}, g^{d_1})^2$. Under mild conditions,

$$\sigma_n^2 = \frac{1}{J} \sum_{j=1}^{J} P_{n,j}^1 \left\{ D^*(d_{nj}, Q_{nj}^{d_{nj}*}, g_{nj}) \right\}^2$$

consistently estimates σ_0^2. Under the consistency of σ_n^2 and the conditions of Theorem 22.4, an asymptotically valid 95% confidence interval for $\tilde{\psi}_{0n}$ is given by

$$\left[\tilde{\psi}_n^* \pm \frac{\sigma_n}{\sqrt{n}} \right]. \tag{22.7}$$

22.3.3 Statistical Inference for the True Optimal Rule ψ_0

Suppose now that we are interested in estimating the mean outcome under the optimal rule d_0 rather than the data-adaptive parameter $\tilde{\psi}_{0n}$. Note that

$$\sqrt{n}\left(\tilde{\psi}_n^* - \psi_0\right) = \sqrt{n}\left(\tilde{\psi}_n^* - \tilde{\psi}_{0n}\right) + \sqrt{n}\left(\tilde{\psi}_{0n} - \psi_0\right)$$

$$= \sqrt{n}\left(\tilde{\psi}_n^* - \tilde{\psi}_{0n}\right) + \frac{\sqrt{n}}{J} \sum_{j=1}^{J} \left[\Psi_{d_{nj}}(P_0) - \psi_0\right].$$

If $J^{-1} \sum_{j=1}^{J} \left[\Psi_{d_{nj}}(P_0) - \psi_0 \right]$, then by Slutsky's theorem the left-hand side has the same normal limit as $\sqrt{n} \left(\tilde{\psi}_n^* - \tilde{\psi}_{0n} \right)$ provided the conditions of Theorem 22.4 hold. Furthermore, as J is fixed as $n \to \infty$, $J^{-1} \sum_{j=1}^{J} \left[\Psi_{d_{nj}}(P_0) - \psi_0 \right] = o_P(n^{-1/2})$ if

$$\Psi_{d_{nj}}(P_0) - \psi_0 = o_P(n^{-1/2}) \text{ for each } j. \tag{22.8}$$

To analyze $\Psi_{d_{nj}}(P_0) - \psi_0$, we will assume that the user estimates $\bar{Q}_{b,10}$ and $\bar{Q}_{b,20}$ using $\bar{Q}_{b,1nj}$ and $\bar{Q}_{b,2nj}$, and then subsequently uses the plug-in estimators of the format described in Theorem 22.1. Data-adaptive estimators of $\bar{Q}_{b,10}$ and $\bar{Q}_{b,20}$ were previously described in Luedtke and van der Laan (2016b). While we do not require that d_{nj} result from a plug-in estimator, this is the estimation scheme we will focus on analyzing here. Given that the main result needed to show (22.8) for the plug-in estimator is analytic in nature, we focus on a general Q with corresponding blip functions $\bar{Q}_{b,1}$, $\bar{Q}_{b,2}$ and optimal rule plug-in estimates $d_{Q,A(0)}$, $d_{Q,A(1)}$. One can then apply this result directly to our fold-specific estimator.

The following result is proved in Sect. 22.5.

Lemma 22.1. *Recall the definitions of $\bar{Q}_{b,20}$ and $\bar{Q}_{b,10}$ in Theorem 22.1. We can represent $\Psi(P_0) = E_0 Y_{d_0}$ as follows:*

$$\Psi(P_0) = E_0 Y_{(0,1),(0,1)} + E_0 \left[d_{0,A(1)}((0,1), V_{(0,1)}(1)) \bar{Q}_{b,20}((0,1), V_{(0,1)}(1)) \right]$$
$$+ E_0 d_{0,A(0)}(V(0)) \bar{Q}_{b,10}(V(0)).$$

where $V_{(0,1)}(1)$ is drawn under the g-computation distribution for which treatment $(0,1)$ is given at the first time point.

It follows that

$$R_2(Q, Q_0) = E_0 (d_{Q,A(0)} - d_{0,A(0)})(V(0)) \bar{Q}_{b,10}(V(0))$$
$$+ E_0 (d_{Q,A(1)} - d_{0,A(1)})((0,1), V_{(0,1)}(1)) \bar{Q}_{b,20}((0,1), V_{(0,1)}(1))$$
$$\equiv R_{2,A(0)}(Q, Q_0) + R_{2,A(1)}(Q, Q_0).$$

We will be able to attain a fast rate on $R_2(Q, Q_0)$ under margin assumptions. We start with the assumption that we use to bound $R_{2,A(0)}(Q, Q_0)$. Suppose there exist positive constants C_1, β_1 such that, for all $t > 0$,

$$P_0 \left\{ 0 < \left| \bar{Q}_{b,10}(V(0)) \right| \leq t \right\} \leq C_1 t^{\beta_1}. \tag{MA1}$$

The above assumption requires that the blip function at the first time point does not concentrate too much mass near (but not at) the decision boundary (zero). The assumption is different from the exceptional law condition, since that condition requires that this blip function places no mass exactly at the decision boundary. For β_1 and β_2 small, this is a weak assumption, though it may not attain the rates of convergence needed to satisfy (22.8). These assumptions hold for $\beta_1 = 1$ if that the blip functions applied to the data have bounded Lebesgue density in a neighborhood of zero.

We make a similar assumption on $\bar{Q}_{b,20}$. In particular, we assume there exists some C_2, β_2 such that, for all $t > 0$,

$$P_0 \left\{ 0 < \left| \bar{Q}_{b,20}((0,1), V_{(0,1)}(1)) \right| \le t \right\} \le C_2 t^{\beta_2}. \tag{MA2}$$

We now show that (MA1) and (MA2) give a β_1, β_2-specific upper bound of $R_2(Q, Q_0)$ by the distance of $\bar{Q}_{b,1}$ and $\bar{Q}_{b,2}$ from $\bar{Q}_{b,10}$ and $\bar{Q}_{b,20}$.

Theorem 22.5. *If (MA1) holds for some $C_1, \beta_1 > 0$, then, for some constant $C > 0$,*

$$\left| R_{2,A(0)}(Q, Q_0) \right| \le C \min \left\{ \left\| \bar{Q}_{b,1} - \bar{Q}_{b,10} \right\|_{2,P_0}^{2(1+\beta_1)/(2+\beta_1)} , \left\| \bar{Q}_{b,1} - \bar{Q}_{b,10} \right\|_{\infty,P_0}^{1+\beta_1} \right\}. \tag{22.9}$$

If (MA2) holds for some $C_2, \beta_2 > 0$, then, for some constant $C > 0$,

$$\left| R_{2,A(1)}(Q, Q_0) \right| \le C \min \left\{ \left\| \bar{Q}_{b,2} - \bar{Q}_{b,20} \right\|_{2,P_{0,(0,1)}}^{2(1+\beta_1)/(2+\beta_1)} , \left\| \bar{Q}_{b,1} - \bar{Q}_{b,10} \right\|_{P_{0,(0,1)}}^{1+\beta_1} \right\}, \tag{22.10}$$

where $P_{0,(0,1)}$ represents the static intervention specific g-computation distribution where treatment $(0,1)$ is given at the first time point.

The above analytic result is useful for evaluating the plausibility of (22.8) when d_{nj} is estimated using a plug-in estimator. In a parametric model, one could typically estimate $\bar{Q}_{b,10}$ and $\bar{Q}_{b,20}$ at $n^{-1/2}$ rates for both the $L_2(P_0)$ and the supremum norms presented above. Hence, if the margin conditions hold with $\beta_1 = \beta_2 = 1$, the supremum norm result yields n^{-1} rates on each $\Psi_{d_{nj}}(P_0) - \psi_0$. In practice we of course do not expect to be able to correctly specify a parametric model. Rather, we would use data-adaptive estimators for the blip functions, such as super learning, to make correct specification of the estimators more likely (Luedtke and van der Laan 2016b). Under smoothness assumptions on the blip functions, one can ensure nearly parametric rates on the $L_2(P_0)$ norm using smoothing. These rates can, under enough smoothness, achieve the $o_P(n^{-3/8}$ rate required by the above theorem at $\beta_1 = \beta_2 = 1$ to show that $\Psi_{d_{nj}}(P_0) - \psi_0 = o_P(n^{-1/2})$. Nonetheless, in general we may not expect such a fast rate to hold. If this fast rate does not hold, then one can still achieve inference for the data-adaptive parameter $\tilde{\psi}_{0n}$. If the fast rate does hold, as may be possible if V is low-dimensional, then the implication is that, under the conditions of Theorem 22.4 and the consistency of σ_n^2, the confidence interval presented in (22.7) is asymptotically valid for both $\tilde{\psi}_{0n}$ and ψ_0.

22.4 Discussion

This chapter investigated semiparametric statistical inference for the mean outcome under the V-optimal rule and statistical inference for the data-adaptive target parameter defined as the mean outcome under a data adaptively determined V-optimal rule (treating the latter as given). We proved a surprising and useful result stating that the mean outcome under the V-optimal rule is represented by a statistical parameter whose pathwise derivative is identical to what it would have been if the unknown rule had been treated as known, under the condition that the data is generated by a nonexceptional law. As a consequence, the efficient influence curve is immediately known, and any of the efficient estimators for the mean outcome under a given rule can be applied at the estimated rule. In particular, we demonstrate a CV-TMLE, and present asymptotic linearity results. However, the dependence of the statistical target parameter on the unknown rule affects the second-order terms of the CV-TMLE, and, as a consequence, the asymptotic linearity of the CV-TMLE requires that a second-order difference between the estimated rule and the V-optimal rule converges to zero at a rate faster than $1/\sqrt{n}$. While this can be expected to hold for rules that are only a function of one continuous score (such as a biomarker), only strong smoothness assumptions will guarantee this when V is moderate-to-high dimensional, so that, even in an RCT, we cannot expect valid statistical inference for such V-optimal rules.

To account for this challenge, we also described estimation of the average of sample split specific data-adaptive target parameters, as in general proposed in Hubbard et al. (2016). Specifically, our data-adaptive target parameter is defined as an average across J sample splits in training and validation sample of the mean outcome under the dynamic treatment fitted on the training sample. We presented a CV-TMLE of this data-adaptive target parameter, and we established an asymptotic linearity theorem that does not require that the estimated rule be consistent for the optimal rule, let alone at a particular rate. We showed that statistical inference for this data-adaptive target parameter does not rely on the convergence rate of our estimated rule to the optimal rule, and in fact only requires that the data adaptively fitted rule converges to some (possibly suboptimal) fixed rule. As a consequence, in a sequential RCT, this method provides valid asymptotic statistical inference under very mild conditions, the primary of which is that the estimated rule converges to some (possibly suboptimal) fixed rule.

Drawing inferences concerning optimal treatment strategies is an important topic that will hopefully help guide future health policy decisions. We believe that working with a large semiparametric model is desirable because it helps to ensure that the projected health benefits from implementing an estimated treatment strategy are not due to bias from a misspecified model. The CV-TMLEs presented in this chapter have many desirable statistical properties and allow one to get estimates and make inference in this large model.

22.5 Proofs

Proof (Theorem 22.1). Let $V_d = (V(0), V_d(1))$. For a rule in $\mathcal{D}$, we have

$$E_{P_d} Y_d = E_{P_d} E_{P_d}(Y_d \mid V_d)$$
$$= E_{V_d} \left(E(Y_{a(0),a(1)} \mid V_{a(0)}) I(a(1) = d_{A(1)}(a(0), V_{a(0)}(1))) I(a(0) = d_{A(0)}(V(0))) \right).$$

For each value of $a(0)$, $V_{a(0)} = (V(0), V_{a(0)}(1))$ and $d_{A(0)}(V(0))$, the inner conditional expectation is maximized over $d_{A(1)}(a(0), V_{a(0)}(1))$ by $d_{0,A(1)}$ as presented in the theorem, where we used that $V(1)$ includes $V(0)$. This proves that $d_{0,A(1)}$ is indeed the optimal rule for assignment of $A(1)$. Suppose now that $V(1)$ does not include $V(0)$, but the stated assumption holds. Then the optimal rule $d_{0,A(1)}$ that is restricted to be a function of $(V(0), V(1), A(0))$ is given by $I(\bar{Q}_{b,20}(A(0), V(0), V(1)) > 0)$, where

$$\bar{Q}_{b,20}(a(0), v(0), v(1)) =$$
$$E_0(Y_{a(0),A(1)=(1,1)} - Y_{a(0),A(1)=(0,1)} \mid V_{a(0)}(1) = v(1), V(0) = v(0)).$$

However, by assumption, the latter function only depends on $(a(0), v(0), v(1))$ through $(a(0), v(1))$, and equals $\bar{Q}_{b,20}(a(0), v(1))$. Thus, we now still have that $d_{0,A(1)}(V) = (I(\bar{Q}_{b,20}(A(0), V(1)) > 0), 1)$, and, in fact, it is now also an optimal rule among the larger class of rules that are allowed to use $V(0)$ as well.

Given we found $d_{0,A(1)}$, it remains to determine the rule $d_{0,A(0)}$ that maximizes

$$E_{V_d}\left(E_P(Y_{a(0),d_{0,A(1)}} \mid V_{a(0)})I(a(0) = d_{A(0)}(V(0)))\right)$$
$$= E_0 E(Y_{a(0),d_{0,A(1)}} \mid V(0))I(a(0) = d_{A(0)}(V(0))),$$

where we used the iterative conditional expectation rule, taking the conditional expectation of $V_{a(0)}$, given $V(0)$. This last expression is maximized over $d_{A(0)}$ by $d_{0,A(0)}$ as presented in the theorem. This completes the proof.

Proof (Theorem 22.3). By the definition of R_{1d} we have

$$P_0 D^*(Q, g) = P_0 D^*(d_Q, Q, g) = \Psi_{d_Q}(Q_0^{d_Q}) - \Psi_{d_Q}(Q^{d_Q}) + R_{1d_Q}(Q^{d_Q}, Q_0^{d_Q}, g, g_0)$$
$$= \Psi_{d_0}(Q_0^{d_0}) - \Psi_{d_Q}(Q^{d_Q}) + \{\Psi_{d_Q}(Q_0^{d_Q}) - \Psi_{d_0}(Q_0^{d_0})\} + R_{1d_Q}(Q^{d_Q}, Q_0^{d_Q}, g, g_0)$$
$$= \Psi(Q_0) - \Psi(Q) + R_2(Q, Q_0) + R_{1d_Q}(Q^{d_Q}, Q_0^{d_Q}, g, g_0).$$

Proof (Theorem 22.4). For all $j = 1, \ldots, J$, we have that:

$$\Psi_{d_{nj}}(Q_{nj}^{d_{nj}*}) - \Psi_{d_{nj}}(Q_0^{d_{nj}*}) = -P_0 D^*(d_{nj}, Q_{nj}^{d_{nj}*}, g_{nj})$$
$$+ R_{1d_{nj}}(Q_{nj}^{d_{nj}*}, Q_0^{d_{nj}*}, g_{nj}, g_0)$$

Summing over j and using (22.6) gives:

$$\tilde{\psi}_n^* - \tilde{\psi}_{0n} = \frac{1}{J} \sum_{j=1}^{J} \left((P_{n,j}^1 - P_0)D^*(d_{nj}, Q_{nj}^{d_{nj}*}, g_{nj}) + R_{1d_{nj}}(Q_{nj}^{d_{nj}*}, Q_0^{d_{nj}*}, g_{nj}, g_0)\right).$$

We also have that:

$$\frac{1}{J} \sum_{j=1}^{J} (P_{n,j}^1 - P_0)\left(D^*(d_{nj}, Q_{nj}^{d_{nj}*}, g_{nj}) - D^*(d_1, Q^{d_1}, g)\right) = o_{P_0}(n^{-1/2}).$$

The above follows from the first by applying the law of total expectation conditional on the training sample, and then noting that each $\hat{Q}^*(P_{n,B_n}^0, \epsilon_n)$ only relies on P_{n,B_n}^0

through the finite dimensional parameter ϵ_n. Because GLM-based parametric classes easily satisfy an entropy integral condition (van der Vaart and Wellner 1996), the consistency assumption on $D^*(d_{nj}, Q_{nj}^{d_{nj}*}, g_{nj})$ shows that the above is second order. We refer the reader to Zheng and van der Laan (2010) for a detailed proof of the above result for general cross-validation schemes, including J-fold cross-validation.

It follows that:

$$\tilde{\psi}_n^* - \tilde{\psi}_{0n} = (P_n - P_0)D^*(d_1, Q^{d_1}, g)$$

$$+ \frac{1}{J}\sum_{j=1}^{J} R_{1d_{nj}}(Q_{nj}^{d_{nj}*}, Q_0^{d_{nj}*}, g_{nj}, g_0) + o_{P_0}(n^{-1/2}).$$

Finally, note that $\frac{1}{J}\sum_{j=1}^{J} R_{1d_{nj}}(Q_{nj}^{d_{nj}*}, Q_0^{d_{nj}*}, g_{nj}, g_0)$ is $o_P(n^{-1/2})$ by the last assumption of the theorem.

Proof (Lemma 22.1). For a point treatment data structure $O = (L(0), A(0), Y)$ and binary treatment $A(0)$, we have for a rule $V \to d(V)$, $E_0 Y_d = E_0 Y_0 + E_0 d(V)\bar{Q}_0(V))$ with $\bar{Q}_0(V) = E_0[Y_1 - Y_0 \mid V]$. This identity is applied twice in the following derivation:

$$\Psi(P_0) = E_0 Y_{(0,1),d_{0,A(1)}} + E_0 d_{0,A(0)}(V(0))\bar{Q}_{b,10}(V(0))$$
$$= E_0 E_0[Y_{(0,1),d_{0,A(1)}} \mid V_{(0,1)}(1)] + E_0 d_{0,A(0)}(V(0))\bar{Q}_{b,10}(V(0))$$
$$= E_0 E_0[Y_{(0,1),(0,1)} \mid V_{(0,1)}(1)] + E_0 I(\bar{Q}_{b,20}((0,1), V_{(0,1)}(1)) > 0)\bar{Q}_{b,20}(0, V_{(0,1)}(1))$$
$$+ E_0 d_{0,A(0)}(V(0))\bar{Q}_{b,10}(V(0))$$
$$= E_0 E_0[Y_{(0,1),(0,1)} \mid V_{(0,1)}(1)] + E_0 d_{0,A(1)}((0,1), V_{(0,1)}(1))\bar{Q}_{b,20}(0, V_{(0,1)}(1))$$
$$+ E_0 d_{0,A(0)}(V(0))\bar{Q}_{b,10}(V(0))$$
$$= E_0 Y_{(0,1),(0,1)} + E_0 d_{0,A(1)}((0,1), V_{(0,1)}(1))\bar{Q}_{b,20}(0, V_{(0,1)}(1))$$
$$+ E_0 d_{0,A(0)}(V(0))\bar{Q}_{b,10}(V(0)).$$

Proof (Theorem 22.5). In this proof we will omit the dependence of $d_{0,A(0)}$, $d_{Q,A(0)}$, $\bar{Q}_{b,10}$, and $\bar{Q}_{b,1}$ on $V(0)$ in the notation. This part of the proof mimics the proof of Lemma 5.2 in Audibert and Tsybakov (2007). For any $t > 0$,

$$|R_{2,A(0)}(Q, Q_0)| = E_0[|\bar{Q}_{b,10}|I(d_{0,A(0)} \neq d_{Q,A(0)})]$$
$$= E_0[|\bar{Q}_{b,10}|I(d_{0,A(0)} \neq d_{Q,A(0)})I(0 < |\bar{Q}_{b,10}| \leq t)]$$
$$+ E_0[|\bar{Q}_{b,10}|I(d_{0,A(0)} \neq d_{Q,A(0)})I(|\bar{Q}_{b,10}| > t)]$$
$$\leq E_0[|\bar{Q}_{b,1} - \bar{Q}_{b,10}|I(0 < |\bar{Q}_{b,10}| \leq t)]$$
$$+ E_0[|\bar{Q}_{b,1} - \bar{Q}_{b,10}|I(|\bar{Q}_{b,1} - \bar{Q}_{b,10}| > t)]$$
$$\leq \left\|\bar{Q}_{b,1} - \bar{Q}_{b,10}\right\|_{2,P_0} \Pr(0 < |\bar{Q}_{b,10}| \leq t)^{1/2} + \frac{\left\|\bar{Q}_{b,1} - \bar{Q}_{b,10}\right\|_{2,P_0}^2}{t}$$
$$\leq \left\|\bar{Q}_{b,1} - \bar{Q}_{b,10}\right\|_{2,P_0} C_0^{1/2} t^{\beta_1/2} + \frac{\left\|\bar{Q}_{b,1} - \bar{Q}_{b,10}\right\|_{2,P_0}^2}{t},$$

where the first inequality holds because $d_{0,A(0)} \neq d_{Q,A(0)}$ implies that $|\bar{Q}_{b,1} - \bar{Q}_{b,10}| > |\bar{Q}_{b,10}|$, the second inequality holds by the Cauchy-Schwarz and Markov inequalities, and the third inequality holds by (MA1). The first result follows by optimizing over t to find that the upper bound is minimized when $t = C' \left\| \bar{Q}_{b,1} - \bar{Q}_{b,10} \right\|_{2,P_0}^{2(1+\beta_1)/(2+\beta_1)}$ for a constant C' that depends on C_0 and β_1.

We now establish the supremum-norm result. Note that

$$
\begin{aligned}
|R_{2,A(0)}(Q, Q_0)| &= E_0 \left| I(d_{Q,A(0)} \neq d_{0,A(0)}) \bar{Q}_{b,10} \right| \\
&\leq E_0 \left[I(0 < |\bar{Q}_{b,10}| \leq |\bar{Q}_{b,1} - \bar{Q}_{b,10}|) |\bar{Q}_{b,10}| \right] \\
&\leq E_0 \left[I\left(0 < |\bar{Q}_{b,10}| \leq \left\| \bar{Q}_{b,1} - \bar{Q}_{b,10} \right\|_{\infty,P_0} \right) |\bar{Q}_{b,10}| \right] \\
&\leq \left\| \bar{Q}_{b,1} - \bar{Q}_{b,10} \right\|_{\infty,P_0} \Pr \left(0 < |\bar{Q}_{b,10}| \leq \left\| \bar{Q}_{b,1} - \bar{Q}_{b,10} \right\|_{\infty,P_0} \right).
\end{aligned}
$$

By (MA1), $|\Psi_{d_{Q,A(0)}}(P_0) - \Psi_{d_{0,A(0)}}(P_0)| \leq C_1 \left\| \bar{Q}_{b,1} - \bar{Q}_{b,10} \right\|_{\infty,P_0}^{1+\beta_1}$. Combining the two results yields (22.9). The proof of (22.10) is analogous and so is omitted.

22.6 CV-TMLE for the Mean Outcome Under Data-Adaptive V-Optimal Rule

Let $\hat{d} : \mathcal{M} \to \mathcal{D}$ be an estimator of the V-optimal rule d_0. Firstly, without loss of generality we can assume that $Y \in [0, 1]$. Denote the realizations of B_n with $j = 1, \ldots, J$, and let $d_{nj} \equiv \hat{d}(P_{n,j}^0)$ denote the estimated rule on training sample j. Let

$$(a(0), \bar{l}(1)) \mapsto E_{nj}[Y | \bar{A}(1) = d_{nj}(a(0), v), \bar{L}(1) = \bar{l}(1)] \qquad (22.11)$$

represent an initial estimate of $E_0[Y \mid \bar{A}(1) = d_{nj}(A(0), V), \bar{L}(1)]$ based on the training sample j. Similarly, let g_{nj} represent the estimated intervention mechanism based on this training sample $P_{n,j}^0$, $j = 1, \ldots, J$. Consider the fluctuation submodel

$$
\begin{aligned}
\text{logit } E_{nj}^{(\epsilon_2)} \left[Y | \bar{A}(1) = d_{nj}(A(0), V), \bar{L}(1) \right] &= \text{logit } E_{nj} \left[Y | \bar{A}(1) = d_{nj}(A(0), V), \bar{L}(1) \right] \\
&\quad + \epsilon_2 H_2(g_{nj})(O)
\end{aligned}
$$

where

$$H_2(g_{nj})(O) = \frac{I(\bar{A}(1) = d_{nj}(A(0), V(1)))}{\prod_{l=0}^{1} g_{nj,A(l)}(O)}.$$

Note that the fluctuation ϵ_2 does not rely on j. Let

$$\epsilon_{2n} = \arg \min_{\epsilon_2} \frac{1}{J} \sum_{j=1}^{J} P_{n,j}^1 \tilde{\phi}(E_{nj}^{(\epsilon_2)}),$$

where $E_{nj}^{(\epsilon_2)}$ refers to the represents the fluctuated function in (22.11) and

$$-\tilde{\phi}(f)(o) = y \log f(o) + (1 - y) \log (1 - f(o)). \tag{22.12}$$

for all $f : O \rightarrow (0, 1)$. For each $i = 1, \ldots, n$, let $j(i) \in \{1, \ldots, J\}$ represent the value of B_n for which element i is in the validation set. The fluctuation ϵ_{2n} can be obtained by fitting a univariate logistic regression of $(y_i : i = 1, \ldots, n)$ on $(H_2(g_{nj(i)})(o_i) : i = 1, \ldots, n)$ using

$$\left(\text{logit } E_{nj(i)} \left[Y|\bar{A}(1) = d_{nj(i)}(a(0)_i, v_i), \bar{L}(1) = \bar{l}(1)_i\right] : i = 1, \ldots, n\right)$$

as offset. Thus each observation i is paired with nuisance parameters are fit on the training sample that does not contain observation i. This defines a targeted estimate

$$E_{nj}^* \left[Y|\bar{A}(1) = d_{nj}(A(0), V), \bar{L}(1)\right] \equiv E_{nj}^{(\epsilon_{2n})} \left[Y|\bar{A}(1) = d_{nj}(A(0), V), \bar{L}(1)\right] \tag{22.13}$$

of $E_0[Y \mid \bar{A}(1) = d_{nj}(A(0), V), \bar{L}(1)]$. We note that this targeted estimate only depends on P_n through the training sample $P_{n,j}^0$ and the one-dimensional ϵ_{2n}.

We now aim to get a targeted estimate of $E_0[Y_{d_{nj}}|L(0)]$. We can obtain an estimate

$$(a_1(0), l(0)) \mapsto E_{nj}\left[E_{nj}\left[Y \mid \bar{A}(1) = d_{nj}(A(0), V), \bar{L}(1)\right]\middle|A(0) = (a_1(0), 1), L(0) = l(0)\right] \tag{22.14}$$

by regressing $E_{nj}\left[Y \mid \bar{A}(1) = d_{nj}(A(0)_i, V_i), \bar{L}(1)_i\right]$ against $A(0)_i, L(0)_i$ for all of the observations i in training sample j. For an estimate $E_{nj}[Y_{d_{nj}}|L(0)]$ of $E_0[Y_{d_{nj}}|L(0)]$, we can use the regression function above but with $a(0)$ fixed to $d_{nj,A(0)}(v(0))$.

Consider the fluctuation submodel

$$\text{logit } E_{nj}^{(\epsilon_1)} \left[Y_{d_{nj}} \mid L(0)\right] = \text{logit } E_{nj} \left[Y_{d_{nj}} \mid L(0)\right] + \epsilon H_1(g_{nj})(O),$$

where

$$H_1(g_{nj})(O) = \frac{I(A(0) = d_{nj,A(0)}(V(0)))}{g_{nj,A(0)}(O)}.$$

Again the fluctuation ϵ_1 does not rely on j. Let

$$\epsilon_{1n} = \arg\min_{\epsilon_1} \frac{1}{J} \sum_{j=1}^{J} P_{n,j}^1 \tilde{\phi}(E_{nj}^{(\epsilon_1)}),$$

where $\tilde{\phi}$ is defined in (22.12). For each $i = 1, \ldots, n$, again let $j(i) \in \{1, \ldots, J\}$ represent the value of B_n for which element i is in the validation set. The fluctuation ϵ_{1n} can be obtained by fitting a univariate logistic regression of

$$\left(E_{nj(i)}^* \left[Y|\bar{A}(1) = d_{nj(i)}(a(0)_i, v_i), \bar{l}(1)_i\right] : i = 1, \ldots, n\right)$$

on $(H_1(g_{nj(i)})(o_i) : i = 1, \ldots, n)$ using

$$\left(\text{logit } E_{nj(i)}\left[Y_{d_{nj(i)}}|L(0) = l(0)_i\right] : i = 1, \ldots, n\right)$$

as offset. This defines a targeted estimate

$$E_{nj}^*\left[Y_{d_{nj}}|L(0)\right] \equiv E_{nj}^{(\epsilon_{1n})}\left[Y_{d_{nj}}|L(0)\right] \qquad (22.15)$$

of $E_0[Y_{d_{nj}}|L(0)]$. We note that this targeted estimate only depends on P_n through the training sample $P_{n,j}^0$ and the one-dimensional ϵ_{1n}.

Let $Q_{L(0),nj}$ be the empirical distribution of $L(0)_i$ for the validation sample $P_{n,j}^1$. For all $j = 1, \ldots, J$, let $\bar{Q}_{nj}^{d_{nj}*}$ be the parameter mapping representing the collection containing $Q_{L(0),nj}$ and the targeted regressions in (22.13) and (22.15). This defines an estimator $\psi_{nj}^* = P_{n,j}^1 \bar{Q}_{b,1nj}^*$ of $\psi_{d_{nj}0} = \Psi_{d_{nj}}(P_0)$ for each $j = 1, \ldots, J$. The cross-validated TMLE is now defined as $\psi_n^* = \frac{1}{J}\sum_{j=1}^J \psi_{nj}^*$. This CV-TMLE solves the cross-validated efficient influence curve equation:

$$\frac{1}{J}\sum_{j=1}^J P_{n,j}^1 D^*(d_{nj}, Q_{nj}^{d_{nj}*}, g_{nj}) = 0.$$

Further, each $Q_{nj}^{d_{nj}*}$ only relies on $P_{n,j}^1$ through the univariate parameters ϵ_{1n} and ϵ_{2n}. This will allow us to use the entropy integral arguments presented in Zheng and van der Laan (2010) that show that no restrictive empirical process conditions are needed on the initial estimates in (22.11) and (22.14).

The only modification relative to the original CV-TMLE presented in Zheng and van der Laan (2010) is that in the above description we change our target on each training sample into the training sample specific target parameter implied by the fit $\hat{d}(P_{n,B_n}^0)$ on the training sample, while in the original CV-TMLE formulation, the target would still be $\Psi_{d_0}(P_0)$. With this minor twist, the (same) CV-TMLE is now used to target the average of training sample specific target parameters averaged across the J training samples.

22.7 Notes and Further Reading

Examples of multiple time-point dynamic treatment regimes are given in Lavori and Dawson (2000, 2008); Murphy (2005); Rosthøj et al. (2006); Thall et al. (2002); Wagner et al. (2001) ranging from rules that change the dose of a drug, change or augment the treatment, to making a decision on when to start a new treatment, in response to the history of the subject. For an excellent overview on dynamic treatments we refer to Chakraborty and Moodie (2013).

We define the optimal dynamic multiple time-point treatment regime as the rule that maximizes the mean outcome under the dynamic treatment, where the candidate rules are restricted to only respond to a user-supplied subset of the baseline and intermediate covariates. The literature on Q-learning shows that we can describe the optimal dynamic treatment among *all* dynamic treatments in a sequential manner (Murphy 2003; Robins 2004; Murphy 2005). The optimal rule can be learned through fitting the likelihood and then calculating the optimal rule under this fit of the likelihood. This approach can be implemented with maximum likelihood estimation based on parametric models. It has been noted (e.g., Robins 2004) that the estimator of the parameters of one of the regressions (except the first one) when using parametric regression models is a nonsmooth function of the estimator of the parameters of the previous regression, and that this results in nonregularity of the estimators of the parameter vector. This raises challenges for obtaining statistical inference, even when assuming that these parametric regression models are correctly specified. Chakraborty and Moodie (2013) discuss various approaches and advances that aim to resolve this delicate issue such as inverting hypothesis testing (Robins 2004), establishing nonnormal limit distributions of the estimators (Laber et al. 2014a), or using the m out of n bootstrap (Chakraborty et al. 2014). The proof of the fast rate for the estimate of the optimal rule provided in Theorem 22.5 is similar to the proofs of the fast classification rates obtained in Audibert and Tsybakov (2007). It was presented for single time point optimal treatment rules in van der Laan and Luedtke (2015).

Murphy (2003) and Robins (2004) develop structural nested mean models tailored to optimal dynamic treatments. These models assume a parametric model for the "blip function" defined as the additive effect of a blip in current treatment on a counterfactual outcome, conditional on the observed past, in the counterfactual world in which future treatment is assigned optimally. Statistical inference for the parameters of the blip function proceeds accordingly, but Robins (2004) points out the irregularity of the estimator, resulting in some serious challenges for statistical inference as referenced above. Structural nested mean models have also been generalized to blip functions that condition on a (counterfactual) subset of the past, thereby allowing the learning of optimal rules that are restricted to only using this subset of the past (Robins 2004 and Section 6.5 in van der Laan and Robins 2003).

Each of the above referenced approaches for learning an optimal dynamic treatment that also aims to provide statistical inference relies on parametric assumptions: obviously, Q-learning based on parametric models, but also the structural nested mean model rely on parametric models for the blip function. As a consequence, even in a SMART, the statistical inference for the optimal dynamic treatment heavily relies on assumptions that are generally believed to be false, and will thus be expected to be biased. To avoid these biases, in this chapter we defined our model as nonparametric, beyond possible restrictions on the treatment/censoring mechanism.

Chapter 23
Optimal Individualized Treatments Under Limited Resources

Alexander R. Luedtke and Mark J. van der Laan

In this chapter, we consider a resource constraint under which there is a maximum proportion of the population that can be treated. Given this constraint, we develop a root-n rate estimator for the optimal resource-constrained (R-C) value and corresponding confidence intervals. We show that our estimator is efficient among all regular and asymptotically linear estimators in our nonparametric model $\mathcal{M}$ under conditions. When the baseline covariates are continuous and the resource constraint is active, i.e., when the optimal R-C value is less than the optimal unconstrained value, these conditions are more reasonable than the nonexceptional law assumption needed for regular estimation of the optimal unconstrained value discussed in Chap. 22.

23.1 Optimal Resource-Constrained Rule and Value

Suppose we observe n i.i.d. draws from a single time point data structure $(W, A, Y) \sim P_0$, where the vector of covariates W has support $\mathcal{W}$, the treatment A has support $\{0, 1\}$, and the outcome Y has support in the closed unit interval. Little generality is lost with the bound on Y, given that any continuous outcome bounded can be rescaled to the unit interval via a linear transformation. Our statistical model is nonparametric, beyond possible knowledge of the treatment mechanism, i.e., the

A. R. Luedtke (✉)
Vaccine and Infectious Disease Division, Fred Hutchinson Cancer Research Center, 1100 Fairview Ave, Seattle, WA 98109, USA
e-mail: aluedtke@fredhutch.org

M. J. van der Laan
Division of Biostatistics and Department of Statistics, University of California, Berkeley, 101 Haviland Hall, #7358, Berkeley, CA 94720, USA
e-mail: laan@berkeley.edu

© Springer International Publishing AG 2018
M.J. van der Laan, S. Rose, *Targeted Learning in Data Science*,
Springer Series in Statistics, https://doi.org/10.1007/978-3-319-65304-4_23

probability of treatment given covariates. Suppose that the treatment resource is limited so that at most a $\kappa \in (0, 1)$ proportion of the population can receive the treatment $A = 1$. A deterministic treatment rule $\tilde{d}$ takes as input a covariate vector $w \in \mathcal{W}$ and outputs a binary treatment decision $\tilde{d}(w)$. The stochastic treatment rules considered in this chapter are maps from $\mathcal{U} \times \mathcal{W}$ to $\{0, 1\}$, where $\mathcal{U}$ is the support of some random variable $U \sim P_U$. If d is a stochastic rule and $u \in \mathcal{U}$ is fixed, then $d(u, \cdot)$ represents a deterministic treatment rule. Throughout this chapter, we will let U be drawn independently of all draws from P_0. We will be consistent with the use of the tilde to represent deterministic rules and lack of tilde to represent stochastic rules so that throughout $\tilde{d} : \mathcal{W} \to \mathbb{R}$ and $d : \mathcal{U} \times \mathcal{W} \to \mathbb{R}$.

For a distribution P, let $\bar{Q}_P(a, w) \triangleq E_P[Y|A = a, W = w]$. For notational convenience, we let $\bar{Q}_0 \triangleq \bar{Q}_{P_0}$. Let $\tilde{d}$ be a deterministic treatment regime. For a distribution P, let $\tilde{\Psi}_{\tilde{d}} \triangleq E_{P_0}[\bar{Q}_P(\tilde{d}(V), W)]$ represent the value of $\tilde{d}$. Under causal assumptions, this quantity is equal to the counterfactual mean outcome if, possibly contrary to fact, the rule $\tilde{d}$ were implemented in the population (Robins 1986). The optimal R-C deterministic regime at P is defined as the deterministic regime $\tilde{d}$ that solves the optimization problem

$$\text{Maximize } \Psi_{\tilde{d}}(P) \text{ subject to } E_0[\tilde{d}(W)] \le \kappa. \tag{23.1}$$

For a stochastic regime d, let $\Psi_d(P) \triangleq E_{P_U}[\tilde{\Psi}_{d(U,\cdot)}(P)]$ represent the value of d. Under causal assumptions, this quantity is equal to the counterfactual mean outcome if, possibly contrary to fact, the stochastic rule d were implemented in the population (see Díaz and van der Laan 2013b for a similar identification result). The optimal R-C stochastic regime at P is defined as the stochastic treatment regime d that solves the optimization problem

$$\text{Maximize } \Psi_d(P) \text{ subject to } E_{P_U \times P}[d(U, W)] \le \kappa. \tag{23.2}$$

We call the optimal value under a R-C stochastic regime $\Psi(P)$. Because any deterministic regime can be written as a stochastic regime that does not rely on the stochastic mechanism U, we have that $\Psi(P) \ge \tilde{\Psi}(P)$. Let S_P represent the survival function of the blip function $\bar{Q}_{b,P}(\cdot) \triangleq \bar{Q}_P(1, \cdot) - \bar{Q}_P(0, \cdot)$, i.e. $\tau \mapsto P(\bar{Q}_{b,P}(W) > \tau)$. Let

$$\eta_P \triangleq \inf \{\tau : S_P(\tau) \le \kappa\}$$
$$\tau_P \triangleq \max \{\eta_P, 0\}. \tag{23.3}$$

For notational convenience we let $S_0 \triangleq S_{P_0}$, $\eta_0 \triangleq \eta_{P_0}$, and $\tau_0 \triangleq \tau_{P_0}$.

Define the deterministic treatment rule $\tilde{d}_P$ as $w \mapsto I(\bar{Q}_{b,P}(w) > \tau_P)$, and for notational convenience let $\tilde{d}_0 \triangleq \tilde{d}_{P_0}$. We have the following result.

Theorem 23.1. *If $P(\bar{Q}_{b,P}(W) = \tau_P) = 0$, then the $\tilde{d}_P$ is an optimal deterministic rule satisfying the resource constraint, i.e. $\Psi_{\tilde{d}_P}(P)$ attains the maximum described in (23.1).*

One can in fact show that $\tilde{d}_P$ is the P almost surely unique optimal deterministic regime under the stated condition. We do not treat the case where $P(\bar{Q}_{b,P}(W) = \tau_P) > 0$ for deterministic regimes, since in this case (23.1) is a more challenging problem: for discrete W with positive treatment effect in all strata, (23.1) is a special case of the 0–1 knapsack problem, which is NP-hard, though is considered one of the easier problems in this class (Karp 1972; Korte and Vygen 2012). In the knapsack problem, one has a collection of items, each with a value and a weight. Given a knapsack that can only carry a limited weight, the objective is to choose which items to bring so as to maximize the value of the items in the knapsack while respecting the weight restriction. Considering the optimization problem over stochastic rather than deterministic regimes yields a fractional knapsack problem, which is known to be solvable in polynomial time (Dantzig 1957; Korte and Vygen 2012). The fractional knapsack problem differs from the 0–1 knapsack problem in that one can pack partial items, with the value of the partial items proportional to the fraction of the item packed.

Define the stochastic treatment rule d_P by its distribution with respect to a random variable drawn from P_U:

$$P_U\left(d_P(U, w) = 1\right) = \begin{cases} \frac{\kappa - S_P(\tau_P)}{\Pr_P(\bar{Q}_{b,P}(W) = \tau_P)}, & \text{if } \bar{Q}_{b,P}(w) = \tau_P \text{ and } \tau_P > 0 \\ I(\bar{Q}_{b,P}(w) > \tau_P), & \text{otherwise.} \end{cases}$$

If $\Pr_P(\bar{Q}_{b,P}(W) = \tau_P) = 0$, then the first case occurs with probability zero, and this the division by this quantity will not prove problematic. We will let $d_0 \triangleq d_{P_0}$. Note that $\tilde{d}_P(W)$ and $d_P(U, W)$ are $P_U \times P$ almost surely equal if $P(\bar{Q}_{b,P}(W) = \tau_P) = 0$ or if $\tau_P \leq 0$, and thus have the same value in these settings. It is easy to show that

$$E_{P_U \times P}[d_P(U, W)] = \kappa \text{ if } \tau_P > 0. \tag{23.4}$$

The following theorem establishes the optimality of the stochastic rule d_P in a resource-limited setting.

Theorem 23.2. *The maximum in (23.2) is attained at $d = d_P$, i.e. d_P is an optimal stochastic rule.*

Note that the above theorem does not claim that d_P is the unique optimal stochastic regime. For discrete W, the above theorem is an immediate consequence of the discussion of the knapsack problem in Dantzig (1957).

In this chapter we focus on the value of the optimal stochastic rule. Nonetheless, the techniques that we present in this chapter will only yield valid inference in the case where the data are generated according to a distribution P_0 for which $P_0(\bar{Q}_{b,0}(W) = \tau_0) = 0$. This is analogous to assuming a nonexceptional law in settings where resources are not limited (Robins 2004; Luedtke and van der Laan 2016a), though we note that for continuous covariates W this assumption is much more likely if $\tau_0 > 0$. It seems unlikely that the treatment effect in some positive probability stratum of covariates will concentrate on some arbitrary (determined by the constraint κ) value τ_0.

23.2 Estimating the Optimal Resource-Constrained Value

We now present an estimation strategy for the optimal R-C rule. The upcoming sections justify this strategy and suggest that it will perform well for a wide variety of data generating distributions. The estimation strategy proceeds as follows:

1. Obtain estimates $\bar{Q}_n^0$, $\bar{Q}_{b,n}$, and g_n of $\bar{Q}_0$, $\bar{Q}_{b,0}$, and g_0 using any desired estimation strategy that respects the fact that Y is bounded in the unit interval.
2. Estimate the marginal distribution of W with the corresponding empirical distribution.
3. Estimate S_0 with the plug-in estimator S_n given by $\tau \mapsto \frac{1}{n} \sum_{i=1}^n I\left(\bar{Q}_{b,n}(w_i) > \tau\right)$.
4. Estimate η_0 with the plug-in estimator $\eta_n \triangleq \inf\{\tau : S_n(\tau) \le \kappa\}$.
5. Estimate τ_0 with the plug-in estimator given by $\tau_n \triangleq \max\{\eta_n, 0\}$.
6. Estimate d_0 with the plug-in estimator d_n with distribution

$$P_U(d_n(U, w) = 1) = \begin{cases} \frac{\kappa - S_n(\tau_n)}{\Pr_{P_n}(\bar{Q}_{b,n}(W) = \tau_n)}, & \text{if } \bar{Q}_{b,n}(w) = \tau_n \text{ and } \tau_n > 0 \\ I(\bar{Q}_{b,n}(w) > \tau_n), & \text{otherwise.} \end{cases}$$

7. Run a TMLE for the parameter $\Psi_{d_n}(P_0)$:

    (a) For $\tilde{a} \in \{0, 1\}$, define $H_n^*(a, w) \triangleq \frac{P_U(d_n(U,w)=a)}{g_n(a|w)}$. Run a univariate logistic regression using:

    $$\text{Outcome: } (y_i : i = 1, \ldots, n)$$
    $$\text{Offset: } \left(\text{logit } \bar{Q}_n^0(a_i, w_i) : i = 1, \ldots, n\right)$$
    $$\text{Covariate: } (H_n^*(a_i, w_i) : i = 1, \ldots, n).$$

    Let ϵ_n represent the estimate of the coefficient for the covariate, i.e.

    $$\epsilon_n \triangleq \arg\max \epsilon \in \mathbb{R} \frac{1}{n} \sum_{i=1}^n \left[\bar{Q}_n^\epsilon(a_i, w_i) \log y_i + \left(1 - \bar{Q}_n^\epsilon(a_i, w_i)\right) \log(1 - y_i)\right],$$

    where $\bar{Q}_n^\epsilon(a, w) \triangleq \text{logit}^{-1}\left(\text{logit } \bar{Q}_n^0(a, w) + \epsilon H_n^*(a, w)\right)$.
    (b) Define $\bar{Q}_n^* \triangleq \bar{Q}_n^{\epsilon_n}$.
    (c) Estimate $\Psi_{d_n}(P_0)$ using the plug-in estimator given by

    $$\Psi_{d_n}(P_n^*) \triangleq \frac{1}{n} \sum_{i=1}^n \sum_{a=0}^1 \bar{Q}_n^*(a, w_i) P_U(d_n(U, w_i) = a).$$

We use $\Psi_{d_n}(P_n^*)$ as our estimate of $\Psi(P_0)$. We will denote this estimator $\hat{\Psi}$, where we have defined $\hat{\Psi}$ so that $\hat{\Psi}(P_n) = \Psi_{d_n}(P_n^*)$. Note that we have used a TMLE for the data-dependent parameter $\Psi_{d_n}(P_0)$, which represents the value under a *stochastic* intervention d_n. Nonetheless, we assume that $P_0(\bar{Q}_{b,0}(W) = \tau_0) = 0$ for many of the results pertaining to our estimator $\hat{\Psi}$, i.e., we assume that the optimal R-C rule is deterministic. We view estimating the value under a stochastic rather than deterministic intervention as worthwhile because one can give conditions under which the above estimator is (root-n) consistent for $\Psi(P_0)$ at all laws P_0, even if nonnegligible bias invalidates standard Wald-type confidence intervals for the parameter of interest at laws P_0 for which $P_0(\bar{Q}_{b,0}(W) = \tau_0) > 0$.

We will use P_n^* to denote any distribution for which $\bar{Q}_{P_n^*} = \bar{Q}_n^*$, $g_{P_n^*} = g_n$, and P_n^* has the marginal empirical distribution of W for the marginal distribution of W. We note that such a distribution P_n^* exists provided that $\bar{Q}_n^*$ and g_n fall in the parameter spaces of $P \mapsto \bar{Q}_P(W)$ and $P \mapsto g_P$, respectively. In practice we recommend estimating $\bar{Q}_0$ and $\bar{Q}_{b,0}$ using an ensemble method such as super learning to make an optimal bias-variance trade-off (or, more generally, minimize cross-validated risk) between a mix of parametric models and data-adaptive regression algorithms (van der Laan et al. 2007; Luedtke and van der Laan 2016b). If the treatment mechanism g_0 is unknown then we recommend using similar data-adaptive approaches to obtain the estimate g_n. If g_0 is known (as in a randomized controlled trial without missingness), then one can either take $g_n = g_0$ or estimate g_0 using a correctly specified parametric model, which we expect to increase the efficiency of estimators when the $\bar{Q}_0$ part of the likelihood is misspecified (van der Laan and Robins 2003; van der Laan and Luedtke 2015).

There is typically little downside to using data-adaptive approaches to estimate the needed portions of the likelihood, though we do give a formal empirical process condition in Sect. 23.4 that describes exactly how data adaptive these estimators can be. If one is concerned about the data adaptivity of the estimators of the needed portions of the likelihood, then one can consider a cross-validated TMLE approach such as that presented in van der Laan and Luedtke (2015). This approach makes no restrictions on the data adaptivity of the estimators of $\bar{Q}_0$, $\bar{Q}_{b,0}$, or g_0.

We now outline the main results of this chapter, which hold under appropriate consistency and regularity conditions.

- Asymptotic linearity of $\hat{\Psi}$:

$$\hat{\Psi}(P_n) - \Psi(P_0) = \frac{1}{n} \sum_{i=1}^{n} D_0(O_i) + o_{P_0}(n^{-1/2}),$$

with D_0 a known function of P_0.
- $\hat{\Psi}$ is an asymptotically efficient estimate of $\Psi(P_0)$.
- One can obtain a consistent estimate σ_n^2 for the variance of $D_0(O)$. An asymptotically valid 95% confidence intervals for $\Psi(P_0)$ given by $\hat{\Psi}(P_n) \pm 1.96\sigma_n/\sqrt{n}$.

The upcoming sections give the consistency and regularity conditions that imply the above results.

23.3 Canonical Gradient of the Optimal Resource-Constrained Value

The pathwise derivative of Ψ will provide a key ingredient for analyzing the asymptotic properties of our estimator. We refer the reader to Pfanzagl (1990) and Bickel et al. (1997b) for an overview of the crucial role that the pathwise derivative plays in semiparametric efficiency theory. We remind the reader that an estimator $\hat{\Phi}$ is an asymptotically linear estimator of a parameter $\Phi(P_0)$ with influence curve IC_{P_0} provided that

$$\hat{\Phi}(P_n) - \Phi(P_0) = \frac{1}{n} \sum_{i=1}^{n} IC_{P_0}(O_i) + o_{P_0}(n^{-1/2}).$$

If Φ is pathwise differentiable with canonical gradient IC_{P_0}, then $\hat{\Phi}$ is RAL and asymptotically efficient (minimum variance) among all such RAL estimators of $\Phi(P_0)$ (Pfanzagl 1990; Bickel et al. 1997b).

For $o \in O$, a deterministic rule $\tilde{d}$, and a real number τ, define

$$D_1(\tilde{d}, P)(o) \triangleq \frac{I(a = \tilde{d}(w))}{g_P(a|w)} \left(y - \bar{Q}_P(a, w)\right)$$

$$D_2(\tilde{d}, P)(o) \triangleq \bar{Q}_P(\tilde{d}(w), w) - E_P \bar{Q}_P(\tilde{d}(W), W),$$

where $g_P(a|W) \triangleq P(A = a|W)$. We will let $g_0 \triangleq g_{P_0}$. We note that $D_1(\tilde{d}, P) + D_2(\tilde{d}, P)$ is the efficient influence curve of the parameter $\Psi_{\tilde{d}}(P)$.

Let d be some stochastic rule. The canonical gradient of Ψ_d is given by

$$IC_d(P)(o) \triangleq E_{P_U}[D_1(d(U, w), P)(o) + D_2(d(U, w), P)(o)].$$

Define

$$D(d, \tau, P)(o) \triangleq IC_d(P)(o) - \tau \left(E_{P_U}[d(U, w)] - \kappa\right).$$

For ease of reference, let $D_0 \triangleq D(d_0, \tau_0, P_0)$. The upcoming theorem makes use of the following assumptions.

(C1) g_0 satisfies the positivity assumption: $P_0(0 < g_0(1|W) < 1) = 1$.
(C2) $\bar{Q}_{b,0}(W)$ has density f_0 at η_0, and $0 < f_0(\eta_0) < \infty$.
(C3) S_0 is continuous in a neighborhood of η_0.
(C4) $P_0(\bar{Q}_{b,0}(W) = \tau) = 0$ for all τ in a neighborhood of τ_0.

We now present the canonical gradient of the optimal R-C value.

Theorem 23.3. *Suppose 23.3 through 23.3. Then Ψ is pathwise differentiable at P_0 with canonical gradient D_0.*

Note that 23.3 implies that $P_0(\bar{Q}_{b,0}(W) = \tau_0) = 0$. Thus d_0 is (almost surely) deterministic and the expectation over P_U in the definition of D_0 is superfluous. Nonetheless, this representation will prove useful when we seek to show that our estimator solves the empirical estimating equation defined by an estimate of $D(d_0, \tau_0, P_0)$.

When the resource constraint is active, i.e., $\tau_0 > 0$, the above theorem shows that Ψ has an additional component over the optimal value parameter when no resource constraints are present (van der Laan and Luedtke 2015). The additional component is $\tau_0 \times (E_{P_U}[d_0(U, w)] - \kappa)$, and is the portion of the derivative that relies on the fact that d_0 is estimated and falls on the edge of the parameter space. We note that it is possible that the variance of $D_0(O)$ is greater than the variance of $IC_{d_0}(P_0)(O)$. If $\tau_0 = 0$ then these two variances are the same, so suppose $\tau_0 > 0$. Then, provided that $P_0(\bar{Q}_{b,0}(W) = \tau_0) = 0$, we have that

$$\mathrm{Var}_{P_0}(D_0(O)) - \mathrm{Var}_{P_0}(IC_{d_0}(P_0))$$
$$= \tau_0 \kappa (1 - \kappa)\left(\tau_0 - 2E_0\left[\bar{Q}_0(1, W)\middle| \tilde{d}_0(W) = 1\right] + 2E_0\left[\bar{Q}_0(0, W)\middle| \tilde{d}_0(W) = 0\right]\right).$$

For any $\kappa \in (0, 1)$, it is possible to exhibit a distribution P_0 that satisfies the conditions of Theorem 23.3 and for which $\mathrm{Var}_{P_0}(D_0(O)) > \mathrm{Var}_{P_0}(IC_{d_0}(P_0)(O))$. Perhaps more surprisingly, it is also possible to exhibit a distribution P_0 that satisfies the conditions of Theorem 23.3 and for which $\mathrm{Var}_{P_0}(D_0(O)) < \mathrm{Var}_{P_0}(IC_{d_0}(P_0)(O))$. We omit further the discussion here because the focus of this chapter is on considering the estimating the value from the optimization problem (23.2), rather than discussing how this procedure relates to the estimation of other parameters.

23.4 Inference for $\Psi(P_0)$

We now show that $\hat{\Psi}$ is an asymptotically linear estimator for $\Psi(P_0)$ with influence curve D_0 provided our estimates of the needed parts of P_0 satisfy consistency and regularity conditions. For any distributions P and P_0 satisfying positivity, stochastic intervention d, and real number τ, define the second-order remainder terms:

$$R_{10}(d, P) \triangleq E_{P_U \times P_0}\left[\left(1 - \frac{g_0(d|W)}{g(d|W)}\right)\left(\bar{Q}_P(d, W) - \bar{Q}_0(d, W)\right)\right]$$
$$R_{20}(d) \triangleq E_{P_U \times P_0}\left[(d - d_0)(\bar{Q}_{b,0}(W) - \tau_0)\right].$$

Above the reliance of d and d_0 on (U, W) is omitted in the notation. Let $R_0(d, P) \triangleq R_{10}(d, P) + R_{20}(d)$. The upcoming theorem makes use of the following assumptions.

(C5) g_0 satisfies the strong positivity assumption: $P_0(\delta < g_0(1|W) < 1 - \delta) = 1$ for some $\delta > 0$.
(C6) g_n satisfies the strong positivity assumption for a fixed $\delta > 0$ with probability approaching 1: there exists some $\delta > 0$ such that, with probability approaching 1, $P_0(\delta < g_n(1|W) < 1 - \delta) = 1$.
(C7) $R_0(d_n, P_n^*) = o_{P_0}(n^{-1/2})$.
(C8) $E_0\left[(D(d_n, \tau_0, P_n^*)(O) - D_0(O))^2\right] = o_{P_0}(1)$.
(C9) $D(d_n, \tau_0, P_n^*)$ belongs to a P_0-Donsker class $\mathcal{D}$ with probability approaching 1.
(C10) $\frac{1}{n}\sum_{i=1}^n D(d_n, \tau_0, P_n^*)(O_i) = o_{P_0}(n^{-1/2})$.

We note that the τ_0 in the final condition above only enters the expression in the sum as a multiplicative constant in front of $-E_{P_U}[d(U, w_i)] - \kappa$.

Theorem 23.4 ($\hat{\Psi}$ Is Asymptotically Linear). *Suppose 23.3 through 23.4. Then $\hat{\Psi}$ is a RAL estimator of $\Psi(P_0)$ with influence curve D_0, i.e.*

$$\hat{\Psi}(P_n) - \Psi(P_0) = \frac{1}{n} \sum_{i=1}^{n} D_0(O_i) + o_{P_0}(n^{-1/2}).$$

Further, $\hat{\Psi}$ is efficient among all such RAL estimators of $\Psi(P_0)$.

Let $\sigma_0^2 \triangleq \text{Var}_{P_0}(D_0)$. By the central limit theorem, $\sqrt{n}\left(\hat{\Psi}(P_n) - \Psi(P_0)\right)$ converges in distribution to a $N(0, \sigma_0^2)$ distribution. Let $\sigma_n^2 \triangleq \frac{1}{n} \sum_{i=1}^{n} D(d_n, \tau_n, P_n^*)(O_i)^2$ be an estimate of σ_0^2. We now give the following lemma, which gives sufficient conditions for the consistency of τ_n for τ_0.

Lemma 23.1 (Consistency of τ_n). *Suppose 23.3 and 23.3. Also suppose $\bar{Q}_{b,n}$ is consistent for $\bar{Q}_{b,0}$ in $L^1(P_0)$ and that the estimate $\bar{Q}_{b,n}$ belongs to a P_0 Glivenko Cantelli class with probability approaching 1. Then $\tau_n \to \tau_0$ in probability.*

It is easy to verify that conditions similar to those of Theorem 23.4, combined with the convergence of τ_n to τ_0 as considered in the above lemma, imply that $\sigma_n \to \sigma_0$ in probability. Under these conditions, an asymptotically valid two-sided $1 - \alpha$ confidence interval is given by

$$\hat{\Psi}(P_n) \pm z_{1-\alpha/2} \frac{\sigma_n}{\sqrt{n}},$$

where $z_{1-\alpha/2}$ denotes the $1 - \alpha/2$ quantile of a $N(0, 1)$ random variable.

23.5 Discussion of Theorem 23.4 Conditions

Conditions 23.3 and 23.3. These are standard conditions used when attempting to estimate the κ-quantile η_0, defined in (23.3). Provided good estimation of $\bar{Q}_{b,0}$, these conditions ensure that gathering a large amount of data will enable one to get a good estimate of the κ-quantile of the random variable $\bar{Q}_{b,0}$. See Lemma 23.1 for an indication of what is meant by "good estimation" of $\bar{Q}_{b,0}$. It seems reasonable to expect that these conditions will hold when W is continuous and $\eta_0 \neq 0$, since we are assuming that $\bar{Q}_{b,0}$ is not degenerate at the arbitrary (determined by κ) point η_0.

Condition 23.3. If $\tau_0 > 0$, then 23.3 is implied by 23.3. If $\tau_0 = 0$, then 23.3 is like assuming a nonexceptional law, i.e. that the probability of a there being no treatment effect in a stratum of W is zero. Because τ_0 is not known from the outset, we require something slightly stronger, namely that the probability of *any specific* small treatment effect is zero in a stratum of W is zero. Note that this condition does *not* prohibit the treatment effect from being small, e.g. $P_0(|\bar{Q}_{b,0}(W)| < \tau) > 0$ for all $\tau > 0$, but rather it prohibits there existing a sequence $\tau_m \downarrow 0$ with the property that $P_0(\bar{Q}_{b,0}(W) = \tau_m) > 0$ infinitely often. Thus this condition does not really seem any stronger than assuming a nonexceptional law. If one is concerned about such exceptional laws then we suggest adapting the methods in Luedtke and van der Laan (2016b) to the R-C setting.

Condition 23.4. This condition assumes that people from each stratum of co-variates have a reasonable (at least a $\delta > 0$) probability of treatment.

Condition 23.4. This condition requires that our estimates of g_0 respect the fact that each stratum of covariates has a reasonable probability of treatment.

Condition 23.4. This condition is satisfied if $R_{10}(d_n, P_n^*) = o_{P_0}(n^{-1/2})$ and $R_{20}(d_n) = o_{P_0}(n^{-1/2})$. The term $R_{10}(d_n, P_n^*)$ takes the form of a typical double robust term that is small if either g_0 or $\bar{Q}_0$ is estimated well, and is second order, i.e., one might hope that $R_{10}(d_n, P_n^*) = o_{P_0}(n^{-1/2})$, if both g_0 and $\bar{Q}_0$ are estimated well. One can upper bound this remainder with a product of the $L^2(P_0)$ rates of convergence of these two quantities using the Cauchy-Schwarz inequality. If g_0 is known, then one can take $g_n = g_0$ and this term is zero.

Ensuring that $R_{20}(d_n) = o_{P_0}(n^{-1/2})$ requires a little more work but will still prove to be a reasonable condition. We will use the following margin assumption for some $C_0, \alpha > 0$:

$$P_0\left(0 < |\bar{Q}_{b,0} - \tau_0| \le t\right) \le C_0 t^\alpha \text{ for all } t > 0, \tag{23.5}$$

This margin assumption is analogous to that used in Audibert and Tsybakov (2007). The following result relates the rate of convergence of $R_{20}(d_n)$ to the rate at which $\bar{Q}_{b,n} - \tau_n$ converges to $\bar{Q}_{b,0} - \tau_0$.

Theorem 23.5. *If (23.5) holds for some $C_0, \alpha > 0$, then, for some constant $C > 0$,*

$$|R_{20}(d_n)| \le C \min\left\{\left\|(\bar{Q}_{b,n} - \tau_n) - (\bar{Q}_{b,0} - \tau_0)\right\|_{2,P_0}^{2(1+\alpha)/(2+\alpha)}, \left\|(\bar{Q}_{b,n} - \tau_n) - (\bar{Q}_{b,0} - \tau_0)\right\|_{\infty,P_0}^{1+\alpha}\right\}.$$

The proof of this result is analogous to Theorem 22.5 and is omitted. If S_0 has a finite derivative at τ_0, as is given by (23.3), then one can take $\alpha = 1$. The above theorem then implies that $R_{20}(d_n) = o_{P_0}(n^{-1/2})$ if either $\left\|(\bar{Q}_{b,n} - \tau_n) - (\bar{Q}_{b,0} - \tau_0)\right\|_{2,P_0}$ is $o_{P_0}(n^{-3/8})$ or $\left\|(\bar{Q}_{b,n} - \tau_n) - (\bar{Q}_{b,0} - \tau_0)\right\|_{\infty,P_0}$ is $o_{P_0}(n^{-1/4})$.

Condition 23.4. This is a mild consistency condition that is implied by the $L^2(P_0)$ consistency of d_n, g_n, and $\bar{Q}_n^*$ to d_0, g_0, and $\bar{Q}_0$. We note that the consistency of the initial (unfluctuated) estimate $\bar{Q}_n^0$ for $\bar{Q}_0$ will imply the consistency of $\bar{Q}_n^*$ to $\bar{Q}_0$ given 23.4, since in this case $\epsilon_n \to 0$ in probability, and thus $\left\|\bar{Q}_n^* - \bar{Q}_n^0\right\|_{2,P_0} \to 0$ in probability.

Condition 23.4. This condition places restrictions on how data adaptive the estimators of d_0, g_0, and $\bar{Q}_0$ can be. We refer the reader to Section 2.10 of van der Vaart and Wellner (1996) for conditions under which the estimates of d_0, g_0, and $\bar{Q}_0$ belonging to Donsker classes implies that $D(d_n, \tau_0, P_n^*)$ belongs to a Donsker class. This condition was avoided for estimating the value function using a cross-validated TMLE in van der Laan and Luedtke (2015), and using this technique will allow one to avoid the condition here as well.

Condition 23.4. Using the notation $Pf = \int f(o)dP(o)$ for any distribution P and function $f : O \to \mathbb{R}$, we have that

$$P_n D(d_n, \tau_0, P_n^*) = P_n D_1(d_n, P_n^*) + P_n D_2(d_n, P_n^*)$$

$$- \tau_0 \left(\frac{1}{n} \sum_{i=1}^{n} E_{P_U}[d_n(U, w_i)] - \kappa \right).$$

The first term is zero by the fluctuation step of the TMLE algorithm and the second term on the right is zero because P_n^* uses the empirical distribution of W for the marginal distribution of W. If $\tau_0 = 0$ then clearly the third term is zero, so suppose $\tau_0 > 0$. Combining (23.4) and the fact that d_n is a substitution estimator shows that the third term is 0 with probability approaching 1 provided that $\tau_n > 0$ with probability approaching 1. This will of course occur if $\tau_n \to \tau_0 > 0$ in probability, for which Lemma 23.1 gives sufficient conditions.

23.6 Discussion

We considered the problem of estimating the optimal resource-constrained value. Under causal assumptions, this parameter can be identified with the maximum attainable population mean outcome under individualized treatment rules that rely on measured covariates, subject to the constraint that a maximum proportion κ of the population can be treated. We also provided an explicit expression for an optimal stochastic rule under the resource constraint.

Additionally, we derived the canonical gradient of the optimal R-C value under the key assumption that the treatment effect is not exactly equal to τ_0 in some stratum of covariates that occurs with positive probability. The canonical gradient plays a key role in developing asymptotically linear estimators. We found that the canonical gradient of the optimal R-C value has an additional component when compared to the canonical gradient of the optimal unconstrained value when the resource constraint is active, i.e., when $\tau_0 > 0$.

We presented a TMLE for the optimal R-C value. This estimator was designed to solve the empirical mean of an estimate of the canonical gradient. This quickly yielded conditions under which our estimator is RAL, and efficient among all such RAL estimators. All of these results rely on the condition that the treatment effect is not exactly equal to τ_0 for positive probability stratum of covariates. This assumption is more plausible than the typical nonexceptional law assumption when the covariates are continuous and the constraint is active because it may be unlikely that the treatment effect concentrates on an arbitrary (determined by κ) $\tau_0 > 0$. We note that this pseudo-nonexceptional law assumption has implied that the optimal stochastic rule is almost surely equal to the optimal deterministic rule.

Some resource constraints encountered in practice may not be of the form $E[d(U, W)]$ less than or equal to κ. For example, the cost of distributing the treatment to people may vary based on the values of the covariates. If $c : \mathcal{W} \to [0, \infty)$ is a cost function, then this constraint may take the form $E[c(W)d(U, W)] \le \kappa$. The optimal rule takes the form $(u, w) \mapsto I(\bar{Q}_{b,0}(w) > \tau c(w))$ for w for which $\bar{Q}_{b,0}(w) \ne \tau_0 c(w)$ or $c(w) = 0$, and randomly distributes the remaining resources uniformly among all remaining w. We leave further consideration of this more general resource constraint problem to future work.

We have not considered the ethical considerations associated with allocating limiting resources to a population. The debate over the appropriate means to distribute limited treatment resources to a population is ongoing (see, e.g., Brock and Wikler 2009; Macklin and Cowan 2012; Singh 2013, for examples in the treatment of HIV/AIDS). Clearly any investigator needs to consider the ethical issues associated with certain resource allocation schemes. Our method is optimal in a particular utilitarian sense (maximizing the expected population mean outcome with respect to an outcome of interest) and yields a treatment strategy that treats individuals who are expected to benefit most from treatment in terms of our outcome of interest. One must be careful to ensure that the outcome of interest truly captures the most important public health implications. Unlike in unconstrained individualized medicine, inappropriately prescribing treatment to a stratum will also have implications for individuals outside of that stratum, namely for the individuals who do not receive treatment due to its lack of availability. We leave further ethical considerations to experts on the matter. It will be interesting to see if there are settings in which it is possible to transform the outcome or add constraints to the optimization problem so that the statistical problem considered in this chapter adheres to the ethical guidelines in those settings.

We have looked to generalize previous works in estimating the value of an optimal individualized treatment regime to the case where the treatment resource is a limited resource, i.e., where it is not possible to treat the entire population. The results in this chapter should allow for the application of optimal personalized treatment strategies to many new problems of interest. This chapter includes content, with permission, from Luedtke and van der Laan (2016c).

23.7 Proofs

Proofs for Sect. 23.1. We first state a simple lemma.

Lemma 23.2. *For a distribution P and a stochastic rule d, we have the following representation for Ψ_d:*

$$\Psi_d(P) \triangleq E_{P_U \times P}\left[d(U, W)\bar{Q}_{b,P}(W)\right] + E_P[\bar{Q}_P(0, W)].$$

Proof (Lemma 23.2). We have that

$$
\begin{aligned}
\Psi_d(P) &= E_{P_U \times P}[d(U, W)\bar{Q}_P(1, W)] + E_{P_U \times P}[(1 - d(U, W))\bar{Q}_P(0, W)] \\
&= E_{P_U \times P}[d(U, W)(\bar{Q}_P(1, W) - \bar{Q}_P(0, W))] + E_P[\bar{Q}_P(0, W)] \\
&= E_{P_U \times P}[d(U, W)\bar{Q}_{b,P}(W)] + E_P[\bar{Q}_P(0, W)].
\end{aligned}
$$

Proof (Theorem 23.1). This result will be a consequence of Theorem 23.2. If $P(\bar{Q}_{b,0}(W) = \tau_P) = 0$, then $d_P(U, W)$ is $P_U \times P$ almost surely equal to $\tilde{d}_P(W)$, and thus $\Psi_{\tilde{d}_P}(P) = \Psi_{d_P}(P)$. Thus $(u, w) \mapsto \tilde{d}_P(w)$ is an optimal stochastic regime. Because the class of deterministic regimes is a subset of the class of stochastic regimes, $\tilde{d}_P$ is an optimal deterministic regime.

Proof (Theorem 23.2). Let d be some stochastic treatment rule that satisfies the resource constraint. For $(b, c) \in \{0, 1\}^2$, define $B_{bc} \triangleq \{(u, w) : d_P(u, w) = b, d(u, w) = c\}$. Note that

$$
\begin{aligned}
\Psi_{d_P}(P) - \Psi_d(P) &= E_{P_U \times P}\left[(d_P(U, W) - d(U, W))\,\bar{Q}_{b,0}(W)\right] \\
&= E_{P_U \times P}\left[\bar{Q}_{b,0}(W)I((U, W) \in B_{10})\right] - E_{P_U \times P}\left[\bar{Q}_{b,0}(W)I((U, W) \in B_{01})\right] \quad (23.6)
\end{aligned}
$$

The $\bar{Q}_{b,0}(W)$ in the first term in (23.6) can be upper bounded by τ_P, and in the second term can be lower bounded by τ_P. Below we use $\mathrm{Pr}_{P_U \times P}(\text{Event})$ to denote the probability distribution of Event under the product distribution $P_U \times P$. Thus,

$$
\begin{aligned}
\Psi_{d_P}(P) - \Psi_d(P) &\geq \tau_P\left[\mathrm{Pr}_{P_U \times P}((U, W) \in B_{10}) - \mathrm{Pr}_{P_U \times P}((U, W) \in B_{01})\right] \\
&= \tau_P\left[\mathrm{Pr}_{P_U \times P}((U, W) \in B_{10} \cup B_{11}) - \mathrm{Pr}_{P_U \times P}((U, W) \in B_{01} \cup B_{11})\right] \\
&= \tau_P\left(E_{P_U \times P}[d_P(U, W)] - E_{P_U \times P}[d(U, W)]\right).
\end{aligned}
$$

If $\tau_P = 0$ then the final line is zero. Otherwise, $E_{P_U \times P}[d_P(U, W)] = \kappa$ by (23.4). Because d satisfies the resource constraint, $E_{P_U \times P}[d(U, W)] \leq \kappa$ and thus the final line above is at least zero. Thus $\Psi_{d_P}(P) - \Psi_d(P) \geq 0$ for all τ_P. Because d was arbitrary, d_P is an optimal stochastic rule.

Proofs for Sect. 23.3

Proof (Theorem 23.3). The pathwise derivative of $\Psi(Q)$ is defined as $\frac{d}{d\epsilon}\Psi(Q(\epsilon))\big|_{\epsilon=0}$ along paths $\{P_\epsilon : \epsilon\} \subset \mathcal{M}$. In particular, these paths are chosen so that

$$dQ_{W,\epsilon} = (1 + \epsilon H_W(W))dQ_W,$$

$$\text{where } EH_W(W) = 0 \text{ and } C_W \triangleq \sup_w |H_W(w)| < \infty;$$

$$dQ_{Y,\epsilon}(Y \mid A, W) = (1 + \epsilon H_Y(Y \mid A, W))dQ_Y(Y \mid A, W),$$

$$\text{where } E(H_Y \mid A, W) = 0 \text{ and } \sup_{w,a,y} |H_Y(y \mid a, w)| < \infty.$$

The parameter Ψ is not sensitive to fluctuations of $g_0(a|w) = P_0(a|w)$, and thus we do not need to fluctuate this portion of the likelihood. Let $\bar{Q}_{b,\epsilon} \triangleq \bar{Q}_{b,P_\epsilon}$, $\bar{Q}_\epsilon \triangleq \bar{Q}_{P_\epsilon}$, $d_\epsilon \triangleq d_{P_\epsilon}$, $\eta_\epsilon \triangleq \eta_{P_\epsilon}$, $\tau_\epsilon \triangleq \tau_{P_\epsilon}$, and $S_\epsilon \triangleq S_{P_\epsilon}$. First note that

$$\bar{Q}_{b,\epsilon}(w) = \bar{Q}_{b,0}(w) + \epsilon h_\epsilon(w) \tag{23.7}$$

for an h_ϵ with

$$\sup_{|\epsilon|<1} \sup_w |h_\epsilon(w)| \triangleq C_1 < \infty. \tag{23.8}$$

Note that 23.3 implies that d_0 is (almost surely) deterministic, i.e. $d_0(U, \cdot)$ is almost surely a fixed function. Let $\tilde{d}$ represent the deterministic rule $w \mapsto I(\bar{Q}_{b,0}(w) > 0)$ to which $d(u, \cdot)$ is (almost surely) equal for all u. By Lemma 23.2,

$$\begin{aligned}
\Psi(P_\epsilon) - \Psi(P_0) &= \int_w \left(E_{P_U}[d_\epsilon(U, W)] - \tilde{d}_0(W) \right) \bar{Q}_{b,\epsilon} dQ_{W,\epsilon} \\
&\quad + \int_w \tilde{d}_0(W) \left(\bar{Q}_{b,\epsilon} dQ_{W,\epsilon} - \bar{Q}_{b,0} dQ_{W,0} \right) + E_{P_\epsilon} \bar{Q}_\epsilon(0, W) \\
&\quad - E_0 \bar{Q}_0(0, W) \\
&= \int_w \left(E_{P_U}[d_\epsilon(U, W)] - \tilde{d}_0(W) \right) \left(\bar{Q}_{b,\epsilon} - \tau_0 \right) dQ_{W,\epsilon} \\
&\quad + \tau_0 \int_w \left(E_{P_U}[d_\epsilon(U, W)] dQ_{W,\epsilon} - \tilde{d}_0(W) dQ_{W,0} \right) \\
&\quad - \tau_0 \int_w \tilde{d}_0(W) \left(dQ_{W,\epsilon} - dQ_{W,0} \right) + \left[\Psi_{d_0}(P_\epsilon) - \Psi_{d_0}(P_0) \right]. \tag{23.9}
\end{aligned}$$

Dividing the fourth term by ϵ and taking the limit as $\epsilon \to 0$ gives the pathwise derivative of the mean outcome under the rule that treats d_0 as known. The third term can be written as $-\epsilon\tau_0 \int_w \tilde{d}_0(W) H_W dQ_{W,0}$, and thus the pathwise derivative of this term is $-\int_w \tau_0 \tilde{d}_0(W) H_W dQ_{W,0}$. If $\tau_0 > 0$, then $E_{P_U \times P_0}[\tilde{d}_0(W)] = \kappa$. The pathwise derivative of this term is zero if $\tau_0 = 0$. Thus, for all τ_0,

$$\lim_{\epsilon \to 0} -\frac{1}{\epsilon} \tau_0 \int_w \tilde{d}_0(W) \left(dQ_{W,\epsilon} - dQ_{W,0} \right) = \int_w \left(-\tau_0(\tilde{d}_0(w) - \kappa) \right) H_W(w) dQ_{W,0}(w).$$

Thus the third term in (23.9) generates the $w \mapsto -\tau_0(\tilde{d}_0(w) - \kappa)$ portion of the canonical gradient, or equivalently $w \mapsto -\tau_0(E_{P_U}[d_0(U, w)] - \kappa)$. The remainder of this proof is used to show that the first two terms in (23.9) are $o(\epsilon)$.

Step 1: $\eta_\epsilon \to \eta_0$

We refer the reader to (23.3) for a definition of the quantile $P \mapsto \eta_P$. This is a consequence of the continuity of S_0 in a neighborhood of η_0. For $\gamma > 0$,

$$|\eta_\epsilon - \eta_0| > \gamma \text{ implies that } S_\epsilon(\eta_0 - \gamma) \le \kappa \text{ or } S_\epsilon(\eta_0 + \gamma) > \kappa. \tag{23.10}$$

For positive constants C_1 and C_W,

$$S_\epsilon(\eta_0 - \gamma) \ge (1 - C_W|\epsilon|)P_0\left(\bar{Q}_{b,\epsilon} > \eta_0 - \gamma\right) \ge (1 - C_W|\epsilon|)S_0(\eta_0 - \gamma + C_1|\epsilon|).$$

Fix $\gamma > 0$ small enough so that S_0 is continuous at $\eta_0 - \gamma$. In this case we have that $S_0(\eta_0 - \gamma + C_1|\epsilon|) \to S_0(\eta_0 - \gamma)$ as $\epsilon \to 0$. By the infimum in the definition of η_0, we know that $S_0(\eta_0 - \gamma) > \kappa$. Thus $S_\epsilon(\eta_0 - \gamma) > \kappa$ for all $|\epsilon|$ small enough.

Similarly, $S_\epsilon(\eta_0 + \gamma) \le (1 + C_W|\epsilon|)S_0(\eta_0 + \gamma - C_1|\epsilon|)$. Fix $\gamma > 0$ small enough so that S_0 is continuous at $\eta_0 + \gamma$. Then $S_0(\eta_0 + \gamma - C_1|\epsilon|) \to S_0(\eta_0 + \gamma)$ as $\epsilon \to 0$. Condition 23.3 implies the uniqueness of the κ-quantile of $\bar{Q}_{b,0}$, and thus that $S_0(\eta_0 + \gamma) < \kappa$. It follows that $S_\epsilon(\eta_0 + \gamma) < \kappa$ for all $|\epsilon|$ small enough. Combining $S_\epsilon(\eta_0 - \gamma) > \kappa$ and $S_\epsilon(\eta_0 + \gamma) < \kappa$ for all ϵ close to zero with (23.10) shows that $\eta_\epsilon \to \eta_0$ as $\epsilon \to 0$.

Step 2: Second Term of (23.9) Is 0 Eventually

If $\tau_0 = 0$ then the result is immediate, so suppose $\tau_0 > 0$. By the previous step, $\eta_\epsilon \to \eta_0$, which implies that $\tau_\epsilon \to \tau_0 > 0$ by the continuity of the max function. It follows that $\tau_\epsilon > 0$ for ϵ large enough. By (23.4), $\Pr_{P_U \times P_\epsilon}(d_\epsilon(U, W) = 1) = \kappa$ for all sufficiently small $|\epsilon|$ and $P_0(\tilde{d}_0(W) = 1) = \kappa$. Thus the second term of (23.9) is 0 for all $|\epsilon|$ small enough.

Step 3: $\tau_\epsilon - \tau_0 = O(\epsilon)$

Note that $\kappa < S_\epsilon(\eta_\epsilon - |\epsilon|) \le (1 + C_W|\epsilon|)S_0(\eta_\epsilon - (1 + C_1)|\epsilon|)$. A Taylor expansion of S_0 about η_0 shows that

$$\begin{aligned}
\kappa &< (1 + C_W|\epsilon|)\left(S_0(\eta_0) + (\eta_\epsilon - \eta_0 - (1 + C_1)|\epsilon|)(-f_0(\eta_0) + o(1))\right) \\
&= \kappa + (\eta_\epsilon - \eta_0 - (1 + C_1)|\epsilon|)(-f_0(\eta_0) + o(1)) + O(\epsilon) \\
&= \kappa - (\eta_\epsilon - \eta_0)f_0(\eta_0) + o(\eta_\epsilon - \eta_0) + O(\epsilon).
\end{aligned} \tag{23.11}$$

The fact that $f_0(\eta_0) \in (0, \infty)$ shows that $\eta_\epsilon - \eta_0$ is bounded above by some $O(\epsilon)$ sequence. Similarly, $\kappa \ge S_\epsilon(\eta_\epsilon + |\epsilon|) \ge (1 - C_W|\epsilon|)S_0(\eta_\epsilon + (1 + C_1)|\epsilon|)$. Hence,

$$\begin{aligned}
\kappa &\ge (1 - C_W|\epsilon|)\left(S_0(\eta_0) + (\eta_\epsilon - \eta_0 + (1 + C_1)|\epsilon|)(-f_0(\eta_0) + o(1))\right) \\
&= \kappa - (\eta_\epsilon - \eta_0)f_0(\eta_0) + o(\eta_\epsilon - \eta_0) + O(\epsilon).
\end{aligned}$$

It follows that $\eta_\epsilon - \eta_0$ is bounded below by some $O(\epsilon)$ sequence. Combining these two bounds shows that $\eta_\epsilon - \eta_0 = O(\epsilon)$, which immediately implies that $\tau_\epsilon - \tau_0 = \max\{O(\epsilon), 0\} = O(\epsilon)$.

Step 4: First Term of (23.9) Is $O(\epsilon)$

We know that

$$\bar{Q}_{b,0}(W) - \tau_0 + O(\epsilon) \le \bar{Q}_{b,\epsilon}(W) - \tau_\epsilon \le \bar{Q}_{b,0}(W) - \tau_0 + O(\epsilon).$$

By 23.3, it follows that there exists some $\delta > 0$ such that $\sup_{|\epsilon|<\delta} P_0(\bar{Q}_{b,\epsilon}(W) = \tau_\epsilon) = 0$. By the absolute continuity of $Q_{W,\epsilon}$ with respect to $Q_{W,0}$, $\sup_{|\epsilon|<\delta} \mathrm{Pr}_{P_\epsilon}(\bar{Q}_{b,\epsilon}(W) = \tau_\epsilon) = 0$. It follows that, for all small enough $|\epsilon|$ and almost all u, $d_\epsilon(u,w) = I(\bar{Q}_{b,\epsilon}(w) > \tau_\epsilon)$. Hence,

$$\int_w \left(E_{P_U}[d_\epsilon(U,W)] - d_0(W) \right) \left(\bar{Q}_{b,\epsilon} - \tau_0 \right) dQ_{W,\epsilon}$$

$$= \left| \int_w \left(I(\bar{Q}_{b,\epsilon} > \tau_\epsilon) - I(\bar{Q}_{b,0} > \tau_0) \right) \left(\bar{Q}_{b,\epsilon} - \tau_0 \right) dQ_{W,\epsilon} \right|$$

$$\le \int_w \left| I(\bar{Q}_{b,\epsilon} > \tau_\epsilon) - I(\bar{Q}_{b,0} > \tau_0) \right| \left(\left| \bar{Q}_{b,0} - \tau_0 \right| + C_1|\epsilon| \right) dQ_{W,\epsilon}$$

$$\le \int_w I(|\bar{Q}_{b,0} - \tau_0| \le |\bar{Q}_{b,0} - \tau_0 - \bar{Q}_{b,\epsilon} + \tau_\epsilon|) \left(\left| \bar{Q}_{b,0} - \tau_0 \right| + C_1|\epsilon| \right) dQ_{W,\epsilon}$$

$$= \int_w I(0 < |\bar{Q}_{b,0} - \tau_0| \le |\bar{Q}_{b,0} - \tau_0 - \bar{Q}_{b,\epsilon} + \tau_\epsilon|) \left(\left| \bar{Q}_{b,0} - \tau_0 \right| + C_1|\epsilon| \right) dQ_{W,\epsilon}$$

$$\le O(\epsilon) \int_w I(0 < |\bar{Q}_{b,0} - \tau_0| \le O(\epsilon)) dQ_{W,\epsilon}$$

$$\le O(\epsilon)(1 + C_W|\epsilon|) P_0 \left(0 < |\bar{Q}_{b,0} - \tau_0| \le O(\epsilon) \right),$$

where the penultimate inequality holds by Step 3 and (23.7). The last line above is $o(\epsilon)$ because $Pr(0 < X \le \epsilon) \to 0$ as $\epsilon \to 0$ for any random variable X. Thus dividing the left-hand side above by ϵ and taking the limit as $\epsilon \to 0$ yields zero.

Proofs for Sect. 23.4. We give the following lemma before proving Theorem 23.4.

Lemma 23.3. *Let P_0 and P be distributions that satisfy the positivity assumption and for which Y is bounded in probability. Let d be some stochastic treatment rule and τ be some real number. We have that $\Psi_d(P) - \Psi(P_0) = -E_0[D(d, \tau_0, P)(O)] + R_0(d, P)$.*

Proof (Lemma 23.3). Note that

$$\Psi_d(P) - \Psi(P_0) + E_0[D(d, \tau_0, P)(O)]$$

$$= \Psi_d(P) - \Psi_d(P_0) + \sum_{j=1}^{2} E_{P_U \times P_0}[D_j(d(U, \cdot), P)(O)]$$

$$+ \Psi_d(P_0) - \Psi_{d_0}(P_0) - \tau_0 E_{P_U \times P_0}[d(U, W) - \kappa].$$

Standard calculations show that the first term on the right is equal to $R_{10}(d, P)$ (van der Laan and Robins 2003). If $\tau_0 > 0$, then (23.4) shows that $\tau_0 E_{P_U \times P_0}[d - \kappa] = \tau_0 E_{P_U \times P_0}[d - d_0]$. If $\tau_0 = 0$, then obviously $\tau_0 E_{P_U \times P_0}[d - \kappa] = \tau_0 E_{P_U \times P_0}[d - d_0]$. Lemma 23.2 shows that $\Psi_d(P_0) - \Psi_{d_0}(P_0) = E_{P_U \times P_0}[(d - d_0)\bar{Q}_{b,0}]$. Thus the second line above is equal to $R_{20}(d)$.

Proof (Theorem 23.4). We make use of empirical process theory notation in this proof so that $Pf = E_P[f(O)]$ for a distribution P and function f. We have that

$$\hat{\Psi}(P_n) - \Psi(P_0)$$
$$= -P_0 D(d_n, \tau_0, P_n^*) + R_0(d_n, P_n^*) \qquad \text{(by Lemma 23.3)}$$
$$= (P_n - P_0)D(d_n, \tau_0, P_n^*) + R_0(d_n, P_n^*) + o_{P_0}(n^{-1/2}) \qquad \text{(by Condition 23.4)}$$
$$= (P_n - P_0)D_0 + (P_n - P_0)(D(d_n, \tau_0, P_n^*) - D_0) + R_0(d_n, P_n^*).$$

The middle term on the last line is $o_{P_0}(n^{-1/2})$ by 23.4, 23.4, 23.4, and 23.4 (van der Vaart and Wellner 1996), and the third term is $o_{P_0}(n^{-1/2})$ by 23.4. This yields the asymptotic linearity result. Proposition 1 in Section 3.3 of Bickel et al. (1997b) yields the claim about regularity and asymptotic efficiency when conditions 23.3, 23.3, 23.3, and 23.4 hold (see Theorem 23.3).

Proof (Lemma 23.1). We will show that $\eta_n \to \eta_0$ in probability, and then the consistency of τ_n follows by the continuous mapping theorem. By 23.3, there exists an open interval N containing η_0 on which S_0 is continuous. Fix $\eta \in N$. Because $\bar{Q}_{b,n}$ belongs to a Glivenko-Cantelli class with probability approaching 1, we have that

$$|S_n(\eta) - S_0(\eta)| = \left| P_n I(\bar{Q}_{b,n} > \eta) - P_0 I(\bar{Q}_{b,0} > \eta) \right|$$
$$\leq \left| P_0 \left(I(\bar{Q}_{b,n} > \eta) - I(\bar{Q}_{b,0} > \eta) \right) \right| + \left| (P_n - P_0)I(\bar{Q}_{b,n} > \eta) \right|$$
$$\leq \underbrace{\left| P_0 \left(I(\bar{Q}_{b,n} > \eta) - I(\bar{Q}_{b,0} > \eta) \right) \right|}_{\triangleq T_n(\eta)} + o_{P_0}(1), \qquad (23.12)$$

where we use the notation $Pf = E_P[f(O)]$ for any distribution P and function f. Let $Z_n(\eta)(w) \triangleq \left(I(\bar{Q}_{b,n}(w) > \eta) - I(\bar{Q}_{b,0}(w) > \eta) \right)^2$. The following display holds for all $q > 0$:

$$T_n(\eta) \leq P_0 Z_n(\eta)$$
$$= P_0 Z_n(\eta) I(|\bar{Q}_{b,0} - \eta| > q) + P_0 Z_n(\eta) I(|\bar{Q}_{b,0} - \eta| \leq q)$$
$$= P_0 Z_n(\eta) I(|\bar{Q}_{b,0} - \eta| > q) + P_0 Z_n(\eta) I(0 < |\bar{Q}_{b,0} - \eta| \leq q) \qquad (23.13)$$
$$\leq P_0 Z_n(\eta) I(|\bar{Q}_{b,n} - \bar{Q}_{b,0}| > q) + P_0 Z_n(\eta) I(0 < |\bar{Q}_{b,0} - \eta| \leq q) \qquad (23.14)$$
$$\leq P_0 \left(|\bar{Q}_{b,n} - \bar{Q}_{b,0}| > q \right) + P_0 \left(0 < |\bar{Q}_{b,0} - \eta| \leq q \right)$$
$$\leq \frac{P_0 |\bar{Q}_{b,n} - \bar{Q}_{b,0}|}{q} + P_0 \left(0 < |\bar{Q}_{b,0} - \eta| \leq q \right).$$

Above (23.13) holds because 23.3 implies that $P_0(\bar{Q}_{b,0} = \eta) = 0$, (23.14) holds because $Z_n(\eta) = 1$ implies that $|\bar{Q}_{b,n} - \bar{Q}_{b,0}| \geq |\bar{Q}_{b,0} - \eta|$, and the final inequality holds by Markov's inequality. The lemma assumes that $E_0|\bar{Q}_{b,n} - \bar{Q}_{b,0}| = o_{P_0}(1)$, and thus we can choose a sequence $q_n \downarrow 0$ such that

$$T_n(\eta) \leq P_0\left(0 < |\bar{Q}_{b,0} - \eta| \leq q_n\right) + o_{P_0}(1).$$

To see that the first term on the right is $o(1)$, note that $P_0(\bar{Q}_{b,0} = \eta) = 0$ combined with the continuity of S_0 on N yield that, for n large enough,

$$P_0\left(0 < |\bar{Q}_{b,0} - \eta| \leq q_n\right) = S_0(-q_n + \eta) - S_0(q_n + \eta).$$

The right-hand side is $o(1)$, and thus $T_n(\eta) = o_{P_0}(1)$. Plugging this into (23.12) shows that $S_n(\eta) \to S_0(\eta)$ in probability. Recall that $\eta \in N$ was arbitrary.

Fix $\gamma > 0$. For γ small enough, $\eta_0 - \gamma$ and $\eta_0 + \gamma$ are contained in N. Thus $S_n(\eta_0 - \gamma) \to S_0(\eta_0 - \gamma)$ and $S_n(\eta_0 + \gamma) \to S_0(\eta_0 + \gamma)$ in probability. Further, $S_0(\eta_0 - \gamma) > \kappa$ by the definition of η_0 and $S_0(\eta_0 + \gamma) < \kappa$ by Condition 23.3. It follows that, with probability approaching 1, $S_n(\eta_0 - \gamma) > \kappa$ and $S_n(\eta_0 + \gamma) < \kappa$. But $|\eta_n - \eta_0| > \gamma$ implies that $S_n(\eta_0 - \gamma) \leq \kappa$ or $S_n(\eta_0 + \gamma) > \kappa$, and thus $|\eta_n - \eta_0| \leq \gamma$ with probability approaching 1. Thus $\eta_n \to \eta_0$ in probability, and $\tau_n \to \tau_0$ by the continuous mapping theorem.

Chapter 24
Targeting a Simple Statistical Bandit Problem

Antoine Chambaz, Wenjing Zheng, and Mark J. van der Laan

Statistical Challenge. An infinite sequence of independent and identically distributed (i.i.d.) random variables $(W_n, Y_n(0), Y_n(1))_{n \geq 1}$ drawn from a common law Q_0 is to be sequentially and partially disclosed during the course of a controlled experiment. The first component, W_n, describes the nth context in which we will have to carry out one action out of two, denoted $a = 0$ and $a = 1$. The second and third components, $Y_n(0)$ and $Y_n(1)$, are the rewards that actions $a = 0$ and $a = 1$ would grant. The set $\mathcal{W}$ of contexts may be high-dimensional. The rewards take their values in $]0, 1[$.

The controlled experiment will unfold as follows. Sequentially, we will be informed of the new context W_n. We will then carry out a randomized action $A_n \in \{0, 1\}$ with probability either $g_n(1|W_n)$ or $g_n(0|W_n) \equiv 1 - g_n(1|W_n)$ to go for either action $a = 1$ or action $a = 0$, where $g_n(\cdot|W_n)$ will be determined by us based on observations $O_1, \ldots, O_{n-1}$ accrued so far during the course of the experiment. We will then be granted reward $Y_n \equiv Y_n(A_n)$ corresponding to the action undertaken, the alternative

A. Chambaz (✉)
MAP5 (UMR CNRS 8145), Université Paris Descartes, 45 rue des Saints-Pères, 75270
Paris cedex 06, France
e-mail: antoine.chambaz@parisdescartes.fr

W. Zheng
Netflix, 100 Winchester Circle, Los Gatos, CA 95032, USA
e-mail: wzheng@netflix.com

M. J. van der Laan
Division of Biostatistics and Department of Statistics, University of California, Berkeley,
101 Haviland Hall, #7358, Berkeley, CA 94720, USA
e-mail: laan@berkeley.edu

© Springer International Publishing AG 2018
M.J. van der Laan, S. Rose, *Targeted Learning in Data Science*,
Springer Series in Statistics, https://doi.org/10.1007/978-3-319-65304-4_24

reward being kept undisclosed, hence the nth observation $O_n \equiv (W_n, A_n, Y_n)$. This setting is one of the simplest bandits settings in the machine learning literature, hence the expression "simple bandit problem" in the title of this chapter.

Our objective justifies why the expression actually reads "simple *statistical* bandit problem". Indeed, it consists in inferring the optimal rule

$$r_0(W) \equiv \arg\max_{a=0,1} E_{Q_0}(Y(a)|W)$$

(by convention, $r_0(W) = 1$ if equality occurs) with r_n and the mean reward under r_0,

$$\psi_0 \equiv E_{Q_0}(Y(r_0(W))),$$

trying to get a narrow confidence interval (CI) for ψ_0 and a sense of how well we sequentially determined our actions through the estimation of the following regret:

$$\mathcal{R}_n \equiv \frac{1}{n} \sum_{i=1}^{n} Y_i(r_n(W_i)) - Y_i.$$

Regret is one the most central notion in the bandits literature. Viewed here as a data-adaptive parameter, $\mathcal{R}_n$ compares the actual average of the rewards granted at step n, $n^{-1} \sum_{i=1}^{n} Y_i$, with the counterfactual average of the rewards we would have been be granted at step n if we had constantly used r_n from the start of the experiment to decide which action to carry out at the n successive steps, $n^{-1} \sum_{i=1}^{n} Y_i(r_n(W_i))$. We emphasize that the former average is known to us but the latter is not. It may occur indeed that $A_i \neq r_n(W_i)$ for some $1 \leq i \leq n$, in which case $Y_i(r_n(W_i))$ is the reward that was kept secret from us at step i. If all actions A_i coincide with $r_n(W_i)$ ($1 \leq i \leq n$), a very unlikely event, then $\mathcal{R}_n = 0$. In general, $n\mathcal{R}_n$ equals

$$\sum_{\substack{1 \leq i \leq n \\ A_i \neq r_n(W_i)}} Y_i(1 - A_i) - Y_i(A_i).$$

This alternative expression shows that $n\mathcal{R}_n$ is the counterfactual sum of the differences between the two possible rewards at each step i where the randomized action A_i differs from the optimal action $r_n(W_i)$ according to the estimate of the optimal rule at step n. Since the optimal action is that which has the larger conditional mean given the context, as opposed to that action which grants the larger reward, it is not guaranteed that $\mathcal{R}_n$ is nonnegative.

Inference of data-adaptive parameters are at the core of the present chapter. We will derive CIs for ψ_0 and for $\mathcal{R}_n$, the first data-adaptive parameter we introduced, from a targeted minimum loss estimator (TMLE, which also stands for targeted minimum loss estimation) of the second data-adaptive parameter

$$\psi_{r_n,0} \equiv E_{Q_0}(Y(r_n(W))),$$

the mean reward under r_n, thus justifying entirely the title of the chapter. There is much more to $\psi_{r_n,0}$ than being a convenient proxy for the inference of ψ_0. In fact, we may argue that $\psi_{r_n,0}$ is more interesting than ψ_0 itself because it is the mean reward under rule r_n that we know and can use concretely. The same reasoning motivates our choice of regret $\mathcal{R}_n$ instead of its counterpart with r_0 substituted for r_n.

Precision Medicine. The general story told so far can be cast in the context of precision medicine. Precision medicine is the burgeoning field whose general focus is on identifying which treatments and preventions will be effective for which patients based on genetic, environmental, and lifestyle factors. To do so, imagine that each random variable $(W_n, Y_n(0), Y_n(1))$ corresponds with a patient: W_n is a set of baseline covariates measured on her; $Y_n(0)$ is the potential outcome of treatment $a = 0$ and $Y_n(1)$ is the potential outcome of treatment $a = 1$. A rule defines an individualized treatment strategy in which treatment assignment for a patient is based on her measured baseline covariates. Only one potential outcome is observed, Y_n, which corresponds with the assigned treatment A_n. Rule r_0 is optimal in the sense that it maximizes the mean value of the outcome of the assigned treatment.

In this context, the targeted sequential elaboration of the design and the inference procedure developed on top of it are driven by two goals. The first one is to increase the robustness and efficiency of statistical inference through the construction of targeted, narrower confidence intervals. This appeals to the investigators of the study. The second goal is to increase the likelihood that each patient enrolled in the study be assigned that treatment which is more favorable to her according to data accrued so far. This appeals to the patients enrolled in it and their doctors. To understand why, let us contrast a trial based on our design with a traditional RCT. In a RCT, clinicians must admit to each potential patient that it is not known which of the treatments would be best for her, thereby potentially eroding their relationship. In addition, they should believe that the treatments are equivalent with respect to potential patient benefit, a situation many of them find uncomfortable (Stanley 2007). These two disadvantages would be respectively considerably diminished and irrelevant in a trial based on our design, at the cost of a more complex implementation. Moreover, one may expect a gain in compliance.

Quick Review of Literature. Chakraborty and Moodie (2013) present an excellent unified overview on the estimation of optimal rules. Their focus is on dynamic rules, which actually prescribe successive actions at successive time points based on time-dependent contexts. The estimation of the optimal rule from i.i.d. observations has been studied extensively, with a recent interest in the use of machine learning algorithms to reach this goal (Qian and Murphy 2011; Zhao et al. 2012, 2015; Zhang et al. 2012a,b; Rubin and van der Laan 2012; Luedtke and van der Laan 2016b). The estimation of the mean reward under the optimal rule is more challenging. Zhao et al. (2012, 2015) use their theoretical risk bounds evaluating the statistical performance of the estimator of the optimal rule as measures of statistical performance of the resulting estimators of the mean reward under the optimal rule. However, this approach does not yield CIs.

Constructing CIs for the mean reward under the optimal rule is known to be more difficult when there exists a stratum of context where no action dominates the other (if action means treatment, no treatment is neither beneficial nor harmful) (Robins 2004). In this so called "exceptional" case, the definition of the optimal rule has to be disambiguated. Assuming nonexceptionality, Zhang et al. (2012a) derive CIs for the mean reward under the (sub)optimal rule defined as the optimal rule over a parametric class of candidate rules. van der Laan and Luedtke (2015) derive CIs for the actual mean reward under the optimal rule. In the more general case where exceptionality can occur, different approaches have been considered (Chakraborty et al. 2014; Goldberg et al. 2014; Laber et al. 2014b; Luedtke and van der Laan 2016a). Here, we focus on the nonexceptional case under a companion margin assumption (Mammen and Tsybakov 1999).

We already unveiled that our pivotal TMLE is actually conceived as an estimator of the mean reward under the current estimate of the optimal rule. Worthy of interest on its own, this data-adaptive statistical parameter (or similar ones) has also been considered in Chakraborty et al. (2014); Laber et al. (2014a,b); van der Laan and Luedtke (2015); Luedtke and van der Laan (2016a). Our main result is a central limit theorem (CLT), which enables the construction of various CIs. The analysis (for the proofs that we omit here, see the full-blown Chambaz et al. 2016) builds upon previous studies on the construction and statistical analysis of targeted, covariate-adjusted, response-adaptive trials also based on TMLE (Chambaz and van der Laan 2014; Zheng et al. 2015; Chambaz et al. 2015). The asymptotic variance in the CLT takes the form of the variance of an efficient influence curve at a limiting distribution. This allows to discuss the efficiency of inference. One of the cornerstones of the theoretical study is a new maximal inequality for martingales wrt the uniform entropy integral. Proved by decoupling (de la Peña and Giné 1999), symmetrization and chaining, it is used to control several empirical processes indexed by random functions.

24.1 Sampling Strategy and TMLE

Let us introduce some notation. We let $\bar{Q}_{0,Y}$ and $\bar{q}_{0,Y}$ respectively denote the true conditional expectation $\bar{Q}_{0,Y}(a, W) \equiv E_{Q_0}(Y(a)|W)$ (for $a = 0, 1$) and related "blip function" $\bar{q}_{0,Y}(W) \equiv \bar{Q}_{0,Y}(1, W) - \bar{Q}_{0,Y}(0, W)$. More generally, every (measurable) function $\bar{Q}_Y$ from $\{0, 1\} \times \mathcal{W}$ to $]0, 1[$ is associated with its blip function $\bar{q}_Y(W) \equiv \bar{Q}_Y(1, W) - \bar{Q}_Y(0, W)$. Thus,

$$r_0(W) = \arg\max_{a=0,1} \bar{Q}_{0,Y}(a, W) = I\{\bar{q}_{0,Y}(W) \geq 0\} \equiv R(\bar{Q}_{0,Y})(W) \qquad (24.1)$$

(recall that $r_0(W) = 1$ if equality occurs), ψ_0 equals $E_{Q_0}(\bar{Q}_{0,Y}(r_0(W), W))$ and $\psi_{r_n,0}$ equals $E_{Q_0}(\bar{Q}_{0,Y}(r_n(W), W))$.

The adaptive sampling strategy and TMLE rely on a working model $\bar{Q}_Y$ and loss function L_Y for $\bar{Q}_{0,Y}$ that we determine prior to starting the controlled experiment. Requirements on the complexity of $\bar{Q}_Y$ will be given in Sect. 24.2. They also rely on a nondecreasing, Lipschitz function G from $[-1, 1]$ to $[0, 1]$ such that $G(0) = 1/2$ and, for some fixed and small real numbers $p, \xi > 0$, $|x| > \xi$ implies $G(x) = p$ if $x < 0$ and $G(x) = (1 - p)$ if $x > 0$. Function G is a smooth approximation to the mapping $x \mapsto I\{x \geq 0\}$ from $[-1, 1]$ to $\{0, 1\}$. It will be used to define an approximation g_{n+1} to r_n, both derived from an estimator $\bar{q}_{n,Y}$ of the blip function $\bar{q}_{0,Y}$, see (24.3).

24.1.1 Sampling Strategy

The first n_0 randomized actions $A_1, \ldots, A_{n_0}$ are drawn from the Bernoulli distribution with parameter $1/2$. In other words, we set $g_i = g^b$ for $i = 1, \ldots, n_0$ where $g^b(1|W) = 1 - g^b(0|W) \equiv 1/2$, thus giving equiprobable chance to each action to be carried out as long as deemed necessary to start estimating $\bar{Q}_{0,Y}$ from the accrued observations. Suppose now that $O_1, \ldots, O_{n-1}$ have been observed. Explaining how the next observation is obtained will complete the description of the sampling strategy.

We estimate $\bar{Q}_{0,Y}$ with

$$\bar{Q}_{n,Y} \in \arg\min_{\bar{Q}_Y \in \bar{\mathcal{Q}}_Y} \frac{1}{n-1} \sum_{i=1}^{n-1} L_Y(\bar{Q}_Y)(O_i) \frac{g^b(A_i|W_i)}{g_i(A_i|W_i)}. \tag{24.2}$$

The weights $g^b(A_i|W_i)/g_i(A_i|W_i)$ $(i = 1, \ldots, n)$ compensate for the fact that our observations are not identically distributed. We associate the above estimator with its blip function $\bar{q}_{n,Y}$ and rule $r_n \equiv R(\bar{Q}_{n,Y})(W) \equiv I\{\bar{q}_{n,Y}(W) \geq 0\}$. They are substitution estimators of $\bar{q}_{0,Y}$ and r_0. We now define

$$g_{n+1}(1|W) = 1 - g_{n+1}(0|W) \equiv G(\bar{q}_{n,Y})(W), \tag{24.3}$$

and thus are in a position to sample O_{n+1}: we request the disclosure of W_{n+1}, draw A_{n+1} from the Bernoulli distribution with parameter $g_{n+1}(1|W_{n+1})$, carry out action A_{n+1}, are granted reward $Y_{n+1} = Y_{n+1}(A_{n+1})$ and form $O_{n+1} \equiv (W_{n+1}, A_{n+1}, Y_{n+1})$.

The randomized action A_{n+1} rarely differs from the deterministic action $r_n(W_{n+1})$ in the sense that

$$|g_{n+1}(1|W_{n+1}) - r_n(W_{n+1})| I\{|\bar{q}_{n,Y}(W_{n+1})| > \xi\} = p : \tag{24.4}$$

if $|\bar{q}_{n,Y}(W_{n+1})|$ is sufficiently away from 0, meaning that we confidently believe that one action is superior to the other, then A_{n+1} equals $r_n(W_{n+1})$ with (large) probability $(1 - p)$. On the contrary, if $|\bar{q}_{n,Y}(W_{n+1})|$ is small, meaning that it is unclear whether an action is superior to the other or not, then the probability that A_{n+1} be equal $r_n(W_{n+1})$ lies between $(1 - t)$ and $1/2$, and is continuously closer to $1/2$ as $|\bar{q}_{n,Y}(W_{n+1})|$ gets closer to 0.

24.1.2 TMLE

The initial substitution estimator of $\psi_{r_n,0}$,

$$\psi_n^0 \equiv \frac{1}{n} \sum_{i=1}^{n} \bar{Q}_{n,Y}(r_n(W_i), W_i),$$

may fail to be $\sqrt{n}$-consistent and must therefore be enhanced. Fortunately, we can rely on TMLE. Indeed, just like any mapping $\Psi_\rho : P_{Q,g} \mapsto E_Q(Y(\rho(W)))$ with a fixed rule ρ from $\mathcal{W}$ to $\{0, 1\}$, the data-adaptive Ψ_{r_n} is pathwise differentiable from the nonparametric set of all possible data-generating distributions $P_{Q,g}$ of $O \equiv (W, A, Y)$ with g bounded away from 0 to $[0, 1]$ (van der Laan and Luedtke 2015; Luedtke and van der Laan 2016a). Its efficient influence curve at $P_{Q,g}$ is $\Delta_{r_n}(Q, g)$ where, for every rule $\rho : \mathcal{W} \to \{0, 1\}$, $\Delta_\rho(Q, g)$ is characterized by

$$\Delta_\rho(Q, g)(O) = (Y - \bar{Q}_Y(\rho(W), W)) \frac{I\{A = \rho(W)\}}{g(A|W)} + \bar{Q}_Y(\rho(W), W) - \Psi_\rho(P_{Q,g}). \quad (24.5)$$

We let ℓ denote the quasi negative-log-likelihood loss function, which is characterized by

$$-\ell(\bar{Q}_Y)(O) \equiv Y \log(\bar{Q}_Y(A, W)) + (1 - Y) \log(1 - \bar{Q}_Y(A, W)),$$

and introduce the one-dimensional regression model through $\bar{Q}_{n,Y}$ given by

$$\text{logit}\left(\bar{Q}_{n,Y}(\epsilon)(A, W)\right) \equiv \text{logit}\left(\bar{Q}_{n,Y}(A, W)\right) + \epsilon \frac{I\{A = r_n(W)\}}{g_n(A|W)}$$

for all $\epsilon \in \mathbb{R}$. It is tailored to the estimation of $\psi_{r_n,0} = \Psi_{r_n}(P_{Q_0,g_n})$ in the sense that $\frac{\partial}{\partial \epsilon} \ell(\bar{Q}_{n,Y}(\epsilon))(O)|_{\epsilon=0}$ equals $-(Y - \bar{Q}_{n,Y}(A, W))I\{A = r_n(W)\}/g_n(A|W)$. In words, its generalized score equals the component of $\Delta_{r_n}(Q_n, g_n)$ which is orthogonal to the set of P_{Q_n,g_n}-square-integrable and centered functions of W. Here, Q_n denotes any distribution of $(W, Y(0), Y(1))$ such that $E_{Q_n}(Y(a)|W) = \bar{Q}_{n,Y}(a, W)$ for each $a = 0, 1$, Q_n-almost surely.

The optimal fluctuation parameter is

$$\epsilon_n \in \arg\min_{\epsilon \in \mathbb{R}} \frac{1}{n} \sum_{i=1}^{n} \ell(\bar{Q}_{n,Y}(\epsilon))(O_i) \frac{g_n(A_i|W_i)}{g_i(A_i|W_i)}.$$

Setting $\bar{Q}_{n,Y}^* \equiv \bar{Q}_{n,Y}(\epsilon_n)$, the TMLE of $\psi_{r_n,0}$ finally writes

$$\psi_n^* \equiv \frac{1}{n} \sum_{i=1}^{n} \bar{Q}_{n,Y}^*(r_n(W_i), W_i).$$

24.2 Convergence of Sampling Strategy and Asymptotic Normality of TMLE

We must choose the working model $\bar{Q}_Y$ and loss function L_Y for $\bar{Q}_{0,Y}$ in such a way that $\bar{Q}_Y$ and the subsequent working models $L_Y(\bar{Q}_Y) \equiv \{L(\bar{Q}_Y) : Q_Y \in \bar{Q}_Y\}$ and $R(\bar{Q}_Y) \equiv \{R(\bar{Q}_Y) : Q_Y \in \bar{Q}_Y\}$ be reasonably large/complex relative to a measure of complexity central to the theory of empirical processes (van der Vaart and Wellner 1996). Specifically, we must choose them so that $\bar{Q}_Y$, $L_Y(\bar{Q}_Y)$, $R(\bar{Q}_Y)$ be separable (countable would be sufficient) and that each admit a finite uniform entropy integral with respect to an envelope function (van der Vaart and Wellner 1996, Sections 2.5.1 and 2.6).

Introduce the norm $\| \cdot \|_{Q_0}$ characterized by $\|f\|_{Q_0}^2 \equiv E_{P_{Q_0,g^b}}(f^2(O))$. We will assume that $\bar{Q}_Y$ satisfies the following assumption:

A1. There exists $\bar{Q}_{1,Y} \in \bar{Q}_Y$ such that $\bar{Q}_Y \mapsto E_{P_{Q_0,g^b}}(L_Y(\bar{Q}_Y)(O))$ from $\bar{Q}_Y$ to $\mathbb{R}$ is minimized at $\bar{Q}_{1,Y}$. Moreover, $\bar{Q}_{1,Y}$ is well-separated in the sense that, for all $\delta > 0$,

$$E_{P_{Q_0,g^b}}\left(L_Y(\bar{Q}_{1,Y})(O)\right) < \inf\left\{E_{P_{Q_0,g^b}}\left(L_Y(\bar{Q}_Y)(O)\right) : \bar{Q}_Y \in \bar{Q}_Y, \|\bar{Q}_Y - \bar{Q}_{1,Y}\|_{Q_0} \geq \delta\right\}.$$

Finally, $\bar{q}_{1,Y} = \bar{q}_{0,Y}$.

The most stringent condition is the equality of the blip functions.

Our second assumption concerns the fluctuation/targeting step in the construction of the TMLE. Let g_0 be given by

$$g_0(1|W) = 1 - g_0(0|W) \equiv G(\bar{q}_{0,Y}(W)). \tag{24.6}$$

Just like g_n is an approximation to r_n, see (24.3), (24.4) and its comment, g_0 is an approximation to the optimal rule r_0. We will soon see that g_0 is the limit of g_n. For every rule $\rho : W \to \{0, 1\}$, consider the one-dimensional regression model through $\bar{Q}_{1,Y}$ characterized by

$$\text{logit}\left(\bar{Q}_{1,Y,\rho}(\epsilon)(A, W)\right) \equiv \text{logit}\left(\bar{Q}_{1,Y}(A, W) + \epsilon \frac{I\{A = \rho(W)\}}{g_0(A|W)}\right) \tag{24.7}$$

for all $\epsilon \in \mathbb{R}$. We will assume that:

A2. For every rule $\rho : W \to \{0, 1\}$, there exists a unique $\epsilon_0(\rho) \in \mathbb{R}$ which minimizes the real-valued mapping $\epsilon \mapsto E_{P_{Q_0,g_0}}\left(\ell(\bar{Q}_{1,Y,\rho}(\epsilon))(O)\right)$ over $\mathbb{R}$.

The third and last assumption concerns Q_0:

A3. The conditional distributions of $Y(0)$ and $Y(1)$ given W under Q_0 is not degenerated. Moreover, there exist $\gamma_1, \gamma_2 > 0$ such that, for all $t \geq 0$,

$$P_{Q_0}\left(0 \leq |\bar{q}_{0,Y}(W)| \leq t\right) \leq \gamma_1 t^{\gamma_2}. \tag{24.8}$$

Taking $t = 0$ in (24.8) yields $\bar{q}_{0,Y}(W) = 0$ with probability zero under Q_0. In words, the optimal action $r_0(W)$ is defined without ambiguity Q_0-almost surely. In the terminology of (Robins 2004), Q_0 is nonexceptional. More generally, (24.8) for $t > 0$ is known as a margin assumption. Inspired from the seminal article (Mammen and Tsybakov 1999), **A3** formalizes a tractable concentration of $\bar{q}_{0,Y}(W)$ around 0, where our inference task is the most challenging.

We may now state our results. According to the first proposition, the sampling strategy nicely converges as n tends to infinity:

Proposition 24.1. *Under* **A1**, **A2** *and* **A3**, *it holds that* $\|\bar{Q}_{n,Y} - \bar{Q}_{1,Y}\|_{Q_0}$, $\|\bar{q}_{n,Y} - \bar{q}_{0,Y}\|_{Q_0}$, $\|r_n - r_0\|_{Q_0}$, $\|g_n - g_0\|_{Q_0}$ *and the nonnegative data-adaptive parameter* $\psi_0 - \psi_{r_n,0}$ *all converge in probability to zero as* n *tends to infinity.*

The second proposition establishes the asymptotic normality of $\sqrt{n}(\psi_n^* - \psi_{r_n,0})$. Let us introduce $\bar{Q}_{1,Y}^* \equiv \bar{Q}_{1,Y,r_0}(\epsilon_0(r_0))$ (see (24.7) and **A2**), D_1^* given by

$$D_1^*(O) \equiv (Y - \bar{Q}_{1,Y}^*(A, W))\frac{I\{A = r_0(W)\}}{g_0(A|W)} + \bar{Q}_{1,Y}^*(r_0(W), W) - \psi_0, \qquad (24.9)$$

and $\sigma_1^2 \equiv E_{P_{Q_0,g_0}}\left(D_1^*(O)^2\right)$. Analogously, recalling the definition of $\bar{Q}_{n,Y}^* \equiv \bar{Q}_{n,Y}(\epsilon_n)$, let us define

$$D_{ni}^*(O_i) \equiv (Y_i - \bar{Q}_{n,Y}^*(A_i, W_i))\frac{I\{A_i = r_n(W_i)\}}{g_i(A_i|W_i)} + \bar{Q}_{n,Y}^*(r_n(W_i), W_i) - \psi_n^* \quad (\text{each } 1 \leq i \leq n)$$

then $\sigma_n^2 \equiv n^{-1} \sum_{i=1}^n D_{ni}^*(O_i)^2$.

Proposition 24.2. *Under* **A1**, **A2** *and* **A3**, ψ_n^* *consistently estimates* $\psi_{r_n,0}$ *hence* ψ_0 *as well by Proposition 24.1. Moreover,* σ_n^2 *consistently estimates* σ_1^2, *which is positive, and* $\sqrt{n/\sigma_n^2}(\psi_n^* - \psi_{r_n,0})$ *converges in law to the standard normal distribution as* n *tends to infinity.*

Obviously, the larger is γ_2 from **A3**, the less concentrated is $\bar{q}_{0,Y}(W)$ around zero under Q_0, the less difficult is our inference task. If we assume that $\gamma_2 \geq 1$ and that the rate of convergence of $\bar{q}_{n,Y}$ to $\bar{q}_{0,Y}$ is sufficiently fast, then a first corollary to Proposition 24.2 shows that $\sqrt{n}(\psi_n^* - \psi_0)$ is also asymptotically normal. Introduce $\gamma_3 \equiv \frac{1}{4} + \frac{1}{2(1+\gamma_2)}$.

Corollary 24.1. *Under* **A1**, **A2** *and* **A3**, *if* $\gamma_2 \geq 1$ *hence* $\gamma_3 \in (\frac{1}{4}, \frac{1}{2}]$ *and if* $n^{\gamma_3}\|\bar{q}_{n,Y} - \bar{q}_{0,Y}\|_{Q_0}$ *converges in probability to zero as* n *tends to infinity, then the data-adaptive parameter* $\sqrt{n}(\psi_{r_n,0} - \psi_0)$ *converges in probability to zero as* n *tends to infinity. Therefore,* $\sqrt{n/\sigma_n^2}(\psi_n^* - \psi_0)$ *converges in law to the standard normal distribution as* n *tends to infinity.*

The proofs of Propositions 24.1, 24.2 and Corollary 24.1 rely on arguments typical of empirical processes theory and the analysis of TMLEs (Chambaz et al. 2016). The underlying martingale structure of the empirical process proves again a nice extension to an i.i.d. structure.

Let Q_1^* be any distribution of $(W, Y(0), Y(1))$ such that W has the same distribution under Q_0 and Q_1^* and $E_{Q_1^*}(Y(a)|W) = \bar{Q}_{1,Y}^*(a, W)$ for each $a = 0, 1$, Q_0-almost surely. The influence function D_1^* in (24.9) equals $\Delta_{r_0}(Q_1^*, g_0)$, the efficient influence curve of Ψ_{r_0} at $P_{Q_1^*, g_0}$ (24.5). Consequently, $\sigma_1^2 = E_{P_{Q_0, g_0}} \left(\Delta_{r_0}(Q_1^*, g_0)(O)^2 \right)$. If $\bar{Q}_{1,Y} = \bar{Q}_{0,Y}$ (a stronger condition than equality $\bar{q}_{1,Y} = \bar{q}_{0,Y}$ in **A1**), then $\bar{Q}_{1,Y}^* = \bar{Q}_{0,Y}$ (because $\epsilon_0(r_0)$ from **A2** equals zero) hence $\sigma_1^2 = E_{P_{Q_0, g_0}} \left(\Delta_{r_0}(Q_0, g_0)(O)^2 \right)$: the asymptotic variance of $\sqrt{n}(\psi_n^* - \psi_{r_n,0})$ coincides with the generalized Cramér-Rao lower bound for the asymptotic variance of any regular and asymptotically linear estimator of $\Psi_{r_0}(P_{Q_0, g_0}) = \psi_0$ when sampling independently from P_{Q_0, g_0} (Luedtke and van der Laan 2016a). Otherwise, the discrepancy between σ_1^2 and $E_{P_{Q_0, g_0}} \left(\Delta_{r_0}(Q_0, g_0)(O)^2 \right)$ will vary subtly depending on that between $\bar{Q}_{1,Y}$ and $\bar{Q}_{0,Y}$, hence in particular on our working model $\bar{Q}_Y$.

24.3 Confidence Intervals

Set a confidence level $\alpha \in]0, 1/2[$ and let $\xi_{1-\alpha/2}$ be the corresponding $(1 - \alpha/2)$-quantile of the standard normal distribution. By Proposition 24.2 and Corollary 24.1, the TMLE can be used to construct CIs for the data-adaptive parameter $\psi_{r_n,0}$ or ψ_0 itself, as stated in this second corollary to Proposition 24.2:

Corollary 24.2. *Under the assumptions of Proposition 24.2,*

$$\left[\psi_n^* \pm \xi_{1-\alpha/2} \frac{\sigma_n}{\sqrt{n}} \right] \qquad (24.10)$$

contains $\psi_{r_n,0}$ with probability tending to $(1 - \alpha)$ as n tends to infinity. Moreover, under the stronger assumptions of Corollary 24.1, the above CI also contains ψ_0 with probability tending to $(1 - \alpha)$ as n tends to infinity.

Deriving a CI for $\mathcal{R}_n$ is not as immediate because of its counterfactual nature. We need to introduce a new assumption:

A4. There exist an infinite sequence $(U_n)_{n \geq 1}$ of i.i.d. random variables independent from $(W_n)_{n \geq 1}$ and taking values in $\mathcal{U}$ and a deterministic (measurable) function $\bar{\mathbb{Q}}_{0,Y}$ mapping $\{0, 1\} \times \mathcal{U} \times \mathcal{W}$ to $]0, 1[$ such that $Y_n(a) = \bar{\mathbb{Q}}_{0,Y}(a, U_n, W_n)$ for all $n \geq 1$ and both $a = 0, 1$.

With **A4**, we frame the present discussion in the context of nonparametric structural equations models (Pearl 2009a). The notation $\bar{\mathbb{Q}}_{0,Y}$ is justified by the equalities

$$\bar{Q}_{0,Y}(a, W_n) = E_{Q_0}(Y_n(a)|W_n) = E_{Q_0}(\bar{\mathbb{Q}}_{0,Y}(a, U_n, W_n)|W_n)$$

showing that, for each $n \geq 1$ and $a = 0, 1$, the conditional mean of $Y_n(a)$ given W_n is obtained by averaging out U_n from $\bar{\mathbb{Q}}_{0,Y}(a, U_n, W_n)$ conditionally on W_n.

Introduce

$$
s_1^2 \equiv E_{P_{Q_0,g_0}} \left(\left(D_1^*(O) + \psi_0 - \bar{Q}_{0,Y}(r_0(W),W) \right)^2 \right),
$$

$$
s_n^2 \equiv \frac{1}{n} \sum_{i=1}^n \left(D_{ni}^*(O_i) + \psi_n^0 - \bar{Q}_{n,Y}(r_n(W_i),W_i) \right)^2 .
$$

The latter is an empirical counterpart to and estimator of the former. We may now state the last result of this chapter, which exhibits a conservative CI for $\mathcal{R}_n$:

Proposition 24.3. *Under* **A1**, **A2**, **A3** *and* **A4**, s_n^2 *consistently estimates* s_1^2, *which is positive. Moreover,*

$$
\left[\psi_n^* - \frac{1}{n} \sum_{i=1}^n Y_i \pm \xi_{1-\alpha/2} \frac{s_n}{\sqrt{n}} \right] \tag{24.11}
$$

contains $\mathcal{R}_n$ *with probability converging to* $(1 - \alpha') \geq (1 - \alpha)$ *as n tends to infinity.*

The proof of Proposition 24.3 unfolds as follows. Pretending, contrary to facts, that U_n is also observed at each step though not used to define the TMLE, which is thus the same as before, we adapt the proof of Proposition 24.2 to obtain a similar CLT. The normalization factor involved now depends on $U_1, \ldots, U_n$ as well. We straightforwardly derive from it a CI for $\mathcal{R}_n$ whose width λ_n depends on $U_1, \ldots, U_n$ too. Fortunately, we can prove that the width of the CI in (24.11) is always larger than λ_n. Since it is free of $U_1, \ldots, U_n$, this yields the desired result. This clever scheme of proof draws its inspiration from Balzer et al. (2016c).

24.4 Simulation

Under Q_0, the baseline covariate W decomposes as $W \equiv (U, V) \in [0,1] \times \{1,2,3\}$, where U and V are independent random variables respectively drawn from the uniform distributions on $[0,1]$ and $\{1,2,3\}$. Moreover, $Y(0)$ and $Y(1)$ are conditionally drawn given W from Beta distributions with a constant variance set to 0.01 and means $\bar{Q}_{0,Y}(0,W)$ and $\bar{Q}_{0,Y}(1,W)$ satisfying $\bar{q}_{0,Y}(W) = \bar{Q}_{0,Y}(1,W) - \bar{Q}_{0,Y}(0,W) \equiv \frac{9}{8}\left(U^2 - \frac{5}{2}U + \frac{2}{3} \right) + \frac{3\sqrt{V}}{4\sqrt{3}}I\{U \geq \frac{1}{4}\lfloor\frac{V+3}{3}\rfloor\}$ and $\bar{Q}_{0,Y}(1,W) + \bar{Q}_{0,Y}(0,W) \equiv \frac{4}{5} + \frac{1}{3\sqrt{V}}\left(\cos\left(\frac{\pi}{2}\frac{4U}{V}\right)I\{4U \leq V\} + \sin\left(\frac{\pi}{2}\frac{4U-V}{4-V}\right)I\{4U > V\} - \frac{1}{2} \right)$.

The conditional means $\bar{Q}_{0,Y}(0,\cdot)$, $\bar{Q}_{0,Y}(1,\cdot)$ and associated blip function $\bar{q}_{0,Y}$ are represented in Fig. 24.2 (left plots). We compute the numerical values of the following parameters: $\psi_0 \approx 0.5570$ (mean reward under optimal rule r_0); $\mathrm{Var}_{P_{Q_0,g^b}} \Delta(Q_0, g^b)(O) \approx 0.1812^2$ (variance under P_{Q_0,g^b} of the efficient influence curve of Ψ at P_{Q_0,g^b}, i.e., under Q_0 with equiprobability of carrying out action $a = 1$ or $a = 0$); $\mathrm{Var}_{P_{Q_0,g_0}} \Delta(Q_0, g_0)(O) \approx 0.1548^2$ (variance under P_{Q_0,g_0} of the efficient influence curve of Ψ at P_{Q_0,g_0}, i.e., under Q_0 and the approximation g_0 to r_0); and $\mathrm{Var}_{P_{Q_0,r_0}} \Delta(Q_0, r_0)(O) \approx 0.1512^2$ (variance under P_{Q_0,r_0} of the efficient influence curve of Ψ at P_{Q_0,r_0}, i.e., under Q_0 and r_0).

We set $p = 10\%$, $\xi = 1\%$ and choose G characterized over $[-1, 1]$ by $G(x) \equiv pI\{x \leq -\xi\} + \left(-\frac{1/2-p}{2\xi^3}x^3 + \frac{1/2-p}{2\xi/3}x + \frac{1}{2}\right)I\{-\xi \leq x \leq \xi\} + (1-p)I\{x \geq \xi\}$. Reducing p to 5% did not change the results significantly (not shown). Working model $\bar{Q}_Y$ consists of functions $\bar{Q}_{Y\beta}$ mapping $\{0, 1\} \times \mathcal{W}$ to $[0, 1]$ such that, for each $a = 0, 1$ and $v \in \{1, 2, 3\}$, logit $\bar{Q}_{Y\beta}(a, (U, v))$ is a linear combination of $1, U, U^2, \ldots, U^5$ and $I\{\frac{j-1}{10} \leq U < \frac{j}{10}\}$ $(1 \leq j \leq 10)$. The resulting global parameter β belongs to $\mathbb{R}^{96}$. Neither $\bar{Q}_{0,Y}$ nor $\bar{q}_{0,Y}$ belongs to $\bar{Q}_Y$ or $\{\bar{q}_{Y\beta} : \bar{Q}_{Y\beta} \in \bar{Q}_Y\}$. However, expit$(\bar{q}_{Y,0})$ does belong to the latter working model.

The targeting steps were performed when sample size is a multiple of 25, at least 200 and no more than 1000, when the experiment is stopped. Working model $\bar{Q}_Y$ was fitted with respect to the negative log-likelihood loss function ℓ using penalized regression imposing (data-adaptive) upper bounds on the ℓ^1- and ℓ^2-norms of parameter β (via penalization), hence the search for a sparse optimal parameter. We repeated $N = 1000$ times, independently, the strategy described in Sect. 24.1. Each time a targeting step was performed, we constructed the CIs of Corollary 24.2 and Proposition 24.3, with a nominal coverage set to $(1 - \alpha) = 95\%$ for each of them. The simulation study was conducted in R (R Development Core Team 2016), using the package `tsml.cara.rct` (Chambaz 2016).

Results. Figures 24.1 and 24.2 illustrate a typical realization. Figure 24.2 represents $\bar{Q}_{0,Y}$, $\bar{q}_{0,Y}$ and their estimators $\bar{Q}_{n,Y}$, $\bar{q}_{n,Y}$ at final sample size $n = 1000$. The top plot of Fig. 24.1 shows the 95%-CI I_n in (24.10) at every sample size n where a CI is derived. By Corollary 24.2, the probability of the event "$\psi_{r_n,0} \in I_n$" is more likely to be close to 95% than the probability of the event "$\psi_0 \in I_n$" in the sense that the latter property requires that the rate of convergence of $\bar{q}_{n,Y}$ to $\bar{q}_{0,Y}$ be sufficiently fast. Nevertheless, we observe on this realization that each I_n contains both its corresponding data-adaptive parameter $\psi_{r_n,0}$ (cross) and ψ_0 (horizontal black line). Moreover, the difference between the length of I_n and that of the vertical segment joining the two curves of the same nuance of darker gray at sample size n gets smaller as n grows. This indicates that the variance of ψ_n^* gets closer to the optimal variance $\text{Var}_{P_{Q_0,r_0}} \Delta(Q_0, r_0)(O)$ as n grows. The bottom plot of Fig. 24.1 shows the actual value of $\mathcal{R}_n$ (cross) and 95%-CI in (24.11) at every sample size n where a CI is derived. We observe on this realization that the regrets are all positive, a fact that was not granted. Moreover, each CI contains its corresponding data-adaptive parameter $\mathcal{R}_n$.

We can evaluate if our 95%-CIs achieve their nominal 95%-coverage. To do so, we carry out binomials tests. By construction, the empirical number of CIs which cover $\psi_{r_n,0}$ is a random variable drawn from a Binomial distribution with parameters (N, π). We choose to test the null "$\pi \geq 95\%$" against its one-sided alternative "$\pi < 95\%$". A large p-value is interpreted as the absence of empirical evidence supporting that the CI does not achieve its nominal coverage. We do the same for ψ_0 and $\mathcal{R}_n$, *mutatis mutandis*.

Instead of reporting $3 \times 33 = 99$ empirical proportions of coverage and related p-values, we simply plot the logarithms of the p-values of the tests evaluating the coverage of $\psi_{r_n,0}$ and ψ_0, see Fig. 24.3. Overall, the gray curve dominates the black one, indicating that empirical coverage tends to be higher for $\psi_{r_n,0}$ (it ranges between

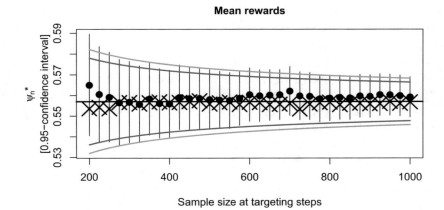

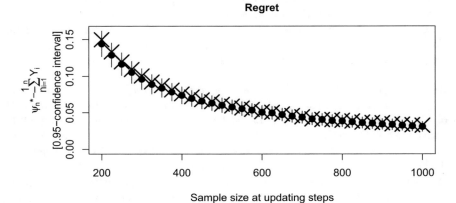

Fig. 24.1 Illustrating the data-adaptive inference of the optimal rule, its mean reward and the re-lated regret (see also Fig. 24.2). *Top plot.* The *black horizontal line* represents the value of the mean reward under the optimal rule, ψ_0. The *gray curves* represent the mapping $n \mapsto \psi_0 \pm \xi_{97.5\%}\sigma_k/\sqrt{n}$ ($k = 1, 2$), where $\sigma_1 \approx 0.1512$ is the square root of $\mathrm{Var}_{P_{Q_0,r_0}} \Delta(Q_0, r_0)(O)$ (*darker gray*) and $\sigma_2 \approx 0.1812$ is the square root of $\mathrm{Var}_{P_{Q_0,g^b}} \Delta(Q_0, g^b)(O)$ (*lighter gray*). Thus, at a given sample size n, the length of the vertical segment joining the two darker gray curves equals the length of a CI based on a regular, asymptotically efficient estimator of ψ_0. The *crosses* represent the successive values of the data-adaptive parameters $\psi_{r_n,0}$. The *black dots* represent the successive values of ψ_n^*, and the vertical segments centered at them represent the successive 95%-CIs for $\psi_{r_n,0}$ and, under additional assumptions, for ψ_0 as well. *Bottom plot.* The *crosses* represent the successive values of regret $\mathcal{R}_n$. The *black dots* represent the successive values of $\psi_n^* - n^{-1}\sum_{i=1}^n Y_i$, and the vertical segments represent the successive 95%-CIs for $\mathcal{R}_n$

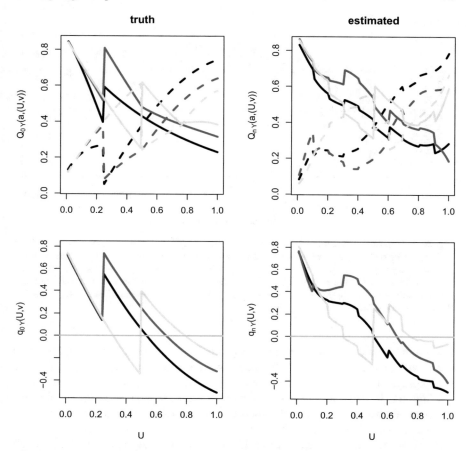

Fig. 24.2 Illustrating the data-adaptive inference of the optimal rule, its mean reward and the related regret through the representation of the conditional mean $\bar{Q}_{0,Y}$, blip function $\bar{q}_{0,Y}$ and their estimators (see also Fig. 24.1). *Top left plot*: The *solid curves* represent $U \mapsto \bar{Q}_{0,Y}(1,(U,v))$ for $v = 1$ (in *black*, lowest value in 1), $v = 2$ (in *dark gray*, middle value in 1) and $v = 3$ (in *light gray*, largest value in 1). The *dashed curves* represent $U \mapsto \bar{Q}_{0,Y}(0,(U,v))$ for $v = 1$ (in *black*, largest value in 1), $v = 2$ (in *dark gray*, middle value in 1) and $v = 3$ (in *light gray*, smallest value in 1). *Bottom left plot*: The *curves* represent $U \mapsto \bar{q}_{0,Y}(U,v)$ for $v = 1$ (in *black*, smallest value in 1), $v = 2$ (in *dark gray*, middle value in 1) and $v = 3$ (in *light gray*, largest value in 1). *Right plots*. Counterparts to the *left plots*, where $\bar{Q}_{0,Y}$ and $\bar{q}_{0,Y}$ are replaced with $\bar{Q}_{n,Y}$ and $\bar{q}_{n,Y}$ for $n = 1000$, the final sample size

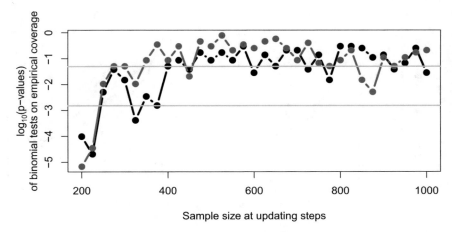

Fig. 24.3 Empirical evaluation of the coverage of the CIs. The *curves* represent the logarithms of *p*-values of binomial tests of adequate coverage (null) *vs.* inadequate coverage (alternative). A large *p*-value is interpreted as the absence of empirical evidence supporting that the related CI does not achieve its nominal coverage of 95%. The *black curve* corresponds with CIs for $\psi_{r_n,0}$, and the gray with CIs for ψ_0. The *gray horizontal lines* show the threshold of 5% (*top*) and the Bonferonni-corrected threshold of 5/33% (*bottom*)

0.917 and 0.955 with an average of 0.940) than for ψ_0 (it ranges between 0.919 and 0.946 with an average of 0.937). This does not come as a surprise, as argued in the first paragraph of this section. Moreover, a majority of the *p*-values are larger than 5% (top gray horizontal line), and even more of them are larger than the Bonferroni-corrected threshold of 5/33%. Furthermore, the smallest *p*-values correspond to sample sizes $n = 200$ and $n = 225$, where inference is based on little information. As for the coverage of $\mathcal{R}_n$, it is far above the nominal 95%-coverage, ranging between 0.951 and 0.990 with an average of 0.997. This does not come as a surprise either since the CIs for $\mathcal{R}_n$ are conservative by construction.

24.5 Conclusion (on a Twist)

We acknowledged that assuming the equality $\bar{q}_{1,Y} = \bar{q}_{0,Y}$ in **A1** is a stringent condition. It happens that the equality is mandatory only in the context of Corollary 24.1, which provides sufficient conditions for the TMLE to estimate ψ_0, the mean reward under r_0. Yet we argued that we are more interested in the data-adaptive parameter $\psi_{r_n,0}$, the mean reward under r_n, than in ψ_0. What can be said then without assuming $\bar{q}_{1,Y} = \bar{q}_{0,Y}$?

Let **A1*** be assumption **A1** deprived of its condition $\bar{q}_{1,Y} = \bar{q}_{0,Y}$. In light of (24.1) and (24.6), let rule r_1 and its approximation g_1 be given by $r_1(W) \equiv I\{\bar{q}_{1,Y}(W) \geq 0\}$ and $g_1(1|W) = 1 - g_1(0|W) \equiv G(\bar{q}_{1,Y}(W))$. Introduce

$$\psi_1 \equiv E_{Q_0}\left(Y(r_1(W))\right),$$

the mean reward under rule r_1. Now, let **A2*** be assumption **A2** with $\epsilon \mapsto E_{P_{Q_0, g_1}}\left(\ell(\bar{Q}'_{1,Y,\rho}(\epsilon))(O)\right)$ substituted for $\epsilon \mapsto E_{P_{Q_0, g_0}}\left(\ell(\bar{Q}_{1,Y,\rho}(\epsilon))(O)\right)$, where $\bar{Q}'_{1,Y,\rho}(\epsilon)$ is defined as in (24.7) using g_1 in lieu of g_0. Introduce $\bar{Q}'^*_{1,Y,r_1} \equiv \bar{Q}'_{1,Y,r_1}(\epsilon_0(r_1))$ and, in light of (24.9), D'^*_1 given by

$$D'^*_1(O) \equiv (Y - \bar{Q}'^*_{1,Y}(A, W))\frac{I\{A = r_1(W)\}}{g_1(A|W)} + \bar{Q}'^*_{1,Y}(r_1(W), W) - \psi_1,$$

then $\Sigma^2_1 \equiv E_{P_{Q_0, g_1}}(D'^*_1(O)^2)$. Finally, consider the following counterpart to **A3**:

A3*. The conditional distributions of $Y(0)$ and $Y(1)$ given W under Q_0 is not degenerated. Moreover, there exist $\gamma_1, \gamma_2 > 0$ such that, for all $t \geq 0$,

$$P_{Q_0}\left(0 \leq |\bar{q}_{1,Y}(W)| \leq t\right) \leq \gamma_1 t^{\gamma_2}. \tag{24.12}$$

In addition, the ratio $|\bar{q}_{0,Y}/\bar{q}_{1,Y}|$ can be defined and its (essential) supremum is finite.

The margin condition in **A3*** now concerns the limit blip function $\bar{q}_{1,Y}$. The true blip function $\bar{q}_{0,Y}$ needs not take positive values Q_0-almost surely anymore. As for the constraint on the ratio $|\bar{q}_{0,Y}/\bar{q}_{1,Y}|$ (which is obviously met when $\bar{q}_{1,Y} = \bar{q}_{0,Y}$), we could simply enforce it by choosing $\bar{Q}_Y$ in such a way that $|\bar{q}_Y| \geq \delta > 0$ for all $\bar{Q}_Y \in \bar{\mathcal{Q}}_Y$. We may now state the final result of this chapter.

Proposition 24.4. *Under* **A1***, **A2*** *and* **A3***, *it holds that* $\|\bar{Q}_{n,Y} - \bar{Q}_{1,Y}\|_{Q_0}$, $\|\bar{q}_{n,Y} - \bar{q}_{1,Y}\|_{Q_0}$, $\|r_n - r_1\|_{Q_0}$, $\|g_n - g_1\|_{Q_0}$ *and the data-adaptive parameter* $\psi_1 - \psi_{r_n,0}$ *all converge in probability to zero as n tends to infinity. Furthermore,* ψ^*_n *consistently estimates* $\psi_{r_n,0}$ *hence* ψ_1 *as well. It does so in such a way that* $\sqrt{n/\sigma^2_n}(\psi^*_n - \psi_{r_n,0})$ *converges in law to the standard normal distribution as n tends to infinity, where* σ^2_n *consistently estimates the positive* Σ^2_1.

Therefore, under the assumptions of Proposition 24.4, the CI defined in (24.10) still contains $\psi_{r_n,0}$ with probability tending to $(1 - \alpha)$ as n tends to infinity. The most important result of the chapter is thus preserved without assuming that the limit blip function and the true one coincide.

Acknowledgement Antoine Chambaz and Wenjing Zheng acknowledge the support of the French Agence Nationale de la Recherche (ANR), under grant ANR-13-BS01-0005 (project SPADRO). Antoine Chambaz also acknowledges that this research has been conducted as part of the project Labex MME-DII (ANR11-LBX-0023-01).

Part VIII
Special Topics

Chapter 25
CV-TMLE for Nonpathwise Differentiable Target Parameters

Mark J. van der Laan, Aurélien Bibaut, and Alexander R. Luedtke

TMLE has been developed for the construction of efficient substitution estimators of pathwise differentiable target parameters. Many parameters are nonpathwise differentiable such as a density or regression curves at a single point in a nonparametric model. In these cases one often uses a specific estimator under a specific smoothness assumptions for which it is possible to establish a limit distribution and thereby provide statistical inference. However, such estimators do not adapt to the true unknown smoothness of the data density and, as a consequence, can be easily outperformed by an adaptive estimator that is able to adapt to the underlying true smoothness.

In this chapter, we present a fully adaptive estimator that converges at an adaptive optimal rate implied by the underlying unknown smoothness of the true data density, while still providing formal statistical inference. Our estimator utilizes CV-TMLE for a data adaptively selected smooth approximation of the nonpathwise differentiable target parameter, and thereby integrates efficiency theory and the state-of-art in data-adaptive estimation through super learning.

M. J. van der Laan (✉)
Division of Biostatistics and Department of Statistics, University of California, Berkeley, 101 Haviland Hall, #7358, Berkeley, CA 94720, USA
e-mail: laan@berkeley.edu

A. Bibaut
Division of Biostatistics, University of California, Berkeley, 101 Haviland Hall, #7358, Berkeley, CA 94720, USA
e-mail: aurelien.bibaut@berkeley.edu

A. R. Luedtke
Vaccine and Infectious Disease Division, Fred Hutchinson Cancer Research Center, 1100 Fairview Ave, Seattle, WA 98109, USA
e-mail: aluedtke@fredhutch.org

© Springer International Publishing AG 2018
M.J. van der Laan, S. Rose, *Targeted Learning in Data Science*,
Springer Series in Statistics, https://doi.org/10.1007/978-3-319-65304-4_25

25.1 Definition of the Statistical Estimation Problem

Let $O_1, \ldots, O_n$ be n i.i.d. copies of a random variable O with probability distribution P_0 known to be an element of a specified statistical model $\mathcal{M}$. Let $\Psi : \mathcal{M} \to \mathbf{R}$ be the target parameter of interest, and consider the case that $\mathcal{M}$ is an infinite dimensional model. Suppose that this target parameter is *not* pathwise differentiable. We will assume that it can be represented as a point-wise evaluation $\Psi(P) = \Psi_1(P)(x_0)$ of a function-valued parameter $\Psi_1(P) : (a, b) \subset \mathbf{R}^d \to \mathbf{R}$ for an interval (a, b) containing x_0. For notational convenience, we denote $\Psi_1(P)(x)$ with $\Psi(P)(x)$ again. In this chapter we focus on estimation and inference for such a nonpathwise differentiable target parameter.

Let's first consider some examples of such target parameters.

Example I: Let $O = (W, Y) \sim P_0$, and let $\mathcal{M}$ be a nonparametric model. Consider the case that W is continuous valued. Let the target parameter be defined as $\Psi(P) = E_P(Y \mid W = w_0)$ for a given w_0. Note that indeed $\Psi(P) = \Psi_1(P)(w_0)$ for the function valued parameter $\Psi_1(P)(W) = E_P(Y \mid W)$. In order to formally define $E_P(Y \mid W = w_0)$ at a given point w_0, we assume that for each $P \in \mathcal{M}$, $w \to E_P(Y \mid W = w)$ is continuous at w_0 and that we select a version of $E_P(Y \mid W)$ in the equivalence class that is continuous at w_0. In the following examples, we avoid discussing the technical conditions needed to formally define $\Psi(P)(x_0)$.

Example II: Let $O = (W, A, Y) \sim P_0$ and let $\mathcal{M}$ be a model which only makes assumptions on the conditional density of A, given W. Consider the case that $A \in [0, 1]$ is continuous valued, and assume that the conditional density of A, given W, is positive at a_0 for P_0-almost all W. The target parameter is defined as $\Psi(P) = E_P E_P(Y \mid A = a_0, W)$, where a_0 is a given value. We have that $\Psi(P) = \Psi_1(P)(a) \equiv E_P E_P(Y \mid A = a, W)$. Note that $\Psi(P)$ represents the estimand of the counterfactual mean EY_{a_0} if $Y = Y_A$ and A is independent of $(Y_a : a)$, conditional on W.

Example III: Let $O = (L(0), A(0), L(1), A(1), Y)$ be a longitudinal data structure with two intervention nodes $A(0), A(1)$. Suppose that both $A(0), A(1)$ are continuous valued with values in $[0, 1]$. Consider a statistical model $\mathcal{M}$ which only makes assumptions on the conditional distribution of $A(j)$, given $\bar{L}(j), \bar{A}(j - 1)$, $j = 0, 1$, where we use the convention that $\bar{A}(-1)$ is empty. In particular, we assume that the conditional density of $A(j)$, given $\bar{L}(j), \bar{A}(j - 1)$, is positive at $a_0(j)$, P_0-a.e., $j = 0, 1$. Let $\Psi(P) = \Psi_1(P)(a_0)$, where

$$\Psi_1(P)(a) \equiv E_P E_P(E_P(Y \mid \bar{L}(1), A(0) = a(0), A(1) = a(1)) \mid L(0), A(0) = a(0)).$$

Under the Neyman-Rubin causal model, sequential randomization, this represents the estimand for the counterfactual mean $EY_{a_0(0), a_0(1)}$.

Example IV: Let $O \sim P_0$ be a continuous valued random variable and let $\mathcal{M}$ be a nonparametric model. Our target parameter is $\Psi(P) = \Psi_1(P)(x_0) \equiv p(x_0)$ at a given point x_0, where $p = dP/d\mu$ is the density of P w.r.t. Lebesgue measure μ.

Example V: Let $O = (W, A, \tilde{T} = \min(T, C), \Delta = I(T \leq C)) \sim P_0$, where W is a vector of baseline covariates, $A \in [0, 1]$ is a continuous valued treatment, T is a continuous valued survival time of interest, and C is the time at which failure time T is right-censored. Let $A_c(t) = I(\tilde{T} \leq t, \Delta = 0)$ be the indicator process that jumps from 0 to 1 when an observed censoring event occurs. Similarly, let $N(t) = I(\tilde{T} \leq t, \Delta = 1)$. The statistical model $\mathcal{M}$ only makes assumptions about the conditional distribution of treatment A, given W, and the conditional distribution of $A_c(t)$, given $\bar{A}_c(t-), W, A$. Our target parameter is $\Psi(P) = \Psi_1(P)(a_0) \equiv E_P S(t \mid A = a_0, W)$, where $S(t \mid A, W) = \exp(-\Lambda(t \mid A, W))$ and $d\Lambda(t \mid A, W) = E(dN(t) \mid A, W, \tilde{T} \geq t)$ is the intensity of the counting process N w.r.t. history $(W, A, \bar{N}(t-), \bar{A}_c(t-))$ and conditional of still being at risk for a jump. If we assume that C and T are independent given A, W, then $\Psi(P) = E_P P(T > t \mid A = a_0, W)$. In addition, if we also assume a causal model $T = T_A$, where T_a is a treatment specific counterfactual survival time, and that A is randomized conditional on W, then the latter represents the counterfactual treatment specific survival curve $P(T_{a_0} > t)$ at time t.

Example VI: Let $O = (C, \Delta = I(T \leq C)) \sim P_0$ be a marginal current status data structure on a time until onset T, where C is the monitoring time at which the status $I(T \leq C)$ is measured. Our target parameter is $\Psi(P) = \Psi_1(P)(t_0) \equiv E(\Delta \mid C = t_0)$. If we assume that T and C are independent, then $\Psi(P) = P(T \leq t_0)$ is the cumulative distribution function of T.

Example VII: Let $O = (W, A, C, \Delta = I(T \leq C)) \sim P_0$, where W are baseline covariates, $A \in \{0, 1\}$ is a binary treatment, C is a monitoring time and T is a time until onset of a disease such as cancer or an infection. Thus, this is the current status data structure extended to include baseline covariates and a treatment indicator. Consider a statistical model $\mathcal{M}$ that only makes assumptions on the conditional distribution of (A, C), given W. The target parameter is defined as $\Psi_a(P) = \Psi_{1,a}(P)(t_0) \equiv E_P E_P(\Delta \mid C = t_0, A = a, W)$ for $a \in \{0, 1\}$. If we assume that C is independent of T, given A, W, then $\Psi_a(P) = E_P P(T \leq t_0 \mid A = a, W)$. In addition, if we also assume the causal model and the randomization assumption, then the latter equals the counterfactual cumulative distribution function $P(T_a \leq t_0)$ at time t. Specifically, one might be interested in the causal contrast $P(T_1 \leq t_0) - P(T_0 \leq t_0) = \Psi_1(P) - \Psi_0(P)$.

It is clear that we can construct an endless list of interesting examples of nonpathwise differentiable target parameters. Essentially, for every data structure and realistic statistical model there are a variety of nonpathwise differentiable statistical target parameters that answer questions of interest.

At an intuitive level, a pathwise differentiable target parameter is a parameter that is a smooth function of the data density $p_0(o)$ across o in a set O_1 that has positive P_0-probability. One might refer to O_1 as the "support" of the target parameter. Since for such a pathwise differentiable target parameter all observations that fall in the set O_1 provide information about the target parameter, it is typically possible to construct an estimator ψ_n of $\Psi(P_0)$ that is asymptotically linear, so that it converges at rate $n^{-1/2}$, even for nonparametric models $\mathcal{M}$. On the other hand, any target parameter on a nonparametric model that is defined by the data density at a point or the data density over a set O_1 with P_0-measure zero is only able to utilize observations close to this support of the target parameter, and each of these observations will result in a bias of the estimator. The latter amount of bias will depend on the underlying smoothness of p_0. As a result, one is only able to use the observations in a local neighborhood of its support, and an estimator needs to carefully select the size of the neighborhood in order to trade-off bias and variance. The optimal trade-off of bias and variance will depend on the underlying smoothness of the data density at the support set and the choice of estimator. Either way, the optimal rate of convergence will be slower than $n^{-1/2}$.

25.2 Approximating Our Target Parameter by a Family of Pathwise Differentiable Target Parameters

We will approximate $\Psi(P)$ with a kernel smoother of $\Psi(P)(x)$ over x in a neighborhood of x_0 with bandwidth h. Specifically, let

$$\Psi_h(P) = \int K_{h,x_0}(x)\Psi(P)(x)dx,$$

where $K_{h,x_0}(x) = h^{-d}K((x - x_0)/h)$, $K(x) = K(x_1,\ldots,x_d)$ is a kernel satisfying $\int K(x)dx = 1$. For simplicity, we use the same bandwidth h for all components so that this family of parameters is indexed by a scalar, and we assume that the kernel $K(x) = \prod_{j=1}^d K^*(x_j)$ is a product kernel defined by a univariate kernel K^*. We will refer to a kernel K as a J-orthogonal kernel if $\int K^*(s)ds = 1$ and $\int K^*(s)s^j ds = 0$, $j = 1,\ldots,J$. If K is a J-orthogonal kernel, then it is orthogonal to any polynomial of degree smaller or equal than J:

$$\int K(x) \prod_{j=1}^d x_j^{s_j}dx = 0 \text{ if } s_j \in \{0,\ldots,J\}, j = 1,\ldots,d.$$

Sometimes we will denote $\Psi_h(P)$ with $\Psi_{h,J}(P)$ to indicate that it uses a $(J - 1)$-orthogonal kernel. Notice that

$$\Psi_h(P) = \int_z K(z)\Psi(P)(x_0 + hz)dz.$$

At a minimum we assume that $x \to \Psi(P)(x)$ is continuous at x_0 so that

$$\Psi(P) = \lim_{h \to 0} \Psi_h(P).$$

Let $\mathcal{M}(J) = \{P \in \mathcal{M} : x \to \Psi(P)(x) \text{ is } J\text{-times continuously differentiable at } x_0\}$. The following lemma proves that if $P_0 \in \mathcal{M}(J)$, then $\Psi_{h,J}(P_0) - \Psi(P_0) = O(h^J)$.

Lemma 25.1. *Assume $P_0 \in \mathcal{M}(J_0)$ for an integer $J_0 \geq 1$. Let*

$$\Psi_{h,J}(P) = \int_x K_{h,x_0}(x)\Psi(P)(x)dx$$

be the parameter approximation of $\Psi(P) = \Psi(P)(x_0)$ using a $J-1$-orthogonal kernel with bandwidth h centered at x_0. Let $J_0^* = \min(J_0, J)$. Let

$$B_0(J_0^*) \equiv \sum_{\{m \in \{0,\ldots,J_0^*\}^d : \sum_l m_l = J_0^*\}} \psi_0^m(x_0) \int_y K(y) \frac{\prod_l y_l^{m_l}}{\prod_l m_l!} dy.$$

Then,

$$h^{-J_0^*}(\Psi_{h,J_0^*}(P_0) - \Psi(P_0)) \to B_0(J_0^*) \text{ as } h \to 0.$$

Proof. For notational convenience, let J denote J_0^* in this proof. We have

$$\Psi_{h,J}(P_0) - \Psi(P_0) = \int_x K_{h,x_0}(x)\{\Psi(P_0)(x) - \Psi(P_0)(x_0)\}dx$$

$$= \int_x h^{-d} K((x-x_0)/h)\{\psi_0(x) - \psi_0(x_0)\}dx$$

$$= \int_y K(y)\{\psi_0(x_0 + hy) - \psi_0(x_0)\}dy$$

$$= \int_y K(y) \sum_{j=1}^{J-1} \sum_{m,\sum_l m_l = j} \frac{h^j \prod_l y_l^{m_l}}{\prod_l m_l!} \psi_0^m(x_0)dy$$

$$+ h^J \int_y K(y) \sum_{m,\sum_l m_l = J} \frac{\prod_l y_l^{m_l}}{\prod_l m_l!} \psi_0^m(x_0)dy + o(h^J)$$

$$= h^J \int_y K(y) \sum_{m,\sum_l m_l = J} \frac{\prod_l y_l^{m_l}}{\prod_l m_l!} \psi_0^m(x_0)dy + o(h^J)$$

$$\equiv h^J B_0(J) + o(h^J).$$

Thus,

$$h^{-J}(\Psi_{h,J}(P_0) - \Psi(P_0)) = B_0(J) + o(1).$$

This completes the proof. $\square$

Efficient Influence Curve. Let $D_h^*(P)$ be the canonical gradient of Ψ_h at $P \in \mathcal{M}$. It is assumed that

$$D_h^*(P) = D(K_{h,x_0}, P), \text{ where } K \to D(K, P) \text{ is linear.}$$

In our examples, we have that $O = (X, O_1)$ and

$$D_h^*(P)(O) = K_{h,x_0}(X)D_1(P)(O) + \int_x K_{h,x_0}(x)D_2(P)(x, O_1)dx$$

for some $D_1(P), D_2(P) \in L^2(P)$, where $\int K_{h,x_0}(x)D_2(P)(x, O_1)dx = \int K(y)D_2(P)$
$(x_0 + hy, O_1)dy$. In addition, it is assumed that

$$\sup_{P \in \mathcal{M}, o} | D_h^*(P) | (o) < Ch^{-d} \tag{25.1}$$

for a finite constant $C < \infty$ and the supremum over o is over a support of P_0. One
might refer to the latter assumption as a positivity assumption.

Second-Order Taylor Expansion $\Psi_h(P)$. Because Ψ_h is pathwise differentiable,

$$\Psi_h(P) - \Psi_h(P_0) = (P - P_0)D_h^*(P) + R_h(P, P_0),$$

where $R_h(P, P_0)$ is a second-order remainder, and we can use that $PD_h^*(P) = 0$. We
will assume that this second-order remainder $R_h(P, P_0)$ has the following structure:

$$R_h(P, P_0) = \int_x K_{h,x_0}(x)R_x(P, P_0)dx \tag{25.2}$$

for some second-order remainder term $R_x(P, P_0)$. This structure is essentially implied by $\Psi_h(P)$ being a kernel average of $\Psi(P)(x)$ around x_0. Indeed this assumption
holds in all our examples.

A typical form of the second-order remainder $R_x(P, P_0)$ is given by

$$R_x(P, P_0) = \int_y (H_1(P) - H_1(P_0))(x, y)(H_2(P) - H_2(P_0))(x, y)H_3(P, P_0)(x, y)d\mu(y),$$

where $H_1(P), H_2(P)$ are statistical parameters that are functions of O through X and
another subvector Y. If $H_1(P) \neq H_2(P)$ and these two parameters are variation independent, then this represents a so-called double robust structure since $R_x(P, P_0) = 0$
if either $H_1(P) = H_1(P_0)$ or $H_2(P) = H_2(P_0)$. We note that $R_x(P, P_0)$ typically involves differences of the P and P_0-density or regression at a single point x, while
the other variables (i.e., Y) are integrated out. For pathwise differentiable parameters, one will always have that the second-order remainder involves integrating over
all variables, so that the second-order remainder can be bounded by L^2-norms of
$H_1(P) - H_1(P_0)$ and $H_2(P) - H_2(P_0)$. Such bounding is not possible for $R_x(P, P_0)$,
but is possible for $R_h(P, P_0)$. One could view $R_x(P, P_0)$ as a second-order remainder
corresponding with the second-order Taylor expansion of the nonpathwise differentiable parameter $\Psi(P)(x)$.

Even though our general approach does not rely on any specific structure of
$\Psi_h(P)$ beyond that it approximates $\Psi(P)$ and that it is pathwise differentiable, the
above structure allows us to fully understand the dependence of both the efficient influence curve and the second-order remainder on the tuning parameter h. This will
allow us to establish asymptotic normality of our proposed estimator of ψ_0 under
concrete assumptions.

25.3 CV-TMLE of h-Specific Target Parameter Approximation

Let $\{P_{\epsilon,h} : \epsilon\}$ be a local least favorable parametric submodel through P so that the linear span of its score $\frac{d}{d\epsilon} \log dP_{\epsilon,h}/dP\big|_{\epsilon=0}$ includes $D_h^*(P)$. Let $p_{\epsilon,h} = dP_{\epsilon,h}/dP$. For example, this could be a one-dimensional universal least favorable submodel $p_{\epsilon,h} = p \exp\left(\int_0^\epsilon D_h^*(P_e)de\right)$ in which case we even have $\frac{d}{d\epsilon} \log dP_{\epsilon,h}/dP = D_h^*(P_{\epsilon,h})$ for all ϵ. We could also select the one-dimensional universal least favorable submodel $\{P_{\epsilon,h} : \epsilon \geq 0\}$ for targeting a multivariate target parameter (or more generally, solving a multivariate vector of desired equations $P_n \bar{D}_h(P) = 0$) defined by

$$\frac{dP_{\epsilon,h}}{dP} = \exp\left(\int_0^\epsilon \frac{\{P_n \bar{D}_h(P_{e,h})\}^\top}{\| P_n \bar{D}_h(P_{e,h}) \|} \bar{D}_h(P_{e,h}) de\right),$$

where $\bar{D}_h(P)$ should include as one of its components $D_h^*(P)$, and $\| \cdot \|$ is the standard Euclidean norm. In this case, we have $\frac{d}{d\epsilon} P_n \log p_{\epsilon,h} = \| P_n \bar{D}_h(P_{\epsilon,h}) \|$.

25.3.1 CV-TMLE of $\Psi_h(P_0)$

Let $\mathbf{P}_n = \hat{P}(P_n)$ be an initial estimator of P_0. Using our notation for cross-validation, let P_{n,B_n}^1, P_{n,B_n}^0 be the empirical probability distribution of the validation sample $\{i : B_n(i) = 1\}$ and training sample $\{i : B_n(i) = 0\}$, respectively, implied by a random split $B_n \in \{0,1\}^n$. Let $\mathbf{P}_{n,B_n} = \hat{P}(P_{n,B_n}^0)$ be the initial estimator of P_0 applied to the training sample. Let $\{\mathbf{P}_{n,B_n,\epsilon,h} : \epsilon\}$ be our least favorable submodel through $\mathbf{P}_{n,B_n}$. Let ϵ_n be the MLE:

$$\epsilon_n = \arg\max_\epsilon E_{B_n} P_{n,B_n}^1 \log \mathbf{p}_{n,B_n,\epsilon,h}.$$

This defines TMLE updates $\mathbf{P}_{n,B_n,h}^* = \mathbf{P}_{n,B_n,\epsilon_n,h}$ for each possible realization of B_n. The CV-TMLE of $\Psi_h(P_0)$ is defined by

$$\psi_{n,h}^* \equiv E_{B_n} \Psi_h(P_{n,B_n,h}^*).$$

If we select one of the two universal least favorable submodels defined above, then we have

$$E_{B_n} P_{n,B_n}^1 D_h^*(P_{n,B_n,h}^*) = 0.$$

For our analysis of the CV-TMLE, we only need that the left-hand side is $o_P(r_{n,h_n})$ for a rate r_{n,h_n} that converges to zero at faster rate than the rate of convergence of ψ_{n,h_n}^* using our proposed bandwidth h_n, so that is represents a second-order term. For example, if we use a local least favorable submodel and the initial estimator $\hat{P}(P_n)$ converges fast enough to P_0, then this will typically be the case. One could also use a local least favorable submodel and simply iterate the TMLE-updating step until $r_{n,h}$ is small enough.

25.3.2 Asymptotic Normality of CV-TMLE

The following theorem establishes asymptotic normality of the CV-TMLE $\psi^*_{n,h_n} = E_{B_n} \Psi_{h_n}(\mathbf{P}^*_{n,B_n,h_n})$ of $\Psi_{h_n}(P_0)$ along a sequence $h_n \to 0$ without making smoothness assumptions on P_0.

Theorem 25.1. *We make the following assumptions:*

- *Let $r(n)$ be a rate converging to zero such that for some $P_1 \in M$ with $R_h(P_1, P_0) = 0$ for all $h > 0$*

$$\| D^*_h(\mathbf{P}^*_{n,B_n,h}) - D^*_h(P_1) \|_{P_0} = O_P(h^{-d/2} r(n)). \tag{25.3}$$

- *Let $h_n \to_p 0$, $r(n)\{1 - \log(h_n^{d/2} r(n))\} \to 0$, and*

$$\lim_{h \to 0} h^d P_0 \{D^*_h(P_1)\}^2 = \sigma_0^2. \tag{25.4}$$

- *There exists a deterministic sequence $\tilde{h}_n$ so that $h_n/\tilde{h}_n \to_p 1$.*
- *Consider the class of functions $\mathcal{F}_n = \{D(K_{1,h,x_0}, P_{n,B_n,\epsilon,h}) - D(K_{1,h,x_0}, P_1) : h, \epsilon\}$ and let F_n be its envelope. Assume that*

$$\sup_\Lambda N(\delta \mid F_n \mid, \mathcal{F}_n, L^2(\Lambda)) = O(\delta^{-p}) \text{ for some integer } p > 0.$$

-

$$\sqrt{nh_n^d} E_{B_n} \int_y K(y) R_{x_0+h_n y}(\mathbf{P}^*_{n,B_n,h_n}, P_0) dy = o_P(1). \tag{25.5}$$

Then,

$$\sqrt{nh_n^d} \left(E_{B_n} \Psi_{h_n}(\mathbf{P}^*_{n,B_n,h_n}) - \Psi_{h_n}(P_0) \right) = \sqrt{nh_n^d}(P_n - P_0) D^*_{h_n}(P_1) + o_P(1)$$
$$+ \sqrt{nh_n^d} E_{B_n} \int_y K(y) R_{x_0+h_n y}(\mathbf{P}^*_{n,B_n,h_n}, P_0) dy,$$

and

$$\sqrt{nh_n^d} \left(E_{B_n} \Psi_{h_n}(\mathbf{P}^*_{n,B_n,h_n}) - \Psi_{h_n}(P_0) \right) \Rightarrow_d N(0, \sigma_0^2).$$

Recall $D^*_h(P) = D(K_{h,x_0}, P)$, where $K \to D(K, P)$ is linear in K. For example, $O = (X, O_1)$ and $D^*_h(P) = K_{h,x_0} D_1(P) + \int K_{h,x_0}(X) D_2(P)(x, O_1) dx$, where $\int K_{h,x_0}(x) D_2(P)(x, O_1) dx = \int K(y) D_2(P)(x_0 + hy, O_1) dy$. In this case, the left-hand side of (25.3) is typically dominated by $\| K_{h,x_0} D_1(\mathbf{P}^*_{n,B_n,h}) - K_{h,x_0} D_1(P_1) \|_{P_0}$ and $r(n)$ can be chosen so that

$$\left(\int_y K^2(y) E_0(\{D_1(\mathbf{P}^*_{n,B_n,h_n}) - D_1(P_1)\}^2 \mid X = x_0 + h_n y) p_{0,X}(x_0 + h_n y) dy \right)^{0.5} = O_P(r(n)).$$

The simplest scenario for $R_{20}(P_1, P_0) = 0$ is that $P_1 = P_0$. In many problems we have a double robustness structure so that

$$R_x(P, P_0) = \int_y (H_1(P) - H_1(P_0))(x, y)(H_2(P) - H_2(P_0))(x, y)H_3(P, P_0)(x, y)d\mu(y).$$

In that case, we will have that $R_x(P, P_1)$ equals zero if either $H_1(P_1) = H_1(P_0)$ or $H_2(P_1) = H_2(P_0)$. For example, consider the scenario that $H_2(P_1) = H_2(P_0)$. In that case, we have the following decomposition:

$$R_x(\mathbf{P}_n, P_0) = \int_y (H_1(\mathbf{P}_n) - H_1(P_1))(H_2(\mathbf{P}_n) - H_2(P_0))(x, y)H_3(\mathbf{P}_n, P_0)(x, y)d\mu(y)$$

$$+ \int_y (H_1(P_1) - H_1(P_0))(H_2(\mathbf{P}_n) - H_2(P_0))(x, y)H_3(\mathbf{P}_n, P_0)(x, y)d\mu(y).$$

Thus, the condition (25.9) would be satisfied if these two terms on the right-hand side are $o_P((nh^d)^{-0.5})$. So this would require that $H_2(\mathbf{P}_n)$ converges to $H_2(\mathbf{P}_0)$ at a faster rate than $(nh^d)^{-0.5}$.

Proof of Theorem 25.1. We have

$$\sqrt{nh_n^d}\left(E_{B_n}\Psi_{h_n}(\mathbf{P}^*_{n,B_n,h_n}) - \Psi_{h_n}(P_0)\right) = \sqrt{nh^d}E_{B_n}(P^1_{n,B_n} - P_0)D^*_{h_n}(\mathbf{P}^*_{n,B_n,h_n})$$

$$+ \sqrt{nh_n^d}E_{B_n}R_{h_n}(\mathbf{P}^*_{n,B_n,h_n}, P_0).$$

Consider the leading empirical process term. Let $f_{n,B_n} \equiv D^*_{h_n}(\mathbf{P}^*_{n,B_n,h_n}) - D^*_{h_n}(P_1))$. Let $K_{1,h,x_0}(x) = K((x - x_0)/h)$ so that $K_{h,x_0} = h^{-d}K_{1,h,x_0}$. Notice that $f_{n,B_n} = h_n^{-d}g_{n,B_n}$, where $g_{n,B_n} = D(K_{1,h_n,x_0}, \mathbf{P}^*_{n,B_n,h_n}) - D(K_{1,h_n,x_0}, P_1)$. Let $p = \sum_{i=1}^n B_n(i)/n$. We have

$$\sqrt{nh_n^d}E_{B_n}(P^1_{n,B_n} - P_0)(D^*_{h_n}(\mathbf{P}^*_{n,B_n,h_n}) - D^*_{h_n}(P_1)) = \sqrt{nh_n^d}E_{B_n}(P^1_{n,B_n} - P_0)f_{n,B_n}$$

$$= \sqrt{nh_n^d h_n^{-d}}E_{B_n}(P^1_{n,B_n} - P_0)g_{n,B_n}$$

$$= p^{-0.5}h_n^{-0.5d}\sqrt{np}E_{B_n}(P^1_{n,B_n} - P_0)g_{n,B_n}.$$

For a given B_n, conditional on P^0_{n,B_n}, we apply Lemma 25.2 below to $\sqrt{np}(P^1_{n,B_n} - P_0)g_{n,B_n}$, where g_{n,B_n} is only random through $\theta_n = (\epsilon_n, h_n)$. For notational convenience, denote this function g_{n,B_n} with g_{n,θ_n}. We have that $g_{n,\theta_n} \in \mathcal{F}_n = \{g_{n,\theta} : \theta\}$ so that indeed $\mathcal{F}_n$ is finite dimensional as in Lemma 25.2. In addition, any function in $\mathcal{F}_n$ is bounded by a universal constant C due to our assumption that $D^*_{h_n}(P)$ is bounded by Ch_n^{-d} universally in o and $P \in \mathcal{M}$. Thus, the envelope F_n of $\mathcal{F}_n$ satisfies $\| F_n \|_{P_0} < C < \infty$ for some $C < \infty$. By assumption, we have $\| g_{n,\theta_n} \|_{P_0} = O_P(h_n^{d/2}r(n))$. Application of Lemma 25.2 proves that $\sqrt{np}(P^1_{n,B_n} - P_0)g_{n,\theta_n} = O_P(h_n^{d/2}r(n)\{1 - \log(h_n^{d/2}r(n))\})$. Since B_n has only a finite number of realization, this proves that $\sqrt{np}E_{B_n}(P^1_{n,B_n} - P_0)g_{n,B_n} = O_P(h_n^{d/2}r(n)\{1 - \log(h_n^{d/2}r(n))\})$. This shows that

$$p^{-0.5}h_n^{-0.5d}\sqrt{np}E_{B_n}(P_{n,B_n}^1 - P_0)g_{n,B_n} = O_P(r(n)\{1 - \log(h_n^{d/2}r(n))\}).$$

By assumption, the right-hand side is $o_P(1)$. Thus, we have shown that

$$\sqrt{nh_n^d}E_{B_n}(P_{n,B_n}^1 - P_0)(D_{h_n}^*(\mathbf{P}_{n,B_n,h_n}^*) - D_{h_n}^*(P_1)) = o_P(r(n)\{1 - \log(h_n^{d/2}r(n))\}) = o_P(1).$$

This now yields

$$\sqrt{nh_n^d}\left(E_{B_n}\Psi_{h_n}(\mathbf{P}_{n,B_n,h_n}^*) - \Psi_{h_n}(P_0)\right) = \sqrt{nh_n^d}E_{B_n}(P_{n,B_n}^1 - P_0)D_{h_n}^*(P_1) + o_P(1)$$
$$+ \sqrt{nh_n^d}E_{B_n}\int_x K_{h_n,x_0}(x)R_x(\mathbf{P}_{n,B_n,h_n}^*, P_0)dx.$$

By the Lindeberg Theorem for triangular arrays, using assumption (25.4), it follows that for a deterministic sequence $\tilde{h}_n$,

$$\sqrt{n\tilde{h}_n^d}E_{B_n}(P_{n,B_n}^1 - P_0)D_{\tilde{h}_n}^*(P_1) \Rightarrow_d N(0, \sigma_0^2).$$

By assumption, there exists a deterministic sequence $\tilde{h}_n$ so that $h_n/\tilde{h}_n \to_p 1$. We have

$$\sqrt{nh_n^d}E_{B_n}(P_{n,B_n}^1 - P_0)D_{h_n}^*(P_1) - \sqrt{n\tilde{h}_n^d}E_{B_n}(P_{n,B_n}^1 - P_0)D_{\tilde{h}_n}^*(P_1)$$
$$= n^{0.5}\{\sqrt{h_n^d} - \sqrt{\tilde{h}_n^d}\}E_{B_n}(P_{n,B_n}^1 - P_0)D_{\tilde{h}_n}^*(P_1)$$
$$+ \sqrt{nh_n^d}E_{B_n}(P_{n,B_n}^1 - P_0)\{D_{h_n}^*(P_1) - D_{\tilde{h}_n}^*(P_1)\}.$$

The first term can be written as $\sqrt{h_n^d/\tilde{h}_n^d} - 1$ times $(n\tilde{h}_n^d)^{0.5}E_{B_n}(P_{n,B_n}^1 - P_0)D_{\tilde{h}_n}^*(P_1)$. Since the scalar converges to zero and the empirical process term converges to $N(0, \sigma_0^2)$, it follows that the first term is $o_P(1)$. The second term can be analyzed as the above empirical process term and is thus $o_P(1)$. Let's now consider the second-order remainder term $\sqrt{nh_n^d}E_{B_n}R_{h_n}(\mathbf{P}_{n,B_n,h_n}^*, P_0)$. It equals

$$\sqrt{nh_n^d}E_{B_n}\int_y K(y)R_{x_0+hy}(\mathbf{P}_{n,B_n,h}^*, P_0)dy.$$

By assumption (25.5) the latter is $o_P(1)$. This proves the theorem. $\square$

The proof used the following lemma.

Lemma 25.2. *Let $G_n(f) = \sqrt{n}(P_n - P_0)f$ be a standardized empirical mean of $f(O_i)$. Let $f_{n,\epsilon_n} \in \mathcal{F}_n = \{f_{n,\epsilon} : \epsilon\}$ where ϵ varies over a bounded set in $\mathbf{R}^p$ and $f_{n,\epsilon}$ is a nonrandom function (i.e., not based on data $O_1, \ldots, O_n$). Suppose that $\| f_{n,\epsilon_n} \|_{P_0} = o_P(r_n)$ for a rate r_n satisfying $r_n \log r_n^{-1} \to 0$, and $r_n n^{0.5} \to \infty$. Suppose that the envelope F_n of $\mathcal{F}_n$ satisfies $\| F_n \|_{P_0} \le M_n$. Assume that*

$$\sup_\Lambda N(\delta \mid F_n \mid, \mathcal{F}_n, L^2(\Lambda)) = O(\delta^{-p}) \text{ for some integer } p > 0.$$

Then,

$$E_0 \mid G_n(f_{n,\epsilon_n}) \mid = O\left(M_n r_n (1 + \log r_n^{-1})\right).$$

Thus, if $r_n = o(1)$ and $M_n < M < \infty$, then $G_n(f_{n,\epsilon_n}) = o_P(1)$.

Proof. For notational convenience, let's denote f_{n,ϵ_n} with f_n. We apply the Theorem in van der Vaart and Wellner (2011) providing us with

$$E_0 \mid G_n(f_n) \mid \leq J(\delta_n, \mathcal{F}_n)\left(1 + \frac{J(\delta_n, \mathcal{F}_n)}{\delta_n^2 n^{0.5} \parallel F_n \parallel_{P_0}}\right) \parallel F_n \parallel_{P_0}, \qquad (25.6)$$

where we can select $\delta_n = r_n$. Using the bound ϵ^{-p} on the uniform covering number, it follows that $J(\delta_n, \mathcal{F}_n) = -p^{0.5} \int_0^{\delta_n} (\log \epsilon)^{0.5} d\epsilon$. We can conservatively bound $(\log \epsilon)^{0.5}$ by $\log \epsilon$, and use that $\int_0^{\delta_n} \log \epsilon d\epsilon = \delta_n - \delta \log \delta_n = \delta(1 + \log(\delta_n^{-1}))$. This shows that $J(\delta, \mathcal{F}_n) \leq \delta_n(1 + \log \delta_n^{-1})$. If $J(\delta_n, \mathcal{F}_n) = O(\delta_n^2 n^{0.5})$, then the leading term in (25.6) is given by $J(\delta_n, \mathcal{F}_n) \parallel F_n \parallel_{P_0}$. Using the above bound for $J(\delta_n, \mathcal{F}_n)$, it follows that this holds if $\delta_n(1 + \log \delta_n^{-1}) = O(\delta_n^2 n^{0.5})$, or equivalently, $\delta_n(1 + \log \delta_n^{-1}) = O(\delta_n n^{0.5})$. By assumption we have $\delta_n n^{0.5} \to \infty$ and $\delta_n \log \delta_n^{-1} \to 0$, so that this always holds. This results in the following bound:

$$E_0 \mid G_n(f_n) \mid = O(M_n r_n (1 + \log r_n^{-1})),$$

which equals the stated bound. $\square$

25.3.3 Asymptotic Normality of CV-TMLE as an Estimator of ψ_0

Let $P_0 \in \mathcal{M}(J_0)$. The following theorem is an immediate consequence of $(nh^d)^{0.5}(\psi_n^* - \psi_{h0}) \Rightarrow_d N(0, \sigma_0^2)$ and $h^{-J_0}(\psi_{h0} - \psi_0) \to B_0(J_0)$, which shows that $\psi_n^* - \psi_0 \approx N(h^{J_0} B_0(J_0), \sigma_0^2/(nh^d))$. Balancing the square of the bias and the variance yields the optimal rate $h_n^* = n^{-1/(2J_0+d)}$, while slight undersmoothing yields a mean zero $N(0, \sigma_0^2)$ limit distribution.

Theorem 25.2. *Suppose that we use a J-orthogonal kernel K and assume the conditions of Theorem 25.1. Let $P_0 \in \mathcal{M}(J_0)$ for an unknown degree of smoothness $J_0 \geq 1$. Let $J_0^* = \min(J_0, J)$. The optimal rate for the bandwidth is given by $h_n^* = n^{-1/(2J_0^*+d)}$. If we select this rate, then we obtain*

$$n^{-J_0^*/(2J_0^*+d)}(\psi_n^* - \psi_0) \Rightarrow_d N(B_0(J_0^*), \sigma_0^2).$$

If we select h_n such that $h_n/n^{-1/(2J_0^*+d)} \to 0$ (i.e., if we undersmooth), then we obtain

$$\sqrt{nh_n^d}(\psi_n^* - \psi_0) \Rightarrow_d N(0, \sigma_0^2).$$

If we select such an undersmoothed h_n, then an asymptotically valid 0.95-*confidence interval for ψ_0 is given by*

$$\psi_n^* \pm 1.96\sigma_n/(nh_n^d)^{0.5},$$

where σ_n^2 is a consistent estimator of σ_0^2.

If it is known that $J_0 \geq J$, then the above optimal rate h_n^* is known and identified by $J_0^* = J$. However, without that knowledge, this optimal rate h_n^* is unknown. Even when $J_0 \geq J$, regarding finite sample performance, one wants to select ch_n^* for a well chosen c, which would be unknown. Therefore, for either case it is important that we propose a data-adaptive selector for h. This will be proposed in the next section.

25.4 A Data-Adaptive Selector of the Smoothing Bandwidth

Let h_n be a solution of the equation

$$\frac{d}{dh_n}\psi_{n,h_n}^* + C_n\frac{d}{dh_n}\left\{\sigma_n/(nh_n^d)^{0.5}\right\} = 0 \qquad (25.7)$$

for some finite or slowly converging sequence C_n such as $C_n = C\log n$ or $-C\log n$ for some $C > 0$. The choice of sign of C_n depends on $h \to \psi_{n,h}^*$ being increasing or decreasing for h close to 0. In the unlikely case that $\psi_{n,h}$ (and possibly $\psi_{0,h}$) is constant in h for local neighborhood $(0, a)$ of 0, we would select $h_n = a$. In practice, we expect that (25.7) only has a solution for either a positive or negative C_n so that this choice will naturally follow. We allow C_n to slowly converge to infinity in order to guarantee the existence of a solution of (25.7), while the optimal adaptive rate will only be achieved if $C_n = O_P(1)$. Therefore, in practice, we recommend to select C_n as the smallest C_n with $| C_n |\geq 1.96$ for which a solution is found.

One interesting choice is $C_n = Z_{1-\alpha_n}$ or $C = -Z_{1-\alpha_n}$ for some sequence α_n, where $Z(1 - \alpha)$ is the $1 - \alpha$-quantile of the standard normal distribution. For example, $\alpha_n = 1/n$ corresponds with C_n behaving as $\log n$, but if possible one should select α at a fixed level such as 0.05. In this manner, one can view h_n as the value that minimizes the upper bound or maximizes the lower bound of the $1 - \alpha_n$-confidence interval for ψ_0, which provides the following intuition of this selector h_n. Suppose that an investigation of the behavior of $\psi_{n,h}$ as a function of $h > 0$ at 0 (away from pure noise) demonstrates an increasing trend as h approximates 0. In that case, one might be willing to assume that $\psi_0 > \psi_{0,h}$ for all $h > 0$, suggesting that we want to maximize $\psi_{n,h}^*$. However, as h approximates zero, the noise of $\psi_{n,h}^*$ starts to dominate the signal $\psi_{0,h}$. On the other hand, one is $(1 - \alpha_n)100\%$-confident that $\psi_{0,h} > \psi_{n,h}^* - Z_{1-\alpha_n}\sigma_n/(nh^d)^{0.5}$, suggesting that we should instead maximize $h \to \psi_{n,h}^* - Z_{1-\alpha_n}\sigma_n/(nh^d)^{0.5}$, and thus define

$$h_n = \arg\max_h \psi_{n,h}^* - Z_{1-\alpha_n}\sigma_n/(nh^d)^{0.5}.$$

The latter definition of h_n solves (25.7). Alternatively, if one determines that $\psi_{0,h}$ is decreasing as h approximates zero, one wants to minimize the upper bound $\psi^*_{n,h}$ + $Z_{1-\alpha_n}\sigma_n/(nh^d)^{0.5}$:

$$h_n = \arg\min_h \psi^*_{n,h} + Z_{1-\alpha_n}\sigma_n/(nh^d)^{0.5}.$$

To simplify our analysis of our proposed selector h_n for the CV-TMLE, we will define h_n as the solution of (25.7) with ψ^*_{n,h_n} replaced by a cross-validated one-step estimator:

$$\psi^1_{n,h} = E_{B_n}\Psi_h(\mathbf{P}_{n,B_n}) + E_{B_n}P^1_{n,B_n}D^*_h(\mathbf{P}_{n,B_n}), \tag{25.8}$$

where the initial estimator $\mathbf{P}_{n,B_n}$ does not depend on h. The fact that $D^*_h(\mathbf{P}_{n,B_n})$ only depends on h through K_{h,x_0} makes the conditions easier, although we expect the theorem for h_n below also applies to the selector (25.7) using the CV-TMLE itself (which also depends on h through the targeted estimator $\mathbf{P}^*_{n,B_n}$). For many target parameters we have $D^*_h(P) = D_h(P) - \Psi_h(P)$, in which case this cross-validated one-step estimator reduces to

$$\psi^1_{n,h} = E_{B_n}P^1_{n,B_n}D_h(\mathbf{P}_{n,B_n}) = E_{B_n}P^1_{n,B_n}D(K_{h,x_0},\mathbf{P}_{n,B_n}),$$

while the CV-TMLE allows the same representation but now with $\mathbf{P}_{n,B_n}$ replaced by $\mathbf{P}^*_{n,B_n}$.

Theorem 25.3. *Consider the definition (25.7) of h_n with ψ^*_{n,h_n} replaced by the CV-one-step estimator ψ^1_{n,h_n} (25.8). We define*

$$K_{1,h,x_0} = h\frac{d}{dh}K_{h,x_0}$$
$$e_{1,n,B_n} \equiv D^*_h(K_{1,h,x_0},\mathbf{P}_{n,B_n}) - D^*_h(K_{1,h,x_0},P_1)$$
$$r_1(n) = \| e_{1,n,B_n} \|_{P_0}$$
$$Z_n = \sqrt{n}(P_n - P_0)h^{-d/2}D^*_h(K_{1,h,x_0},P_1)$$
$$\sigma^2_{10} \equiv \lim_{h\to 0} VAR\left(h^{d/2}D^*_h(K_{1,h,x_0},P_1)\right),$$

where P_1 is the limit of $\mathbf{P}^*_{n,B_n}$, and the right-hand side limit is assumed to exist.

Assumptions. *Let $P_0 \in M(J_0)$ for some unknown integer $J_0 \geq 1$. and let $h_n \to_p 0$ be a solution of (25.7). Let $J^*_0 = \min(J_0, J)$, where $J - 1$ is the degree of our kernel K. Let $r(n)$ be such that for some $P_1 \in M$ with $R_{20}(P_1, P_0) = 0$*

$$\| D^*_h(K_{1,h_n,x_0},\mathbf{P}_{n,B_n}) - D^*_h(K_{1,h_n,x_0},P_1) \|_{P_0} = O_P(h_n^{-d/2}r(n)).$$

Let $r(n)\log(h_n^{-d/2}r(n)) \to_p 0$. Suppose that there exists a deterministic sequence $\tilde{h}_n$ so that $h_n/\tilde{h}_n \to_p 1$. Consider the class of functions $\mathcal{F}_n = \{D(K_{1,h,x_0},\mathbf{P}_{n,B_n}) - D(K_{1,h,x_0},P_1) : h, \epsilon\}$ and let F_n be its envelope. Assume that

$$\sup_\Lambda N(\delta \mid F_n \mid, \mathcal{F}_n, L^2(\Lambda)) = O(\delta^{-p}) \text{ for some integer } p > 0.$$

In addition, assume

$$\sqrt{nh_n^d}E_{B_n}\int_x K_{1,x_0,h_n}(x)R_x(\mathbf{P}_{n,B_n},P_0)dx = o_P(1). \tag{25.9}$$

Then,

$$-h_n^{J_0^*}J_0B_0(J_0) + o(h_n^{J_0^*}) = (nh_n^d)^{-0.5}(Z_n + 0.5dC_n\sigma_n + o_P(1)),$$

where $Z_n = O_P(1)$. As a consequence, h_n behaves as $(C_n^2/n)^{-1/(2J_0^*+d)}$. Specifically, if $C_n = O_P(1)$, then h_n converges to zero at the optimal rate $n^{-1/(2J_0^*+d)}$.

In the special case that $\frac{d}{dh}\psi_{0,h} = 0$ for $h \in (0,a)$ for some $a > 0$, we only need that $h_n \to_p a*$ for some $a* \in (0,a]$. This scenario corresponds with the case that ψ_0 is actually pathwise differentiable, in which case the CV-TMLE ψ_{n,h_n}^* converges at rate $n^{-0.5}$ and will in fact be asymptotically efficient estimator of ψ_0.

The function K_{1,h,x_0} happens to behave in the same way as K_{h,x_0} as a function of h. Specifically, $EK_{1,h,x_0}^2(X)$ behaves as $EK_{h,x_0}^2(X)$, and similarly the variance of $D(K_{1,h,x_0},P)$ behaves as the variance of $D(K_{h,x_0},P)$ as a function of h, and is thus $O(h^{-d})$. To understand this, consider $d = 2$ so that $K(x_1,x_2) = K^*(x_1)K^*(x_2)$ is a bivariate product kernel. We have

$$K_{1,h,x_0}(X) = -dK_{h,x_0}(X) - h^{-d}K^{*\prime}((x_1-x_{01})/h)(x_1-x_{01})/hK^*((x_2-x_{02})/h)$$
$$-h^{-d}K^{*\prime}((x_2-x_{02})/h)(x_2-x_{02})/hK^*((x_1-x_{01})/h).$$

The first term is obviously analogue to $K_{h,x_0}(X)$. The second and third term combined can be written as $h_{-d}K_1((x-x_0)/h)$ for a multivariate product kernel $K_1(x_1,x_2) = K^{*\prime}(x_1)x_1K^*(x_2) + K^{*\prime}(x_2)x_2K(x_1)$. We have

$$E(h^{-d}K_1((X-x_0)/h))^2 = h^{-d}\int_y K_1^2(y)p_{0,X}(x_0+hy)dy.$$

This proves that indeed the kernel K_{1,h,x_0} behaves in the same manner as K_{h,x_0} as a function of h. As a result we have $\mathrm{Var}D(K_{1,h,x_0},P) = O(h^{-d})$.

Proof of Theorem 25.3. Rearranging (25.7) shows that

$$\frac{d}{dh_n}\Psi_{n,h_n}^1 = -0.5dC_n\sigma_n n^{-1/2}h_n^{-d/2-1}. \tag{25.10}$$

We also have

$$\Psi_{n,h_n}^1 - \Psi_{h_n}(P_0) = E_{B_n}(\Psi_{h_n}(\mathbf{P}_{n,B_n}) - \Psi_{h_n}(P_0)) + E_{B_n}P_{n,B_n}^1 D_{h_n}^*(\mathbf{P}_{n,B_n})$$
$$= E_{B_n}(\Psi_{h_n}(\mathbf{P}_{n,B_n}) - \Psi_{h_n}(P_0)) + E_{B_n}(P_{n,B_n}^1 - P_0)D_{h_n}^*(\mathbf{P}_{n,B_n})$$
$$+ E_{B_n}P_0 D_{h_n}^*(\mathbf{P}_{n,B_n})$$
$$= E_{B_n}(P_{n,B_n}^1 - P_0)D_{h_n}^*(\mathbf{P}_{n,B_n}) + E_{B_n}R_{20,h_n}(\mathbf{P}_{n,B_n},P_0).$$

We now take the derivative w.r.t. h_n on the left-hand and right-hand side, and apply (25.10). This yields

$$-\frac{d}{dh_n}\Psi_{h_n}(P_0) = E_{B_n}(P^1_{n,B_n} - P_0)D\left(\frac{d}{dh_n}K_{h_n,x_0}, \mathbf{P}_{n,B_n}\right)$$

$$+E_{B_n}\int_x \frac{d}{dh_n}K_{h,x_0}(x)R_x(\mathbf{P}_{n,B_n}, P_0)dx + 0.5dC_n\sigma_n n^{-1/2}h_n^{-d/2-1}.$$

We now write

$$\frac{d}{dh_n}K_{h_n,x_0} = h_n^{-1}\left(h_n\frac{d}{dh_n}K_{h_n,x_0}\right) \equiv h_n^{-1}K_{1,h_n,x_0},$$

where $K_{1,h,x_0} = h\frac{d}{dh}K_{h,x_0}$. Then, we can represent the equation above as follows:

$$-\frac{d}{dh_n}\Psi_{h_n}(P_0) = h_n^{-1}E_{B_n}(P^1_{n,B_n} - P_0)D(K_{1,h_n,x_0}, \mathbf{P}_{n,B_n})$$

$$+h_n^{-1}E_{B_n}\int_x K_{1,h_n,x_0}(x)R_x(\mathbf{P}_{n,B_n}, P_0)dx + 0.5dC_n\sigma_n n^{-1/2}h_n^{-d/2-1}.$$

As shown above, we have $\mathrm{Var}D(K_{1,h,x_0}, P) = O(h^{-d})$. Thus, we can analyze the empirical process term in the same manner as we analyzed the empirical process term $E_{B_n}(P^1_{n,B_n} - P_0)D^*_{h_n}(\mathbf{P}^*_{n,B_n,h_n})$ in the CV-TMLE (Theorem 25.1). Therefore, by Lemma 25.2, under the stated conditions,

$$\sqrt{nh_n^d}E_{B_n}(P^1_{n,B_n} - P_0)D^*_{h_n}(\mathbf{P}_{n,B_n}) = \sqrt{nh_n^d}(P_n - P_0)h_n^{-d/2}D^*_{h_n}(K_{1,h_n,x_0}, \mathbf{P}_1) + o_P(1),$$

and the right-hand side converges in distribution to $N(0, \sigma_{10}^2)$. In other words,

$$\sqrt{nh_n^d}E_{B_n}(P^1_{n,B_n} - P_0)D^*_{h_n}(\mathbf{P}_{n,B_n}) = Z_n + o_P(1),$$

where $Z_n \Rightarrow_d N(0, \sigma_{01}^2)$.

Combined with assumption (25.9) this latter result yields:

$$-\frac{d}{dh_n}\Psi_{h_n}(P_0) = h_n^{-1}(nh_n^d)^{-0.5}(Z_n + o_P(1)) + 0.5dC_n\sigma_n n^{-1/2}h_n^{-d/2-1}.$$

We now apply Lemma 25.3 below to obtain the expression for $\frac{d}{dh_n}\Psi_{h_n}(P_0)$. This yields

$$-h_n^{J_0^*-1}J_0B_0(J_0^*) + o(h_n^{J_0^*-1}) = h_n^{-1}(nh_n^d)^{-0.5}(Z_n + o_P(1)) + 0.5dC_n\sigma_n n^{-1/2}h_n^{-d/2-1}.$$

Thus,

$$-h_n^{J_0^*}J_0^*B_0(J_0^*) + o(h_n^{J_0^*}) = (nh_n^d)^{-0.5}(Z_n + 0.5dC_n\sigma_n + o_P(1)).$$

In the special case that $C_n = O_P(1)$, this corresponds with $h_n^{2J_0^*} = (nh_n^d)^{-1}$, giving the optimal rate $h_n = n^{-1/(2J_0^*+d)}$. This completes the proof. □

Lemma 25.3. *Suppose that $P_0 \in \mathcal{M}(J_0)$ and the degree $J-1$ of the kernel K satisfies $J \geq J_0$*

Then,

$$\frac{d}{dh}\Psi_{h,J_0}(P_0) = h^{J_0-1}J_0 B_0(J_0) + o(h^{J_0-1}).$$

Proof. We have

$$\begin{aligned}
\frac{d}{dh}\Psi_h(P_0) &= \frac{d}{dh}\int_x K_{h,x_0}(x)\Psi(P_0)(x)dx \\
&= \frac{d}{dh}\int_y K(y)\psi_0(x_0+hy)dy \\
&= \frac{d}{dh}\int_y K(y)\left\{\psi_0(x_0) + \sum_{j=1}^{J_0-1}\sum_{m,\sum_l m_l=j}\frac{h^j \prod_l y_l^{m_l}}{\prod_l m_l!}\psi_0^m(x_0)\right\} \\
&\quad + \frac{d}{dh}\int_y h^{J_0}K(y)\sum_{m,\sum_l m_l=J_0}\frac{\prod_l y_l^{m_l}}{\prod_l m_l!}\psi_0^m(x_0) + o(h^{J_0}) \\
&= J_0 h^{J_0-1}\int_y K(y)\sum_{m,\sum_l m_l=J_0}\frac{\prod_l y_l^{m_l}}{\prod_l m_l!}\psi_0^m(x_0) + o(h^{J_0-1}) \\
&= J_0 h^{J_0-1}B_0(J_0) + o(h^{J_0-1}). \quad\square
\end{aligned}$$

Remark. If C_n converges to infinity, then our bandwidth h_n oversmooths relative to the optimal rate, and, as a consequence, we will not have that $(nh_n^d)^{0.5}(\psi_{h,n}^* - \psi_0)$ converges to a normal distribution. We would still have that $(nh_n^d)^{0.5}(\psi_{h,n}^* - \psi_{h_n,0}) \Rightarrow_d N(0,\sigma_0^2)$, but the bias is now larger than the standard error. Therefore, in that case, we want to shrink down this choice h_n. Specifically, if we choose $h_n^1 = C_n^{-2/(2J_0^*+d)}h_n$, then h_n^1 would follow the optimal rate. However, this choice of scaling is unknown due to J_0 being unknown.

25.5 Generalization of Result for Data-Adaptive Bandwidth Selector

Let $\Psi : \mathcal{M} \to \mathbf{R}$ be our target parameter of interest, which is not pathwise differentiable. Let $\Psi_h : \mathcal{M} \to \mathbf{R}$, $h \in \mathbf{R}_{\geq 0}$, be a family of pathwise differentiable parameters approximating Ψ in the sense that $\{\Psi_h(P_0) \in \mathbf{R} : h \in \mathbf{R}_{\geq 0}\}$ represent a family of pathwise differentiable approximations of $\Psi(P_0) = \lim_{h\to 0}\Psi_h(P_0)$. In this chapter, we assumed that this family was generated by a kernel smoother, but in this section we allow general definitions of this family of target parameters. Let $R_{20,h}(P,P_0)$ be defined by

$$\Psi_h(P) - \Psi_h(P_0) = -P_0 D_h^*(P) + R_{20,h}(P,P_0),$$

where $D_h^*(P)$ is the canonical gradient of $\Psi_h : \mathcal{M} \to \mathbf{R}$ at P. Suppose that $D_h^*(P_0) = h^{-\alpha_0} \tilde{D}_h(P_0)$ for some $\tilde{D}_h(P_0)$, where $E_{P_0}\{\tilde{D}_h(P_0)\}^2 \to \sigma_0^2 > 0$ for some σ_0^2 as $h \to 0$: i.e., the variance of the canonical gradient behaves as $h^{-2\alpha_0}$ for some $\alpha_0 > 0$. In addition, regarding the bias of this family, let $\Psi_h(P_0) - \Psi(P_0) = h^{J_0} B_0(J_0) + o(h^{J_0})$ for some $J_0 > 0$ and term $B_0(J_0)$. We allow that both α_0 and J_0 are unknown and can depend on the true data distribution P_0. In the previous sections based on the kernel-smooth approximations of Ψ based on a d-dimensional kernel, $\alpha_0 = 2d$ was known, while J_0 represented the unknown underlying smoothness of $\Psi(P_0)$ at x_0. Let $\sigma_{0h}^2 \equiv E_{P_0}\{D_h^*(P_0)\}^2$ be the variance of the canonical gradient. By the above assumption, we have that $\sigma_{0h}^2 \approx h^{-2\alpha_0}\sigma_0^2$. Let σ_{nh}^2 be a consistent estimator of σ_{0h}^2 so that $\sigma_{nh}^2/\sigma_{0h}^2 \to 1$ in probability as $n \to \infty$. Let $r_h(n) \equiv (n/\sigma_{hn}^2)^{0.5}$ represent the standardizing rate for the CV-TMLE of $\Psi_h(P_0)$. Note that $r_h(n)$ behaves as $h^{\alpha_0}n^{0.5}$. Let ψ_{nh}^* be a CV-TMLE and assume that under appropriate conditions and any sequence $h_n \to 0$

$$r_{h_n}(n)(\psi_{nh_n}^* - \psi_{0h_n}) \Rightarrow_d N(0, 1) \text{ as } n \to \infty.$$

This can be established in an analogue manner as in our analysis of the CV-TMLE in the previous section. Thus, ψ_{nh}^* has a variance that behaves as $h^{-2\alpha_0}n^{-1}$ and a bias that behaves as h^{J_0} so that an optimal rate for h w.r.t. MSE is given by

$$h_n^* = n^{-1/2(J_0+\alpha_0)}.$$

We now want to propose a data-adaptive selector h_n and establish that it behaves as h_n^*, analogue to our proposal and analysis in the previous section. As in the previous section, we will base the selector h_n on a one-step estimator:

$$\psi_{nh}^1 = E_{B_n} \Psi_h(\mathbf{P}_{n,B_n}) + E_{B_n} P_{n,B_n}^1 D_h^*(\mathbf{P}_{n,B_n}),$$

where $\mathbf{P}_{n,B_n} \in \mathcal{M}$ is an estimator of P_0 based on the training sample $\{O_i : B_n(i) = 0\}$. Let h_n be a solution of

$$\frac{d}{dh_n}\psi_{nh_n}^1 + C_n \frac{d}{dh_n}\sigma_{nh_n}n^{-0.5} = 0 \text{ for some user supplied sequence } C_n.$$

We refer to the previous section for a discussion regarding the existence of a solution and the choice C_n, where C_n is either bounded uniformly in n or converges very slowly to infinity.

Theorem 25.4. *We make the following assumptions:*

- $\frac{d}{dh}D_h^*(P) = h^{-1}\tilde{D}_{1h}(P)$ *for some* $\tilde{D}_{1h}(P)$, *where* $h^{2\alpha_0}P_0\{\tilde{D}_{1h}(P_0)\}^2 \to \sigma_{10}^2$ *for some* $\sigma_{10}^2 > 0$ *as* $h \to 0$. *In other words,* $\tilde{D}_{1h}(P_0)$ *behaves as* $D_h^*(P_0)$ *as a function of* h.
- *for any sequence* $h \to 0$, *we have*

$$E_{B_n}(P_{n,B_n}^1 - P_0)\tilde{D}_{1h_n}(\mathbf{P}_{n,B_n}) = Z_n/(n^{0.5}h_n^{\alpha_0})+o_P(n^{-0.5}h_n^{-\alpha_0}), \text{ where } Z_n \Rightarrow_d N(0, \sigma_{01}^2).$$

-

$$h_n E_{B_n}\frac{d}{dh}R_{20,h_n}(\mathbf{P}_{n,B_n}, P_0) = o_P(n^{-0.5}h_n^{-\alpha_0}).$$

-
$$-\frac{d}{dh}\sigma_{nh} = h^{-(\alpha_0+1)}c_{nh} + o_P(h^{-(\alpha_0+1)}),$$

where $c_{nh} > 0$ is bounded away from zero and infinity uniformly in $h \in (0, \delta)$ for some $\delta > 0$.

- for any sequence $h \to 0$, we have

$$-\frac{d}{dh}\Psi_h(P_0) \sim c(J_0, h)h^{J_0-1} + o(h^{J_0-1}),$$

where $c(J_0, h)$ is bounded away from zero and infinity uniformly in $h \in (0, \delta)$ for some $\delta > 0$.

Then,

$$c(J_0, h_n)h_n^{J_0} + o(h_n^{J_0}) = n^{-0.5}h_n^{-\alpha_0}\{Z_n - c_{n,h_n}C_n + o_P(1)\}.$$

This proves that if $\lim\sup_n C_n < \infty$, then h_n achieves the optimal rate by balancing the square bias with the variance, and, in general,

$$h_n \sim n^{-1/2(J_0+\alpha_0)}C_n^{1/(J_0+\alpha_0)}.$$

Since $D_h^*(P_0) = h^{-\alpha_0}\tilde{D}_h(P_0)$ where $P_0\{\tilde{D}_h(P_0)\}^2 \to \sigma_0^2$, the third bullet assumption on $\frac{d}{dh}\sigma_{nh}$ is reasonable. Similarly, since $\Psi_h(P_0) - \Psi(P_0) = h^{J_0}B_0(J_0) + o(h^{J_0})$, the fourth bullet assumption is reasonable as well.

Proof. We will analyze this data-adaptive selector h_n, analogue to the proof of Theorem 25.3. We have

$$\psi_{nh_n}^1 - \Psi_{h_n}(P_0) = E_{B_n}(P_{n,B_n}^1 - P_0)D_{h_n}^*(\mathbf{P}_{n,B_n}) + E_{B_n}R_{20,h_n}(\mathbf{P}_{n,B_n}, P_0).$$

Taking the derivative w.r.t. h at $h = h_n$ and using the definition of h_n yields

$$C_n\frac{d}{dh_n}\sigma_{nh_n}/n^{0.5} - \frac{d}{dh_n}\Psi_{h_n}(P_0) = E_{B_n}(P_{n,B_n}^1 - P_0)\frac{d}{dh_n}D_{h_n}^*(\mathbf{P}_{n,B_n})$$

$$+ E_{B_n}\frac{d}{dh_n}R_{20,h_n}(\mathbf{P}_{n,B_n}, P_0).$$

We assumed that $\frac{d}{dh}D_h^*(P) = h^{-1}\tilde{D}_{1h}(P)$, where the $h^{2\alpha_0}P\{\tilde{D}_{1h}(P)\}^2 \to \sigma_1^2$ as $h \to 0$. We assume that under regularity conditions (analogue to the conditions needed in the proof of the asymptotic normality of the one-step estimator ψ_{nh}^1),

$$E_{B_n}(P_{n,B_n}^1 - P_0)\tilde{D}_{1h_n}(\mathbf{P}_{n,B_n}) = Z_n/(n^{0.5}h_n^{\alpha_0}) + o_P(1), \text{ where } Z_n \Rightarrow_d N(0, \sigma_{01}^2).$$

In addition, we assumed that

$$h_nE_{B_n}\frac{d}{dh_n}R_{20,h_n}(\mathbf{P}_{n,B_n}, P_0) = o_P(n^{-0.5}h_n^{-\alpha_0}).$$

This yields

$$-\frac{d}{dh_n}\Psi_{h_n}(P_0) = h_n^{-1}Z_n n^{-0.5}h_n^{-\alpha_0} + h_n^{-1}o_P(n^{-0.5}h_n^{-\alpha_0}) - C_n n^{-0.5}\frac{d}{dh_n}\sigma_{nh_n}.$$

We also assumed that

$$-\frac{d}{dh}\sigma_{nh} = h^{-(\alpha_0+1)}c_{nh},$$

for a sequence $c_{nh} > 0$ bounded away from 0 and ∞. This now yields

$$-\frac{d}{dh_n}\Psi_{h_n}(P_0) = Z_n n^{-0.5}h_n^{-(\alpha_0+1)} + o_P(n^{-0.5}h_n^{-(\alpha_0+1)}) - C_n n^{-0.5}c_{nh_n}h_n^{-(\alpha_0+1)}.$$

We assumed that

$$-\frac{d}{dh}\Psi_h(P_0) \sim c(J_0, h)h^{J_0-1} + o(h^{J_0-1}),$$

where $c(J_0, h)$ is bounded away from zero and infinity uniformly in $h \in (0, \delta)$ for some $\delta > 0$. Imputing this in the last equality yields

$$c(J_0, h_n)h_n^{J_0} + o(h_n^{J_0}) = n^{-0.5}h_n^{-\alpha_0}\{Z_n - c_{n,h_n}C_n + o_P(1)\}.$$

This proves that if $\limsup_n C_n < \infty$, then h_n balances the square bias with the variance, and, in general,

$$h_n \sim n^{-1/2(J_0+\alpha_0)}C_n^{1/(J_0+\alpha_0)}.\quad\square$$

25.5.1 Selecting among Different Classes of Pathwise Differentiable Approximations of Target Parameter

Let $\{\Psi_{h,j}(P) : h\}$ is a family of target parameters approximating $\Psi(P)$ with tuning parameter $h > 0$, indexed by a discrete choice $j = 1, \ldots, J$. For example, $\Psi_{h,j}(P)$ could be our kernel smoothed approximation of $\Psi(P)$ in which the kernel is chosen to be orthogonal to polynomials up to degree j, as analyzed in the previous sections. In the previous sections, we a priori selected j large, but in practice the practical performance could differ substantially for different choices of j. Let $\psi^*_{h,j,n}$ be a CV-TMLE of $\Psi_{h,j}(P_0)$ for any h, j. Let $\sigma^2_{h,j,n}$ be an estimator of the variance of the canonical gradient $D^*_{h,j}(P_0)$ of $\Psi_{h,j}$ at P_0 so that for a sequence $h_n \to 0$

$$(\sigma^{-1}_{h_n,j,n}n^{0.5})(\psi^*_{h_n,j,n} - \Psi_{h_n,j}(P_0)) \Rightarrow_d N(0, 1).$$

Let $h_n(j)$ be a data-adaptive selector of the type proposed above and suppose that the regularity conditions hold so that $h_n(j)$ optimally trades-off the variance and bias of $\psi^*_{h,j,n}$ w.r.t. ψ_0, $j = 1, \ldots, J$. Let $\psi^*_{j,n} \equiv \psi^*_{h_n(j),j,n}$ and $\sigma_{j,n} = \sigma_{h_n(j),j,n}$. How does one select among different approximation strategies indexed by j? Since each $h_n(j)$ is chosen so that the square of the bias of $\psi^*_{j,n}$ is of the same order as the variance of $\psi_{j,n}^*$, for all these estimators $\psi_{j,n}^*$ we have that the MSE is of the same order as $\sigma_{j,n}^2/n$. Therefore, a sensible selector j_n of j is defined as the minimizer of the estimated variance of the estimator:

$$j_n \equiv \arg\min_j \sigma_{j,n}^2.$$

This choice guarantees that we select with probability tending to 1 the family for which $\psi_{j,n}^*$ converges at the fastest rate to $\Psi(P_0)$. This can be viewed as a generalization of the empirical efficiency maximization approach for pathwise differentiable target parameters (Rubin and van der Laan 2008).

For the sake of statistical inference, we would select a slightly under-smoothed version of $h_n(j)$ for each j, so that asymptotic bias can be ignored (e.g., the constant C_n in the definition of the data-adaptive selector $h_n(j)$ is chosen to converge to infinity at a slow rate such as $\log n$). Let $h_n(j)$ denote such a choice. Regarding statistical inference, just selecting the confidence interval based on an undersmoothed version of $\psi_{j_n,n}^*$ as if j_n is a priori-specified might result in too optimistic inference due to the fact that j_n is chosen to minimize the width of the confidence interval. We propose the following adjustment of the confidence interval. We have

$$Z_n(j) \equiv \sigma_{j,n}^{-1} n^{0.5}(\psi_{j,n}^* - \Psi_{h_n(j),j}(P_0)) = \frac{1}{\sqrt{n}} \sum_{i=1}^n \sigma_{j,n}^{-1} D_{h_n(j),j}^*(P_0) + o_P(1), \ j = 1, \dots, J.$$

Therefore, under regularity conditions, one has that $Z_n = (Z_n(j) : j) \Rightarrow N(0, \Sigma_0)$, where

$$\Sigma_0(j_1, j_2) = \lim_{n\to\infty} P_0 \sigma_{j_1,n}^{-1} \sigma_{j_2,n}^{-1} D_{h_n(j_1),j_1}^*(P_0) D_{h_n(j_2),j_2}^*(P_0)$$

is the asymptotic correlation of the standardized versions of the estimators $\psi_{j_1,n}^*$ and $\psi_{j_2,n}^*$. Here $\sigma_{0,h,j}^2 \equiv P_0\left\{D_{h,j}^*(P_0)\right\}^2$ denotes the true variance of $D_{h,j}^*(P_0)$, and we implicitly assumed that the covariance limit exists as a limit almost everywhere. Note that $\Sigma_0(j, j) = 1$, $j = 1, \dots, J$. Let $q_{0.95}$ be the 0.95-quantile of the max of the absolute value of $N(0, \Sigma_0)$. It follows that $\psi_{j,n}^* \pm q_{0.95}\sigma_{j,n}/n^{0.5}$, $j = 1, \dots, J$, is a simultaneous asymptotic 0.95-confidence interval for ψ_0. Therefore $\psi_{j_n,n}^* \pm q_{0.95}\sigma_{j,n}/n^{0.5}$ is an 0.95-confidence interval for ψ_0. We can estimate $q_{0.95}$ consistently by replacing Σ_0 in its definition by a consistent estimator Σ_n.

25.6 Example: Estimation of a Univariate Density at a Point

Let $O \sim P_0$, $\mathcal{M}$ be nonparametric, and $\Psi(P_0) = p_0(x_0)$ for a given point x_0. We define $\Psi_h(P) = \int_t K_{h,x_0}(t)p(t)dt$, where K is a $J - 1$-orthogonal kernel and $K_{h,x_0}(t) = h^{-1}K((t - x_0)/h)$. The efficient influence curve of Ψ_h at P is given by $D_h^*(P) = K_{h,x_0}(O) - \Psi_h(P)$. It satisfies the identity $\Psi_h(P) - \Psi_h(P_0) = -P_0 D_h^*(P)$ so that $R_{20,h}(P, P_0) = 0$. Let $\mathbf{p}_n$ be an initial estimator of p_0. Let $B_n \in \{0, 1\}^n$ be a cross-validation scheme, and $\mathbf{p}_{n,B_n}$ be the initial density estimator applied to the training sample $\{O_i : B_n(i) = 0\}$. Consider the local least favorable submodel through p at $\epsilon = 0$ given by

$$\mathbf{p}_{\epsilon,h}(o) = (1 + \epsilon D_h^*(P)(o))p(o).$$

One could either use this local least favorable submodel, or the universal least favorable submodel

$$\mathbf{p}_{\epsilon,h}(o) = p(o) \exp\left(\int_0^\epsilon D_h^*(p_e)(o)de\right).$$

Let $\mathbf{p}_{n,B_n,\epsilon,h}$ be this least favorable submodel through $\mathbf{p}_{n,B_n}$. Let

$$\epsilon_n = \arg\min_\epsilon E_{B_n} P_{n,B_n} L(\mathbf{p}_{n,B_n,\epsilon,h}),$$

where $L(p)(o) = -\log p(o)$ is the log-likelihood loss. This defines $\mathbf{p}_{n,B_n,h}^* = \mathbf{p}_{n,B_n,\epsilon_n,h}$ and thereby the CV-TMLE $\psi_{h,n}^* = E_{B_n} \Psi_h(\mathbf{P}_{n,B_n}^*)$ of $\Psi_h(P_0)$. Let h_n be defined as a solution of

$$\frac{d}{dh_n}\psi_{h_n,n} + C_n\sigma_n\frac{d}{dh_n}(nh_n)^{-0.5} = 0, \tag{25.11}$$

where we define $\psi_{h,n}$ as the CV-one-step estimator:

$$\psi_{h,n} = E_{B_n} P_{n,B_n}^1 K_{h,x_0} = P_n K_{h,x_0} = \frac{1}{nh}\sum_{i=1}^n K((x_0 - O_i)/h).$$

As in the description of our method, we can set $C_n \in \{-Z_{1-\alpha_n}, Z_{1-\alpha_n}\}$ for some sequence α_n. In addition, $\sigma_n^2 = \tilde{p}_n(x_0) \int K(y)^2 dy$ is a consistent estimator of $\sigma_0^2 = p_0(x_0) \int K(y)^2 dy$. Here $\tilde{p}_n(x_0)$ can be replaced by any consistent estimator, such as our CV-TMLE $\psi_{h_n^0,n}^*$ for a certain bandwidth selector h_n^0 for which $nh_n^0 \to \infty$. Note that, since K_{h,x_0} does not depend on nuisance parameters estimated on the training sample, the CV-one-step estimator used in the bandwidth selector h_n reduces to a standard kernel density estimator at x_0. Our proposed CV-TMLE is given by $\psi_n^* = \psi_{h_n,n}^*$.

Let's now apply Theorem 25.1. Suppose that $p_0(x_0) > 0$ so that it is also bounded away from zero on a small neighborhood of t_0. Let's first consider condition (25.3). We have

$$\| D_h^*(\mathbf{P}_n) - D_h^*(P_0) \|_{P_0} = | \Psi_h(\mathbf{P}_n) - \Psi_h(P_0) |$$

$$= | \int_x h^{-1} K((x - x_0)/h)(\mathbf{p}_n(x) - p_0(x))dx |$$

$$= | \int K(y)(\mathbf{p}_n(x_0 + hy) - p_0(x_0 + hy))dy |$$

$$= O\left(h^{-1} \int_{x_0-h}^{x_0+h} | (\mathbf{p}_n - p_0)(x) | \, dx\right).$$

For example, if $\mathbf{p}_n - \mathbf{p}_0$ converges uniformly to zero over a small interval around 0 at rate $r_1(n)$, then it follows that the right-hand side is $O_P(r_1(n))$. We can also bound it more conservatively with Cauchy-Schwarz inequality: $h^{-1} \int I_{(x_0-h,x_0+h)} \mid \mathbf{p}_n - p_0 \mid dx \leq O\left(h^{-0.5} \left(\int (\mathbf{p}_n - p_0)^2 dx\right)^{0.5}\right)$, where the integral is over a fixed interval containing x_0. In that case, it would suffice to have that $\| \mathbf{p}_n - p_0 \|_{t_0,P_0} = O_P(r(n))$ for some polynomial rate so that we obtain the bound $O_P(h^{-0.5} r(n))$. Here we defined $\| p - p_0 \|_{x_0,P_0}^2 = \int_{x_0-a}^{x_0+a} (p - p_0)^2 dP_0$ for a fixed $a > 0$. This verifies the first condition (25.3). Note

$$hP_0 D_h^*(p_0)^2 = hP_0 \{K_{h,x_0} - \Psi_h(P_0)\}^2 = hP_0 K_{h,x_0}^2 + O(h) = \int K(y)^2 p_0(x_0+hy) dy + O(h)$$

which converges to $\sigma_0^2 = p_0(x_0) \int K(y)^2 dy$. Condition (25.5) automatically holds since $R_{20,h} = 0$. Thus, we can now apply Theorem 25.1, which proves that

$$\sqrt{nh}(\psi_{h,n}^* - \psi_{h,0}) = \sqrt{nh}(P_n - P_0)D_h^*(P_0) + o_P(1) \Rightarrow_d N(0, \sigma_0^2).$$

Assume that p_0 is J_0-times continuously differentiable at x_0 and let $J_0^* = \min(J_0, J)$. Then, we have $h^{-J_0^*}(\psi_{h0} - \psi_0) \rightarrow B_0(J_0^*)$, where

$$B_0(J_0^*) \equiv \psi_0^{J_0^*}(x_0) \int_y K(y) \frac{y^{J_0^*}}{J_0^*!} dy. \tag{25.12}$$

Application of Theorem 25.2 proves that if we select an optimal rate $h_n^* = n^{-1/(2J_0^*+1)}$, then

$$n^{-J_0^*/(2J_0^*+1)}(\psi_{h_n,n}^* - \psi_0) \Rightarrow_d N(B_0(J_0^*), \sigma_0^2).$$

Since the conditions of Theorem 25.2 are the same as of Theorem 25.1 with K_{h,x_0} replaced by $K_{1,h,x_0} = h\frac{d}{dh}K_{h,x_0}$ (and we showed that the latter is a similarly behaved kernel), the verification of the conditions is the same as above. Application of Theorem 25.3 proves that $h_n \sim (C_n^2/n)^{-1/(2J_0^*+1)}$, which achieves the optimal rate if $C_n = O_P(1)$.

This proves the following theorem.

Theorem 25.5. *Consider the CV-TMLE $\psi_{h,n}^*$ defined above and the bandwidth selector h_n (25.11). Consider also the definition $B_0(J)$ (25.12) and σ_0^2. We make the following assumptions:*

- $p_0(x_0) > 0$.
- *p_0 is J_0-times continuously differentiable at x_0 for an unknown J_0.*
- *$\| \mathbf{p}_n - p_0 \|_{x_0,P_0} = O_P(r(n))$ for some polynomial rate (i.e., n^{-p} for some $p > 0$), where $\| p - p_0 \|_{x_0,P_0}^2 = \int_{x_0-a}^{x_0+a} (p - p_0)^2 dP_0$ for a fixed $a > 0$.*

Let $J_0^* = \min(J_0, J)$. We have that $h_n \sim (C_n^2/n)^{-1/(2J_0^*+1)}$, and, if $C_n = O_P(1)$, then

$$(nh_n^d)^{0.5}(\psi_{h_n,n}^* - \psi_0) \Rightarrow_d N(B_0(J_0^*), \sigma_0^2).$$

Suppose $C_n = O_P(1)$ and let $h_n^* = h_n/\log n$ be an undersmoothed version of h_n. Let $\psi_{h_n^*,n}^*$ be the corresponding CV-TMLE. Then,

$$(nh_n^*)^{-0.5}(\psi_{h_n^*,n}^* - \psi_0) \Rightarrow N(0, \sigma_0^2).$$

In particular,

$$\psi_{h_n^*,n}^* \pm 1.96\sigma_n/(nh_n^*)^{0.5}$$

is an asymptotic 95%-confidence interval for ψ_0.

25.7 Example: Causal Dose Response Curve Estimation at a Point

We refer to Kennedy et al. (2015) for a presentation of an estimator and inference for the causal dose response curve. The estimator in Kennedy et al. (2015) was based on an augmented IPCW-estimator instead of a CV-TMLE and used the more general local polynomial smoothing instead of our kernel smoothing.

Let $O = (W, A, Y) \sim P_0$, where $Y \in \{0, 1\}$ and $A \in [0, 1]$ is continuous valued. Let $\mathcal{M}$ be a statistical model for P_0 that only makes assumptions on the conditional density $g_0(a \mid W)$ of A, given W. Let $\bar{Q}_0(A, W) = P_0(Y = 1 \mid A, W)$, and let $Q_{W,0}$ be the probability distribution of W. The target parameter is defined as $\Psi(P) = E_P E_P(Y \mid A = a_0, W)$, where a_0 is a given value. Let $\Psi(P)(a) \equiv E_P E_P(Y \mid A = a, W)$. Let $\Psi_h(P) = \int K_{h,a_0}(a)\Psi(P)(a)da$ for a bandwidth h and $J - 1$-orthogonal kernel K. We can also represent this target parameter as $\Psi_h(Q_W, \bar{Q})$. The efficient influence curve of Ψ_h at P is given by

$$D_h^*(P) = \frac{K_{h,a_0}(A)}{g(A \mid W)}(Y - \bar{Q}(A, W)) + \int_a K_{h,a_0}(a)\bar{Q}(a, W)da - \Psi_h(P)$$
$$\equiv D_h(\bar{Q}, g) - \Psi_h(P).$$

We can also represent this efficient influence curve as $D_h^*(\bar{Q}, g, \psi_h)$. We have

$$\Psi_h(Q_W, \bar{Q}) - \Psi_h(Q_{W,0}, \bar{Q}_0) = -P_0 D_h^*(P) + R_{20,h}(P, P_0),$$

where

$$R_{20,h}(P, P_0) = E_{P_0} \int_a K_{h,a_0}(a)(\bar{Q} - \bar{Q}_0)(a, W)\frac{(g - g_0)(a \mid W)}{g(a \mid W)}da.$$

Given a cross-validation scheme $B_n \in \{0, 1\}^n$, let $Q_{W,n,B_n}, g_{n,B_n}, \bar{Q}_{n,B_n}$ be initial estimators applied to the training sample $\{O_i : B_n(i) = 0\}$. Let

$$\text{Logit}\bar{Q}_{n,B_n,\epsilon,h} = \text{Logit}\bar{Q}_{n,B_n} + \epsilon C(h, a_0, g_{n,B_n}),$$

where $C(h, a_0, g)(A, W) = K_{h,a_0}(A)/g(A \mid W)$. Let

$$\epsilon_n = \arg\min_\epsilon E_{B_n} P^1_{n,B_n} L(\bar{Q}_{n,B_n,\epsilon,h}),$$

where $L(\bar{Q})(O) = -Y \log \bar{Q}(A, W) - (1 - Y) \log(1 - \bar{Q}(A, W))$ is the log-likelihood loss for $\bar{Q}_0$. Let $\bar{Q}^*_{n,B_n,h} = \bar{Q}_{n,B_n,\epsilon_n,h}$.

We could now also target Q_{W,n,B_n} with a least favorable submodel with score $D^*_{h,W}(Q_W, \bar{Q}) = \int_a K_{h,a_0}(a)\bar{Q}(a, W)da - \Psi_h(Q_W, \bar{Q})$ so that we have

$$E_{B_n} P^1_{n,B_n} D^*_{h,W}(Q^*_{W,n,B_n,h}, \bar{Q}^*_{n,B_n,h}) = 0.$$

For example, we could define the universal least favorable model

$$dQ_{W,n,B_n,\epsilon,h} = dQ_{W,n,B_n} \exp\left(\int_0^\epsilon D^*_{h,W}(Q_{W,n,B_n,x,h}, \bar{Q}^*_{n,B_n,h})dx\right),$$

and $\epsilon_n = \arg\min_\epsilon E_{B_n} P^1_{n,B_n} L_1(Q_{W,n,B_n,\epsilon,h})$, where $L(Q_W) = -\log q_W$ is the log-likelihood loss. The targeted version of Q_{W,n,B_n} is then defined as $Q^*_{W,n,B_n,h} = Q_{W,n,B_n,\epsilon_n,h}$. Indeed, we then have

$$E_{B_n} P^1_{n,B_n} D^*_{h,W}(Q^*_{W,n,B_n,h}, \bar{Q}^*_{n,B_n,h}) = 0.$$

The CV-TMLE of $\Psi_h(P_0)$ is then defined as $E_{B_n} \Psi_h(Q^*_{n,B_n,h})$. where $Q^*_{n,B_n,h} = (Q^*_{W,n,B_n,h}, \bar{Q}^*_{n,B_n,h})$. However, note that solving the latter equation implies

$$E_{B_n} \Psi_h(Q^*_{W,n,B_n,h}, \bar{Q}^*_{n,B_n,h}) = E_{B_n} P^1_{n,B_n} \int_a K_{h,a_0}(a)\bar{Q}^*_{n,B_n,h}(a, \cdot)da.$$

Thus, the targeting of Q_{W,n,B_n} ended up redefining the evaluation of the target parameter. Therefore, the targeting step Q_{W,n,B_n} and defining the CV-TMLE as $E_{B_n} \Psi_h(Q^*_{n,B_n,h})$ is equivalent with not targeting Q_{W,n,B_n} but just defining the CV-TMLE as

$$\psi^*_{n,h} = E_{B_n} P^1_{n,B_n} \int_a K_{h,a_0}(a)\bar{Q}^*_{n,B_n,h}(a, \cdot). \tag{25.13}$$

Note that $D^*_h(Q^*_{W,n,B_n,h}, \bar{Q}^*_{n,B_n,h}, g_{n,B_n}) = D^*_h(\bar{Q}^*_{n,B_n,h}, g_{n,B_n}, \psi^*_{n,h})$ so that the CV-TMLE solves

$$E_{B_n} P^1_{n,B_n} D^*_h(\bar{Q}^*_{n,B_n,h}, g_{n,B_n}, \psi^*_{n,h}) = 0.$$

We define the bandwidth selector h_n as the solution of

$$\frac{d}{dh_n}\psi_{n,h_n} + C_n\sigma_n\frac{d}{dh_n}(nh_n)^{-0.5} = 0, \tag{25.14}$$

where C_n is the smallest negative or positive constant for which a solution exists. For example, if possible, we select $C_n \in \{-1.96, 1.96\}$. Here σ_n^2 is a consistent estimator of

$$\sigma_0^2 = \int K^2(y) dy \int_w \frac{\bar{Q}_0(1 - \bar{Q}_0)(a_0, w)}{g_0(a_0 \mid w)} dQ_{W,0}(w). \tag{25.15}$$

Here $\psi_{n,h}$ is the CV-one-step estimator (a cross-validated version of the estimator in Kennedy et al. 2015):

$$\psi_{n,h} = E_{B_n} P_{n,B_n}^1 D_h(\bar{Q}_{n,B_n,h}^*, g_{n,B_n}).$$

The proposed CV-TMLE of ψ_0 is now defined as

$$E_{B_n} \Psi_{h_n}(Q_{n,B_n,h_n}^*) = E_{B_n} P_{n,B_n}^1 \int_a K_{h_n,a_0}(a) \bar{Q}_{n,B_n,h_n}^*(a, \cdot) da.$$

We now apply Theorem 25.1. We assume that $g_0(a_0 \mid W) > \delta > 0$ a.e. for some $\delta > 0$. Using Cauchy-Schwarz inequality, we can bound $\| D_h^*(Q_{n,B_n,h_n}^*, g_{n,B_n}) - D_h^*(Q_0, g_0) \|_{P_0}$ by the sum of $h^{-0.5} \| \bar{Q}_{n,B_n,h_n}^* - \bar{Q}_0 \|_{a_0,P_0}$ and $h^{-0.5} \| g_{n,B_n} - g_0 \|_{a_0,P_0}$, where

$$\| f \|_{a_0,P_0}^2 = \int_w \int_y K^2(y) f(a_0 + h_1 y, w)^2 dy dP_0(w)$$

for some fixed (arbitrarily small) $h_1 > 0$. We assume that $\max(\| g_{n,B_n} - g_0 \|_{a_0,P_0}, \| \bar{Q}_{n,B_n,h_n}^* - \bar{Q}_0 \|_{a_0,P_0}) = O_P(r(n))$ for some polynomial rate $r(n)$. It also follows that $h P_0 \{ D_h^*(P_0) \}^2 \to \sigma_0^2$ as $h \to 0$. It remains to show the second-order term condition:

$$(nh)^{0.5} E_{P_0} \int_y K(y) (\bar{Q}_{n,B_n,h_n}^* - \bar{Q}_0)(a_0 + h_n ya, W) \frac{(g_{n,B_n} - g_0)(a_0 + h_n y \mid W)}{g_{n,B_n}(a_0 + h_n y \mid W)} dy = o_P(1).$$

By using the Cauchy-Schwarz inequality, it suffices to show

$$(nh_n)^{0.5} \| \bar{Q}_{n,B_n,h_n}^* - \bar{Q}_0 \|_{a_0,P_0} \| g_{n,B_n} - g_0 \|_{a_0,P_0} = o_P(1).$$

We will assume this to hold. We can now apply Theorem 25.1, which shows that

$$\sqrt{nh_n}(\psi_{h_n,n}^* - \psi_{h_n,0}) = \sqrt{nh_n}(P_n - P_0) D_{h_n}^*(P_0) + o_P(1) \Rightarrow_d N(0, \sigma_0^2).$$

Suppose that $a \to \psi_0(a) = E_{P_0} \bar{Q}_0(a, W)$ is J_0-times continuously differentiable at $a = a_0$. Then, we have $h^{-J_0}(\psi_{h0} - \psi_0) \to B_0(J_0)$, where

$$B_0(J_0) \equiv \int_y K(y) \frac{y^{J_0}}{J_0!} dy \psi_0^{J_0}(a_0). \tag{25.16}$$

Let $J_0^* = \min(J_0, J)$. Application of Theorem 25.2 proves that if we select an optimal rate $h_n = n^{-1/(2J_0^*+1)}$, then

$$n^{-J_0^*/(2J_0^*+1)}(\psi_{h_n,n}^* - \psi_0) \Rightarrow_d N(B_0(J_0^*), \sigma_0^2).$$

Since the conditions of Theorem 25.2 are the same as of Theorem 25.1 with K_{h,a_0} replaced by $K_{1,h,a_0} = h\frac{d}{dh}K_{h,a_0}$ (and we showed that the latter is a similarly behaved kernel), the verification of the conditions is the same as above. Application of Theorem 25.3 proves that if $C_n = O_P(1)$, then $h_n \sim n^{-1/(2J_0^*+1)}$. This proves the following theorem.

Theorem 25.6. *Consider the CV-TMLE $\psi_{h,n}^*$ (25.13) defined above and the bandwidth selector h_n (25.14). Consider also the definition $B_0(J)$ (25.16), σ_0^2 (25.15), and $\| f \|_{a_0,P_0} = \sqrt{\int_w \int_y K^2(y)f(a_0 + h_1 y, w)dy dQ_{W,0}(w)}$ for some (arbitrarily small) $h_1 > 0$. We make the following assumptions:*

- $g_0(a_0 \mid W) > \delta > 0$ *a.e. for some $\delta > 0$.*
- $a \to \psi_0(a) = E_{P_0}\bar{Q}_0(a, W)$ *is J_0-times continuously differentiable at a_0 for an unknown J_0.*
- $\max(\| g_{n,B_n} - g_0 \|_{a_0,P_0}, \| \bar{Q}_{n,B_n,h_n}^* - \bar{Q}_0 \|_{a_0,P_0}) = O_P(r(n))$ *for some polynomial rate $r(n)$.*
-
$$(nh)^{0.5} \| \bar{Q}_{n,B_n,h_n}^* - \bar{Q}_0 \|_{a_0,P_0}\| g_{n,B_n} - g_0 \|_{a_0,P_0} = o_P(1).$$

- $C_n = O_P(1)$.

Let $J_0^* = \min(J_0, J)$. We have that $h_n \sim n^{-1/(2J_0^*+1)}$, and

$$n^{-J_0^*/(2J_0^*+1)}(\psi_{h_n,n}^* - \psi_0) \Rightarrow_d N(B_0(J_0^*), \sigma_0^2).$$

Let $h_n^* = h_n / \log n$ be an undersmoothed version of h_n, and let $\psi_{h_n^*,n}^*$ be the corresponding CV-TMLE. Then,

$$(nh_n^*)^{-0.5}(\psi_{h_n^*,n}^* - \psi_0) \Rightarrow N(0, \sigma_0^2).$$

In particular,

$$\psi_{h_n^*,n}^* \pm 1.96\sigma_n/(nh_n^*)^{0.5}$$

is an asymptotic 95%-confidence interval for ψ_0.

25.8 Notes and Further Reading

Rosenblatt (1956) introduced kernel density estimation, and Parzen (1962) proved its asymptotic normality. Nadaraya (1964) and Watson (1964) introduced the kernel regression estimator. Einmahl and Mason (2000) studied the convergence of kernel estimators when the bandwidth remains random even as the sample size grows. Einmahl and Mason (2005) derived stronger distributional results than we have presented in this chapter, namely confidence bands for kernel regression estimators that allow the bandwidth to remain random. For a further overview of developments in the study of kernel density estimators, see Wied and Weißbach (2012).

For our causal example, estimating the mean outcome under a continuous point treatment was previously studied by Kennedy et al. (2015). Neugebauer and van der Laan (2007) defines continuous exposure marginal structural working models as projections of the true dose-response curve onto a finite dimensional space. Unlike the parameters discussed in this chapter, the parameters indexing the projection are pathwise differentiable and, under some regularity conditions, can be estimated at a root-n rate. Rosenblum and van der Laan (2010a) describes a TMLE for the parameters indexing these working models.

Chapter 26
Higher-Order Targeted Loss-Based Estimation

Marco Carone, Iván Díaz, and Mark J. van der Laan

The objective of this chapter is to describe how the TMLE framework can be generalized to explicitly utilize higher-order rather than first-order asymptotic representations. The practical significance of this is to provide guidelines for constructing estimators that have sound behavior in finite samples and are asymptotically efficient under less restrictive conditions.

The construction of TMLEs often depends upon certain first-order asymptotic representations. For these representations to be useful, the resulting second-order remainder term must tend to zero in probability faster than $n^{-1/2}$. This ensures that the first-order approximation suffices to guide the construction of estimators and to study their asymptotic limit theory. To satisfy this condition, it must be possible to construct an estimator of the data-generating distribution, or any relevant portions thereof, that converges sufficiently fast. For example, when the density of the data-generating distribution is directly involved in the target parameter, a density estimator converging in a suitable norm at a rate faster than $n^{-1/4}$ is often required to guarantee that the second-order remainder term is negligible. In many settings

M. Carone (✉)
Department of Biostatistics, University of Washington, F-644 Health Sciences Building, 1705 NE Pacific Street, Seattle, WA 98195, USA
e-mail: mcarone@uw.edu

I. Díaz
Division of Biostatistics and Epidemiology, Department of Healthcare Policy and Research, Weill Cornell Medical College, Cornell University, 402 East 67th Street, New York, NY 10065, USA
e-mail: ild2005@med.cornell.edu

M. J. van der Laan
Division of Biostatistics and Department of Statistics, University of California, Berkeley, 101 Haviland Hall, #7358, Berkeley, CA 94720, USA
e-mail: laan@berkeley.edu

© Springer International Publishing AG 2018
M.J. van der Laan, S. Rose, *Targeted Learning in Data Science*,
Springer Series in Statistics, https://doi.org/10.1007/978-3-319-65304-4_26

however, it is rather implausible for such a condition to hold, particularly when the data unit vector is high-dimensional. The remainder term will then itself contribute to the first-order asymptotic behavior of the estimator and derail it from asymptotic linearity.

It is then natural to consider the construction of estimators based on higher-order asymptotic expansions, allowing us to account for additional analytic terms and instead require that a higher-order remainder term be asymptotically negligible. Minimal rate conditions would then be significantly relaxed. For example, if an rth order expansion exists and could be utilized, an estimator of the density function converging at a rate faster than $n^{-1/[2(r+1)]}$ would suffice for the resulting remainder term to be negligible. More importantly, even when such higher-order expansions do not exist, approximate higher-order expansions can generally be constructed, leading to concrete rate gains. In the latter case however, describing the resulting minimal rate conditions in generality is more difficult since these can be somewhat context-dependent.

In this chapter, we propose a novel second-order TMLE of a second-order pathwise-differentiable target parameter, based on n independent draws from an unknown element of a given semiparametric model. We will refer to this estimator as a 2-TMLE, with 1-TMLE corresponding to the usual first-order TMLE. Analogously to the large-sample properties of the usual TMLE requiring the asymptotic negligibility of a second-order remainder term, the asymptotic normality and efficiency of this 2-TMLE will rely on the asymptotic negligibility of a third-order remainder term. As an illustration of its construction, we develop a 2-TMLE of a point intervention g-computation parameter based on observing n independent and identically distributed copies of a random vector consisting of baseline covariates, a binary treatment and a final binary outcome. While for simplicity of presentation and brevity we exclusively focus on 2-TMLE in this chapter, the techniques are readily generalized to an arbitrary order. Further details regarding this generalization can be found in a technical report (Carone et al. 2014).

We provide a general template for constructing a 2-TMLE and for establishing its asymptotic efficiency. We show that this 2-TMLE can be constructed within the standard framework of TMLE, notably by augmenting the least favorable parametric submodel used in the standard TMLE with an additional parameter. We also demonstrate that the statistical inference of this 2-TMLE can be based on a second-order Taylor expansion, possibly providing some finite-sample improvements in the construction of confidence intervals.

In many problems, the target parameter is only first-order pathwise-differentiable and a second-order gradient does not exist. As a solution, we propose a 2-TMLE based on an approximate second-order gradient and present a corresponding theory. This additional approximation yields a supplementary bias term for the resulting 2-TMLE, referred to as the *representation error* in Robins et al. (2009), and the asymptotic linearity and efficiency theorem of the 2-TMLE requires this bias term to be asymptotically negligible as well. Control of this bias term can be carried out within the standard TMLE framework at no risk of losing the desirable properties of the first-order TMLE. In our example, we demonstrate that, for an appropriate choice of kernel and bandwidth rate, and provided the data-generating distribution satisfies certain smoothness conditions, the asymptotic efficiency of the 2-TMLE still only relies on the asymptotic negligibility of a third-order remainder term. We also propose a data-adaptive bandwidth selector based on the C-TMLE framework, essentially selecting the bandwidth maximizing bias reduction in the targeting step of the TMLE. We illustrate the use of 2-TMLE by describing how it can be implemented to infer about a point intervention g-computation parameter and we report numerical results from its use, all of which are discussed in great detail in Díaz et al. (2016).

26.1 Overview of Higher-Order TMLE

Suppose that n independent draws $O_1, \ldots, O_n$ are obtained from $P_0 \in \mathcal{M}$, where the statistical model $\mathcal{M}$ refers to the set of all possible probability distributions for the prototypical data structure O compatible with the available knowledge about P_0. Let P_n be the empirical distribution of $O_1, \ldots, O_n$ and define $O := \cup_{P \in \mathcal{M}} \text{supp}(P)$ with supp(P) denoting the support of P. Let $\Psi : \mathcal{M} \to \mathbb{R}$ be the target parameter mapping and suppose we are interested in inferring about the true parameter value $\psi_0 := \Psi(P_0)$ from the observed data. For simplicity, we focus on univariate target parameters in this article but all the developments herein can immediately be extended to Euclidean-valued target parameters with image in $\mathbb{R}^d$ as well.

26.1.1 TMLE

We focus here on cases where the parameter Ψ is sufficiently smooth in P in an appropriate sense. To be precise, denote by $L_0^2(P)$ the Hilbert space of square-integrable real-valued functions defined on the support of P and with mean zero under P, endowed with inner product $\langle h_1, h_2 \rangle_P := \int h_1(o) h_2(o) dP(o)$. Let $T_{\mathcal{M}}(P)$ be the tangent space of $\mathcal{M}$ at $P \in \mathcal{M}$, defined as the closure of the linear span of all scores of regular parametric submodels through P—this is a subspace of $L_0^2(P)$. If there exists an element $D^{(1)}(P) \in L_0^2(P)$ such that for any $P_1, P_2 \in \mathcal{M}$ the linearization

$$\Psi(P_2) - \Psi(P_1) = (P_2 - P_1)D^{(1)}(P_2) + R_2(P_1, P_2)$$
$$= -P_1 D^{(1)}(P_2) + R_2(P_1, P_2)$$

holds with $R_2(P_1, P_2)/\Delta(P_1, P_2) \to 0$ as $\Delta(P_1, P_2) \to 0$ for some distance Δ on $\mathcal{M}$, then Ψ is said to be *strongly differentiable* over $\mathcal{M}$ and any such element $D^{(1)}(P)$ is called a first-order gradient of Ψ at P (Pfanzagl 1982). Here, for any function f and (possibly signed) measure P, we write Pf to denote $\int f(o) dP(o)$. From hereon, we will informally refer to $R_2(P_1, P_2)$ as a second-order remainder.

The unique first-order gradient $D^{(1)*}(P)$ in $T_{\mathcal{M}}(P)$ is referred to as both the canonical first-order gradient and the efficient influence function. The latter terminology reflects the fact that a regular estimator of ψ_0 is asymptotically efficient if and only if it is asymptotically linear with influence function given by $D^{(1)*}(P_0)$ (Bickel et al. 1997b). Formally, a regular estimator ψ_n of ψ_0 is asymptotically efficient if and only if

$$\psi_n - \psi_0 = (P_n - P_0)D^{(1)*}(P_0) + o_P(n^{-1/2}) .$$

TMLE provides a general template for constructing asymptotically efficient substitution estimators $\psi_n^* := \Psi(P_n^*)$ of ψ_0, whose asymptotic efficiency is in part a consequence of the property

$$P_n D^{(1)*}(P_n^*) = \frac{1}{n} \sum_{i=1}^{n} D^{(1)*}(P_n^*)(O_i) = 0 , \tag{26.1}$$

and the first-order expansion

$$\Psi(P) - \Psi(P_0) = (P - P_0)D^{(1)*}(P) + R_2(P, P_0)$$
$$= -P_0 D^{(1)*}(P) + R_2(P, P_0) \tag{26.2}$$

following from the strong differentiability of Ψ over $\mathcal{M}$. Identity (26.1) can be combined with (26.2) evaluated at $P = P_n^*$ to yield that

$$\Psi(P_n^*) - \Psi(P_0) = (P_n - P_0)D^{(1)*}(P_n^*) + R_2(P_n^*, P_0) . \tag{26.3}$$

This forms the basis of any theorem establishing the asymptotic linearity and efficiency of a TMLE. Provided it can be shown that

1. $D^{(1)*}(P_n^*)$ falls in a P_0-Donsker class with probability tending to one,
2. $P_0 \left[D^{(1)*}(P_n^*) - D^{(1)*}(P_0) \right]^2$ tends to zero in probability, and
3. $R_2(P_n^*, P_0)$ tends to zero in probability faster than $n^{-1/2}$,

results from empirical process theory (e.g., van der Vaart and Wellner 1996) imply the desired asymptotic efficiency, notably that

$$\psi_n^* - \psi_0 = (P_n - P_0)D^{(1)*}(P_0) + o_P(n^{-1/2}) .$$

To construct an estimator P_n^* solving (26.1), TMLE relies on an initial estimator $P_{n,0}$ of P_0, obtained, for example, using super learning (van der Laan and Dudoit 2003; van der Vaart et al. 2006; van der Laan et al. 2007; Polley et al. 2011) for optimal performance, and a so-called least favorable parametric submodel $\{P_{n,0}(\epsilon) : \epsilon\} \subset \mathcal{M}$ such that

1. $P_{n,0}(0) = P_{n,0}$, and
2. $D^{(1)*}(P_{n,0})$ lies in the closure of the linear span of all scores of ϵ at $\epsilon = 0$.

A single update of TMLE is given by $P_{n,1} := P_{n,0}(\epsilon_n^0)$, where

$$\epsilon_n^0 := \arg\max_\epsilon P_n \log p_{n,0}(\epsilon)$$

is the MLE in the parametric submodel constructed and $p_{n,0}(\epsilon) := dP_{n,0}(\epsilon)/d\mu$ is the Radon-Nikodym derivative of $P_{n,0}$ with respect to some dominating measure μ. Subsequently, the least favorable submodel previously constructed using $P_{n,0}$ is constructed using $P_{n,1}$ and the process is repeated to map $P_{n,1}$ into a further update $P_{n,2}$. This process is repeated until convergence. The TMLE of P_0 is then defined as the limit of this iterative algorithm, say $P_n^* \in \mathcal{M}$, while the TMLE of ψ_0 is the corresponding substitution estimator $\psi_n^* := \Psi(P_n^*)$. At each step j for which ϵ_n^j is an interior point, the MLE ϵ_n^j of ϵ solves the score equation $0 = \frac{\partial}{\partial\epsilon} P_n \log p_{n,j}(\epsilon)\big|_{\epsilon=\epsilon_n^j}$, where $p_{n,j}(\epsilon) := dP_{n,j}(\epsilon)/d\mu$ is the Radon-Nikodym derivative of $P_{n,j}$ relative to μ. Therefore, we obtain that

$$P_n D^{(1)*}(P_n^*) = \frac{\partial}{\partial\epsilon} P_n \log p_n^*(\epsilon)\bigg|_{\epsilon=0} = 0$$

since the submodel $\{P_n^*(\epsilon) : \epsilon\}$ has score $D^{(1)*}(P_n^*)$ for ϵ at $\epsilon = 0$, where p_n^* is the Radon-Nikodym derivative of P_n^* relative to μ.

26.1.2 Extensions of TMLE

The above iterative algorithm relies on the log-likelihood loss, which directly motivates the nomenclature *targeted maximum likelihood estimator*. However, provided an appropriate loss function can be identified, the same idea can be applied to any representation of the target parameter as $\Psi(P) = \Psi_1(Q)$ with $Q := Q(P)$, where $Q(P)$ represents the relevant portion of P for the sake of computing $\Psi(P)$, provided $D^{(1)*}(P) := D^{(1)*}(Q, g)$ can be written in terms of Q and a nuisance parameter $g := g(P)$. This generalized algorithm, referred to as *targeted minimum loss-based estimation*, requires an initial estimator $(Q_{n,0}, g_n)$ of (Q_0, g_0), a loss function $(o, Q) \mapsto L(Q)(o)$ such that $Q_0 = \arg\min_{Q \in Q} P_0 L(Q)$ with $Q := \{Q(P) : P \in \mathcal{M}\}$, a least favorable submodel $\{Q_{n,0}(\epsilon) : \epsilon\} \subset Q$ which generally depends on g_n and is such that components of the generalized score

$$\frac{\partial}{\partial\epsilon} L(Q_{n,0}(\epsilon))\bigg|_{\epsilon=0}$$

span $D^{(1)*}(Q_{n,0}, g_n)$. An updating scheme directly analogous to that described above is then defined, where minimization of the empirical risk is performed instead of maximization of the log-likelihood—this yields a targeted estimator Q_n^* and corresponding targeted minimum loss-based estimator $\Psi_1(Q_n^*)$ of ψ_0 such that

$$P_n D^{(1)*}(Q_n^*, g_n) = 0 \, ,$$

an integral requirement for asymptotic linearity and efficiency.

A simple but key observation is that, for the sake of ascribing additional properties to the TMLE, a targeted estimate g_n^* of g_0 can be constructed to solve specified equations and Q_n^* can easily be tailored to solve additional equations of interest. This is accomplished notably by incorporating additional parameters in the least favorable parametric submodel through $(Q_{n,0}, g_n)$. This generality of TMLE has been utilized in several instances before (e.g., van der Laan and Rubin 2006; Rubin and van der Laan 2011; van der Laan and Rose 2011; Gruber and van der Laan 2012b; Lendle et al. 2013; van der Laan 2014b) and plays a fundamental role in the construction of a higher-order TMLE, as described in this article.

26.1.3 Second-Order Asymptotic Expansions

The first-order expansion (26.2), which forms the basis of TMLE, is a direct consequence of the first-order strong differentiability of the parameter mapping Ψ over $\mathcal{M}$. If the asymptotic negligibility of the resulting second-order remainder term $R_2(P_n^*, P_0)$ is implausible, it may be sensible to instead utilize a second-order expansion. The associated remainder term of such an expansion will generally be asymptotically negligible under weaker conditions.

Denote by $L_0^{2*}(P^2)$ the Hilbert space of square-integrable real-valued functions defined on the support of P^2, symmetric in its two arguments, and satisfying that

$$\int f(o_1, o)dP(o_1) = \int f(o, o_2)dP(o_2) = 0$$

for P-almost every o, equipped with the inner product

$$\langle f_1, f_2 \rangle_{P^2} := P^2(f_1 f_2) = \int f_1(o_1, o_2) f_2(o_1, o_2) dP(o_1) dP(o_2).$$

If for any $P_1, P_2 \in \mathcal{M}$ the representation

$$\Psi(P_2) - \Psi(P_1) = (P_2 - P_1)D^{(1)}(P_2) + \frac{1}{2}(P_2 - P_1)^2 D^{(2)}(P_2) + R_3(P_1, P_2)$$

$$= -P_1 D^{(1)}(P_2) - \frac{1}{2}P_1^2 D^{(2)}(P_2) + R_2(P_1, P_2) \qquad (26.4)$$

holds for some element $D^{(2)}(P_2) \in L_0^{2*}(P_2^2)$ and a third-order remainder term $R_3(P_1, P_2)$ such that $R_3(P_1, P_2)/\Delta^2(P_1, P_2) \to 0$ as $\Delta(P_1, P_2) \to 0$ for some distance Δ on $\mathcal{M}$, then Ψ is said to be *second-order strongly differentiable* over $\mathcal{M}$. Any such element $D^{(2)}(P)$ is called a second-order gradient of Ψ at P (Pfanzagl 1985).

In order to construct a 2-TMLE, it will be necessary to identify a gradient $D^{(2)}(P) \in L_0^{2*}(P^2)$ such that either $o_1 \mapsto D^{(2)}(P)(o_1, o)$ or $o_2 \mapsto D^{(2)}(P)(o, o_2)$ lies in $T_{\mathcal{M}}(P)$ for each $o \in O$. Any such element will be referred to as a second-order partial canonical gradient. The notion of second-order scores, tangent spaces and canonical gradients have been defined and studied before (see, e.g., Pfanzagl 1985; Robins et al. 2009; van der Vaart 2014). These concepts are useful for the sake of obtaining a second-order partial canonical gradient since, in particular, any second-order canonical gradient is also a second-order partial canonical gradient. In Carone et al. (2014), we provide a self-contained overview of these concepts, including an account of the link between strong differentiability and pathwise differentiability. We also provide a constructive approach for obtaining a second-order gradient and deriving the second-order canonical gradient.

Denoting by f° the symmetrization

$$(o_1, o_2) \mapsto f^\circ(o_1, o_2) := \frac{f(o_1, o_2) + f(o_2, o_1)}{2}$$

of a given function $(o_1, o_2) \mapsto f(o_1, o_2) \in \mathbb{R}$, we note that $P^2 f = P^2 f^\circ$ for any measure $P \in \mathcal{M}$. As will become clear below, since in the construction of a 2-TMLE the second-order partial canonical gradient $D^{(2)}(P)$ appears only via empirical moments of the form

$$P_n^2 D^{(2)}(P)$$

for $P \in \mathcal{M}$, any element $D_+^{(2)}(P) \in L_0^2(P^2)$ such that $P_n^2 D_+^{(2)}(P) = P_n^2 D^{(2)}(P)$ for each $P \in \mathcal{M}$ could be used in practice. In view of this observation and to simplify our presentation, hereafter we allow abuse of definition and refer as second-order gradient or canonical gradient any element of $L_0^2(P^2)$ whose symmetrization is a second-order partial canonical gradient or canonical gradient of Ψ at $P \in \mathcal{M}$.

26.1.4 Construction of a 2-TMLE

One of the main conditions in the proof of asymptotic efficiency for the TMLE is that the second-order term $R_2(P_n^*, P_0)$ must be $o_P(n^{-1/2})$. In this article, our goal is to construct a TMLE in which this second-order remainder condition is replaced by a third-order remainder condition, requiring instead that $R_3(P_n^*, P_0)$ is $o_P(n^{-1/2})$. This will facilitate the construction of estimators known to be asymptotically linear and efficient under less stringent and therefore more realistic conditions on rates of convergence of the initial estimators used in TMLE.

To achieve this, it is necessary to include additional targeted fitting in the construction of P_n^* based on a higher-order asymptotic expansion of Ψ around P_0. If Ψ is second-order strongly differentiable at P_0 in the sense that (26.4) holds at $P_1 = P_0$, then for any $P_n^* \in \mathcal{M}$ such that (26.1) holds and any second-order partial canonical gradient $D^{(2)}(P)$, it is possible to write

$$\Psi(P_n^*) - \Psi(P_0) = (P_n - P_0)D^{(1)*}(P_n^*) - \frac{1}{2}P_0^2 D^{(2)}(P_n^*) + R_3(P_n^*, P_0) .$$

In view of this, we will arrange that the TMLE not only solves $P_n D^{(1)*}(P_n^*) = 0$ but also the V-statistic equation

$$0 = P_n^2 D^{(2)}(P_n^*) = \frac{1}{n^2} \sum_{i=1}^{n} \sum_{j=1}^{n} D^{(2)}(P_n^*)(O_i, O_j)$$

by including additional parameters in the least favorable parametric submodel. This allows us to obtain that

$$\Psi(P_n^*) - \Psi(P_0) = (P_n - P_0) D^{(1)*}(P_n^*) + \frac{1}{2}(P_n^2 - P_0^2) D^{(2)}(P_n^*) + R_3(P_n^*, P_0) \,.$$

Lemma 26.1 provides conditions under which $(P_n^2 - P_0^2) D^{(2)}(P_n^*)$ converges in probability to zero at a rate faster than $n^{-1/2}$. Asymptotic efficiency is then established as above but with the condition $R_2(P_n^*, P_0) = o_P(n^{-1/2})$ replaced by $R_3(P_n^*, P_0) = o_P(n^{-1/2})$.

Our proposed 2-TMLE is based on a remarkably simple observation: if we denote

$$\bar{D}_n^{(2)}(P_n^*)(O_i) := \frac{1}{n} \sum_{j=1}^{n} D^{(2)}(P_n^*)(O_i, O_j) \,,$$

then $\bar{D}_n^{(2)}(P_n^*)$ can itself be perceived as a score at P_n^*—here, we suppose without loss of generality that $o_1 \mapsto D^{(2)}(P)(o_1, o)$ is in $T_{\mathcal{M}}(P)$ for each $o \in O$. Furthermore, we can write

$$P_n^2 D^{(2)}(P_n^*) = \frac{1}{n} \sum_{i=1}^{n} \bar{D}_n^{(2)}(P_n^*)(O_i) = P_n \bar{D}_n^{(2)}(P_n^*) \,.$$

As a consequence, we may simply augment our least favorable parametric submodel $\{P(\epsilon_1) : \epsilon_1\}$ with score $D^{(1)*}(P)$ at $\epsilon_1 = 0$ in the 1-TMLE to also include a parameter ϵ_2 such that the expanded least favorable submodel $\{P(\epsilon_1, \epsilon_2) : \epsilon_1, \epsilon_2\}$ also generates the score $\bar{D}_n^{(2)}(P)$ at $(\epsilon_1, \epsilon_2) = (0, 0)$. In this manner, the resulting 2-TMLE P_n^* solves the collection of equations

$$\begin{aligned} 0 &= P_n D^{(1)*}(P_n^*) \\ 0 &= P_n \bar{D}^{(2)}(P_n^*) = P_n^2 D^{(2)}(P_n^*) \,. \end{aligned}$$

Thus, the proposed 2-TMLE is no more than a usual 1-TMLE with least favorable submodel extended in a particular manner.

It is important to note that the construction above would not be possible with just any second-order gradient: for arranging a TMLE construction, it is critical that one of its coordinate projections uniformly lie in the tangent space $T_{\mathcal{M}}(P)$. This property generally applies to second-order canonical gradients, which can therefore typically be relied upon for defining a proper 2-TMLE.

26.1.5 Insufficiently Differentiable Target Parameters

As has been noted in Robins et al. (2009), most statistical parameters of interest are unfortunately not smooth enough as functions of the data-generating distribution to allow the required expansion for $R_2(P_n^*, P_0)$. These second-order remainder terms often involve squared differences between a density estimator and the true density. As a consequence, while $R_2(P_n^*, P_0)$ can often itself be expanded into second-order differences between P_n^* and P_0, these second-order terms cannot be represented as the expectation under P_0 of a second-order gradient $D^{(2)}(P_n^*)$. This is a significant hurdle for any method aiming to carry out higher-order bias reduction; in particular, it is a challenge we must face when constructing a higher-order TMLE.

In such cases, we will search for a surrogate $D_h^{(2)}$, indexed by a smoothing parameter h, such that

$$R_2(P_n^*, P_0) = -\frac{1}{2} \lim_{h \to 0} P_0^2 D_h^{(2)}(P_n^*) + R_3(P_n^*, P_0)$$

for some third-order remainder $R_3(P_n^*, P_0)$, thus leading to the representation

$$R_2(P_n^*, P_0) = -\frac{1}{2} P_0^2 D_h^{(2)}(P_n^*) + B_n(h) + R_3(P_n^*, P_0) .$$

Here, the representation error $B_n(h) := [P_0^2 D_h^{(2)}(P_n^*) - \lim_{h \to 0} P_0^2 D_h^{(2)}(P_n^*)]/2$ quantifies the bias resulting from this approximation of a second-order gradient. We may therefore write

$$\Psi(P_n^*) - \Psi(P_0) = (P_n - P_0)D^{(1)*}(P_n^*) - \frac{1}{2} P_0^2 D_h^{(2)}(P_n^*) + B_n(h) + R_3(P_n^*, P_0) .$$

In order to preserve the validity of our general proof of asymptotic efficiency of the 2-TMLE above, we will require that $B_n(h)$ tend to zero faster than $n^{-1/2}$ for an appropriately chosen tuning parameter $h = h_n$ under appropriate smoothness conditions for P_0. When using kernel smoothing to construct approximate second-order gradients, choosing a bandwidth sequence that tends to zero sufficiently fast and utilizing higher-order kernels that leverage potential smoothness in the underlying distribution P_0 will often suffice to ensure that $B_n(h_n)$ is asymptotically negligible.

Because the least favorable submodel used has scores at $\epsilon = 0$ that span both $D^{(1)*}(P)$ and $\bar{D}_{h_n}^{(2)}(P)$, the construction of our proposed 2-TMLE guarantees that

$$P_n D^{(1)*}(P_n^*) = P_n^2 D_{h_n}^{(2)}(P_n^*) = 0$$

for some value h_n of the tuning parameter. In practice, selection of the tuning parameter h_n requires great care. On one hand, to ensure that

$$(P_n^2 - P_0^2)D_{h_n}^{(2)}(P_n^*)$$

converges to zero in probability faster than $n^{-1/2}$, h_n must generally tend to zero slowly enough. On the other hand, for the representation error to be asymptotically negligible, it is required that h_n tend to zero quickly enough. This constitutes the primary theoretical challenge this 2-TMLE must contend with as a result of the target parameter failing to be second-order pathwise differentiable: namely, selection of h_n necessarily involves a careful balance to ensure that

$$\frac{1}{2}(P_n^2 - P_0^2)D_{h_n}^{(2)}(P_n^*) + B_n(h_n) = o_P(n^{-1/2}) .$$

This involves a sensible trade-off between control of the V-statistic term and the representation error.

Of course, this theoretical challenge translates directly into a fundamental practical challenge. Indeed, one algorithm within a large collection of candidate 2-TMLE algorithms, each indexed by the corresponding choice of h, must be chosen in practice. Optimal rates for h_n can often be derived in particular applications, but these are primarily of theoretical interest and generally provide little or no practical guidance regarding the selection of h_n. Fortunately, this is precisely the kind of challenge TMLE can very naturally handle, notably by adjudicating the quality of a particular tuning value h based on the gain in fit resulting from the ensuing parametric TMLE updating step. This is accomplished formally within the C-TMLE framework previously referenced and discussed later.

26.2 Inference Using Higher-Order TMLE

We assume, in this section, that the target parameter is second-order strongly differentiable at P_0 and that (26.4) holds at $P_1 = P_0$ for a given second-order partial canonical gradient $D^{(2)}(P)$.

26.2.1 Asymptotic Linearity and Efficiency

Under prescribed conditions, a 2-TMLE will be asymptotically linear and efficient irrespective of the particular second-order partial canonical gradient selected. A precise enumeration of these conditions are given in the below theorem.

Theorem 26.1. *Suppose that the target parameter Ψ admits the second-order expansion $\Psi(P) - \Psi(P_0) = -P_0 D^{(1)*}(P) - \frac{1}{2}P_0^2 D^{(2)}(P) + R_3(P, P_0)$, and that $P_n^* \in \mathcal{M}$ satisfies the equations*

$$P_n D^{(1)*}(P_n^*) = 0 \ \text{ and } \ P_n^2 D^{(2)}(P_n^*) = P_n \bar{D}_n^{(2)}(P_n^*) = 0$$

with $\bar{D}_n^{(2)}(P_n^*)(o) := \frac{1}{n}\sum_{i=1}^n D^{(2)}(P_n^*)(o, O_i)$. Then, provided that

1. there exists a P_0-Donsker class $\mathcal{F}$ such that $D^{(1)*}(P_n^*)$ is in $\mathcal{F}$ with probability tending to one, and $P_0 \left[D^{(1)*}(P_n^*) - D^{(1)*}(P_0) \right]^2 = o_P(1)$;

2. $(P_n^2 - P_0^2)D^{(2)}(P_n^*) = o_P(n^{-1/2})$;

3. $R_3(P_n^*, P_0) = o_P(n^{-1/2})$;

ψ_n^* is an asymptotically linear estimator of ψ_0 with influence function $D^{(1)*}(P_0)$. It is thus also asymptotically efficient.

Verification of condition 2 in this theorem may seem particularly daunting. The following lemma, which relies upon the concept of *uniform sectional variation norm* (Gill et al. 1995), provides a set of conditions that suffice to establish condition 2 and may be easier to verify in practice. For a given function $f : \mathbb{R}^d \to \mathbb{R}$, the uniform sectional variation norm $\|f\|_v^*$ of f is defined as $\|f\|_v^* := \sup_s \sup_{x_s} \int |f(dx_s, x_{-s})|$, where the supremum is over all possible sections of f and $\int |f(dx_s, x_{-s})|$ represents the variation norm of the section $x_s \mapsto f(x_s, x_{-s})$. Here, the latter section is defined by a given subset $s \subseteq \{1, \ldots, d\}$, $x_s := (x_j : j \in s)$ and $x_{-s} := (x_j : j \notin s)$.

Lemma 26.1. *Provided it can be established that*

1. *the mappings* $o \mapsto \int D^{(2)}(P_n^*)(o, o_2)dP_0(o_2)$ *and* $o \mapsto \int D^{(2)}(P_n^*)(o_1, o)dP_0(o_1)$
   are contained in some fixed P_0-Donsker class $\mathcal{G}$ with probability tending to one;

2. $\int \left[\int D^{(2)}(P_n^*)(o_1, o_2)dP_0(o_1) \right]^2 dP_0(o_2)$ *and* $\int \left[\int D^{(2)}(P_n^*)(o_1, o_2)dP_0(o_2) \right]^2 dP_0(o_1)$ *both tend to zero in probability;*

3. $n^{-1/2}\|D^{(2)}(P_n^*)\|_v^* = o_P(1)$,

then condition 2 of Theorem 1 holds.

26.2.2 Constructing Confidence Intervals

The same techniques for constructing confidence intervals based on the usual TMLE can be utilized in the context of a higher-order TMLE. Since a 2-TMLE ψ_n^* is asymptotically linear with influence function $D^{(1)*}(P_0)$, it follows that $n^{1/2}(\psi_n^* - \psi_0)$ converges in law to a normal variate with mean zero and variance $\sigma_0^2 := P_0[D^{(1)*}(P_0)]^2$. This suggests that the Wald-type interval

$$\left(\psi_n^* - z_{1-\alpha/2}\sigma_n n^{-1/2}, \ \psi_n^* + z_{1-\alpha/2}\sigma_n n^{-1/2} \right), \tag{26.5}$$

where $\sigma_n^2 := P_n[D^{(1)*}(P_n^*)]^2$ and z_β is the β-quantile of the standard normal distribution, has asymptotic coverage level $(1 - \alpha)$. TMLE procedures of different order exhibit identical first-order behavior, although inclusion of higher-order terms will generally guarantee such behavior holds under a wider range of scenarios. The interval (26.5) can therefore be utilized with any higher-order targeted estimator P_n^* of P_0.

The above approach provides asymptotically correct inference. However, it does not explicitly utilize the higher-order expansion upon which a 2-TMLE is constructed. It is plausible that confidence intervals with improved finite-sample performance may be obtained by incorporating higher-order terms from this expansion. For this purpose, a simple bootstrap approach can be devised based on the fact that, provided a random sample $O_1^{\#}, \ldots, O_n^{\#} \sim P_n^{\circ}$ for some consistent estimator P_n° of P_0, the conditional distribution of the bootstrapped statistic $Z_n^{\#}$, defined as

$$n^{-1/2} \sum_{i=1}^{n} \left[D^{(1)*}(P_n^*)(O_i^{\#}) - P_n D^{(1)*}(P_n^*) \right]$$

$$+ \frac{n^{-3/2}}{2} \sum_{i=1}^{n} \sum_{j=1}^{n} \left[D^{(2)}(P_n^*)(O_i^{\#}, O_j^{\#}) - P_n^2 D^{(2)}(P_n^*) \right],$$

given P_n and the distribution of $n^{1/2}(\psi_n - \psi_0)$ approximate each other arbitrarily well for large n and for almost every P_n. Obvious choices for P_n° include, for example, the empirical measure P_n and the targeted estimator P_n^*. This suggests the confidence interval

$$\left(\psi_n^* - q_{1-\alpha/2,n} n^{-1/2}, \ \psi_n^* - q_{\alpha/2,n} n^{-1/2} \right)$$

with $q_{\beta,n}$ the β-quantile of the conditional distribution of $Z_n^{\#}$ given P_n. Of course, these quantiles can be estimated arbitrarily well by simulation. This confidence interval is easy to implement and takes into account the second-order variability explained by the V-statistic process implied by the second-order partial canonical gradient $D^{(2)}(P)$.

26.2.3 Implementing a Higher-Order TMLE

The practical implementation of a higher-order TMLE is no more difficult than a regular TMLE since the former can indeed be seen as an example of the latter. The additional effort involved, in reality, lies in the computation of higher-order partial canonical gradients, or of suitable approximations to such if need be, as discussed in the next section. Below, we describe the implementation of a second-order targeted minimum loss-based estimator.

Given independent variates $O_1, O_2, \ldots, O_n$ distributed according to $P_0 \in \mathcal{M}$, we wish to estimate $\psi_0 := \Psi(P_0)$ for a given target parameter $\Psi : \mathcal{M} \to \mathbb{R}$ of interest. Suppose that the parameter $P \mapsto Q(P)$ satisfies that

1. $\Psi = \Psi_1 \circ Q$ for some mapping $\Psi_1 : Q(\mathcal{M}) \to \mathbb{R}$, and
2. $Q_0 = \arg\min_{Q \in Q(\mathcal{M})} P_0 L(Q)$ for some loss function $(o, Q) \mapsto L(Q)(o)$.

Here, Q_0 represents the true summary $Q(P_0)$. With slight abuse of notation, for the sake of notational convenience, we take $\Psi(Q)$ to mean $\Psi_1(Q)$ hereafter. Several parametrizations may be possible and it will generally be preferable to choose the least complex such parametrization for which we can find an appropriate loss function and parametric fluctuation submodels with scores at Q spanning the appropriate empirical gradients, as described below.

Let $g = g(P)$ be a nuisance parameter and write $g_0 := g(P_0)$. Suppose that the first-order canonical gradient $D^{(1)*}(P)$ of Ψ can be represented as $D^{(1)*}(Q(P), g(P))$ and that $D^{(2)}(P) := D^{(2)}(Q(P), g(P))$ is any associated second-order partial canonical gradient of Ψ. Similar abuse of notation is tolerated here as well. Suppose, as before, that $o_1 \mapsto D^{(2)}(P)(o_1, o_2)$ lies in $T_{\mathcal{M}}(P)$ for each o_2. Then, setting

$$\bar{D}_n^{(2)}(Q, g)(o) := \frac{1}{n} \sum_{j=1}^{n} D^{(2)}(P)(o, O_j) ,$$

we have that $\bar{D}_n^{(2)}(Q, g) \in T_{\mathcal{M}}(P)$ and will play the role of a second-order score at P.

Let $Q(Q, g) := \{Q_g(\epsilon) : \epsilon\} \subset Q(\mathcal{M})$ be a second-order least favorable submodel, in the sense that $Q_g(0) = Q$ and both $D^{(1)*}(Q, g)$ and $\bar{D}_n^{(2)}(Q, g)$ lie in the closure of the linear span

$$\left\{ z^T \frac{\partial}{\partial \epsilon} L(Q_g(\epsilon)) \Big|_{\epsilon=0} : z \in \mathbb{R}^p \right\}$$

of the generalized score vector at $\epsilon = 0$, where p denotes the dimension of ϵ.

Suppose that an initial estimator $(Q_{n,0}, g_n)$ of (Q_0, g_0) is at our disposal. As with a regular TMLE, a 2-TMLE will generally be an iterative procedure, though analytic convergence after a single step can be demonstrated in important examples. A 2-TMLE updating step will be identical to that of a regular TMLE, except for the use of a second-order rather than first-order least favorable submodel. Specifically, given the current estimate $Q_{n,m}$ of Q_0, the updated estimate $Q_{n,m+1}$ is defined as

$$Q_{n,m+1} := \arg\min_{Q \in Q(Q_{n,m}, g_n)} P_n L(Q) .$$

Alternatively, setting $\epsilon_n^m := \arg\min_\epsilon P_n L(Q_{n,m}(\epsilon))$, we can express the updated estimate of Q_0 as $Q_{n,m+1} := Q_{n,m,g_n}(\epsilon_n^m)$. This iterative updating step will generally be repeated until convergence, adjudicated by ϵ_n^m being sufficiently close to zero. Denoting by Q_n^* the limit of this iterative procedure, the 2-TMLE of ψ_0 is given by $\psi_n^* := \Psi(Q_n^*)$. The desirable asymptotic properties of ψ_n^* are in large part a consequence of the fact that

$$P_n D^{(1)*}(Q_n^*, g_n) = P_n \bar{D}_n^{(2)}(Q_n^*, g_n) = 0$$

by construction.

26.3 Inference Using Approximate Second-Order Gradients

The theory of higher-order TMLE outlined above applies to settings wherein the parameter is higher-order pathwise differentiable. However, as indicated before, in realistic models, many parameters of interest are not smooth enough as functionals of the data-generating distribution to admit even a second-order gradient. Nonetheless, a useful higher-order theory can still be developed with the introduction of approximate higher-order partial canonical gradients.

26.3.1 Asymptotic Linearity and Efficiency

The following theorem, which is a direct generalization of Theorem 26.1 allowing for the lack of existence of a second-order partial canonical gradient, illustrates this in the case of the 2-TMLE.

Theorem 26.2. *Suppose that for each h the element $D_h^{(2)}(P) \in L_0^2(P^2)$ satisfies that*

$$o_1 \mapsto D_h^{(2)}(P)(o_1, o)$$

lies in $T_{\mathcal{M}}(P)$ for each $o \in O$, that the target parameter Ψ admits the second-order expansion $\Psi(P) - \Psi(P_0) = -P_0 D^{(1)*}(P) - \frac{1}{2} \lim_{h \to 0} P_0^2 D_h^{(2)}(P) + R_3(P, P_0)$, and that $P_n^* \in \mathcal{M}$ satisfies the equations

$$P_n D^{(1)*}(P_n^*) = 0 \quad and \quad P_n^2 D_{h_n}^{(2)}(P_n^*) = P_n \bar{D}_{h_n,n}^{(2)}(P_n^*) = 0$$

with $\bar{D}_{h_n,n}^{(2)}(P_n^*)(o) := \frac{1}{n} \sum_{i=1}^{n} D_{h_n}^{(2)}(P_n^*)(o, O_i)$. Then, provided that

1. *there exists a P_0-Donsker class $\mathcal{F}$ such that $D^{(1)*}(P_n^*)$ is in $\mathcal{F}$ with probability tending to one, and $P_0 \left[D^{(1)*}(P_n^*) - D^{(1)*}(P_0) \right]^2 = o_P(1)$;*
2. *$(P_n^2 - P_0^2) D_{h_n}^{(2)}(P_n^*) = o_P(n^{-1/2})$;*
3. *$B_n(h_n) := \frac{1}{2} \left[P_0^2 D_{h_n}^{(2)}(P_n^*) - \lim_{h \to 0} P_0^2 D_h^{(2)}(P_n^*) \right] = o_P(n^{-1/2})$;*
4. *$R_3(P_n^*, P_0) = o_P(n^{-1/2})$;*

ψ_n^* is an asymptotically linear estimator of ψ_0 with influence function $D^{(1)*}(P_0)$. It is thus also asymptotically efficient.

Sufficient conditions for establishing the validity of condition 2 in the above theorem are identical to those discussed in Lemma 26.1 upon replacing the second-order partial canonical gradient by its approximation. On one hand, h_n must not tend to zero too quickly to ensure appropriate control of the uniform sectional variation norm. On the other hand, h_n must tend to zero quickly enough for the representation error to be asymptotically negligible. A careful trade-off must therefore be achieved.

26.3.2 Implementation and Selection of Tuning Parameter

A 2-TMLE procedure yields an estimate $P_n^* \in \mathcal{M}$ of P_0 such that

$$P_n D^{(1)*}(P_n^*) = P_n \bar{D}_{h_n,n}^{(2)}(P_n^*) = 0 \,,$$

where $\bar{D}^{(2)}_{h_n,n}(P^*_n)$ is defined similarly as $\bar{D}^{(2)}_n(P^*_n)$ but with $D^{(2)}$ replaced by $D^{(2)}_{h_n}$. Thus, a 2-TMLE procedure can be defined in exactly the same way as in Sect. 26.2.3, where the scores from second-order terms are replaced by their corresponding approximation. As such, except for the selection of an appropriate tuning parameter value h_n, the algorithm is no more difficult to implement than when second-order gradients exist. Of course, the key practical challenge lies in determining an appropriate value for h_n.

The choice of h in this problem determines the least favorable parametric submodel used in the TMLE algorithm. Thus, the selection of h can be seen as completely analogous to the selection of an estimator of the treatment or censoring mechanism, principal determinants of the first-order least favorable submodel, in the problem of estimating a g-computation parameter when all potential confounders have been recorded or when the available data are subject to censoring. C-TMLE, as described, thoroughly discussed and implemented in van der Laan and Gruber (2010), Gruber and van der Laan (2010a), Stitelman and van der Laan (2010), van der Laan and Rose (2011), Wang et al. (2011a), Gruber and van der Laan (2012b), and Chap. 10 of this book, provides a principled solution to this problem. This approach can readily be utilized here as well.

C-TMLE consists of a TMLE algorithm that automatically selects among a collection of candidate least favorable parametric submodels in its updating step. For each candidate submodel, the resulting TMLE algorithm yields a decrease in empirical risk; the magnitude of this change can serve as a criterion for adjudicating the value of this particular submodel. C-TMLE uses precisely this criterion for data-adaptively building or selecting a least favorable parametric submodel, and thereby a corresponding TMLE. Under certain regularity conditions, the resulting estimator is asymptotically linear and efficient provided at least one of the candidate submodels allows for complete bias reduction asymptotically.

Specifically, denoting by $(Q_{n,0}, g_n)$ an initial estimate of (Q_0, g_0) and letting h index a candidate second-order least favorable submodel determined by $D^{(1)*}(Q_{n,0}, g_n)$ and $D^{(2)}_h(Q_{n,0}, g_n)$, a C-TMLE solution would first select the h-value for which the possible decrease in the empirical risk $P_n L(Q_{h,n,0}(\epsilon))$ along the corresponding h-specific parametric submodel $\{Q_{h,n,0}(\epsilon) : \epsilon\}$ through $Q_{n,0}$ is maximal and perform a single updating step along this submodel. In other words, h^0_n would be selected as the minimizer of

$$h \mapsto P_n L(Q_{h,n,0}(\epsilon^0_n(h)))$$

with $\epsilon^0_n(h) := \arg\min_\epsilon P_n L(Q_{h,n,0}(\epsilon))$ for each h and the first update would then be $Q_{n,1} := Q_{h^0_n,n,0}(\epsilon^0_n(h^0_n))$. The next targeting step would be carried out similarly, with subsequent choices of h constrained to be no larger than h^0_n. Specifically, letting h^1_n denote the minimizer of $h \mapsto P_n L(Q_{h,n,1}(\epsilon^1_n(h)))$ over the interval $[0, h^0_n]$, where $\epsilon^1_n(h) := \arg\min_\epsilon P_n L(Q_{h,n,1}(\epsilon))$ for each h, the updated estimate of Q_0 would then be $Q_{n,2} := Q_{h^1_n,n,1}(\epsilon^1_n(h^1_n))$. This iterative process would proceed until the decrease in empirical risk from an additional step is no longer significant according to some pre-specified criterion, such as BIC or some cross-validation risk, for example. Denoting the final estimate of Q_0 and the final value of h_n at convergence by Q^*_n and h^*_n, respectively, the TMLE $\Psi(Q_n^*)$ could be directly used, or the 2-TMLE based on the initial estimate $Q_{n,0}$ and final tuning parameter choice $h = h_n^*$ could be constructed anew.

In the scheme proposed above, the performance of a particular h-value is adjudicated on the basis of the decrease in empirical risk resulting from a single targeting step. While this seems sensible in settings where convergence of the TMLE updating process occurs analytically in a single step, in other settings, it may not be an optimal way to proceed. As an alternative, it would be possible to carry out a fully iterated TMLE until convergence for every choice of h, to select the optimal h-value on this basis, and then, to repeat until overall convergence. Of course, this variant of the algorithm would be potentially much more computationally intensive.

The approach described above is used to fully automate the selection of the tuning parameter required in the setting of a 2-TMLE based on an approximate second-order partial canonical gradient. The same principle could be used to construct more involved collaborative TMLE algorithms that not only select tuning parameters but also other choices that define the approximate second-order partial canonical gradient.

26.4 Illustration: Estimation of a g-Computation Parameter

We provide a concrete example of the above algorithm by developing a 2-TMLE in the context of estimation of $\psi_0 := \Psi(P_0)$ using n independent copies of $O = (W, A, Y) \sim P_0$, where $\Psi(P) = E_P E_P(Y \mid A = 1, W)$ is the point intervention g-computation parameter. Under untestable causal assumptions, this parameter corresponds to the mean of the counterfactual outcome corresponding to the point intervention $A = 1$.

Denoting by $o := (w, a, y)$ a possible realization of O, and writing $\bar{g}(w) := P(A = 1 \mid W = w)$, $\bar{Q}(w) := E_P(Y \mid A = 1, W = w)$ and Q_W is the density of the marginal distribution of W with respect to an appropriate counting measure, the first-order canonical gradient is known in this case to be

$$P \mapsto D^{(1)*}(P) := D_Y^{(1)*}(P) + D_W^{(1)*}(P) ,$$

where we have defined pointwise

$$D_Y^{(1)*}(P)(o) := H^{(1)}(\bar{g})(a, w)\left[y - \bar{Q}(w)\right] \text{ and } D_W^{(1)*}(P)(o) := \bar{Q}(w) - \Psi(P) ,$$

with $H^{(1)}(\bar{g})(a, w) := a/\bar{g}(w)$ and $\bar{g}(w) := P(A = 1 \mid W = w)$. With this definition, we find that $\Psi(P) - \Psi(P_0) = -P_0 D^{(1)*}(P) + R_2(P, P_0)$ with

$$R_2(P, P_0) := P_0\left[\left(\frac{\bar{g} - \bar{g}_0}{\bar{g}}\right)(\bar{Q} - \bar{Q}_0)\right],$$

where $\bar{g}_0$ and $\bar{Q}_0$ denote $\bar{g}$ and $\bar{Q}$, respectively, under P_0. Our goal is to find a representation $R_2(P, P_0) = -\frac{1}{2}P_0^2 D^{(2)}(P) + R_3(P, P_0)$. Below, we deal separately with the cases wherein W has finite versus infinite support.

26.4.1 Case I: Finite Support

Suppose that W has finite support under each considered P. It can be shown that the second-order canonical gradient at P and at (o_1, o_2), with $o_1 := (w_1, a_1, y_1)$ and $o_2 := (w_2, a_2, y_2)$, is given by the symmetrization of

$$D^{(2)}(P)(o_1, o_2) := H^{(2)}(P)(w_1, a_1, w_2, a_2)[y_1 - \bar{Q}(w_1)] , \tag{26.6}$$

where

$$H^{(2)}(P)(w_1, a_1, w_2, a_2) := \frac{2a_1 I(w_1 = w_2)}{\bar{g}(w_1)Q_W(w_1)}\left[1 - \frac{a_2}{\bar{g}(w_1)}\right].$$

It is not difficult to directly verify that indeed

$$\frac{1}{2}P_0^2 D^{(2)}(P) = -P_0\left[\left(\frac{\bar{g} - \bar{g}_0}{\bar{g}}\right)(\bar{Q} - \bar{Q}_0)\right] + R_3(P, P_0) , \tag{26.7}$$

where $R_3(P, P_0)$ is given by

$$P_0\left[\left(1 - \frac{\bar{g}_0 Q_{W,0}}{\bar{g}Q_W}\right)\left(\frac{\bar{g} - \bar{g}_0}{\bar{g}}\right)(\bar{Q} - \bar{Q}_0)\right].$$

Thus, it holds that $\Psi(P) - \Psi(P_0) = -P_0 D^{(1)*}(P) - \frac{1}{2}P_0 D^{(2)}(P) + R_3(P, P_0)$ for a third-order term $R_3(P, P_0)$, as desired. Since W is finitely supported, the event $\{W_1 = W_2\}$ occurs with positive probability and thus $D^{(2)}(P)$ is not degenerate.

In this case, the target parameter $\Psi(P)$ depends on P via $Q(P) = (\bar{Q}(P), Q_W(P))$. It is straightforward to verify that the loss function $L(Q) := L_1(\bar{Q}) + L_2(Q_W)$, where we define

$$L_1(\bar{Q})(o) := -a\left[y \log \bar{Q}(w) + (1 - y) \log(1 - \bar{Q}(w))\right] \text{ and}$$
$$L_2(Q_W)(o) := -\log Q_W(w)$$

pointwise for each $o = (w, a, y)$, is such that $(\bar{Q}_0, Q_{W,0}) := (\bar{Q}(P_0), Q_W(P_0))$ indeed minimizes the true risk $P_0 L(Q)$.

Setting $g(P) := \bar{g}$, both $D^{(1)*}(P)$ and $D^{(2)}(P)$ depend on P through $(Q(P), g(P))$, and the components of $(Q(P), g(P))$ are variationally independent of each other since they involve orthogonal portions of the likelihood. This provides greater flexibility in constructing appropriate fluctuation submodels. Defining

$$\bar{D}_n^{(2)}(P)(o) := \bar{H}_n^{(2)}(P)(w, a)\left[y - \bar{Q}(w)\right]$$

with $\bar{H}_n^{(2)}(P)(w, a) := \frac{1}{n} \sum_{j=1}^{n} H^{(2)}(P)(w, a, W_j, A_j)$, we note that $\bar{D}_n^{(2)}(P) \in T_{\mathcal{M}}(P)$ and in fact is a score of the conditional distribution of Y given A and W. Furthermore, we observe that

$$P_n \bar{D}_n^{(2)}(P) = P_n^2 D^{(2)}(P) .$$

We also note that, at $Q_W = Q_{W,n}$, the empirical distribution of W,

$$\bar{H}_n^{(2)}(P)(w, a) = \frac{2a}{\bar{g}(w)} \left[1 - \frac{\bar{g}_{n,NP}(w)}{\bar{g}(w)} \right],$$

where $\bar{g}_{n,NP}(w) := \sum_{i=1}^{n} I(A_i = 1, W_i = w) / \sum_{i=1}^{n} I(W_i = w)$ is the nonparametric maximum likelihood estimator of $\bar{g}_0(w)$.

We can readily verify that $P_n D_W^{(1)*}(P) = 0$ if $Q_W(P) = Q_{W,n}$. Thus, if the initial estimator of $Q_{W,0}$ is taken to be $Q_{W,n}$, to achieve our objective of solving the requisite first- and second-order estimating equations, it will suffice to produce an estimate $\bar{Q}_n^*$ of $\bar{Q}_0$ satisfying

$$P_n D_Y^{(1)*}(\bar{Q}_n^*, Q_{W,n}, \bar{g}_n) = P_n \bar{D}_n^{(2)}(\bar{Q}_n^*, Q_{W,n}, \bar{g}_n) = 0 \qquad (26.8)$$

with $\bar{g}_n$ our initial estimate of $\bar{g}_0$. Given any particular $\bar{Q}$, Q_W and $\bar{g}$, the submodel determined by

$$\bar{Q}_{\bar{g}, Q_W}(\epsilon) := \text{expit} \left[\text{logit}(\bar{Q}) + \epsilon_1 H^{(1)}(\bar{g}) + \epsilon_2 \bar{H}_n^{(2)}(\bar{g}, Q_W) \right]$$

with $\epsilon := (\epsilon_1, \epsilon_2)$ satisfies that $\bar{Q}_{\bar{g}, Q_W}(0) = \bar{Q}$ as well as

$$\left. \frac{\partial}{\partial \epsilon_1} L(\bar{Q}_{\bar{g}, Q_W}(\epsilon), Q_W)(o) \right|_{\epsilon=0} = H^{(1)}(\bar{g})(o) \left[y - \bar{Q}(w) \right] = D_Y^{(1)*}(P)(o) ,$$

$$\left. \frac{\partial}{\partial \epsilon_2} L(\bar{Q}_{\bar{g}, Q_W}(\epsilon), Q_W)(o) \right|_{\epsilon=0} = \bar{H}_n^{(2)}(\bar{g}, Q_W)(o) \left[y - \bar{Q}(w) \right] = \bar{D}_n^{(2)}(P)(o) .$$

As such, given an initial estimate $(\bar{Q}_{n,0}, \bar{g}_n)$ of $(\bar{Q}_0, \bar{g}_0)$, a first updated estimate $\bar{Q}_{n,1}$ of $\bar{Q}_0$ is obtained by selecting the minimizer of the empirical risk $P_n L_1(\bar{Q})$ with $\bar{Q}$ ranging over the parametric submodel determined by

$$\bar{Q}_{n,0}(\epsilon) := \text{expit} \left[\text{logit}(\bar{Q}_{n,0}) + \epsilon_1 H^{(1)}(\bar{g}_n) + \epsilon_2 \bar{H}_n^{(2)}(\bar{g}_n, Q_{W,n}) \right]$$

and $-\infty < \epsilon_1, \epsilon_2 < +\infty$. The optimal values of ϵ_1 and ϵ_2 can be readily obtained as estimated regression coefficients in the fit of a logistic regression model with outcome Y_i, covariates $H^{(1)}(\bar{g}_n)(O_i) = A_i / \bar{g}_n(W_i)$ and

$$\bar{H}_n^{(2)}(\bar{g}_n, Q_{W,n})(O_i) = \frac{2A_i}{\bar{g}_n(W_i)} \left[1 - \frac{\bar{g}_{n,NP}(W_i)}{\bar{g}_n(W_i)} \right],$$

and offset $\text{logit}(\bar{Q}_{n,0}(W_i))$ restricted to the subset of data points for which $A_i = 1$. It is not difficult to see that any further attempt to update $\bar{Q}_{n,1}$ by considering the second-order least favorable submodel through it will not produce any change. As such, in this case, the algorithm terminates in a single step, so that $\bar{Q}_n^* = \bar{Q}_{n,1}$ and (26.8) must then hold. The resulting 2-TMLE of ψ_0 is finally given by

$$\Psi(\bar{Q}_n^*, Q_{W,n}) = \frac{1}{n} \sum_{i=1}^{n} \bar{Q}_n^*(W_i) .$$

This estimator allows concrete theoretical gains relative to the 1-TMLE described previously and preliminary simulation results suggest these indeed translate well into practical gains, as reported below.

In the algorithm described above, there was no need to update the estimate of $Q_{W,0}$ at all. This resulted from the selection of the NPMLE $Q_{W,n}$ as initial estimator and of the log-likelihood loss for Q_W. Had any of these two choices differed, it would generally have been necessary to iteratively update the estimate of $Q_{W,0}$ using an appropriate fluctuation submodel to ensure that the resulting targeted estimate $Q_{W,n}^*$ satisfy that

$$P_n D_W^{(1)*}(\bar{Q}_n^*, Q_{W,n}^*) = 0 ,$$

leading then to the 2-TMLE of ψ_0 given by $\sum_w \bar{Q}_n^*(w) Q_{W,n}^*(w)$.

26.4.2 Case II: Infinite Support

Suppose now that W has infinite support under each considered P. In such case, the g-computation parameter generally does not admit a second-order gradient in the nonparametric model we have been considering. An approximate second-order canonical gradient can nonetheless be considered, leading to a well-defined 2-TMLE, as we now describe.

Suppose that $W := (W(1), \ldots, W(d))$ and W_j is real-valued. Defining pointwise

$$D_h^{(2)}(\bar{Q}, \bar{g}, Q_W)(o_1, o_2) := H_h^{(2)}(\bar{g}, Q_W)(w_1, a_1, w_2, a_2)[y_1 - \bar{Q}(w_1)] ,$$

where we have also defined

$$H_h^{(2)}(\bar{g}, Q_W)(w_1, a_1, w_2, a_2) := \frac{1}{h^d} K\left(\frac{w_1 - w_2}{h}\right) \frac{2a_1}{\bar{g}(w_1)Q_W(w_1)} \left[1 - \frac{a_2}{\bar{g}(w_1)}\right]$$

with K a compactly-supported multivariate kernel function and h a positive bandwidth, $D_h^{(2)}$ is seen to be a kernel approximation of the second-order partial canonical gradient $D^{(2)}$ defined in the previous subsection. For each w, we henceforth denote $h^{-d}K(h^{-1}w)$ by $K_h(w)$. Lemma 1 of Díaz et al. (2016) establishes conditions under which the bias term arising in a TMLE due to the use of this approximate second-order canonical gradient is asymptotically negligible.

Computation of the g-computation parameter does not in principle require an estimate of the density function of W — indeed, an estimate of the distribution function of W suffices. Nonetheless, the second-order targeting process explicitly requires this density function, as is apparent from the form of the approximate second-order canonical gradient. Thus, to construct a 2-TMLE, an estimate of the density of W must first be obtained. At least two approaches can be considered to tackle this issue:

1. a fixed smooth estimate $Q_{W,n}^{\circ}$ of $Q_{W,0}$ is used whenever required in the second-order targeting process to yield a targeted estimate of $\bar{Q}_n^*$ but final computation of the targeted estimate of ψ_0 is based on the empirical measure $Q_{W,n}$;
2. the same smooth estimate $Q_{W,n}^{\circ}$ of $Q_{W,0}$ is used both in the second-order targeting process and in the computation of the targeted estimate of ψ_0.

The second option requires a substantial amount of additional work since in order for the smooth estimator $Q_{W,n}^{\circ}$ to be appropriate for the sake of constructing the targeted estimate of ψ_0 it must itself be targeted. As such, it will need to be iteratively updated along with estimates of $\bar{Q}_0$. This issue is circumvented in the first option since the empirical distribution $Q_{W,n}$ is a NPMLE and therefore already solves the relevant score equation. For this reason, we recommend the first option in practice and restrict our attention to this option alone.

Consider the 2-TMLE using the empirical distribution $Q_{W,n}$ as initial estimator of $Q_{W,0}$ in the TMLE algorithm but using a fixed smooth estimator $Q_{W,n}^{\circ}$ in the quantity

$$H_h^{(2)}(\bar{g}_n, Q_{W,n}^{\circ})$$

used to construct the logistic regression-based least favorable submodel through a current estimate $\bar{Q}_{n,m}$ of $\bar{Q}_0$. This is then precisely an example of the 2-TMLE of Sect. 26.4. The implementation of this algorithm is particularly simple because a single step of targeting suffices to achieve analytic convergence. To perform this single updating step, the maximum likelihood estimate ϵ_n of $\epsilon = (\epsilon_1, \epsilon_2)$ in the logistic regression model

$$\bar{Q}_{n,0,h}(\epsilon) := \text{expit}\left[\text{logit}(\bar{Q}_{n,0}) + \epsilon_1 H^{(1)}(\bar{g}_n) + \epsilon_2 \bar{H}_{h,n}^{(2)}(\bar{g}_n, Q_{W,n}^{\circ})\right]$$

is obtained, with $H^{(1)}(\bar{g}_n)(o) = a/\bar{g}_n(w)$ and

$$\bar{H}_{h,n}^{(2)}(\bar{g}_n, Q_{W,n}^{\circ})(o) = \frac{1}{n}\sum_{j=1}^{n}\frac{1}{h^d}K\left(\frac{w - W_j}{h}\right)\frac{2a}{\bar{g}_n(w)Q_{W,n}^{\circ}(w)}\left[1 - \frac{A_j}{\bar{g}_n(w)}\right]$$

$$= \frac{2a}{\bar{g}_n(w)Q_{W,n}^{\circ}(w)}\left[\frac{1}{n}\sum_j K_h(w - W_j) - \frac{\frac{1}{n}\sum_j K_h(w - W_j)A_j}{\bar{g}_n(w)}\right].$$

If the kernel estimate $Q_{W,n}^{\circ}(w) := \frac{1}{n}\sum_{i=1}^{n} K_h(w - W_i)$ is used, it is easy to verify that the simplification

$$\bar{H}_{h,n}^{(2)}(\bar{g}_n, Q_{W,n}^{\circ})(o) = \frac{2a}{\bar{g}_n(w)}\left[1 - \frac{\bar{g}_{n,h}(w)}{\bar{g}_n(w)}\right]$$

ensues, where $\bar{g}_{n,h}(w) := \sum_j K_h(w - W_j)A_j / \sum_j K_h(w - W_j)$ is the Nadaraya-Watson estimator of $\bar{g}_0(w)$ indexed by bandwidth h.

Given the targeted estimate $\bar{Q}^*_{n,h_n} := \bar{Q}_{n,0,h_n}(\epsilon_n)$ of $\bar{Q}_0$ based on this single-step procedure, the resulting 2-TMLE of ψ_0 is $\psi^*_n := \Psi(\bar{Q}^*_{n,h_n}, Q_{W,n}) = \frac{1}{n} \sum_{i=1}^n \bar{Q}^*_{n,h_n}(W_i)$. Below, we will make reference to the third-order remainder term R_3 defined as

$$R_3(P, P_0) := P_0 \left[\left(1 - \frac{Q_{W,0}\bar{g}_0}{Q_W \bar{g}} \right) \left(\frac{\bar{g} - \bar{g}_0}{\bar{g}} \right) (\bar{Q} - \bar{Q}_0) \right],$$

as before. The conditions under which ψ^*_n is an asymptotically linear and efficient estimator of ψ_0 are given in Theorem 1 of Díaz et al. (2016). The theorem includes the conditions that:

1. The kernel function K is $2d$-times differentiable and $h_n^{2d} n \to +\infty$, and
2. $R_3(P^*_n, P_0) = o_P(n^{-1/2})$ and $\|\bar{Q}^*_n - \bar{Q}_0\| h_n^{m_0+1} = o_P(n^{-1/2})$,

where m_0 is such that both $\bar{g}_0$ and $\bar{Q}_0$ are $(m_0 + 1)$-times continuously differentiable. As is evident from the above conditions, and according to our discussion in Sect. 26.1.5, the convergence of the bandwidth h_n plays a critical role in the asymptotic behavior of the 2-TMLE described.

We remark that when K is indeed taken to be a tensor product of uniform kernels over $(-0.5, +0.5)$, $m_0 = 1$ and condition 2 becomes more restrictive. While for $d \le 2$, there does indeed exist a rate h_n for which both conditions are true, this is not the case if $d > 2$. Therefore, even though higher-order kernels increase the variation norm and thereby lead to more stringent conditions on h_n, when the vector of potential confounders includes more than two components, it is necessary to use higher-order kernels that fully exploit the underlying smoothness of the distribution of (A, W) in order to control the representation error.

A 2-TMLE will generally be asymptotically linear and efficient in a larger model compared to a corresponding 1-TMLE. On one hand, it is generally true that whenever a 1-TMLE is efficient, so will be a 2-TMLE. This means that a 2-TMLE operates in a safe haven wherein we expect not to damage an asymptotically efficient 1-TMLE by performing the additional targeting required to construct a 2-TMLE. On the other hand, 2-TMLE will be efficient in many instances in which 1-TMLE is not. As an illustration, suppose that W is a univariate random variable with a sufficiently smooth density function. Suppose also that $\bar{g}_0$ is smooth enough so that a univariate second-order kernel smoother produces an optimal estimator of $\bar{g}_0$. In this case, efficiency of a 1-TMLE requires that $\bar{Q}_n$ tends to Q_0 at a rate faster than $n^{-1/10}$. In contrast, a corresponding 2-TMLE built upon a second-order canonical gradient approximated using an optimal second-order kernel smoother will be efficient provided that $\bar{Q}_n$ is consistent for $\bar{Q}_0$, irrespective of the actual rate of convergence. This problem is exacerbated further if W has several components. For example, if W is five-dimensional, a 1-TMLE requires that $\bar{Q}_n$ tend to $\bar{Q}_0$ faster than $n^{-5/18}$, whereas the corresponding 2-TMLE based on a third-order kernel-smoothed approximation of the second-order canonical gradient requires that $\bar{Q}_n$ tend to $\bar{Q}_0$ faster than $n^{-1/5}$.

While the latter is achievable using an optimal second-order kernel smoother, the former is not, and without further smoothness assumptions on $\bar{Q}_0$, a 1-TMLE will generally not be efficient.

We may hope that systematic dimension reduction may be performed by replacing $K_h(w_1 - w_2)$ in the definition of $D_h^{(2)}$ by a kernel-based discrepancy based on the propensity score at w_1 and w_2, as discussed in Díaz et al. (2016). This can be accomplished if $\bar{g}_0$ is known, in which case $K_h(\bar{g}_0(w_1) - \bar{g}_0(w_2))$ can be used to define a dimension-reduced approximate second-order partial canonical gradient without sacrificing the order of the remainder in the associated expansion. Replacing $\bar{g}_0$ by $\bar{g}$ in this kernel discrepancy unfortunately introduces a second-order term in the remainder, thereby invalidating the theoretical justification for using such a second-order partial canonical gradient. However, this does not preclude the possibility that finite-sample benefits might be derived from taking this path. Indeed, in the following section we present numerical results from a simulation study illustrating the gains of this estimator, which we denote 1^*-TMLE, for a particular data-generating mechanism.

26.4.3 Numerical Results

We present a simulation study illustrating the performance of the 1^*-TMLE and 2-TMLE compared to the 1-TMLE, using covariate dimension $d = 3$ and sample sizes $n \in \{500, 1000, 2000, 10,000\}$. This simulation study was originally presented in Díaz et al. (2016). For each sample size $n \in \{500, 1000, 2000, 10,000\}$, we simulated 1000 datasets from the joint distribution implied by the conditional distributions

$$W_1 \sim \text{Beta}(2, 2)$$
$$W_2 \mid W_1 = w_1 \sim \text{Beta}(2w_1, 2)$$
$$W_3 \mid W_1 = w_1, W_2 = w_2 \sim \text{Beta}(2w_1, 2w_2)$$
$$A \mid W = (w_1, w_2, w_3) \sim \text{Bernoulli}(\text{expit}(1 + 0.12w_1 + 0.1w_2 + 0.5w_3))$$
$$Y \mid A = 1, W = (w_1, w_2, w_3) \sim \text{Bernoulli}(\text{expit}(-4 + 0.2w_1 + 0.3w_2 + 0.5 \exp(w_3))) ,$$

where Bernoulli(π) denotes the Bernoulli distribution with success probability π, expit represents the function $u \mapsto \exp(u)/[1 + \exp(u)]$, and Beta($a, b$) denotes the Beta distribution with parameters a and b.

Estimation of $\bar{Q}_0$ and g_0 was carried out as follows. For each dataset, we first fitted correctly-specified parametric models. Then, for a perturbation parameter p, we multiplied the linear predictor of $\bar{Q}_n$ by a random variable with distribution $U(1 - n^{-p}, 1)$, and subtracted a Gaussian random variable with mean $3 \times n^{-p}$ and standard deviation n^{-p}. Analogously, for a perturbation parameter q, we multiplied the linear predictor of g_n by a random variable $U(1 - n^{-q}, 1)$, and subtracted a Gaussian random variable with mean $3 \times n^{-q}$ and standard deviation n^{-q}. We varied the values of p and q each in $\{0.01, 0.02, 0.05, 0.1, 0.2, 0.5\}$.

The above perturbation of the MLE in a correctly-specified parametric model is carried out to obtain initial estimators that have varying convergence rates. For example, suppose that $g_{n,MLE}$ denotes the MLE of g_0 in the correct parametric model, and let $g_{n,MLE,q}$ denote the perturbed estimator. Let U_n and V_n be random variables distributed according to $U(1 - n^{-q}, 1)$ and $N(-3n^{-q}, n^{-2q})$, respectively. Substituting $g_{n,MLE,q} := U_n g_{n,MLE} + V_n$ into $\|\hat{g}_q^{MLE} - g_0\|_{P_0}^2$ yields

$$\|g_{n,MLE,q} - g_0\|_{P_0}^2 \leq \|U_n(g_{n,MLE,q} - g_0)\|_{P_0}^2 + \|g_0(U_n - 1)\|_{P_0}^2 + \|V_n\|_{P_0}^2$$
$$= O_P(n^{-1} + n^{-2q}).$$

Consider now different values of q. For example, $q = 0.5$ yields the parametric consistency rate $\|g_{n,MLE,q} - g_0\|_{P_0}^2 = O_P(n^{-1})$, whereas $q = 0$ results in inconsistency.

We compute a 1-TMLE, $1^\star$-TMLE and a 2-TMLE for each initial estimator $(\hat{Q}, \hat{g})$ obtained through this perturbation. We compare the performance of the estimators through their bias inflated by a factor $n^{1/2}$, relative variance compared to the nonparametric efficiency bound, and the coverage probability of a 95% confidence interval based on the true variance. Using the true variance allows us to isolate the performance of the estimator itself from the estimator of its variance. The true sampling variance, bias and coverage probabilities are approximated through empirical means across the 1000 simulation runs.

Table 26.1 shows the relative variance (rVar, defined as n times the variance divided by the efficiency bound), the absolute bias inflated by $n^{1/2}$, and the coverage probability of a 95% confidence interval using the true variance of the estimators for selected values of the perturbation parameter (p, q). Figure 26.1 shows the absolute bias of each estimator multiplied by $n^{1/2}$, and Fig. 26.2 shows the coverage probability of a 95% confidence interval for all values of (p, q) used in the simulation.

We notice that for certain slow convergence rates, e.g., $(p, q) = (0.01, 0.01)$ or $(p, q) = (0.1, 0.01)$, all the estimators have very large bias. In contrast, for some other slow convergence rates, e.g., $(p, q) = (0.01, 0.1)$, the absolute bias scaled by $n^{1/2}$ of the 1-TMLE diverges very fast in comparison to the 2-TMLE and $1^\star$-TMLE. The improvement in large-sample absolute bias of the proposed estimators comes at the price of increased variance in certain small sample scenarios, e.g., $n \leq 2000$, when, for example, the outcome model converges at a fast enough rate but the missingness mechanism does not, e.g., $(p, q) = (0.5, 0.1)$. In this case, the 1-TMLE has lower variance than its competitors. This advantage of the first-order TMLE disappears asymptotically, as predicted by theory.

In terms of coverage, the improvement obtained with use of the $1^\star$-TMLE and the 2-TMLE for small values of both p and q is noteworthy. As an example, when $n = 2000$ and $(p, q) = (0.01, 0.1)$, the estimated coverage probability is 0, 0.9 and 0.8 for the 1-TMLE, the $1^\star$-TMLE and the 2-TMLE, respectively. This simulation illustrates the potential for dramatic improvement obtained by use of the $1^\star$-TMLE and the 2-TMLE, which nevertheless comes at the cost of over-coverage in small sample sizes when convergence rates are fast enough, e.g., $n \leq 2000$ and $(p, q) = (0.5, 0.1)$.

Table 26.1 Summary of estimator performance as a function of sample size and convergence rate of initial estimators of $\bar{Q}_0$ and g_0

| | | | 1-TMLE | | | | 1*-TMLE | | | | 2-TMLE | | | |
| | | | n | | | | n | | | | n | | | |
	p	q	500	1000	2000	10,000	500	1000	2000	10,000	500	1000	2000	10,000
		0.01	3.02	4.34	6.14	13.56	1.39	1.97	2.77	5.98	0.47	1.00	2.69	6.40
	0.01	0.10	1.93	2.55	3.27	5.65	0.18	0.22	0.32	0.41	1.47	1.95	0.61	0.73
		0.50	0.07	0.05	0.04	0.03	0.09	0.09	0.12	0.15	1.71	2.05	0.67	0.63
		0.01	1.33	1.77	2.31	4.26	0.63	0.84	1.17	2.22	0.03	0.28	1.08	2.28
$n^{1/2}$|Bias|	0.10	0.10	0.87	1.05	1.25	1.70	0.08	0.11	0.14	0.17	0.96	1.02	0.22	0.12
		0.50	0.01	0.02	0.01	0.03	0.07	0.05	0.07	0.04	0.87	0.90	0.21	0.23
		0.01	0.09	0.08	0.08	0.07	0.00	0.01	0.03	0.03	0.02	0.02	0.04	0.02
	0.50	0.10	0.04	0.03	0.03	0.00	0.00	0.07	0.19	0.01	0.02	0.09	0.11	0.03
		0.50	0.01	0.00	0.01	0.01	0.02	0.00	0.01	0.01	0.15	0.08	0.01	0.03
		0.01	1.60	1.59	1.57	1.60	3.22	2.13	2.15	2.11	2.58	2.87	1.89	2.09
	0.01	0.10	1.73	1.72	1.56	1.46	1.05	1.09	1.02	0.99	2.14	1.97	1.27	1.17
		0.50	1.06	1.08	1.03	1.01	1.10	1.08	0.97	1.07	2.15	2.10	1.31	1.17
		0.01	1.10	1.07	1.06	1.04	1.13	1.26	1.14	1.25	1.71	1.58	1.31	1.19
rVar	0.10	0.10	1.21	1.20	1.09	1.04	0.95	0.99	0.99	0.92	1.75	1.78	1.17	1.06
		0.50	0.97	0.98	1.02	0.98	1.03	1.01	1.01	1.01	2.17	1.96	1.19	1.04
		0.01	1.02	1.04	1.01	1.01	1.04	1.00	1.05	1.03	1.02	1.02	0.99	0.99
	0.50	0.10	1.04	0.95	0.97	0.97	1.14	6.10	17.17	1.02	3.58	9.19	16.22	0.84
		0.50	0.99	0.98	1.01	0.95	1.10	0.93	1.00	0.93	1.37	1.23	1.10	0.97
		0.01	0.01	0.00	0.00	0.00	0.76	0.23	0.03	0.00	0.91	0.78	0.03	0.00
	0.01	0.10	0.19	0.03	0.00	0.00	0.93	0.92	0.90	0.86	0.49	0.22	0.80	0.73
		0.50	0.94	0.95	0.94	0.94	0.93	0.95	0.95	0.94	0.38	0.22	0.80	0.79
		0.01	0.30	0.08	0.01	0.00	0.79	0.69	0.42	0.03	0.95	0.93	0.54	0.02
Cov. P	0.10	0.10	0.66	0.53	0.35	0.10	0.95	0.94	0.94	0.93	0.70	0.67	0.93	0.94
		0.50	0.95	0.95	0.95	0.94	0.94	0.95	0.94	0.94	0.79	0.77	0.92	0.93
		0.01	0.95	0.95	0.95	0.95	0.95	0.95	0.96	0.96	0.95	0.95	0.94	0.96
	0.50	0.10	0.94	0.94	0.95	0.95	0.95	0.99	0.98	0.95	0.99	0.98	0.98	0.95
		0.50	0.95	0.95	0.95	0.95	0.94	0.95	0.95	0.96	0.93	0.95	0.95	0.95

The 2-TMLE has poorer performance in terms of bias than the 1-TMLE and the 1*-TMLE for small samples whenever one of the models converges at a fast enough rate, i.e., either p or q equal to 0.5, which translates into lower coverage probabilities. The problem dissipates somewhat as n increases. While this suggests that caution may be required when using the 2-TMLE in small samples, we note that achieving either p or q equal to 0.5 in practice corresponds to using a correctly-specified parametric model, a nearly impossible feat in most applications.

In Figs. 26.1 and 26.2, we note that there is a region of slow convergence rates in which the proposed estimators outperform the standard first-order TMLE. In addition, as seen in Fig. 26.1, we observe a small advantage of the 2-TMLE over the 1*-TMLE in terms of bias. Generally though, we do not see any practical advantage

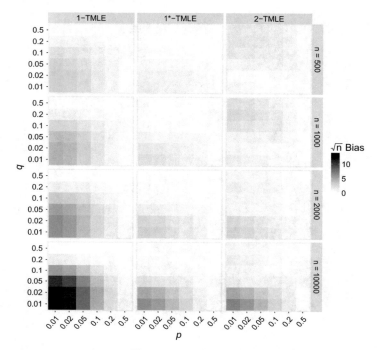

Fig. 26.1 Absolute bias scaled by $n^{1/2}$ as a function of sample size and convergence rate of initial estimators of $\bar{Q}_0$ and g_0

of the 2-TMLE over the 1*-TMLE. In fact, the 1*-TMLE performs better than the 2-TMLE for small samples and outperforms the 1-TMLE in all sample sizes, with the caveat of increased variance in certain scenarios, as discussed above.

26.5 Concluding Remarks

If the target parameter is higher-order pathwise differentiable, it seems desirable to implement a higher-order TMLE using appropriate higher-order gradients. Compared to its first-order counterpart, a higher-order TMLE will be an asymptotically efficient substitution estimator under weaker conditions. This article provides a template describing how to construct such a second-order TMLE, and a higher-order TMLE of arbitrary order can be implemented in similar fashion. Its implementation is identical to that of a regular TMLE, save for the use of a higher-dimensional least favorable parametric submodel that generates a set of scores defined by the gradients of all orders.

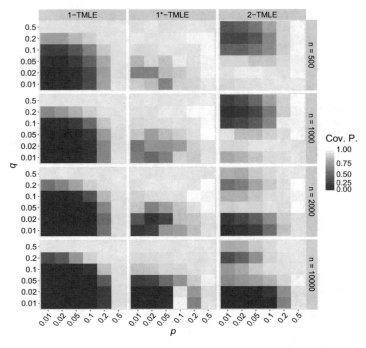

Fig. 26.2 Coverage probability of confidence intervals as a function of sample size and convergence rate of initial estimators of $\bar{Q}_0$ and g_0

To tackle target parameters that are not higher-order pathwise differentiable, we make use of approximate higher-order gradients in this chapter. This approximation strategy impacts only the higher-order bias reduction performed by the higher-order gradients. Since such a higher-order TMLE is tailored to eliminate the higher-order terms in the asymptotic expansions for the TMLE, a meaningful bias reduction relative to a standard first-order TMLE can be expected in practice if higher-order gradients are reasonably approximated. Unfortunately, these approximations rely on possibly high-dimensional smoothing and one may wonder whether we have simply replaced a difficult problem by another difficult problem. Fortunately, approximation-based higher-order TMLEs operate in a safe haven in which the desirable properties of the first-order TMLE are preserved. Adding extra parameters to the least-favorable submodel always improve the estimator asymptotically and can be very reasonably expected not to substantially harm the estimator in finite samples.

In practice, the higher-order TMLE, based on actual or approximate higher-order partial canonical gradients, should aim to achieve as much bias reduction as possible with the higher-order least favorable submodel. In this chapter, we provided theoretically sound and practicable building blocks for achieving this. Since any higher-order TMLE is a substitution estimator, the finite-sample variability induced by the need to fit a higher-order least favorable submodel is controlled by global bounds on the model and target parameter mapping. This is likely even more crucial in higher-order procedures given the need to correct a greater number of terms. Even more importantly, the seemingly daunting task of selecting tuning parameters and other potential inputs of the algorithm can be seamlessly overcome using an implementation of C-TMLE. The latter provides concrete tools for data adaptively fine-tuning the selection of a higher-order least favorable submodel with the objective of maximizing its effectiveness in reducing bias.

As highlighted in this chapter, the second-order TMLE in our illustrative example provides a concrete demonstration of a gain relative to a first-order TMLE: it is asymptotically linear and efficient in a significantly larger statistical model than an analogue first-order TMLE. This is directly parallel to the advances made in the seminal work of Robins et al. (2009) in the context of the one-step estimators. Another advantage of including higher-order partial canonical gradients into the TMLE framework as carried out in this work is that they yield contributions to the influence function of the higher-order TMLE that can be directly incorporated in the construction of confidence intervals, thereby possibly leading to finite-sample performance improvements.

Finally, since the Donsker class conditions imposed in our theorems can be restrictive in some settings, it is certainly of interest to develop a higher-order cross-validated TMLE. The latter would use a cross-validated version of the empirical risk in the iterative updating procedure and could be shown to lead to an asymptotically linear and efficient estimator even when such Donsker class conditions fail to hold. Such a development would be a direct extension of the work of Zheng and van der Laan (2010) in the first-order case, particularly since, as we have illustrated, a higher-order TMLE can be framed as a first-order TMLE with an augmented least-favorable submodel.

26.6 Notes and Further Reading

The building blocks of this generalization were set several decades ago, notably in the works of J. Pfanzagl (Pfanzagl 1985), wherein the notion of higher-order gradients was introduced and higher-order expansions of finite-dimensional parameters over arbitrary model spaces were formalized. In this work, we seek to utilize higher-order expansions to enable and guide the construction of regular and asymptotically linear estimators of statistical parameters in rich infinite-dimensional models. This is the perspective that motivated the seminal contributions of J. Robins, L. Li, E. Tchetgen and A. van der Vaart (e.g., Robins et al. 2008a, 2009; Li et al. 2011; van der Vaart 2014); these authors are the first to have provided a rigorous framework for precisely addressing this problem. The first exposition is that of Robins et al. (2008a), where the focus resides primarily on the use of higher-order gradients to derive optimal estimators in settings where regular estimation is not possible. Subsequent works are concerned with the development of a higher-order analogue of the one-step estimator introduced early on in Levit (1975), Ibragimov and Khasminskii (1981), Pfanzagl (1982) and Bickel (1982), for example. The general approach is thus to identify the dominating terms of an asymptotic expansion for a naive plug-in estimator and to perform, accordingly, an additive correction on this naive estimator. In Robins et al. (2009), the authors carefully establish the required foundations of a second-order extension of this approach and illustrate its use in the context of two problems, namely that of estimating the integral of the square of a density function and of estimating the mean response in missing data models. In Li et al. (2011), the authors study in great detail the problem of inferring a treatment effect using this approach and also establish minimax rate results for certain situations in which regular inference is not possible.

An excellent review of the general higher-order extension of the one-step estimator is provided in van der Vaart (2014). The one-step estimator is simple and easy to describe. However, in finite samples, it is vulnerable to decreased performance since it does not include any safeguard ensuring that the additive correction performed on the naive plug-in estimator does not drive the estimator near and possibly beyond the natural bounds of the parameter space. Rather than performing post-hoc bias correction in the parameter space, as does the one-step procedure, the TMLE framework, which we focus on in this chapter, provides guidelines for constructing an estimator of the underlying data-generating distribution, or whichever portion of it is needed to compute the parameter of interest, such that the resulting plug-in estimator enjoys regular and asymptotically linear asymptotic behavior. As such, the correction step is performed in the model space rather than in the parameter space. The appeal of devising such an approach, even decades before its formal development, was highlighted in Pfanzagl (1982). Since the resulting estimator automatically satisfies bounds on the parameter space, it never produces nonsensical output, such as a probability estimate outside the unit interval, and can outperform in finite samples asymptotically equivalent estimators that do not have a plug-in form. This issue arises when comparing first-order TMLE and one-step estimators, but is likely of even greater importance in higher-order inference due to the increased complexity of the correction process.

As is highlighted by Robins et al. (2009) and discussed in this chapter, higher-order gradients do not exist for several statistical parameters of interest. Nonetheless, approximate higher-order gradients can be used to produce regular estimators. In practice, the notion of approximate higher-order gradients necessarily involves the selection of certain tuning parameters. This requirement is discussed in Robins et al. (2009) but practical guidelines are neither provided nor appear particularly easy to develop. An advantage of using TMLE in this context, as we discuss later, is that the selection of such tuning parameters can be effortlessly embedded in the framework of collaborative TMLE, hereafter referred to as C-TMLE. Thus, the practical complications associated with the need for carefully-tuned approximations in the context of inference based on higher-order expansions can be tackled readily. Furthermore, as will be discussed, the framework of TMLE also provides useful tools that aim to guarantee that the resulting higher-order estimator will not behave worse than the usual first-order TMLE.

Chapter 27
Sensitivity Analysis

Iván Díaz, Alexander R. Luedtke, and Mark J. van der Laan

Causal inference problems are often tackled through the study of parameters of the distribution of a sequence of counterfactual variables, that represent the outcome in a hypothetical world where an intervention is enforced. The fundamental problem of causal inference is that, for a given individual, we only observe one such counterfactual outcome: the outcome under the treatment level actually assigned. Therefore, it is necessary to make certain untestable assumptions to *identify* the distribution of the missing counterfactual outcomes from the distribution of the observed data. One common such assumption is that the treatment mechanism does not depend on unmeasured factors that are causally related to the outcome. This assumption is often referred to as nonignorability of treatment assignment (Rubin 1976) or the (sequential) randomization assumption (van der Laan and Robins 2003).

> If the identifiability assumptions are not defensible in light of current scientific knowledge, the counterfactual distribution cannot be estimated from observed data. A useful approach in these cases is to use sensitivity analyses, which may allow investigators to understand how their analysis would change under varying degrees of violations to the assumptions.

I. Díaz (✉)
Division of Biostatistics and Epidemiology, Department of Healthcare Policy and Research, Weill Cornell Medical College, Cornell University, 402 East 67th Street, New York, NY 10065, USA
e-mail: ild2005@med.cornell.edu

A. R. Luedtke
Vaccine and Infectious Disease Division, Fred Hutchinson Cancer Research Center, 1100 Fairview Ave, Seattle, WA 98109, USA
e-mail: aluedtke@fredhutch.org

M. J. van der Laan
Division of Biostatistics and Department of Statistics, University of California, Berkeley, 101 Haviland Hall, #7358, Berkeley, CA 94720, USA
e-mail: laan@berkeley.edu

© Springer International Publishing AG 2018
M.J. van der Laan, S. Rose, *Targeted Learning in Data Science*,
Springer Series in Statistics, https://doi.org/10.1007/978-3-319-65304-4_27

There is an extensive literature on sensitivity analysis for causal inference. One of the most common approaches was developed by Rotnitzky et al. (1998); Scharfstein et al. (1999); Robins et al. (1999); Robins (1999); Rotnitzky et al. (2001); Scharfstein and Robins (2002). The approach they proposed assumes that the treatment mechanism distribution, conditional on the counterfactual outcome, follows a distribution in a parametric model, which is indexed by an unidentifiable parameter α. Their sensitivity analysis consists in estimating the treatment effect under various hypothesized levels of α. Subject-matter scientific knowledge is then used to judge the plausibility of each α value, and the estimates of the causal parameter corresponding to plausible values for α are used as the conclusions of the study.

In Díaz and van der Laan (2013b) and Luedtke et al. (2015b), we discuss the above approach, and highlight some of its limitations. The method of the above paragraph relies on the capability of the researcher to correctly pose a parametric model for the relation between the counterfactual outcome and the treatment assignment mechanism. The use of parametric models in this step poses at least three threats to the validity of the conclusions. First, parametric models have long been recognized to be inadequate to describe the complex relations arising in medium to large dimensional data (see e.g., Starmans 2011). Second, even if the parametric model is used only as an approximation, the adequacy of the approximation is unverifiable, because counterfactual variables are unobserved. A third problem, common to all parametric models, is that α does not have an intelligible interpretation under model misspecification. As a result, the pillar of the sensitivity analysis collapses because the plausibility of certain α values cannot be judged based on auxiliary scientific knowledge.

In this chapter, we present a sensitivity analysis that overcomes the above issues. The approach presented here is closely related to the partial identification literature, and can be summarized as follows. We decompose the bias in two main components: identification bias, defined as the difference between the observed data parameter and the causal parameter; and statistical bias, defined as the difference between the repeated sampling expectation of the estimator and the observed data parameter. We discuss ways of dealing with each source of bias. For dealing with identification bias, we propose to use *nonparametric* bounds on the causal effects, often indexed by some *interpretable* Condition δ. For reducing the statistical bias, we propose to use the targeted learning framework.

27.1 The Problem

Throughout the chapter, we use the following examples to illustrate the relevant concepts.

Example 27.1. Díaz and van der Laan (2013b) present a study demonstrating the effectiveness of a Chagas disease treatment. Causal conclusions were previously thought impossible for this application due to the inevitable informative dropout resulting from the disease's long (30 year) incubation period. This long incubation period also renders randomized studies prohibitively expensive, so that researchers can only use observational data to evaluate the efficacy of treatment. The original analysis for this problem proceeds as follows. A thorough review of the literature about the Chagas disease yielded 19 studies comprising about 520 patients. A standard meta analysis of these studies was deemed inappropriate because: (a) many studies did not have a control group; (b) there is very limited information about treatment allocation in all the manuscripts; (c) none of the studies reported baseline/confounder information (e.g., baseline health status, exposure to the vector, etc.); and (d) many of the studies presented large numbers of lost to follow-up and drop-outs. In terms of treatment assignment, the previous points imply that these studies cannot be considered randomized, and that there may be a considerable amount of unmeasured confounding (e.g., treatment was allocated according to unmeasured baseline status of the infection). Additionally, since patients may drop out of the study as a consequence of the worsening or improving of their health condition, loss to follow-up and drop out were often related to the unobserved outcome.

Example 27.2. Luedtke et al. (2015b) present an analysis of data from the Western Collaborative Group Study (WCGS) (Rosenman et al. 1964, 1975). The WCGS was a prospective cohort study designed to learn about the effect of binary personality type (Type A or B) on coronary heart disease (CHD) within an eight and a half year period. The data is publicly available through the `epitools` package in R (Aragon 2012). The authors focus on the effect of smoking status (yes or no) on CHD. This example is particularly useful for testing a sensitivity analysis method because the causal link between smoking and CHD is well-established in the literature. A similar decision was made by Ding and VanderWeele (2015) when evaluating their sensitivity analysis procedure on a historical data set exploring the effect of smoking on lung cancer. The covariates used as potential confounders of the relation between smoking status at baseline and CHD are: age, height, weight, and an indicator of Type A personality. The outcome we consider is the presence of CHD within 8 1/2 years of baseline.

Let us now formally define the inference problem. For simplicity, we focus on a cross-sectional study with a potentially missing outcome. Let A denote a binary treatment variable, let Y denote a continuous or binary outcome, observed only when $\Delta = 1$, and let W denote a vector of observed pre-treatment covariates. Let $O = (W, A, \Delta, \Delta Y)$ represent a random variable with distribution P_0, and let $O_1, \ldots, O_n$ denote a sample of n i.i.d. observations of O. We assume $P_0 \in \mathcal{M}$, where $\mathcal{M}$ is the nonparametric model defined as all continuous densities on O with respect to a dominating measure v. Examples 27.1 and 27.2 are particular cases of this setup. In Example 27.1, no covariates W are observed. In Example 27.2, $P_0(\Delta = 1 \mid A = a, W = w) = 1$ with probability one over draws of (A, W), so that there are no missing outcomes.

In Examples 27.1 and 27.2 above, we are interested in estimating causal parameters, which are defined in terms of the distribution of the counterfactual outcomes $Y_a : a \in \{0, 1\}$. These outcomes represent the outcome observed in a hypothetical world in which $P_0(A = a, \Delta = 1) = 1$. Let us now examine the parameters and identifiability assumptions in our illustrating examples.

Example 27.1 (continued). The causal parameter of interest is the effect of treatment among the treated, defined as $\psi_{\text{causal}} = E(Y_1 - Y_0 \mid A = 1)$. Under the assumptions that

1. $Y = \Delta(AY_1 + (1 - A)Y_0)$ *(consistency)*; and
2. $(Y_0, Y_1) \perp\!\!\!\perp (A, \Delta)$ *(randomization and missing completely at random)*;
3. $P_0(A = a, \Delta = 1) > 0$ for $a \in \{0, 1\}$ *(positivity)*.

identification of ψ_{causal} may be obtained as

$$E(Y_1 - Y_0 \mid A = 1) = E(Y \mid \Delta = 1, A = 1) - E(Y \mid \Delta = 1, A = 0), \qquad (27.1)$$

where the right-hand side parameter is estimable from the observed data. Randomization and missing at random state that the treatment and missingness mechanisms are independent of the counterfactual outcomes. Since treatment is not randomized and drop-out is believed to be associated to the severity of the disease, assumption (2) is likely violated.

Example 27.2 (continued). The causal parameter in this problem is the average treatment effect given by $\psi_{\text{causal}} = E(Y_1 - Y_0)$. The assumptions required are the standard assumptions for making causal inferences from observational data, namely:

1. $Y = AY_1 + (1 - A)Y_0$ *(consistency)*;
2. $(Y_0, Y_1) \perp\!\!\!\perp A \mid W$ *(randomization)*; and
3. $P_0(A = a \mid W = w) > 0$ for $a \in \{0, 1\}$ with probability one over draws of W *(positivity)*.

Under these assumptions, we can write $E(Y_a) = E\{E(Y \mid A = a, W)\}$, where the right hand side is a parameter depending only on the distribution of the observed data $O = (W, A, Y)$.

27.2 Sensitivity Analysis

When the necessary identifiability conditions do not hold, as may be the case in our illustrating examples, a sensitivity analysis may be used to estimate the causal effect under various degrees of violation to the identifiability assumptions. We propose a sensitivity analysis which can be carried out in the following steps.

Step 1: Define a Statistical Parameter of Interest. In contrast to causal parameters, whose computation requires the distribution of unobserved counterfactuals

$Y_a : a \in \{0, 1\}$, statistical parameters are defined as parameters whose true value can be computed solely based on the distribution P_0. We denote our statistical parameter of interest with ψ_0. Here, ψ_0 may be given by the parameter that would have identified ψ_{causal} had the identifiability assumptions been correct, though this is not necessary, as we will see in Example 27.1.

Let ψ_n denote an estimator of ψ_0, and denote $\psi_1 = E(\psi_n)$, where the expectation is taken over draws of $O_1, \ldots, O_n$. If ψ_n is used to estimate ψ_{causal}, its bias may be decomposed in terms of *statistical bias* and *identification bias* as follows:

$$E(\psi_n) - \psi_{\text{causal}} = \underbrace{\psi_1 - \psi_0}_{\text{statistical bias}} + \underbrace{\psi_0 - \psi_{\text{causal}}}_{\text{identification bias}} .$$

Statistical bias arises because the statistical parameter, ψ_0, is incorrectly estimated. Identification bias arises because the identifiability assumptions are not met. Unlike identification bias, statistical bias may be corrected through better estimation methods or increased the sample sizes. Other names found in the literature for identification bias include confounding bias and selection bias. We favor the former term because it encompasses a wider variety of problems.

Step 2: Estimate the Statistical Parameter, Remove Statistical Bias. Due to their frequent misspecification, parametric working models are generally inadequate for reducing statistical bias. Instead, we encourage the use of targeted learning. In particular, TMLE provides estimators with the property that

$$\sqrt{n}\,(\psi_n - \psi_0) = \frac{1}{\sqrt{n}} \sum_{i=1}^{n} IC_{\hat{\psi}}(O_i) + o_P(1), \tag{27.2}$$

where $IC_{\hat{\psi}}$ is a mean zero, finite variance function, known as the influence function. Result (27.2) is known as *asymptotic linearity* of ψ_n. Given an estimate $\hat{\sigma}^2$ of the variance of $IC_{\hat{\psi}}(O)$, an asymptotically valid $(1 - \alpha)100\%$ Wald-type confidence interval for ψ_0 is given by

$$(L_n, U_n) \equiv \psi_n \pm z_{\alpha/2} \frac{\hat{\sigma}}{\sqrt{n}}, \tag{27.3}$$

where z_α is the $1 - \alpha$ percentile of a standard normal distribution.

Step 3: Find Bounds on the Identification Bias. Suppose that we know that $L_{\text{causal}}(\delta) \leq \psi_{\text{causal}} - \psi_0 \leq U_{\text{causal}}(\delta)$, where these bounds depend on some Condition δ, and may depend on P_0. In addition, δ must be interpretable in the entire nonparametric model, so that experts' knowledge and scientific literature can be used to postulate plausible values. Methods developed in the literature of partial identification can be of great help in this step. Below we illustrate the construction of the bounds in Examples 27.1 and 27.2.

Step 4: Compute Uncertainty Interval for the Causal Parameter. Assume that (L_n, U_n) is a valid confidence interval for ψ_0. This is achieved if there is no statistical bias, if (27.2) holds, and if $\hat{\sigma}^2$ is a consistent estimator of the variance of $IC_{\hat{\psi}}$. Given known bounds $(L_{\text{causal}}(\delta), U_{\text{causal}}(\delta))$ on the identification bias, we know that

$$(L_n + L_{\text{causal}}(\delta), U_n + U_{\text{causal}}(\delta)) \tag{27.4}$$

is an asymptotically valid $(1 - \alpha)100\%$ uncertainty interval for ψ_{causal} under Condition δ. If $L_{\text{causal}}(\delta)$ or $U_{\text{causal}}(\delta)$ depend on P_0, they must be estimated and the uncertainty interval must be modified accordingly. We develop such a modification below in Sect. 27.3.

Let us now illustrate these steps in our examples.

Example 27.1 (continued). In this example, we use the statistical parameter

$$\psi_0 = E_0(\Delta Y \mid A = 1) - E(Y \mid \Delta = 1, A = 0).$$

Note that this is *not* equal to (27.1), the parameter that would have identified ψ_{causal} if the identifiability conditions were met. The reasons to choose this parameter instead of (27.1) will become apparent momentarily. Consistent and asymptotically linear estimation of ψ_0 is trivially achieved by the nonparametric estimator

$$\psi_n = \frac{\sum_i^n A_i \Delta_i Y_i}{\sum_i^n A_i} - \frac{\sum_i^n (1 - A_i) \Delta_i Y_i}{\sum_i^n (1 - A_i) \Delta_i},$$

which uses empirical means to estimate the expectations involved in the definition of ψ_0. A straightforward application of the delta method yields (27.2) with

$$IC_{\hat{\psi}} = \frac{A}{E(A)} \{\Delta Y - E(\Delta Y \mid A = 1)\} - \frac{(1 - A)\Delta}{E\{(1 - A)\Delta\}} \{Y - E(Y \mid \Delta = 1, A = 0)\}.$$

A Wald-type confidence interval as in (27.3) may be constructed with $\hat{\sigma}^2$ equal to the empirical variance of $IC_{\hat{\psi},n}(O)$, where $IC_{\hat{\psi},n}(O)$ is equal to $IC_{\hat{\psi}}(O)$ with the unknown expectations replaced by their empirical counterparts. The main goal of this study is to establish treatment efficacy. Thus, we focus on $L_{\text{causal}}(\delta)$ and find it as follows. First, note that $E(\Delta Y \mid A = 1)$ is a conservative estimate of $E(Y_1 \mid A = 1)$. That is:

$$E(\Delta Y \mid A = 1) = E(\Delta Y_1 \mid A = 1) \leq E(Y_1 \mid A = 1).$$

It follows that

$$\psi_{\text{causal}} - \psi_0 = E(Y_1 \mid A = 1) - E(\Delta Y \mid A = 1) + E(Y \mid \Delta = 1, A = 0) - E(Y_0 \mid A = 1)$$
$$\geq E(Y \mid \Delta = 1, A = 0) - E(Y_0 \mid A = 1)$$

Thus, under Condition $\delta \leq E(Y_0 \mid A = 1)$, the lower bound is equal to $L_{\text{causal}}(\delta) = E(Y \mid \Delta = 1, A = 0) - \delta$. In this example, knowledge about the plausibility of the Condition $\delta \leq E(Y_0 \mid A = 1)$ can be obtained as follows. The parameter $E(Y_0 \mid A = 1)$ is the probability of cure for the treated population had they not been treated (i.e., it is the probability of a "spontaneous" cure). The literature in the Chagas disease provides auxiliary information that such probability is usually very low. This illustrates the advantage of defining the Condition δ nonparametrically: parametric definitions may have led to uninterpretable parameters whose value cannot be learned from auxiliary knowledge. Note also that the quantity $E(Y \mid \Delta = 1, A = 0)$ in $L_{\text{causal}}(\delta)$ is unknown. In this example, the effect magnitude $\psi_n = 0.47$ is large enough to allow us to conservatively set $E(Y \mid \Delta = 1, A = 0) = 0$. In other applications this practice would likely lead to bounds that are too wide to be informative. With a standard deviation estimated at $\hat{\sigma} = 0.02$, we obtain a one-sided 95% uncertainty interval for ψ_{causal} of $(0.4384 - \delta, \infty)$, assuming $\delta \leq E(Y_0 \mid A = 1)$. Thus, auxiliary knowledge that the probability of spontaneous cure among the treated $E(Y_0 \mid A = 1)$ is smaller than 0.4384 would suffice to reject the null hypothesis $H_0 : \psi_{\text{causal}} \leq 0$ (negative or no treatment effect) in favor of $H_1 : \psi_{\text{causal}} > 0$.

Example 27.2 (continued). In contrast to the previous example, here we define the causal parameter as

$$\psi_0 = E\{E(Y \mid A = 1, W) - E(Y \mid A = 0, W)\},$$

which is the parameter that identifies ψ_{causal} under conditions (a)–(c). Double robust estimating equation and targeted minimum loss based estimators for ψ_0 have been presented in van der Laan and Robins (2003) and van der Laan and Rose (2011), respectively. When both the outcome regression and treatment mechanism are estimated consistently and at a fast enough rate, these estimators have influence curve

$$IC_{\hat{\psi}}(O) = \frac{2A - 1}{P_0(A|W)} \{Y - E(Y|A, W)\} + E(Y|A, W) - \psi_0.$$

The assumption that both the outcome regression and treatment mechanism are estimated well enough can be unfeasible in practice. A recent work presents an estimator which is asymptotically linear with valid inference when only one of these objects is estimated well (van der Laan 2014b). This approach can be integrated into a sensitivity analysis, but we omit such discussion here.

We now find intervals $(L_{\text{causal}}(\delta), U_{\text{causal}}(\delta))$ that contain ψ_{causal} under a Condition δ. Unlike the previous example, the bounds are such that, if the identifiability assumptions are met, $L_{\text{causal}}(\delta) = U_{\text{causal}}(\delta) = 0$. We start by developing bounds on $E(Y_1)$, and then extend them to $\psi_{\text{causal}} = E(Y_1 - Y_0)$. Note that

$$E(Y_1) = E\{E(Y_1|A = 0, W)P_0(A = 0|W) + E(Y|A = 1, W)P_0(A = 1|W)\}. \quad (27.5)$$

Suppose that, with probability 1 over draws of the covariate W and some $\delta > 0$,

$$-\delta E(Y|A = 1, W) \leq E(Y_1|A = 0, W) - E(Y|A = 1, W) \leq \delta\{1 - E(Y|A = 1, W)\}.$$

Consider the lower bound. Given that $E(Y|A = 1, W) > 0$, the above says that $\frac{E(Y_1|A=0,W)}{E(Y|A=1,W)} \geq 1 - \delta$ for some $\delta > 0$, i.e. that the relative risk of $Y_1 = 1$ among untreated versus treated people is at least $1-\delta$ in every strata of covariates. Similarly, the above says that $\frac{E(1-Y_1|A=0,W)}{E(1-Y|A=1,W)} \geq 1 - \delta$, which is a similar relative risk bound but now for the risk of $Y_1 = 0$. At $\delta = 1$ the above condition reproduces the known bounds on the outcome. For rare (or highly common) outcomes it is worth using a two-dimensional δ parameter in the above expression, one for the lower bound and one for the upper, but for simplicity we do not explore this here. Plugging the above bound into (27.5) yields that

$$-\delta E\{E(Y|A = 1, W)P_0(A = 0|W)\} \leq$$
$$E(Y_1) - E\{E(Y \mid A = 1, W)\} \leq$$
$$\delta E\{E(1 - Y|A = 1, W)P_0(A = 0|W)\}. \quad (27.6)$$

Let us now return to bounds on ψ_{causal}. We use a bivariate sensitivity parameter $\delta = (\delta^-, \delta^+)$, where δ^- and δ^+ fall in $(0, 1)$. Condition δ is satisfied when the following two inequalities hold with probability 1 over draws of the covariate W:

$$-\delta^- E(Y|A = 0, W) \leq E(Y_0|A = 1, W) - E(Y|A = 0, W) \leq \delta^+ E(1 - Y|A = 0, W)$$
$$-\delta^- E(Y|A = 1, W) \leq E(Y_1|A = 0, W) - E(Y|A = 1, W) \leq \delta^+ E(1 - Y|A = 1, W).$$

A straightforward extension of (27.6) shows that, under Condition δ,

$$L_{\text{causal}}(\delta) = -E\{\delta^- E(Y|A = 1, W)P_0(A = 0|W) + \delta^+ E(1 - Y|A = 0, W)P_0(A = 1|W)\}$$
$$U_{\text{causal}}(\delta) = E\{\delta^+ E(1 - Y|A = 1, W)P_0(A = 0|W) + \delta^- E(Y|A = 0, W)P_0(A = 1|W)\}. \quad (27.7)$$

Below we refer to individuals who smoke in the observed population as "natural smokers" and people who do not smoke in the observed population as "natural non-smokers". We also refer to the risk ratio as the RR. Condition δ implies the following four inequalities within each stratum of the covariates:

1. $\frac{E(Y_1|A=0,W)}{E(Y|A=1,W)} \geq 1 - \delta^-$: The RR for a natural nonsmoker versus a natural smoker having a CHD event if, contrary to fact, everyone is intervened upon to be a smoker is at most $1 - \delta^-$.
2. $\frac{E(Y_0|A=1,W)}{E(Y|A=0,W)} \geq 1 - \delta^-$: The RR for a natural smoker versus a natural nonsmoker having a CHD event if, contrary to fact, everyone is intervened upon not to smoke is at most $1 - \delta^-$.
3. $\frac{E(1-Y_1|A=0,W)}{E(1-Y|A=1,W)} \geq 1 - \delta^+$: The RR for a natural nonsmoker versus a natural smoker *not* having a CHD event if, contrary to fact, everyone is intervened upon to be a smoker is at most $1 - \delta^+$.
4. $\frac{E(1-Y_0|A=1,W)}{E(1-Y|A=0,W)} \geq 1 - \delta^+$: The RR for a natural smoker versus a natural nonsmoker *not* having a CHD event if, contrary to fact, everyone is intervened upon not to smoke is at most $1 - \delta^+$.

The validity of the lower bound $L_{\text{causal}}(\delta)$ only relies on 1 and 4. Under 1, natural nonsmokers are not too protected from CHD events by some unmeasured cause. Under 4, smokers are not too inclined towards CHD events by some unmeasured cause. The prevalence of coronary events is low in our data set ($257/3154 \approx 0.08$), so large values of δ^- should be more plausible than large values of δ^+. Since the interval $(L_{\text{causal}}(\delta), U_{\text{causal}}(\delta))$ depends on P_0, it must be estimated, and the uncertainty in its estimation must be incorporated in the interval (27.4). We show such modification in the following section.

27.3 Bounds on the Causal Bias Are Unknown

Suppose that $(L_{\text{causal}}(\delta), U_{\text{causal}}(\delta))$ depends on the observed data distribution P_0. This holds in our Example 27.2, see (27.6). Note that $L_{\text{causal}}(\delta)$ and $U_{\text{causal}}(\delta)$ are now parameters which take as input an observed data distribution and output a number. Thus, provided $L_{\text{causal}}(\delta)$ is pathwise differentiable, it is reasonable to expect that, for an influence curve $IC_{\hat{\ell}}$, we can develop an asymptotically linear estimators of $L_{\text{causal}}(\delta)$ with the property that

$$\sqrt{n}\,(\ell_{n,\text{causal}} - L_{\text{causal}}(\delta)) = \frac{1}{\sqrt{n}} \sum_{i=1}^{n} IC_{\hat{\ell}}(O_i) + o_P(1)$$

If $U_{\text{causal}}(\delta)$ is sufficiently smooth, then we would expect to be able to develop an estimator so that the same expression above holds with $u_{n,\text{causal}}$, $U_{\text{causal}}(\delta)$, and $IC_{\hat{\ell}}$ replaced by $u_{n,\text{causal}}$, $U_{\text{causal}}(\delta)$, and $IC_{\hat{u}}$, respectively. Combining the above and (27.2) yields

$$\sqrt{n}\,\{\psi_n + \ell_{n,\text{causal}} - (\psi_0 + L_{\text{causal}}(\delta))\} = \frac{1}{\sqrt{n}} \sum_{i=1}^{n} \left\{ IC_{\hat{\psi}}(O_i) + IC_{\hat{\ell}}(O_i) \right\} + o_P(1).$$

The right hand side converges to a mean zero normal distribution with variance equal to the variance of the sum of influence curves on the right. An analogous argument yields that $\psi_0 + U_{\text{causal}}(\delta)$ has influence curve $IC_{\hat{\psi}} + IC_{\hat{u}}$. Hence the joint distribution of these two influence curves applied to $O \sim P_0$ converges to a multivariate normal distribution with mean zero and covariance matrix given by that of the two-dimensional random variable with coordinates $IC_{\hat{\psi}}(O) + IC_{\hat{\ell}}(O)$ and $IC_{\hat{\psi}}(O) + IC_{\hat{u}}(O)$. Given a consistent estimate Σ_n of the covariance matrix, one can take Monte Carlo draws $(Z_L^1, Z_U^1), \ldots, (Z_L^m, Z_U^m)$ from the $N(0, \Sigma_n)$ distribution. Given these draws, one can then choose s_n to be the 95% quantile of $\max\{Z_L^k, -Z_U^k\}$ among the observations $k \in \{1, \ldots, m\}$. In that case, the uncertainty interval

$$\left(\psi_n + \ell_{n,\text{causal}} - \frac{s_n}{\sqrt{n}}, \psi_n + u_{n,\text{causal}} + \frac{s_n}{\sqrt{n}} \right) \tag{27.8}$$

will contain the causal parameter with probability approaching 0.95 under Condition δ. One could alternatively replace s_n in the lower bound with a $s_{n,L}$ and s_n in the upper bound with $s_{n,U}$ and choose an empirically valid uncertainty interval (in the

Monte Carlo draws) which minimizes $s_{n,U} + s_{n,L}$. This may be beneficial when, e.g., the lower bound is significantly easier to estimate than the upper bound and one wants to make the interval lower bound tighter to reflect this.

We now illustrate this in Example 27.2.

Example 27.2 (continued). Let us first estimate ψ_0 using the TMLE as presented in Chap. 7 of van der Laan and Rose (2011). We then estimate the interval $(L_{\text{causal}}(\delta), U_{\text{causal}}(\delta))$ resulting from (27.7) using a TMLE algorithm described in Web Appendix A. Given the broader focus of this paper, we omit a theoretical analysis of the asymptotic properties of this estimator, though refer the reader to van der Laan and Rose (2011) for a general template for how to analyze such an estimator. We use 2.5×10^4 draws from a bivariate normal distribution to implement the uncertainty bound estimation procedure described in Sect. 27.3.

Code to replicate the analysis is available in the original research article (Luedtke et al. 2015b). Figure 27.1 shows how the lower bound is impacted by different choices of δ. Consider $\delta^+ = 0.02$, and suppose that the probability of not having a CHD event is at most 0.3 within all strata of covariates (according to our estimate of $E(1 - Y|A = 0, W)$, the maximum probability of not having a CHD event is approximately 0.25). In this case 4 is satisfied provided $E(Y_0|A = 1, W0 \le E(Y|A = 1, W) + 0.015$, so that within any stratum of covariates natural smokers could have an at most a 1.5% higher additive heart attack risk than natural nonsmokers if an intervention had set everyone to be nonsmokers at baseline. For the lower bound on the average treatment effect to remain positive, we then need that δ^- is no more than approximately 0.4. Inequality 2 is irrelevant for the lower bound on the average treatment effect, so we focus on 1. This says that if we intervened in the population to make everyone a smoker then, within each stratum of covariates, the relative risk of a heart attack between natural nonsmokers and natural smokers is no less than 60%. Figure 27.2 provides similar insights for the lower bounds, but also allows

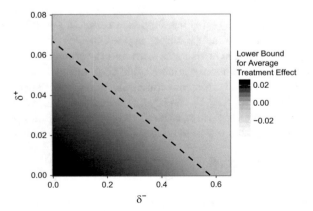

Fig. 27.1 95% uncertainty bounds for the average treatment effect of smoking on CHD events at several δ values

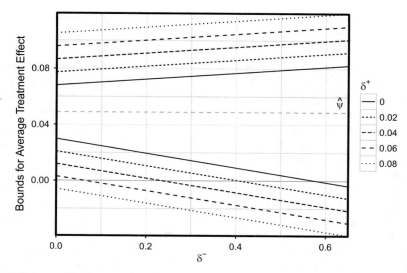

Fig. 27.2 Lower 97.5% uncertainty bound for the average treatment effect of smoking on CHD events at a continuum of δ values

one to visualize the upper bound for the average treatment effect under different choices of δ. From these choices of sensitivity parameters, it appears unlikely that the average treatment effect of smoking on CHD events within eight and a half years will be larger than 0.12 in this population.

27.4 Notes and Further Reading

The contents of this chapter are based on previous work by Díaz and van der Laan (2013b) and Luedtke et al. (2015b); the reader interested in further discussion and details is encouraged to consult these sources. We stress that subject matter experts should be able to judge the plausibility of each value of δ, the sensitivity parameter, based solely on background scientific knowledge. We also stress the importance of using statistically valid methods to estimate ψ_0. In particular, we encourage the use of targeted learning. The standard practice of posing unjustified parametric assumptions in this stage would in general increase the statistical bias, and would therefore invalidate the conclusions of the sensitivity analysis.

We now discuss some methods related to the approach discussed in this chapter. Rosenbaum and Rubin (1983a) consider the estimation of a parametric model for the average treatment effect if one assumes that adjusting for an unmeasured confounder would make the parameter identifiable. VanderWeele and Arah (2011) improve these results by giving general formulas for the bias of the statistical parameter for the causal parameter under this same unmeasured confounder assumption. These approaches yield an interpretable sensitivity parameter under sometimes strong simplifying assumptions, such as that this unmeasured confounder is binary. Obtaining uncertainty bounds for the causal parameter requires obtaining confidence bounds for a statistical parameter that adjusts for measured confounders, and thus requires modern statistical approaches for these confidence bounds to be valid. Similar approaches were used to get bounds for interaction parameters (VanderWeele et al. 2012a) and for direct effects (VanderWeele 2010). The bounds in Ding and VanderWeele (2015) are defined using this approach, but require far fewer parameters than the earlier methods when the unmeasured confounder is not binary.

These sensitivity analysis procedures are related to the partial identification literature which develops bounds on the difference between the causal and statistical parameter that hold under very weak assumptions (if any), such as bounds on the outcome (Manski 1990, 2003; Horowitz and Manski 2000; MacLehose et al. 2005). Though the bounds resulting from these analyses are convincing when informative, in many cases they can be too conservative to be informative about even the sign of an effect. In the context of our Example 27.2, Manski (2003) considers bounds on causal parameters which reduce to using the known bounds on the outcome in great detail. That work also gives bounds under other assumptions, such as monotonicity assumptions, which may help inspire other choices of Condition δ.

Horowitz and Manski (2000) used an approach similar to that of Sect. 27.3 to get an uncertainty region for the bounds in the partial identifiability context. Woutersen (2006) considered how to develop such an uncertainty region given a asymptotically linear estimators of the upper and lower bound in partial identifiability problems. Both of these works actually consider a refined procedure which guarantees coverage for the parameter ψ_{causal}, rather than the entire region $(\psi_0 + L_{\text{causal}}(\delta), \psi_0 + U_{\text{causal}}(\delta))$ known to contain ψ_{causal}, with probability approaching 0.95. Such refinements will be analogous for sensitivity analyses, but we do not explore them here. The uncertainty region we presented in Sect. 27.3 is similar in spirit to those presented in Vansteelandt et al. (2006), but does not require the specification of an implausible semiparametric model for the identification bias.

Chapter 28
Targeted Bootstrap

Jeremy Coyle and Mark J. van der Laan

The bootstrap is used to obtain statistical inference (confidence intervals, hypothesis tests) in a wide variety of settings (Efron and Tibshirani 1993; Davison and Hinkley 1997). Bootstrap-based confidence intervals have been shown in some settings to have higher-order accuracy compared to Wald-style intervals based on the normal approximation (Hall 1988, 1992; DiCiccio and Romano 1988). For this reason it has been widely adopted as a method for generating inference in a range of contexts, not all of which have theoretical support. One setting in which it fails to work in the manner it is typically applied is in the framework of targeted learning. We describe the reasons for this failure in detail and present a solution in the form of a targeted bootstrap, designed to be consistent for the first two moments of the sampling distribution.

Suppose we want to estimate a particular pathwise differentiable parameter using a targeted learning approach. The typical workflow is to obtain initial estimates for relevant factors of the likelihood using super learner, and then generate a targeted estimate using TMLE. By using super learner and TMLE, we can generate correct inference for our parameter of interest without assuming that the likelihood can be modeled by simple parametric models. Relying on the fact that TMLE is an asymptotically linear estimator, we can use the normal approximation to generate Wald-style confidence intervals where the standard error is based on the in-

J. Coyle (✉)
Division of Biostatistics, University of California, Berkeley, 101 Haviland Hall, #7358, Berkeley, CA 94720, USA
e-mail: jrcoyle@berkeley.edu

M. J. van der Laan
Division of Biostatistics and Department of Statistics, University of California, Berkeley, 101 Haviland Hall, #7358, Berkeley, CA 94720, USA
e-mail: laan@berkeley.edu

© Springer International Publishing AG 2018
M.J. van der Laan, S. Rose, *Targeted Learning in Data Science*,
Springer Series in Statistics, https://doi.org/10.1007/978-3-319-65304-4_28

fluence curve. These confidence intervals are first-order accurate. It is tempting to instead obtain higher-order correct confidence intervals by applying the nonparametric bootstrap. However, in the case of TMLE with initial estimates obtained via the super learner algorithm, naïve application of the nonparametric bootstrap is not justified and which we will show to have poor performance, because super learner and therefore TMLE behaves differently on nonparametric bootstrap samples than it does on samples from the true data generating distribution. It is therefore important to develop a bootstrap method that works in the context of targeted learning.

We illustrate the reason for this difference in super learner's behavior, and present a solution in the form of the targeted bootstrap, a novel model based bootstrap that samples from a distribution targeted to the parameter of interest and to the asymptotic variance of its influence curve. In the process, we outline a TMLE that targets both a parameter of interest and its asymptotic variance. This TMLE can be used to generate another Wald-style confidence interval, by directly using the targeted estimate of the asymptotic variance. Additionally, it can be used to generate a confidence of interval for the asymptotic variance itself. We demonstrate the practical performance of the targeted bootstrap confidence intervals relative to the Wald-type confidence intervals as well as confidence intervals generated by other bootstrap approaches.

28.1 Problem Statement

Suppose that we observe n independent and identically distributed copies of O with probability distribution P_0 known to be an element of the statistical model $\mathcal{M}$. In addition, assume we are concerned with statistical inference of the target parameter value $\psi_0 = \Psi(P_0)$ for a given parameter mapping $\Psi : \mathcal{M} \to \mathbf{R}$. Consider a given estimator $\hat{\Psi} : \mathcal{M}_{np} \to \mathbf{R}$ that maps an empirical distribution P_n of $O_1, \ldots, O_n$ into an estimate of ψ_0, and assume that this estimator $\psi_n = \hat{\Psi}(P_n)$ is asymptotically linear at P_0 with influence curve $O \to D(P_0)(O)$ at P_0, so that we can write:

$$\psi_n - \psi_0 = (P_n - P_0)D(P_0) + o_P(1/\sqrt{n}).$$

In that case, we have that $\sqrt{n}(\psi_n - \psi_0)$ converges in distribution to a normal distribution $N(0, \Sigma^2(P_0))$, where $\Sigma^2 : \mathcal{M} \to \mathbf{R}$ is defined by $\Sigma^2(P) = PD(P)^2$ as the variance of the influence curve $D(P)$ under P.

We wish to estimate a confidence interval for ψ_n. A one-sided confidence interval is defined by a quantity $\psi_{n,[\alpha]}$ such that $P_0(\psi_0 < \psi_{n,[\alpha]}) = \alpha$. Two sided confidence intervals are typically equal-tailed intervals, having the same error in each tail: $P_0(\psi_0 < \psi_{n,[\alpha/2]}) = P_0(\psi_0 > \psi_{n,[1-\alpha/2]}) = \alpha/2$. These can be constructed using a pair of one-sided intervals. A one-sided Wald confidence interval can be generated using the asymptotic normality discussed above: defining variance estimator $\hat{\sigma}_n^2 = \Sigma^2(P_n)$, the endpoint is $\psi_{n,[\alpha],\text{Wald}} = \psi_n - n^{-1/2}\hat{\sigma}_n\phi^{-1}(1-\alpha)$, where $\phi^{-1}(1-\alpha)$

is the $1 - \alpha$th quantile of the standard normal distribution. Similarly, a two-sided Wald confidence interval is given by $\psi_n \pm \phi^{-1(1-\alpha/2)}n^{-1/2}\hat{\sigma}_n$. This approach ignores the remainder term $o_P(1/\sqrt{n})$ and is therefore said to be first order correct.

To provide a concrete motivating example, suppose we observe n i.i.d. observations of $O = (W, A, Y) \sim P_0$, for baseline covariates W, treatment $A \in \{0, 1\}$, and outcome $Y \in \{0, 1\}$, and suppose that $\mathcal{M}$ is the nonparametric model, making no assumptions about the distribution from which O is sampled. The target parameter $\Psi : \mathcal{M} \to \mathbb{R}$, is a treatment-specific mean defined as $\Psi(P) = E_P E_P(Y \mid A = 1, W)$. Let $\bar{Q}(P)(W) = E_P(Y \mid A = 1, W)$ and $\bar{g}(P)(W) = E_P(A \mid W)$.

28.2 TMLE

As previously discussed elsewhere in this book, TMLE produces asymptotically linear substitution estimators of target parameters. TMLE fluctuates an initial estimate of the target parameter, resulting in an estimate which makes the correct bias-variance trade-off. TMLEs are asymptotically linear with a known influence curve, even when the components of the likelihood are estimated using data-adaptive methods (like these).

28.2.1 TMLE for Treatment Specific Mean

The efficient influence curve of Ψ at P is given by:

$$D^*(P)(O) = \frac{A}{\bar{g}(W)}(Y - \bar{Q}(W)) + \bar{Q}(W) - \Psi(P)$$

(van der Laan and Robins 2003). Note that $\Psi(P)$ only depends on P through $\bar{Q}(P)$ and the probability distribution $Q_W(P)$ of W. Let $Q(P) = (Q_W(P), \bar{Q}(P))$ and let $Q(\mathcal{M}) = \{Q(P) : P \in \mathcal{M}\}$ be its model space. We will also denote the target parameter as $\Psi : Q(\mathcal{M}) \to \mathbf{R}$ as a mapping that maps a Q in the parameter space $Q(\mathcal{M})$ into a numeric value, abusing notation by using the same notation Ψ for this mapping. Similarly, we will also denote $D^*(P)$ with $D^*(Q, G)$. The efficient influence curve $D^*(P)$ satisfies the expansion $\Psi(P) - \Psi(P_0) = -P_0 D^*(P) + R_\psi(P, P_0)$, where

$$R_\psi(P, P_0) = P_0 \frac{\bar{g} - \bar{g}_0}{\bar{g}}(\bar{Q} - \bar{Q}_0).$$

Let $\psi_n^* = \Psi(Q_n^*)$ be a TMLE of ψ_0 so that it is asymptotically linear at P_0 with influence curve $D^*(P_0)$. This TMLE can be defined by letting $\bar{Q}_n^0$ being an initial estimator of $\bar{Q}_0$, $\bar{g}_n$ an estimator of g_0, $L(\bar{Q})(O) = -I(A = 1)(Y \log \bar{Q}(W) + (1 - Y) \log(1 - \bar{Q}(W))$ being the log-likelihood loss function for $\bar{Q}_0$, the submodel $\text{Logit}\bar{Q}_n^0(\epsilon)) = \text{Logit}\bar{Q}_n^0 + \epsilon H(\bar{g}_n)$ with $H(\bar{g}_n) = A/\bar{g}_n(W)$, $\bar{Q}_n^1 = \bar{Q}_n^0(\epsilon_n^0)$ with

$\epsilon_n^0 = \arg\min_\epsilon P_n L(\bar{Q}_n^0(\epsilon))$, and $\psi_n^* = \Psi(Q_n^1)$, where $Q_n^1 = (\bar{Q}_n^1, Q_{W,n})$ and $Q_{W,n}$ is the empirical distribution of $W_1, \ldots, W_n$. Let P_n^* be a probability distribution compatible with Q_n^*.

28.2.2 TMLE of the Variance of the Influence Curve

Let $O \sim P_0 \in \mathcal{M}$, and we have two target parameters $\Psi : \mathcal{M} \to \mathbf{R}$ and $\Sigma^2 : \mathcal{M} \to \mathbf{R}$. We are given an estimator ψ_n^* that is asymptotically linear at P_0 with influence curve $D(P_0)$. For simplicity, we will consider the case that $\psi_n^* = \Psi(P_n^*)$ is an efficient targeted maximum likelihood estimator so that $D(P) = D^*(P)$ and $D^*(P)$ is the efficient influence curve of Ψ at P. In this case, $\Sigma^2(P) = P\{D^*(P)\}^2$. Let $D_\Sigma^*(P)$ be the efficient influence curve of Σ^2 at P. Suppose that $\Sigma^2(P) = \Sigma_1^2(Q_\Sigma(P))$ for some parameter $Q_\Sigma(P)$ that can be defined by minimizing the risk of a loss function $L_\Sigma(Q_\Sigma)$ so that $Q_\Sigma(P) = \arg\min_{Q_\Sigma} PL_\Sigma(Q)$. In addition, we assume that $D_\Sigma^*(P)$ only depends on P through $Q_{\Sigma(P)}$ and some other parameter $g_\Sigma(P)$. For notational convenience, we will denote these to alternative representations of the asymptotic variance parameter and its efficient influence curve with $\Sigma^2(Q_\Sigma)$ and $D_\Sigma^*(Q_\Sigma(P), g_\Sigma(P))$ respectively.

We now develop a TMLE of $\Sigma^2(P_0)$ as follows. First, let $Q_{\Sigma,n}^0$ be an initial estimator of $Q_\Sigma(P_0)$, which could be based on the super learner ensemble algorithm using the loss function $L_\Sigma()$. Similarly, let $g_{\Sigma,n}$ be an estimator of $g_{\Sigma,0}$. Set $k = 0$. Consider now a submodel $\{Q_{\Sigma,n}^k(\epsilon \mid g_{\Sigma,n}) : \epsilon\} \subset Q_\Sigma(\mathcal{M})$ so that the linear span of the components of the generalized score $\frac{d}{d\epsilon} L_\Sigma(Q_{\Sigma,n}^k(\epsilon \mid g_{\Sigma,n}))$ at $\epsilon = 0$ spans $D_\Sigma^*(Q_{\Sigma,n}^k, g_{\Sigma,n})$. Define $\epsilon_n^k = \arg\min_\epsilon P_n L_\Sigma(Q_{\Sigma,n}^k(\epsilon \mid g_{\Sigma,n}))$ as the MLE and define the update $Q_{\Sigma,n}^{k+1} = Q_{\Sigma,n}^k(\epsilon_n^k \mid g_{\Sigma,n})$. We iterate this updating process until convergence at which step K we have $\epsilon_n^K \approx 0$. We denote this final update with $Q_{\Sigma,n}^*$ and we call that the TMLE of $Q_\Sigma(P_0)$, while $\Sigma^2(Q_{\Sigma,n}^*)$ is the TMLE of the asymptotic variance $\Sigma^2(Q_0)$ of the TMLE ψ_n^* of ψ_0. Let $\tilde{P}_n^*$ be a probability distribution compatible with $Q_{\Sigma,n}^*$.

Application to the Treatment Specific Mean. The asymptotic variance of $\sqrt{n}(\psi_n^* - \psi_0)$ is given by:

$$\Sigma^2(P_0) = E_{P_0}\{D^*(P_0)\}^2$$
$$= Q_{0,W}\left(\frac{\bar{Q}_0(1 - \bar{Q}_0)}{\bar{g}_0} + (\bar{Q}_0 - Q_{0,w}\bar{Q}_0)^2\right)$$

The following lemma presents its efficient influence curve $D_\Sigma^*(P)$.

Lemma 28.1. *The efficient influence curve $D_\Sigma^*(P)$ of Σ^2 at P is given by:*

$$D_\Sigma^*(P)(W, A, Y) = D_{\Sigma^2, Q_W}(P)(W) + D_{\Sigma^2, \bar{Q}}(P)(O) + D_{\Sigma^2, \bar{g}}(P)(O),$$

where

$$D_{\Sigma^2, Q_W}(P)(W) = \frac{\bar{Q}(1 - \bar{Q})}{\bar{g}}(W) - Q_W \frac{\bar{Q}(1 - \bar{Q})}{\bar{g}}$$
$$+ (\bar{Q}(W) - \Psi(Q))^2 - Q_W(\bar{Q} - \Psi(Q))^2$$

$$D_{\Sigma^2, \bar{Q}}(P)(O) = \frac{I(A = 1)}{\bar{g}(W)}\left(\frac{1 - 2\bar{Q}(W)}{\bar{g}(W)} + 2(\bar{Q}(W) - \Psi(Q))\right)(Y - \bar{Q}(W))$$

$$D_{\Sigma^2, \bar{g}}(P)(O) = -\frac{\bar{Q}(1 - \bar{Q})(W)}{\bar{g}^2(W)}(A - \bar{g}(W)).$$

This allows us to develop a TMLE $\Sigma^2(Q_{W,n}, \bar{Q}_n^*, \bar{g}_n^*)$ of $\Sigma^2(Q_{W,0}, \bar{Q}_0, \bar{g}_0)$. Define the clever covariates:

$$C_Y(\bar{g}, Q)(A, W) \equiv \frac{I(A = 1)}{\bar{g}(W)}\left(\frac{1 - 2\bar{Q}(W)}{\bar{g}(W)} + 2(\bar{Q}(W) - \Psi(Q))\right)$$

$$C_A(\bar{g}, \bar{Q})(W) \equiv \frac{\bar{Q}(1 - \bar{Q})(W)}{\bar{g}^2(W)}.$$

Let $Q_n^0 = (Q_{W,n}, \bar{Q}_n^0)$ for an initial estimator $\bar{Q}_n^0$ of $\bar{Q}_0$, where $Q_{W,n}$ is the empirical distribution which will not be changed by the TMLE algorithm. Let $k = 0$. Consider the submodels

$$\text{Logit}\bar{Q}_n^k(\epsilon_1) = \text{Logit}\bar{Q}_n^k + \epsilon_1 C_Y(\bar{g}_n^k, Q_n^k)$$
$$\text{Logit}\bar{g}_n^k(\epsilon_2) = \text{Logit}\bar{g}_n^k + \epsilon_1 C_A(\bar{g}_n^k, \bar{Q}_n^k).$$

In addition, consider the log-likelihood loss functions

$$L_1(\bar{Q}) = -I(A = 1)\left\{Y \log \bar{Q}(W) + (1 - Y) \log(1 - \bar{Q}(W))\right\}$$
$$L_2(\bar{g}) = -\left\{A \log \bar{g}(W) + (1 - A) \log(1 - \bar{g}(W))\right\}.$$

Define the MLEs $\epsilon_{1n}^k = \arg\min_\epsilon P_n L_1(\bar{Q}_n^k(\epsilon))$ and $\epsilon_{2n}^k = \arg\min_\epsilon P_n L_2(\bar{g}_n^k(\epsilon))$. This defines the first step update $\bar{Q}_n^{k+1} = \bar{Q}_n^k(\epsilon_{1n}^k)$ and $\bar{g}_n^{k+1} = \bar{g}_n^k(\epsilon_{2n}^k)$. Now set $k = k + 1$ and iterate this process until convergence defined by $(\epsilon_{1n}^*, \epsilon_{2n}^*)$ being close enough to $(0, 0)$. Let $\bar{g}_n^*, \bar{Q}_n^*$ denote these limits of this TMLE procedure, and let $Q_n^* = (Q_{W,n}, \bar{Q}_n^*)$. The TMLE of $\Sigma^2(P_0)$ is given by $\Sigma^2(\tilde{P}_n^*)$ where $\tilde{P}_n^*$ is defined by $(Q_{W,n}, \bar{Q}_n^*, \bar{g}_n^*)$. We note that at $(\epsilon_{1n}^*, \epsilon_{2n}^*) = (0, 0)$, we have

$$0 = P_n D_{\Sigma^2}(\tilde{P}_n^*) = 0,$$

and if the algorithm stops earlier at step K, and one defines $\tilde{P}_n^* = P_n^K$, one just needs to make sure that

$$P_n D_{\Sigma^2}(\tilde{P}_n^*) = o_P(1/\sqrt{n}).$$

28.2.3 Joint TMLE of Both the Target Parameter and Its Asymptotic Variance

We could also define a TMLE targeting both parameters Ψ and Σ^2. This is defined exactly as above, but now using a submodel $\{P_n^k(\epsilon) : \epsilon\} \subset \mathcal{M}$ that has a score $\frac{d}{d\epsilon} L(P_n^k(\epsilon))$ at $\epsilon = 0$ whose components span both efficient influence curves $(D_\psi^*(P), D_\Sigma^*(P))$. In this manner, one obtains a TMLE $\tilde{P}_n^*$ that solves $P_n D_\psi^*(\tilde{P}_n^*) = P_n D_\Sigma^*(\tilde{P}_n^*) = 0$ and, under regularity conditions, yields an asymptotically efficient estimator of both ψ_0 and σ_0^2. In this case our TMLE of ψ_0 could just be $\Psi(\tilde{P}_n^*)$: so in this special case, we have $P_n^* = \tilde{P}_n^*$. In particular, we could estimate both ψ_0 and σ_0^2 with a bivariate TMLE $(\Psi(\tilde{P}_n^*), \Sigma^2(\tilde{P}_n^*))$ where $\tilde{P}_n^*$ is a TMLE that targets both ψ_0 and $\sigma_0^2 = \Sigma^2(P_0)$. In this case, $P_n^* = \tilde{P}_n^*$ and thus $\psi_n^* = \Psi(\tilde{P}_n^*)$, $\sigma_n^* = \Sigma^2(\tilde{P}_n^*)$. This TMLE can be defined as the above iterative TMLE of $\Sigma^2(P_0)$ but now using the augmented submodel:

$$\text{Logit}\bar{Q}_n^k(\epsilon_1) = \text{Logit}\bar{Q}_n^k + \epsilon_0 H(\bar{g}_n^k) + \epsilon_1 C_Y(\bar{g}_n^k, Q_n^k),$$

where $H(\bar{g})(A, W) = I(A = 1)/\bar{g}(W)$. Conditions under which $(\Psi(\tilde{P}_n^*), \Sigma^2(\tilde{P}_n^*))$ is an asymptotically efficient estimator of $(\Psi(P_0), \Sigma^2(P_0))$ are essentially the same as needed for efficiency of $\Psi(P_0)$.

28.3 Super Learner

TMLE requires initial estimates of factors of the likelihood. For the treatment-specific mean example we need estimates of $\bar{Q}(A, W)$ and $\bar{g}(W)$. In the targeted learning framework, these factors are typically estimated with super learner, discussed earlier in Chap. 3 and elsewhere. For the purposes of this chapter, we applied discrete super learner (also referred to as the cross-validation selector), which selects the risk-minimizing individual algorithm.

Loss based estimation allows us to objectively evaluate the quality of estimates and select amongst competing estimators based on this evaluation. Super learner is a particular implementation of this framework, and understanding this framework is important to understanding the behavior of super learner on nonparametric bootstrap samples. In the context of super learner, we will refer to estimators of the factors of the likelihood as "learners". This exposition will focus on the example of learning an estimate of $\bar{Q}_0(P_0) = E_{P_0}[Y|A, W]$, but it applies equally well to other factors of the likelihood. Here, $\bar{Q}(P)$ indicates an estimate of $\bar{Q}$ based on P. Consider the problem of selecting an estimate $\bar{Q}$ from a class of possible distributions $\bar{\mathbf{Q}}$. In the context of discrete super learner, this becomes selecting from a set of candidate learners: $\bar{Q}_k : k = 1, \ldots, K$.

The key ingredient in this framework is a loss function, $L(\bar{Q}, O)$, that describes the severity of the difference between a value predicted by a learner and the true observed value. For example, the squared error loss: $L(\bar{Q}, O) = (\bar{Q} - Y)^2$. This leads to the risk, which is the expected value of the loss with respect to distribution P: $\theta(\bar{Q}, P) = PL(\bar{Q}, O) = E_{P_0}[L(\bar{Q}, O)]$. Evaluated at the truth, P_0, we get the true risk $\theta_0(\bar{Q}) = \theta(\bar{Q}, P_0)$, which provides a criteria by which to select a learner: $\bar{Q}_0 = \arg\min_{\bar{Q} \in \bar{\mathbf{Q}}} \theta_0(\bar{Q}) = \bar{Q}(P_0)$, the learner we want is the one that minimizes the true risk. Crucially, this should be equal to the parameter we're trying to estimate, here $\bar{Q}_0(A, W)$. The value of this risk at the minimizer is called the optimal risk, or the irreducible error: $\theta_0(\bar{Q}_0) = \min_{\bar{Q} \in \bar{\mathbf{Q}}} \theta_0(\bar{Q})$. Then, in the context of discrete super learner, we can define oracle selector as $\tilde{k}_n = \arg\min_{k \in \{1, \dots, K_n\}} \theta_0(\bar{Q}_k)$, which selects the learner that minimizes true risk amongst set of candidates. This is the learner we would select if we knew P_0

The empirical or resubstitution risk estimate, $\hat{\theta}_{P_n}(\bar{Q}(P_n)) = \theta(\bar{Q}(P_n), P_n)$, estimates the risk on the same dataset used to train the learner. This is known to be optimistic (biased downwards) in most circumstances, with the optimism increasing as a function of model complexity. Therefore, the resubstitution selector $\arg\min_{k \in \{1, \dots, K_n\}} \hat{\theta}_{P_n}(\bar{Q}_k(P_n))$ selects learners which "overfit" the data, selecting learners which are unnecessarily complex, and therefore have a higher risk than models which make the correct bias-variance trade-off. Hastie et al. (2001) discusses this phenomenon in more detail.

Cross-validation allows more accurate risk estimates that are not biased towards more complex models. Our formulation relies on a split vector $B_n \in \{0, 1\}^n$, which divides the data into two sets, a training set $(O_i : B_n(i) = 0)$, with the empirical distribution P^0_{n,B_n}, and a validation set $(O_i : B_n(i) = 1)$, with the empirical distribution P^1_{n,B_n}. Averaging over the distribution of B_n, yields a cross-validated risk estimate: $\hat{\theta}_{CV}(\bar{Q}) = E_{B_n} \theta(\bar{Q}(P^0_{n,B_n}), P^1_{n,B_n})$. This yields a cross-validated selector $\hat{k}_n = \arg\min_{k \in \{1, \dots, K_n\}} \theta_{CV}(\bar{Q})$, which selects the learner that minimizes the cross-validated risk estimate. Because cross-validation uses separate data for training and risk estimation for each split vector B_n, it has an important oracle property.

Under appropriate conditions the cross-validation selector will do asymptotically as well as the oracle selector in terms of a risk difference ratio:

$$\frac{\theta_0(\bar{Q}_{\hat{k}}) - \theta_0(\bar{Q}_0)}{\theta_0(\bar{Q}_{\tilde{k}}) - \theta_0(\bar{Q}_0)} \xrightarrow{P} 1. \tag{28.1}$$

That is, the ratio of the risk difference between the cross-validation selector and the optimal risk and the risk difference between the oracle selector and the optimal risk approaches 1 in probability. Conditions and proofs for this result are given in Dudoit and van der Laan (2005); van der Laan and Dudoit (2003); van der Vaart et al. (2006). It is through this property that discrete super learner does asymptotically as well as the best of its candidate learners. We will soon describe how this property fails for nonparametric bootstrap samples.

28.4 Bootstrap

Before presenting our generalization of the bootstrap, we briefly review the bootstrap theoretical framework. The key idea of the bootstrap is as follows: we wish to estimate the sampling distribution $G(x) = P(\hat{\Psi}(P_n) \leq x)$ of an estimator $\hat{\Psi} : \mathcal{M}_{NP} \to \Psi$, where $\hat{\Psi}(P_n)$ is viewed as a random variable through the random P_n. If the estimator is asymptotically linear, then this sampling distribution could be approximated with a normal distribution so that it suffices to estimate its first and second moment. It is difficult to estimate this distribution directly because we only observe one sample from P_0, and therefore only one realization of ψ_n. However, we can draw B repeated samples of size n from some estimate of P_0 and apply our estimator to those samples. Denoting such a sample $O_1^{\#}, \ldots, O_n^{\#}$ and the empirical distribution corresponding to that sample $P_n^{\#}$ and estimate $\psi_n^{\#} = \Psi(P_n^{\#})$, we can obtain an estimate of the desired sampling distribution:

$$\hat{G}(x) = \frac{1}{B} \sum_{i=1}^{B} I(\psi_{n,i}^{\#} \leq x)$$

28.4.1 Nonparametric Bootstrap

The nonparametric bootstrap applies this approach by sampling from the empirical distribution, P_n. This approach has been demonstrated to be an effective tool for estimating the sampling distribution in a wide range of settings. However, the nonparametric bootstrap is not universally appropriate for sampling distribution estimation. Because P_n is a discrete distribution, repeated sampling from it will create "copied" observations—bootstrap samples will have more than one identical observation in a sample. Bickel et al. (1997a) notes that the bootstrap can fail if the estimator is sensitive to ties in the dataset. One example of a class of estimators that may be sensitive to ties are those that use cross-validation to make a bias-variance trade-off. If cross-validation is applied to a nonparametric bootstrap sample, duplicate observations have the potential to appear in both the training and testing portions of a given sample split. This creates an issue for estimators that rely on cross-validation. Hall (1992) specifically notes issue of ties for cross-validation based model selection.

The severity of this problem is determined by how many copied observations we can expect. For a bootstrap sample of size n, and validation proportion p_{B_n}, the probability of a validation observation having a copy in the training sample is given by

$$p(\text{copy}) = 1 - \left(1 - \frac{1}{n}\right)^{(1-p_{B_n})n} \approx 1 - e^{-(1-p_{B_n})}$$

For ten-fold cross-validation $p_{B_n} = 0.1$, so we can expect $\approx 59\%$ of validation observations to also be in the training sample. An average of 59% of a cross-validated risk estimate on a bootstrap sample is therefore something like a resubstitution risk estimate, thus having suboptimal properties. One ad-hoc solution to the problem of duplicate observations for cross-validation is to do "clustered" cross-validation where a cluster is defined as a set of identical bootstrap observations, and then split the clusters between training and validation. This way, no observation will appear in both the training and testing sets. Although we lack rigorous justification for this approach, it was evaluated in the simulation study below. It appears in the results as "NP Bootstrap + CVID".

28.4.2 Model-Based Bootstrap

In contrast, the parametric bootstrap draws samples from an estimate of P_0 based on an assumed parametric model: $P_{n,\beta}$. The parametric bootstrap can be generalized to a "model-based" bootstrap that using semi- or nonparametric estimates of factors of the likelihood. In the context of our treatment-specific mean example, this means using estimates of $P(Y|A, W)$ and $\bar{g}$. If Y is binary, as is the case in our simulation, $P(Y = 1|A, W) = E(Y|A, W) = \bar{Q}(A, W)$. Otherwise, we need an estimate of the conditional distribution of $\epsilon(A, W)$, given A, W, where $Y = E(Y|A, W) + \epsilon(A, W)$. To be explicit, observations are drawn from an estimate $\tilde{P}_n = Q_{W,n} g_n Q_{Y,n}$ according to the algorithm below.

Algorithm. *Model-Based Bootstrap*

$\star$ Sample $W^\#$ from the empirical distribution of W: $Q_{W,n}$,
$\star$ Using $W^\#$, sample $A^\#$ from $g_n(\cdot \mid W^\#)$,
$\star$ Using $A^\#$ and $W^\#$, sample $Y^\#$ from $Q_{Y,n}(\cdot \mid (A^\#, W^\#))$.

The targeted bootstrap, described in the next section, is a particular model-based bootstrap using estimates of $\bar{Q}_n^* \bar{g}_n^*$ targeted to ensure correct asymptotic performance.

28.4.3 Targeted Bootstrap

The idea of targeted bootstrap is to construct a TMLE $\tilde{P}_n^*$ so that $\Sigma^2(\tilde{P}_n^*)$ is a TMLE of $\sigma_0^2 = \Sigma^2(P_0)$. Then we know that under regularity conditions, $\Sigma^2(\tilde{P}_n^*)$ is an asymptotically linear and efficient estimator of σ_0^2 at P_0 so that we can construct a confidence interval for σ_0^2. In addition, since it is a substitution estimator of σ_0^2 it is often more reliable in finite samples resulting in potential finite sample improvements in coverage of the confidence interval. In addition, we will show that under appropriate regularity conditions, due to the consistency of $\Sigma^2(\tilde{P}_n^*)$, the bootstrap distribution of $\sqrt{n}(\hat{\Psi}(P_{n,\#}) - \Psi(\tilde{P}_n^*))$ based on sampling $O_1^\#, \ldots, O_n^\# \sim_{iid} \tilde{P}_n^*$, given almost every $(P_n : n)$, $\sqrt{n}(\hat{\Psi}(P_n) - \psi_0)$ converges to the desired limit distribution $N(0, \sigma_0^2)$, even when $\tilde{P}_n^*$ is misspecified. Thus, we can show that the $\tilde{P}_n^*$-bootstrap is consistent almost everywhere for the purpose of estimating the limit distribution of $\sqrt{n}(\psi_n - \psi_0)$, but the bootstrap has the advantage of also obtaining an estimate of the finite sample sampling distribution of the estimator under this bootstrap distribution. Normally, the consistency of a model based bootstrap that samples from an estimate $\tilde{P}_n^*$ of P_0 relies on consistency of the density of $\tilde{P}_n^*$ as an estimator of the density of P_0. In this case, however, the consistency of the bootstrap only relies on the conditions under which the TMLE $\Sigma^2(\tilde{P}_n^*)$ is a consistent estimator of $\Sigma^2(P_0)$. This in turn can allow that parts of P_0 are inconsistently estimated within $\tilde{P}_n^*$. We refer to this bootstrap as the *targeted bootstrap*.

The TMLE of $\Sigma^2(P_0)$ is typically represented as $\Sigma^2(Q_{\Sigma,n}^*)$ for a smaller parameter $P \rightarrow Q_\Sigma(P)$ utilizing a possible nuisance parameter estimator $g_{\Sigma,n}^*$ of a $g_\Sigma(P_0)$. As a consequence, $\tilde{P}_n^*$ can be defined as any distribution for which $Q_\Sigma(\tilde{P}_n^*) = Q_n^*$, without affecting the consistency of the targeted bootstrap. This demonstrates that the targeted bootstrap is robust to certain types of model misspecification. For the best finite sample performance in estimating the actual sample sampling distribution of $\hat{\Psi}(P_n)$, it might still be helpful that also the remaining parts of P_0, beyond Q_0, are well approximated; however that contribution will be asymptotically negligible.

28.4.4 Bootstrap Confidence Intervals

Once a bootstrap sampling distribution is obtained, a number of methods have been proposed to generate confidence interval endpoints from them. Hall (1988) presents a framework by which to evaluate bootstrap confidence intervals. We follow that approach here. We are interested in studying the *accuracy* of various confidence intervals. For a given confidence interval endpoint, $\psi_{n,[\alpha]}$, we say that it's jth order accurate if we can write $P_0(\psi_0 < \psi_{n,[\alpha]}) = \alpha + O(n^{-j/2})$. Coverage probability of a one-sided interval is closely related to its accuracy. We also discuss the coverage error of a two-sided confidence interval.

The most general theoretical results for bootstrap confidence interval accuracy for the nonparametric bootstrap come from the smooth functions setting of Hall (1988). This setting is for parameters that can be written as $f(P_0Y)$ where Y is a vector generated from a set of transformations of O, (i.e. $h^j(O)$), and where f is a smooth function. This setting accommodates many common parameters such as means and other moments but also leaves out other common parameters like quantiles. Notably, it does not include the treatment-specific mean or other kinds of targeted learning parameters. Below we present some bootstrap confidence interval methods and state the relevant theoretical results in this setting.

Bootstrap Wald Interval. A Wald interval using the bootstrap estimate of variance:

$$\hat{\sigma}^2_{n,\text{boot}} = \frac{1}{B} \sum_{i=1}^{B} \left(\Psi(P_{n,\#,i}) - \bar{\Psi}(P_{n,\#,i}) \right)^2$$

with $\bar{\Psi}(P_{n,\#,i}) = \frac{1}{B} \sum_{i=1}^{B} \Psi(P_{n,\#,i})$. As before:

$$\psi_{n,[\alpha],\text{Wald}} = \psi_n - n^{-1/2} \hat{\sigma}_{n,boot} \phi^{-1}(1 - \alpha)$$

The Wald interval method is first order accurate in the smooth functions setting (Hall 1988).

Percentile Interval. Efron's percentile interval directly using the α quantile of $\hat{G}(x)$:

$$\psi_{n,[\alpha],\text{Percentile}} = \hat{G}^{-1}(\alpha)$$

The percentile interval is also first order accurate in the smooth functions setting (Hall 1988).

Bootstrap-t Interval. The bootstrap-t interval can be thought of as an improvement to the Wald-style interval. It relies on the following "studentized" distribution function.

$$K(x) = P\left(\frac{n^{1/2}(\hat{\psi}_n - \psi_0)}{\hat{\sigma}_n} < x \right)$$

The bootstrap estimate of this distribution is as follows:

$$\hat{K}(x) = \frac{1}{B} \sum_{i=1}^{B} I\left(\frac{n^{1/2}(\hat{\psi}_n^{\#} - \hat{\psi}_n)}{\hat{\sigma}_n} < x \right)$$

Defining $\hat{y}_\alpha = \hat{K}^{-1}(\alpha)$ as the estimate of the α quantile of this distribution, we modify the Wald interval as follows:

$$\psi_{n,[\alpha],\text{bootstrap-}t} = \psi_n + n^{-1/2} \hat{\sigma}_n \hat{y}_\alpha$$

A commonly cited drawback of this method is it requires a reliable estimate of σ (Hall 1988). However, in our setting we have access to estimates of σ both from influence curves and targeted estimates of variance. In our simulation study (below), we used the influence curve variance estimate except in the case of the targeted and joint targeted bootstraps, where we used the targeted estimate. The bootstrap-t interval is second-order accurate in the smooth functions setting (Hall 1988).

BC_a **Interval.** The BC_a (bias-corrected, accelerated) interval, first presented in Efron (1987), accounts for bias and skew in a sampling distribution when forming a confidence interval. Its development was motivated by the practice of employing monotone transformations to normalize the sampling distribution of an estimator. It depends on two additional parameters. The bias constant z_0 captures the bias in the sampling distribution, while the acceleration constant a captures the skewness of the sampling distribution.

Given both of these quantities, the BC_a defines a new quantile to look up:

$$\beta_{z_0,a,\alpha} = \Phi\left(z_0 + \frac{z_0 + z_\alpha}{1 - a(z_0 + z_\alpha)}\right)$$

$$\psi_{n,[\alpha],BC_a} = \hat{G}^{-1}(\beta_{z_0,a,\alpha}),$$

where $\Phi(x)$ is the standard normal distribution and $z_\alpha \equiv \Phi^{-1}(\alpha)$ is its α quantile. To generate this interval in practice, we require estimates of z_0 and a. We estimate z_0 as the normal quantile for the proportion of the bootstrap estimates that fall below the original sample estimate:

$$\hat{z}_0 = \Phi^{-1}\left[\hat{G}(\hat{\psi}_n)\right].$$

We use our knowledge of the influence function to estimate the acceleration constant a from the original sample:

$$\hat{a} = \frac{\sum_{i=1}^n D(O_i)^3}{6\left(\sum_{i=1}^n D(O_i)^2\right)^{3/2}}.$$

The BC_a interval is also second order accurate in the smooth functions setting (Hall 1988).

28.5 Simulation

To evaluate the practical performance of the targeted bootstrap, we simulate data from the following P_0:

$$W_1 \sim U(-1, 1),$$
$$W_2 \sim U(-1, 1),$$
$$W^* = W_2 - W_1,$$
$$A|W \sim \text{Bernoulli}\left(\text{inv.logit}(-0.5W^*)\right),$$
$$Y|A, W \sim \text{Bernoulli}\left(\text{inv.logit}(A(1 - 0.5W^* + sin(W^*)))\right).$$

Positivity, $(\bar{g}(P_0)(W) = P_0(A = 1|W) > 0$, is met: $0.26 < \bar{g}(P_0)(W) < 0.74$. Samples of size $n = 1000$ were generated for each of $B = 1000$ Monte Carlo simulations.

In our simulation, we estimated $\bar{Q}(P_0)(W) = E[Y|A = 1, W]$ using kernel regression with bandwidth selected by 10-fold cross-validation (i.e., the discrete super learner). We estimate $\bar{g}(P_0)(W)$ using a correctly specified logistic regression. For each simulation iteration, we estimated Q, and fit three TMLEs: a TMLE for the treatment-specific mean, a TMLE for its asymptotic variance, and a joint TMLE for both the treatment-specific mean and its asymptotic variance. After fitting the TMLEs, we generated 1000 repeated bootstrap samples from five different methods: nonparametric bootstrap, clustered nonparametric bootstrap, model-based bootstrap based on the initial super learner fit, the targeted bootstrap sampling from the TMLE distribution targeting the asymptotic variance, and the joint targeted bootstrap sampling from the joint targeted TMLE distribution.

The three TMLEs fit to the simulated dataset generated different confidence interval estimates: one Wald-style interval based on the influence curve from the first TMLE, and direct estimates of the variance for the remaining two TMLEs. For the five bootstrap methods, we estimated intervals for all four methods described in Sect. 28.4.4. We evaluated coverage and interval lengths for all estimated confidence intervals. We also compared the performance of the super learner on full samples and samples from all the bootstrap approaches. To evaluate the performance of super learner on bootstrap samples, we compared which bandwidths were selected on the various sample types, as well as the risk difference ratios for those selections.

Results. As described above, super learner behaves differently on nonparametric bootstrap samples than on full samples, behaving more like a resubstitution estimator. While on full samples, the super learner (cross-validation) often selects bandwidths close to those selected by the oracle selector (minimizing the true risk), on nonparametric bootstrap samples, super learner most often selects the lowest available bandwidth, over-fitting the data. On other kinds of bootstrap samples, including targeted bootstrap, super learner behaves more like it does on full samples, suggesting that these bootstrap methods don't have the same problem. This difference in the selection behavior impacts the performance of the resulting super learner in terms of the risk difference ratio (Eq. (28.1)). This can be seen in Fig. 28.1. Again, other bootstrap samples behave more like full samples in terms of the risk difference ratio.

Figures 28.2, 28.3, and 28.4 show how super learner performance impacts confidence interval performance for the resulting TMLE estimate. In general, the over-fit super learner being used in TMLE on nonparametric bootstrap samples is more variable than the well-fit super learner being used in full samples. Therefore, nonparametric bootstrap confidence intervals are unnecessarily long and over-cover. The effect of this over-coverage on length is modest at $n = 1000$, with the Wald intervals estimated from the nonparametric bootstraps are on average just 4% longer than the standard influence-curve based confidence intervals. At smaller sample sizes, the effect is more severe: nonparametric bootstrap intervals are 21% longer than influence curve intervals at $n = 250$, and 39% longer at $n = 100$. This substantial increase in length will negatively impact the power of nonparametric bootstrap confidence intervals. In our simulation, the set of bandwidths from which super learner could

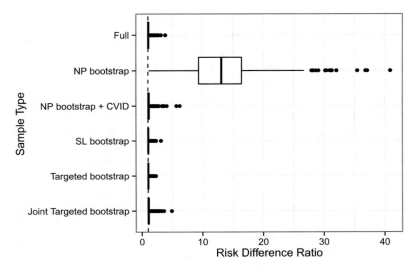

Fig. 28.1 Risk difference ratio (defined in Eq. (28.1)) of super learner on different sample types

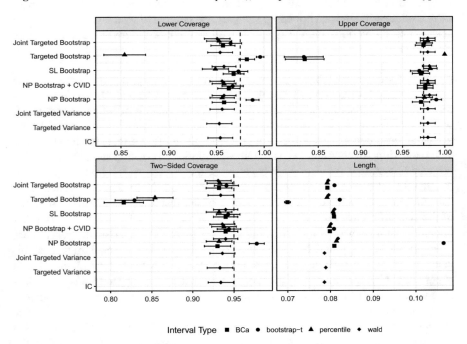

Fig. 28.2 Confidence interval coverage and length for $n = 1000$

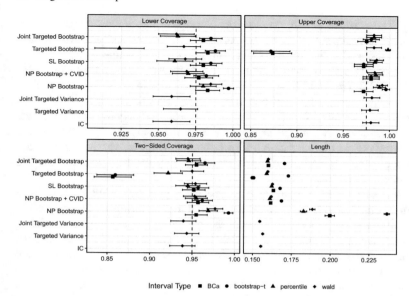

Fig. 28.3 Confidence interval coverage and length for $n = 250$

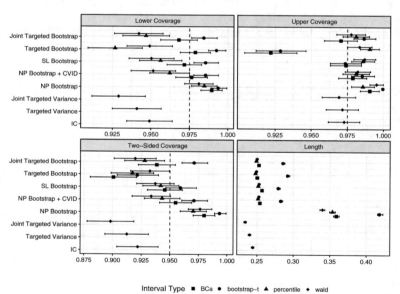

Fig. 28.4 Confidence interval coverage and length for $n = 100$

select was fixed with respect to sample size. We expect that, if smaller bandwidths had been available, super learner on nonparametric bootstrap samples would have chosen them, increasing the impact of over-fitting on larger sample sizes.

These figures also show the importance of a bootstrap that jointly targets both the parameter of interest and its asymptotic variance. For interval types other than Wald, the (variance-only) targeted bootstrap intervals have very poor coverage. This is because these intervals are not centered on the treatment-specific mean estimate from the full dataset, and are instead centered on the average estimate from the bootstrap intervals. In the case of targeted bootstrap samples, these estimates are biased, because the targeted bootstrap is targeting only the variance, and not the actual parameter of interest.

Figure 28.4 shows that at small sample sizes, the asymptotic Wald intervals have lower than nominal coverage, with all methods under-covering by at least 2.5%. Small sample sizes such as this are where the bootstrap has the most potential to improve upon asymptotic confidence intervals. At larger sample sizes, the second-order terms become relatively unimportant. However, even at this small sample size, asymptotic intervals are only modestly anti-conservative in this simulation. This may be due to the fact that even at $n = 100$, our simulated sampling distribution is already very close to normal.

Focusing only on the joint targeted bootstrap, we can compare the performance of different bootstrap confidence interval types. Figure 28.5 shows this comparison. At modest sample sizes, Bootstrap-t intervals over-cover and are longer than other interval types. The other bootstrap methods generate intervals of similar length. Of

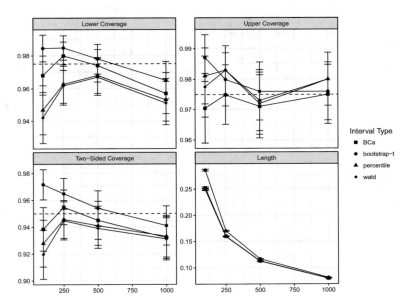

Fig. 28.5 Joint targeted bootstrap interval performance comparison

the three, BC_a has the closest to nominal coverage over the range of sample sizes tested. Therefore, it is recommended that this interval type be used with the Joint targeted bootstrap going forward.

28.6 Conclusion

We have shown the effectiveness of the targeted bootstrap for estimating properties of the sampling distribution theoretically and through simulation results. Our simulation illustrates the problems of applying the nonparametric bootstrap to a TMLE estimate with initial estimates based on super learner. Specifically, ties in nonparametric bootstrap samples sabotage the sample splitting that occurs in cross-validation, causing cross-validating risk estimates to behave more like resubstitution estimates. This leads super learner to select over-fit models on nonparametric bootstrap samples. By sampling from a continuous distribution estimate, and one that is targeted to the parameters of interest, targeted bootstrap does not create the ties that break cross-validation, and therefore generates confidence intervals with acceptable performance. We have demonstrated the superiority of the targeted bootstrap over the nonparametric bootstrap in the context of targeted learning.

Additional work is necessary to further explore the issue of bootstrap confidence intervals for targeted learning. A simulation study with a continuous outcome variable, especially one with a skewed error distribution, would be interesting in several ways. First, it would validate the targeted bootstrap approach for continuous outcomes, which theory tells us should be consistent even when the error distribution is misspecified. Secondly, it would allow us to investigate the magnitude of second-order terms in a setting with a sampling distribution that might be more skewed at smaller sample sizes. Another extension would be to investigate additional bootstrap confidence interval types, especially the tilted and automatic percentile interval types (DiCiccio and Romano 1990).

We also want to highlight another type of bootstrap that is asymptotically consistent in great generality. In this bootstrap method one estimates the sampling distribution of the TMLE with the sampling distribution, under sampling from the empirical probability distribution P_n, of the TMLE *that fixes the initial estimator and only recomputes the TMLE update step in the TMLE algorithm*. It follows that this is a consistent bootstrap method as long as the TMLE itself is asymptotically efficient. Of course, this type of bootstrap fails to pick up second-order terms due to the estimation of the nuisance parameters.

Chapter 29
Targeted Learning Using Adaptive Survey Sampling

Antoine Chambaz, Emilien Joly, and Xavier Mary

Consider the following situation: we wish to build a confidence interval (CI) for a real-valued pathwise differentiable parameter Ψ evaluated at a law P_0, $\psi_0 \equiv \Psi(P_0)$, from a data set $O_1, \ldots, O_N$ of independent random variables drawn from P_0 but, as is often the case nowadays, N is so large that we will not be able to use all data. To overcome this computational hurdle, we decide *(a)* to select n among N observations randomly with unequal inclusion probabilities and *(b)* to adapt TMLE from the smaller data set that results from the selection.

Our analysis is asymptotic: we assume that N goes to infinity and that n goes to infinity as N does, in such a way that the ratio n/N go to 0. The selection of n among N observations will be the random outcome of a survey sampling design. From now on, we assume that each observation O_i is summarized by V_i, a low-dimensional random variable, and that $V_1, \ldots, V_N$ are all observed. We will draw advantage from $V_1, \ldots, V_N$ to adjust the probability that each O_i be sampled.

First explored in Bertail et al. (2016a), our approach is an alternative to the so called "online targeted learning" developed by van der Laan and Lendle (2014). It bears many similarities with inverse probability of censoring (IPCW) TMLE in two-stage designs (Rose and van der Laan 2011), but extends it in three directions. First, the random selection process is conditional on $(V_1, \ldots, V_N)$ and carried out collectively rather than individually. This induces dependence between the selected

A. Chambaz (✉)
MAP5 (UMR CNRS 8145), Université Paris Descartes, 45 rue des Saints-Pères, 75270 Paris cedex 06, France
e-mail: antoine.chambaz@parisdescartes.fr

E. Joly · X. Mary
Modal'X, Université Paris Nanterre, 200 av de la République, 92000 Nanterre, France
e-mail: emilien.joly@gmail.com; xmary@parisnanterre.fr

© Springer International Publishing AG 2018
M.J. van der Laan, S. Rose, *Targeted Learning in Data Science*,
Springer Series in Statistics, https://doi.org/10.1007/978-3-319-65304-4_29

observations. Second, the inclusion probabilities of $O_1, \ldots, O_N$ converge to 0 as N goes to infinity. This is convenient from a computational viewpoint, since sampling $n = o(N)$ observations out of $O_1, \ldots, O_N$ would be difficult otherwise. Third, the approach naturally lends itself to a multiple-stage sampling version, allowing to optimize the inclusion probabilities in order to meet an objective such as minimizing the asymptotic variance of the targeted minimum loss estimator (TMLE).

We will develop two examples of survey sampling designs: Sampford's and determinantal sampling designs. Also known as rejective sampling design based on Poisson sampling with unequal inclusion probabilities, Sampford's sampling design is a particular case of sampling without replacement (Hanif and Brewer 1980). It has been thoroughly studied since the publication of the seminal articles (Hajek 1964; Sampford 1967). Recently introduced in sampling theory by Loonis and Mary (2015), determinantal sampling design benefits from a rich literature on determinantal point processes (Macchi 1975; Lyons 2003; Hough et al. 2006). The methodology will be illustrated with an example, that of the inference of a nonparametric variable importance measure of a continuous exposure (Chambaz et al. 2012). A simulation study will contribute to assessing the performance of the procedure in this setting.

29.1 Template for Targeted Inference by Survey Sampling

This section presents a template for carrying out targeted inference from large data sets by survey sampling. Throughout the chapter, we denote $\mu f \equiv \int f d\mu$ and $\|f\|_{2,\mu} \equiv (\mu f^2)^{1/2}$ for any measure μ and function f (measurable and integrable with respect to μ).

29.1.1 Retrieving the Observations by Survey Sampling

As explained in introduction, the first step of the inference procedure is the random selection without replacement of n among N observations. The survey sample size n is set beforehand. Down to earth computational considerations typically drive its choice. We assume that both n and N go to infinity and that n/N go to 0. How n depends on N may or may not need to be described more precisely. The results of this chapter could be extended to the case that n is random and satisfies these two conditions almost surely (with respect to the law of the sampling design; more details to follow).

The random selection of n among N observations takes the form of a vector $\eta \equiv (\eta_1, \ldots, \eta_N)$ of binary random variables where, for each $1 \leq i \leq N$, O_i is selected if and only if η_i equals 1. The conditional joint distribution of η given $O_1, \ldots, O_N$ is the survey sampling design. By construction, it coincides with the conditional joint distribution of η given the summary measures $V_1, \ldots, V_n$ which, contrary to $O_1, \ldots, O_N$, are all observed at the beginning of the study.

We denote P^s a generic conditional joint distribution of η given $O_1, \ldots, O_n$ (the superscript "s" stands for "survey"). The first order inclusion probabilities are the (conditional marginal) probabilities $\pi_i \equiv P^s(\eta_i = 1)$ for $1 \leq i \leq N$. In case they are equal, the sampling design is said equally weighted. The second-order inclusion probabilities are the (conditional joint) probabilities $\pi_{ij} \equiv P^s(\eta_i = 1, \eta_j = 1) = P^s(\eta_i \eta_j = 1)$ for $1 \leq i \neq j \leq N$. The Horvitz-Thompson empirical measure

$$P_n^{\mathrm{HT}} \equiv \frac{1}{N} \sum_{i=1}^{N} \frac{\eta_i}{\pi_i} \mathrm{Dirac}(O_i) \qquad (29.1)$$

takes up the role that the empirical measure $P_N \equiv N^{-1} \sum_{i=1}^{N} \mathrm{Dirac}(O_i)$ would play if we had access to it. The former is an unbiased estimator of the latter in the sense that, for any function f of O drawn from P_0,

$$\mathrm{E}_{P^s} \left[P_n^{\mathrm{HT}} f \right] = P_N f.$$

A CLT may hold for $\sqrt{n}(P_n^{\mathrm{HT}} - P_N)f$ (conditionally on $O_1, \ldots, O_N$). Whether it does or not notably depends on the asymptotic behavior of $\mathrm{Var}_{P^s} \left[P_n^{\mathrm{HT}} f \right]$. In general, it holds that

$$\mathrm{Var}_{P^s} \left[P_n^{\mathrm{HT}} f \right] = \frac{1}{N^2} \sum_{i=1}^{N} \left(\frac{1}{\pi_i} - 1 \right) f^2(O_i) + \frac{1}{N^2} \sum_{1 \leq i \neq j \leq N} \left(\frac{\pi_{ij}}{\pi_i \pi_j} - 1 \right) f(O_i) f(O_j).$$

$$(29.2)$$

Sections 29.2 and 29.3 focus on the specific examples of the Sampford and determinantal survey sampling designs. The choice of the sampling design affects the limit variance of the TMLE. For the time being, we do not characterize further the survey sampling design.

29.1.2 CLT on the TMLE and Resulting Confidence Intervals

Constructing the TMLE. Let $\mathcal{M}$ be a statistical model for P_0. Let O denote a set where O drawn from any $P \in \mathcal{M}$ takes its values and, for each $P \in \mathcal{M}$, let $L_0^2(P)$ be the set of measurable functions f on O satisfying $Pf = 0$ and Pf^2 finite. We consider the case that Ψ, viewed as a functional from $\mathcal{M}$ to $\mathbb{R}$, is pathwise differentiable in the following sense. For a given loss function $L : \mathcal{M} \times O \rightarrow \mathbb{R}$ there exist, for each $P \in \mathcal{M}$, a subset $\mathcal{S}(P)$ of $L_0^2(P)$ and an influence function $D(P) \in L_0^2(P)$ such that, for all $s \in \mathcal{S}(P)$, there exists a submodel $\{P_{s,\epsilon} : \epsilon \in (-c, c)\}$ of $\mathcal{M}$ with generalized score s (i.e., for P-almost all $o \in O$, $\epsilon \mapsto L(P_{s,\epsilon}, o)$ is differentiable at $\epsilon = 0$ with a derivative equal to $s(o)$) such that $\epsilon \mapsto \Psi(P_{s,\epsilon})$ is differentiable at $\epsilon = 0$ with a derivative equal to $PsD(P)$. For instance, in the example of Sect. 29.4, $\mathcal{M}$ is the nonparametric model of all possible distributions of O which meet some positivity constraints; L is the negative log-likelihood function; $S(P)$ is the subset of all elements of $L_0^2(P)$ which are not identically 0 and have finite supremum norm; for each $s \in S(P)$, the submodel can be characterized by defining $P_{s,\epsilon} \ll P$ with $dP_{s,\epsilon}/dP \equiv 1 - \epsilon s$ (take $c \equiv \|s\|_\infty^{-1}$). In this setting, s is the classical score of submodel $\{P_{s,\epsilon} : \epsilon \in (-c, c)\}$ and $D(P)$ is the efficient influence curve of Ψ at P.

Suppose that we have constructed $P_n^* \in \mathcal{M}$ targeted to ψ_0 in the sense that

$$P_n^{\mathrm{HT}} D(P_n^*) = o_P(1/\sqrt{n}). \tag{29.3}$$

The TMLE is the substitution estimator $\psi_n^* \equiv \Psi(P_n^*)$.

Central Limit Theorem and CIs. The CLT hinges on three assumptions. Let h be a real-valued function over O to be determined later. We suppose the existence of a class $\mathcal{F}$ of (measurable) real-valued functions over O such that

A1. The empirical process $\sqrt{n}(P_n^{\mathrm{HT}} - P_0)$ converges in law in $\ell^\infty(\mathcal{F})$ to a zero-mean Gaussian process with covariance function Σ_h.

A2. With P_0-probability tending to one, $D(P_n^*) \in \mathcal{F}$, and there exists $f_1 \in \mathcal{F}$ such that $\|D(P_n^*) - f_1\|_{2,P_0} = o_P(1)$. Moreover, one knows a conservative estimator σ_n^2 of $\sigma_1^2 \equiv \Sigma_h(f_1, f_1)$.

Under **A1** and **A2**,

$$\hat{\sigma}_n^2 \equiv P_n^{\mathrm{HT}} D(P_n^*)^2 h^{-1} \tag{29.4}$$

consistently estimates σ_1^2. In practice, however, the estimation of σ_1^2 may be more difficult. Requesting a conservative estimator is suboptimal, but grants some flexibility. The third assumption guarantees that a second-order term in an expansion of $\psi_n^* = \Psi(P_n^*)$ around P_0 is indeed of second order:

A3. It holds that $\psi_0 - \psi_n^* - P_0 D(P_n^*) = o_P(1/\sqrt{n})$.

Assumptions **A1**, **A2** and **A3** are slight variations on the assumptions typically made in the asymptotic analysis of TMLEs. They allow to derive the following CLT, whose proof is sketched in Sect. 29.6.1. Set $\alpha \in (0, 1)$ and denote $\xi_{1-\alpha/2}$ the $(1 - \alpha/2)$-quantile of the standard normal distribution.

Proposition 29.1. *Under* **A1**, **A2** *and* **A3**, $\sqrt{n}(\psi_n^* - \psi_0)$ *converges in law to the centered Gaussian distribution with variance* σ_1^2. *Consequently,*

$$\left[\psi_n^* \pm \xi_{1-\alpha/2} \sqrt{\frac{\sigma_n^2}{n}} \right] \tag{29.5}$$

is a confidence interval for ψ_0 *with asymptotic coverage no less than* $(1 - \alpha)$.

Resorting to survey sampling thus makes it possible to construct a CI for ψ_0. It is through σ_n^2 that the width of CI (29.5) depends on the survey sampling design and more precisely on the covariance function in **A1**. In this regard, Sects. 29.2 and 29.3 will show that all survey sampling designs are not equal. In particular, simple random sampling (selecting n among N observations without replacement and with equal weights) is suboptimal.

29.2 Survey Sampling Designs and Assumption A1

This section introduces two examples survey sampling designs and discusses **A1** in their respective contexts.

29.2.1 Sampford's Survey Sampling Design

Denote $\mathcal{V}$ the space where V drawn from P_0 takes its values and let h be a (measurable) function mapping $\mathcal{V}$ to $\mathbb{R}_+$, chosen by us in such a way that h be bounded away from 0 and $P_0 h = 1$. For each $1 \leq i \leq N$, define

$$p_i \equiv \frac{nh(V_i)}{N}.$$

Let P^{sP} be characterized by the fact that, under P^{sP}, η is distributed from the conditional law of $(\varepsilon_1, \ldots, \varepsilon_N)$ given $\sum_{i=1}^N \varepsilon_i = n$ when $\varepsilon_1, \ldots, \varepsilon_N$ are independently sampled from Bernoulli laws with parameters $p_1, \ldots, p_N$, respectively (we recall that this statement is conditional on $O_1, \ldots, O_N$). This survey sampling design is an instance of Sampford's survey sampling design. It is also called rejective sampling design based on Poisson (hence the superscript "P" in P^{sP}) sampling with unequal inclusion probabilities (unequal as soon as h is not constant).

By Bertail et al. (2016a), Theorem 2 which builds upon Bertail et al. (2017), assumption **A1** is met when using Sampford's survey sampling design P^{sP} provided that $\mathcal{F}$, the class introduced in Sect. 29.1.2, is not too complex: this is the message of Proposition 29.2 below.

Proposition 29.2. *Assume that $\mathcal{F}$ is separable (for instance, countable), that it admits an envelope function such that the corresponding uniform entropy integral be finite (see Condition (2.1.7) in van der Vaart and Wellner 1996), and that $P_0 f^2 h^{-1}$ is finite for all $f \in \mathcal{F}$. Then **A1** holds when using Sampford's survey sampling design P^{sP} with a covariance function Σ_h^P given by $\Sigma_h^P(f, g) \equiv P_0 f g h^{-1}$.*

The conclusions of Proposition 29.2 still hold under the same conditions when substituting

$$\frac{1}{N} \sum_{i=1}^N \frac{\eta_i}{p_i} \mathrm{Dirac}(O_i) = \frac{1}{n} \sum_{i=1}^N \frac{\eta_i}{h(V_i)} \mathrm{Dirac}(O_i) \qquad (29.6)$$

for P_n^{HT}. It is thus unnecessary to compute the first order inclusion probabilities of Sampford's survey sampling design, which differ from $p_1, \ldots, p_N$ when h is not constant, and the targeting of $P_n^* \in \mathcal{M}$ to ψ_0 can be achieved by ensuring $n^{-1} \sum_{i=1}^N \eta_i D(P_n^*)(O_i) h^{-1}(V_i) = o_P(1/\sqrt{n})$ instead of (29.3).

29.2.2 Determinantal Survey Sampling Design

A Minimalist Introduction. Determinantal survey sampling designs are built upon determinantal point processes. Let K be a $N \times N$ Hermitian matrix whose eigenvalues belong to $[0, 1]$. It happens that the set of equalities: for all $I \subset \{1, \ldots, N\}$,

$$\sum_{I' \supset I} P^s(I') = \det(K_{|I}), \qquad (29.7)$$

uniquely characterizes the determinantal survey sampling design P^{sK}, a probability measure on the powerset of $\{1, \ldots, N\}$. Here, $K_{|I}$ denotes the Hermitian matrix derived from K by keeping only its rows and columns indexed by the elements of I.

The first and second-order inclusion probabilities of P^{sK} characterized by (29.7) are easily derived from the entries of K: for all $1 \leq i \neq j \leq N$, if holds that $\pi_i = \det(K_{|\{i\}}) = K_{ii}$ and $\pi_{ij} = \det(K_{|\{i,j\}}) = K_{ii} \times K_{jj} - |K_{ij}|^2$. Furthermore, draws from P^{sK} are of fixed size if and only if the eigenvalues of K belong to $\{0, 1\}$, in which case K is a projection matrix and the fixed size equals the trace of K. From now on, we focus on this case.

If the first order inclusion probabilities are all positive then, for any bounded function f of O drawn from P_0, $P_n^{\mathrm{HT}} f - \mathrm{E}_{P^{sK}} \left[P_n^{\mathrm{HT}} f \right]$ satisfies a concentration inequality (Pemantle and Peres 2014, Theorem 3.1; see also (29.20) in our Sect. 29.6.3). Moreover, if f meets the so called Soshnikov conditions (29.17), (29.18) and (29.19), then $\sqrt{n}(P_n^{\mathrm{HT}} - P_N)f$ satisfies a CLT (Soshnikov 2000). These two remarkable properties are the building blocks of Proposition 29.3 below. Let $\mathcal{F}'$ be defined as $\mathcal{F}$ deprived of its elements which depend on O through V only.

Proposition 29.3. *Assume that $\mathcal{F}'$ is countable, uniformly bounded, and that its bracketing entropy with respect to the supremum norm is finite (see the condition preceding Condition (2.1.7) in van der Vaart and Wellner 1996). Assume moreover that, for every $f \in \mathcal{F}'$, $n \operatorname{Var}_{P^{sK}} \left[P_n^{\mathrm{HT}} f \right] > 0$ converges in P_0-probability to a positive number and f meets the Soshnikov conditions (29.17), (29.18), (29.19) P_0-almost surely. Then A1 holds with $\mathcal{F}'$ substituted for $\mathcal{F}$ when using any fixed-size determinantal survey sampling design P^{sK}, provided that its first order inclusion probabilities are bounded away from 0 uniformly in N. The covariance function is defined as a limit with no closed-form expression in general.*

The message of Proposition 29.3 is the following: if $\mathcal{F}'$ is not too complex, and if n/N goes to 0 sufficiently slowly, then A1 is met with $\mathcal{F}'$ substituted for $\mathcal{F}$ when using most determinantal survey sampling designs P^{sK}. The proof of Proposition 29.3 is sketched in Sect. 29.6.3. The condition on the ratio n/N is included implicitly in the assumption that the elements of $\mathcal{F}'$ satisfy the Soshnikov conditions P_0-almost surely. We elaborate further on this issue in Proposition 29.4 below.

We wish to follow the same strategy as in Sect. 29.2.1, i.e., to define possibly unequal first order inclusion probabilities depending on $V_1, \ldots, V_N$. There exists an algorithm to both construct and sample from a fixed-size determinantal survey sampling design with given first order inclusion probabilities (Loonis and Mary 2015). Unfortunately, its computational burden is considerable for both tasks in general, especially in the context of large data sets (N large). In addition, the second set of conditions on $\mathcal{F}'$ (and not $\mathcal{F}$) in Proposition 29.3 would typically be very demanding for the yielded determinantal survey sampling design. Moreover, computing the limit variance of the TMLE resulting from its use would be difficult, and its inference would typically be achieved through the use of a very conservative estimator. These difficulties can be overcome by focusing on V-stratified determinantal survey sampling designs equally weighted on each V-stratum.

V-Stratified Determinantal Sampling Equally Weighted on Each V-Stratum.
We now consider the case that V drawn from P_0 takes finitely many different values. To alleviate notation, we assume without loss of generality that $\mathcal{V} \equiv \{1, \ldots, v\}$ and that $O_1, \ldots, O_N$ are ordered by values of $V_1, \ldots, V_N$. Let h be a function mapping $\mathcal{V}$ to $\mathbb{R}_+^*$ such that $P_N h = N^{-1} \sum_{i=1}^N h(V_i) = 1$. We will hide and neglect notation-wise the dependency of h on $V_1, \ldots, V_N$ due to the normalization $P_N h = 1$. In the limit, h does not depend on the summary measures anymore: by the strong law of large numbers, $P_N h$ converges P_0-almost surely to $P_0 h$, revealing that condition $P_N h = 1$ is similar to its counterpart $P_0 h = 1$ from Sect. 29.2.1. For each $1 \le i \le N$, define

$$\pi_i \equiv \frac{nh(V_i)}{N}.$$

Similar to the proportions $p_1, \ldots, p_N$ used in Sect. 29.2.1 to characterize a Sampford survey sampling design, $\pi_1, \ldots, \pi_N$ are the exact (as opposed to approximate) first order inclusion probabilities that we choose for our determinantal survey sampling design. Its complete characterization now boils down to elaborating a $N \times N$ Hermitian matrix Π with $\pi_1, \ldots, \pi_N$ as diagonal elements and eigenvalues in $\{0, 1\}$. Since $\sum_{i=1}^N \pi_i = n$, the resulting determinantal survey sampling design will be of fixed size n.

For simplicity, we elaborate Π under the form of a block matrix with zero matrices as off-diagonal blocks and make each of the v diagonal blocks be a projection matrix featuring the prescribed diagonal elements. This last step is easy provided that $n_v \equiv \sum_{i=1}^N \pi_i I\{V_i = v\}$ is an integer dividing $N_v \equiv \sum_{i=1}^N I\{V_i = v\}$. In that case, the projection matrix can be a block matrix consisting of n_v^2 square matrices of size $N_v/n_v \times N_v/n_v$, with zero off-diagonal blocks and diagonal blocks having all their entries equal to n_v^{-1}. Otherwise, we may rely on an algorithm to derive the desired projection matrix.

The determinantal survey sampling design $P^{s\Pi}$ encoded by Π (hence the superscript "Π") is said V-stratified and equally weighted on each V-stratum. It randomly selects a deterministic number n_v of observations from the stratum where $V = v$, for each $1 \le v \le v$. Sampling from it makes it possible to derive the next result, proven in Sect. 29.6.2: for any function f of O drawn from P_0,

$$\mathrm{E}_{P_0}\left[\mathrm{Var}_{P^{s\mathit{\Pi}}}\left[P_n^{\mathrm{HT}}f\right]\right] = \frac{1}{n}\mathrm{E}_{P_0}\left[\mathrm{Var}_{P_0}\left[f(O)|V\right]h^{-1}(V)\right] - \frac{1}{N}\mathrm{E}_{P_0}\left[\mathrm{Var}_{P_0}\left[f(O)|V\right]\right].$$

$$(29.8)$$

Equality (29.8) is instrumental in deriving the following corollary to Proposition 29.3, whose proof is sketched in Sect. 29.6.3.

Proposition 29.4. *Let us impose that n is chosen in such a way that $N/n = o((N^2/n)^\epsilon)$ for all $\epsilon > 0$. This is the case if $n \equiv N/\log^a(N)$ for some $a > 0$, for instance. Assume that $\mathcal{F}$ is separable (for instance, countable) and that its bracketing entropy with respect to the supremum norm is finite (see the condition preceding Condition (2.1.7) in van der Vaart and Wellner 1996). Then* **A1** *holds when using the V-stratified and equally weighted on each V-stratum determinantal survey sampling design $P^{s\mathit{\Pi}}$ with a covariance function $\Sigma_h^{\mathit{\Pi}}$ given by*

$$\Sigma_h^{\mathit{\Pi}}(f,g) = \mathrm{E}_{P_0}\left[\mathrm{Cov}_{P_0}\left[f(O)g(O)|V\right]h^{-1}(V)\right].$$

Note that $\Sigma_h^{\mathit{\Pi}}(f,f) = 0$ for every $f \in \mathcal{F}$ which depends on O through V only. In fact, for such a function, $\sqrt{n}(P_n^{\mathrm{HT}} - P_0)f = \sqrt{n}(P_N - P_0)f = O_P(\sqrt{n/N}) = o_P(1)$. Moreover, for every $f \in \mathcal{F}$, combining (29.8) and equality $\mathrm{E}_{P^{s\mathit{\Pi}}}\left[P_n^{\mathrm{HT}}f\right] = P_N f$ readily implies

$$\mathrm{Var}_{P_0 P^{s\mathit{\Pi}}}\left[\sqrt{n}(P_n^{\mathrm{HT}} - P_0)f\right] = \Sigma_h^{\mathit{\Pi}}(f,f) + \frac{n}{N}\left(\mathrm{Var}_{P_0}\left[f(O)\right] - \mathrm{E}_{P_0}\left[\mathrm{Var}_{P_0}\left[f(O)|V\right]\right]\right).$$

$$(29.9)$$

Proved in Sect. 29.6.2, (29.9) relates the exact variance of $\sqrt{n}(P_n^{\mathrm{HT}} - P_0)f$ with the limit variance $\Sigma_h^{\mathit{\Pi}}(f,f)$, showing that their difference is upper-bounded by a $O(n/N) = o(1)$-expression.

It is an open question to determine whether or not the extra condition on how n depends on N could be relaxed or even given up by proving directly a functional CLT for $\sqrt{n}(P_n^{\mathrm{HT}} - P_0)$. By "directly", we mean without building up on functional CLTs conditional on the observations, and managing to go around the Soshnikov conditions. This route was followed to prove (Bertail et al. 2016a, Theorem 2). Sobolev classes are known to have finite bracketing entropy with respect to the supremum norm (van der Vaart 1998, Example 19.10). The fact that the bracketing entropy is meant relative to the supremum norm instead of the $L^2(P_0)$-norm is a little frustrating, though. Indeed, a bracketing entropy condition relative to the latter would have allowed a larger variety of classes. The supremum norm comes from the concentration inequality (Pemantle and Peres 2014, Theorem 3.1). Perhaps the aforementioned direct proof might also allow to replace it with the $L^2(P_0)$-norm.

The covariance functions Σ_h^{P} and $\Sigma_h^{\mathit{\Pi}}$ in Propositions 29.2 and Proposition 29.4 differ. In particular, for every $f \in \mathcal{F}$,

$$\Sigma_h^{\mathrm{P}}(f,f) = \mathrm{E}_{P_0}\left[\mathrm{E}_{P_0}\left[f^2(O)|V\right]h^{-1}(V)\right] \geq \mathrm{E}_{P_0}\left[\mathrm{Var}_{P_0}\left[f(O)|V\right]h^{-1}(V)\right] = \Sigma_h^{\mathit{\Pi}}(f,f)$$

$$(29.10)$$

(using the same h on both sides of (29.10) is allowed because, in the limit, condition $P_N h = 1$ is similar to condition $P_0 h = 1$). Consequently, $P^{s\mathit{\Pi}}$ is more efficient than P^{sP} when $\mathcal{V}$ is finite in the sense that whichever function h^P is used to define P^{sP}, it is always possible to choose function h^{Π} to define $P^{s\Pi}$ in such a way that $\Sigma^P_{h^P}(f, f) \geq \Sigma^{\Pi}_{h^{\Pi}}(f, f)$ for every $f \in \mathcal{F}$.

29.3 Optimizing the Survey Sampling Designs

This section discusses the optimization of functions h^P and h^{Π} used to define the first order inclusion probabilities of the survey sampling designs P^{sP} and $P^{s\Pi}$ that we developed in Sects. 29.2.1 and 29.2.2. The optimization is relative to the asymptotic variance of the TMLE, $\Sigma^P_h(f_1, f_1)$ or $\Sigma^{\Pi}_h(f_1, f_1)$, respectively. In light of (29.10), let f_2^P and f_2^{Π} be the functions from $\mathcal{V}$ to $\mathbb{R}_+$ given by

$$f_2^P(V) \equiv \sqrt{\mathrm{E}_{P_0}\left[f_1^2(O)|V\right]} \quad \text{and} \quad f_2^{\Pi}(V) \equiv \sqrt{\mathrm{Var}_{P_0}\left[f_1(O)|V\right]}. \qquad (29.11)$$

Then (29.10) shows in particular that $\Sigma^P_h(f_1, f_1) = P_0(f_2^P)^2 h^{-1}$ is always larger than $\Sigma^{\Pi}_h(f_1, f_1) = P_0(f_2^{\Pi})^2 h^{-1}$. Now, with f_2 equal to either f_2^P or f_2^{Π}, the Cauchy-Schwarz inequality yields that

$$P_0(f_2)^2 h^{-1} \times P_0 h \geq (P_0 f_2)^2,$$

where equality holds if and only if h is proportional to f_2.

In the case of P^{sP}, h satisfies $P_0 h = 1$. Therefore, the optimal h and corresponding optimal asymptotic variance of the TMLE are

$$h^P \equiv f_2^P / P_0 f_2^P \quad \text{and} \quad \Sigma^P_{h^P}(f_1, f_1) = \left(P_0 f_2^P\right)^2. \qquad (29.12)$$

In the case of $P^{s\Pi}$, h satisfies $P_N h = 1$ and $P_0 h = 1$ in the limit. By analogy with (29.12), the optimal h and corresponding optimal asymptotic variance of the TMLE are

$$h^{\Pi} \equiv f_2^{\Pi} / P_0 f_2^{\Pi} \quad \text{and} \quad \Sigma^{\Pi}_{h^{\Pi}}(f_1, f_1) = \left(P_0 f_2^{\Pi}\right)^2. \qquad (29.13)$$

29.4 Example: Variable Importance of a Continuous Exposure

We illustrate our template for survey sampling targeted learning with the inference of a variable importance measure of a continuous exposure. In this example, the ith observation O_i writes $(W_i, A_i, Y_i) \in O \equiv \mathcal{W} \times \mathcal{A} \times [0, 1]$. Here, $W_i \in \mathcal{W}$ is the ith context, $A_i \in \mathcal{A}$ is the ith exposure and $Y_i \in [0, 1]$ is the ith outcome. Exposures take their values in $\mathcal{A} \ni 0$, a bounded subset of $\mathbb{R}$ containing 0, which serves as a reference level of exposure. Typically, in biostatistics or epidemiology, W_i could be the baseline covariate describing the ith subject, A_i could describe her assignment (e.g., a dose-level) or her level of exposure, and Y_i could quantify her biological response.

29.4.1 Preliminaries

For each probability measure P on O equipped with a Borel σ-field, we let $g_P(0|W) \equiv P(A = 0|W)$ be the conditional probability that the exposure equal the reference value 0 and $Q_P(A, W) \equiv \mathrm{E}_P[Y|A, W]$ be the conditional expectation of the response given exposure and context. We assume that $P_0(A \neq 0|W)$ is positive $P_{0,W}$-almost surely and that there exists a constant $c(P_0) > 0$ such that $P_0(A = 0|W) \geq c(P_0) \, P_{0,W}$-almost surely. Introduced in Chambaz et al. (2012) and Chambaz and Neuvial (2015), the true parameter of interest is

$$\psi_0^c \equiv \arg\min_{\beta \in \mathbb{R}} \mathrm{E}_{P_0}\left[\left(Y - \mathrm{E}_{P_0}[Y|A = 0, W] - \beta A\right)^2\right]$$

$$= \arg\min_{\beta \in \mathbb{R}} \mathrm{E}_{P_0}\left[\left(\mathrm{E}_{P_0}[Y|A, W] - \mathrm{E}_{P_0}[Y|A = 0, W] - \beta A\right)^2\right]$$

(the superscript "c" stands for "continuous").

In the context of this example, $\mathcal{M}$ stands for the nonparametric set of all probability measures P on O equipped with a Borel σ-field such that there exists a constant $c(P) > 0$ guaranteeing that $P(A \neq 0|W) > 0$ and $g_P(0|W) \geq c(P) \, P_W$-almost surely. In particular, $P_0 \in \mathcal{M}$ by the above assumption. We view ψ_0^c as the value at P_0 of the functional Ψ^c characterized over $\mathcal{M}$ by

$$\Psi^c(P) \equiv \arg\min_{\beta \in \mathbb{R}} \mathrm{E}_P\left[\left(\mathrm{E}_P[Y|A, W] - \mathrm{E}_P[Y|A = 0, W] - \beta A\right)^2\right].$$

Set arbitrarily $P \in \mathcal{M}$ and define $\mu_P(W) \equiv \mathrm{E}_P[A|W]$ and $\zeta^2(P) \equiv \mathrm{E}_P\left[A^2\right]$. By Chambaz et al. (2012), Proposition 1 it holds that

$$\Psi^c(P) = \frac{\mathrm{E}_P\left[A(Q_P(A, W) - Q_P(0, W))\right]}{\mathrm{E}_P\left[A^2\right]}.$$

Moreover, Ψ^c is pathwise differentiable at P with influence curve $D^c(P) \equiv D_1^c(P) + D_2^c(P) \in L_0^2(P)$ given by

$$\zeta^2(P)D_1^c(P)(O) \equiv A\left(Q_P(A, W) - Q_P(0, W) - A\Psi^c(P)\right),$$

$$\zeta^2(P)D_2^c(P)(O) \equiv (Y - Q_P(A, W))\left(A - \frac{\mu_P(W)I\{A = 0\}}{g_P(0|W)}\right).$$

Specifically, for all bounded $s \in L_0^2(P) \setminus \{0\}$, if we define $P_{s,\epsilon} \ll P$ by setting $dP_{s,\epsilon} \equiv 1 + \epsilon s$ for all $|\epsilon| < c$ with $c \equiv \|s\|_\infty^{-1}$, then $\{P_{s,\epsilon} : \epsilon \in (-c, c)\}$ is a submodel of

$\mathcal{M}$ with score s and $\epsilon \mapsto \Psi^c(P_{s,\epsilon})$ is differentiable at $\epsilon = 0$ with derivative $PsD^c(P)$. Because the closure of the linear span of the set of scores is $L_0^2(P)$ itself, it happens that $D^c(P)$ is the efficient influence curve at P. Let now $\mathcal{R}^c : \mathcal{M}^2 \to \mathbb{R}$ be given by

$$\mathcal{R}^c(P, P') \equiv \Psi^c(P') - \Psi^c(P) - (P' - P)D^c(P)$$
$$\equiv \Psi^c(P') - \Psi^c(P) - P'D^c(P).$$

In light of **A3**, $\psi_0 - \psi_n^* - P_0 D^c(P_n^*) = \mathcal{R}^c(P_n^*, P_0)$. The last step of the proof of Chambaz et al. (2012), Proposition 1 shows that, for every $P, P' \in \mathcal{M}$,

$$\mathcal{R}^c(P, P') = \left(1 - \frac{\zeta^2(P')}{\zeta^2(P)}\right)(\Psi^c(P') - \Psi^c(P))$$
$$+ \frac{1}{\zeta^2(P)}P'\left((Q_{P'}(0, \cdot) - Q_P(0, \cdot))\left(\mu_{P'} - \mu_P \frac{g_{P'}(0|\cdot)}{g_P(0|\cdot)}\right)\right). \quad (29.14)$$

We use this equality to derive an easy to interpret sufficient condition for **A3** to hold.

29.4.2 Construction of the TMLE

Let $\mathcal{Q}^w$, $\mathcal{M}^w$ and $\mathcal{G}^w$ be three user-supplied classes of functions mapping $\mathcal{A} \times \mathcal{W}$, $\mathcal{W}$ and $\mathcal{W}$ to $[0, 1]$, respectively. First, we estimate Q_{P_0}, μ_{P_0} and g_{P_0} with Q_n, μ_n and g_n built upon P_n^{HT}, $\mathcal{Q}^w$, $\mathcal{M}^w$ and $\mathcal{G}^w$. For instance, one can simply minimize (weighted) empirical risks and define

$$Q_n \equiv \arg\min_{Q \in \mathcal{Q}^w} P_n^{\mathrm{HT}} \ell(Y, Q(A, W)), \qquad \mu_n \equiv \arg\min_{\mu \in \mathcal{M}^w} P_n^{\mathrm{HT}} \ell(A, \mu(W)),$$

$$g_n \equiv \arg\min_{g \in \mathcal{G}^w} P_n^{\mathrm{HT}} \ell(I\{A = 0\}, g(0|W))$$

with ℓ the logistic loss function given by

$$-\ell(u, v) \equiv u \log(v) + (1 - u) \log(1 - v)$$

(all $u, v \in [0, 1]$, with convention $\log(0) \equiv -\infty$ and $0 \log(0) \equiv 0$). Alternatively, one could prefer minimizing cross-validated (weighted) empirical risks. One then should keep in mind that the observations are dependent, because of the selection process by survey sampling. Second, we estimate the marginal distribution $P_{0,W}$ of W under P_0 with $P_{n,W}^{\mathrm{HT}}/P_n^{\mathrm{HT}}I$, where $P_{n,W}^{\mathrm{HT}}$ is defined as in (29.1) with W_i substituted for O_i, and the real-valued parameter $\zeta^2(P_0)$ with $\zeta^2(P_n^{\mathrm{HT}}/P_n^{\mathrm{HT}}I)$. Note that $P_{n,W}^{\mathrm{HT}}/P_n^{\mathrm{HT}}I$ and $P_n^{\mathrm{HT}}/P_n^{\mathrm{HT}}I$, the empirical measures $P_{n,W}^{\mathrm{HT}}$ and P_n^{HT} renormalized by $P_n^{\mathrm{HT}}I = N^{-1}\sum_{i=1}^N \eta_i/\pi_i$, are the nonparametric maximum likelihood estimators of the distributions $P_{0,W}$ and P_0. Furthermore, the renormalization factor converges to 1 at $\sqrt{n}$-rate by **A1**.

Let P_n^0 be a measure such that $Q_{P_n^0} = Q_n$, $\mu_{P_n^0} = \mu_n$, $g_{P_n^0} = g_n$, $\zeta^2(P_n^0) = \zeta^2(P_n^{HT}/P_n^{HT}I)$, $P_{n,W}^0 = P_{n,W}^{HT}/P_n^{HT}I$, and from which we can sample A conditionally on W. Picking up such a P_n^0 is an easy technical task, see (Chambaz et al. 2012,
Lemma 5) for a computationally efficient choice. Then the initial estimator $\Psi^b(P_n^0)$
of ψ_0^b can be computed with high accuracy by Monte-Carlo. It suffices to sample
a large number B (say $B = 10^7$) of independent $(A^{(b)}, W^{(b)})$ by (a) sampling $W^{(b)}$
from $P_{n,W}^0 = P_{n,W}^{HT}/P_n^{HT}I$ then (b) sampling $A^{(b)}$ from the conditional distribution of
A given $W = W^{(b)}$ under P_n^0 repeatedly for $b = 1, \ldots, B$ and to make the approximation

$$\Psi^c(P_n^0) \approx \frac{B^{-1} \sum_{b=1}^{B} A^{(b)}(Q_n(A^{(b)}, W^{(b)}) - Q_n(0, W^{(b)}))}{\zeta^2(P_n^0)}.$$

We now target the inference procedure and bend P_n^0 into P_n^* satisfying (29.3)
with D^c substituted for D. We proceed iteratively. Suppose that P_n^k has been constructed for some $k \geq 0$. We fluctuate P_n^k with the one-dimensional parametric model
$\{P_n^k(\epsilon) : |\epsilon| \leq c(P_n^k)/2\|D^c(P_n^k)\|_\infty\}$ characterized by $dP_n^k(\epsilon)/dP_n^k \equiv 1 + \epsilon D^c(P_n^k)$.
Lemma 1 in Chambaz et al. (2012) shows how $Q_{P_n^k(\epsilon)}$, $\mu_{P_n^k(\epsilon)}$, $g_{P_n^k(\epsilon)}$, $\zeta^2(P_n^k(\epsilon))$ and
$P_{n,W}^k(\epsilon)$ depart from their counterparts at $\epsilon = 0$. The optimal move along the fluctu-
ation is indexed by

$$\epsilon_n^k \equiv \arg\max_\epsilon P_n^{HT} \log\left(1 + \epsilon D^c(P_n^k)\right),$$

i.e., the maximum (weighted) likelihood estimator of ϵ. Note that the random function $\epsilon \mapsto P_n^{HT} \log(1 + \epsilon D^c(P_n^k))$ is strictly concave. The optimal move results in the
$(k + 1)$-th update of P_n^0, $P_n^{k+1} \equiv P_n^k(\epsilon_n^k)$.

There is no guarantee that a P_n^{k+1} will coincide with its predecessor P_n^k. We assume that the iterative updating procedure converges (in k) in the sense that, for k_n
large enough, $P_n^{HT} D^c(P_n^{k_n}) = o_P(1/\sqrt{n})$. We set $P_n^* \equiv P_n^{k_n}$. It is actually possible to
come up with a one-step updating procedure (i.e., an updating procedure such that
$P_n^k = P_n^{k+1}$ for all $k \geq 1$) by relying on universally least favorable models (van der
Laan and Gruber 2016). We adopt this multistep updating procedure for simplicity.

We can assume without loss of generality that we can sample A conditionally
on W from P_n^*. The final estimator is computed with high accuracy like $\Psi^c(P_n^0)$
previously: with $Q_n^* \equiv Q_{P_n^*}$, we sample B independent $(A^{(b)}, W^{(b)})$ by (a) sampling
$W^{(b)}$ from $P_{n,W}^*$ then (b) sampling $A^{(b)}$ from the conditional distribution of A given
$W = W^{(b)}$ under P_n^* repeatedly for $b = 1, \ldots, B$ and make the approximation

$$\psi_n^* \equiv \Psi^c(P_n^*) \approx \frac{B^{-1} \sum_{b=1}^{B} A^{(b)}(Q_n^*(A^{(b)}, W^{(b)}) - Q_n^*(0, W^{(b)}))}{\zeta^2(P_n^*)}.$$

To conclude this section, we use (29.14) to derive an easy to interpret alternative
condition to **A3**. If $\zeta^2(P_n^*) = \zeta^2(P_0) + o_P(1)$, then

$$P_0\left((Q_{P_0}(0,\cdot) - Q_{P_n^*}(0,\cdot))\left(\mu_{P_0} - \mu_{P_n^*}\frac{g_{P_0}(0|\cdot)}{g_{P_n^*}(0|\cdot)}\right)\right) = o_P(1/\sqrt{n})$$

can be substituted for **A3**. Through the product, we draw advantage of the synergistic convergences of $Q_{P_n^*}(0,\cdot)$ to $Q_{P_0}(0,\cdot)$ and $(\mu_{P_n^*}, g_{P_n^*})$ to (μ_{P_0}, g_{P_0}) (by the Cauchy-Schwarz inequality for example).

29.5 Simulation

We consider three data-generating distributions $P_{0,1}$, $P_{0,2}$ and $P_{0,3}$ of a data-structure $O = (W, A, Y)$. The three distributions differ only in terms of the conditional mean and variance of Y given (A, W). Specifically, $O = (W, A, Y)$ drawn from $P_{0,j}$ ($j = 1, 2, 3$) is such that

- $W \equiv (V, W_1, W_2)$ with $P_{0,j}(V = 1) = 1/6$, $P_{0,j}(V = 2) = 1/3$, $P_{0,j}(V = 3) = 1/2$ and, conditionally on V, (W_1, W_2) is a Gaussian random vector with mean $(0, 0)$ and variance $\left(\begin{smallmatrix} 1 & -0.2 \\ -0.2 & 1 \end{smallmatrix}\right)$ (if $V = 1$), $(1, 1/2)$ and $\left(\begin{smallmatrix} 0.5 & 0.1 \\ 0.1 & 0.5 \end{smallmatrix}\right)$ (if $V = 2$), $(1/2, 1)$ and $\left(\begin{smallmatrix} 1 & 0 \\ 0 & 1 \end{smallmatrix}\right)$ (if $V = 3$);
- conditionally on W, $A = 0$ with probability 80% if $W_1 \geq 1.1$ and $W_2 \geq 0.8$ and 10% otherwise; moreover, conditionally on W and $A \neq 0$, $3A - 1$ is drawn from the χ^2-distribution with one degree of freedom and noncentrality parameter $\sqrt{(W_1 - 1.1)^2 + (W_2 - 0.8)^2}$;
- conditionally on (W, A), Y is a Gaussian random variable with mean

  - $A(W_1 + W_2)/6 + W_1 + W_2/4 + \exp((W_1 + W_2)/10)$ for $j = 1, 2$,
  - $A(W_1 + W_2)/6 + W_1 + W_2/4 + \exp((W_1 + W_2)/10) + 3AV$ for $j = 3$,

  and standard deviation

  - 2 (if $V = 1$), 1.5 (if $V = 2$) and 1 (if $V = 3$) for $j = 1$,
  - 9 (if $V = 1$), 4 (if $V = 2$) and 1 (if $V = 3$) for $j = 2, 3$.

The true parameters equal approximately $\Psi^c(P_{0,1}) = \Psi^c(P_{0,2}) = 0.1201$ and $\Psi^c(P_{0,3}) = 6.9456$.

For $B = 10^3$ and each $j = 1, 2, 3$, we repeat independently the following steps:

1. simulate a data set of $N = 10^7$ independent observations drawn from $P_{0,j}$;
2. extract $n_0 \equiv 10^3$ observations from the data set by simple random sampling (SRS, which is identical to P^{sP} with $h_0 \equiv 1$), and based on these observations:

   (a) apply the procedure described in Sect. 29.4 and retrieve $f_{n_0,1} \equiv D^c(P_{n_0}^{k_{n_0}})$;
   (b) regress $f_{n_0,1}(O)$ and $f_{n_0,1}(O)^2$ on V, call $f_{n_0,2}^P$ the square root of the resulting estimate of f_2^P and $f_{n_0,2}^{II}$ the square root of the resulting estimate of f_2^{II}, see (29.11);

(c) estimate the marginal distribution of V, estimate $P_0 f^{\mathrm{P}}_{n_0,2}$ with $\pi_{n_0,2}$ and set $h^{\mathrm{P}}_{n_0} \equiv f^{\mathrm{P}}_{n_0,2}/\pi_{n_0,2}$, $h^{\mathit{\Pi}}_{n_0} \equiv f^{\mathit{\Pi}}_{n_0,2}/P_N f^{\mathit{\Pi}}_{n_0,2}$, see (29.12) and (29.13);

3. for each n in $\{10^3, 5 \times 10^3, 10^4\}$, successively, and for each survey sampling design among SRS, $P^{s\mathrm{P}}$ with $h^{\mathrm{P}}_{n_0}$ and $P^{s\mathit{\Pi}}$ with $P^{\mathit{\Pi}}_{n_0}$, extract a subsample of n observations from the data set (deprived of the observations extracted in step 2) and, based on these observations, apply the procedure described in Sect. 29.4. We use $\hat{\sigma}^2_n$ given in (29.4) to estimate σ^2_1, although we are not sure in advance that it is a conservative estimator.

We thus obtain $27 \times B$ estimates and their respective CIs.

To give an idea of what are $h^{\mathrm{P}}_{n_0}$ and $h^{\mathit{\Pi}}_{n_0}$ in each scenario, we report their averages across the B simulation studies under $P_{0,1}$, $P_{0,2}$ and $P_{0,3}$:

- under $P_{0,1}$, we expect similar $h^{\mathrm{P}}_{n_0}$ and $h^{\mathit{\Pi}}_{n_0}$, and do get that they are approximately equal (on average) to $(h_1(1), h_1(2), h_1(3)) \approx (2.10, 0.83, 0.75)$;
- under $P_{0,2}$, we also expect similar $h^{\mathrm{P}}_{n_0}$ and $h^{\mathit{\Pi}}_{n_0}$, and do get that they are approximately equal (on average) to $(h_1(1), h_1(2), h_1(3)) \approx (3.39, 0.83, 0.32)$;
- under $P_{0,3}$, we do not expect similar $h^{\mathrm{P}}_{n_0}$ and $h^{\mathit{\Pi}}_{n_0}$, and get that they are approximately equal (on average) to $(h_1(1), h_1(2), h_1(3)) \approx (2.93, 0.66, 0.58)$ and $(h_1(1), h_1(2), h_1(3)) \approx (2.97, 0.68, 0.56)$, respectively (although small, the differences are significant).

Applying the TMLE procedure is straightforward in R package `tmle.npvi` (Chambaz and Neuvial 2016, 2015). Note, however, that it is necessary to compute $\hat{\sigma}^2_n$. Specifically, we fine-tune the TMLE procedure by setting `iter` (the maximum number of iterations of the targeting step) to 7 and `stoppingCriteria` to `list(mic=0.01, div=0.001, psi=0.05)`. Moreover, we use the default `flavor` called `"learning"`, thus notably rely on parametric linear models for the estimation of the infinite-dimensional parameters Q_{P_0}, μ_{P_0} and g_{P_0} and their fluctuation. We refer the interested reader to the package's manual and vignette for details. The Sampford sampling method (Sampford 1967) implements $P^{s\mathrm{P}}$. However, when the ratio n/N is close to 0 or 1 (here, when n/N differs from 10^{-3}), this acceptance-rejection algorithm typically takes too much time to succeed. To circumvent the issue, we approximate $P^{s\mathrm{P}}$ with a Pareto sampling (see Algorithm 2 in Bondesson et al. 2006, Section 5). We implement $P^{s\mathit{\Pi}}$ as described in Sect. 29.2.2, with minor changes to account for the fact that for some $1 \leq v \leq 3$, $\sum^N_{i=1} K_{ii} I\{V_i = v\}$ may not be an integer or may not divide $\sum^N_{i=1} I\{V_i = v\}$.

The results are summarized in Table 29.1. We focus on the empirical coverage, empirical variance and mean of the estimated variance of the TMLE. All empirical coverages are larger than 95% but one (equal to 94%). In each case, the mean of estimated variances is larger than the corresponding empirical variance, revealing that we achieve the conservative estimation of σ^2_1. Regarding the variances, we observe that $P^{s\mathrm{P}}$ and $P^{s\mathit{\Pi}}$ perform similarly and provide slightly better results than SRS under $P_{0,1}$. This is in line with what was expected, due to the contrast induced by the conditional standard deviation of Y given (A, W) under $P_{0,1}$. Under $P_{0,2}$, we

Table 29.1 Summarizing the results of the simulation study

		SRS			P^{sP} with $h_{n_0}^P$			P^{sII} with $h_{n_0}^{II}$	
		n			n			n	
	10^3	5×10^3	10^4	10^3	5×10^3	10^4	10^3	5×10^3	10^4
$P_{0,1}$ Empirical coverage	96.2%	98.9%	99.2%	98.1%	98.6%	99.4%	97.8%	99.2%	99.3%
Empirical variance	09	08	07	07	06	06	07	06	06
Estimated variance	13	14	14	11	11	11	11	11	11
$P_{0,2}$ Empirical coverage	94.0%	98.9%	99.2%	98.9%	99.9%	99.1%	98.4%	99.4%	99.3%
Empirical variance	129	104	102	44	41	44	49	42	42
Estimated variance	171	196	200	85	86	87	86	85	86
$P_{0,3}$ Empirical coverage	95.6%	98.8%	97.8%	97.8%	97.9%	97.1%	98.5%	98.3%	96.3%
Empirical variance	157	134	168	85	91	116	81	85	104
Estimated variance	216	242	245	130	133	135	124	128	127

The top, middle and bottom groups of rows correspond to simulations under $P_{0,1}$, $P_{0,2}$ and $P_{0,3}$. Each of them reports the empirical coverage of the CIs ($B^{-1} \sum_{b=1}^{B} I\{\Psi^c(P_{0,j}) \in I_{n,b}\}$), n times the empirical variance of the estimators ($n[B^{-1} \sum_{b=1}^{B} \psi_{n,b}^{*2} - (B^{-1} \sum_{b=1}^{B} \psi_{n,b}^{*})^2]$) and empirical mean of n times the estimated variance of the estimators ($B^{-1} \sum_{b=1}^{B} \hat{\sigma}_{n,b}^2$), for every subsample size n and for each survey sampling design

observe that P^{sP} and P^{sII} perform similarly and provide significantly better results than SRS. This too is in line with what was expected, due to the contrast induced by the conditional standard deviation of Y given (A, W), which is stronger under $P_{0,2}$ than under $P_{0,1}$. Finally, under $P_{0,3}$, we observe that P^{sP} performs better than SRS and that P^{sII} performs even slightly better than P^{sP}. This again is in line with what was expected, due to the contrast induced by the conditional standard deviation of Y given (A, W) (same as under $P_{0,2}$) and to the different conditional means of Y given (A, W) under $P_{0,3}$ and $P_{0,2}$.

29.6 Elements of Proof

For every $f \in \mathcal{F}$, let $\bar{f}, \overline{f^2}$ be given by $\bar{f}(V) \equiv E_{P_0}[f(O)|V]$, $\overline{f^2}(V) \equiv E_{P_0}[f^2(O)|V]$. Note that $\overline{f^2}(V) - \bar{f}^2(V) = \text{Var}_{P_0}[f(O)|V]$. For every $1 \leq v \leq \nu$, let $\ell_1, \ldots, \ell_\nu$ and $I_1, \ldots, I_\nu$ be given by $\ell_v(V) \equiv I\{V = v\}$ and $I_v \equiv \{1 \leq i \leq N : V_i = v\}$.

29.6.1 Proof of Proposition 29.1

Combining (29.3) and **A3** yields that

$$\sqrt{n}(\psi_n^* - \psi_0) = \sqrt{n}(P_n^{HT} - P_0)D(P_n^*) + o_P(1)$$
$$= \sqrt{n}(P_n^{HT} - P_0)f_1 + \sqrt{n}(P_n^{HT} - P_0)(D(P_n^*) - f_1) + o_P(1),$$

where $f_1 \in \mathcal{F}$ is introduced in **A2**. By **A1**, the first RHS term in the above equation converges in distribution to the centered Gaussian distribution with variance σ_1^2. Moreover, by a classical argument of empirical processes theory (van der Vaart 1998, Lemma 19.24), **A1** and the convergence of $D(P_n^*)$ to f_1 in **A2** imply that the second RHS term converges to 0 in probability. This completes the sketch of proof.

29.6.2 Proof of Eqs. (29.8) and (29.9)

By construction of $P^{s\Pi}$, the number of observations sampled from each V-stratum is deterministic. In other words, it holds for each $1 \leq v \leq \nu$ that $\mathrm{Var}_{P^{s\Pi}}\left[P_n^{\mathrm{HT}} \ell_v\right] = 0$. In light of (29.2), this is equivalent to

$$\sum_{i \in I_v} \left(\frac{1}{\Pi_{ii}} - 1\right) = \sum_{i \neq j \in I_v} \frac{|\Pi_{ij}|^2}{\Pi_{ii}\Pi_{jj}} \tag{29.15}$$

for each $1 \leq v \leq \nu$.

Now, since $V_i \neq V_j$ implies $\Pi_{ij} = 0$ by construction, (29.2) rewrites

$$N^2 \mathrm{Var}_{P^{s\Pi}}\left[P_n^{\mathrm{HT}} f\right] = \sum_{i=1}^{N} \left(\frac{1}{\Pi_{ii}} - 1\right) f^2(O_i) - \sum_{1 \leq i \neq j \leq N} |\Pi_{ij}|^2 \frac{f(O_i)}{\Pi_{ii}} \frac{f(O_j)}{\Pi_{jj}}$$

$$= \sum_{v=1}^{\nu} \sum_{i \in I_v} \left(\frac{1}{\Pi_{ii}} - 1\right) f^2(O_i) - \sum_{v=1}^{\nu} \sum_{i \neq j \in I_v} \frac{|\Pi_{ij}|^2}{\Pi_{ii}\Pi_{jj}} f(O_i)f(O_j).$$

Because $O_1, \ldots, O_N$ are conditionally independent given $(V_1, \ldots, V_N)$ and since each factor $|\Pi_{ij}|^2/\Pi_{ii}\Pi_{jj}$ is deterministic given $i, j \in I_v$, the previous equality and (29.15) then imply

$$N^2 \mathrm{E}_{P_0}\left[\mathrm{Var}_{P^{s\Pi}}\left[P_n^{\mathrm{HT}} f\right]\right] = \mathrm{E}_{P_0}\left[\sum_{v=1}^{\nu} \overline{f^2}(v) \sum_{i \in I_v}\left(\frac{1}{\Pi_{ii}} - 1\right) - \sum_{v=1}^{\nu} \bar{f}^2(v) \sum_{i \neq j \in I_v} \frac{|\Pi_{ij}|^2}{\Pi_{ii}\Pi_{jj}}\right]$$

$$= \sum_{v=1}^{\nu} \left(\overline{f^2}(v) - \bar{f}^2(v)\right) \mathrm{E}_{P_0}\left[\sum_{i \in I_v}\left(\frac{1}{\Pi_{ii}} - 1\right)\right]. \tag{29.16}$$

For each $1 \leq v \leq \nu$,

$$\mathrm{E}_{P_0}\left[\sum_{i \in I_v}\left(\frac{1}{\Pi_{ii}} - 1\right)\right] = \left(\frac{N}{nh(v)} - 1\right)\mathrm{E}_{P_0}\left[\mathrm{card}(I_v)\right] = \left(\frac{N}{nh(v)} - 1\right)NP_0(V = v).$$

Therefore, (29.16) yields

$$E_{P_0}\left[\text{Var}_{P^{sII}}\left[P_n^{HT}f\right]\right] = \frac{1}{n}\sum_{v=1}^{v}\left(\overline{f^2}(v) - \bar{f}^2(v)\right)h^{-1}(v)P_0(V=v)$$

$$-\frac{1}{N}\sum_{v=1}^{v}\left(\overline{f^2}(v) - \bar{f}^2(v)\right)P_0(V=v)$$

$$= \frac{1}{n}E_{P_0}\left[\text{Var}_{P_0}\left[f(O)|V\right]h^{-1}(V)\right] - \frac{1}{N}E_{P_0}\left[\text{Var}_{P_0}\left[f(O)|V\right]\right],$$

as stated in (29.8). We now turn to (29.9). Since $E_{P^{sII}}\left[P_n^{HT}f\right] = P_N f$, it holds that

$$\text{Var}_{P_0 P^{sII}}\left[\sqrt{n}(P_n^{HT} - P_0)f\right] = nE_{P_0}\left[E_{P^{sII}}\left[(P_n^{HT}f)^2\right]\right] - n\left(E_{P_0 P^{sII}}\left[P_n^{HT}f\right]\right)^2$$

$$= nE_{P_0}\left[\text{Var}_{P^{sII}}\left[P_n^{HT}f\right]\right] + n\,\text{Var}_{P_0}\left[P_N f\right]$$

$$= n\Sigma_h^{II}(f,f) + \frac{n}{N}\left(\text{Var}_{P_0}\left[f(O)\right] - E_{P_0}\left[\text{Var}_{P_0}\left[f(O)|V\right]\right]\right),$$

where the last equality follows from (29.8). This completes the proof.

29.6.3 Proof of Proposition 29.3

Let us first state the so called Soshnikov conditions (Soshnikov 2000). A function f of O drawn from P_0 meets them if

$$N^2\,\text{Var}_{P^{sK}}\left[P_n^{HT}f\right] \text{ goes to infinity,} \tag{29.17}$$

$$\max_{1\le i\le N} K_{ii}^{-1}f(O_i) = o\left(N^2\,\text{Var}_{P^{sK}}\left[P_n^{HT}f\right]\right)^{\epsilon} \text{ for all } \epsilon > 0, \tag{29.18}$$

$$NE_{P^{sK}}\left[P_n^{HT}|f|\right] = O\left(N^2\,\text{Var}_{P^{sK}}\left[P_n^{HT}f\right]\right)^{\delta} \text{ for some } \delta > 0. \tag{29.19}$$

Conditions (29.17), (29.18) and (29.19) are expressed conditionally on a trajectory $(O_i)_{i\ge 1}$ of mutually independent random variables drawn from P_0. We denote $\Omega(f)$ the set of trajectories for which they are met. By assumption, $P_0(\Omega(f)) = 1$ for all $f \in \mathcal{F}'$. It is worth emphasizing that this assumption may implicitly require that the ratio n/N go to zero sufficiently slowly, as evident in the sketch of proof of Proposition 29.4. Since $\mathcal{F}'$ is countable, $\Omega \equiv \cap_{f\in\mathcal{F}'}\Omega(f)$ also satisfies $P_0(\Omega) = 1$.

Set $f \in \mathcal{F}'$ and define $Z_N(f) \equiv (\text{Var}_{P^s}\left[P_n^{HT}f\right])^{-1/2}(P_n^{HT} - P_0)f$. On Ω, the characteristic function $t \mapsto E_{P^{sK}}\left[e^{itZ_N(f)}\right]$ converges pointwise to $t \mapsto e^{-t^2/2}$. Therefore, $t \mapsto E_{P_0}\left[E_{P^{sK}}\left[e^{itZ_N(f)}\right]I\{\Omega\}\right]$ also does. Since $P_0(\Omega) = 1$, this implies the convergence in distribution of $Z_N(f)$ to the standard normal law hence, by Slutsky's lemma, that of $\sqrt{n}(P_n^{\mathrm{HT}} - P_0)f$ to the centered Gaussian law with a variance equal to the limit in probability of $n \, \mathrm{Var}_{P^{sK}} \left[P_n^{\mathrm{HT}} f \right]$. The asymptotic tightness of $\sqrt{n}(P_n^{\mathrm{HT}} - P_0)f$ follows. Finally, applying the Cramér-Wold device yields the convergence to a centered multivariate Gaussian law of all marginals $\sqrt{n}(P_n^{\mathrm{HT}} - P_0)(f_1, \ldots, f_M)$ with $f_1, \ldots, f_M \in \mathcal{F}'$.

The second step of the proof hinges on the following concentration inequality (Pemantle and Peres 2014, Theorem 3.1): if $C(f) \equiv \max_{1 \leq i \leq N} |K_{ii}^{-1} f(O_i)|$ then, for all $t > 0$,

$$P^{sK} \left[|(P_n^{\mathrm{HT}} - P_0)f| \geq t \right] \leq 2 \exp\left(-nt^2/8C(f)^2\right). \tag{29.20}$$

This statement is conditional on $O_1, \ldots, O_N$. Note that there exists a deterministic upper-bound to all $C(f)$s because $\mathcal{F}'$ is uniformly bounded and because the first order inclusion probabilities are bounded away from 0 uniformly in N. We go from the convergence of all marginals to **A1** by developing a so called chaining argument typical of empirical processes theory. The argument builds upon (29.20) and the assumed finiteness of the bracketing entropy of $\mathcal{F}'$ with respect to the supremum norm. This completes the sketch of the proof.

29.6.4 Proof of Proposition 29.4

Consider $f \in \mathcal{F} \setminus \mathcal{F}'$, a function of O drawn from P_0 which depends on V only. It holds that

$$P_n^{\mathrm{HT}} f = \frac{1}{n} \sum_{v=1}^{\nu} f(v) h^{-1}(v) n_v,$$

where $n_v = \sum_{i=1}^{N} \eta_i \ell_v(V_i) = nh(v)N_v/N$ with $N_v \equiv \sum_{i=1}^{N} \ell_v(V_i)$ (each $1 \leq v \leq \nu$). Therefore, the above display rewrites $P_n^{\mathrm{HT}} f = P_N f$, hence $\mathrm{Var}_{P^{sII}} \left[P_n^{\mathrm{HT}} f \right] = 0$. Moreover, the CLT for bounded, independent and identically distributed observations implies $\sqrt{n}(P_n^{\mathrm{HT}} - P_0)f = \sqrt{n/N} \times \sqrt{N}(P_N - P_0)f = O_P(\sqrt{n/N}) = o_P(1)$.

Consider now $f \in \mathcal{F}'$. We wish to prove that f meets the Soshnikov conditions and that $n \, \mathrm{Var}_{P^{sII}} \left[P_n^{\mathrm{HT}} f \right]$ converges in P_0-probability to $\Sigma_h^{II}(f, f)$, which is positive because $f \in \mathcal{F}'$. When relying on P^{sII}, the LHS expression in (29.18) rewrites $\max_{1 \leq i \leq N} Nf(O_i)/nh(V_i)$ and is clearly upper-bounded by a constant times N/n. As for the LHS of (29.19), it equals $\sum_{i=1}^{N} |f(O_i)|$ and is thus clearly upper-bounded by a constant times N. Let us now turn to $\mathrm{Var}_{P^{sII}} \left[P_n^{HT} f \right]$. By construction of P^{sII}, the variance decomposes as the sum of the variances over each V-stratum, each of them being a quadratic form in sub-Gaussian, independent and identically distributed random variables conditionally on $(V_1, \ldots, V_N)$. Because quadratic forms of independent sub-Gaussian random variables are known to concentrate exponentially fast around their expectations (see the Hanson-Wright concentration inequality in Rudelson and Vershynin 2013), $\mathrm{Var}_{P^{sII}} \left[P_n^{HT} f \right]$ concentrates around its expectation (29.8). Consequently, $N^2 \, \mathrm{Var}_{P^{sII}} \left[P_n^{HT} f \right]$ is of order N^2/n. It is then clear that

$N/n = o((N^2/n)^{\epsilon})$ for all $\epsilon > 0$ ensures that f meets the Soshnikov conditions. This holds for instance if $n \equiv N/\log^a(N)$ for some $a > 0$. Finally, the concentration of $\mathrm{Var}_{P^s}\left[P_n^{\mathrm{HT}}f\right]$ around its expectation also yields the convergence of $n\,\mathrm{Var}_{P^s}\left[P_n^{\mathrm{HT}}f\right]$ to $\Sigma_h^{\Pi}(f,f)$ in P_0-probability. At this point, we have shown that $\sqrt{n}(P_n^{\mathrm{HT}} - P_0)f$ converges in distribution to the centered Gaussian law with variance $\Sigma_h^{\Pi}(f,f)$. The rest of the proof is similar to the end of the proof of Proposition 29.3. This completes the sketch of proof.

Acknowledgement The authors acknowledge the support of the French Agence Nationale de la Recherche (ANR), under grant ANR-13-BS01-0005 (project SPADRO) and that this research has been conducted as part of the project Labex MME-DII (ANR11-LBX-0023-01).

Chapter 30
The Predicament of Truth: On Statistics, Causality, Physics, and the Philosophy of Science

Richard J. C. M. Starmans

30.1 Statistics and the Fragility of Truth

On the one hand, the first part of this essay's title may seem a little querulous and ill-omened, a prelude to a litany of complaints or to sorrowful, grief-stricken pondering on the deterioration of the quest for truth. Undeniably, it suggests a further decline of civilization as we know it, in accordance with Oswald Spengler's pessimistic anticipations on the lifespan of civilizations, published in *Der Untergang des Abendlandes* in 1918 at the end of the First World War. The concept of truth has been essential in the history of ideas and characteristic and distinguishing for the human condition. Even if one adheres to Protagoras' *homo-mensura-principle*, be it in a mitigated or radical manner, people cannot exist, survive or function without proclaiming, stipulating, conjecturing, establishing or cherishing a notion of truth, underlying and motivating their thoughts, words and deeds. As such it has been ubiquitous in religion, metaphysics, epistemology, science, politics and everyday life. From a historical-philosophical point of view the concept of truth is *pivotal* in epistemology; it precedes, subsumes or—at the least—it is *presupposed* by concepts like knowledge, rationality, objectivity, *causality*, justification, inference and many more. At the same time truth may easily be denied, distorted, declared obsolete, or conveniently modified and relativized on behalf of self-interest, religion, political ideology, freedom, stakeholders interests, public health, national security, climate, the Will of the People, et cetera. This fragility of truth may be noticeable in politics, journalism (whether phrased as disinformation, alleged truisms or fake news), on social media, in historiography or—horribile dictu—even in philosophy and modern science.

R. J. C. M. Starmans (✉)
Department of Information and Computing Sciences, Buys Ballot Laboratory, Universiteit Utrecht, Princetonplein 5, De Uithof, 3584 CC Utrecht, The Netherlands
e-mail: starmans@cs.uu.nl

© Springer International Publishing AG 2018
M.J. van der Laan, S. Rose, *Targeted Learning in Data Science*,
Springer Series in Statistics, https://doi.org/10.1007/978-3-319-65304-4_30

On the other hand, dyed-in-the-wool statisticians cannot be that astounded or appalled with the depicted view, as they find themselves standing trial ever since the dawn of their discipline for having a problematic relation with truth. Mark Twain's notorious adagio that "there are lies, damned lies and statistics" or George Box's emblematic dictum or maxim that "all models are wrong, but some are useful" have been quoted frequently for a variety of purposes. Darrell Hoff's booklet *How to Lie with Statistics* (1954) sold more than a million copies worldwide and even gave rise to a new genre on statistical fallacies, caveats and pitfalls. Renowned textbooks on the history of statistics like Stigler (1986, 1999, 2016), Porter (1986, 1995), Krüger et al. (1987), Hacking (1975, 1990), Gigerenzer (1989) and many more, abundantly show that suspicion and mistrust have been pervasive throughout the development of statistics: in science, philosophy and popular culture.

For example, in his monograph *The Taming of Chance* (1990), Ian Hacking depicted a resilient backlash against statistics in nineteenth century, amongst divergent leading intellectuals, including Fjodor M. Dostoyevsky, August Comte and Friedrich Nietzsche, for (alleged) capitalizing on utility, numerical abstractions, averages and mediocracy, rather than pursuing truth, genuine knowledge and understanding, focusing on essences and embracing the dignity of individual human beings. Despite blatant progress in the field of statistics ever since and the undisputed fact that nearly all sciences have taken a probabilistic turn in the last century, skepticism endures in science, politics and popular culture; also today the dialogue between ethics and statistics often proceeds in a strenuous and cumbersome fashion (Starmans 2012a,b,c).

Of course these brief citations and references reflect only a few aspects of the relation between truth and statistics. We will refrain here from the question to what extent the profession is responsible itself for the depicted view in popular culture. Rather than this we will make three observations, restraining ourselves to current research in mathematical statistics.

First, it is clear that from a historical-philosophical point of view, the paradigm of TMLE, among other things, aims to rethink or rather restore the connection between truth and models. It re-establishes the concept of a statistical model in a prudent and parsimonious way, allowing humans to include only their true, realistic knowledge (e.g., data are randomized, representing independent and identically distributed observations of a random variable) in the model. Rather than assuming misspecified parametric or highly restrictive semiparametric statistical models, TMLE defines the statistical estimation problem in terms of nonparametric or semiparametric statistical models, that represent realistic knowledge, i.e., knowledge in terms of the experiment that actually generated the data the hand. The model must contain the (optimal approximation of the) true probability distribution and the targeted parameter of interest is defined as a function of this true distribution. To this aim TMLE "reassigns to the very concept of estimation, canonical as it has always been in statistical inference, the leading role in any theory or approach to learning of data" (Starmans 2011).

Secondly, current practice in statistical data analysis is simultaneously paradoxical and problematic. On the one hand, textbooks wrongly suggest a united field with clear foundations, well-founded techniques, narrating a traditional *whig history*, with emerging, cumulating and justified knowledge. On the other hand, in reality we observe a plethora of different, sometimes *incompatible techniques and principles*, which require a delicate process of *eclecticism* to employ. These techniques and principles have often been detached from their historical background and derived from sometimes incompatible or discordant positions, including those of Karl Pearson, Udny Yule, Ronald Fisher, Jerzy Neyman, Egon Pearson, Jimmy Savage, Bruno de Finetti and John Tukey (Starmans 2011).

Thirdly, recent developments in data science, with divergent powerful machine learning techniques, including deep learning, are likely to deteriorate the situation even more. A reconciliation of statistics and algorithmic data analysis seems mandatory and in fact one of the guiding principles of TMLE (van der Laan and Rose 2011; van der Laan and Starmans 2014) is to achieve this (Starmans 2016a). The third section of this essay will develop the latter issue a little further with respect to the *anti-statistical* and *anti-epistemic* stance, currently gaining popularity in data science.

30.2 Truth in Epistemology and Methodology

To account for this predicament of truth or rather to understand its current situation one could consider taking a historical-philosophical stance. Rather than exploring its many appearances, we confine ourselves to the history of ideas in general and more particularly to philosophy (especially metaphysics an epistemology) and modern science. In doing so, we observe throughout the ages the essential tension between the holy grail of *true knowledge*, pursued by humble and altruistic "ancillae veritatis" or servants of truth, and those who believe all this to be utopic, due to the nature of the concept of truth, be it ill-defined or ill-conceived, or simply because the human condition prevents us from achieving it.

For instance, the proclaimed tension can easily be traced back to the ancient Greeks, starting with Plato's notion of *episteme* or real, certain knowledge as *justified true belief* and with Aristotle's axiomatic-deductive knowledge ideal and syllogistic logic based on *truth-preserving* inferences. Episteme was considered to be the opposite of *doxa*, opinion, *uncertainty* and irregularity. These ideals were challenged and defied almost instantly by the rhetorical sophist movement, including the aforementioned Protagoras and by the cynic school of Diogenes, but most notoriously and profoundly by the *Sceptics*. Especially Pyrrho of Elis, whose deliberations were analyzed and transmitted by Sextus Empiricus in his *Principles of Pyrrhonism* in the second or third century, paved the way for criticizing claims regarding truth. Skepticism, be it in many disguises and forms, would persist as a wave through the ages, proceeding via Michel de Montaigne's sixteenth century skeptical humanism, Rene Descartes' methodological doubt in the seventeenth century, David Hume in the eighteenth century up to Karl Popper in the twentieth century and contemporary philosophy. The expounded tension has lasted to this very day and as such the ideal has kept its significance, also in the era of big data and data science. However, as philosophy evolved and science progressed through the ages, the preoccupations with truth *continued and advanced* along at least two distinct lines, that will be briefly expounded here: first *in epistemology*/philosophy of science, secondly in research methodology.

The philosophy of science has shown many developments that influenced or shaped the current status of truth and in this short essay we will only highlight a few significant ones. First, the primordial question whether truth is a property of reality, a state, process, fact, thought or *proposition/statement/theory/model* has led to a variety of conceptualizations or definitions of truth, developed and scrutinized in epistemology. These include "rigid," mathematical versions like the *correspondence theory of truth* and Alfred Tarski's related semantic *theory of truth*, "moderate" versions like the *coherence theory of truth*, or more or less "relativistic" versions like *the consensus theory*, *conventionalist* or *pragmatic* approaches.

Indeed, it is often claimed that a genuine mathematically precise concept of truth is only viable in logic and model theory, not in empirical science. Mathematical logic is *truth-functional*; key concepts like *entailment*, *inference* and *validity* are understood as such and the criteria of *soundness* and *completeness* guide and guarantee the correctness of the inferential process. If a formula can be derived it must be true (soundness), if it is true it should be derivable (completeness). This strict correspondence theory of *syntactic derivability* and *semantical truth* as dual notions found little resonance in the epistemology of the empirical sciences, although it could be argued that some renowned applications in methodology can be traced back to it.

For example, in epidemiology *sensitivity* and *specificity*, and derived notions including *likelihood ratios of positive and negative tests* define post-test probability in a similar way. In information retrieval and pattern recognition *precision* and *recall* are defined accordingly. The former expresses the idea that if a document is retrieved (derived) it should be relevant (true). Reversely, recall expresses that if a text is relevant we want it to be retrieved. Obviously, these are all *probabilistic* concepts, employed in research methodology, rather than being the backbone of epistemology itself. Later, this topic will be addressed a little more comprehensively.

Secondly, as of the beginning of the twentieth century in the Anglo-Saxon *analytical tradition* the concept of truth has been connected to *meaning*, by logical positivists (verification-criterium, inductive confirmation) like Moritz Schlick and Rudolph Carnap and later associated with *falsificationalism* (predominantly in Poppers critical rationalism). However, all these distinguished scholars focused on the *context of justification* rather than on the *context of discovery* and as such they were predominantly aimed at a rational reconstruction of the empirical sciences and used probabilistic notions to achieve this. They didn't strive for developing a constructive research methodology. What's more, truth was actually studied in a wider context of epistemic issues like *objectivity, rationality, the structure of scientific theories, value-freedom, uniformity of science*, etc.

The underlying concepts of truth were often coherence-based or pragmatic. The latter notion was particularly advocated by philosophers like Charles S. Peirce,

William James, John Dewey and W.V.O. Quine, all pioneers of pragmatism, up to now a dominant movement in the philosophy of science, in order to make the old fashioned notion more suitable and viable for contemporary epistemic issues. Logical positivistic research on the structure of scientific theories led to both *syntactic* approaches (Carnap) and *semantic* approaches, the latter depicting *theories as classes of models* (Suppes, Van Fraassen, Stegmüller). This distinction was only loosely based on the "dual" logical correspondence theory. As a result the semantic approach can best be affiliated with a coherence conception of truth and it appeared to be a precursor to current positions like those of George Box and the physicist John von Neumann, which is actually the problem to be addressed in the next section.

A third development in the philosophy of science that challenged the ideal of true knowledge was the rise of influential historical and sociological studies, like those of Thomas Kuhn, Paul Feyerabend and Richard Rorty, marxist studies by the Frankfurter Schule, externalistic studies published by Robert K. Merton, Bruno Latour, the "Strong Program" by David Bloor and Barry Barnes and more recently social constructivism by Samuel Pickering. For example, Kuhn developed his famous paradigm theory with discontinuity rather than continuous progress, with incommensurability of concepts and methods used along the different paradigms, with theory-laden observations, et cetera in *The Structure of Scientific Revolutions* (1962). Paul Feyerabend shocked the research community with his anarchistic ideas and "anything goes" justification in *Against Method* (1975). In his book *Philosophy and the Mirror of Nature* (1979) Richard Rorty proclaimed the end of representation, or the idea that scientific theories in one way or the other mimic nature. Bruno Latour, one of the founding fathers of network-actor theory conducted detailed studies on how experimental scientist actually are doing science in laboratories, by experimenting, discussing, negotiating, interacting with stakeholders, funding organizations and dealing with conventions and conflicts. His *Science in Action: How to Follow Scientists and Engineers through Society* (1987) is a modern classic in this field and influenced much research in *science and technology studies*.

Although all these scholars aimed at the entire scientific enterprise in a broader sense, including objectivity, rationality, its alleged or disputed value-free status, this led to diverse sorts of *relativism*, which also affected the notion of truth. Whether caused by externalist historical studies, by contemporary skepticism, cultural relativism or social constructivism, the concept of truth eroded and this typifies the situation today. Although truth is the key concept in epistemology, it paradoxically is often used implicitly, tacitly as a primitive or self-evident term and employed if convenient to the user, whereas the real research focus in epistemology is on less fundamental notions, as has been mentioned before. As a result the constellation of issues involving the concept of truth transpire in an indirect fashion and figurate persistently in present-day topics in the philosophy of science. Not only do they encompass or involve traditional themes from normative philosophy (on ethics, values, integrity, moral responsibility, informed consent, code of conducts, et cetera), but the concept of truth especially emerges, figurates or is presumed in canonical epistemological issues, such as the *scientific realism debate, (Bayesian) confirmation theory, the structure of scientific theories*, and last but not least the triptych "explanations, causality and laws," which notions are often studied or put into relation to each other.

The second line along which thinking about truth advanced, concerns developments in research practice and methodology; put roughly, the concept of truth has been operationalized, reshaped or rather supplemented by notions like validity, reliability, soundness, correctness or accuracy. For example, in research methodology a plethora of validity concepts has been deployed, such as internal and external validity, construct validity, statistical conclusion validity, concurrent validity, a list which can easily be extended. The same applies to reliability, that has many connotations which are prerequisites or rudiments to truth, its representation and "measurement." In addition many more or less domain-specific measures or standards of quality have been developed, including the aforementioned concepts like sensitivity, specificity, recall, precision, and many more. Important as both developments are, they have not prevented the above mentioned and depicted view and -all too often- misplaced authoritative appeals on some notion of truth are made, rather exploiting its fragility than safeguarding it. Ironically, on the one hand, the concept seems obsolete, it has been circumvented and disregarded, on the other hand it seems that we appeal to it regularly, require and presuppose it. Here, we will show how this small section gives rise to some important aspects, that shape or at least illuminate the predicament of truth.

30.3 Eroded Models, Von Neumann and the End of Theory

The Hungarian physicist John von Neumann was not only a pioneer of computer science and founder of game theory, but similar to colleagues like Einstein and Heisenberg also actively engaged in the relationship between philosophy and science. A proof of his epistemological preoccupations can be found in his article "Method in the Physical Sciences," which appeared in 1955 and was reprinted in 2000 in *The Neumann Compendium*. In an often quoted passage, the physicist leaves little space for misunderstanding his vision: "The sciences do not try to explain, they hardly even try to interpret, they mainly make models. By a model is meant a mathematical construct which, with the addition of certain verbal interpretations, describes observed phenomena. The justification of such a mathematical construct is solely and precisely that it is expected to work—that is, correctly to describe phenomena from a reasonably wide area."

Amongst some scientists, this short passage may cause little disturbance or confusion (Starmans 2011a). The perceived view seems at the most partly compatible with experiences from everyday practice. The postulated omnipresence of models in science will be endorsed by any physicist, biologist, social scientist, economist, climate scientist or engineer. In addition, models are also manifest out of science, in professional practice and in daily life. However, from an epistemic point of view, Von Neumann's quote is problematic in various respects.

First of all, he advocates a virtually complete reduction of science to model formation. Models do not offer an interpretative framework or an additional and heuristic value in theorizing, but they are the essence of the scientific enterprise. Both appear more or less equivalent. The extent to which this proposition is considered problematic depends on the status and scope of the model concept used. Again, Von Neumann is very explicit. First of all, he considers models in science primarily as mathematical constructions with a verbal explanation. This comment seems to be easy to reprove today, as science and technology uses not only symbolic mathematical models, but also plenty of tangible, physical, analog models, or graphic or digital simulation models. Moreover, Von Neumann indicates the justification for using models solely in utilitarian terms; declaration of a (underlying) reality no longer seems to be a goal, neither the search for truth. This embodies both extreme pragmatism, and a hardly concealed and necessary *instrumentalism* and *antirealism*. It also provides an insight into a conception of science in which explaining (whether causally or otherwise) is no longer a major research function and truth has become an isolated and even useless concept, comparable with the view of the already quoted statistician George Box.

More importantly these deliberations may be seen an prelude or gain—a fortiori—significance by recent developments within big data and data science, which have even more far-reaching consequences for science philosophy and methodology. For example, one can think of a notorious, but frequently quoted article by science journalist Chris Anderson, who proclaimed the end of scientific methodology and knowledge as in 2008. Its "The End of Theory: The Data Deluge Makes the Scientific Method Obsolete" (Anderson 2008) is equally important for philosophers as for methodologists and statisticians. A small anthology of Anderson's sometimes rather unobtrusive comments may suffice.

First of all, the author states that companies like Google are actually sifting through the most measured era since dawn of civilization, "treating this massive corpus as a laboratory of the human condition," establishing a society inhabited by the "children of the Petabyte Age." Clearly, the author does not shy away from any pathos. He is not a futurologist, but unquestionably states that the outlined situation is more about the status quo than a far prospect. According to Anderson in the era of data science information processing is no longer a matter of simple low dimensional classifications, natural laws and theories, but of what he calls *dimensionally agnostic statistics*. It demands a essentially different methodology, that finally will do justice to the data as something that can be "visualized in its totality." To this aim we have "to view data mathematically first and establish a context for it later."

The author shows himself to be a genuine apologist of the new doctrine. According to Anderson, Google's success is based on the conviction that we don't understand why this page is better than that one, and that we actually should not bother: a full reliance on the statistics will do and "no semantic or causal analysis is required." As far as the author concerns, this includes virtually any aspect or dimension of human action: "Out with every theory of human behavior, from linguistics to sociology. Forget taxonomy, ontology, and psychology. Who knows why people do what they do? The point is they do it, and we can track and measure it with unprecedented fidelity. With enough data, the numbers speak for themselves." Indeed, our entire concept of science as hypothesizing, modeling and testing seems rapidly becoming obsolete.

Anderson asserts that the field of physics seems to be a good example. First, Newtonian laws appeared to be rude, useful, but—at the atomic level—wrong approximations of truth. Then in the beginning of the twentieth century, quantum mechanics provided yet another, admittedly improved model, but still flawed no less than a misrepresentation or distortion of a more complex reality underlying the phenomena. After this, biology finds itself standing trial. Canonical high school accounts of "dominant" and "recessive" genes based on a Mendelian genetics appear to be an even more rude oversimplification of reality than classical mechanics. In fact "the discovery of gene-protein interactions and other aspects of epigenetics has challenged the view of DNA as destiny and even introduced evidence that environment can influence inheritable traits, something once considered a genetic impossibility." Even worse, an increase of real knowledge on biological phenomena, will inevitably decrease the possibility to come up with a model that can explain the phenomena. The children of the Petabyte Age will perceive and understand the world in an entirely different way. Scientific progress does no longer require any mechanistic explanation, causal models will be overtaken by correlational patterns, grand unified theories will no longer be needed or strived for.

A year later, Google researchers Alon Havely, Peter Norvig and Fernando Pereira published "The Unreasonable Effectiveness of Data." When it comes to people and not elementary particles, simple mathematical formulas and elegant theories are of limited use, the authors say. The same applies to parametric statistical models. To understand and predict people's behavior, you can rely more on many petabytes of rough, unlisted, unstructured, if necessary, distorted data. These are first represented by efficient data structures/architectures and then forged by intelligent pattern-recognizing algorithms into knowledge. In Starmans (2016b, 2017b) we have tried to do more justice to the vision of the aforementioned authors, but this global representation of their ideas suffices to illustrate the situation in which both statistics and the "scientific method" are, and how the underlying philosophy seems to come under pressure.

In addition, the views of Anderson and Havely cum suis grossly can be found almost everywhere in the popular and professional literature on big data and data science, admittedly sometimes in a slightly more mitigated fashion. Obviously, this does not imply the absence of opposite and more nuanced positions. This also applies to Google (Starmans 2015a) and especially to contemporary debates about the risks of AI and, more generally, about the moral dimension in AI and data science. For example, the recent success of *Weapons of Math Destruction; How Big Data Increases Inequality and Threatens Democracy*, published in 2016 by mathematician and converted data science sceptic Cathy O'Neil, is illustrative in this respect. Be that as it may, the developments we depicted in this section suggest an *anti-epistemic* and *anti-statistical* stance, that seems to offer nothing but leaking prospects for both statistics and epistemology, which might encourage someone to consider the issue of joining forces. This, however appears to be far from trivial.

30.4 Physics, Statistics and the Philosophy of Science

A century ago, the relationship between physics and philosophy was still relatively harmonious. This in spite of fundamental revolutions like the theory of relativity and quantum mechanics, which strongly undermined the traditional, comfortable world-view and made the ancient and familiar philosophical concepts of time, space and causality problematic. Physicists like Einstein, Heisenberg, Bohr and Schrödinger were highly philosophically oriented, and many of them showed their interest in word and writing. Philosophers like Russell, Carnap, Reichenbach and Schlick closely monitored the physical developments in their era. In those day some even believed in the necessity or possibility of interaction and cross-fertilisation or, at least, in a peaceful coexistence. It would appear that now these days are gone.

The physicist and Nobel Prize laureate Richard Feynmann was quoted frequently for decades ago because of his contention that "Philosophy of science is about as useful to scientists as ornithology is to birds." Steven Weinberg, who also won the Nobel Prize for Physics, wrote in 1993 *The Unreasonable Ineffectiveness of Philosophy*, unmistakably with a wink to Eugene Wigner admiration for "the unreasonable effectiveness of mathematics." In Weinberg (1993) he devoted to the problem a separate chapter, which was decorated with the prominent title "Against Philosophy." The astronomer Stephen Hawking once proclaimed the death of philosophy, and more recently, the famous physicist Laurence Kraus also broke into the debate. In 2015, Weinberg again invoked uproar with *To Explain the World: The Discovery of Modern Science*. As a result, he had to face a firm reprimand by historian Stephen Shapin in his review "Why Scientists Should Not Write History" (2015), published in *The Wall Street Journal*. The list can be expanded effortlessly, and annotated with numerous historical examples varying from Goethe's denial and repudiation of Newtons Optics to the vicious Einstein-Bergson controversy on the nature of time in the beginning of the twentieth century. However, also less polemically oriented physicists usually admit to an eliminative reductionism, which eliminates many philosophical concepts and "qualia" or gives them a specific abstract or mere mathematical description. These include a variety of notions such as space, time, movement, causality, intentionality, teleology, but also meaning, mind, free will, consciousness, personal identity and usually all metaphysical or religiously related ideas.

Fortunately, the relationship between statistics and science philosophy is not characterized by a similar animosity or a time-consuming polemic debate. Nevertheless, the conversation between them has been somewhat difficult for decades. Although the genealogy of this problem has a number of factors, we restrict ourselves to one of its crucial aspects: since its origins around 1925, science philosophy has made a radical distinction between the *context of discovery* and the *context or justification*, that was briefly referred to in Sect. 30.2. In her initial objective, inspired by logical positivism, to provide a rational reconstruction of (unity) science and to formulate corresponding demarcation criteria, philosophy of science focused on the context of justification rather than the practice of research and a constructive methodology. Although many science philosophers almost immediately recognized the relevance of the probabilistic revolution, paradoxically, a separation emerged that has continued to exist. As will be outlined in the next section, the "quest for truth" gained a layered structure, a triptych consisting of (empirical) disciplines, (research) methodology and epistemology/science philosophy. Many debates from the history of ideas and contemporary bottlenecks and challenges in the relationship between science and philosophy can be expressed in terms of these "layers" and their mutual relationships. This also applies to the current relationship between statistics and philosophy.

All this can be observed in the work of the American Bayesian statistician Andrew Gelman. In his coauthored article "Philosophy and the Practice of Bayesian Statistics" (Gelman and Shalizi 2013), they emphasize the eminent importance of philosophy for both statistical research and statistical practice. The authors argue that in this regard there is frequent "freewheeling" and dilettantism of certain statisticians, among whom he also generously counts himself. Although in the aforementioned article they confine themselves to figureheads of the philosophy of science such as Popper and Kuhn, they encourage their colleagues to seek a connection to more recent philosophical research.

At the same time they go one step further. Gelman and Shalizi (2013) aims to develop a new philosophy for Bayesian statistics, which is empirically adequate and better accounts for existing statistical practice. The authors state that it is customary in today's statistics to relate or even to anchor statistical approaches or assumptions in philosophical paradigms or positions. They refer in particular to inductivism and the hypothetical-deductive model (HD model) as conceptions of science. Next, they epitomize various achievements of "classical" statistics as manifestations of hypothetical-deductive reasoning; the significance tests of Ronald Fisher, hypothesis testing as developed by Egon Pearson and Jerzy Neyman, and finally the theory of confidence intervals, of which Neyman was also the founder.

According to Gelman and Shalizi (2013), Bayesian statistics is widely regarded as an instantiation of inductivism in statistics. Both theories are presented as opposing, not to say conflicting visions. As an impetus to a new empirically appropriate philosophy for Bayesian statistics, the authors criticize the previously outlined field and claim that Bayesian statistics should be regarded as an extremely successful HD model and not as more (or less) inductive than the above-mentioned manifestations of "classic" statistics. Additionally they assert that the most successful applications of Bayesian reasoning exceed the standard view, in which reasoning is perceived as updating a pre-distribution into a posteriori distribution, using a likelihood function based on new data, after which the "degree of faith" should be adjusted accordingly. It is rather a matter of developing various models (model fitting), which are then evaluated (model checking) and subjected to a series of rigorous tests, after which the strongest model remains. Because, according to the authors (almost), all models are still "wrong," that is, do not contain the probability distribution by which the available data are generated. Rather, there is a trial and error approach, which historically speaking is associated with Popper, but which finds in the recent literature its most prominent advocate in Mayo (1996, 2010).

However, Gelman and Shalizi (2013) does not note that inductivism and the HD model are not at all "claire et distinct" and can hardly be regarded as opposing conceptions of science, playing a role within statistics as they suggest (Starmans 2013). The distinction rather reflects the wrestling of thinkers with the reasoning and "reversals" in their search for knowledge. Induction in particular has taken a long process of development, has different appearances and aspects that are strongly linked to the wise positions of the people involved. This applies to Aristotle, Bacon, Hume, Kant and Mill, but also to Popper and the Bayesians. The concept has undergone remarkable shifts. A strict indication, such as naive inductivism as a logic or discovery, will no longer be advocated by anyone. A moderate and "sophisticated" version, as underpinned by many confirmation theories, will be experienced problematic by only a few. The HD model cannot claim a comparable respectable history and should not be considered a simple and successful repair attempt or alternative for induction. It's too general and has in fact always been linked with induction, even in the naive version of Francis Bacon. As a method, it does not contain significant new elements or insights, does not solve the problems of induction and faces various historical and systematic objections, including the famous Quine-Duhem these, which roughly put claims that it is not possible to develop a "crucial test" for an separate more or less isolated hypothesis. Indeed, the hypothesis is part of a network of other (auxiliary) hypotheses, initial conditions and ceteris paribus clauses, so it is not clear what has now been precisely tested or categorized.

In any case, when categorizing a person or idea in either of these "camps," the deceitfulness of the fallacy of wrong opposition and the strawmans fallacy looks inexorable. Moreover, the time of comprehensive all-encompassing epistemological theories has long been over and consequently the search for or postulation of a one-on-one correspondence between a statistical approach and a philosophical paradigm is less obvious. Despite noted weaknesses, the pursuit of alliances between statistical methodology and epistemology in Gelman and Shalizi (2013) is highly relevant and stimulating. What's more, their efforts to bridge the gap do not remain unnoticed, partly because the authors are not just focusing on initiated peers, but try to reach a wider audience and Gelman is active on many websites through blogs and comments. Its relevance is substantial, it exceeds the realms of statistics in a strict sense and by no means statistics is solely to blame for the current situation. Rather than this, it is to be considered a manifestation of a more general problem, as will be made apparent in the next section.

30.5 The Triptych of True Knowledge

The search for true or reliable knowledge and the validation of the employed methods and (tentative or ultimate) findings are closely linked in the history of ideas. In the many expressions that ultimately form the written results of the above-mentioned efforts, with any good will, a layered structure can be recognized, which can be described by means of a simple "architecture." In Starmans (2017d) it is argued that also the quest for truth has a layered structure: of empirical sciences (I), research methodology (II), and epistemology/philosophy of science (III).

The most concrete and specific layer can be situated within the individual subject disciplines (I), where for convenience the empirical sciences will function as a paradigmatic example. The practitioners of a specific field of science possess the *domain knowledge* and try to solve *knowledge problems* in this domain. In strict terms, they "produce" the knowledge. To that end, the researchers describe and model aspects of reality and seek explanations, which ideally form part of a universally possible, ideally unifying theory, which is both declarative and predictive successful. In view of this, the researchers make observations, collect data, conduct experiments, investigate causal relationships, try to formulate (nature) laws, build models and test hypotheses.

Typically, most knowledge problems from a specific knowledge domain, whether law, medicine, business, natural science or social science are *causal*. They presuppose *cause-effect relations* in the reality underlying the empirical domain, or at least establish causal connotations expressed in or underlying the aforementioned theory. The research problems are causal, the research questions are causal, the argumentation schemes are causal, in fact the entire natural language in which these knowledge problems are expressed, whether by layman or experts are causal. Whether we are dealing with everyday language discourse, a practical or highly theoretical problem, it is hard to imagine a research problem that is not causal. So causal talk is *ubiquitous* and *immanent*. If one seeks to know, to explain, to model or to understand something, causality emerges and appears to be interwoven with the domain knowledge, the level of practice and existing theories, the nature of the problem and the aims of the discipline.

At level I, causality is not only obvious and immanent, but in a way *context-sensitive*, depending on the habits, peculiarities and idiosyncrasies of the field, be it medicine, law, natural science, social science of business. In fact, as we will outline in the next two sections, this context-sensitivity has also been visible in the development of the concept in the history of philosophy. Be that as it may, an essential aspect of this above described, more or less problematic "standard image" of science, of course, concerns the subtle relationships (or suspected allegations and interactions) between empiric and theory. The observations, the data, the "special" and the concrete versus the laws, the hypotheses, the "general" and the abstract.

Epistemology largely exists in grace of the postulated or experienced field of tension between both. Knowledge acquisition in general and the related reasoning or inferences in particular always require *reversal* or *inverse* of patterns of reasoning. The "direction" alternates from the special to the general and vice versa, from the concrete to the abstract and vice versa, from the finite to the infinite and vice versa, from data to hypotheses/theory and vice versa. And above all: from causes to effects and vice versa. In fact, the Bayesian approach that Gelman and Shalizi (2013) tried to justify from an epistemic point of view is often credited for modeling this envisaged aspect in a convenient and cogent way. Needless to say that these intuitions and insights on causality are not only manifest in empirical science, but a fortiori relevant in engineering disciplines or *design science*, where solutions are to be designed for *action problems* rather than for knowledge problems and the scientific process is guided by a *design cycle* rather than an *empirical cycle* (Wieringa 2014).

In performing this research, researchers can fall back on a second layer, the methodological (II). This is more abstract, less specific and provides the formal methods and representations, the techniques and tools that help the researcher to actually conduct the research, including the aforementioned inverses of reasoning patterns, and validate the results. Reliable and valid conclusions require the bridging of the suspected gap between empirical and theory. Sometimes, the methods are generic, such as statistical methods, data-analytical algorithms (machine learning, data mining) or methods of research design. Sometimes (subject) specific methods and techniques are required, such as measuring instruments, experimental environments, simulation models or verification tools.

Regarding the question of causality at this stage we could only say that it is the methodologists duty to help the researcher answering his research questions, including its causal connotations, allusions or talk, and that progress in (natural) science has depended and still depends on how successful the alliance was/has been in answering the causal questions, *rephrasing* or formalizing the causal questions or perhaps *avoiding* or circumventing them. For example, this is what happens in (experimental) methodology through a range of statistical techniques, including, (canonical) correlation, partial correlation, analysis of contingency tables and association measures, analysis of variance, logistic and linear regression analysis, path models and structural comparison models. But also factor analysis, principle component analysis or discriminant analysis and even time series forecasting and survival analysis have partly causal connotations or at least allow for causal interpretations. Also today, these methods dominate experimental and observational studies and form the basis for causal statements in many empirical disciplines.

In a strictly pragmatic and scientistic sense, the quest for the foundation and justification of knowledge could be considered as complete. However, the history of science expounds that usually an appeal is made to a third layer. The architecture is complemented by epistemology/general philosophy of science (III). In particular, in disciplines that have just experienced one or more foundational crises and have come to some degree of maturity this is manifest. Researchers explicitly seek connection to and anchoring in major themes of epistemology, the identified concepts of science, the cherished or dismantled knowledge ideals. Themes such as valid and reliable knowledge, knowledge acquisition, truth, reality, statements, causality, explanation and natural laws, the structure of scientific theories, rationality and scientific progress, and unification or the unity of the sciences, enter into the philosophy of such a discipline.

Now that the time of comprehensive universal theories in the philosophy of science is over, in addition to the *general philosophy of science*, there is more and more space for *a special philosophy of science* (focusing on specific problems in e.g., mathematics, physics, biology, economics, social sciences, neurosciences or computer science). The obvious intended interaction can be achieved from at least two core questions. To what extent can we understand developments within a special science area from generic insights, theories and concepts provided by the general philosophy of science? And vice versa: what contribution can a special field of science provide for further development of the existing theory formation in general knowledge and science studies?

In this short essay we have to confine ourselves to the observation that either way, this simple architecture has at least three relationships: between I and II, between II and III and between I and III. Many debates in the history of ideas and contemporary bottlenecks in the relationship between science and philosophy can be derived from these relationships and many problems can *be understood and accounted for* from this perspective (Starmans 2017b,d). For example, this section showed that the relation between I and III is highly problematic in physics, Gelman and Shalizi (2013) philosophical turn concerned primarily the relation between II and III, and as we will see in the next sections current debates on representing causality are related to the relation between I and II or I and III, respectively. Many aspects should remain unattended here, but at this point it should be emphasized that the attempt in Gelman and Shalizi (2013) to relate II–III is appraisable, but here love is a reciprocal notion and definitely must come from both sides.

One of the big problems of epistemology/general philosophy of science is that as far as statistics is concerned epistemology sticks to the old distinction between context of discovery and justification and does not try to capture or address developments in statistics (with a few exceptions in Bayesian statistics). This appears to be also immanent in the literature on causality that is immensely and unfortunately partitioned along the aforementioned lines with hardly any cross-fertilization. A recent study (Phyllis and Russo 2016) shows some interest in research practice and certainly is an interesting attempt to bridge the gap, but recent developments (on causality) in both statistical science, including TMLE and algorithmic data analysis are largely ignored. Some aspects of this problem will be sketched in the next section.

30.6 Some Roots and Aspects of Causality

Although truth is the fundamental notion underlying the entire scientific enterprise, it has already been stated that the concept does not seem to play a key role in epistemology or research methodology anymore. It is often used informally, not even introduced syncategorematically, and mainly emerges or figurates in debates related to less fundamental issues. As stated before, we will here show how it manifests itself in the context of an equally persistent topic in the philosophy and science, the complicated concept of *causality* and intertwined notions like *causation, cause-effect relations, causal inference, reasoning* or *modeling*. Instead of trying to demarcate these very notions on historical grounds, stipulatively or along methodological lines, we will use causality as the over-arching term covering the constellation of aforementioned notions. Also concepts like physical necessity, logical necessity, unavoidability, sufficiency, determinism, regularity or their antonyms like chance, probability or irregularity constitute the concept of causality, or are historically linked to it.

Still, unlike truth causality has many applications and shows new lines of research in the philosophy of science, it is dominant in research on methodology and in a way even "en vogue," especially in the last two decades due to developments in computer science and AI. Causality appears to be an ancient and a fundamental notion, an unavoidable aspect of the "homo mensura." It seems first and foremost an intuitive concept, a common sense notion or natural category (in the Aristotelian or Kantian sense of the word), which we constantly use to be *adaptive* in the struggle for life, to explain our experiences and to *understand* ourselves, our place in cosmos and the contingencies of human existence. As such, causality is omnipresent in everyday language use. Explicitly, it is subsumed in many connectives, and numerous "causal" verbs, including causing, inducing, affecting, triggering, initiating, leading to, effecting, influencing, producing, resulting in, et cetera. In a sense all transitive verbs are essentially causal. Some of these express physical necessities or contingencies, others presuppose a subject or acting agent, which deliberately and intentionally causes a change in an object. As such causality was already manifest long before the dawn of philosophy or science.

Ever since the concept shows many faces, connotations and interpretations and has evolved over the centuries. It has proven to be chameleonic in the sense that it adapts itself to the philosophical context or theory where it has been utilized or developed. As we will see it appears to be highly *context-sensitive* and as such has been associated with logical necessity, physical necessity, contingence, coincidence, chance, teleology and determinism and shows different faces in divergent positions like *realism, idealism, empiricism, instrumentalism or rationalism*. There is a remarkable consistency in employing the term causality through the ages, although conceptual changes went on and on.

More importantly, it invokes confusion and dispute up to this very day in philosophy in science, statistics and data science. First it appears that even the most elementary and bare expression "X causes Y" opens a Pandora-box of problems and subtleties. The concept covers quite different ideas and approaches. We observe *deterministic versus probabilistic* stances, *difference-making* approaches, *regularity based* methods, *counterfactual* approaches, *interventionist* approaches, *mechanistic* theories, *type or token dilemma*, the alleged gap between *causes and reasons*, accounts at the *individual or group level, physical or mentalist* interpretations, *mono versus multiple* causality, *qualitative or quantitative* accounts, appearance in *discrete and continuous processes, in observational and experimental* settings, *direct and indirect* causality, et cetera. This list with approaches or interpretations, which are clearly neither mutually exclusive, nor totally exhaustive can easily be extended (Tacq 2011; Williamson 2009) Detached from their historical and philosophical background they provide us with a kaleidoscopic picture, allowing only a pluralist, disunified account, especially in view of the regrettable triptych we outlined earlier. A few historical details may elucidate the issue a little.

The British philosopher and mathematician Alfred North Whitehead (1861–1947) is still celebrated by philosophers today, because of his criticism of materialism in science, but especially as a protagonist and herald of process philosophy. In a Western tradition that is heavily influenced by Parmenides work, in which the immutability of being as a ontological principle is postulated, by Platonic time and space less ideas or forms, and by Aristotles substance thinking and essentialism, Whitehead expressed an different view. He rather acted in the footsteps of the process philosopher avant-la-lettre Heraclitus of Ephesus, by developing a "philosophy of organism," where reality is understood as a major process of continuous "actual occasions." "Material" objects are actual patterns within these current events, which can be both objective and subjective, take information of and pass on to other events. In addition, both experience and creativity are leading, constituent principles. In this way, the events are also *causally* active, involving the whole world event as a continuous process; *variation and change are primary* and not derived from or reducible to static states. Understanding reality means understanding the *process, procedure and mechanism*, concepts still recognizable in contemporary thoughts on causality.

Due to this combination of ideas and for no longer treating variation and change in a pejorative way, Whitehead is related to such diverse thinkers and contemporaries as the French Philosopher Henri Bergson, the biologists Wallace and Darwin and the statisticians Galton and Pearson (Starmans 2011a). The field of logic, on the other hand, recognizes Whitehead today primarily as the co-author of Bertrand Russell's monumental *Principia Mathematica*, which appeared between 1910 and 1913 and formed the pinnacle of the logical stance in the philosophy of mathematics.

Although Whitehead's versatility and profoundness is out of order, ironically his most cited statement is of a completely different order. In *Process and Reality* from 1929 he concluded that "The safest general characterization of the European philosophical tradition is that it consists of a series of footnotes to Plato." One can interpret this bold statement polemically and sneeringly, or consider it as a gag emphasizing the timeless character of Plato's philosophy. One can also try to find a canon of great and still topical themes from Western philosophy and then check to what extent they are rooted in or inspired by Plato's work. In conducting such an exercise, one inevitably strikes the persistent and complex problem of causality. Although the atomists and the Stoa already explored the concepts of necessity and determinism, it could be argued that it was indeed Plato, who included the beginning of the theory of causality in his dialogue Phaedo with a number of highly problematic fragments.

It was, on the other hand, his student Aristotle, who offered a much more systematic and influential dissertation in the Analytica Posteriora and the Physics with his doctrine of four causae (aitiai). To know a thing means to know its aitia. A full *explanation* and understanding of a particular entity requires answering four questions regarding the causa materialis ("what is it made of"), causa efficiens ("who made it or initiated the making"), causa finalis ("what is it made for") and causa formalis ("what is it that makes this entity what it is, constitutes its essence and not something else"). This doctrine was at the heart of his metaphysics and epistemology, led to many interpretations and modifications in theology, philosophy and science until the seventeenth century, making Aristotle the founder of the theory of causality.

Since then, issues surrounding causality dominated metaphysics and epistemology, until the nineteenth century, among other things in the work of such diverse thinkers as Aquinas, Bacon, Descartes, Galilei, Spinoza, Hobbes, Newton, Leibniz,

Hume and Kant, Stuart Mill and C.S. Peirce. The context-sensitive and chameleonic character of the notion is manifest in the work of all aforementioned thinkers, and many more; sometimes *explicitly* in the sense that they theorized about the concept, sometimes *implicitly* in the sense that they used it, as a primitive term or otherwise took it for granted. In fact, causality was *reinvented* time after time again, influenced by and *closely tied to the metaphysical and epistemic views* of the relevant thinker. Since in philosophy theories or views are seldom falsified and replaced, many conflicting ideas endured, evolving in nowadays pluralistic view on causality in epistemology. Indeed also today, philosophers scrutinize the ontological and epistemic aspects of the causal relationship and the question which assumptions and other conditions are required for meaningful causal statements and reasoning. As stated before the interactions and interdependence between causality, scientific statements and laws/laws of nature still constitute an important area or focus in the general philosophy of science.

Acknowledging the remarkable evolution that led to the concept of causality in the history of ideas, but without pretending in any way to pursue a detailed historical-wise overview, we begin the conceptual analysis in one way or another by dragging the concept into the classical "philosophical triangle" built up by the concepts of *reality, mind/ideas* and *language*, and the subtle interaction between these concepts. The questions raised here have dominated western metaphysics and epistemology over the past two thousand years. *Is there an objective, independent world (realism) or is reality built up by the mind (idealism, constructivism). Is that world knowledgeable and if so how? Can we make true statements about this world and how can we represent them? And what is truth?*

Focusing on the question of causality and cause-effect relationships, these questions can be specified as follows. Based on *reality*: does causality exist and if so, does it exist in a static way (as a state, fact or factual situation) or in a dynamic way (as a physical process with a production character)? Is it an action initiated by a causal agent, who intentionally performs the action from a certain point of view? Should that process be regarded as a sequence of uninterrupted steps, a kind of chain of causes to the ultimate effect? Is this process necessary, deterministic or stochastic c.q. indefinite? Does it take place continuously or discretely/stepwise? Does that process also require an action mechanism? Can there be multiple causes and can they influence (reinforce, subvert or cancel out) each other? Should we check part of reality or do an intervention to establish causality?

Starting from *the mind*: is causality a construction of the mind, a projection of experiences and expectations on reality (Hume), a characteristic of human ability and a condition for perceiving (Kant) presupposing a dualism of the real noumenal world and the world of phenomena? Can we perceive it, hypothetize or know it with our senses and cognition? And starting with *the language*: can we represent or operationalize it in natural language, in logical formalism (temporal logic, process logic action logic, situation calculus), in mathematical language and if so with deterministic functions, systems (linear) equations or probabilities, etcetera? Are probabilities sufficient or do we need a language such as Pearls do-calculus? Do we need inferential statistics to represent and understand it and if so how should we interpret the results? Through possible worlds semantics or via counterfactuals? Should we quantify, measure and then determine mean averages as operationalization of causality or causal effect (such as potential outcome approaches assume)? Can we indicate it in terms of necessary and sufficient conditions (as in INUS conditions) and if so, are those conditions finitely axiomatizable or in a finite list perceptible? In doing so, can we circumvent the frame-problem, quantification problem or ramification problem that have been notorious in AI? Is a unifying theory of causality that integrates the many different aspects and connotations possible? The multiplicity of positions has several historical roots and seems a disadvantage for the further development of the concept in research methodology, as will be outlined in the next section.

30.7 Elimination, Dualism, the Probabilistic Revolution, and Unification

The obvious role of causality in human behavior and philosophy has not prevented the concept from running into trouble throughout the course of the history of ideas. In the eighteenth century, David Hume famously challenged the (metaphysical) status of causality based on his concept empiricism. According to Hume causality does not exist in reality, there is only a constant regular sequence of phenomena, in our thoughts; due to natural habit they are associated with physical necessity in our minds, and projected on reality, thus explaining our "intuition" of necessity.

In the nineteenth century, the German physicist/philosopher Ernst Mach and the English philosopher/statistician Karl Pearson manifested themselves as prominent anti-causalists. The main critic, however, appeared in the twentieth century. It was the aforementioned Bertrand Russell who published some notes in The Proceedings of the Aristotelian Society in 1913, entitled "On the Notion of Cause" and said, "The law of causality, I believe, as much as passes by philosophers, is a Relic of a bygone age, surviving, like the monarchy, only because it is erroneously supposed to do no harm." Now, this attitude to causality is not unique. Over time, many concepts and concepts and related philosophical themes have been discredited or have at least given a specific abstract or mathematical term. This applies, inter alia, to notions such as space, time, movement, intentionality, but also to concepts such as meaning, sense of mind, mind, free will, consciousness and personal identity (Starmans 2011).

Put roughly this is due to current scientism. According to some scientists and naturalist philosophers, all philosophical problems will ultimately be unraveled and unraveled by science. If the problem is well defined then it will be analyzed and ultimately resolved. If it is not well defined, it is identified as a pseudo-problem or as meaningless and pushed aside. I will evaporate in the fullness of time. Philosophical reflection can at most lead to some kind of pre-scientific theories that can have some explanatory power or practical utility, but eventually they will be replaced by true scientific knowledge. The concepts and concepts that play a role in that philosophical reflection will usually also have to vanish.

This tendency to purify science and her language from metaphysical concepts, common sense notions, natural categories and everyday experiences has a notorious high point in philosopher and neuroscientist Paul Churchland's views, who wanted to break radically with a tradition sometimes pejoratively called "folk psychology." People try to understand, explain and predict the behavior of oneself and others in terms of causal relevant factors, such as motives, intentions, beliefs and obligations. Churchland advocates a radical eliminative materialism about these propositional attitudes, and argues that "folk psychology" including the notion of consciousness, the human mind and its internal processes is completely inaccurate (Starmans 2011). Likewise, he deplores the preoccupations of philosophers with language and its postulated crucial significance for thinking. According to Churchland, developments in neurosciences will lead to the elimination of these "maladies," which he considers equally relevant to science as the eighteenth century phlogiston theory is relevant to modern chemistry, or medieval views on witchcraft are relevant to contemporary psychology. The extreme vision of Churchland does not in any way form the "communis opinion" among philosophers, but is not a new phenomenon either. It fits into a prolonged and impressive development in Western idea history that began with the pre-Socratics and reached a peak in today's naturalistic/physicalist epistemology. As a result, the current worldview has lost much of its intuitive appeal. On the one hand, there is the everyday, familiar life of the phenomena, leading to experiences (perceptions, impressions), representations and ideas and her (postulated) concrete, material objects. On the other hand, there is the scientific worldview with its abstract, often mathematical models, representations of the "real" world, that is supposed to hide behind these experiences and which is supposed to *cause* them and to *explain* them.

This type of dualism is already widely found in the Greek atomism of Demokritos, Leukippos and Epicurus and became manifest in a dominant way in the works of Descartes and John Locke in the seventeenth century. They distinguished *primary* characteristics inherent in matter and form, size, location, movement of indivisible particles, and *secondary* characteristics such as colors, scents, sounds that actually only exist in the human mind. The gap between both is apparently greater than ever before. Not only are these phenomena as perceived to us not a reliable basis for scientific theories, but the intuitive concepts and natural categories do not seem to be in accordance with the underlying mechanisms, abstract principles and laws of the "real" world, as described *by the language* and the nomenclature of science.

In fact, the entire evolution of the notion of causality must be situated and understood in view of this tension. First of all, causality also came to the forefront of attempts for elimination, precisely because of its prominent place in metaphysics. When a philosophical position was criticized or a speculative theory was dismantled, the accompanying notion of causality was also convicted forthwith. A salient example of this is Aristotle himself, who introduced the causa formalis and materialis to underpin his hulemorphism, which connects form and matter. In addition, the causa finalis was crucial to his views on teleology/intentionality in nature and cosmos.

Then in thirteenth century scholastics, the concept was reinterpreted and adjusted predominantly by Aquinos, in order to make it suitable for his famous integration of Aristotelianism and Christianity. But then the physics and metaphysics of Aristotle were discredited in the seventeenth century, criticized by Bacon, Descartes, Hobbes and Huygens. As a result only causa efficiens was deemed to be acceptable. This was done in such a rigid and overhasty way, that it still effects the problem of causality today. However, Bacon also made the notion of causality subordinate to his empiricism, and assumed some kind of "causative forms" that could be derived directly from the observations. Descartes, in turn, embraced a strict mechanistic philosophy, in which the causal effect of production could only be explained by colliding particles, direct contact and instantaneous effects, and no "actio in distans" was allowed, which of course was pivotal in the successful anticartesian physics of Newton. For Spinoza, causality was, above all, a logical notion. But all these visions on causality of Bacon, Descartes and Spinoza, which did not allow any mathematical description of reality, were no longer applicable after the successes of classical mechanics and the consequent rise of Laplace's determinism.

After Hume's skepticism, rather based on his concept-empiricism than on a metaphysical position, it was in the eighteenth century, predominantly Kant who sought to restore the notion and strived for a reconciliation between metaphysics and natural science by considering causality as synthetic a priori knowledge as a basic category. This constitutes a condition for observations and scientific knowledge. However with the anti-Kantian philosophy of logical positivists in the twentieth century, this conception of causality was also argued to be problematic, not the least because of the successes of quantum mechanics, in which causality was no longer considered a Kantian building block of reality.

This list can easily be extended. The history cogently shows how time after time causality ran into trouble due to its metaphysical roots or its intrinsic *connections with specific philosophical positions*. As stated before, the concept has been reinvented time after time again, but since many positions were not actually falsified and replaced, the ideas remained evolving in nowadays *pluralistic view on causality*. So the preoccupations of contemporary philosophers with the principle of causality and the attempts to capture the essence of the causal relationship are by no means merely historical. Aspects of their considerations and interpretations still resemble modern, sophisticated approaches of causality and even in computational causal modeling techniques.

The rule of regularity, which reflects Hume's perception of sequence, proximity and constant convergence, the counterfactual approach (also Hume), the aspect of continuity and mechanism of action (Descartes) or the intervention interpretation (Stuart Mill) are just a few examples. For striving to bridge the gap between levels II and III especially Stuart Mill should be credited. When philosophers in the nineteenth century started thinking about experimental research, they needed a new constructive notion of causality, which could be detached from both the old metaphysics and the laplacian determinism. Attempts to anchor causality in a more pragmatic and experimental context can be found in the work of John Stuart Mill, who considered causality from the whole of circumstances that had to be known, controlled or manipulated to establish causal relationships or a causal effect. His *System of Logic* (1843) contains the famous "Five Methods of Mill" and was actually a methodological handbook avant la lettre, in which the author tried to bridge the gap between abstract epistemology and actual science practice, especially on thinking about causality. Illustrious predecessors like Aristotle with his "Organon," Bacon with his "Novum Organum," and Arnauld and Nicole with their "Logic of Port Royal" were definitely less successful. However, the real improvement would require a probabilistic revolution, incorporating probability theory and statistics. Nowadays nearly all approaches of causality are probabilistic and without the probabilistic revolution the concept would never have gained a place in research methodology.

Thinking about variation and change would be accelerated later in the nineteenth century, especially due to the evolutionary theory of Darwin and Wallace, which of course had no mathematical model for evolution and no adequate mechanism of inheritance at all. It was Francis Galton, who, as an inadmissible advocate for mathematization of science, realized that variation indeed allowed for a mathematical treatment, without pejorative manifestations of abnormalities, error functions, etc. Galton, who first developed the conception of "regression-to-the-mean" was not so much concerned with Quetelets averages, but focused on the deviant, the special and the individual, although he didn't know exactly how small long-term changes could suffice to bring about the wealth and variety of phenomena and evolution.

His pupil Pearson, on the other hand, corrected all this by not identifying the variation in "errors" but in the phenomena themselves (encoded in data), and by distinguishing different (classes of) distributions. He saw that many phenomena were not normal, but skewed and could be classified using four parameters (mean, standard deviation, skewness and curvature). The variability in nature manifested itself in a point cloud of measurements, and Pearson sought the best fitting model, the function that best suited the data. This was rather a parsimonious description of the phenomena, than an indication of a causal underlying mechanism.

In the first place, he gave probabilities a full place in science and saw the world at a level of abstraction, where data, data variation, data-generating mechanisms and parameters of the distributions rather code and build reality, rather than represent or depict an alleged physical world. Probability distributions with associated parameters are the objects of science. The worldview, though, lost some its direct comprehensibility, but paradoxically, the Pearsonian reality as statistical distribution was "observable," close to the data, and above all notable. It was a description of the actual data, a large but finite subset of the collection of all possible measurements, available only in an "ideal" situation (Starmans 2011). With Galtons and Pearson statistical approach to variation and change and Fisher's later synthesis of evolution and mendelian genetics, a major transformation in thinking about uncertainty was a fact. Here we must confine ourselves to the statement that the dualist position of Fisher benefited thinking about causality, whereas the monist position of Pearson made causality obsolete. So far causality has survived Churchlands eliminativism, Von Neumann's eroded models and Andersons pondering on the end of theory, paradoxically *despite and due* to the fact that it appeared into certain forms of dualism which are still courant and needed. In fact causality did even better than that. As an antipode or perhaps even as an antidote, we must refer to the computational turn that was made in causality thinking at the end of the twentieth century. As a result, scientists gained more grip on this former tricky concept.

In 2012, American AI scientist Judea Pearl won the most important scientific prize in computer science, the Turing Award for his work on probabilistic networks and causal modeling. Published in 2000 and revised in 2009, *Causality: Models, Reasoning and Inference* can be considered as a milestone in modern thinking about causality (Pearl 2009a). What's more Pearl considers it a unifying concept in science. In fact, in his introduction he considers "causal relationships to be the fundamental building blocks both of reality and of human understanding of that reality." There can be no such thing as reductionism to probability, statistics or algorithmic. On the contrary, Pearl knows "no greater impediment to scientific progress than the prevailing practice of focusing all of our mathematical resources on probabilistic and statistical inferences" precisely because in this tradition, according to Pearl, causal notions and nomenclatures are banned from the formal scientific language.

He therefore expresses the hope that after the twentieth century of probability, the twenty-first century will show a new era of causality. Pearl pursues unification and in these endeavors his framework of structural equation modeling combined with probabilistic networks plays a crucial role. In fact, his method partly finds its roots in the works of Sewall Wright, who established almost 100 years ago his approach with path coefficient, systems of linear equations, modeling the relationships between the different endogenous and exogenous variables, direct and indirect effects etcetera. Pearl rejects the "symmetric" mathematical language and introduces his famous "do-operator" to prevent confounding and related problems such as Simpsons paradox. It goes without saying that rather anti-statistical position both enriched and complicated the view on the role of causality and its manifestations in statistics, data science and research methodology.

30.8 Conclusion

In this essay we have tried to give shape to the suggested predicament of truth, a concept that is part of the homo mensura and pivotal in epistemology in the sense that it is presupposed by less fundamental notions, including the concept of causality, which unmistakably is more fashionable. We highlighted several aspects of truth, among other things its fragile character in the big evil world, its problematic relation with statistics throughout the development of the discipline, varying from backlash in nineteenth century popular culture to nowadays distrust and skepticism in a period that nearly all sciences have taken a probabilistic turn. We also depicted the erosion of the concept of models, nowadays eclecticism in statistical practice, although textbooks wrongly suggest a united field, and, last but not least, the upcoming anti-statistical stance, currently gaining popularity in data science.

It was shown that some aspects of TMLE do address these issues in a systematic way. Furthermore, we outlined how this anti-statistical stance may turn into an anti-epistemic stance, which appeared to be more encompassing than just "the end of theory." It appeared that a reconciliation between epistemology and research methodology is in many ways problematic and also that the solution is unlikely to be found in the philosophy of science alone. Starting with classical skepticism, shifting truth-definitions, logical positivistic preoccupation with the distinction between context of discovery and context of justification, externalist historical studies, cultural relativism and social constructivism, all positions challenged the rationality and objectivity of science or at the least strengthened the decline of the concept of truth or proclaiming its redundancy. Unsurprisingly research methodology has different, mainly pragmatic priorities. Subsequently, we showed how attempts to counter this development by bridging the gap between statistics and philosophy of science are so far problematic. We generalized this by outlining a regrettable triptych in the quest for truth with often unnecessary demarcations between levels I, II and III. Because the constellation of issues involving the concept of truth nowadays often transpire in an indirect fashion and figurate persistently in contemporary topics in the philosophy of science we focused on the concept of causality from a historical-philosophical perspective. Causal notions and nomenclature continue to play a crucial role in the language of science. It is dominant at levels I, II and III, but the burgeoning literature on causality cogently illustrates the unfortunate depicted compartmentalization. Causality has a long tradition in philosophy, underwent many conceptual changes, but has always been context-sensitive, closely linked to epistemic or meta-physical positions. As such it was reinvented time after time again, influenced by and closely tied to the metaphysical and epistemic views of the relevant thinker. Since in philosophy theories or views are seldom falsified and replaced, many conflicting ideas endured, evolving in nowadays pluralistic view on causality, which does not seem to allow for any sophisticated unification.

This obviously poses challenges for epistemology, many philosophical issues have not even been listed here, let alone solved. The fact is that many old metaphysical and epistemic connotations still play their part. However, as pointed out in Starmans (2016a, 2017a) any attempt to solve the philosophical and practical problems surrounding causality, should take into account its current state, which is largely determined by some important transformations in the history of ideas. These include, the probabilistic revolution of Karl Pearson and Ronald Fisher without which the concept of causality would have no status in research methodology whatsoever, in addition the Copenhagen interpretation of quantum mechanics of Niels Bohr, that has decisively influenced the thinking about causality, but also communication theory of Claude Shannon whose ideas and velvet revolution on information as data in context and flux, surpassed materialism, making information no longer an epiphenomena or derived notion, which appears to be particularly of interest in view of the themes addressed in this essay. Opposed to strong protagonists like Pearl we observe a tendency to eliminate the term throughout history, including Pearson, Mach, Russell, Churchland, Anderson and Havely. Yet, Google has also taken a different position for example in the project of the "automated statistician," (Starmans 2015b) arguing that precisely with huge complex models analyzing automatic thousands of variables, people can only hold confidence in the statistical conclusions and in decisions based when understanding the mechanism, the arguments, including a causal structure and cause-effect relationships. Again, it turns out that the classical themes explanations, causality and (natural) laws should not be treated separately here. In addition it must be noticed that in view of the backlash against statistics, its cumbersome relation with ethics, the appeal to the moral dimension and the homo mensura, these days also the "dark side" of data science is rapidly gaining attention, for example in the before mentioned recent book (O'Neil 2016).

As pointed out by van der Laan and Rose (2011), in TMLE the issue of causality is treated differently. Causality is subordinate to *truth* and *inference*, there is a strict distinction between a purely statistical target parameter and a causal effect. In order to assign a causal *interpretation* to a statistical parameter under the Neyman-Rubin causal network several assumptions are required, such as positivity (i.e. every entity should have a nonzero probability of receiving either exposure within the strata), the assumption of no confounders and the so-called stable unit treatment value assumption (i.e., the exposure status of an entity should not affect the potential outcomes of the other entities and the exposure level should be the same for all individuals exposed at that level). The focus is primarily on causal *estimation*, on clearly formulating the causal question of interest and estimating well-defined statistical parameters, guided by the research question and using flexible algorithms that make minimal assumptions with respect to functional form, enabling proper adjustment for confounding.

Many subtle points are only slightly addressed or had to remain untouched here. For example, acknowledging the social and external aspect of the scientific enterprise does not imply slipping into relativism or social constructivism, even in experimental physics as has been pointed out by the physicist Peter Galison on many occasions (Galison 1987; Starmans 2017c). The same applies to the problematic relation between philosophy and statistics, the lack of an integrated unified approach in data-analysis and the necessity to rebut the depicted "new vision" on science. All in all, it could be contended that the aforementioned issues regarding truth, causality, data science and the triptych of knowledge and their problematic relations are not only big challenges for epistemology and statistics, but a first step towards a genuine philosophy of data science as well.

Appendix A
Foundations

A.1 Data-Adaptive Target Parameters

We present a proof of asymptotic linearity of the CV-TMLE of a data-adaptive target parameter. During this proof we encounter various key assumptions that must be met. After having carried out this proof we collect these assumptions and provide the formal theorem. Recall the notation of our chapter on data-adaptive target parameters.

A.1.1 Statistical Inference Based on the CV-TMLE

Let's now proceed with the analysis of this CV-TMLE $\psi_n^* = \frac{1}{V} \sum_{v=1}^{V} \Psi_{\hat{d}(P_{n,v^c})}(Q_{n,v^c}(\epsilon_n))$ of $\psi_{0n} = \frac{1}{V} \sum_{v=1}^{V} \Psi_{\hat{d}(P_{n,v^c})}(Q_0)$. The identity $\Psi_d(Q) - \Psi_d(Q_0) = -P_0 D_d^*(Q, G) + R_{0,d}(Q, G, Q_0, G_0)$ for the target parameter Ψ_d and its canonical gradient $D_d^*(Q, G)$ immediately translates into the following key identity for the CV-TMLE:

$$\frac{1}{V} \sum_{v=1}^{V} P_0 D_{\hat{d}(P_{n,v^c})}^*(Q_{n,v^c}(\epsilon_n), g_{n,v^c}) = \frac{1}{V} \sum_{v=1}^{V} \Psi_{\hat{d}(P_{n,v^c})}(P_0) - \psi_n^*$$
$$+ \frac{1}{V} \sum_{v=1}^{V} R_{0,\hat{d}(P_{n,v^c})}(Q_{n,v^c}(\epsilon_n), Q_0, g_{n,v^c}, g_0).$$

Combined with the cross-validated empirical mean of the efficient influence curve equation $\frac{1}{V} \sum_v P_{n,v} D_{\hat{d}(P_{n,v^c})}^*(Q_{n,v^c}(\epsilon_n), g_{n,v^c}) = 0$, this establishes the following identity:

$$\psi_n^* - \psi_{0,n} = \frac{1}{V} \sum_{v=1}^{V} (P_{n,v} - P_0) D_{\hat{d}(P_{n,v^c})}^*(Q_{n,v^c}(\epsilon_n), g_{n,v^c})$$
$$+ \frac{1}{V} \sum_{v=1}^{V} R_{0,\hat{d}(P_{n,v^c})}(Q_{n,v^c}(\epsilon_n), Q_0, g_{n,v^c}, g_0).$$

Regarding the empirical process term we have the following lemma.

© Springer International Publishing AG 2018
M.J. van der Laan, S. Rose, *Targeted Learning in Data Science*,
Springer Series in Statistics, https://doi.org/10.1007/978-3-319-65304-4

Lemma A.1. *Assume that the supremum norm of* $D^*_{\hat{d}(P_{n,v^c})}(Q_{n,v^c}(\epsilon_n), g_{n,v^c}))$ *is bounded by some $M < \infty$ with probability tending to 1, and that*

$$P_0\{D^*_{\hat{d}(P_{n,v^c})}(Q_{n,v^c}(\epsilon_n), g_{n,v^c}) - D^*_{d_0}(Q^{d_0}, g^{d_0})\}^2 \to 0 \text{ in probability.}$$

Then,

$$\frac{1}{V}\sum_{v=1}^{V}(P_{n,v} - P_0)D^*_{\hat{d}(P_{n,v^c})}(Q_{n,v^c}(\epsilon_n), g_{n,v^c}) = (P_n - P_0)D^*_{d_0}(Q^{d_0}, g^{d_0})$$
$$+ o_P(1/\sqrt{n}).$$

Thus, under this very mild consistency condition, we have

$$\psi^*_n - \psi_{0n} = (P_n - P_0)D^*_{d_0}(Q^{d_0}, g^{d_0}) + o_P(1/\sqrt{n})$$
$$+ \frac{1}{V}\sum_{v=1}^{V}R_{0,\hat{d}(P_{n,v^c})}(Q_{n,v^c}(\epsilon_n), Q_0, g_{n,v^c}, g_0).$$

Suppose now that $Q^{d_0} = Q_0^{d_0}$ and $g^{d_0} = g_0^{d_0}$, and

$$\frac{1}{V}\sum_{v=1}^{V}R_{0,\hat{d}(P_{n,v^c})}(Q_{n,v^c}(\epsilon_n), Q_0, g_{n,v^c}, g_0) = o_P(1/\sqrt{n}).$$

Then, it follows that

$$\psi^*_n - \psi_{0n} = (P_n - P_0)D^*_{d_0}(Q_0^{d_0}, g_0^{d_0}) + o_P(1/\sqrt{n}),$$

which completes the proof. In general, we assume $g = g_0$, and

$$\frac{1}{V}\sum_{v=1}^{V}R_{0,\hat{d}(P_{n,v^c})}(Q_{n,v^c}(\epsilon_n), Q_0, g_{n,v^c}, g_0) - \frac{1}{V}\sum_{v=1}^{V}R_{0,\hat{d}(P_{n,v^c})}(Q, Q_0, g_{n,v^c}, g_0)$$
$$= o_P(1/\sqrt{n}).$$

In many applications, due to linearity of $(Q - Q_0) \to R_{0,d}(Q, Q_0, g, g_0)$, this difference is represented by an integral involving the product of a difference $Q_{n,v^c}(\epsilon_n) - Q$ and a difference $\hat{g}(P_{n,v^c}) - g_0$. In that case, this assumption corresponds with a second-order term being $o_P(1/\sqrt{n})$, where the second-order term can typically be bounded by an L^2-norm of a difference between Q^*_{n,v^c} and Q times an L^2-norm of a difference between $\hat{g}(P_{n,v^c})$ and g_0. In addition, in this case where we only assume $g = g_0$, we also need to assume the following asymptotic linearity condition on $\hat{g}$:

$$\frac{1}{V}\sum_{v=1}^{V}R_{0,\hat{d}(P_{n,v^c})}(Q, Q_0, \hat{g}(P_{n,v^c}), g_0) = (P_n - P_0)D_g(P_0) + o_P(1/\sqrt{n}).$$

Then, we can conclude:

$$\psi^*_n - \psi_{0n} = (P_n - P_0)\{D^*_{d_0}(Q, g_0) + D_g(P_0)\} + o_P(1/\sqrt{n}).$$

This proves the following theorem.

Theorem A.1. *Let $\mathcal{D}$ be an index set for a collection of target parameters, and for each $d \in \mathcal{D}$, we have a statistical target parameters $\Psi_d : \mathcal{M} \to \mathbb{R}$. Let $\hat{d} : \mathcal{M}_{NP} \to \mathcal{D}$ be an estimator that maps an empirical distribution into estimate of a desired index $d_0 \in \mathcal{D}$, and thereby a choice of target parameter. Consider a sample-split random vector V, and for a split $V = v$, let P_{n,v^c} be the empirical distribution of parameter-generating sample and $P_{n,v}$ be the empirical distribution of the validation/estimating sample. The data-adaptive target parameter is defined as follows:*

$$\psi_{0,n} = \frac{1}{V} \sum_{v=1}^{V} \Psi_{\hat{d}(P_{n,v^c})}(P_0).$$

For each target parameter Ψ_d, let $D_d^*(P_0)$ be its efficient influence curve at P_0. Assume that $\Psi_d(P_0) = \Psi_d(Q_0^d)$ only depends on P_0 through a parameter Q_0^d, and assume that $D_d^*(P_0) = D_d^*(Q_0^d, g_0^d)$ depends on P_0 through Q_0^d and a nuisance parameter g_0^d. Define a second-order term $R_{0,d}()$ as follows:

$$P_0 D_d^*(Q^d, g^d) = \Psi_d(P_0) - \Psi_d(Q^d) + R_{0,d}(Q^d, Q_0^d, g^d, g_0^d).$$

Let $(Q^d, O) \to L^d(Q^d)(O)$ be a valid loss function for Q_0^d so that

$$Q_0^d = \arg\min_{Q^d} P_0 L^d(Q^d),$$

and let $\{Q^d(\epsilon) : \epsilon\}$ be a submodel through Q at $\epsilon = 0$ with a univariate or multivariate parameter ϵ so that the linear span of the generalized score includes the efficient influence curve:

$$D_d^*(Q^d, g^d) \in \langle \frac{d}{d\epsilon} L^d(Q^d(\epsilon)) \Big|_{\epsilon=0} \rangle.$$

Let $\{\hat{Q}^d(\epsilon) : \epsilon\}$ be this submodel through $\hat{Q}^d$, using $\hat{g}^d$. For notational convenience, we use the notation $\hat{Q}(P_{n,v^c}) = \hat{Q}^{\hat{d}(P_{n,v^c})}(P_{n,v^c})$, and similarly, we define $\hat{g}(P_{n,v^c}) = \hat{g}^{\hat{d}(P_{n,v^c})}(P_{n,v^c})$. For each split $V = v$, we define the corresponding updates $Q_{n,v^c}^* \equiv \hat{Q}(P_{n,v^c})(\epsilon_n)$. Let ϵ_n be computed so that it solves/satisfies the following equation:

$$\frac{1}{V} \sum_{v=1}^{V} P_{n,v} D_{\hat{d}(P_{n,v^c})}^*(Q_{n,v^c}(\epsilon_n), \hat{g}(P_{n,v^c})) = o_P(1/\sqrt{n}). \tag{A.1}$$

The proposed estimator of $\psi_{0,n}$ is given by

$$\psi_n^* \equiv \frac{1}{V} \sum_{v=1}^{V} \Psi_{\hat{d}(P_{n,v^c})}(Q_{n,v^c}^*).$$

Assume that the supremum norm of $D_{\hat{d}(P_{n,v^c})}^*(Q_{n,v^c}^*, \hat{g}(P_{n,v^c}))$ is bounded by some $M < \infty$ with probability tending to 1, and that

$$P_0 \{D_{\hat{d}(P_{n,v^c})}^*(Q_{n,v^c}^*, \hat{g}(P_{n,v^c})) - D_{d_0}^*(Q^{d_0}, g^{d_0})\}^2 \to 0 \text{ in probability.}$$

Then,

$$\psi_n^* - \psi_{0n} = (P_n - P_0)D_{d_0}^*(Q^{d_0}, g^{d_0}) + o_P(1/\sqrt{n})$$
$$+ \frac{1}{V}\sum_{v=1}^{V} R_{0,\hat{d}(P_{n,v^c})}(Q_{n,v^c}^*, Q_0, \hat{g}(P_{n,v^c}), g_0).$$

We assume $g = g_0;$

$$\frac{1}{V}\sum_{v=1}^{V} R_{0,\hat{d}(P_{n,v^c})}(Q_{n,v^c}^*, Q_0, \hat{g}(P_{n,v^c}), g_0) - \frac{1}{V}\sum_{v=1}^{V} R_{0,\hat{d}(P_{n,v^c})}(Q, Q_0, \hat{g}(P_{n,v^c}), g_0)$$
$$= o_P(1/\sqrt{n});$$

and the following asymptotic linearity condition on $\hat{g}$:

$$\frac{1}{V}\sum_{v=1}^{V} R_{0,\hat{d}(P_{n,v^c})}(Q, Q_0, \hat{g}(P_{n,v^c}), g_0) = (P_n - P_0)D_g(P_0) + o_P(1/\sqrt{n}).$$

Then,

$$\psi_n^* - \psi_{0,n} = (P_n - P_0)\{D_{d_0}^*(Q, g_0) + D_g(P_0)\} + o_P(1/\sqrt{n}).$$

A.2 Mediation Analysis

We establish three fundamental results for Chap. 17. Firstly, we prove the identification result for the natural direct effect, which is defined as a contrast of two
counterfactual mean outcomes under a static intervention on A_t and an unknown
stochastic intervention on Z_t. This identification result is just a consequence of (1)
the identification of the counterfactual mean outcome under a given intervention
from the g-computation formula and (2) the identification of the conditional distribution $\bar{\Gamma}_{1,t}^{a'}$ of the counterfactual mediator representing the unknown stochastic
intervention (Robins 1986). Secondly, we prove that our claimed efficient influence
curve is indeed the canonical gradient of the pathwise derivative of our statistical
target parameter. We prove this by (1) specifying a rich class of parametric submodels through P; (2) defining the tangent space (i.e., the closure of the linear span of
all the scores generated by this class of submodels) as an orthogonal sum of tangent
spaces of the different factors of the likelihood; (3) expressing the pathwise derivative of the statistical target parameter along a parametric submodel as a covariance
of the claimed efficient influence curve and the score of the submodel; (4) showing
that the claimed efficient influence curve is an element of the tangent space. Our
proof applies to any statistical model for which the conditional distributions of A_t
are modeled while all other conditional distributions of the nodes that makes up the
longitudinal data structure are kept locally nonparametric. Thirdly, we establish the
claimed double robustness of the efficient influence curve.

A.2.1 Proof of Lemma 17.1: Identifiability Result

For $X_t \in (R_t, L_t)$, we also denote $X_t\left(\mathbf{a}, \left(\bar{\Gamma}^{\mathbf{a}'}_{1,s}, \mathbf{z}_{s,t}\right)\right)$ the counterfactual covariate generated at time t by intervening to set $\mathbf{A}_t = \mathbf{a}_t, \mathbf{Z}_{1,s} \sim \bar{\Gamma}^{\mathbf{a}'}_{1,s}, \mathbf{Z}_{s,t} = \mathbf{z}_{s,t}$.

$$E(Y_\tau(\mathbf{a}, \bar{\Gamma}^{\mathbf{a}'})) = \sum_{r,l,z} y_\tau p[L_0 = l_0]$$

$$\times \prod_t \Big\{ p\Big[R_t(\mathbf{a}, \bar{\Gamma}^{\mathbf{a}'}) = r_t \mid L_0 = l_0, \mathbf{R}_{t-1}(\mathbf{a}, \bar{\Gamma}^{\mathbf{a}'}) = \mathbf{r}_{t-1}, \bar{\Gamma}^{\mathbf{a}'}_{t-1} = \mathbf{z}_{t-1}, \mathbf{L}_{t-1}(\mathbf{a}, \bar{\Gamma}^{\mathbf{a}'}) = \mathbf{l}_{t-1}\Big]$$

$$\times p\Big[\Gamma^{\mathbf{a}'}_t = z_t \mid L_0 = l_0, \mathbf{R}_t(\mathbf{a}, \bar{\Gamma}^{\mathbf{a}'}) = \mathbf{r}_t, \bar{\Gamma}^{\mathbf{a}'}_{t-1} = \mathbf{z}_{t-1}, \mathbf{L}_{t-1}(\mathbf{a}, \bar{\Gamma}^{\mathbf{a}'}) = \mathbf{l}_{t-1}\Big]$$

$$\times p\Big[L_t(\mathbf{a}, \bar{\Gamma}^{\mathbf{a}'}) = l_t \mid L_0 = l_0, \mathbf{R}_t(\mathbf{a}, \bar{\Gamma}^{\mathbf{a}'}) = \mathbf{r}_t, \bar{\Gamma}_t(\mathbf{a}') = \mathbf{z}_t, \mathbf{L}_{t-1}(\mathbf{a}, \bar{\Gamma}^{\mathbf{a}'}) = \mathbf{l}_{t-1}\Big]\Big\}.$$

By definition of $\Gamma^{\mathbf{a}'}_t$, $p\Big[\Gamma^{\mathbf{a}'}_t = z_t \mid \mathbf{R}_t(\mathbf{a}, \bar{\Gamma}^{\mathbf{a}'} = \mathbf{r}_t, \bar{\Gamma}^{\mathbf{a}'}_{t-1} = \mathbf{z}_{t-1}, \mathbf{L}_{t-1}(\mathbf{a}, \bar{\Gamma}^{\mathbf{a}'}) = \mathbf{l}_{t-1}\Big] \equiv p\left(Z_t(\mathbf{a}') = z_t \mid \mathbf{R}_t(\mathbf{a}') = \mathbf{r}_t, \mathbf{Z}_{t-1}(\mathbf{a}') = \mathbf{z}_{t-1}, \mathbf{L}_{t-1}(\mathbf{a}') = \mathbf{l}_{t-1}\right)$. This quantity arises in traditional longitudinal total causal effect problems where one sets $\mathbf{A} = \mathbf{a}'$, and measure $\mathbf{Z}$ and $\mathbf{L}$ (Robins 1986). Therefore, under A1, it is identifiable as the conditional probability $p_0\left(Z_t = z_t \mid \mathbf{A}_t = \mathbf{a}'_t, \mathbf{Z}_{t-1} = \mathbf{z}_{t-1}, \mathbf{L}_{t-1} = \mathbf{l}_{t-1}\right)$ from the observed data distribution. To identify the conditional probabilities of $R_t(\mathbf{a}, \bar{\Gamma}^{\mathbf{a}'})$ and $L_t(\mathbf{a}, \bar{\Gamma}^{\mathbf{a}'})$we demonstrate the steps for the first two, the results for the subsequent covariates can be induced thereafter. Firstly

$$p\Big[R_1(\mathbf{a}, \bar{\Gamma}^{\mathbf{a}'}) = r_1 \mid L_0 = l_0\Big] = p[R_1(a_1) = r_1 \mid L_0 = l_0]$$
$$= p[R_1(a_1) = l_1 \mid A_1 = a_1, L_0 = l_0] = p_0[R_1 = r_1 \mid A_1 = a_1, L_0 = l_0].$$

The first equality is by definition of the counterfactuals $R_1(\mathbf{a}, \bar{\Gamma}^{\mathbf{a}'})$. The second equality is due to the assumption A2 that given L_0, A_1 is independent of $R_1(\mathbf{a})$. The last equality follows from consistency. Next,

$$p\Big[L_1(\mathbf{a}, \bar{\Gamma}^{\mathbf{a}'}) = l_1 \mid L_0 = l_0, R_1(\mathbf{a}, \bar{\Gamma}^{\mathbf{a}'}) = r_1, \Gamma^{\mathbf{a}'}_1 = z_1\Big]$$
$$= p\Big[L_1(a_1, z_1) = l_1 \mid L_0 = l_0, R_1(\mathbf{a}, \bar{\Gamma}^{\mathbf{a}'}) = r_1, \Gamma^{\mathbf{a}'}_1 = z_1\Big]$$
$$= p\Big[L_1(a_1, z_1) = l_1 \mid L_0 = l_0, R_1(\mathbf{a}, \bar{\Gamma}^{\mathbf{a}'}) = r_1\Big]$$
$$= p[L_1(a_1, z_1) = l_1 \mid L_0 = l_0, R_1(a_1) = r_1]$$
$$= p_0[L_1 = l_1 \mid L_0 = l_0, A_1 = a_1, R_1 = r_1, Z_1 = z_1]$$

The first equality is by definition of the counterfactuals $L_1(\mathbf{a}, \bar{\Gamma}^{\mathbf{a}'})$ and $L_1(\mathbf{a}, \mathbf{z})$. The second equality is due to the fact that in our ideal experiment conditional on $L_0 = l_0$ and $R_1(\mathbf{a}, \bar{\Gamma}^{\mathbf{a}'}) = R_1(a_1) = r_1$, Z_1 is a random draw from the distribution $\Gamma^{\mathbf{a}'}_1(\cdot \mid l_0, r_1)$, and does not affect the covariates $L_1(a_1, z_1)$, whose value only depend on $R_1(a_1) = r_1$ and l_0. The last equality follows from the usual argument of sequential randomization under static interventions on (A, Z) by applying assumptions A2, A3. The positivity assumptions in A4 assure that the conditional probabilities in the identifying expression (17.5) are well defined.

A.2.2 Proof of Theorem 17.1

For any $P \in \mathcal{M}$, we recall the likelihood decomposition in (17.1):

$$p(O) = p(L_0)$$

$$\times \prod_{t=1}^{\tau} \Big(p(A_t \mid \mathbf{A}_{t-1}, \mathbf{R}_{t-1}, \mathbf{Z}_{t-1}, \mathbf{L}_{t-1}) p(R_t \mid \mathbf{A}_t, \mathbf{R}_{t-1}, \mathbf{Z}_{t-1}, \mathbf{L}_{t-1})$$

$$\times p(Z_t \mid \mathbf{A}_t, \mathbf{R}_t, \mathbf{Z}_{t-1}, \mathbf{L}_{t-1}) p(L_t \mid \mathbf{A}_t, \mathbf{R}_t, \mathbf{Z}_t, \mathbf{L}_{t-1}) \Big).$$

For $O_j \in \{L_0, A_t, R_t, Z_t, L_t : t = 1, \ldots, \tau\}$, let P_{O_j} denote the conditional probability of $P_{O_j}\big(O_j \mid Pa(O_j)\big)$. Let $L^2(P)$ denote the Hilbert space of mean zero functions of O, endowed with the covariance operator. Consider a rich class of one-dimensional parametric submodels $P(\epsilon)$ that are generated by only fluctuating P_{O_j}. Under our model, no restrictions are imposed on the conditional probabilities P_{O_j}. As a result, given any function $S_{O_j} \in L^2(P)$ of $(O_j, Pa(O_j))$ with finite variance and $E_P(S_{O_j}(O_j, Pa(O_j)) \mid Pa(O_j)) = 0$, the fluctuation $P_{O_j}(\epsilon) = (1 + \epsilon S_{O_j}(O_j, Pa(O_j)))P_{O_j}$ is a valid one-dimensional submodel with score S_{O_j}. Therefore, the tangent subspaces corresponding to fluctuations of each P_{O_j} are given by

$$T(P_{L_0}) = \{S_{L_0}(L_0) : E_P(S_{L_0}) = 0\}$$
$$T(P_{A_t}) = \{S_{A_t}(A_t, \mathbf{A}_{t-1}, \mathbf{R}_{t-1}, \mathbf{Z}_{t-1}, \mathbf{L}_{t-1}) : E_P(S_{A_t} \mid \mathbf{A}_{t-1}, \mathbf{R}_{t-1}, \mathbf{Z}_{t-1}, \mathbf{L}_{t-1}) = 0\}$$
$$T(P_{R_t}) = \{S_{R_t}(R_t, \mathbf{A}_t, \mathbf{R}_{t-1}, \mathbf{Z}_{t-1}, \mathbf{L}_{t-1}) : E_P(S_{R_t} \mid \mathbf{A}_t, \mathbf{R}_{t-1}, \mathbf{Z}_{t-1}, \mathbf{L}_{t-1}) = 0\}$$
$$T(P_{Z_t}) = \{S_{Z_t}(Z_t, \mathbf{A}_t, \mathbf{R}_t, \mathbf{Z}_{t-1}, \mathbf{L}_{t-1}) : E_P(S_{Z_t} \mid \mathbf{A}_t, \mathbf{R}_t, \mathbf{Z}_{t-1}, \mathbf{L}_{t-1}) = 0\}$$
$$T(P_{L_t}) = \{S_{L_t}(L_t, \mathbf{A}_t, \mathbf{R}_t, \mathbf{Z}_t, \mathbf{L}_{t-1}) : E_P(S_{L_t} \mid \mathbf{A}_t, \mathbf{Z}_t, \mathbf{R}_t, \mathbf{L}_{t-1}) = 0\}.$$

Due to the factorization in (17.1), $T(P_{O_i})$ is orthogonal to $T(P_{O_j})$ for $O_i \neq O_j$. Moreover, the tangent space $T(P)$, corresponding to fluctuations of the entire likelihood, is given by the orthogonal sum of these tangent subspaces, i.e. $T(P) = \bigoplus_j T(P_{O_j})$, and any score $S(O) \in T(P)$ can be decomposed as $\sum_j S_{O_j}(O)$.

Under this generous definition of the tangent subspaces, any function $S(O)$ that has zero mean and finite variance under P is contained in $T(P)$. This implies in particular that any gradient for the pathwise derivative of $\Psi^{a,a'}(\cdot)$ is contained in $T(P)$, and is thus in fact the canonical gradient. Therefore, it suffices to show that $D^{*,a,a'}(\cdot)$ in (17.9) is a gradient for the pathwise derivative of $\Psi^{a,a'}(\cdot)$. Indeed, for any $S(O) = \sum_j S_{O_j}(O) \in T(P)$, let $P^S(\epsilon)$ denote the fluctuation of P with score S. Under appropriate regularity conditions, the pathwise derivative at P can be expressed as

$$\frac{d}{d\epsilon} \Psi^{a,a'}(P^S(\epsilon)) \mid_{\epsilon=0}$$

$$= \frac{d}{d\epsilon}\Big|_{\epsilon=0} \sum_{r,z,l} y \Big\{ [(1 + \epsilon S_{L_0})p_{L_0}](l_0)$$

$$\times \prod_{t=1}^{\tau} \Big([(1 + \epsilon S_{R_t}) p_{R_t}] (r_t \mid \mathbf{a}_t, \mathbf{r}_{t-1} \mathbf{z}_{t-1}, \mathbf{l}_{t-1})$$

$$\times [(1 + \epsilon S_{Z_t}) p_{Z_t}] (z_t \mid \mathbf{a}'_t, \mathbf{r}_t, \mathbf{z}_{t-1}, \mathbf{l}_{t-1}) [(1 + \epsilon S_{L_t}) p_{L_t}] (l_t \mid \mathbf{a}_t, \mathbf{r}_t, \mathbf{z}_t, \mathbf{l}_{t-1}) \Big) \Big\}$$

$$= \sum_{\mathbf{r},\mathbf{z},\mathbf{l}} y \left(p_{L_0}(l_0) \mathbf{P}_R(\mathbf{r} \mid \mathbf{a}, \mathbf{z}, \mathbf{l}) \mathbf{P}_Z(\mathbf{z} \mid \mathbf{a}', \mathbf{r}, \mathbf{l}) \mathbf{P}_L(\mathbf{l} \mid \mathbf{a}, \mathbf{r}, \mathbf{z}) \sum_{t=1}^{\tau} S_{L_t} \right) \tag{A.2}$$

$$+ \sum_{\mathbf{r},\mathbf{z},\mathbf{l}} y \left(p_{L_0}(l_0) \mathbf{P}_R(\mathbf{r} \mid \mathbf{a}, \mathbf{z}, \mathbf{l}) \mathbf{P}_Z(\mathbf{z} \mid \mathbf{a}', \mathbf{r}, \mathbf{l}) \mathbf{P}_L(\mathbf{l} \mid \mathbf{a}, \mathbf{r}, \mathbf{z}) \sum_{t=1}^{\tau} S_{Z_t} \right) \tag{A.3}$$

$$+ \sum_{\mathbf{r},\mathbf{z},\mathbf{l}} y \left(p_{L_0}(l_0) \mathbf{P}_R(\mathbf{r} \mid \mathbf{a}, \mathbf{z}, \mathbf{l}) \mathbf{P}_Z(\mathbf{z} \mid \mathbf{a}', \mathbf{r}, \mathbf{l}) \mathbf{P}_L(\mathbf{l} \mid \mathbf{a}, \mathbf{r}, \mathbf{z}) \sum_{t=1}^{\tau} S_{R_t} \right) \tag{A.4}$$

$$+ \sum_{\mathbf{r},\mathbf{z},\mathbf{l}} y \left(p_{L_0}(l_0) \mathbf{P}_R(\mathbf{r} \mid \mathbf{a}, \mathbf{z}, \mathbf{l}) \mathbf{P}_Z(\mathbf{z} \mid \mathbf{a}', \mathbf{r}, \mathbf{l}) \mathbf{P}_L(\mathbf{l} \mid \mathbf{a}, \mathbf{r}, \mathbf{z}) S_{L_0} \right), \tag{A.5}$$

where $\mathbf{P}_R(\mathbf{r} \mid \mathbf{a}, \mathbf{z}, \mathbf{l}) \equiv \prod_{t=1}^{\tau} p_R(r_t \mid \mathbf{A}_t = \mathbf{a}_t, \mathbf{r}_{t-1}, \mathbf{z}_{t-1}, \mathbf{l}_{t-1})$, analogously for $\mathbf{P}_L$, and $\mathbf{P}_Z(\mathbf{z} \mid \mathbf{a}, \mathbf{r}, \mathbf{l}) \equiv \prod_{t=1}^{\tau} p_Z(z_t \mid \mathbf{A}_t = \mathbf{a}'_t, \mathbf{r}_t, \mathbf{z}_{t-1}, \mathbf{l}_{t-1})$.

Note firstly that by definition of D_t^L and S_{L_t}, for every $t = 1, \ldots, t_0$,

$$E_P \left(D_{L_t}^{\mathbf{a},\mathbf{a}'}(\mathbf{R}_t, \mathbf{Z}_t, \mathbf{L}_t) S_{L_t}(\mathbf{A}_t, \mathbf{R}_t, \mathbf{Z}_t, \mathbf{L}_t) \right)$$
$$= \sum_{\mathbf{r},\mathbf{z},\mathbf{l}} y \ p_{L_0}(l_0) \mathbf{P}_R(\mathbf{r} \mid \mathbf{a}, \mathbf{z}, \mathbf{l}) \mathbf{P}_Z(\mathbf{z} \mid \mathbf{a}', \mathbf{r}, \mathbf{l}) \mathbf{P}_L(\mathbf{l} \mid \mathbf{a}, \mathbf{r}, \mathbf{z}) S_{L_t}.$$

Moreover, $D_{L_t}^{\mathbf{a},\mathbf{a}'}(P)(\mathbf{R}_t, \mathbf{Z}_t, \mathbf{L}_t)$ satisfies $E_P \left(D_{L_t}^{\mathbf{a},\mathbf{a}'}(P)(\mathbf{R}_t, \mathbf{Z}_t, \mathbf{L}_t) \mid \mathbf{R}_t, \mathbf{Z}_t, \mathbf{L}_{t-1} \right) = 0.$
Therefore $D_{L_t}^{\mathbf{a},\mathbf{a}'}(P) \in T(P_{L_t|Pa(L_t)})$ by the definition of these tangent subspaces. It thus follows from the orthogonal decomposition of $T(P)$ that (P)

$$E_P \left\{ D_{L_t}^{\mathbf{a},\mathbf{a}'}(P) \times S_{L_t} \right\} = E_P \left\{ D_{L_t}^{\mathbf{a},\mathbf{a}'}(P) \left(S_{L_0} + \sum_{t=1}^{\tau} S_{A_t} + S_{Z_t} + S_{L_t} \right) \right\}.$$

By similar arguments, (A.3) can be written as

$$E_P \left\{ D_{Z_t}^{\mathbf{a},\mathbf{a}'}(P) \times S_{Z_t} \right\} = E_P \left\{ D_{Z_t}^{\mathbf{a},\mathbf{a}'}(P) \left(S_{L_0} + \sum_{t=1}^{\tau} S_{A_t} + S_{Z_t} + S_{L_t} \right) \right\},$$

(A.4) can be written as

$$E_P \left\{ D_{R_t}^{\mathbf{a},\mathbf{a}'}(P) \times S_{R_t} \right\} = E_P \left\{ D_{R_t}^{\mathbf{a},\mathbf{a}'}(P) \left(S_{L_0} + \sum_{t=1}^{\tau} S_{A_t} + S_{Z_t} + S_{L_t} \right) \right\}.$$

and (A.5) can be written as

$$E_P\left\{D_{L_0}^{\mathbf{a},\mathbf{a}'}(P)\times S_{L_0}\right\}=E_P\left\{D_{L_0}^{\mathbf{a},\mathbf{a}'}(P)\left(S_{L_0}+\sum_{t=1}^{\tau}S_{A_t}+S_{Z_t}+S_{L_t}\right)\right\}.$$

Combining these results, one concludes that

$$\frac{d}{d\epsilon}\,\psi^{\mathbf{a},\mathbf{a}'}(P^S(\epsilon))\mid_{\epsilon=0}=$$

$$E_P\left\{\left(D_{L_0}^{\mathbf{a},\mathbf{a}'}(P)+\sum_{t=1}^{\tau}D_{R_t}^{\mathbf{a},\mathbf{a}'}(P)+D_{Z_t}^{\mathbf{a},\mathbf{a}'}(P)+D_{L_t}^{\mathbf{a},\mathbf{a}'}(P)\right)S\right\}$$

Therefore, $D^{*,\mathbf{a},\mathbf{a}'}(P)\equiv D_{L_0}^{\mathbf{a},\mathbf{a}'}(P)+\sum_{t=1}^{\tau}D_{R_t}^{\mathbf{a},\mathbf{a}'}(P)+D_{Z_t}^{\mathbf{a},\mathbf{a}'}(P)+D_{L_t}^{\mathbf{a},\mathbf{a}'}(P)$ is a gradient for the pathwise derivative of $\psi^{\mathbf{a},\mathbf{a}'}$ at P. As discussed above, under the nonparametric model, $D^{*,\mathbf{a},\mathbf{a}'}(P)$ is in fact the canonical gradient.

Double Robustness of Efficient Influence Curve. The first case of the robustness condition is trivial. In the second case, correct p_L and p_R yield that $E_{P_0}D_{R_t}^{\mathbf{a},\mathbf{a}'}(P)=0$ and $E_{P_0}D_{L_t}^{\mathbf{a},\mathbf{a}'}(P)=0$; correct p_L, p_R and p_A produce a telescopic sum over t of $D_{Z_t}^{\mathbf{a},\mathbf{a}'}$. Specifically

$$E_{P_0}\sum_t D_{Z_t}^{\mathbf{a},\mathbf{a}'}(P)=$$

$$\sum_{\mathbf{l}_{\tau-1},\mathbf{r},\mathbf{z}}\bar{Q}_{L_\tau}(P_0)^{\mathbf{a},\mathbf{a}'}(\mathbf{r},\mathbf{z},\mathbf{l}_{\tau-1})p_{0,L_0}(l_0)\mathbf{P}_{0,R}(\mathbf{r}\mid\mathbf{a},\mathbf{z},\mathbf{l}_{\tau-1})\mathbf{P}_{0,Z}(\mathbf{z}\mid\mathbf{a}',\mathbf{r},\mathbf{l}_{\tau-1})\mathbf{P}_{0,L}(\mathbf{l}_{\tau-1}\mid\mathbf{a},\mathbf{r},\mathbf{z})$$

$$-\,E_{P_0}\bar{Q}_{Z_1}^{\mathbf{a},\mathbf{a}'}(\mathbf{R}_1,L_0)$$

$$=\psi^{\mathbf{a},\mathbf{a}'}(P_0)-E_{P_0}\bar{Q}_{Z_1}^{\mathbf{a},\mathbf{a}'}(\mathbf{R}_1,L_0)$$

On the other hand, $E_{P_0}D_{L_0}^{\mathbf{a},\mathbf{a}'}(P)=E_{P_0}\sum_{r_1}p_{0,R}(r_1\mid l_0,a_1)\bar{Q}_{Z_1}^{\mathbf{a},\mathbf{a}'}(r_1,L_0)-\psi^{\mathbf{a},\mathbf{a}'}(P)$. Therefore, $E_{P_0}D^{*,\mathbf{a},\mathbf{a}'}(P)=0$ implies that

$$0=E_{P_0}D^{*,\mathbf{a},\mathbf{a}'}(P)=E_{P_0}\sum_t D_{Z_t}^{\mathbf{a},\mathbf{a}'}(P)+E_{P_0}D_{L_0}^{\mathbf{a},\mathbf{a}'}(P)$$

$$=\psi^{\mathbf{a},\mathbf{a}'}(P_0)-E_{P_0}\bar{Q}_{Z_1}^{\mathbf{a},\mathbf{a}'}(\mathbf{R}_1,L_0)+E_{P_0}\sum_{r_1}p_{0,R}(r_1\mid l_0,a_1)\bar{Q}_{Z_1}^{\mathbf{a},\mathbf{a}'}(r_1,L_0)-\psi^{\mathbf{a},\mathbf{a}'}(P)$$

$$=\psi^{\mathbf{a},\mathbf{a}'}(P_0)-\psi^{\mathbf{a},\mathbf{a}'}(P).$$

Similarly, in the third case, correct p_Z yields $P_0D_t^{Z_t,\mathbf{a},\mathbf{a}'}(P)=0$; the correct p_A and correct p_Z will produce the desired telescopic sums over t of $E_{P_0}D_{L_t}^{\mathbf{a},\mathbf{a}'}(P)+E_{P_0}D_{R_t}^{\mathbf{a},\mathbf{a}'}(P)$, leaving

$$0=E_{P_0}D^{*,\mathbf{a},\mathbf{a}'}(P)=\sum_t\left\{E_{P_0}D_{L_t}^{\mathbf{a},\mathbf{a}'}(P)+E_{P_0}D_{R_t}^{\mathbf{a},\mathbf{a}'}(P)\right\}+E_{P_0}D_{L_0}^{\mathbf{a},\mathbf{a}'}(P)$$

$$=\psi^{\mathbf{a},\mathbf{a}'}(P_0)-E_{P_0}\bar{Q}_{R_1}^{\mathbf{a},\mathbf{a}'}(L_0)+E_{P_0}E_{P_0}\bar{Q}_{R_1}^{\mathbf{a},\mathbf{a}'}(L_0)-\psi^{\mathbf{a},\mathbf{a}'}(P)$$

$$=\psi^{\mathbf{a},\mathbf{a}'}(P_0)-\psi^{\mathbf{a},\mathbf{a}'}(P).$$

A.3 Online Super Learning

In this section, we will start out with showing that the online cross-validated risk minus its desired target is a discrete martingale, and present a theorem from the literature that provides an exponential inequality for the tail probability of such discrete martingale. Subsequently, we present the proof of Theorem 18.1. The proof of Theorem 18.2 is much easier and is therefore omitted. In the final section, we provide a succinct review of the literature on stochastic gradient descent algorithms for online estimation.

A.3.1 Online Cross-Validated Risk Minus Online Cross-Validated True Risk Is a Discrete Martingale

The difference between the online cross-validated risk and the online cross-validated true risk (minimized by oracle selector) can be written as a martingale as follows:

$$(n - n_l + 1)\{R_{CV,n}(\hat{\Psi}_k) - \tilde{R}_{CV,n}(\hat{\Psi}_k)\} =$$
$$\sum_{t_0=n_l+1}^{n}\{L(\hat{\Psi}_k(P_{t_0-1}))(Z(t_0), O(t_0)) - L(\psi_0)(Z(t_0), O(t_0))\}$$
$$- \sum_{t_0=n_l+1}^{n} P_{\theta_0,t_0,Z(t_0)}\{L(\hat{\Psi}_k(P_{t_0-1})) - L(\psi_0)\}$$
$$= \sum_{t_0=n_l+1}^{n}\{f(t_0, \bar{O}(t_0 - 1), O(t_0)) - E_0(f(t_0, \bar{O}(t_0 - 1), O(t_0)) \mid \bar{O}(t_0 - 1))\}$$
$$\equiv M_n(f),$$

where

$$f(t_0, \bar{O}(t_0 - 1), O(t_0)) = L(\hat{\Psi}_k(P_{t_0-1}))(Z(t_0), O(t_0)) - L(\psi_0).$$

Clearly, for $k < n$, $E_0(M_n(f) \mid \bar{O}(k)) = M_k(f)$, which proves that $(M_n(f) : n = n_l + 1, \ldots)$ is a discrete martingale in n.

A.3.2 Martingale Exponential Inequality for Tail Probability

In order to establish an oracle inequality for the online cross-validation selector based on data $O(1), \ldots, O(n)$, we need an exponential inequality for tail-probabilities of Martingale sums $M_n(f)$. For that purpose, we refer to Theorem 8 (page 40) in Chambaz and van der Laan (2011a) for the following exponential inequality for Martingales established by van Handel (2009).

Theorem A.2 (Proposition A2 in van Handel 2009). *For the sake of this theorem, let* $M_n(f) = \sum_{i=1}^{n} f(i, O(i), \bar{O}(i - 1)) - P_{\theta_0,i,Z(i)}f$, $P_{\theta_0,i,Z(i)}$ *denoting the conditional probability distribution of* $O(i)$, *given* $Z(i)$, *and let* $\mathcal{F}$ *be a set of such functions* f. *Fix* $K > 0$ *and define, for all* $f \in \mathcal{F}$, $n \geq 1$,

$$\tilde{R}_{n,K}(f) = \frac{2K^2}{n} \sum_{i=1}^{n} P_{\theta_0,i,Z(i)}\phi\left(\frac{|f|}{K}\right),$$

where $\phi(x) \equiv \exp(x) - x - 1$. There exists a universal constant $C > 0$ (e.g, $C = 100$ works) such that, for any $n \geq 1$, $R > 0$,

$$P\left(\sup_{f \in \mathcal{F}} I(\tilde{R}_{n,K}(f) \leq R)\frac{M_n(f)}{n} \geq x\right) \leq 2\exp\left\{-\frac{nx^2}{C^2(c_1+1)R}\right\}$$

for any $x, c_0, c_1 > 0$ satisfying $c_0^2 \geq C^2(c_1 + 1)$ and

$$\frac{c_0}{\sqrt{n}}\int_0^{\sqrt{R}}\sqrt{H(\mathcal{F}, \|\cdot\|_\infty, \epsilon)}d\epsilon \leq x \leq \frac{c_1 R}{K}.$$

Here $H(\mathcal{F}, \|\cdot\|_\infty, \epsilon) = \log(1 + N(\mathcal{F}, \|\cdot\|_\infty, \epsilon))$ is the so called entropy function w.r.t. supremum norm and $N(\mathcal{F}, \|\cdot\|_\infty, \epsilon)$ is the covering number defined as the number of balls with radius ϵ that is needed to cover $\mathcal{F}$.

For specified c_0 and c_1, satisfying $c_0^2 \geq C^2(c_1 + 1)$, R, for x larger than $c_0 E/\sqrt{n}$ and smaller than $c_1 R/K$, the above exponential inequality applies, where $E = \int_0^{\sqrt{R}}\sqrt{H(\mathcal{F}, \|\cdot\|_\infty, \epsilon)}d\epsilon$. On this interval of x-values we have $x \leq c_1 R/K$, which implies $c_1 \geq xK/R$. Therefore, we can restate the above result as follows: For a specified R, c_0, c_1 satisfying $c_0^2 \geq C^2(c_1 + 1)$, and $x \in (c_0/\sqrt{n}E, c_1 R/K)$, we have,

$$P\left(\sup_{f \in \mathcal{F}} I(\tilde{R}_{n,K}(f) \leq R)\frac{M_n(f)}{n} \geq x\right) \leq 2\exp\left\{-\frac{nx^2}{C^2(Kx+R)}\right\}.$$

In words, one can conclude that the above inequality shows that for x of the order $1/\sqrt{n}$, the tail probability behaves as $\exp(-nx^2)$, while for large x, it behaves as $\exp(-nx)$.

Specifically, for a single f, we obtain the following corollary.

Corollary A.1. *For any $c_0, c_1 \geq 0$ satisfying $c_0^2 \geq C^2(c_1 + 1)$ and $x \in (c_0/\sqrt{n}\sqrt{R}, c_1 R/K)$, we have*

$$P\left(I(\tilde{R}_{n,K}(f) \leq R)\frac{M_n(f)}{n} \geq x\right) \leq 2\exp\left\{-\frac{nx^2}{C^2(Kx+R)}\right\}. \tag{A.6}$$

In our proof f plays the role of $L(\psi) - L(\psi_0)$. Regarding bounding $\tilde{R}_{n,K}(f)$, note also that if $\|f\|_\infty < C$ is uniformly bounded, then $\tilde{R}_{n,K}(f)$ is also bounded by a constant depending on C. In our proof for quadratic loss functions we need to bound $\tilde{R}_{n,K}(f)$ in terms of $\frac{1}{n}\sum_{i=1}^n P_{\theta_0,i,Z(i)}f$. For that purpose we will use the following lemma.

Lemma A.2. *Let $L^0(\hat{\Psi})(\bar{O}(i)) = L(\hat{\Psi}(P_{i-1}))(Z(i), O(i)) - L(\psi_0)(Z(i), O(i))$. Suppose that with probability 1, $\sup_{\psi \in \Psi}|L^0(\psi)(Z(i), O(i))| < M_1 < \infty$, and*

$$\sup_{\psi \in \Psi}\frac{P_{\theta_0,i,Z(i)}\{L^0(\psi)\}^2}{P_{\theta_0,i,Z(i)}L^0(\psi)} \leq M_2 < \infty.$$

Then,

$$\tilde{R}_{n,K}(L^0(\hat{\Psi})) = \frac{2K^2}{n} \sum_{i=1}^{n} P_{\theta_0,i,Z(i)} \phi\left(\frac{\mid L^0(\hat{\Psi}(P_{i-1})) \mid}{K}\right)$$

$$\leq 2M_2(1/2K^2 + 1/6M_1K\exp(M_1/K))\frac{1}{n}\sum_{i=1}^{n}P_{\theta_0,i,Z(i)}L^0(\hat{\Psi}(P_{i-1})).$$

Proof. A third order tailor expansion for $\exp(x)$ yields $\phi(x) = x^2/2! + \exp(\xi(x))x^3/3!$ for some $\xi(x)$. This can be bounded by $x^2(1/2 + 1/6\exp(M_1/K)M_1/K)$ by using that $\mid x \mid < M_1/K$. As a consequence, we can bound $P_{\theta_0,i,Z(i)}\phi(\mid L^0(\psi) \mid /K)$ by $(1/2 + 1/6M_1/K\exp(M_1/K))P_{\theta_0,i,Z(i)}\{L^0(\psi)\}^2$, which, by assumption, can thus be bounded by $M_2(1/2 + 1/6M_1/K\exp(M_1/K))P_{\theta_0,i,Z(i)}L^0(\psi)$. This proves the lemma. $\square$

A.3.3 Proof of Theorem 18.1

For notational convenience, we let $n = (n - n_l + 1)$ and let the sum over t_0 run from 1 to n. We have

$$0 \leq d_{0n}(\hat{\Psi}_{k_n}, \psi_0)$$

$$= \frac{1}{n}\sum_{t_0} P_{\theta_0,t_0,Z(t_0)}\{L(\hat{\Psi}_{k_n}(P_{t_0-1})) - L(\psi_0)\}$$

$$-(1+\delta)\frac{1}{n}\sum_{t_0}\{L(\hat{\Psi}_{k_n}(P_{t_0-1})) - L(\psi_0)\}(O(t_0), Z(t_0))$$

$$+(1+\delta)\frac{1}{n}\sum_{t_0}\{L(\hat{\Psi}_{k_n}(P_{t_0-1})) - L(\psi_0)\}(O(t_0), Z(t_0))$$

$$\leq \frac{1}{n}\sum_{t_0} P_{\theta_0,t_0,Z(t_0)}\{L(\hat{\Psi}_{k_n}(P_{t_0-1})) - L(\psi_0)\}$$

$$-(1+\delta)\frac{1}{n}\sum_{t_0}\{L(\hat{\Psi}_{k_n}(P_{t_0-1})) - L(\psi_0)\}(O(t_0), Z(t_0))$$

$$+(1+\delta)\frac{1}{n}\sum_{t_0}\{L(\hat{\Psi}_{\bar{k}_n}(P_{t_0-1})) - L(\psi_0)\}(O(t_0), Z(t_0))$$

$$= \frac{1}{n}\sum_{t_0} P_{\theta_0,t_0,Z(t_0)}\{L(\hat{\Psi}_{k_n}(P_{t_0-1})) - L(\psi_0)\}$$

$$-(1+\delta)\frac{1}{n}\sum_{t_0}\{L(\hat{\Psi}_{k_n}(P_{t_0-1})) - L(\psi_0)\}$$

$$+(1+\delta)\frac{1}{n}\sum_{t_0}\{L(\hat{\Psi}_{\bar{k}_n}(P_{t_0-1})) - L(\psi_0)\}$$

$$-(1 + 2\delta)\frac{1}{n}\sum_{t_0} P_{\theta_0,t_0,Z(t_0)}\{L(\hat{\Psi}_{\tilde{k}_n}(P_{t_0-1})) - L(\psi_0)\}$$

$$+(1 + 2\delta)\frac{1}{n}\sum_{t_0} P_{\theta_0,t_0,Z(t_0)}\{L(\hat{\Psi}_{\tilde{k}_n}(P_{t_0-1})) - L(\psi_0)\}.$$

Denote the sum of the first two terms in the last expression by R_{n,k_n} and the sum of the third and fourth term by $T_{n,\tilde{k}_n}$; the last term is the benchmark $(1+2\delta)d_{0n}(\hat{\Psi}_{\tilde{k}_n}, \psi_0)$. Hence, we have

$$0 \leq d_{0n}(\hat{\Psi}_{k_n}, \psi_0) \leq (1 + 2\delta)d_{0n}(\hat{\Psi}_{\tilde{k}_n}, \psi_0) + R_{n,k_n} + T_{n,\tilde{k}_n} \qquad (A.7)$$

Rewriting $R_{n,k}$ (and $T_{n,k}$) as a Martingale. For notational convenience, we introduce the following notation for the relevant random variables

$$\tilde{H}_k \equiv \frac{1}{n}\sum_{t_0} P_{\theta_0,t_0,Z(t_0)}\{L(\hat{\Psi}_k(P_{t_0-1})) - L(\psi_0)\}$$

$$\bar{H}_k \equiv \frac{1}{n}\sum_{t_0}\{L(\hat{\Psi}_k(P_{t_0-1})) - L(\psi_0)\}(O(t_0), Z(t_0)),$$

where, by definition of ψ_0, $\tilde{H}_k \geq 0 \; \forall \; k$. Rewrite $R_{n,k}$ and $T_{n,k}$ as

$$R_{n,k} = (1 + \delta)\left[\tilde{H}_k - \bar{H}_k\right] - \delta\tilde{H}_k$$

and

$$T_{n,k} = (1 + \delta)\left[\bar{H}_k - \tilde{H}_k\right] - \delta\tilde{H}_k.$$

Approximating $R_{n,k}$ (and $T_{n,k}$) with a Negatively Deterministically Shifted Martingale Sum, Up to Negligible Remainder. In order to exploit that a negatively shifted martingale sum has a nice exponential tail behavior, it is important that the random shift $\delta\tilde{H}_k \geq 0$ is replaced by a deterministic shift that is guaranteed larger than a constant we can control. We will now utilize assumption A3 to succeed in that. For a K, we define $\tilde{R}_{n,k} \equiv \frac{2K^2}{n}\sum_{t_0=1}^n P_{\theta_0,t_0,Z(t_0)}\phi\left(\frac{|L^0(\hat{\Psi}_k(P_{t_0-1}))|}{K}\right)$, where $\phi(x) \equiv \exp(x) - x - 1$. By Lemma A.2, we have $\tilde{R}_{n,k} \leq M_2(1/2K^2 + 1/6M_1 K \exp(M_1/K))\tilde{H}_k$. Let's denote this constant with $C_1(M_1, M_2, K)$ so that $\tilde{R}_{n,k} \leq C_1(M_1, M_2, K)\tilde{H}_k$. Define the event $E_{nk} = \{M_{3n}^{-1} < \tilde{H}_k/E_0\tilde{H}_k < M_{3n}\}$, and let $I_{E_{nk}}$ denote the indicator of this event. By assumption A3, we have $P_0^n(I_{E_{nk_n}} = 1) \to 1$, and $P_0^n(I_{E_{n,\tilde{k}_n}} = 1) \to 1$, as $n \to \infty$. This also implies that $P_0^n(\tilde{R}_{n,k_n}/E_0\tilde{H}_{k_n} < C_1 M_{3n}) \to 1$. For notational convenience, let M_{3n} be redefined by $\max(C_1, 1)M_{3n}$. We decompose $R_{n,k}$ as follows:

$$R_{n,k} = (1 + \delta)\left[\tilde{H}_k - \bar{H}_k\right]I_{E_{nk}} + (1 + \delta)\left[\tilde{H}_k - \bar{H}_k\right]I_{E_{nk}^c}$$
$$-\delta\tilde{H}_k I(\tilde{H}_k > M_{3n}^{-1}E_0\tilde{H}_k) - \delta\tilde{H}_k I(\tilde{H}_k < M_{3n}^{-1}E_0\tilde{H}_k)$$
$$= R_{n,k}^* + e_{n,k},$$

where

$$R^*_{n,k} \equiv (1+\delta)\left[\tilde{H}_k - \bar{H}_k\right] I_{E_{n,k}} - \delta\tilde{H}_k I(\tilde{H}_k > M^{-1}_{3n} E_0 \tilde{H}_k)$$
$$e_{n,k} \equiv (1+\delta)\left[\tilde{H}_k - \bar{H}_k\right] I^c_{E_{n,k}} - \delta\tilde{H}_k I(\tilde{H}_k < M^{-1}_{3n} E_0 \tilde{H}_k).$$

Thus, $R_{n,k_n} = R^*_{n,k_n} + e_{n,k_n}$. By assumption A3 we have $P^n_0(|\ e_{n,k_n}\ |= 0) \to 1$, as $n \to \infty$. Similarly,

$$T_{n,k} = T^*_{n,k} + f_{n,k},$$

where

$$T^*_{n,k} \equiv (1+\delta)\left[\bar{H}_k - \tilde{H}_k\right] I_{E_{n,k}} - \delta\tilde{H}_k I(\tilde{H}_k > M^{-1}_{3n} E_0 \tilde{H}_k)$$
$$f_{n,k} \equiv (1+\delta)\left[\bar{H}_k - \tilde{H}_k\right] I_{E_{n,k}} - \delta\tilde{H}_k I(\tilde{H}_k < M^{-1}_{3n} E_0 \tilde{H}_k).$$

By the same argument as used for e_{n,k_n}, we have $P^n_0(|\ f_{n,\tilde{k}}\ |= 0) \to 1$ as $n \to \infty$. Thus, $T_{n,\tilde{k}_n} = T^*_{n,\tilde{k}_n} + f_{n,\tilde{k}_n}$ where $f_{n,\tilde{k}_n}$ equals zero with probability tending to 1.

What We Have and What We Still Need To Do. Let $Z_{n2} = e_{n,k_n} + f_{n,\tilde{k}_n}$ and $Z_{n1} = R^*_{n,k_n} + T^*_{n,\tilde{k}_n}$. We have shown that $d_{0n}(\hat{\Psi}_{k_n}, \psi_0) \leq (1 + 2\delta)d_{0n}(\hat{\Psi}_{\tilde{k}_n}, \psi_0) + Z_{n1} + Z_{n2}$, where $P^n_0(Z_{n2} = 0) \to 1$ as $n \to \infty$. In the sequel, we will show that

$$EZ_{n1} = ER^*_{n,k_n} + ET^*_{n,\tilde{k}_n} \leq C(M_1, M_2, M_{3n}, \delta)(1 + \log(K(n)))/n$$

for some specified $C(M_1, M_2, M_{3n}, \delta) < \infty$, which then completes the proof.

Bounding the Tail Probability of R^*_{n,k_n}

Step 1: Getting a Deterministic Negative Shift. We also define the event $E_{n,k,1} = \{\tilde{H}_k > M^{-1}_{3n} E_0 \tilde{H}_k\}$. Let $s > 0$. We have

$$P^n_0(R^*_{n,k_n} > s) = P^n_0\left(I_{E_{n,k_n}}\{\tilde{H}_{k_n} - \bar{H}_{k_n}\} > \frac{1}{1+\delta}\left\{s + \delta\tilde{H}_{k_n} I_{E_{n,k_n,1}}\right\}\right)$$

$$\leq P^n_0\left(I_{E_{n,k_n}}\{\tilde{H}_{k_n} - \bar{H}_{k_n}\} > \frac{1}{1+\delta}\left\{s + \delta M^{-1}_{3n} E_0 \tilde{H}_k\big|_{k=k_n} I_{E_{n,k_n,1}}\right\}\right),$$

where we used that event $E_{n,k_n,1}$ implies $\tilde{H}_{k_n} \geq M^{-1}_{3n} E_0 \tilde{H}_k\big|_{k=k_n}$, allowing us to replace the random $\tilde{H}_{k_n}$ by this bound that is only random through k_n. Let's denote the event in the last displayed probability by A_n so that the last displayed bound is denoted with $P^n_0(A_n)$. We can write

$$P^n_0(A_n) = P^n_0(A_n \text{ and } I_{E_{n,k_n,1}} = 1) + P^n_0(A_n \text{ and } I_{E_{n,k_n,1}} = 0).$$

Note that if $I_{E_{n,k_n,1}} = 0$, then the right-hand side of the equality equals $\frac{1}{1+\delta}s > 0$, while the left-hand side of inequality in event A_n equals 0. This shows that $P^n_0(A_n \text{ and } I_{E_{n,k_n,1}} = 0) = 0$. This yields the following bound for $P^n_0(R^*_{n,k_n} > s)$:

$$P_0^n(R_{n,k_n}^* > s)$$
$$\leq P_0^n\left(I_{E_{n,k_n}}\{\tilde{H}_{k_n} - \bar{H}_{k_n}\} > \tfrac{1}{1+\delta}\left\{s + \delta M_{3n}^{-1} E_0\tilde{H}_k\big|_{k=k_n}\right\} \text{ and } E_{n,k_n,1} = 1\right)$$
$$\leq P_0^n\left(I_{E_{n,k_n}}\{\tilde{H}_{k_n} - \bar{H}_{k_n}\} > \tfrac{1}{1+\delta}\left\{s + \delta M_{3n}^{-1} E_0\tilde{H}_k\big|_{k=k_n}\right\}\right)$$
$$\leq K(n)\max_k P_0^n\left(I_{E_{n,k}}\{\tilde{H}_k - \bar{H}_k\} > \tfrac{1}{1+\delta}\left\{s + \delta M_{3n}^{-1}E_0\tilde{H}_k\right\}\right).$$

In the last inequality we used that for some collection of random variables $(X(k) : k)$ and constants $(c(k) : k)$ and random index k_n, we have

$$P_0^n(X(k_n) < c(k_n)) \leq P_0^n(X(k) < c(k)) \text{ for at least one } k)$$
$$\leq \sum_{k=1}^{K(n)} P_0^n(X(k) < c(k))$$
$$\leq K(n)\max_k P_0^n(X(k) < c(k)).$$

Similarly, for $T_{n,\bar{k}_n}^*$, we obtain

$$P_0^n(T_{n,\bar{k}_n}^* > s)$$
$$\leq K(n)\max_k P_0^n\left(I_{E_{n,k}}\{\bar{H}_k - \tilde{H}_k\} > \frac{1}{1+\delta}\left\{s + \delta M_{3n}^{-1}E_0\tilde{H}_k\right\}\right).$$

Step 2: Applying the Martingale Exponential Tail Probability. We have that $\bar{H}_k - \tilde{H}_k$ equals a martingale sum $\frac{1}{n}\sum_{t_0} Z_{k,t_0} - E(Z_{k,t_0} \mid \bar{O}(t_0 - 1))$ where

$$Z_{k,t_0} = \{L(\hat{\Psi}_k(P_{t_0-1})) - L(\psi_0)\}(\bar{O}(t_0)).$$

By assumption A1, the random variables Z_{k,t_0} are bounded: $\mid Z_{k,t_0} \mid \leq M_1$ a.s.

We are now ready to apply the Martingale inequality of Theorem A.2, specifically inequality (A.6), to $\tilde{H}_k - \bar{H}_k$ with $R = R_k = M_{3n}E_0\tilde{H}_k$, for each k separately. Due to this choice of R, we obtain a tail probability at $s > 0$ that behaves for s small as $\exp(-cM_{3n}ns)$ instead of the usual $\exp(-cns^2)$. This on its turn will prove that the expectation of the remainder terms R_{n,k_n}^* and $T_{n,\bar{k}_n}^*$ converge at a rate $\log(K(n))/n$ instead of the usual $\log(K(n))/\sqrt{n}$.

For ease of reference, we state here this martingale exponential inequality at a k explicitly:

Lemma A.3. *Let K be set, and*

$$\tilde{R}_{n,k} \equiv \frac{2K^2}{n}\sum_{i=1}^n P_{\theta_0,i,Z(i)}\phi\left(\frac{\mid L^0(\hat{\Psi}_k(P_{i-1})) \mid}{K}\right),$$

where $\phi(x) \equiv \exp(x) - x - 1$. Let $M_{n,k} = \sum_{i=1}^n\{L^0(\hat{\Psi}_k(P_{i-1}))(Z(i), O(i)) - E_0(L^0(\hat{\Psi}_k(P_{i-1})) \mid Z(i))\}$. For any R_k, $c_0, c_1 \geq 0$ satisfying $c_0^2 \geq C^2(c_1 + 1)$ and $\alpha \in (c_0/\sqrt{n}\sqrt{R_k}, c_1R_k/K)$, we have

$$P\left(I(\tilde{R}_{n,k} \leq R_k)\frac{M_{n,k}}{n} \geq \alpha\right) \leq 2\exp\left\{-\frac{n\alpha^2}{C^2(K\alpha + R_k)}\right\}.$$

In order to apply this inequality to the above tail probability for $I(\tilde{R}_{n,k} < R_k)(\bar{H}_k - \tilde{H}_k)$ with $R_k = M_{3n}E_0\tilde{H}_k$ at a given $\alpha(s) = \frac{1}{1+\delta}(s + \delta M_{3n}^{-1}E_0\tilde{H}_k)$, we need to be able to select c_0, c_1 with $c_0^2 \geq C^2(c_1 + 1)$ so that

$$\frac{c_0\sqrt{R_k}}{\sqrt{n}} \leq \frac{1}{1+\delta}\left[s + \delta M_{3n}^{-1}E_0\tilde{H}_k\right] \leq c_1\frac{R_k}{K}. \tag{A.8}$$

Note $M_{3n}^{-1}E_0\tilde{H}_k = M_{3n}^{-2}R_k$, so that we need to apply the inequality at

$$\alpha(s) = 1/(1 + \delta)(s + \delta M_{3n}^{-2}R_k).$$

So we now need to select c_0, c_1 so that this $\alpha(s) \in (c_0 R_k^{0.5}/n^{0.5}, c_1 R_k/K)$. We select $c_0^2 = c_0^2(c_1) = C^2(c_1 + 1)$. Since the martingale process $\bar{H}_k - \tilde{H}_k$ is bounded by $2M_1$, the upper bound is nonexistent if $c_1 R_k/K > 2M_1$. This implies the choice $c_1 = c_1(M_1) = 2M_1 K/R_k$, thereby guaranteeing that there is no upper bound on $\alpha(s)$ for all s. Let $c_0(M_1) = c_0(c_1(M_1))$ be the corresponding choice for c_0. Thus, for any $\alpha \in (c_0(M_1)R_k^{0.5}n^{-0.5}, \infty)$, we have $P_0^n(I(\tilde{R}_{nk} < R_k)(\bar{H}_k - \tilde{H}_k) > \alpha) \leq 2\exp(-n\alpha^2/\{C^2(K\alpha + R_k)\})$.

The left-inequality $\alpha(s) > c_0(M_1)R_k^{0.5}n^{-0.5}$ is equivalent with

$$s > -\delta M_{3n}^{-2}R_k + C^2(c_1(M_1) + 1)n^{-0.5}R_k^{0.5}(1 + \delta). \tag{A.9}$$

The first term on the right-hand side is negative and converges to zero at rate $M_{3n}^{-2}R_k$, while the second term is positive and converges to zero at rate $R_k^{0.5}n^{-0.5}$. By assumption A4, we have

$$\max_k \frac{R_k^{0.5}n^{-0.5}}{M_{3n}^{-2}R_k} \to 0.$$

This implies that for n large enough, we have that the right-hand side of (A.9) is negative, proving that the inequality $\alpha(s) > c_0(M_1)R_k^{0.5}n^{-0.5}$ holds for all $s > 0$. Thus, there exists an n_1 so that for all $n > n_1$, we have for all $s > 0$,

$$P_0^n(I(\tilde{R}_{nk} < R_k)(\bar{H}_k - \tilde{H}_k) > \alpha(s)) \leq 2\exp(-n\alpha(s)^2/\{C^2(K\alpha(s) + R_k)\})$$
$$= 2\exp\left(-C^{-2}\frac{n}{(1+\delta)^2}\frac{(s+\delta M_{3n}^{-2}R_k)^2}{R_k + \frac{K}{(1+\delta)}(s+\delta M_{3n}^{-2}R_k)}\right).$$

Step 3: Understanding the Asymptotic Behavior of Tail Probability. We now note that

$$\frac{(s + \delta M_{3n}^{-2}R_k)^2}{R_k + \frac{K}{(1+\delta)}(s + \delta M_{3n}^{-2}R_k)} = \frac{(s + \delta M_{3n}^{-2}R_k)}{\frac{R_k}{s+\delta M_{3n}^{-2}R_k} + \frac{K}{(1+\delta)}} \geq \frac{(s + \delta M_{3n}^{-2}R_k)}{\frac{M_{3n}^2}{\delta} + K}$$
$$\geq \frac{s}{\frac{M_{3n}^2}{\delta} + K}$$

$$= M_{3n}^{-2} \frac{s}{\delta^{-1} + M_{3n}^{-2} K}$$

$$\geq M_{3n}^{-2} \frac{s}{\delta^{-1} + K}$$

where we use that $M_{3n} > 1$ for all n so that $K M_{3n}^{-2} \leq K$. This shows that, for $s > 0$,

$$P_0^n(R_{n,k_n}^* > s) \leq 2K(n) \exp\left(-\frac{n M_{3n}^{-2}}{c(M_1, M_2, \delta)} s\right),$$

where $c(M_1, M_2, \delta) = 2C^2(1 + \delta)^2 \left(K + \delta^{-1}\right)$.

Bounding the Expectation of R_{n,k_n}^* Based on Our Tail Probability Bounds. Since $ER_{n,k_n}^* \leq \int_0^\infty P_0^n(R_{n,k_n}^* > s) ds$, for each $u > 0$, we have

$$ER_{n,k_n}^* \leq u + \int_u^\infty 2K(n) \exp\left(-\frac{M_{3n}^{-2} n}{c(M_1, M_2, \delta)} s\right) ds.$$

The minimum is attained at $u_n = c(M_1, M_2, \delta) \log(2K(n))/(n M_{3n}^{-2})$ and is given by $c(M_1, M_2, \delta)(\log(2K(n)) + 1)/(n M_{3n}^{-2})$. Thus,

$$ER_{n,k_n}^* \leq c(M_1, M_2, \delta) \frac{1 + \log(2K(n))}{n M_{3n}^{-2}}.$$

Similarly, we obtain his bound for $ET_{n,\tilde{k}_n}$. This proves the theorem under assumption **A4**.

What Happens If Assumption A4 Does Not Hold. Let's now consider the case that assumption A4 fails to hold. Then we have that the leading term $E_0 d_{0n}(\hat{\Psi}_{\tilde{k}_n}, \psi_0) = O(n^{-1} M_{3n}^3)$. Firstly, consider the case that the right-hand side of (A.9) is negative. In that case, we have our desired inequality for $P_0^n(R_{n,k}^* > s)$ for all $s > 0$ provided above. Consider now the case that the right-hand side of (A.9) is positive. Then, we know that

$$R_k^{0.5} < (1 + \delta)\delta^{-1} C^2(c_1(M_1) + 1) M_{3n}^2 n^{-0.5},$$

which implies that the right-hand side of (A.9) is bounded by

$$cM_{3n}^2 n^{-1} \equiv (1 + \delta)\delta^{-1} C^4(c_1(M_1) + 1)^2 M_{3n}^2 n^{-1}.$$

Thus, in this case, we have the desired exponential bound for $P_0^n(I(\tilde{R}_{nk} < R_k)(\bar{H}_k - \tilde{H}_k) > \alpha(s))$ for any $s > cM_{3n}^2 n^{-1}$ for this specified constant $c > 0$.

We now proceed as follows: for any $u > cM_{3n}^2 n^{-1}$, we have

$$E_0 R_{n,k_n}^* = \int_0^u P_0^n(R_{n,k_n}^* > s)ds + \int_u^\infty P_0^n(R_{n,k_n}^* > s)ds$$

$$\leq u + K(n) \max_k \int_u^\infty P_0^n(R_{n,k}^* > s)ds$$

$$\leq u + 2K(n) \int_u^\infty \exp\left(-\frac{nM_{3n}^{-2}}{c(M_1, M_2, \delta)} s\right)ds$$

$$= u + 2K(n)\frac{c(M_1, M_2, \delta)}{nM_{3n}^{-2}} \exp\left(-\frac{nM_{3n}^{-2}}{c(M_1, M_2, \delta)} u\right).$$

The optimal u is given by

$$u^* = \max(cM_{3n}^2 n^{-1}, c(M_1, M_2, \delta)M_{3n}^2 n^{-1} \log(2K(n))).$$

Suppose that $c > c(M_1, M_2, \delta)\{\log 2 + \log K(n)\}$, so that $u^* = cM_{3n}^2 n^{-1}$. Note also that this implies that $K(n) < \exp(cc(M_1, M_2, \delta)^{-1})$. Plugging this u^* in the final expression yields a first term equal to u^* plus a term

$$2K(n)c(M_1, M_2, \delta)^{-1} n^{-1} M_{3n}^2 \exp(-c/c(M_1, M_2, \delta)).$$

Using the bound on $K(n)$ shows that the final expression is $O(M_{3n}^2 n^{-1})$. Suppose now that $c < c(M_1, M_2, \delta)\{\log 2 + \log K(n)\}$, so that $u^* = c(M_1, M_2, \delta)M_{3n}^2 n^{-1}(\log 2 + \log K(n))$. Plugging this u^* in the final expression now shows that the final expression is $O(M_{3n}^2 n^{-1}(1 + \log K(n)))$. Thus, we have shown that in either case, we have that $E_0 R_{n,k_n}^* < C_1 M_{3n}^2 n^{-1}(1 + \log K(n))$ for some universal $C_1 = C_1(M_1, M_2, \delta) < \infty$. The same bounding applies to $E_0 T_{n,\bar{k}_n}^*$. Thus we have shown that if assumption A4 does not hold, then we have

$$d_{0n}(\hat{\Psi}_{k_n}, \psi_0) = o_P(n^{-1}M_{3n}^3) + o_P(n^{-1}M_{3n}^2(1 + \log K(n))).$$

This completes the proof of Theorem 18.1.

A.3.4 Brief Review of Literature on Online Estimation

Consider a parametric model and an i.i.d. regression setting, where we assume that the conditional mean of Y, given W is described by a linear model $\psi_\beta(W) = \beta' W$. This setting has been studied extensively in the online literature: Zinkevich (2003); Crammer et al. (2006); Bottou (2010); Shalev-Shwartz (2011). Let $\psi_0(W) = E_0(Y \mid W) = \beta_0^\top W$ be the parameter of interest. Let $L(\psi_\beta)(Y, W)$ be the squared error loss function, and let $\hat{\beta}_n$ be the least squares estimator defined by the minimizer of the empirical mean of the loss function:

$$\hat{\beta}_n = \arg\min_\beta \frac{1}{n} \sum_{i=1}^{n} L(\psi_\beta)(O(i)) .$$

More generally, we can have that $\{p_\beta : \beta\}$ is a parametric model, $L(\psi_\beta) = -\log p_\beta$ is the log-likelihood loss and $\hat{\beta}_n$ is the maximum likelihood estimator of β_0.

Stochastic gradient descent is an online algorithm aiming to approximate the MLE $\hat{\beta}_n$. Consider the updating step

$$\beta_{t+1} = \beta_t - \gamma_t \Gamma_t \frac{d}{d\beta_t} L(\psi_{\beta,t})(O(t)) \tag{A.10}$$

where γ_t is a scalar step size or learning rate, Γ_t is a $d \times d$ matrix, and $O(t)$ is the observation used at the t-th step (Bottou 2010). This defines an approximation $\tilde{\beta}_n$ of $\hat{\beta}_n$. In first-order SGD Γ_t is some constant times the identity matrix, while other variants replace Γ_t with an appropriate diagonal matrix (e.g., diagonal elements of the estimated inverse Hessian) (Duchi et al. 2011; Zeiler 2012). Second-order SGD accounts for the curvature of the loss function by using a Γ_t that approximates the inverse Hessian (Murata 1998). However, computing and storing an estimate of this matrix is often computationally expensive for high-dimensional d and, though it is optimal, second-order SGD is rarely used in practice. There are many other methods for online optimization that have been used in a variety of contexts (Polyak and Juditsky 1992; Xu 2011), including settings with regularized loss functions, such as the lasso regression and support vector machines (Fu 1998; Langford et al. 2009; Kivinen et al. 2004; Balakrishnan and Madigan 2008; Shalev-Shwartz et al. 2011).

We have

$$d_{0n}(\psi_{\tilde{\beta}_n}, \psi_0) = d_{0n}(\psi_{\beta,0}, \psi_0) + d_{0n}(\psi_{\hat{\beta},n}, \psi_{\beta,0}) + d_{0n}(\psi_{\tilde{\beta}_n}, \psi_{\hat{\beta}_n}) ,$$

where the first two terms are the approximation and estimation error, while the third term is the optimization error incurred by using $\tilde{\beta}_n$ rather than the true minimizer $\hat{\beta}_n$. Existing results in the online learning literature suggest that in big data settings, the estimation and optimization error will be small (Shalev-Shwartz 2011). Thus, the performance of an online estimator will be determined largely by the approximation error. To minimize the approximation error, we utilize the online-super learning framework.

A.4 Online Targeted Learning

In this section, we will derive a first order expansion of the type

$$\Psi_J(P^{N*}) - \Psi_J(P_0^N) = -\frac{1}{N} \sum_{i=1}^{N} P_{0,o(i)} \bar{D}(P^{N*}) + R_{2,N}(\theta_N^*, \theta_0), \tag{A.11}$$

where $P_{0,o(i)}$ denotes the conditional expectation w.r.t. $O(i)$, given $\bar{O}(i-1)$, and $R_{2,N}(\theta_N^*, \theta_0)$ is a second-order remainder. This result is presented in Theorem A.4. In order to establish this first order expansion, we first establish the identity

$$\Psi_J(P^{N*}) - \Psi_J(P_0^N) = -\frac{1}{N}\sum_{i=1}^N P_0^N \bar{D}(P^{N*}) + R_{21}(\theta_N^*, \theta_0),$$

for an explicitly defined second-order remainder $R_{21}(\theta_N^*, \theta_0)$. This is presented in Theorem A.3. Note that this involves an *marginal expectation of the efficient influence curve*, instead of the desired conditional expectation. Subsequently, in the second subsection we study the additional remainder $R_{22,N}(\theta_N^*, \theta_0) = \frac{1}{N}\sum_{i=1}^N (P_{0,o(i)} - P_0^N)\{\bar{D}(\theta_N^*) - \bar{D}(\theta_0)\}$: $R_{2,N}() = R_{21}() + R_{22,N}()$. Subsequently, we show that under conditions on the dependence structure of the time series this term $R_{22,N}$ also represents a second-order remainder.

The identity (A.11) provides the basis for the analysis of the TMLE. Specifically, this identity (A.11) combined with the TMLE θ_N^* solving $0 = \frac{1}{N}\sum_{i=1}^N \bar{D}(\theta_N^*) = 0$ results in the first order expansion:

$$\Psi_J(P^{N*}) - \Psi_J(P_0^N) = \frac{1}{N}\sum_{i=1}^N \{\bar{D}(P^{N*}) - P_{0,o(i)}\bar{D}(P^{N*})\} + R_{2,N}(\theta_N^*, \theta_0).$$

The leading term on the right hand side is a martingale process in θ_N^*, allowing to establish that it converges to a normal limit distribution under weak conditions.

Subsequently, we establish an analogue first order expansion for *online estimation* of the type:

$$-\frac{1}{N}\sum_{i=1}^N P_{0,o(i)}\bar{D}_{g^*}(\theta_{i-1}) = \frac{1}{N}\sum_{i=1}^N \Psi_{g^*}(\theta_{i-1}) - \Psi_{g^*}(\theta_0) + \bar{R}_{21,g^*,N} + \bar{R}_{22,g^*,N}.$$

Again, we derive this result in a two stage fashion, first establishing one in terms of the marginal expectation $P_0^N \bar{D}_{g^*}(\theta_{i-1})$ only involving the remainder $\bar{R}_{21,g^*,N}$. Subsequently, we study the remainder $\bar{R}_{22,g^*,N}$ and note that controlling it will require an additional stationarity assumption.

A.4.1 First Order Expansion of Target Parameter Based on Marginal Expectation of Efficient Influence Curve

Note that

$$P_{g^*,o(t)}f = P_{w(t)}P_{g^*,a(t)}P_{y(t)}f.$$

For notational convenience, we will also denote P_{g^*} with P^*. We have

$$\Psi_{g^*}(P) = \left(\prod_{t=1}^\tau P_{g^*,o(t)}\right)Y(\tau).$$

Theorem A.3. *Let*

$$\Delta_w(P^*_{o(s)}, P^*_{0,o(s)}) = (P_{w(s)} - P_{0,w(s)})P^*_{0,a(s)}P_{0,y(s)},$$
$$\Delta_a(P^*_{o(s)}, P^*_{0,o(s)}) = P_{w(s)}(P^*_{a(s)} - P^*_{0,a(s)})P_{0,y(s)},$$
$$\Delta_y(P^*_{o(s)}, P^*_{0,o(s)}) = P_{w(s)}P_{a(s)}(P_{y(s)} - P_{0,y(s)}),$$
$$\Phi_{x(s)}(x, c_x) = E(Y_{g^*}(\tau) \mid X(s) = x, C_x(s) = c_x).$$

We have

$$-P_0^N D_{g^*}^N(P) = \Psi_{g^*}(P) - \Psi_{g^*}(P_0) + R_{21,g^*}(P, P_0),$$

where $R_{21,g^*}(P, P_0) = R_{21a,g^*}(P, P_0) + R_{21b,g^*}(P, P_0)$, *and*

$$R_{21a,g^*}(\theta, \theta_0) = \sum_{x \in \{w,a,y\}} \sum_{s=1}^{\tau} \left\{ \prod_{t=1}^{s-1} P^*_{o(t)} \right\} \Delta_x(P^*_{o(s)}, P^*_{0,o(s)}) \times \qquad (A.12)$$

$$\left\{ \sum_{j=s+1}^{\tau} \prod_{l=s+1}^{j-1} P^*_{o(l)}(P^*_{o(j)} - P^*_{0,o(j)}) \prod_{l=j+1}^{\tau} P^*_{0,o(l)} \right\} Y(\tau),$$

$$R_{21b,g^*}(P, P_0) = \sum_{x \in \{w,a,y\}} \sum_{s=1}^{\tau} R_{21b,x(s),g^*}(P, P_0),$$

$$R_{21b,x(s),g^*}(P, P_0) = \int_c \int_x \left(\frac{\bar{h}_{c_x,0}}{\bar{h}_{c_x}}(c) - 1 \right) \Phi_{x(s)}(x, c) d(P^*_{x(s)} - P^*_{0,x(s)})(x \mid c) h^*_{c_x(s)}(c) d\mu(c).$$

As a consequence, recalling that $\Psi_J(P) = \frac{1}{J} \sum_{j=1}^{J} \Psi_{g_j^*}(P)$ *and* $D^N(P) = \frac{1}{J} \sum_{j=1}^{J} D_{g_j^*}^N$ *(P), we have*

$$-P_0^N D^N(P) = \Psi_J(P) - \Psi_J(P_0) + R_{21}(\theta, \theta_0),$$

where $R_{21}(\theta, \theta_0) = \frac{1}{J} \sum_{j=1}^{J} R_{21,g_j^*}(\theta, \theta_0)$.

Proof. Define

$$\Psi_{g^*,P,P_0,w(s)}(P^*_{1,w(s)}) \equiv \left\{ \prod_{t=1}^{s-1} P^*_{o(t)} \right\} P_{1,w(s)} P^*_{0,a(s)} P_{0,y(s)} \left\{ \prod_{t=s+1}^{\tau} P^*_{0,o(t)} \right\} Y(\tau),$$

$$\Psi_{g^*,P,P_0,a(s)}(P^*_{1,a(s)}) \equiv \left\{ \prod_{t=1}^{s-1} P^*_{o(t)} \right\} P_{w(s)} P^*_{1,a(s)} P_{0,y(s)} \left\{ \prod_{t=s+1}^{\tau} P^*_{0,o(t)} \right\} Y(\tau),$$

$$\Psi_{g^*,P,P_0,y(s)}(P^*_{1,y(s)}) \equiv \left\{ \prod_{t=1}^{s-1} P^*_{o(t)} \right\} P_{w(s)} P^*_{a(s)} P_{1,y(s)} \left\{ \prod_{t=s+1}^{\tau} P^*_{0,o(t)} \right\} Y(\tau).$$

Note that

$$\Psi_{g^*}(P) - \Psi_{g^*}(P_0) = \sum_{x\in\{w,a,y\}} \sum_{s=1}^{\tau} \left(\left\{ \prod_{t=1}^{s-1} P^*_{o(t)} \right\} \Delta_x(P^*_{o(s)}, P^*_{0,o(s)}) \left\{ \prod_{t=s+1}^{\tau} P^*_{0,o(t)} \right\} \right) Y(\tau)$$

$$= \sum_{x\in\{w,a,y\}} \sum_{s=1}^{\tau} \left\{ \Psi_{g^*,P,P_0,x(s)}(P^*_{x(s)}) - \Psi_{g^*,P,P_0,x(s)}(P^*_{0,x(s)}) \right\}.$$

Define $\Psi_{g^*,P,x(s)} = \Psi_{g^*,P,P,x(s)}$ and also define the second-order term

$$-R_{21a,g^*}(\theta,\theta_0) = \sum_{x\in\{w,a,y\}} \sum_{s=1}^{\tau} \left(\left\{ \prod_{t=1}^{s-1} P^*_{o(t)} \right\} \Delta_x(P^*_{o(s)}, P^*_{0,o(s)}) \left\{ \prod_{t=s+1}^{\tau} P^*_{0,o(t)} \right\} \right) Y(\tau)$$

$$- \sum_{x\in\{w,a,y\}} \sum_{s=1}^{\tau} \left(\left\{ \prod_{t=1}^{s-1} P^*_{o(t)} \right\} \Delta_x(P^*_{o(s)}, P^*_{0,o(s)}) \left\{ \prod_{t=s+1}^{\tau} P^*_{o(t)} \right\} \right) Y(\tau)$$

$$= - \sum_{x\in\{w,a,y\}} \sum_{s=1}^{\tau} \left\{ \prod_{t=1}^{s-1} P^*_{o(t)} \right\} \Delta_x(P^*_{o(s)}, P^*_{0,o(s)}) \times$$

$$\left\{ \sum_{j=s+1}^{\tau} \prod_{l=s+1}^{j-1} P^*_{o(l)}(P^*_{o(j)} - P^*_{0,o(j)}) \prod_{l=j+1}^{\tau} P^*_{0,o(l)} \right\} Y(\tau).$$

We have

$$\Psi_{g^*}(P) - \Psi_{g^*}(P_0) = \sum_{x\in\{w,a,y\}} \sum_{s=1}^{\tau} \left(\Psi_{g^*,P,x(s)}(P^*_{x(s)}) - \Psi_{g^*,P,x(s)}(P^*_{0,x(s)}) \right) + R_{21a,g^*}(\theta,\theta_0).$$

We have

$$D^N_{g^*}(P)(O^N) = \sum_{x\in\{y,a,w\}} \sum_{s=1}^{\tau} \frac{1}{N} \sum_{i=1}^{N} \bar{D}_{x(s)}(P)(X(i), C_x(i)),$$

where $\bar{D}_{x(s)}$ is the component of the efficient influence curve that is a score of the conditional distribution $X(s)$, given $C_x(s)$. Let $D^N_{x(s)}(P) = \frac{1}{N} \sum_{i=1}^{N} \bar{D}_{x(s)}(P)(X(i), C_x(i))$. We will show that

$$- P_0^N D^N_{x(s)}(P) = \Psi_{g^*,P,x(s)}(P^*_{x(s)}) - \Psi_{g^*,P,x(s)}(P^*_{0,x(s)}) + R_{21b,x(s),g^*}(P,P_0). \quad (A.13)$$

This yields the following derivation:

$$-P_0^N D^N_{g^*}(P) = \sum_{x\in\{y,a,w\}} \sum_{s=1}^{\tau} -P_0^N D^N_{x(s)}(P)$$

$$= \sum_{x\in\{y,a,w\}} \sum_{s=1}^{\tau} \left\{ \Psi_{g^*,P,x(s)}(P^*_{x(s)}) - \Psi_{g^*,P,x(s)}(P^*_{0,x(s)}) + R_{21b,x(s),g^*}(P,P_0) \right\}$$

$$= \sum_{x\in\{y,a,w\}} \sum_{s=1}^{\tau} \left\{ \Psi_{g^*,P,x(s)}(P^*_{x(s)}) - \Psi_{g^*,P,x(s)}(P^*_{0,x(s)}) \right\}$$

$$+ \sum_{x\in\{y,a,w\}} \sum_{s=1}^{\tau} R_{21b,x(s),g^*}(P,P_0)$$

$$
= \Psi_{g^*}(P) - \Psi_{g^*}(P_0) - R_{21a,g^*}(\theta, \theta_0) + \sum_{x \in \{y,a,w\}} \sum_{s=1}^{\tau} R_{21b,x(s),g^*}(P, P_0)
$$

$$
\equiv \Psi_{g^*}(P) - \Psi_{g^*}(P_0) - R_{21a,g^*}(\theta, \theta_0) + R_{21b,g^*}(P, P_0),
$$

which proves the theorem. Thus, it remains to establish (A.13).

We have

$$
-P_0^N D_{x(s)}^N(P) = -P_0^N \frac{1}{N} \sum_{i=1}^N \frac{h_{c_x(s)}^*(C_x(i))}{\bar{h}_{c_x}(C_x(i))} \times
$$
$$
\{E(Y_{g^*}(\tau) \mid X(s) = X(i), C_x(s) = C_x(i)) - E(Y_{g^*}(\tau) \mid C_X(s) = C_x(i))\}
$$
$$
= P_0^N \frac{1}{N} \sum_{i=1}^N \frac{h_{c_x(s)}^*(C_x(i))}{\bar{h}_{c_x}(C_x(i))} \int_x \Phi_{x(s)}(x, C_x(i)) d(P_{x(s)}^* - P_{0,x(s)}^*)(x \mid C_x(i))
$$
$$
= \frac{1}{N} \sum_{i=1}^N \int_c \frac{h_{c_x(s)}^*(c)}{\bar{h}_{c_x}(c)} \int_x \Phi_{x(s)}(x, c) d(P_{x(s)}^* - P_{0,x(s)}^*)(x \mid c) h_{c_x(i),0}(c) d\mu(c)
$$
$$
= \int_c \int_x \frac{\bar{h}_{c_x,0}(c)}{\bar{h}_{c_x}(c)} \Phi_{x(s)}(x, c) d(P_{x(s)}^* - P_{0,x(s)}^*)(x \mid c) h_{c_x(s)}^*(c) d\mu(c)
$$
$$
= \int_c \int_x \left(\frac{\bar{h}_{c_x,0}}{\bar{h}_{c_x}}(c) - 1 \right) \Phi_{x(s)}(x, c) d(P_{x(s)}^* - P_{0,x(s)}^*)(x \mid c) h_{c_x(s)}^*(c) d\mu(c)
$$
$$
+ \int_{x,c} \Phi_{x(s)}(x, c) d(P_{x(s)}^* - P_{0,x(s)}^*)(x \mid c) h_{c_x(s)}^*(c) d\mu(c)
$$
$$
\equiv R_{21b,x(s),g^*}(P, P_0) + \Psi_{g^*,P,x(s)}(P_{x(s)}^*) - \Psi_{g^*,P,x(s)}(P_{0,x(s)}^*).
$$

This completes the proof. □

A.4.2 First Order Expansion of Target Parameter Based on Conditional Expectations of Efficient Influence Curve Components

We have $D_{g^*}^N(P)(O^N) = \frac{1}{N} \sum_{i=1}^N \bar{D}_{g^*}(P)(\bar{O}(i))$. The identity above combined with the TMLE solving the efficient influence curve equation $D_{g^*}^N(P_N^*)(O^N) = 0$ shows that

$$
\Psi_{g^*}(P_N^*) - \Psi_{g^*}(P_0) = \frac{1}{N} \sum_{i=1}^N \{\bar{D}_{g^*}(P_N^*)(\bar{O}(i)) - P_0^N \bar{D}_{g^*}\} + R_{21,g^*}(P_N^*, P_0).
$$

Using martingale theory, we will be able to show that $\frac{1}{\sqrt{N}} \sum_{i=1}^N \{\bar{D}_{g^*}(P_N^*)(\bar{O}(i)) - P_{0,o(i)} \bar{D}_{g^*}(P_N^*)\}$ converges to a normal limit distribution under reasonable assumptions. Therefore, we are left with a remainder

$$
R_{22,g^*,N}(P_N^*, P_0) = \frac{1}{N} \sum_{i=1}^N (P_{0,o(i)} - P_0^N) \bar{D}_{g^*}(P_N^*),
$$

which we want to show represents a second-order remainder that can be reasonably assumed to be $o_P(N^{-1/2})$. We have

$$
\bar{D}_{g^*}(P) = \sum_{x \in \{y,a,w\}} \sum_{s=1}^{\tau} \bar{D}_{g^*,x(s)}(P),
$$

where $\bar{D}_{g^*,x(s)}$ is the component of the efficient influence curve that is a score of the conditional distribution $X(s)$, given $C_x(s)$.

Thus, we have

$$R_{22,g^*,N}(P_N^*, P_0) = \sum_{x\in\{y,a,w\}} \sum_{s=1}^{\tau} R_{22,x(s),g^*,N}(P_N^*, P_0),$$

where

$$R_{22,x(s),g^*,N}(\theta, \theta_0) = \frac{1}{N} \sum_{i=1}^{N} (P_{0,o(i)} - P_0^N)\bar{D}_{g^*,x(s)}(P).$$

Therefore, it suffices to study this latter $x(s)$-specific term. Define

$$f_{x(s),g^*}(\theta,\theta_0)(C_x(i)) \equiv \frac{h_{c_x(s)}^*(C_x(i))}{\bar{h}_{c_x}(C_x(i))} \int_x \Phi_{x(s)}(x,C_x(i))d(P_{x(s)}^* - P_{0,x(s)}^*)(x \mid C_x(i)),$$

and note that this is a first order difference between P and P_0. We note that

$$E_{P_0}f_{x(s),g^*}(\theta,\theta_0)(C_x(i)) = \int_x \int_c \frac{\bar{h}_{c_x,0}(c)}{\bar{h}_{c_x}(c)} \Phi_{x(s)}(x,c)d(P_{x(s)}^* - P_{0,x(s)}^*)(x \mid c)h_{c_x(s)}^*(c)d\mu(c).$$

We have

$$R_{22,x(s),g^*,N}(\theta,\theta_0) = \frac{1}{N} \sum_{i=1}^{N} \left\{ f_{x(s),g^*}(\theta,\theta_0)(C_x(i)) - E_0 f_{x(s),g^*}(\theta,\theta_0)(C_x(i)) \right\}.$$

Thus, $R_{22,x(s),g^*,N}(\theta,\theta_0)$ is an average of N mean zero random variables, where each random variable goes to zero as $P \to P_0$. So this is a second-order remainder since it involves a double difference. In combination of the previous Theorem A.5, this results in the following theorem.

Theorem A.4. *Let* $\Phi_{x(s)}(x,c_x) = E(Y_{g^*}(\tau) \mid X(s) = x, C_x(s) = c_x)$. *Recall the representation* $D_{g^*}^N(P) = \frac{1}{N} \sum_{i=1}^{N} \bar{D}_{g^*}(P)(\bar{O}(i))$ *of the efficient influence curve.*
We have

$$-\frac{1}{N} \sum_{i=1}^{N} P_{0,o(i)} \bar{D}_{g^*}(P) = \Psi_{g^*}(P) - \Psi_{g^*}(P_0) + R_{21,g^*}(\theta,\theta_0) + R_{22,g^*,N}(\theta,\theta_0),$$

where $R_{21,g^*}(\theta,\theta_0)$ *is defined in Theorem A.5, and*

$$R_{22,g^*,N}(\theta,\theta_0) = \sum_{x\in\{y,a,w\}} \sum_{s=1}^{\tau} R_{22,x(s),g^*,N}(\theta,\theta_0),$$

$$R_{22,x(s),g^*,N}(\theta,\theta_0) = \frac{1}{N} \sum_{i=1}^{N} (P_{0,o(i)} - P_0^N)\bar{D}_{g^*,x(s)}(P)$$

$$= \frac{1}{N} \sum_{i=1}^{N} f^0_{x(s),g^*}(\theta, \theta_0)(C_x(i)),$$

$$f^0_{x(s),g^*}(\theta, \theta_0)(C_x(i)) = \{f_{x(s),g^*}(\theta, \theta_0)(C_x(i)) - P_0^N f_{x(s),g^*}(\theta, \theta_0)\},$$

$$f_{x(s),g^*}(\theta, \theta_0)(C_x(i)) \equiv \frac{h^*_{c_x(s)}(C_x(i))}{\bar{h}_{c_x}(C_x(i))} \int_x \Phi_{x(s)}(x, C_x(i)) d(P^*_{x(s)} - P^*_{0,x(s)})(x \mid C_x(i)).$$

Thus, $R_{22,x(s),g^*,N}(\theta, \theta_0)$ is an average $X_N = \frac{1}{N} \sum_{i=1}^N f_i$ of N mean zero random variables $f_i = f^0_{x(s),g^*}(\theta, \theta_0)(C_x(i))$ that are functions of $C_x(i)$, where each random variable goes to zero as $P \to P_0$.

Finally, recalling that $\Psi_J(P) = \frac{1}{J} \sum_{j=1}^J \Psi_{g^*_j}(P)$ and $D^N(P) = \frac{1}{J} \sum_{j=1}^J D^N_{g^*_j}(P)$, we have

$$-\frac{1}{N} \sum_{i=1}^N P_{0,o(i)} \bar{D}(P) = \Psi_J(P) - \Psi_J(P_0) + R_{21}(\theta, \theta_0) + R_{22,N}(\theta, \theta_0),$$

where $R_{21}(\theta, \theta_0) = \frac{1}{J} \sum_{j=1}^J R_{21,g^*_j}(\theta, \theta_0)$ and $R_{22,N}(\theta, \theta_0) = \frac{1}{J} \sum_{j=1}^J R_{22,g^*_j,N}(\theta, \theta_0)$.

A.4.3 Discussion of $R_{22,g^*,N}$ Remainder: Finite Memory Assumption

The variance of $R_{22,x(s),g^*,N}(\theta, \theta_0)$ is given by

$$\sigma^2(N) = \frac{2}{N^2} \sum_{i \le j} E_{P_0^N} \left\{ f^0_{x(s),g^*}(\theta, \theta_0)(C_x(i)) f^0_{x(s),g^*}(\theta, \theta_0)(C_x(j)) \right\}.$$

Suppose that at a fixed (nonrandom) sequence $P = P_N$ converging to P_0 we have $P_0^N f_i^2 \le r(N)^2$. Then the variance of X_N (i.e., $R_{22,x(s),g^*,N}(P_N, P_0)$) can be bounded by

$$\sigma^2(N) \le r^2(N) \frac{2}{N^2} \sum_{i \le j} \rho_N(f_i, f_j),$$

where $\rho_N(f_i, f_j)$ is the correlation of f_i and f_j under P_0^N. Thus, to make $X_N = o_P(N^{-1/2})$ we need

$$\frac{2}{N^2} \sum_{i \le j} \rho_N(f_i, f_j) = o\left(\frac{1}{Nr^2(N)}\right).$$

The latter is therefore a necessary assumption in order to have that $R_{21,x(s),g^*,N}(P^*_N, P_0) = o_P(N^{-1/2})$ at an estimator P^*_N for which $P_0^N f_{x(s),g^*}(P^*_N, P_0)^2 = O_P(r(N)^2)$.

If $\rho_N(f_{N,i}, f_{N,j}) = 0$ if $\mid j - i \mid > K(N)$ for some universal (in i, j) $K(N)$, then we obtain

$$\sigma^2(N) \sim pr^2(N)\frac{2}{N^2} \sum_{i \le j, j-i < K(N)} \sim r^2(N)\frac{K(N)}{N}.$$

So in this case, $X_N(f_N) = O_P(r(N)\sqrt{K(N)/N})$. This allows any sequence $K(N)$ converging to infinity slowly enough so that $K(N)/N = o(N^{-0.5})$, i.e., $K(N)$ grows to infinity at a rate slower than $N^{0.5}$.

Instead of correlations between $f_{N,i}$ and $f_{N,j}$ being equal to zero for large enough $|j - i|$, suppose now that $\rho_N(f_i, f_j) = O(|j - i|^{-\delta})$ for some $\delta > 0$. Then we obtain

$$\sigma^2(N) \sim r^2(N)\frac{2}{N^2} \sum_{i \le j} |j - i|^{-\delta} \sim r^2(N)N^{-\delta}.$$

Thus, in this case, we have that $X_N(f) = O_P(r(N)N^{-\delta/2})$. One would thus want that $\delta > 0.5$ so that $r(N) = o(N^{-1/4})$ will suffice.

Let's now consider a situation in which $C_x(i)$ is a summary measure of $\bar{O}(i - 1)$ that converges to a constant as $i \to \infty$. That would correspond with $f_N(C_x(i))$ converging to a constant 0 (recall it has mean zero) as $i \to \infty$, even when we fix N. Specifically, let's assume that $P_0^N f_{i,N}^2 = O(r^2(N)r_1^2(i))$ for some rate $r_1(i) = i^{-\delta}$ for some $\delta > 0$. Then,

$$\sigma^2(N) \sim \frac{2}{N^2} \sum_{i \le j} \sqrt{P_0^N f_{i,N}^2 P_0^N f_{j,N}^2} \sim \frac{1}{N^2} \sum_{i \le j} r(N)^2 r_1(i)r_1(j)$$
$$= O(r^2(N)N^{-2\delta}).$$

Thus, we now have that $X_N(f_N) = O_P(r(N)N^{-\delta})$ so that one wants $\delta > 0.25$. If $C_x(i)$ is some (e.g., maximum likelihood based) estimator of a parameter of the distribution of $\bar{O}(i)$, one might expect a rate of convergence $i^{-0.5}$, so that this assumption would be easily met.

Let's now consider an example for which the dependence of the time series is too strong for the condition to hold. For the sake of concreteness, let's assume $C_x(i) = O(i-1)$ so that $P_0^N f_{N,i} f_{N,j}$ concerns the correlation between $f_N(O_{i-1})$ and $f_N(O_{j-1})$. In addition, assume that $f_N(O(i)) = \sum_{l=1}^{i} Z(l)$, where $Z(l), l = 1, \ldots, N$, are independent mean zero random variables with variance $\sigma^2(l)$. That is, we assume that the time series develops by adding independent increments to the previous value. Now, we have for $i < j$,

$$\rho(f_{N,i}, f_{N,j}) = \sqrt{\frac{\sum_{l=1}^{i} \sigma^2(l)}{\sum_{l=1}^{j} \sigma^2(l)}}.$$

Let's for concreteness, assume that the variance of $Z(l)$ does not shrink to zero as $l \to \infty$, so that we can bound this correlation with $\rho(f_{N,i}, f_{N,j}) = O(\sqrt{i/j})$. In that case, we have

$$\sigma_f^2(N) \sim \frac{2}{N^2} \sum_{i \le j}^{N} \sqrt{i/j} r^2(N) \sim r^2(N)\frac{2}{N^2} \sum_{j=1}^{N} j = O(r^2(N)).$$

So this type of correlation structure is not weak enough to make the variance converge to zero at a faster rate than $1/N$. In this example, each observation in the past contributes proportionally to the realization of $O(i)$, and that is not the kind of long dependence that is allowed. We believe that the above discussion clarifies that $R_{22,g^*,N}(P_N^*, P_0) = o_P(N^{-1/2})$ does go beyond requiring that P_N^* converges at a fast enough rate to P_0, but it also enforces an intrinsic assumption on the dependence structure in the time series.

A.4.4 First Order Expansion for Online Estimation Based on Marginal Expectation of Efficient Influence Curve

We now establish the analogue expansion for online estimation.

Theorem A.5. *Recall* $D_{g^*}^N(P) = \frac{1}{N} \sum_{i=1}^N \bar{D}_{g^*,i}(P)$, *where*

$$\bar{D}_{g^*,i}(P)(O^N) = \sum_{x \in \{y,a,w\}} \sum_{s=1}^{\tau} \bar{D}_{x(s)}(P)(X(i), C_x(i)),$$

$\bar{D}_{x(s)}$ *is the component of the efficient influence curve that is a score of the conditional distribution* $X(s)$, *given* $C_x(s)$.
We have

$$-P_0^N \bar{D}_{g^*,i}(P) = \Psi_{g^*}(P) - \Psi_{g^*}(P_0) + R_{21,g^*,i}(P, P_0),$$

where $R_{21,g^*,i}(P, P_0) = R_{21a,g^*}(\theta, \theta_0) + R_{21b,g^*,i}(P, P_0)$, *and*

$$R_{21a,g^*}(\theta, \theta_0) = \sum_{x \in \{w,a,y\}} \sum_{s=1}^{\tau} \left\{ \prod_{t=1}^{s-1} P_{o(t)}^* \right\} \Delta_x(P_{o(s)}^*, P_{0,o(s)}^*) \times$$

$$\left\{ \sum_{j=s+1}^{\tau} \prod_{l=s+1}^{j-1} P_{o(l)}^* (P_{o(j)}^* - P_{0,o(j)}^*) \prod_{l=j+1}^{\tau} P_{0,o(l)}^* \right\} Y(\tau),$$

$$R_{21b,g^*,i}(P, P_0) = \sum_{x \in \{w,a,y\}} \sum_{s=1}^{\tau} R_{21b,x(s),g^*,i}(P, P_0),$$

$$R_{21b,x(s),g^*,i}(P, P_0) = \int_c \int_x \left(\frac{h_{c_x(i),0}}{h_{c_x}}(c) - 1 \right) \Phi_{x(s)}(x, c) d(P_{x(s)}^* - P_{0,x(s)}^*)(x \mid c) h_{c_x(s)}^*(c) d\mu(c).$$

Specifically, this shows the following identity:

$$-\frac{1}{N} \sum_{i=1}^N P_0^N \bar{D}_{g^*,i}(\theta_{i-1}) = \frac{1}{N} \sum_{i=1}^N \Psi_{g^*}(\theta_{i-1}) - \Psi_{g^*}(\theta_0) + \bar{R}_{21,g^*,N},$$

where

$$\bar{R}_{21,g^*,N} \equiv \bar{R}_{21a,g^*,N} + \bar{R}_{21b,g^*,N}, \tag{A.14}$$

and

$$\bar{R}_{21a,g^*,N} \equiv \frac{1}{N} \sum_{i=1}^{N} R_{21a,g^*}(\theta_{i-1}, \theta_0),$$

$$\bar{R}_{21b,g^*,N} \equiv \frac{1}{N} \sum_{i=1}^{N} R_{21b,g^*,i}(\theta_{i-1}, \theta_0).$$

Proof. In the proof of Theorem A.3 we showed:

$$\Psi_{g^*}(\theta) - \Psi_{g^*}(\theta_0) = \sum_{x \in \{w,a,y\}} \sum_{s=1}^{\tau} \left(\Psi_{g^*,P,x(s)}(P^*_{x(s)}) - \Psi_{g^*,P,x(s)}(P^*_{0,x(s)}) \right) + R_{21a,g^*}(\theta, \theta_0).$$

We will show that

$$-P_0^N \bar{D}_{x(s),i}(P) = \Psi_{g^*,P,x(s)}(P^*_{x(s)}) - \Psi_{g^*,P,x(s)}(P^*_{0,x(s)}) + R_{21b,x(s),g^*,i}(P, P_0). \quad (A.15)$$

This yields the following derivation:

$$-P_0^N \bar{D}_{g^*,i}(P) = \sum_{x \in \{y,a,w\}} \sum_{s=1}^{\tau} -P_0^N \bar{D}_{x(s),i}(P)$$

$$= \sum_{x \in \{y,a,w\}} \sum_{s=1}^{\tau} \left\{ \Psi_{g^*,P,x(s)}(P^*_{x(s)}) - \Psi_{g^*,P,x(s)}(P^*_{0,x(s)}) + R_{21b,x(s),g^*,i}(P, P_0) \right\}$$

$$= \sum_{x \in \{y,a,w\}} \sum_{s=1}^{\tau} \left\{ \Psi_{g^*,P,x(s)}(P^*_{x(s)}) - \Psi_{g^*,P,x(s)}(P^*_{0,x(s)}) \right\}$$

$$+ \sum_{x \in \{y,a,w\}} \sum_{s=1}^{\tau} R_{21b,x(s),g^*,i}(P, P_0)$$

$$= \Psi_{g^*}(P) - \Psi_{g^*}(P_0) - R_{21a,g^*}(\theta, \theta_0) + \sum_{x \in \{y,a,w\}} \sum_{s=1}^{\tau} R_{21b,x(s),g^*,i}(P, P_0)$$

$$\equiv \Psi_{g^*}(P) - \Psi_{g^*}(P_0) - R_{21a,g^*}(\theta, \theta_0) + R_{21b,g^*,i}(P, P_0),$$

which proves the theorem. Thus, it remains to establish (A.15).
We have

$$-P_0^N \bar{D}_{x(s),i}(P) = -P_0^N \frac{h^*_{c_x(s)}(C_x(i))}{h_{c_x}(C_x(i))} \times$$
$$\{E(Y_{g^*}(\tau) \mid X(s) = X(i), C_x(s) = C_x(i)) - E(Y_{g^*}(\tau) \mid C_x(s) = C_x(i))\}$$

$$= P_0^N \frac{h^*_{c_x(s)}(C_x(i))}{h_{c_x}(C_x(i))} \int_x \Phi_{x(s)}(x, C_x(i)) d(P^*_{x(s)} - P^*_{0,x(s)})(x \mid C_x(i))$$

$$= \int_c \frac{h^*_{c_x(s)}(c)}{h_{c_x}(c)} \int_x \Phi_{x(s)}(x, c) d(P^*_{x(s)} - P^*_{0,x(s)})(x \mid c) h_{c_x(i),0}(c) d\mu(c)$$

$$= \int_c \int_x \frac{h_{c_x(i),0}(c)}{h_{c_x}(c)} \Phi_{x(s)}(x, c) d(P^*_{x(s)} - P^*_{0,x(s)})(x \mid c) h^*_{c_x(s)}(c) d\mu(c)$$

$$= \int_c \int_x \left(\frac{h_{c_x(i),0}}{h_{c_x}}(c) - 1 \right) \Phi_{x(s)}(x, c) d(P^*_{x(s)} - P^*_{0,x(s)})(x \mid c) h^*_{c_x(s)}(c) d\mu(c)$$

$$+ \int_{x,c} \Phi_{x(s)}(x, c) d(P^*_{x(s)} - P^*_{0,x(s)})(x \mid c) h^*_{c_x(s)}(c) d\mu(c)$$

$$\equiv R_{21b,x(s),g^*,i}(P, P_0) + \Psi_{g^*,P,x(s)}(P^*_{x(s)}) - \Psi_{g^*,P,x(s)}(P^*_{0,x(s)}).$$

This completes the proof. □

Discussion of Remainder $\bar{R}_{21b,g^*,N}$**: Another Stationarity Assumption.** Despite their resemblance, the remainder terms $\bar{R}_{21b,g^*,N}$ (arising from the analysis of the online one-step estimator and online TMLE estimators) and R_{21b,g^*} (coming from the analysis of the regular TMLE and one-step estimators) are quite different. A closer look at them reveals that $\bar{R}_{21b,g^*,N}$ essentially writes as an average across $1 \le i \le N$ of i-specific products of the form $(1-h_i(\theta_0)/N^{-1} \sum_{j=1}^N h_j(\theta_i)) \times (f(\theta_i)-f(\theta_0))$ whereas R_{21b,g^*} averages over i $(1 - N^{-1} \sum_{j=1}^N h_j(\theta_0)/N^{-1} \sum_{j=1}^N h_j(\theta_N)) \times (f(\theta_N) - f(\theta_0))$. In this coarse but enlightening description, f and h_i map continuous θ to functions, and θ_N converges to θ_0. It appears that $\bar{R}_{21b,g^*,N}$ only benefits from the convergence to zero of $(f(\theta_N) - f(\theta_0))$ whereas R_{21b,g^*} benefits, through a product, from the convergence to zero of *both* $(f(\theta_N) - f(\theta_0))$ *and* $(1 - N^{-1} \sum_{j=1}^N h_j(\theta_0)/N^{-1} \sum_{j=1}^N h_j(\theta_N))$. This suggests that the behavior of $\bar{R}_{21b,g^*,N}$ might be significantly worse than that of R_{21b,g^*}.

A more in depth study of $\bar{R}_{21b,g^*,N}$ will be necessary. Here we observe that, under a strong form of stationarity of the time-series, we can replace the i-specific term in the expression of $\bar{R}_{21b,g^*,N}$ by $(1 - N^{-1} \sum_{j=1}^N h_j(\theta_0)/N^{-1} \sum_{j=1}^N h_j(\theta_i)) \times (f(\theta_i) - f(\theta_0))$, showing that $\bar{R}_{21b,g^*,N}$ now can also benefit from the convergence to zero of *both* factors as i and N go to infinity. Concretely, our assumption that the mechanism for generating $O(t)$, given $\bar{O}(t-1)$, is constant in time t may be insufficient, but can be complemented by the additional assumption that the marginal density of $C_x(i)$ is constant in i for $x = w, a, y$.

A.4.5 First Order Expansion for Online Estimation Based on Conditional Expectation of Efficient Influence Curve

The following theorem is an easy consequence of Theorem A.5.

Theorem A.6. *We have*

$$-\frac{1}{N} \sum_{i=1}^N P_{0,o(i)} \bar{D}_{g^*}(\theta_{i-1}) = \frac{1}{N} \sum_{i=1}^N \Psi_{g^*}(\theta_{i-1}) - \Psi_{g^*}(\theta_0) + \bar{R}_{21,g^*,N} + \bar{R}_{22,g^*,N},$$

where $\bar{R}_{21,g^*,N}$ *is defined in (A.14), and*

$$\bar{R}_{22,g^*,N} \equiv \frac{1}{N} \sum_{i=1}^N (P_{0,o(i)} - P_0^N) \left\{ \bar{D}_{g^*}(\theta_{i-1}) - \bar{D}_{g^*}(\theta_0) \right\}.$$

References

O. Aalen, Nonparametric estimation of partial transition probabilities in multiple decrement models. Ann. Stat. **6**, 534–545 (1978)

A. Abadie, G. Imbens, Simple and bias-corrected matching estimators for average treatment effects. Technical Report 283. NBER Working Paper (2002)

Action to Control Cardiovascular Risk in Diabetes Study Group. Effects of intensive glucose lowering in type 2 diabetes. N. Engl. J. Med. **358**, 2545–2549 (2008)

ADVANCE Collaborative Group, Intensive blood glucose control and vascular outcomes in patients with type 2 diabetes. N. Engl. J. Med. **358**, 2560–2562 (2008)

A. Afifi, S. Azen, *Statistical Analysis: A Computer Oriented Approach*, 2nd edn. (Academic, New York, 1979)

C. Anderson, The end of theory: the data deluge makes the scientific method obsolete. Wired (2008)

T.J. Aragon, epitools: Epidemiology tools (2012). http://cran.r-project.org/package=epitools

S. Aral, D. Walker, Identifying social influence in networks using randomized experiments. IEEE Intell. Syst. **26**(5), 91–96 (2011)

S. Aral, D. Walker, Tie strength, embeddedness, and social influence: a large-scale networked experiment. Manag. Sci. **60**(6), 1352–1370 (2014)

P.M. Aronow, C. Samii, Estimating average causal effects under interference between units. ArXiv e-prints, May (2013)

J.Y. Audibert, A.B. Tsybakov, Fast learning rates for plug-in classifiers. Ann. Stat. **35**(2), 608–633 (2007)

L. Auret, C. Aldrich, Empirical comparison of tree ensemble variable importance measures. Chemom. Intel. Lab. Syst. **105**(2), 157–170 (2011)

P.C. Austin, A. Manca, M. Zwarensteina, D.N. Juurlinka, M.B. Stanbrook, A substantial and confusing variation exists in handling of baseline covariates in randomized controlled trials: a review of trials published in leading medical journals. J. Clin. Epidemiol. **63**, 142–153 (2010)

C. Avin, I. Shpitser, J. Pearl, Identifiability of path-specific effects. Proceedings of International Joint Conference on Artificial Intelligence, 357–363 (2005)

S. Balakrishnan, D. Madigan, Algorithms for sparse linear classifiers in the massive data setting. J. Mach. Learn. Res. **9**, 313–337 (2008)

L. Balzer, M. Petersen, M.J. van der Laan, Adaptive pair-matching in randomized trials with unbiased and efficient effect estimation. Stat. Med. **34**(6), 999–1011 (2015)

L. Balzer, J. Ahern, S. Galea, M.J. van der Laan, Estimating effects with rare outcomes and high dimensional covariates: Knowledge is power. Epidemiol. Methods. **5**(1), 1–18 (2016a)

L. Balzer, M. van der Laan, M. Petersen, the SEARCH Collaboration, Adaptive pre-specification in randomized trials with and without pair-matching. Stat. Med. **35**(25), 4528–4545 (2016b)

L.B. Balzer, M.L. Petersen, M.J. van der Laan, the SEARCH Collaboration, Targeted estimation and inference of the sample average treatment effect in trials with and without pair-matching. Stat. Med. **35**(21), 3717–3732 (2016c)

H. Bang, J.M. Robins, Doubly robust estimation in missing data and causal inference models. Biometrics **61**, 962–972 (2005)

A.-L. Barabási, R. Albert, Emergence of scaling in random networks. Science **286**(5439), 509–512 (1999)

E. Bareinboim, J. Pearl, A general algorithm for deciding transportability of experimental results. J. Causal Inf. **1**(1), 107–134 (2013)

G.W. Basse, E.M. Airoldi, Optimal design of experiments in the presence of network-correlated outcomes. ArXiv e-prints, July (2015)

C. Beck, B. Lu, R. Greevy, nbpMatching: functions for optimal non-bipartite optimal matching (2016). https://CRAN.R-project.org/package=nbpMatching

O. Bembom, M.J. van der Laan, A practical illustration of the importance of realistic individualized treatment rules in causal inference. Electron. J. Stat. **1**, 574–596 (2007)

O. Bembom, M.J. van der Laan, Analyzing sequentially randomized trials based on causal effect models for realistic individualized treatment rules. Stat. Med. **27**, 3689–3716 (2008)

O. Bembom, M.L. Petersen, S.-Y. Rhee, W.J. Fessel, S.E. Sinisi, R.W. Shafer, M.J. van der Laan, Biomarker discovery using targeted maximum likelihood estimation: application to the treatment of antiretroviral resistant HIV infection. Stat. Med. **28**, 152–72 (2009)

J. Benichou, M.H. Gail, Estimates of absolute cause-specific risk in cohort studies. Biometrics **46**, 813–826 (1990)

Y. Benjamini, Y. Hochberg, Controlling the false discovery rate: a practical and powerful approach to multiple testing. J. R. Stat. Soc. Ser. B **57**, 289–300 (1995)

D. Benkeser, M.J. van der Laan, The highly adaptive lasso estimator, in *IEEE International Conference on Data Science and Advanced Analytics*, pp. 689–696 (2016)

D. Benkeser, M. Carone, M.J. van der Laan, P. Gilbert, Doubly-robust nonparametric inference on the average treatment effect. Biometrika. **104**(4), 863–880 (2017a)

D. Benkeser, S.D. Lendle, J. Cheng, M.J. van der Laan, Online cross-validation-based ensemble learning. Stat. Med. **37**(2), 249–260 (2017b)

P. Bertail, A. Chambaz, E. Joly, Practical targeted learning from large data sets by survey sampling. ArXiv e-prints, June (2016)

P. Bertail, E. Chautru, S. Clémençon, Empirical processes in survey sampling with (conditional) Poisson designs. Scand. J. Stat. **44**(1), 97–111 (2017)

P.J. Bickel, On adaptive estimation. Ann. Stat. **10**, 647–671 (1982)

P.J. Bickel, F. Götze, W.R. van Zwet, Resampling fewer than n observations: gains, losses, and remedies for losses. Stat. Sin. **7**(1), 1–31 (1997a)

P.J. Bickel, C.A.J. Klaassen, Y. Ritov, J. Wellner, *Efficient and Adaptive Estimation for Semiparametric Models* (Springer, Berlin, Heidelberg, New York, 1997b)

L. Bondesson, I. Traat, A. Lundqvist, Pareto sampling versus Sampford and conditional Poisson sampling. Scand. J. Stat. Theory Appl. **33**(4), 699–720 (2006)

L. Bottou, Large-scale machine learning with stochastic gradient descent, in *Proceedings of COMPSTAT'2010* (Springer, Berlin, 2010), pp. 177–186

L. Bottou, Stochastic gradient descent tricks, in *Neural Networks: Tricks of the Trade* (Springer, Berlin, 2012), pp. 421–436

J. Bowers, M.M. Fredrickson, C. Panagopoulos, Reasoning about interference between units: a general framework. Polit. Anal. **21**(1), 97–124 (2013)

L. Breiman, Random forests. Mach. Learn. **45**, 5–32 (2001)

L. Breiman, J.H. Friedman, R. Olshen, C.J. Stone, *Classification and Regression Trees* (Chapman & Hall, Boca Raton, 1984)

L. Breiman et al., Statistical modeling: the two cultures (with comments and a rejoinder by the author). Stat. Sci. **16**(3), 199–231 (2001)

D.I. Broadhurst, D.B. Kell, Statistical strategies for avoiding false discoveries in metabolomics and related experiments. Metabolomics **2**(4), 171–196 (2006)

D.W. Brock, D. Wikler, Ethical challenges in long-term funding for HIV/AIDS. Health Aff. **28**(6), 1666–1676 (2009)

J.C. Brooks, Super learner and targeted maximum likelihood estimation for longitudinal data structures with applications to atrial fibrillation. PhD thesis, University of California, Berkeley (2012)

L.E. Cain, J.M. Robins, E. Lanoy, R. Logan, D. Costagliola, M.A. Hernan, When to start treatment? A systematic approach to the comparison of dynamic regimes using observational data. Int. J. Biostat. **6**, Article 18 (2010)

R.M. Califf, D.A. Zarin, J.M. Kramer, R.E. Sherman, L.H. Aberle, and A. Tasneem, Characteristics of clinical trials registered in ClinicalTrials.gov, 2007–2010. J. Am. Med. Assoc. **307**(17), 1838–1847 (2012)

A.C. Cameron, J.B. Gelbach, D.L. Miller, Boostrap-based improvements for inference with clustered errors. Rev. Econ. Stat. **90**(3), 414–427 (2008)

M.J. Campbell, Cluster randomized trials, in *Handbook of Epidemiology*, 2nd edn., ed. by W. Ahrens, I. Pigeot (Springer, Berlin, 2014)

M.J. Campbell, A. Donner, N. Klar, Developments in cluster randomized trials and statistics in medicine. Stat. Med. **26**, 2–19 (2007)

M. Carone, I. Díaz, M.J. van der Laan, Higher-order targeted minimum loss-based estimation. Technical Report, Division of Biostatistics, University of California, Berkeley

B. Chakraborty, E.E. Moodie, *Statistical Methods for Dynamic Treatment Regimes* (Springer, Berlin, Heidelberg, New York, 2013)

B. Chakraborty, E.B. Laber, Y.-Q. Zhao, Inference about the expected performance of a data-driven dynamic treatment regime. Clin. Trials **11**(4), 408–417 (2014)

A. Chambaz, tsml.cara.rct: targeted sequential minimum loss CARA RCT design and inference (2016). https://github.com/achambaz/tsml.cara.rct

A. Chambaz, P. Neuvial, Targeted, integrative search of associations between DNA copy number and gene expression, accounting for DNA methylation. Bioinformatics **31**(18), 3054–3056 (2015)

A. Chambaz, P. Neuvial, Targeted learning of a non-parametric variable importance measure of a continuous exposure (2016). http://CRAN.R-project.org/package=tmle.npvi

A. Chambaz, M.J. van der Laan, Inference in targeted group-sequential covariate-adjusted randomized clinical trials. Scand. J. Stat. **41**(1), 104–140 (2014)

A. Chambaz, M.J. van der Laan, Targeting the optimal design in randomized clinical trials with binary outcomes and no covariate: theoretical study. Int. J. Biostat. **7**(1), Article 10 (2011a)

A. Chambaz, M.J. van der Laan, Targeting the optimal design in randomized clinical trials with binary outcomes and no covariate: simulation study. Int. J. Biostat. **7**(1), Article 11 (2011b)

A. Chambaz, M.J. van der Laan, TMLE in adaptive group sequential covariate-adjusted RCTs, in *Targeted Learning: Causal Inference for Observational and Experimental Data*, ed. by M.J. van der Laan, S. Rose (Springer, Berlin Heidelberg, New York, 2011c)

A. Chambaz, P. Neuvial, M.J. van der Laan, Estimation of a non-parametric variable importance measure of a continuous exposure. Electron. J. Stat. **6**, 1059–1099 (2012)

A. Chambaz, D. Choudat, C. Huber, J.C. Pairon, M.J. van der Laan, Analysis of the effect of occupational exposure to asbestos based on threshold regression modeling of case–control data. Biostatistics **15**(2), 327–340 (2014)

A. Chambaz, M.J. van der Laan, W. Zheng, Targeted covariate-adjusted response-adaptive lasso-based randomized controlled trials, in *Modern Adaptive Randomized Clinical Trials: Statistical, Operational, and Regulatory Aspects*, ed. by A. Sverdlov (CRC Press, Boca Raton, 2015), pp. 345–368

A. Chambaz, W. Zheng, M.J. van der Laan, Targeted sequential design for targeted learning of the optimal treatment rule and its mean reward. Ann Stat. **45**(6), 1–28 (2017)

T. Chen, C. Guestrin, Xgboost: a scalable tree boosting system, in *Proceedings of the 22nd ACM SIGKDD International Conference on Knowledge Discovery and Data Mining* (ACM, New York, 2016), pp. 785–794

O.Y. Chén, C. Crainiceanu, E.L. Ogburn, B.S. Caffo, T.D. Wager, M.A. Lindquist, High-dimensional multivariate mediation with application to neuroimaging data. Biostatistics (2017, in press)

D.S. Choi, Estimation of monotone treatment effects in network experiments. ArXiv e-prints, August (2014)

N.A. Christakis, J.H. Fowler, The spread of obesity in a large social network over 32 years. N. Engl. J. Med. **357**(4), 370–379 (2007)

N.A. Christakis, J.H. Fowler, Social contagion theory: examining dynamic social networks and human behavior. Stat. Med. **32**(4), 556–577 (2013)

W.G. Cochran, Analysis of covariance: its nature and uses. Biometrics **13**, 261–281 (1957)

E. Colantuoni, M. Rosenblum, Leveraging prognostic baseline variables to gain precision in randomized trials. Technical Report 263, Johns Hopkins University, Department of Biostatistics Working Papers (2015)

S.R. Cole, E.A. Stuart, Generalizing evidence from randomized clinical trials to target populations: the ACTG 320 trial. Am. J. Epidemiol. **172**(1), 107–115 (2010)

S.R. Cole, M.A. Hernan, J.M. Robins, K. Anastos, J. Chmiel, R. Detels, C. Ervin, J. Feldman, R. Greenblatt, L. Kingsley, S. Lai, M. Young, M. Cohen, A. Munoz, Effect of highly active antiretroviral therapy on time to acquired immunodeficiency syndrome or death using marginal structural models. Am. J. Epidemiol. **158**(7), 687–694 (2003)

D.R. Cox, P. McCullagh, Some aspects of analysis of covariance. Biometrics **38**(3), 541–561 (1982)

K. Crammer, O. Dekel, J. Keshet, S. Shalev-Shwartz, Y. Singer, Online passive-aggressive algorithms. J. Mach. Learn. Res. **7**, 551–585 (2006)

G.B. Dantzig, Discrete-variable extremum problems. Oper. Res. **5**(2), 266–288 (1957)

A.C. Davison, D.V. Hinkley, *Bootstrap methods and Their Application*. Cambridge Series in Statistical and Probabilistic Mathematics, vol. 1 (Cambridge University Press, Cambridge, Cambridge, 1997)

A.P. Dawid, V. Didelez, Identifying the consequences of dynamic treatment strategies: a decision-theoretic overview. Stat. Surv. **4**, 184–231 (2010)

V.H. de la Peña, E. Giné, Decoupling, in *Probability and its Applications* (Springer, New York, 1999)

L. Denby, C. Mallows, Variations on the histogram. J. Comput. Graph. Stat. **18**(1), 21–31 (2009)

I. Díaz, M. van der Laan, Super learner-based conditional density estimation with application to marginal structural models. Int. J. Biostat. **7**(1), 38 (2011)

I. Díaz, M. van der Laan, Population intervention causal effects based on stochastic interventions. Biometrics **68**(2), 541–549 (2012)

I. Díaz, M.J. van der Laan, Assessing the causal effect of policies: an example using stochastic interventions. Int. J. Biostat. **9**(2), 161–174 (2013a)

I. Díaz, M.J. van der Laan, Sensitivity analysis for causal inference under unmeasured confounding and measurement error problems. Int. J. Biostat. **9**(2), 149–160 (2013b)

I. Díaz, M. Carone, M.J. van der Laan, Second-order inference for the mean of a variable missing at random. Int. J. Biostat. **12**(1), 333–349 (2016)

I. Díaz, A. Hubbard, A. Decker, M. Cohen, Variable importance and prediction methods for longitudinal problems with missing variables. PLoS One **10**(3), e0120031 (2015)

T.J. DiCiccio, J.P. Romano, A review of bootstrap confidence intervals. J. R. Stat. Soc. Ser. B (1988)

T.J. DiCiccio, J.P. Romano, Nonparametric confidence limits by resampling methods and least favorable families. Int. Stat. Rev./Revue Internationale de Statistique **58**(1), 59 (1990)

V. Didelez, A.P. Dawid, S. Geneletti, Direct and indirect effects of sequential treatments, in *Proceedings of the 22nd Annual Conference on Uncertainty in Artificial Intelligence* (2006), pp. 138–146

P. Ding, T. VanderWeele, Sensitivity analysis without assumptions. Epidemiol. **27**(3), 368–377 (2016)

A. Donner, N. Klar, *Design and Analysis of Cluster Randomization Trials in Health Research* (Arnold, London, 2000)

J. Duchi, E. Hazan, Y. Singer, Adaptive subgradient methods for online learning and stochastic optimization. J. Mach. Learn. Res. **12**, 2121–2159 (2011)

W. Duckworth, C. Abraira, T. Moritz, D. Reda, N. Emanuele, P.D. Reaven, F.J. Zieve, J. Marks, S.N. Davis, R. Hayward, S.R. Warren, S. Goldman, M. Mc-Carren, M.E. Vitek, W.G. Henderson, G.D. Huang for the VADT Investigators, Glucose control and vascular complications in veterans with type 2 diabetes. N. Engl. J. Med. **360**, 129–39 (2009a)

W. Duckworth et al., Glucose control and vascular complications in veterans with type 2 diabetes. N. Engl. J. Med. **360**(2), 129–139 (2009b)

S. Dudoit, M.J. van der Laan, Asymptotics of cross-validated risk estimation in estimator selection and performance assessment. Stat. Methodol. **2**(2), 131–154 (2005)

F. Eberhardt, R. Scheines, Interventions and causal inference. Department of Philosophy. Paper 415 (2006)

B. Efron, Better bootstrap confidence intervals. J. Am. Stat. Assoc. **82**(397), 171–185 (1987)

B. Efron, R.J. Tibshirani, *An Introduction to the Bootstrap* (Chapman & Hall, Boca Raton, 1993)

U. Einmahl, D.M. Mason, An empirical process approach to the uniform consistency of kernel-type function estimators. J. Theor. Probab. **13**(1) 1–37 (2000)

U. Einmahl, D.M. Mason, Uniform in bandwidth consistency of kernel-type function estimators. Ann. Stat. **33**(3), 1380–1403 (2005)

European Medicines Agency, Guideline on adjustment for baseline covariates in clinical trials. London, February (2015)

J.P. Fine, R.J. Gray, A proportional hazards model for the subdistribution of a competing risk. J. Am. Stat. Assoc. **94**(446), 496–509 (1999)

M. Finster, M. Wood, The Apgar score has survived the test of time. Anesthesiology **102**(4), 855–857 (2005)

R.A. Fisher, *Statistical Methods for Research Workers*, 4th edn. (Oliver and Boyd Ltd., Edinburgh, 1932)

R.A. Fisher, *The Design of Experiments*, (Oliver and Boyd Ltd, London, 1935)

C.E. Frangakis, T. Qian, Z. Wu, I. Diaz, Deductive derivation and Turing-computerization of semiparametric efficient estimation. Biometrics **71**(4), 867–874 (2015)

L.S. Freedman, M.H. Gail, S.B. Green, D.K. Corle, The COMMIT Research Group, The Efficiency of the matched-pairs design of the community intervention trial for smoking cessation (COMMIT). Control. Clin. Trials **18**(2), 131–139 (1997)

J.H. Friedman, Multivariate adaptive regression splines. Ann. Stat. **19**(1), 1–141 (1991)

J.H. Friedman, Greedy function approximation: a gradient boosting machine. Ann. Stat. **29**, 1189–1232 (2001)

J.H. Friedman, T.J. Hastie, R.J. Tibshirani, Glmnet: lasso and elastic-net regularized generalized linear models (2010). http://CRAN.R-project.org/package=glmnet

K.J. Friston, L. Harrison, W. Penny, Dynamic causal modelling. Neuroimage **19**(4), 1273–1302 (2003)

K. Friston, R. Moran, A.K. Seth, Analysing connectivity with granger causality and dynamic causal modelling. Curr. Opin. Neurobiol. **23**(2), 172–178 (2013)

W.J. Fu, Penalized regressions: the bridge versus the lasso. J. Comput. Graph. Stat. **7**(3), 397–416 (1998)

P. Galison, *How Experiments End* (University of Chicago Press, Chicago, 1987)

J.J. Gaynor, E.J. Feuer, C.C. Tan, D.H. Wu, C.R. Little, D.J. Straus, B.D. Clarkson, M.F. Brennan, On the use of cause-specific failure and conditional failure probabilities: examples from clinical oncology data. J. Am. Stat. Assoc. **88**(422), 400–409 (1993)

A. Gelman, C. Shalizi, Philosophy and the practice of bayesian statistics. Br. J. Math. Stat. Psychol. 66(1), 8–38 (2013)

A. Gelman, Y.-S. Su, M. Yajima, J. Hill, M.G. Pittau, J. Kerman, T. Zheng, Arm: data analysis using regression and multilevel/hierarchical models (2010). http://CRAN.R-project.org/package=arm

H.C. Gerstein et al., Effects of intensive glucose lowering in type 2 diabetes. N. Engl. J. Med. **358**(24), 2545–2559 (2008)

G. Gigerenzer, *The Empire of Chance: How Probability Changed Science and Everyday Life* (Cambridge University Press, Cambridge, 1989)

P.B. Gilbert, Comparison of competing risks failure time methods and time-independent methods for assessing strain variations in vaccine protection. Stat. Med. **19**(22), 3065–3086 (2000)

P.B. Gilbert, S.G. Self, M.A. Ashby, Statistical methods for assessing differential vaccine protection against human immunodeficiency virus types. Biometrics **54**(3), 799–814 (1998)

P.B. Gilbert, S.G. Self, M. Rao, A. Naficy, J. Clemens, Sieve analysis: methods for assessing from vaccine trial data how vaccine efficacy varies with genotypic and phenotypic pathogen variation. J. Clin. Epidemiol. **54**(1), 68–85 (2001)

R.D. Gill, Non- and semiparametric maximum likelihood estimators and the von Mises method (Part 1). Scand. J. Stat. **16**, 91–128 (1989)

R.D. Gill, J.M. Robins, Causal inference in complex longitudinal studies: continuous case. Ann. Stat. **29**(6), 1785–1811 (2001)

R.D. Gill, M.J. van der Laan, J.A. Wellner, Inefficient estimators of the bivariate survival function for three models. Ann. l'Institut Henri Poincaré **31**(3), 545–597 (1995)

Y. Goldberg, R. Song, D. Zeng, M.R. Kosorok, Comment on "Dynamic treatment regimes: technical challenges and applications". Electron. J. Stat. **8**, 1290–1300 (2014)

N. Grambauer, M. Schumacher, J. Beyersmann, Proportional subdistribution hazards modeling offers a summary analysis, even if misspecified. Stat. Med. **29**(7–8), 875–884 (2010)

R. Greevy, B. Lu, J.H. Silber, P. Rosenbaum, Optimal multivariate matching before randomization. Biostatistics **5**(2), 263–275 (2004)

U. Grömping, Variable importance assessment in regression: linear regression versus random forest. Am. Stat. **63**(4) (2009)

H. Grosskurth, F. Mosha, J. Todd, E. Mwijarubi, A. Klokke, K. Senkoro, P. Mayaud, J. Changalucha, A. Nicoll, G. ka-Gina, J. Newell, K. Mugeye, D. Mabey, R. Hayes, Impact of improved treatment of sexually transmitted diseases on HIV infection in rural Tanzania: randomised controlled trial. Lancet **346**(8974), 530–536 (1995)

S. Gruber, M.J. van der Laan, An application of collaborative targeted maximum likelihood estimation in causal inference and genomics. Int. J. Biostat. **6**(1) (2010a)

S. Gruber, M.J. van der Laan, A targeted maximum likelihood estimator of a causal effect on a bounded continuous outcome. Int. J. Biostat. **6**(1), Article 26 (2010b)

S. Gruber, M.J. van der Laan, tmle: an R package for targeted maximum likelihood estimation. J. Stat. Softw. 51(13) (2012a)

S. Gruber, M.J. van der Laan, Targeted minimum loss based estimator that outperforms a given estimator. Int. J. Biostat. **8**(1), (2012b)

I. Hacking, *The Emergence of Probability* (Cambridge University Press, Cambridge, 1975)

I. Hacking, *The Taming of Chance (1990)* (Cambridge University Press, Cambridge, 1990)

D.M. Hafeman, T.J. VanderWeele, Alternative assumptions for the identification of direct and indirect effects. Epidemiology **22**, 753–764 (2010)

J. Hahn, On the role of the propensity score in efficient semiparametric estimation of average treatment effects. Econometrica **2**, 315–331 (1998)

J. Hajek, Asymptotic theory of rejective sampling with varying probabilities from a finite population. Ann. Math. Stat. **35**(4), 1491–1523, 12 (1964)

P Hall, Theoretical comparison of bootstrap confidence intervals. Ann. Stat. **16**, 927–953 (1988)

P. Hall, *The Bootstrap and Edgeworth Expansion*. Springer Series in Statistics (Springer, New York, NY, 1992)

M.E. Halloran, C.J. Struchiner, Causal inference in infectious diseases. Epidemiology **6**(2), 142–151 (1995)

S.M. Hammer, M.E. Sobieszczyk, H. Janes, S.T. Karuna, M.J. Mulligan, D. Grove, B.A. Koblin, S.P. Buchbinder, M.C. Keefer, G.D. Tomaras, Efficacy trial of a DNA/rAd5 HIV-1 preventive vaccine. N. Engl. J. Med. **369**(22), 2083–2092 (2013)

S. Haneuse, A. Rotnitzky, Estimation of the effect of interventions that modify the received treatment. Stat. Med. (2013)

M. Hanif, K.R.W. Brewer, Sampling with unequal probabilities without replacement: a review. International Statistical Review/Revue Internationale de Statistique, pp. 317–335 (1980)

E. Hartman, R. Grieve, R. Ramsahai, J.S. Sekhon, From sample average treatment effect to population average treatment effect on the treated: combining experimental with observational studies to estimate population treatment effects. J. R. Stat. Soc. Ser. A **178**(3), 757–778 (2015)

T. Hastie, gam: generalized additive models (2011) http://CRAN.R-project.org/package=gam

T.J. Hastie, R.J. Tibshirani, J.H. Friedman, *The Elements of Statistical Learning: Data Mining, Inference, and Prediction* (Springer, Berlin Heidelberg New York, 2001)

R.J. Hayes, L.H. Moulton, *Cluster Randomised Trials.* (Chapman & Hall/CRC, Boca Raton, 2009)

M.A. Hearst, S.T Dumais, E. Osman, J. Platt, B. Scholkopf. Support vector machines. IEEE Intell. Syst. Appl. **13**(4), 18–28 (1998)

M.A. Hernan, B.A. Brumback, J.M. Robins, Estimating the causal effect of zidovudine on CD4 count with a marginal structural model for repeated measures. Stat. Med. **21**, 1689–1709 (2002)

M.A. Hernan, B. Brumback, J.M. Robins, Marginal structural models to estimate the causal effect of zidovudine on the survival of HIV-positive men. Epidemiology **11**(5), 561–570 (2000)

M.A. Hernan, E. Lanoy, D. Costagliola, J.M. Robins, Comparison of dynamic treatment regimes via inverse probability weighting. Basic Clin. Pharmacol. **98**, 237–242 (2006)

R. Holiday, What the failed $1m Netflix prize says about business advice. Forbes (2012)

R.R. Holman, S.K. Paul, M.A. Bethel, D.R. Matthews, H.A. Neil, 10-year follow-up of intensive glucose control in type 2 diabetes. N. Engl. J. Med. **359**, 1577–89 (2008)

J.L. Horowitz, C.F. Manski, Nonparametric analysis of randomized experiments with missing covariate and outcome data. J. Am. Stat. Assoc. **95**(449), 77–84 (2000)

D.G. Horvitz, D.J. Thompson, A generalization of sampling without replacement from a finite universe. J. Am. Stat. Assoc. **47**, 663–685 (1952)

J.B. Hough, M. Krishnapur, Y. Peres, B. Virág, Determinantal processes and independence. Probab. Surv. **3**, 206–229 (2006)

F. Hu, W.F. Rosenberger, *The Theory of Response Adaptive Randomization in Clinical Trials* (Wiley, New York, 2006)

A.E. Hubbard, M.J. van der Laan, Mining with inference: data adaptive target parameters, in *Handbook of Big Data*. Chapman-Handbooks-Statistical-Methods, ed. by P. Buhlmann, P. Drineas, M. Kane, M.J. van der Laan (Chapman & Hall/CRC, Boca Raton, 2016)

A.E. Hubbard, I Diaz Munoz, A. Decker, J.B. Holcomb, M.A. Schreiber, E.M. Bulger, K.J. Brasel, E.E. Fox, D.J. del Junco, C.E. Wade et al., Time-dependent prediction and evaluation of variable importance using superlearning in high-dimensional clinical data. J. Trauma-Injury Infect. Crit. Care **75**(1), S53–S60 (2013)

A.E. Hubbard, S. Kherad-Pajouh, M.J. van der Laan, Statistical inference for data adaptive target parameters. Int. J. Biostat. **12**(1), 3–19 (2016)

M.G. Hudgens, M.E. Halloran, Toward causal inference with interference. J. Am. Stat. Assoc. **103**(482), 832–842 (2008)

I.A. Ibragimov, R.Z. Khasminskii, *Statistical Estimation* (Springer, Berlin, 1981)

ICH Harmonised Tripartite Guideline, Statistical principles for clinical trials E9, February (1998)

K. Imai, Variance identification and efficiency analysis in randomized experiments under the matched-pair design. Stat. Med. **27**(24), 4857–4873 (2008)

K. Imai, G. King, C. Nall, The essential role of pair matching in cluster-randomized experiments, with application to the Mexican universal health insurance evaluation. Stat. Sci. **24**(1), 29–53 (2009)

K. Imai, L. Keele, D. Tingley, A general approach to causal mediation analysis. Psychol methods **15**(4), 309–334 (2010a)

K. Imai, L. Keele, T. Yamamoto, Identification, inference and sensitivity analysis for causal mediation effects. Stat. Sci. **25**(1), 51–71 (2010b)

G.W. Imbens, Nonparametric estimation of average treatment effects under exogeneity: a review. Rev. Econ. Stat. **86**(1), 4–29 (2004)

G.W. Imbens, Experimental design for unit and cluster randomized trials. Technical Report. NBER Working Paper (2011)

G. Imbens, D.B. Rubin, *Causal Inference for Statistics, Social, and Biomedical Sciences* (Cambridge University Press, New York, 2015)

J.P. Ioannidis, Why most discovered true associations are inflated. Epidemiology **19**(5), 640–648 (2008)

F. Ismail-Beigi, T. Craven, M.A. Banerji, J. Basile, J. Calles, R.M. Cohen, R. Cuddihy, W.C Cushman, S. Genuth, R.H. Grimm, B.P. Hamilton, B. Hoogwerf, D. Karl, L. Katz, A. Krikorian, P. O'Connor, R. Pop-Busui, U. Schubart, D. Simmons, H. Taylor, A. Thomas, D. Weiss, I. Hramiak for the ACCORD trial group, Effect of intensive treatment of hyperglycaemia on microvascular outcomes in type 2 diabetes: an analysis of the ACCORD randomised trial. Lancet **376**, 419–430 (2010)

Joint National Committee, The fifth report of the joint national committee on detection, evaluation, and treatment of high blood pressure (JNC V). Arch. Intern. Med. **153**(2), 154–183 (1993)

B.C. Kahn, V. Jairath, C.J. Doré, T.P. Morris, The risks and rewards of covariate adjustment in randomized trials: an assessment of 12 outcomes from 8 studies. Trials **15**(139), 1–7 (2014)

R.M. Karp, *Reducibility Among Combinatorial Problems* (Springer, New York, Berlin, Heidelberg, 1972)

S. Keleş, M.J. van der Laan, S. Dudoit, Asymptotically optimal model selection method for regression on censored outcomes. Technical Report, Division of Biostatistics, University of California, Berkeley (2002)

E.H. Kennedy, Z. Ma, M.D. McHugh, D.S. Small, Nonparametric methods for doubly robust estimation of continuous treatment effects. ArXiv e-prints (2015)

R. Kessler, S. Rose, K. Koenen et al., How well can post-traumatic stress disorder be predicted from pre-trauma risk factors? an exploratory study in the who world mental health surveys. World Psychiatry **13**(3), 265–274 (2014)

D. Kibler, D.W. Aha, M.K. Albert, Instance-based prediction of real-valued attributes. Comput. Intell. **5**, 51 (1989)

J. Kivinen, A.J. Smola, R.C. Williamson, Online learning with kernels. IEEE Trans. Signal Process. **52**(8), 2165–2176 (2004)

N. Klar, A. Donner, The merits of matching in community intervention trials: a cautionary tale. Stat. Med. **16**(15), 1753–1764 (1997)

D.C. Knill, A. Pouget, The Bayesian brain: the role of uncertainty in neural coding and computation. Trends Neurosci. **27**(12), 712–719 (2004)

K. Korb, L. Hope, A. Nicholson, K. Axnick, Varieties of causal intervention. in *PRICAI 2004: Trends in Artificial Intelligence*, ed. by C. Zhang, H.W. Guesgen, W.-K. Yeap. Lecture Notes in Computer Science, vol. 3157 (Springer, Berlin, Heidelberg, 2004), pp. 322–331

B. Korte, J. Vygen, *Combinatorial Optimization*, 5th edn. (Springer, Berlin, Heidelberg, New York, 2012)

M.S. Kramer, B. Chalmers, E.D. Hodnett, Z. Sevkovskaya, I. Dzikovich, S. Shapiro, J.P. Collet, I. Vanilovich, I. Mezen, T. Ducruet, G. Shishko, V. Zubovich, D. Mknuik, E. Gluchanina, V. Dombrovskiy, A. Ustinovitch, T. Kot, N. Bogdanovich, L. Ovchinikova, E. Helsing, PROmotion of breastfeeding intervention trial (PROBIT). J. Am. Med. Assoc. **285**(4), 413–420 (2001)

M.S. Kramer, T. Guo, R.W. Platt, S. Shapiro, J.P. Collet, B. Chalmers, E. Hodnett, Z. Sevkovskaya, I. Dzikovich, I. Vanilovich, Breastfeeding and infant growth: biology or bias? Pediatrics **110**(2), 343–347 (2002)

L. Krüger, L. Daston, M. Heidelberger, G. Gigerenzer, M.S. Morgan, *The Probabilistic Revolution.* (MIT Press, Cambridge, 1987)

L. Kunz, S. Rose, D. Spiegelman, S.-L. Normand, Causal inference methods in comparative effectiveness research, in *Methods in Comparative Effectiveness Research*, ed. by C. Gatsonis, S.C. Morton (Chapman & Hall, Boca Raton, 2017)

E.B. Laber, D.J. Lizotte, M. Qian, W.E. Pelham, S.A. Murphy, Dynamic treatment regimes: Technical challenges and applications. Electron. J. Stat. **8**(1), 1225–1272 (2014a)

E.B. Laber, D.J. Lizotte, M. Qian, W.E. Pelham, S.A. Murphy, Rejoinder of "Dynamic treatment regimes: technical challenges and applications". Electron. J. Stat. **8**(1), 1312–1321 (2014b)

J. Langford, L. Li, T. Zhang, Sparse online learning via truncated gradient. J. Mach. Learn. Res. **10**, 777–801 (2009)

P. Lavori, R. Dawson, Adaptive treatment strategies in chronic disease. Annu. Rev. Med. **59**, 443–453 (2008)

P.W. Lavori, R. Dawson, A design for testing clinical strategies: Biased adaptive within-subject randomization. J. R. Stat. Soc. Ser. A **163** 29–38 (2000)

D. Lazer, R. Kennedy, What we can learn from the epic failure of Google flu trends. Wired (2015)

S.D. Lendle, M.J. van der Laan, Identification and efficient estimation of the natural direct effect among the untreated. Technical Report, Division of Biostatistics, University of California, Berkeley (2011)

S.D. Lendle, B. Fireman, M.J. van der Laan, Balancing score adjusted targeted minimum loss-based estimation. Technical Report, Division of Biostatistics, University of California, Berkeley (2013)

S. Lendle, J. Schwab, M.L. Petersen, M.J. van der Laan, ltmle: an R package for implementing targeted minimum loss-based estimation for longitudinal data. J. Stat. Softw. **81**(1) (2017)

B.Y. Levit, On the efficiency of a class of non-parametric estimates. Theory Probab. Appl. **20**(4), 723–740 (1975)

L. Li, E. Tchetgen Tchetgen, A.W. van der Vaart, J.M. Robins, Higher order inference on a treatment effect under low regularity conditions. Stat. Probab. Lett. **81**(7), 821–828 (2011)

A. Liaw, M. Wiener, Classification and regression by randomforest. R News **2**(3), 18–22 (2002)

L. Liu, M.G. Hudgens, Large sample randomization inference of causal effects in the presence of interference. J. Am. Stat. Assoc. **109**(505), 288–301 (2014). ISSN 0162-1459

Z. Liu, T. Stengos, Nonlinearities in cross country growth regressions: a semiparametric approach. J. Appl. Econom. **14**, 527–538 (1999)

V. Loonis, X. Mary, Determinantal sampling designs. ArXiv e-prints, October (2015)

B. Lu, R. Greevy, X. Xu, C. Beck, Optimal nonbipartite matching and its statistical applications. Am. Stat. **65**(1), 21–30 (2011)

A.R. Luedtke, M.J. van der Laan, Statistical inference for the mean outcome under a possibly non-unique optimal treatment strategy. Ann. Stat. **44**(2), 713–742 (2016a)

A.R. Luedtke, M.J. van der Laan, Super-learning of an optimal dynamic treatment rule. Int. J. Biostat. **12**(1), 305–332 (2016b)

A.R. Luedtke, M.J. van der Laan, Optimal individualized treatments in resource-limited settings. Int. J. Biostat. **12**(1), 283–303 (2016c)

A..R Luedtke, M. Carone, M.J. van der Laan, Discussion of deductive derivation and turing-computerization of semiparametric efficient estimation by Frangakis et al. Biometrics **71**(4), 875–879 (2015a)

A.R. Luedtke, I. Díaz, M.J. van der Laan, The statistics of sensitivity analyses. Technical Report, Division of Biostatistics, University of California, Berkeley (2015b)

K. Lum, Limitations of mitigating judicial bias with machine learning. Nat Hum. Behav. **1**, 0141 (2017)

M. Lunn, D. McNeil, Applying Cox regression to competing risks. Biometrics **51**, 524–532 (1995). ISSN 0006-341X

R. Lyons, Determinantal probability measures. Publications Mathématiques de l'Institut des Hautes Études Scientifiques **98**, 167–212 (2003)

R. Lyons, The spread of evidence-poor medicine via flawed social-network analysis. Stat. Politics Policy **2**(1) 1–26 (2010)

O. Macchi, The coincidence approach to stochastic point processes. Adv. Appl. Probab. **7**, 83–122 (1975)

R. Macklin, E. Cowan, Given financial constraints, it would be unethical to divert antiretroviral drugs from treatment to prevention. Health Aff. **31**(7), 1537–1544 (2012)

R.F. MacLehose, S. Kaufman, J.S. Kaufman, C. Poole, Bounding causal effects under uncontrolled confounding using counterfactuals. Epidemiology **16**(4), 548–555 (2005)

E. Mammen, A.B. Tsybakov, Smooth discrimination analysis. Ann. Stat. **27**(6), 1808–1829 (1999)

J.K. Mann, J.R. Balmes, T.A. Bruckner, K.M. Mortimer, H.G. Margolis, B. Pratt, S.K. Hammond, F.W. Lurmann, I.B. Tager, Short-term effects of air pollution on wheeze in asthmatic children in Fresno, California. Environ Health Perspect. **118**(10), 06 (2010)

C.F. Manski, *Partial Identification of Probability Distributions* (Springer, Berlin, Heidelberg, New York, 2003)

C.F. Manski, Nonparametric bounds on treatment effects. Am. Econ. Rev. **80**, 319–323 (1990)

D. Mayo, *Error and the Growth of Experimental Knowledge* (University of Chicago Press, Chicago, 1996)

D. Mayo, *Error and Inference: Recent Exchanges on Experimental Reasoning, Reliability, and the Objectivity and Rationality of Science* (Cambridge, Chicago, 2010)

S. Milborrow, T Hastie, R Tibshirani, Earth: multivariate adaptive regression spline models. R package version 3.2-7 (2014)

T. Mildenberger, Y. Rozenholc, D. Zasada, histogram: Construction of regular and irregular histograms with different options for automatic choice of bins (2009). http://CRAN.R-project.org/package=histogram

E.E.M. Moodie, T.S. Richardson, D.A. Stephens, Demystifying optimal dynamic treatment regimes. Biometrics **63**(2), 447–455 (2007)

K.L. Moore, M.J. van der Laan, Application of time-to-event methods in the assessment of safety in clinical trials, in *Design, Summarization, Analysis & Interpreta-*

tion of Clinical Trials with Time-to-Event Endpoints, ed. by K.E. Peace (Chapman & Hall, Boca Raton, 2009a)

K.L. Moore, M.J. van der Laan, Covariate adjustment in randomized trials with binary outcomes: targeted maximum likelihood estimation. Stat. Med. **28**(1), 39–64 (2009b)

K.L. Moore, M.J. van der Laan, Increasing power in randomized trials with right censored outcomes through covariate adjustment. J. Biopharm. Stat. **19**(6), 1099–1131 (2009c)

K.L. Moore, R. Neugebauer, T. Valappil, M.J. van der Laan, Robust extraction of covariate information to improve estimation efficiency in randomized trials. Stat. Med. **30**(19), 2389–2408 (2011)

N. Murata, A statistical study of on-line learning, in *Online Learning and Neural Networks* (Cambridge University Press, Cambridge, 1998)

S.A. Murphy, Optimal dynamic treatment regimes. J. R. Stat. Soc. Ser. B **65**(2), 331–66 (2003)

S.A. Murphy, An experimental design for the development of adaptive treatment strategies. Stat. Med. **24**, 1455–1481 (2005)

S.A. Murphy, M.J. van der Laan, J.M. Robins, Marginal mean models for dynamic treatment regimens. J. Am. Stat. Assoc. **96**, 1410–1424 (2001)

E.A. Nadaraya, On estimating regression. Theory Probab. Appl. **9**(1), 141–142 (1964)

A.I Naimi, E.E.M. Moodie, N. Auger, J.S. Kaufman, Stochastic mediation contrasts in epidemiologic research: interpregnancy interval and the educational disparity in preterm delivery. Am. J. Epidemiol. **180**(4), 436–445 (2014)

D.M. Nathan, J.B. Buse, M.B. Davidson, E. Ferrannini, R.R. Holman, R. Sherwin, B. Zinman, Medical management of hyperglycemia in type 2 diabetes: a consensus algorithm for the initiation and adjustment of therapy: a consensus statement of the American Diabetes Association and the European Association for the Study of Diabetes. Diab. Care **32**(1), 193–203 (2009)

D.M. Nathan, P. A. Cleary, J.Y. Backlund, S.M. Genuth, J.M. Lachin, T.J. Orchard, P. Raskin, B. Zinman, Diabetes control and complications trial/epidemiology of diabetes interventions and complications (DCCT/EDIC) study research group. Intensive diabetes treatment and cardiovascular disease in patients with type 1 diabetes. N. Engl. J. Med. **22**(353), 2643–2653 (2005)

D.M. Nathan, J.B. Buse, M.B. Davidson, R.J. Heine, R.R. Holman, R. Sherwin, B. Zinman, Management of hyperglycemia in type 2 diabetes: a consensus algorithm for the initiation and adjustment of therapy: a consensus statement from the American Diabetes Association and the European Association for the Study of Diabetes. Diab. Care **29**, 1963–1972 (2006)

NCEP (2002), Third Report of the National Cholesterol Education Program (NCEP) Expert Panel on Detection (2002)

R. Neugebauer, J. Bullard, DSA: data-adaptive estimation with cross-validation and the D/S/A algorithm (2010). http://www.stat.berkeley.edu/~laan/Software/

R. Neugebauer, M.J. van der Laan, Nonparametric causal effects based on marginal structural models. J. Stat. Plann. Infererence **137**(2), 419–434 (2007)

R. Neugebauer, M.J. Silverberg, M.J. van der Laan, Observational study and individualized antiretroviral therapy initiation rules for reducing cancer incidence in HIV-infected patients, chap. 26 (Springer, New York, 2011), pp. 436–456

R. Neugebauer, B. Fireman, J.A. Roy, P.J. O'Connor, J.V. Selby, Dynamic marginal structural modeling to evaluate the comparative effectiveness of more or less aggressive treatment intensification strategies in adults with type 2 diabetes. Pharmacoepidemiol. Drug Saf. 21(Suppl. 2), 99–113 (2012)

R. Neugebauer, B. Fireman, J.A. Roy, P.J. O'Connor, Impact of specific glucose-control strategies on microvascular and macrovascular outcomes in 58,000 adults with type 2 diabetes. Diab. Care 36(11), 3510–3516 (2013)

R. Neugebauer, J. Schmittdiel, M.J. Laan, Targeted learning in real-world comparative effectiveness research with time-varying interventions. Stat. Med. 33(14), 2480–2520 (2014a)

R. Neugebauer, J.A. Schmittdiel, Z. Zhu, J.A. Rassen, J.D. Seeger, S. Schneeweiss, High-dimensional propensity score algorithm in comparative effectiveness research with time-varying interventions. Stat. Med. 34(5), 753–781 (2014b)

R. Neugebauer, J.A. Schmittdiel, M.J. van der Laan, A case study of the impact of data-adaptive versus model-based estimation of the propensity scores on causal inferences from three inverse probability weighting estimators. Int. J. Biostat. 12(1), 131–155 (2016)

J. Neyman, Sur les applications de la theorie des probabilites aux experiences agricoles: Essai des principes (In Polish). English translation by D.M. Dabrowska and T.P. Speed (1990). Stat. Sci. 5, 465–480 (1923)

P.J. O'Connor, F. Ismail-Beigi, Near-normalization of glucose and microvascular diabetes complications: data from ACCORD and ADVANCE. Ther. Adv. Endocrinol. Metab. 2(1), 17–26 (2011)

E.L. Ogburn, T.J. VanderWeele, Vaccines, contagion, and social networks. ArXiv e-prints, March (2014)

E.L. Ogburn, O. Sofrygin, M.J. van der Laan, I. Diaz, Causal inference for social network data with contagion. ArXiv e-prints, October (2017)

B.A. Olken, Pre-analysis plans in economics. Technical report, Massachusetts Institute of Technology Department of Economics (2015)

C. O'Neil, *Weapons of Math Destruction: How Big Data Increases Inequality and Threatens Democracy* (Crown Publishing Group, New York, 2016)

L. Orellana, A. Rotnitzky, J.M. Robins, Dynamic regime marginal structural mean models for estimation of optimal treatment regimes, part I: main content. Int. J. Biostat. 6(2), Article 8 (2010)

E. Parzen, On estimation of a probability density function and mode. Ann. Math. Stat. 33(3), 1065–1076 (1962)

A. Patel, S. MacMahon, J. Chalmers, B. Neal, L. Billot, M. Woodward, M. Marre, M. Cooper, P. Glasziou, D. Grobbee, P. Hamet, S. Harrap, S. Heller, Intensive blood glucose control and vascular outcomes in patients with type 2 diabetes. N. Engl. J. Med. 358(24), 2560–2572 (2008)

J. Pearl, Causal diagrams for empirical research. Biometrika 82, 669–710 (1995)

J. Pearl, Direct and indirect effects, in *Proceedings of the 17th Conference Uncertainty in Artificial Intelligence* (Morgan Kaufmann, San Francisco, 2001)

J. Pearl, *Causality: Models, Reasoning, and Inference*, 2nd edn. (Cambridge, New York, 2009a)

J. Pearl, Myth, confusion, and science in causal analysis. Technical Report R-348, Cognitive Systems Laboratory, Computer Science Department University of California, Los Angeles, Los Angeles, CA, May 2009b

J. Pearl, On the consistency rule in causal inference: axiom, definition, assumption, or theorem? Epidemiology 21(6), 872–875 (2010)

J. Pearl, The mediation formula: a guide to the assessment of causal pathways in nonlinear models, in *Causality: Statistical Perspectives and Applications*, ed. by C. Berzuini, P. Dawid, L. Bernardinelli (Springer, Berlin, 2011)

R. Pemantle, Y. Peres, Concentration of Lipschitz functionals of determinantal and other strong Rayleigh measures. Comb. Probab. Comput. 23(1), 140–160 (2014)

W.D. Penny, K.E. Stephan, A. Mechelli, K.J. Friston, Modelling functional integration: a comparison of structural equation and dynamic causal models. Neuroimage 23, S264–S274 (2004)

G. Peoples, New study from Pandora touts the Pandora effect on music sales. Billboard (2014)

A. Peters, T. Hothorn, ipred: improved predictors (2009) http://CRAN.R-project.org/package=ipred

M. Petersen, J. Schwab, S. Gruber, N. Blaser, M. Schomaker, M.J. van der Laan, Targeted maximum likelihood estimation for dynamic and static longitudinal marginal structural working models. J. Causal Inference 2(2), 147–185 (2014)

M.L. Petersen, E. LeDell, J. Schwab, V. Sarovar, R. Gross, N. Reynolds, J.E. Haberer, K. Goggin, C. Golin, J. Arnsten et al., Super learner analysis of electronic adherence data improves viral prediction and may provide strategies for selective HIV RNA monitoring. J. Acquir. Immune Defic. Syndr. 69(1), 109 (2015)

J. Pfanzagl, *Contributions to a General Asymptotic Statistical Theory* (Springer, Berlin, 1982)

J. Pfanzagl, *Asymptotic Expansions for General Statistical Models*, vol. 31 (Springer, Berlin, 1985)

J. Pfanzagl, *Estimation in Semiparametric Models* (Springer, Berlin, Heidelberg, New York, 1990)

I. Phyllis, F. Russo. *Causality; Philosophical Theory meets Scientific Practice* (Oxford University Press, Oxford, 2016)

M. Pintilie, Analysing and interpreting competing risk data. Stat. Med. 26(6), 1360–1367 (2007)

R. Pirracchio, M.L. Petersen, M.J. van der Laan, Improving propensity score estimators' robustness to model misspecification using super learner. Am. J. Epidemiol. 181(2), 108–119 (2014)

R. Pirracchio, M.L. Petersen, M. Carone, M.R. Rigon, S. Chevret, M.J. van der Laan, Mortality prediction in intensive care units with the super ICU learner algorithm (SICULA): a population-based study. Lancet Respir. Med. 3(1), 42–52 (2015)

R.W. Platt, E.F. Schisterman, S.R. Cole, Time-modified confounding. Am. J. Epidemiol. **170**(6), 687–694 (2009)

S.J. Pocock, S.E. Assmann, L.E. Enos, L.E. Kasten, Subgroup analysis, covariate adjustment, and baseline comparisons in clinical trial reporting: current practice and problems. Stat. Med. **21**, 2917–2930 (2002)

E.C. Polley, M.J. van der Laan, SuperLearner: super learner prediction (2013). http://CRAN.R-project.org/package=SuperLearner

E.C. Polley, M.J. van der Laan, Predicting optimal treatment assignment based on prognostic factors in cancer patients. in *Design, Summarization, Analysis & Interpretation of Clinical Trials with Time-to-Event Endpoints*, ed. by K.E. Peace (Boca Raton, Chapman & Hall, 2009)

E.C. Polley, M.J. van der Laan, Super learner in prediction. Technical Report 266, Division of Biostatistics, University of California, Berkeley (2010)

E.C Polley, S. Rose, M.J. van der Laan, Super-learning, in *Targeted Learning: Causal Inference for Observational and Experimental Data*, ed. by M.J. van der Laan, S. Rose (Springer, Berlin, Heidelberg, New York, 2011)

E.C. Polley, E. LeDell, C. Kennedy, M.J. van der Laan, SuperLearner: super learner prediction (2017). https://github.com/ecpolley/SuperLearner

B.T. Polyak, A.B. Juditsky, Acceleration of stochastic approximation by averaging. SIAM J. Control. Optim. **30**(4), 838–855 (1992)

T.M. Porter, *The Rise of Statistical Thinking* (Princeton University Press, Princeton, 1986)

T.M. Porter, *Trust in Numbers: The Pursuit of Objectivity in Science and Public Life* (Princeton University Press, Princeton, 1995)

K.E Porter, S. Gruber, M.J. van der Laan, J.S. Sekhon, The relative performance of targeted maximum likelihood estimators. Int. J. Biostat. 7(1) (2011)

R.L. Prentice, J.D. Kalbfleisch, A.V. Peterson Jr, N. Flournoy, V.T. Farewell, N.E. Breslow, The analysis of failure times in the presence of competing risks. Biometrics **34**(4), 541–554 (1978)

M. Qian, S.A. Murphy, Performance guarantees for individualized treatment rules. Ann. Stat. **39**(2), 1180–1210 (2011)

R Development Core Team. R: A language and environment for statistical computing. R Foundation for Statistical Computing, Vienna (2016). http://www.R-project.org.

K.K. Ray, S.R. Seshasai, S. Wijesuriya, R. Sivakumaran, S. Nethercott, D. Preiss, S. Erqou, N. Sattar, Effect of intensive control of glucose on cardiovascular outcomes and death in patients with diabetes mellitus: a meta-analysis of randomised controlled trials. Lancet **373**, 1765–72 (2009)

J.M. Robins, A new approach to causal inference in mortality studies with sustained exposure periods–application to control of the healthy worker survivor effect. Math. Modell. **7**, 1393–1512 (1986)

J.M. Robins, Addendum to: "A new approach to causal inference in mortality studies with a sustained exposure period—application to control of the healthy worker survivor effect". Comput. Math. Appl. **14**(9–12), 923–945 (1987)

J.M. Robins, Marginal structural models, in *1997 Proceedings of the American Statistical Association. Section on Bayesian Statistical Science*, pp. 1–10 (1998)

J.M. Robins, Association, causation and marginal structural models. Synthese **121**, 151–179 (1999)

J.M. Robins, Robust estimation in sequentially ignorable missing data and causal inference models, in *Proceedings of the American Statistical Association* (2000)

J.M. Robins, Optimal structural nested models for optimal sequential decisions, in *Proceedings of the Second Seattle Symposium in Biostatistics: Analysis of Correlated Data* (2004)

J.M. Robins, S. Greenland, Identifiability and exchangeability for direct and indirect effects. Epidemiol **3**, 143–155 (1992)

J.M. Robins, Y. Ritov, Toward a curse of dimensionality appropriate (coda) asymptotic theory for semi-parametric models. Stat. Med. **16**, 285–319 (1997)

J.M. Robins, A. Rotnitzky, Recovery of information and adjustment for dependent censoring using surrogate markers, in *AIDS Epidemiology* (Birkhäuser, Basel, 1992)

J.M. Robins, A. Rotnitzky, L.P. Zhao, Estimation of regression coefficients when some regressors are not always observed. J. Am. Stat. Assoc. **89**(427), 846–866 (1994)

J.M. Robins, A. Rotnitzky, D.O. Scharfstein, Sensitivity analysis for selection bias and unmeasured confounding in missing data and causal inference models, in *Statistical Models in Epidemiology, the Environment and Clinical Trials*. IMA Volumes in Mathematics and Its Applications (Springer, Berlin, 1999)

J.M. Robins, M.A. Hernan, B. Brumback, Marginal structural models and causal inference in epidemiology. Epidemiology **11**(5), 550–560 (2000)

J.M. Robins, M.A. Hernán, U. Siebert, Effects of multiple interventions, in *Comparative Quantification of Health Risks: Global and Regional Burden of Disease Attributable to Selected Major Risk Factors*, vol. 1 (World Health Organization, Geneva, 2004), pp. 2191–2230

J.M. Robins, L. Li, E. Tchetgen Tchetgen, A.W. van der Vaart, Higher order influence functions and minimax estimation of nonlinear functionals, in *Probability and Statistics: Essays in Honor of David A. Freedman*, (Institute of Mathematical Statistics, 2008a), pp. 335–421

J.M. Robins, L. Orellana, A. Rotnitzky, Estimation and extrapolation of optimal treatment and testing strategies. Stat. Med. **27**, 4678–4721 (2008b)

J.M. Robins, L. Li, E. Tchetgen Tchetgen, A.W. van der Vaart, Quadratic Semiparametric Von Mises calculus. Metrika **69**(2–3), 227–247 (2009)

M. Rolland, P.T. Edlefsen, B.B. Larsen, S. Tovanabutra, E. Sanders-Buell, T. Hertz, C. Carrico, S. Menis, C.A. Magaret, H. Ahmed, Increased HIV-1 vaccine efficacy against viruses with genetic signatures in Env V2. Nature **490**(7420), 417–420 (2012). ISSN 0028-0836

S. Rose, Mortality risk score prediction in an elderly population using machine learning. Am. J. Epidemiol. **177**(5), 443–452 (2013)

S. Rose, Targeted learning for pre-analysis plans in public health and health policy research. Observational Stud. **1**, 294–306 (2015)

S. Rose, A machine learning framework for plan payment risk adjustment. Health Serv. Res. **51**(6), 2358–2374 (2016)

S. Rose, Robust machine learning variable importance analyses of medical conditions for health care spending. Health Serv. Res. (2018, in press)

S. Rose, S. Bergquist, T. Layton, Computational health economics for identification of unprofitable health care enrollees. Biostatistics **18**(4), 682–694 (2017)

S. Rose, M.J. van der Laan, Simple optimal weighting of cases and controls in case-control studies. Int. J. Biostat. **4**(1), Article 19 (2008)

S. Rose, M.J. van der Laan, Why match? Investigating matched case-control study designs with causal effect estimation. Int. J. Biostat. **5**(1), Article 1 (2009)

S. Rose, M.J. van der Laan, A targeted maximum likelihood estimator for two-stage designs. Int. J. Biostat. **7**(1), Article 17 (2011)

S. Rose, M.J. van der Laan, A double robust approach to causal effects in case-control studies. Am. J. Epidemiol. **179**(6), 663–669 (2014a)

S. Rose, M.J. van der Laan, Rose and van der Laan respond to "Some advantages of RERI". Am. J. Epidemiol. **179**(6), 672–673 (2014b)

P.R. Rosenbaum, D.B. Rubin, Assessing sensitivity to an unobserved binary covariate in an observational study with binary outcome. J. R. Stat. Soc. Ser. B **45**, 212–218 (1983a)

P.R. Rosenbaum, Interference Between Units in Randomized Experiments. J. Am. Stat. Assoc. **102**(477), 191–200 (2007)

P.R. Rosenbaum, D.B. Rubin, The central role of the propensity score in observational studies for causal effects. Biometrika **70**, 41–55 (1983b)

M. Rosenblatt, Remarks on some nonparametric estimates of a density function. Ann. Math. Stat. **27**(3), 832–837 (1956)

M. Rosenblum, M.J. van der Laan, Using regression models to analyze randomized trials: asymptotically valid hypothesis tests despite incorrectly specified models. Biometrics **65**(3), 937–945 (2009)

M. Rosenblum, M.J. van der Laan, Targeted maximum likelihood estimation of the parameter of a marginal structural model. Int. J. Biostat. **6**(2), 19 (2010a)

M. Rosenblum, M.J. van der Laan, Simple, efficient estimators of treatment effects in randomized trials using generalized linear models to leverage baseline variables. Int. J. Biostat. **6**(1), Article 13 (2010b)

M. Rosenblum, S.G. Deeks, M.J. van der Laan, D.R. Bangsberg, The risk of virologic failure decreases with duration of HIV suppression, at greater than 50% adherence to antiretroviral therapy. PLoS ONE **4**(9), e7196 (2009)

R.H. Rosenman, M. Friedman, R. Straus, M. Wurm, R. Kositchek, W. Hahn, N.T. Werthessen, A predictive study of coronary heart disease: the western collaborative group study. J. Am. Med. Assoc. **189**(1), 15–22 (1964)

R.H. Rosenman, R.J. Brand, C.D. Jenkins, M. Friedman, R. Straus, M. Wurm, Coronary heart disease in the western collaborative group study: final follow-up experience of 8 1/2 years. J. Am. Med. Assoc. **233**(8), 872–877 (1975)

B. Rosner, *Fundamentals of Biostatistics*, 5th edn. (Duxbury, Pacific Grove, 1999)

S. Rosthø j, C. Fullwood, R. Henderson, S. Stewart, Estimation of optimal dynamic anticoagulation regimes from observational data: a regret-based approach. Stat. Med. **88**, 4197–4215 (2006)

A. Rotnitzky, D. Scharfstein, S. Ting-Li Su, J. Robins, Methods for conducting sensitivity analysis of trials with potentially nonignorable competing causes of censoring. Biometrics **57**(1), 103–113 (2001)

A. Rotnitzky, J.M. Robins, D.O. Scharfstein, Semiparametric regression for repeated outcomes with nonignorable nonresponse. J. Am. Med. Assoc. **93**(444), 1321–1339 (1998)

Y. Rozenholc, T. Mildenberger, U. Gather, Combining regular and irregular histograms by penalized likelihood. Comput. Stat. Data Anal. **54**(12), 3313–3323 (2010)

D.B. Rubin, Randomization analysis of experimental data: The fisher randomization test comment. J. Am. Stat. Assoc. **75**(371), 591–593 (1980)

D.B. Rubin, Estimating causal effects of treatments in randomized and nonrandomized studies. J. Educ. Psychol. **66**, 688–701 (1974)

D.B. Rubin, Multivariate matching methods that are equal percent bias reducing, II: maximums on bias reduction for fixed sample sizes. Biometrics **32**(1), 121–132 (1976)

D.B. Rubin, Comment: Neyman (1923) and causal inference in experiments and observational studies. Stat. Sci. **5**(4), 472–480 (1990)

D.B. Rubin, *Matched Sampling for Causal Effects* (Cambridge, Cambridge, MA, 2006)

D.B. Rubin, M.J. van der Laan, Empirical efficiency maximization: improved locally efficient covariate adjustment in randomized experiments and survival analysis. Int. J. Biostat. **4**(1), Article 5 (2008)

D.B. Rubin, M.J. van der Laan, Targeted ANCOVA estimator in RCTs, in *Targeted Learning* (Springer, Berlin, 2011), pp. 201–215

D.B. Rubin, M.J. van der Laan, Statistical issues and limitations in personalized medicine research with clinical trials. Int. J. Biostat. **8**(1), Article 1 (2012)

M. Rudelson, R. Vershynin, Hanson-Wright inequality and subGaussian concentration. Electron. Commun. Probab. **18**(82), 1–9 (2013)

M.R. Sampford, On sampling without replacement with unequal probabilities of selection. Biometrika **54**(3–4), 499–513 (1967)

S. Sapp, M.J. van der Laan, K. Page, Targeted estimation of binary variable importance measures with interval-censored outcomes. Int. J. Biostat. **10**(1), 77–97 (2014)

D.O. Scharfstein, J.M. Robins, Estimation of the failure time distribution in the presence of informative censoring. Biometrika **89**(3), 617–634 (2002)

D.O. Scharfstein, A. Rotnitzky, J.M. Robins, Adjusting for nonignorable drop-out using semiparametric nonresponse models, (with discussion and rejoinder). J. Am. Stat. Assoc. **94**, 1096–1120, 1121–1146 (1999)

M.E. Schnitzer, J. Lok, S. Gruber, Variable selection for confounder control, flexible modeling and collaborative targeted minimum loss-based estimation in causal inference. Int. J. Biostat. **12**(1), 97–115 (2016)

M.E. Schnitzer, M.J. van der Laan, E.E.M. Moodie, R.W. Platt, Effect of breast-feeding on gastrointestinal infection in infants: a targeted maximum likelihood approach for clustered longitudinal data. Ann. Appl. Stat. **8**(2), 703–725 (2014)

P. Schochet, Estimators for clustered education RCTs using the Neyman model for causal inference. J. Educ. Behav. Stat. **38**(3), 219–238 (2013)

M.S. Schuler, S. Rose, Targeted maximum likelihood estimation for causal inference in observational studies. Am. J. Epidemiol. **185**(1), 65–73 (2017)

S. Selvaraj, V. Prasad. Characteristics of cluster randomized trials: Are they living up to the randomized trial? JAMA Intern. Med. **173**(23), 313 (2013)

S. Shalev-Shwartz, Online learning and online convex optimization. Found. Trends Mach. Learn. **4**(2), 107–194 (2011)

S. Shalev-Shwartz, Y. Singer, N. Srebro, A. Cotter, Pegasos: primal estimated sub-gradient solver for SVM. Math. Programm. **127**(1), 3–30 (2011)

C.R. Shalizi, A.C. Thomas, Homophily and contagion are generically confounded in observational social network studies. Sociol. Methods Res. **40**(2), 211–239 (2011)

C. Shen, X. Li, L. Li, Inverse probability weighting for covariate adjustment in randomized studies. Stat. Med. **33**, 555–568 (2014)

A. Shrestha, S. Bergquist, E. Montz, S. Rose, Mental health risk adjustment with clinical categories and machine learning. Health Serv. Res. (2018, in press)

J.A. Singh, Antiretroviral resource allocation for HIV prevention. AIDS **27**(6), 863–865 (2013)

S.E. Sinisi, M.J. van der Laan, Deletion/Substitution/Addition algorithm in learning with applications in genomics. Stat. Appl. Genet. Mol. **3**(1), Article 18 (2004)

J.S. Skyler, R. Bergenstal, R.O. Bonow, J. Buse, P. Deedwania, E.A.M. Gale, B.V. Howard, M.S. Kirkman, M. Kosiborod, P. Reaven, R.S. Sherwin, Intensive Glycemic Control and the prevention of cardiovascular events: implications of the accord, advance, and VA diabetes trials: a position statement of the American Diabetes Association and a scientific statement of the American College of Cardiology Foundation and the American Heart Association. Diab. Care **32**, 187–92 (2009)

J.W. Smith, J.E. Everhart, W.C. Dickson, W.C. Knowler, R.S. Johannes, Using the adap learning algorithm to forecast the onset of diabetes mellitus, in *Proceedings of the Annual Symposium on Computer Application in Medical Care* (American Medical Informatics Association, Bethesda, 1988), p. 261

J.M. Snowden, S. Rose, K.M. Mortimer, Implementation of g-computation on a simulated data set: demonstration of a causal inference technique. Am. J. Epidemiol. **173**(7), 731–738 (2011)

M. Sobel, What do randomized studies of housing mobility demonstrate? J. Am. Stat. Assoc. **101**(476), 1398–1407 (2006)

O. Sofrygin, M.J. van der Laan, R. Neugebauer, Simcausal R package: conducting transparent and reproducible simulation studies of causal effect estimation with complex longitudinal data. J. Stat. Softw. 81, 2 (2017)

O. Sofrygin, M.J. van der Laan, tmlenet: targeted maximum likelihood estimation for network data (2015)

O. Sofrygin, M.J. van der Laan, Semi-parametric estimation and inference for the mean outcome of the single time-point intervention in a causally connected population. J. Causal Inference **5**(1), 20160003 (2017)

A. Soshnikov, Gaussian limit for determinantal random point fields. Ann. Probab. **30**(1), 171–187 (2000)

K. Stanley, Design of randomized controlled trials. Circulation **115**, 1164–1169 (2007)

R.J.C.M. Starmans, Models, inference, and truth: probabilistic reasoning in the information era, in *Targeted Learning: Causal Inference for Observational and Experimental Data*, ed. by M. van der Laan, S. Rose (Springer, Berlin, 2011)

R.J.C.M. Starmans, The reality behind the model and the cracks in the mirror of nature (in Dutch), in *Filosofie, Tweemaandelijks Vlaams-Nederlands Tijdschrift, jaargang*, vol. 21 (Garant Publishers, Antwerpen, Apeldoorn, 2011a)

R.J.C.M. Starmans, Ethics and statistics; the progress of a laborious dialogue (in Dutch), in *Filosofie, Tweemaandelijks Vlaams-Nederlands Tijdschrift, jaargang*, vol. 22 (Garant Publishers, Antwerpen, Apeldoorn, 2012a)

R.J.C.M. Starmans, Statistics, discomfort and the human dimension (in Dutch), in *STAtOR*, vol. 13 (2012b)

R.J.C.M. Starmans, The world of values; statistics, evolution and ethics (in Dutch), in *STAtOR*, vol. 13 (2012c)

R.J.C.M. Starmans, Idols and ideals; francis bacon, induction and the hypothetico-deductive model (in Dutch). in *Filosofie, Tweemaandelijks Vlaams-Nederlands Tijdschrift, jaargang*, vol. 23 (Garant Publishers, Antwerpen, Apeldoorn, 2013)

R.J.C.M. Starmans, Between hobbes and turing; george boole and the laws of thinking (in Dutch), in *Filosofie, Tweemaandelijks Vlaams-Nederlands Tijdschrift, jaargang*, vol. 25 (Garant Publishers, Antwerpen, Apeldoorn, 2015a)

R.J.C.M. Starmans, With google toward the automatic statistician (in Dutch), in *STAtOR*, vol. 16 (2015b)

R.J.C.M. Starmans, Shannon; information, entropy and the probabilistic worldview (in Dutch), in *Filosofie Tweemaandelijks Vlaams-Nederlands Tijdschrift, jaargang*, vol. 26 (Garant Publishers, Antwerpen, Apeldoorn, 2016a)

R.J.C.M. Starmans, The advent of data science - some considerations on the unreasonable effectiveness of data, in *Handbook of Big Data - Handbooks of Modern Statistical Methods*, ed. by P. Buhlmann, P. Drineas, M. Kane, M.J. van der Laan (Chapman & Hall/CRC, New York, 2016b)

R.J.C.M. Starmans, From heraclitus to shannon: the velvet revolution of data in context and flux (in Dutch), in *STAtOR*, vol. 18 (2017a)

R.J.C.M. Starmans, The end of theory or the unreasonableness of data (in Dutch), in *Filosofie, Tweemaandelijks Vlaams-Nederlands Tijdschrift, jaargang*, vol. 27 (Garant Publishers, Antwerpen, Apeldoorn, 2017b), p. 2

R.J.C.M. Starmans, The new house of salomon: Peter galison and the empirical tradition (in Dutch), in *Filosofie, Tweemaandelijks Vlaams-Nederlands Tijdschrift, jaargang*, vol. 27 (Garant Publishers, Antwerpen, Apeldoorn, 2017c), p. 4

R.J.C.M. Starmans, The tryptych of the Bayesian paradigm: confirmation, inference and algoritmics, in *Filosofie, Tweemaandelijks Vlaams-Nederlands Tijdschrift, jaargang*, vol. 27 (Garant Publishers, Antwerpen, Apeldoorn, 2017d)

C. Steglich, T.A.B. Snijders, M. Pearson, Dynamic networks and behavior: separating selection from influence. Sociol. Methodol. **40**(1), 329–393 (2010)

S. Stigler, *The History of Statistics: The Measurement of Uncertainty Before 1900* (Harvard University Press, Cambridge, MA, 1986)

S. Stigler, *The History of Statistical Concepts and Methods* (Harvard University Press, Cambridge, MA, 1999)

S. Stigler, *The Seven Pillars of Statistical Wisdom* (Harvard University Press, Cambridge, MA, 2016)

O.M. Stitelman, V. De Gruttola, M.J. van der Laan, A general implementation of TMLE for longitudinal data applied to causal inference in survival analysis. Int. J. Biostat. **8**(1), 1–37 (2012)

O.M. Stitelman, M.J. van der Laan, Collaborative targeted maximum likelihood for time-to-event data. Int. J. Biostat. **6**(1), Article 21 (2010)

O.M. Stitelman, M.J. van der Laan. Targeted maximum likelihood estimation of effect modification parameters in survival analysis. Int. J. Biostat. **7**(1), 1–34 (2011)

O.M. Stitelman, V. De Gruttola, C.W. Wester, M.J. van der Laan, Rcts with time-to-event outcomes and effect modification parameters, in *Targeted Learning: Causal Inference for Observational and Experimental Data*, ed. by M. J. van der Laan, S. Rose (Springer, Berlin, 2011)

C.A. Struthers, J.D. Kalbfleisch, Misspecified proportional hazard models. Biometrika **73**(2), 363–369 (1986)

E.A. Stuart, S.R. Cole, C.P. Bradshaw, P.J. Leaf, The use of propensity scores to assess the generalizability of results from randomized trials. J. R. Stat. Soc. Ser. A **174**(Part 2), 369–386 (2011)

J. Tacq, Causality in qualitative and quantitative research. Qual. Quant. **45**(2), 263–291 (2011)

I. Tager, M. Hollenberg, W. Satariano, Self-reported leisure-time physical activity and measures of cardiorespiratory fitness in an elderly population. Am. J. Epidemiol. **147**, 921–931 (1998)

E.J. Tchetgen Tchetgen, I. Shpitser, Semiparametric theory for causal mediation analysis: efficiency bounds, multiple robustness, and sensitivity analysis. Technical report 130, Biostatistics, Harvard University, June (2011a)

E.J. Tchetgen Tchetgen, I. Shpitser, Semiparametric estimation of models for natural direct and indirect effects. Technical Report 129, Biostatistics, Harvard University, June (2011b)

E.J. Tchetgen Tchetgen, T.J. VanderWeele. On causal inference in the presence of interference. Stat. Methods Med. Res. **21**(1), 55–75 (2012)

P. Thall, H. Sung, E. Estey, Selecting therapeutic strategies based on efficacy and death in multicourse clinical trials. J. Am. Stat. Assoc. **39**, 29–39 (2002)

The Diabetes Control and Complications Trial Research Group, The effect of intensive treatment of diabetes on the development and progression of long-term complications in insulin-dependent diabetes mellitus. N. Engl. J. Med. **329**, 977–86 (1993)

M. Toftager, L.B. Christiansen, P.L. Kristensen, J. Troelsen, Space for physical activity-a multicomponent intervention study: study design and baseline findings from a cluster randomized controlled trial. BMC Public Health **11**, 777 (2011)

P. Toulis, E. Kao, Estimation of causal peer influence effects, in *Proceedings of The 30th International Conference on Machine Learning* (2013), pp. 1489–1497

A.A. Tsiatis, *Semiparametric Theory and Missing Data.* (Springer, Berlin, Heidelberg, New York, 2006)

A.A. Tsiatis, M. Davidian, M. Zhang, X. Lu, Covariate adjustment for two-sample treatment comparisons in randomized clinical trials: a principled yet flexible approach. Stat. Med. **27**, 4658–4677 (2008)

C. Tuglus, M.J. van der Laan, Targeted methods for biomarker discovery, in *Targeted Learning: Causal Inference for Observational and Experimental Data.* ed. by M.J. van der Laan, S. Rose (Springer, Berlin, 2011)

UK Prospective Diabetes Study (UKPDS) Group, Effect of intensive blood-glucose control with metformin on complications in overweight patients with type 2 diabetes (UKPDS 34). Lancet **352**, 854–865 (1998)

M.J. van der Laan, Causal effect models for intention to treat and realistic individualized treatment rules. Technical Report, Division of Biostatistics, University of California, Berkeley (2006a)

M.J. van der Laan, Statistical inference for variable importance. Int. J. Biostat. **2**(1), Article 2 (2006b)

M.J. van der Laan, Estimation based on case-control designs with known prevalence probability. Int. J. Biostat. **4**(1), Article 17 (2008a)

M.J. van der Laan, The construction and analysis of adaptive group sequential designs. Technical Report 232, Division of Biostatistics, University of California, Berkeley (2008b)

M.J. van der Laan, Targeted maximum likelihood based causal inference: Part I. Int. J. Biostat. **6**(2), Article 2 (2010a)

M.J. van der Laan, Targeted maximum likelihood based causal inference: Part II. Int. J. Biostat. **6**(2), Article 3 (2010b)

M.J. van der Laan, Estimation of causal effects of community-based interventions. Technical Report 268, Division of Biostatistics, University of California, Berkeley (2010c)

M.J. van der Laan, Causal inference for networks. Technical Report, Division of Biostatistics, University of California, Berkeley (2012)

M.J. van der Laan, Causal inference for a population of causally connected units. J. Causal Inference **2**(1), 13–74 (2014a)

M.J. van der Laan, Targeted estimation of nuisance parameters to obtain valid statistical inference. Int. J. Biostat. **10**(1), 29–57 (2014b)

M.J. van der Laan, A generally efficient targeted minimum loss based estimator. Int. J. Biostat. **13**(2), 1106–1118 (2017)

M.J. van der Laan, S. Dudoit, Unified cross-validation methodology for selection among estimators and a general cross-validated adaptive epsilon-net estimator:

finite sample oracle inequalities and examples. Technical Report, Division of Biostatistics, University of California, Berkeley (2003)

M.J. van der Laan, S. Gruber, Collaborative double robust penalized targeted maximum likelihood estimation. Int. J. Biostat. **6**(1), Article 17 (2010)

M.J. van der Laan, S. Gruber, Targeted minimum loss based estimation of causal effects of multiple time point interventions. Int. J. Biostat. **8**(1), Article 9 (2012)

M.J. van der Laan, S. Gruber, One-step targeted minimum loss-based estimation based on universal least favorable one-dimensional submodels. Int. J. Biostat. **12**(1), 351–378 (2016)

M.J. van der Laan, S. Lendle, Online targeted learning. Technical Report, Division of Biostatistics, University of California, Berkeley (2014)

M.J. van der Laan, A.R. Luedtke, Targeted learning of an optimal dynamic treatment, and statistical inference for its mean outcome. Technical Report, Division of Biostatistics, University of California, Berkeley

M.J. van der Laan, A.R. Luedtke, Targeted learning of the mean outcome under an optimal dynamic treatment rule. J. Causal Inference **3**(1), 61–95 (2015)

M.J. van der Laan, M.L. Petersen, Causal effect models for realistic individualized treatment and intention to treat rules. Int. J. Biostat. **3**(1), Article 3 (2007)

M.J. van der Laan, M.L. Petersen, Direct effect models. Int. J. Biostat. **4**(1), Article 23 (2008)

M.J. van der Laan, K.S. Pollard, Hybrid clustering of gene expression data with visualization and the bootstrap. J. Stat. Plann. Inference **117**, 275–303 (2003)

M.J. van der Laan, E.C. Polley, A.E. Hubbard, Super learner. Stat. Appl. Genet. Mol. **6**(1), Article 25 (2007)

M.J. van der Laan, J.M. Robins, *Unified Methods for Censored Longitudinal Data and Causality* (Springer, Berlin Heidelberg New York, 2003)

M.J. van der Laan, S. Rose, *Targeted Learning: Causal Inference for Observational and Experimental Data* (Springer, Berlin, Heidelberg, New York, 2011)

M.J. van der Laan, D.B. Rubin, Targeted maximum likelihood learning. Int. J. Biostat. **2**(1), Article 11 (2006)

M.J. van der Laan, R.J.C.M. Starmans, Entering the era of data science: targeted learning and the integration of statistics and computational data analysis. Adv. Stat. 2014, 502678 (2014)

M.J. van der Laan, S. Dudoit, S. Keleş, Asymptotic optimality of likelihood-based cross-validation. Stat. Appl. Genet. Mol. **3**(1), Article 4 (2004)

M.J. van der Laan, S. Dudoit, A.W. van der Vaart. The cross-validated adaptive epsilon-net estimator. Stat. Decis. **24**(3), 373–395 (2006)

M.J. van der Laan, L.B. Balzer, M.L. Petersen, Adaptive matching in randomized trials and observational studies. J. Stat. Res. **46**(2), 113–156 (2013a)

M.J. van der Laan, M. Petersen, W. Zheng, Estimating the effect of a community-based intervention with two communities. J. Causal Inference **1**(1), 83–106 (2013b)

M.J. van der Laan, A.R. Luedtke, I. Díaz, Discussion of identification, estimation and approximation of risk under interventions that depend on the natural value of treatment using observational data, by Jessica Young, Miguel Hernán, and James Robins. Epidemiol Methods **3**(1), 21–31 (2014)

M.J. van der Laan, M. Carone, A.R. Luedtke, Computerizing efficient estimation of a pathwise differentiable target parameter. Technical Report, Division of Biostatistics, University of California, Berkeley (2015)

A.W. van der Vaart, *Asymptotic Statistics* (Cambridge, New York, 1998)

A.W. van der Vaart, Higher order tangent spaces and influence functions. Stat. Sci. **29**(4), 679–686 (2014)

A.W. van der Vaart, J.A. Wellner, *Weak Convergence and Empirical Processes* (Springer, Berlin, Heidelberg, New York, 1996)

A.W. van der Vaart, J.A. Wellner, A local maximal inequality under uniform entropy. Electron. J. Stat. **5**, 192–203 (2011)

A.W. van der Vaart, S. Dudoit, M.J. van der Laan, Oracle inequalities for multi-fold cross-validation. Stat. Decis. **24**(3), 351–371 (2006)

R. van Handel, On the minimal penalty for Markov order estimation. Probab. Theory Relat. Fields **150**, 709–738 (2009)

T.J. VanderWeele, Marginal structural models for the estimation of direct and indirect effects. Epidemiology **20**, 18–26 (2009)

T.J. VanderWeele, Bias formulas for sensitivity analysis for direct and indirect effects. Epidemiology **21**(4), 540 (2010)

T.J VanderWeele, Sensitivity analysis for contagion effects in social networks. Sociol. Methods Res. **40**(2), 240–255 (2011)

T.J. VanderWeele, Inference for influence over multiple degrees of separation on a social network. Stat. Med. **32**(4), 591–596 (2013)

T.J. VanderWeele, W. An, Social networks and causal inference, in *Handbook of Causal Analysis for Social Research* (Springer, Berlin, 2013), pp. 353–374

T.J. VanderWeele, O.A. Arah, Unmeasured confounding for general outcomes, treatments, and confounders: bias formulas for sensitivity analysis. Epidemiology **22**(1), 42 (2011)

T.J. VanderWeele, M.A. Hernán, Causal inference under multiple versions of treatment. J. Causal Inference **1**(1), 1–20 (2013)

T.J. VanderWeele, E.J. Tchetgen Tchetgen, Mediation analysis with time-varying exposures and mediators. J. R. Stat. Soc. Ser. B **79**(3), 917–938 (2017)

T.J. VanderWeele, B. Mukherjee, J. Chen, Sensitivity analysis for interactions under unmeasured confounding. Stat. Med. **31**(22), 2552–2564 (2012a)

T.J. VanderWeele, J.P. Vandenbrouke, E.J. Tchetgen Tchetgen, J.M. Robins, A mapping between interactions and interference: implications for vaccine trials. Epidemiology **23**(3), 285–292 (2012b)

T.J. VanderWeele, E.L. Ogburn, E.J. Tchetgen Tchetgen, Why and when "flawed" social network analyses still yield valid tests of no contagion. Stat. Polit. Policy **3**(1), 2151–2160 (2012c)

T.J. VanderWeele, S. Vansteelandt, J.M. Robins, Effect decomposition in the presence of an exposure-induced mediator-outcome confounder. Epidemiology **25**(2), 300–306 (2014a)

T.J. VanderWeele, E.J. Tchetgen Tchetgen, M.E. Halloran, Interference and sensitivity analysis. Stat. Sci. **29**(4), 687–706 (2014b)

S. Vansteelandt, E. Goetghebeur, M.G. Kenward, G. Molenberghs, Ignorance and uncertainty regions as inferential tools in a sensitivity analysis. Stat. Sin. **16**(3), 953–979 (2006)

W.N. Venables, B.D. Ripley, *Modern Applied Statistics with S*, 4th edn. (Springer, Berlin, Heidelberg, New York, 2002)

T.M. Vogt, J. Elston-Lafata, D. Tolsma, S.M. Greene, The role of research in integrated healthcare systems: the HMO Research Network. Am. J. Manag. Care **10**(9), 643–648 (2004)

E. Wagner, B. Austin, C. Davis, M. Hindmarsh, J. Schaefer, A. Bonomi, Improving chronic illness care: translating evidence into action. Health Aff. **20**, 64–78 (2001)

D. Walker, L. Muchnik, Design of randomized experiments in networks. Proc. IEEE **102**(12), 1940–1951 (2014)

H. Wang, M.J. van der Laan, Dimension reduction with gene expression data using targeted variable importance measurement. BMC Bioinf. **12**(1), 312 (2011)

H. Wang, S. Rose, M.J. van der Laan, Finding quantitative trait loci genes with collaborative targeted maximum likelihood learning. Stat. Probab. Lett. **81**(7), 792–796 (2011a)

H. Wang, S. Rose, M.J. van der Laan. Finding quantitative trait loci genes, in *Targeted Learning: Causal Inference for Observational and Experimental Data*, ed. by M.J. van der Laan, S. Rose (Springer, Berlin Heidelberg, New York, 2011b)

H. Wang, Z. Zhang, S. Rose, M.J. van der Laan, A novel targeted learning methods for quantitative trait Loci mapping. Genetics **198**(4), 1369–1376 (2014)

G.S. Watson, Smooth regression analysis. Sankhyā Indian J. Stat. Ser. A 359–372 (1964)

L. Watson, R. Small, S. Brown, W. Dawson, J. Lumley, Mounting a community-randomized trial: sample size, matching, selection, and randomization issues in PRISM. Control. Clin. Trials **25**(3), 235–250 (2004)

S. Weinberg, *Dreams of a Final Theory: The Scientist's Search for the Ultimate Laws of Nature* (Random House Inc., New York, 1993)

D. Wied, R. Weißbach, Consistency of the kernel density estimator: a survey. Stat. Pap. **53**(1), 1–21 (2012)

R.J. Wieringa, Design Science Methodology for Information Systems and Software Engineering (Springer, New York, 2014)

J. Williamson, Probabilistic theories of causality, in *The Oxford Handbook of Causation*, ed. by H. Beebee, C. Hitchcock, P. Menzies (Oxford University Press, Oxford, 2009), pp. 185–212

P. Wilson, R.B. D'Agostino, D. Levy, A.M. Belanger, H. Silbershatz, W.B. Kannel, Prediction of coronary heart disease using risk factor categories. Circulation **97**(18), 1837–1847 (1998)

T Woutersen, A simple way to calculate confidence intervals for partially identified parameters. Technical Report, Johns Hopkins University (2006)

W. Xu, Towards optimal one pass large scale learning with averaged stochastic gradient descent. ArXiv e-prints, December (2011)

J.G. Young, M.A. Hernán, J.M. Robins, Identification, estimation and approximation of risk under interventions that depend on the natural value of treatment using observational data. Epidemiol. Methods **3**(1), 1–19 (2014)

S. Yuan, H.H. Zhang, M. Davidian, Variable selection for covariate-adjusted semiparametric inference in randomized clinical trials. Stat. Med. **31**, 3789–3804 (2012)

M.D. Zeiler, Adadelta: an adaptive learning rate method. arXiv e-prints, December (2012)

K. Zhang, D.S. Small, Comment: the essential role of pair matching in clusterrandomized experiments, with application to the Mexican universal health insurance evaluation. Stat. Sci. **25**(1), 59–64 (2009)

B. Zhang, A. Tsiatis, M. Davidian, M. Zhang, E. Laber, A robust method for estimating optimal treatment regimes. Biometrics **68**, 1010–1018 (2012a)

B. Zhang, A. Tsiatis, M. Davidian, M. Zhang, E. Laber, Estimating optimal treatment regimes from a classification perspective. Stat **68**(1), 103–114 (2012b)

M. Zhang, A.A. Tsiatis, M. Davidian, Improving efficiency of inferences in randomized clinical trials using auxiliary covariates. Biometrics **64**(3), 707–715 (2008)

T. Zhang, J. Wu, F. Li, B. Caffo, D. Boatman-Reich, A dynamic directional model for effective brain connectivity using electrocorticographic (ECoG) time series. J. Am. Stat. Assoc. **110**(509), 93–106 (2015)

Y. Zhao, D. Zeng, A. Rush, M Kosorok, Estimating individual treatment rules using outcome weighted learning. J. Am. Stat. Assoc. **107**, 1106–1118 (2012)

Y. Zhao, D. Zeng, E.B. Laber, M.R. Kosorok, New statistical learning methods for estimating optimal dynamic treatment regimes. J. Am. Stat. Assoc. **110**(510), 583–598 (2015)

W. Zheng, M.J. van der Laan, Asymptotic theory for cross-validated targeted maximum likelihood estimation. Technical Report, Division of Biostatistics, University of California, Berkeley (2010)

W. Zheng, M.J. van der Laan, Causal mediation in a survival setting with timedependent mediators. Technical Report, Division of Biostatistics, University of California, Berkeley (2012a)

W. Zheng, M.J. van der Laan, Targeted maximum likelihood estimation of natural direct effects. Int. J. Biostat. **8**(1), 1–40 (2012b)

W. Zheng, M.J. van der Laan, Longitudinal mediation analysis with time-varying mediators and exposures, with application to survival outcomes. J. Causal Inference **5**(2), 20160006 (2017)

W. Zheng, A. Chambaz, M.J. van der Laan, Drawing valid targeted inference when covariate-adjusted response-adaptive RCT meets data-adaptive loss-based estimation, with an application to the lasso. Technical Report, Division of Biostatistics, University of California, Berkeley (2015)

M. Zinkevich, Online convex programming and generalized infinitesimal gradient ascent. Proceedings of ICML (2003)

Printed in the United States
By Bookmasters